THE LATTICE BOLTZMANN EQUATION

The Lattice Boltzmann Equation

For Complex States of Flowing Matter

Sauro Succi

Center for Life Nanoscience at La Sapienza,
Istituto Italiano di Tecnologia, 00161, Roma, Italy
Viale Regina Margherita, 295, 00161, Roma, Italy

Istituto Applicazioni del Calcolo, National Research Council of Italy,
via dei Taurini 19, 00185, Roma, Italy

Institute for Applied Computational Science, J. Paulson
School of Engineering and Applied Sciences, Harvard University,
Oxford Street 29, 02138, Cambridge, USA

OXFORD
UNIVERSITY PRESS

OXFORD
UNIVERSITY PRESS

Great Clarendon Street, Oxford, OX2 6DP,
United Kingdom

Oxford University Press is a department of the University of Oxford.
It furthers the University's objective of excellence in research, scholarship,
and education by publishing worldwide. Oxford is a registered trade mark of
Oxford University Press in the UK and in certain other countries

First Edition published in 2018

Impression: 1

Published in the United States of America by Oxford University Press
198 Madison Avenue, New York, NY 10016, United States of America

British Library Cataloguing in Publication Data

Data available

Library of Congress Control Number: 2017943732

ISBN 978–0–19–959235–7

Printed and bound by
CPI Group (UK) Ltd, Croydon, CR0 4YY

*To Claudia and Cate, to the many friends who shine light through my journey.
And to the memory of my parents, up there.*

Preface

Start by doing what is necessary, then do what is possible and suddenly you are doing the impossible.

(Saint Francis of Assisi)

Preface 2018

Since the appearance of my previous book (The lattice Boltzmann equation, Oxford U.P, 2001), Lattice Boltzmann has known an exponential growth of methodologies and applications, especially in connection with the simulation of complex and soft-matter flows. Providing a complete and in-depth account of such burgeoning developments is probably beyond the scope of any self-contained book and by all means beyond the capability, knowledge and time-energy of the present author. As a result, this second book remains more modest in scope: it just provides an account of the major developments, with no aim of completeness. At the same time, it is also more ambitious, because it aims at discussing research items beyond Navier-Stokes hydrodynamics, with special focus on interfaces between fluid dynamics and allied disciplines, such as material science, soft matter and biology. Moreover, it also ventures into the realm of quantum and relativistic fluids.

As with the first book, the only major criticism I have heard of, is lack of self-containedness for readers not trained in physics. A single chapter of kinetic theory, apparently, did not fill the bill. In response to this just criticism, I have considerably expanded the pedagogical part on continuum kinetic theory. To the point that, with some doses of wishful thinking, I hope that this part can be used to stand-alone as an introductory text to the kinetic theory of fluids. In response to significant developments which have occurred in the last decade, I have also expanded the quantum-relativistic part, hoping that this may help capture the attention of the growing community dealing with the fascinating confluence between fluid dynamics, condensed matter and high-energy physics which we are witnessing these days.

This book splits naturally into three main components: the first one deals with basic notions continuum kinetic theory, the fundamentals of early-day lattice kinetic theory, a few applications thereof, and finally a few more advanced topics in lattice kinetic theory. All in all, this first part is basically concerned with *macroscopic* fluid dynamics.

Thes second part deals with lattice kinetic theory for generalized hydrodynamics beyond Navier-Stokes, basically the territory where the physics of *flowing matter* makes contact with its allied disciplines, soft matter in the first place.

Finally, the third part deals with fluids beyond Newtonian mechanics, namely relativistic and quantum fluids. This latter part is less voluminous, but it holds its own place in view of the amazing developments at the frontier between fluid dynamics, condensed matter and high-energy physics, which have taken place in the last decade, holographic fluids and graphene on the frontline. It is still much less developed than the mainstreams covered in the two parts above, but very rich in promise, I believe.

As usual, I tried to keep math at a minimum, but hopefully never below the threshold where equations cannot be traded for "words and pics."

As a result, I am reasonably confident that reading this book is not "like chewing glass," as Sydney Chapman once commented on his wonderful but demanding Chapman-Cowling cornerstone.

I also tried to watch against another danger, wittily described by Sir Winston Churchill: "By its very length, this report defends itself against the risk of being read." Not sure I managed, but I did my best.

For any mistakes, typos and/or inconsistencies which, I'm afraid, are inevitably left in a near-800 pages, single-handed book, I have no chance but appealing to the reader's benevolent understanding and cooperation.

Enjoy!

Acknowledgements

This book collects another fifteen years of intense research on the subject of Lattice Boltzmann (LB). Many things have changed since the early days of LB, when nearly every idea was new, fertile and usually well received by a small but growing and self-supportive community. This second fifteens have witnessed an exponential growth of the LB applications, along with many important methodological extensions.

Despite my best intentions for conciseness, this book nearly triples in size its predecessor, and I'm afraid that errors and infelicities grow way more than linearly with size .... Thus, even though I fully subscribe to Joyce's statement below, I would definitely make my errors zero, if I only could ....

Therefore, I shall again be grateful to any reader kindly willing to signal mistakes, inconsistencies and any sort of imperfections to me.

The near-three decades long LB adventure has got me in touch with a huge number of colleagues all over the world, from whom I invariably learned more than I can possibly express. Unlike Sinatra's regrets, they are "too many to mention", without incurring a serious risk of self-embarrassing omissions.

Yet, I shall make three exceptions: one for those who took the time and pain of reading selected portions of this manuscript, i.e., Hakan Basagaoglu, M. Bernaschi, L. Biferale, Burkhard Duenweg, Giacomo Falcucci, Iliya Karlin, Tony Ladd, Simone Melchionna, Miller Mendoza, Andrea Montessori, M. Sbragaglia and Stefano Ubertini.

The other for the senior colleagues and friends who inspired and shared substantial stretches of the journey, namely Roberto Benzi, Jean-Pierre Boon, Bruce Boghosian, Hudong Chen, Tim Kaxiras, Iliya Karlin, Hans Herrmann and the late Steven Orszag.

The third is for my early wonderful mentors in kinetic theory, the late V. Boffi, V. Molinari and G. Spiga at Bologna University, K. Appert and J. Vaclavik at the Ecole Polytechnique Federale de Lausanne.

I would also like to acknowledge Sonke Adlung, Ania Wronski and Indumadhi Srinivasan for their continued kind (and patient) help throughout this ordeal.

Finally, I would like to thank my wife Claudia and daughter Caterina, for coping with a rather absent and absent-minded husband and father.

Errors are the portals of discovery (J. Joyce)

S. S.
Rome
March 2018

Contents

Part VI Beyond Newtonian Mechanics: Quantum and Relativistic Fluids

Part I

Kinetic Theory of Fluids

The first Part of this Book is entirely devoted to pedagogical material, namely the kinetic theory of fluids.

Since the Lattice Boltzmann method was historically devised to solve the equations of continuum fluid mechanics, in Chapter I we present a rapid survey of such equations and introduce the motivations for a microscopic approach to fluid dynamics, namely the kinetic theory of fluids.

Chapter II presents the basic notions of kinetic theory, with special focus on its mathematical cornerstone: Boltzmann's kinetic equation. In Chapter III we discuss the approach to equilibrium and the attendant notions of entropy and irreversibility. Chapter IV provides an elementary illustration of transport phenomena and their link to the underlying microphysics. In Chapter V we discuss the hydrodynamic limit of Boltzmann's kinetic theory, taking to the Navier-Stokes equations of continuum fluid mechanics, while in Chapter VI we illustrate Grad's formulation of generalized hydrodynamics, beyond the Navier-Stokes picture. Chapter VII deals with extensions of Boltzmann's kinetic theory to the case of dense fluids. Chapter VIII illustrates simplified versions of the Boltzmann's equation, so called Model Boltzmann equations, which have been devised in order to facilitate its analytical and numerical solution. Finally, Chapter IX provides a cursory view of stochastic processes of direct relevance to the kinetic theory of fluids, such as Brownian motion and the associated Langevin equations.

The above material was prompted in response to the only substantial criticism received by the previous book, namely lack of self-containedness for readers with no specific training in statistical physics, and particularly in kinetic theory. This material is meant precisely to those readers, hopefully providing a satisfactory fix to the gap they have kindly highlighted. Those with no urge of filling such gap, can safely skip to Part 2.

1

Why a Kinetic Theory of Fluids?

In this chapter we present the Navier–Stokes equations of fluid mechanics and discuss the main motivations behind the kinetic approach to computational fluid dynamics.

For, sometimes, the longest tour is the shortest way home.

(C.S. Lewis)

1.1 The Navier–Stokes Equation

Fluid flows are a pervasive presence across most branches of human activity and daily life in the first place. Although the basic equations governing the motion of fluid flows have been known for nearly two centuries, since the work of Claude-Louis Navier (1785–1836) and Gabriel Stokes (1819–1903), these equations still set a formidable challenge to our quantitative, and sometimes even qualitative, understanding of the way fluid matter flows in space and time. Climate and meteorological phenomena are among the most popular examples in point, as expressed by Bob Dylan's vivid metaphor "The answer, my friend, is blowing in the wind."

At a first glance, the Navier Stokes equations (NSEs) look relatively harmless. In conservative Eulerian form (1):

$$\begin{cases} \partial_t \rho + \nabla \cdot (\rho \vec{u}) = 0 \\ \partial_t (\rho \vec{u}) + \nabla \cdot (\rho \vec{u}\vec{u} + p) = \nabla \cdot \vec{\vec{\sigma}} + \vec{f}^{ext} \end{cases} \tag{1.1}$$

In (1.1), n and $\rho = nm$ are the fluid number and mass density (m is the mass of the molecules), respectively, p is the fluid pressure, $\vec{u}$ is the fluid velocity and $\vec{f}^{ext}$ is the external force per unit volume. Whenever such force can be derived from a potential, say Φ, the latter adds to the total fluid pressure, $p_{tot} = p + n\Phi$. The shear-stress tensor $\vec{\vec{\sigma}}$ represents dissipative effects, as induced by the deformation of the fluid elements.

The Lattice Boltzmann Equation. Sauro Succi, Oxford University Press (2018).
© Sauro Succi. DOI: 10.1093/oso/9780199592357.001.0001

The first of these two equations is simply a statement of mass conservation, as applied to a small but finite volumlet of fluid. The change in time of the mass in the volumlet is due to the unbalance between the incoming and outgoing mass fluxes. The second equation is basically Newton's equation in reverse, $m\vec{a} = \vec{F}$, as applied to the same volumlet. In this, "small" means much smaller than the typical scale of change of the macroscopic fields, and yet, large enough to allow the neglect molecular fluctuations.

The first term on the left-hand side of the second equation is the change per unit time of the momentum in the fluid volumlet at a fixed position in space (Eulerian representation). The second term on the right-hand side is the flux of momentum across the surface of the volumlet due to the fluid motion itself, namely the inertial forces per unit volume. The third term on the left-hand side is the force per unit volume exerted by the adjacent fluid along the normal surface (pressure gradient). The first term on the right-hand side represents the contact forces tangential to the surface, the main source of strain, hence dissipation, on the fluid.

The NSE represents four equations for eleven unknowns, density, pressure, three components of the velocity field and six components of the (symmetric) stress tensor. Hence, to close the system, they must be supplemented with seven extra-relations.

The first is the *equation of state*, relating pressure to density and temperature

$$p = p(\rho, T) \tag{1.2}$$

For an ideal gas, pressure depends linearly on the density

$$p = n\mathcal{R}T \tag{1.3}$$

where $\mathcal{R}$ is the universal gas constant. For non-ideal gases, the relation is no longer linear due to the contribution of potential energy.

The other six constraints come from the so-called *constitutive relations*, namely an expression of the stress tensor in terms of the gradient of the velocity field.

In the simplest instance, the so-called Newtonian fluids, this relation takes the following linear form:

$$\vec{\vec{\sigma}} = \lambda(\nabla \cdot \vec{u})I + \mu(\nabla\vec{u} + \vec{u}\nabla) \tag{1.4}$$

where μ is the shear viscosity and λ is related to the bulk viscosity, μ_b, via

$$\mu_b = \lambda + \frac{2}{3}\mu. \tag{1.5}$$

On stability grounds, both viscosities must be non-negative, although the bulk viscosity is often taken to be zero (Stoke's hypothesis).

Note that these *constitutive relations* are heuristic in nature; they state that the strain of the fluid resulting from a given applied stress is linearly proportional to the stress itself.

Although fairly reasonable (for small enough strains), this does *not* follow from any fundamental law of Newtonian mechanics. Indeed, constitutive relations are not universal, i.e., they take different forms depending on the specific class of fluid considered.

The simplest case, λ and μ constant, designates so-called Newtonian fluids.[1] For more complex fluids, typically endowed with internal structure, the constitutive relations may become nonlinear, nonlocal in space and time, and non-isotropic as well.

Regardless of the specific form of the constitutive relations, the physical content of the NSE remains quite transparent, basically mass and momentum conservation as applied to a small, yet *finite*, volume of fluid (volumlet).

1.1.1 Elementary Derivation of the Navier–Stokes Equations

Having noted that the continuity equation is simply a statement of mass conservation, while the second NSE is basically Newton's law plus a statement of strain-stress linearity, we now proceed to an elementary derivation of the statements based on simple arguments of continuum mechanics. With reference to the square volumlet of size $\Delta V = \Delta x \Delta y$ depicted in Figures 1.1 and 1.2 (in two dimensions for simplicity), the change of mass in the volumlet over a time interval δt is given by

$$\frac{\delta(\rho \Delta x \Delta y)}{\delta t} = \left[(\rho u_x)\left(x - \frac{\Delta x}{2}, y\right) - (\rho u_x)\left(x + \frac{\Delta x}{2}, y\right) \right] \Delta y \qquad (1.6)$$
$$+ \left[(\rho u_y)\left(x, y - \frac{\Delta y}{2}\right) - (\rho u_y)\left(x, y + \frac{\Delta y}{2}\right) \right] \Delta x$$

where the right-hand side is the sum of the incoming and outgoing mass fluxes.

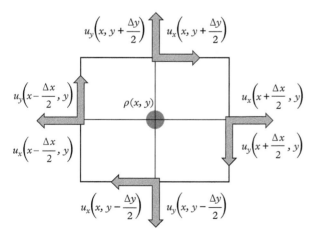

Figure 1.1 *The mass balance illustrating the elementary derivation of the continuity equation. For simplicity, the two-dimensional case is considered.*

---

[1] This is a bit ironic, given that constitutive equations represent precisely the only component of the NSEs which goes beyond a plain transcription of Newtonian physics from the microscopic to the macroscopic level!

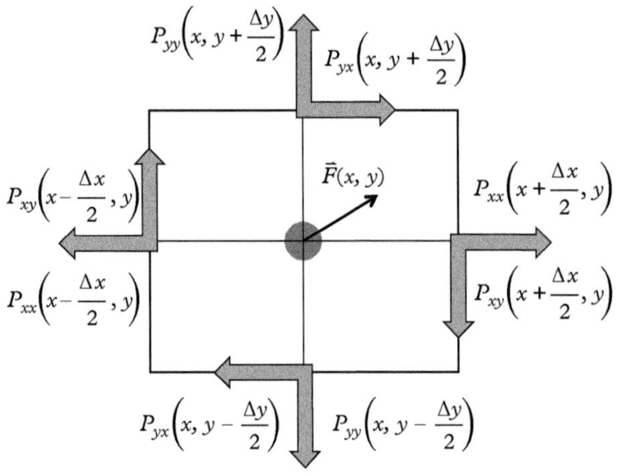

Figure 1.2 *The momentum balance illustrating the elementary derivation of the NSEs.*

Dividing by $\Delta x \Delta y$, and taking the continuum limit $\delta t \to 0$ and $\Delta x \to 0$, we obtain

$$\partial_t \rho = -\partial_x(\rho u_x) - \partial_y(\rho u_y) \tag{1.7}$$

which is precisely the continuity equation.

By the same token, the change of momentum along the x direction in the volumlet over a time interval δt is given by

$$\frac{\delta(\rho u_x \Delta x \Delta y)}{\delta t} = \left[(\rho u_x u_x) \left(x - \frac{\Delta x}{2}, y \right) - (\rho u_x u_x) \left(x + \frac{\Delta x}{2}, y \right) \right] \Delta y \tag{1.8}$$

$$+ \left[(\rho u_x u_y) \left(x, y - \frac{\Delta y}{2} \right) - (\rho u_x u_y) \left(x, y + \frac{\Delta y}{2} \right) \right] \Delta x$$

The first two terms on the right-hand side represent the flux of momentum along x, entering across the left facelet at $x - \Delta x/2$, minus the same flux exiting across the right facelet at $x + \Delta x/2$. The other two terms represent the mass flux along x, entering along y across the top and bottom boundaries, respectively (see Fig. 1.1). Dividing by $\Delta x \Delta y$, and taking the continuum limit $\delta t \to 0$ and $\Delta x \to 0$, delivers

$$\partial_t(\rho u_x) = -\partial_x(\rho u_x^2) - \partial_y(\rho u_x u_y) \tag{1.9}$$

This accounts for the convective term on the left-hand side of the momentum equation.

Pressure and dissipative effects are represented by the contact forces acting on the four surfaces of the two-dimensional volumlet. By summing the forces, i.e., change of momentum per unit time, acting upon the four faces of the volumlet along x, we obtain

$$f_x \Delta x \Delta y = \left[P_{xx}\left(x - \frac{\Delta x}{2}, y\right) - P_{xx}\left(x + \frac{\Delta x}{2}\right) \right] \Delta y$$

$$+ \left[P_{xy}\left(x, y - \frac{\Delta y}{2}, y\right) - P_{xy}\left(x, y + \frac{\Delta y}{2}\right) \right] \Delta x$$

Using the standard definition of the pressure, namely

$$p = P_{xx} - P_{yy} = \Gamma_{zz}$$

and dividing again by the area of the volumlet, we obtain

$$f_x = \partial_x p - \partial_y P_{xy} \tag{1.10}$$

Summing this to (1.9) delivers precisely the x component of the NSEs.

This shows that the NSEs are essentially Newton's equations

$$\frac{d}{dt}(m\Delta N \vec{u}) = \Delta V \vec{f} \tag{1.11}$$

as applied to the finite volumlet of volume ΔV and mass $m\Delta N = \rho \Delta V$.

The forces on the right-hand side of the NSE are most conveniently split into *conservative* and *dissipative* components, respectively.

The former is given by

$$\vec{f}^{con} = -\nabla(p + n\Phi) \tag{1.12}$$

where we have assumed that the external force derives from a potential Φ, i.e., $\vec{F} = -\nabla\Phi$. The overall pressure

$$p_{tot} = p + n\Phi \tag{1.13}$$

plays the role of a generalized potential. The most familiar case is perhaps the one of a fluid in a gravitational field, in which case $\Phi = -mgz$, g being the gravitational accelera-tion and z the elevation of the fluid element (minus sign indicates that lower-lying fluid layers experience higher hydrostatic pressure).

For *perfect* fluids, i.e., (idealized) fluids with strictly zero dissipation, NSE go by the name of Euler equations. Even though Euler fluids represent a very useful idealization, any real fluid is bound to display some form of dissipation, especially in the vicinity of solid walls (we dispense here with quantum effects leading to superfluidity).

According to the NSE, the dissipative force is given by

$$\vec{f}^{dis} = \nabla \cdot \vec{\vec{\sigma}} \tag{1.14}$$

As anticipated, this is *not* implied by Newtonian physics.

Indeed, at the level of continuum mechanics, it cannot be derived from first principles, but must be postulated based on general heuristics guidelines. As we shall see, such first-principle microscopic derivation comes quite naturally at the level of the kinetic theory of fluids.

Notwithstanding the heuristic nature of the dissipative force, the NSEs prove amazingly robust, whence their spectacular success in describing the physics of fluids under an impressively broad range of conditions.

The reason is that they encode very basic principles of Newtonian mechanics, plus a (very reasonable) assumption of linearity between the applied stress and the resulting strain. This immunizes them against the vagaries of the underlying microscopic interactions, a strong manifestation of what statistical physicists use to call by the beautiful name of *Universality*.

1.1.2 Navier–Stokes Equations in Lagrangian Form

The NSEs, as given in eqns (1.1), come in *mathematically* conservative form, in that the time change of density and momentum is driven by the divergence of a corresponding vector or tensor.

For many fluid-dynamic problems it proves expedient to recast them in a form which emphasizes the transport along the material-fluid lines, i.e., the lines whose tangent identifies with the fluid velocity itself.

Using the identities: $\nabla \cdot (\rho \vec{u}) = \vec{u} \cdot \nabla \rho + \rho \nabla \cdot \vec{u}$ and $\nabla \cdot (\rho \vec{u} \vec{u}) = \vec{u} \cdot \nabla(\rho \vec{u}) + \rho \vec{u} \cdot \nabla \vec{u}$, the NSE can be cast in so-called Lagrangian form, namely (2):

$$D_t \rho \equiv \partial_t \rho + \vec{u} \cdot \nabla \rho = -\rho \, \nabla \cdot \vec{u} \tag{1.15}$$

$$D_t \vec{u} \equiv \partial_t \vec{u} + \vec{u} \cdot \nabla \vec{u} = -\frac{\nabla p}{\rho} + \frac{1}{\rho} \nabla \cdot \vec{\vec{\sigma}} \tag{1.16}$$

The Lagrangian representation emphasizes the role of the material derivative D_t; both density and velocity are advected by the velocity itself, with no change along the fluid trajectory. Changes, on the other hand, are described by the terms on the right-hand side.

The qualitative difference between the Eulerian and Lagrangian viewpoints is sketched in Fig. 1.3.

For instance, the continuity equation describes a process whereby the density is carried with no change "on the back" of a fluid element, from position $\vec{x}$ to $\vec{x} + \vec{u}dt$ in a time lapse dt. Changes then occur due to the compressibility term on the right-hand side.

For the case of *incompressible* fluids, characterized by a divergence-free (solenoidal) flow velocity:

$$\nabla \cdot \vec{u} = 0 \tag{1.17}$$

the density looses the status of a space-time dependent field and can be set to a constant, conventionally $\rho = 1$. In this case, the solenoidal condition (1.17) takes the role of a *kinematic constraint* on the velocity field.

Since the material derivative is nothing but the acceleration in a frame moving with the fluid, the Lagrangian form of the Euler equation (zero-dissipative force) best reveals its genuine relation to Newtonian mechanics. The left-hand side is the acceleration, and

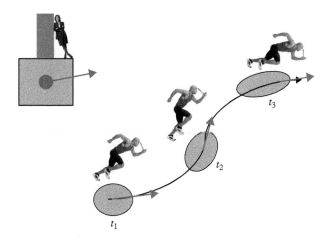

Figure 1.3 *The Eulerian (top-left) and the Lagrangian (bottom-right) approaches to the physics of fluids. In the Eulerian approach, the observer stands still at a given space location and watches the fluid moving in and out of the observing box. In the Lagrangian approach the observer "goes with the flow."*

the right-hand side is the force per unit mass stemming from the generalized potential $p_{tot} = p + n\Phi$.

In fact, leaving only $\partial_t \vec{u}$ on the left-hand side identifies three type of conservative forces per unit mass, namely:

- Inertial: $\frac{\vec{F}^{ine}}{m} = -\vec{u} \cdot \nabla \vec{u}$
- Pressure: $\frac{\vec{F}^{pre}}{m} = -(\nabla p)/\rho$
- External: $\frac{\vec{F}^{ext}}{m} = -n\nabla\Phi$

The inertial forces account for the kinetic energy of the fluid. In the absence of dissipation, the total energy per unit volume

$$h = \rho\frac{u^2}{2} + p + n\Phi \tag{1.18}$$

is conserved along a fluid trajectory. This important statement of conservation is known as Bernoulli theorem, a cornerstone of hydrostatics.

Finally, we note that for an observer living in a local frame moving with the fluid, i.e., $\vec{u} \equiv 0$, the inertial forces disappear altogether. This substantial simplification lies at the heart of an important family of Lagrangian techniques for the numerical solution of the NSEs.

1.1.3 Navier–Stokes Equations in Coordinate Form

For the purpose of deriving hydrodynamics from kinetic theory, a task that shall be undertaken in chapter 5, it proves expedient to cast the NSE in coordinate form, that is

$$\partial_t \rho + \partial_a(\rho u_a) = 0 \qquad (1.19)$$

$$\partial_t(\rho u_a) + \partial_b(\rho u_a u_b + p\delta_{ab} - \sigma_{ab}) = nF_a^{ext} \qquad (1.20)$$

where Latin indices denote the spatial coordinates x, y, z and δ_{ab} is the Kronecker delta. As usual, repeated indices are summed upon.

This form emphasizes the conservative nature of the NSEs, while the coordinate notation lends transparency to the algebraic manipulations required to derive it from the underlying Boltzmann kinetic equation.

In particular, the momentum-flux tensor

$$P_{ab} \equiv \rho u_a u_b + p\delta_{ab} - \sigma_{ab} \qquad (1.21)$$

will be shown to play a key role in the kinetic theory of fluids.

It is left as an exercise for the reader to verify that expressions (1.1) and (1.19–20) are basically the same wine in two different bottles.

1.2 Computational Aspects of the Navier–Stokes Equations

The NSEs rest on the representation of the fluid as continuum media, a pervasive field filling up space, with "no holes" in between. For more than a century, and with special thanks to Ludwig Boltzmann (1844–1906), we know that, at the atomic scale, matter is granular, made up of tiny atoms or molecules of sub-nanometric size. Being so tiny, these atoms are also very many, of the order of the Avogadro number $N_{av} \sim 6 \times 10^{23}$ in a little more than a centimeter cube of water (in fact, about 20 N_{av}). As a result, it would be foolish, and needless as well, to attempt a quantitative description of a macroscopic fluid as collection of Avogadro numbers of atoms! The continuum representation achieves a spectacular compression of information: from Avogadro's number of molecules, to just a fistful of continuum fields, density, velocity and pressure.

Beware, though, that such simplification is subtler than it seems. Indeed, by definition, a continuum field contains an *infinite* amount of information, and consequently this compression is, strictly speaking, an infinite inflation instead!

Of course, whenever analytical techniques are available to deal with such infinities, theory comes into its zenith.

However, as for most nonlinear theories, such glorious achievements are mostly confined to idealized situations.

As a result, the actual import of the previous compression of information is basically dictated by the discrete scale at which numerical solutions can be worked out.

Be that as it may, the continuum picture is the dominant one, to the point that the quantitative study of fluid dynamics is still often taken as a by-name for solving the NSEs.

This has spawned a very successful and still fast-growing discipline known as Computational Fluid Dynamics (CFD), a leading forefront of Computational science. CFD has made tremendous progress over the last half century, with a distinguished tradition tracing back to the famous John von Neumann's (1903–57) 1949 "breaking-the-deadlock" report, where he advocated the use of digital simulation to gain knowledge on a variety of fluid phenomena (3).

Von Neumann's bold vision is well captured by his famous statement: "*[W]hat we cannot control, we shall predict, what we cannot predict, we shall control.*"

In actual facts, such bold vision was depleted by the discovery of chaos and, more specifically, by the phenomenon of turbulence, which is a prominent reason (but surely not the only one) why fluid dynamics cannot be declassified under the rubric of "old science."

Indeed, along with major progress, CFD has also taught us a very down-to-earth lesson: despite their harmless look, the NSEs prove *exceedingly* hard to solve on digital computers (let alone analytics).

As anticipated, one of the main reasons is *turbulence*, the name of nonlinearity when it comes to the physics of fluids.

Let us consider for simplicity the case of *incompressible* flows, for which the continuity equation reduces to the solenoidal condition on the velocity field given by (1.17). For such flows, the density is constant in space and time and, consequently, it can be scaled out from the equations by setting it to the conventional value $\rho = 1$.

The momentum equation then simplifies as follows:

$$\partial_t \vec{u} + \vec{u} \cdot \vec{\nabla} \vec{u} = \nu \Delta \vec{u} - \nabla p \tag{1.22}$$

where $\nu = \mu/\rho$ is the kinematic viscosity.

Turbulence results from the competition between large-scale advection, the term $\vec{u} \cdot \vec{\nabla}\vec{u}$, and small-scale dissipation, the term $\nu \Delta \vec{u}$.

This ratio is measured by the Reynolds number

$$Re = \frac{UL}{\nu} \tag{1.23}$$

This number easily exceeds a million in many daily-life fluid phenomena, such as a car at a standard speed of 100 Km/h.

1.2.1 Why Is the Reynolds Number so Large?

The next natural question is: why is the nonlinearity so overwhelming over dissipation?

The reason is best highlighted by recasting the Reynolds numbers as follows:

$$Re = \frac{U}{c_s} \frac{L}{l_\mu} = \frac{Ma}{Kn} \tag{1.24}$$

an expression also known as the Von Karman relation (after Theodor von Karman, 1981–63).

In (1.24), c_s is the speed of sound and $l_\mu = v/c_s$ is the mean-free path, i.e., the average distance traveled by a molecule before colliding with another molecule. Furthermore,

$$Ma = \frac{U}{c_s} \tag{1.25}$$

is the Mach number (Ernst Mach, 1838–1916), the ratio of fluid to sound speed, and

$$Kn = \frac{l_\mu}{L} \tag{1.26}$$

is the Knudsen number, after the Danish scientist Martin Knudsen (1871–1949).

In air, molecules travel about one-tenth of a micron before colliding, which means that for an ordinary macroscopic object, say a car, featuring $L \sim 1$ m, the ratio L/l_μ is about 10 million: here we are with huge Reynolds numbers!

Indeed, the Mach number U/c_s is typically in the order of unity or less; a car traveling at 100 Km/h features $Ma \sim 0.1$, since the speed of sound in air in standard conditions is about $c_s \sim 300$ m/s.

In summary, the Reynolds number is large because it measures the length of the car in units of the mean-free path! The point is that advection acts at macroscopic scales, while dissipation takes over at much shorter scales, known as the dissipative or Kolmogorov scale, after the famous scaling theory formulated by the Russian mathematician A.N. Kolmogorov (1903–87) back in 1941.

According to this theory, the smallest dynamically active scale in a turbulent flow at a given Reynolds number Re is given by

$$l_d = \frac{L}{Re^{3/4}} \tag{1.27}$$

Thus, the number of degrees of freedom in a turbulent flow of size L is approximately:

$$N_{dof} = \left(\frac{L}{l_d}\right)^3 \sim Re^{9/4} \tag{1.28}$$

For an ordinary flow at $Re = 10^6$, this makes about $N_{dof} \sim 10^{14}$ degrees of freedom, still way less than the some 50 Avogadro's number of molecules contained in a cubic meter of air, and yet far in excess of those affordable even on the most powerful supercomputers (current leading-edge simulations of fluid turbulence are handling of the order of ten billion degrees of freedom).

This, combined with the fact that fluids of practical interest typically move in highly complex geometries (cars, airplanes and the like), makes the ordeal of solving the NSE an extremely difficult one. This is essentially the reason behind the relentless pursuit of innovative computational methods for fluid flows.

The overwhelming majority of CFD methods focus on various discretizations of the NSE, as a set of nonlinear partial differential equations. That is, one starts from the NSE in their continuum form, and devises different methods to represent such equations on a discrete grid, be it fixed in space (Eulerian approach) or moving along with the fluid (Lagrangian approach).

In order for the simulation to be spatially resolved, the grid spacing must be smaller than the Kolmogorov dissipative scale, namely

$$\Delta x < l_d \qquad (1.29)$$

This shows that the actual number of grid points, $N_{grid} = (L/\Delta x)^3$, must necessarily exceed the number of physical degrees of freedom. These bounds are dictated by the basics physics of turbulence; hence they do not depend on the specific discretization procedure. However, they speak clearly for computational requirements of CFD.

As we shall see, the Lattice Boltzmann equation falls in the line of a major paradigmatic shift, first opened up by its forerunner, the Lattice Gas Cellular Automata (LGCA) method.

That is computational fluid dynamics without directly discretizing the NSEs.

Instead of discretizing partial differential equations, the idea is to track the dynamics of a fictitious set of *representative particles*, supporting macroscopic fluid behavior described by the NSEs, as an *emergent* phenomenon.

In other words, one does not solve the emergent equation itself (Navier–Stokes), but the underlying stylized microscopic dynamics instead.

The obvious question is: what's the point of such an indirect approach?

The answer, which we shall spell out in the sequel, is indeed not completely obvious.

1.3   The Benefits of Kinetic Extra Dimensions

The advantage is that, as we shall expound in the course of this book, particle dynamics can be made much simpler than the dynamics of hydrodynamic fields.

For instance, force-free particles move along straight trajectories, while material lines follow the fluid velocity itself, which is typically highly complex field even in a force-free fluid.

Let's expand on the point.

As we shall see, the basic object of Boltzmann's kinetic theory (see Fig. 1.4) is the one-particle distribution function (4): $f(\vec{x}, \vec{v}; t)$, which represents the probability density of finding a molecule at position $\vec{x}$ in space at time t, with a given velocity $\vec{v}$.

In actual practice,

$$\Delta N = f \Delta \vec{x} \Delta \vec{v} \qquad (1.30)$$

is the mean number of particles in the *phase-space* element of volume $\Delta \vec{x} \Delta \vec{v}$. From a mere fluid dynamic perspective, $f(\vec{x}, \vec{v}; t)$ is a highly redundant object, as the information on the molecular velocity has no explicit bearing on the fluid equations, which only depend on space and time. As a result, at first sight, the Boltzmann equation looks like a total overkill when it comes to fluid dynamics, six dimensions: three in ordinary space and three in velocity space, against just three!

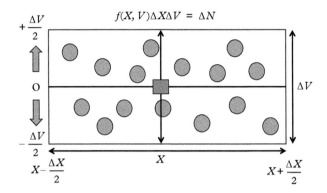

Figure 1.4 *The meaning of the Boltzmann distribution function. Fifteen particles lying in the segment Δx each with its own velocity in the range $-\frac{\Delta v}{2} < v < \frac{\Delta v}{2}$ are represented by fifteen points in the phase-space element $\Delta x \Delta v$. The distribution function is the density of particles in this six-dimensional phase space (two dimensional in the figure).*

1.3.1   Molecular Streaming versus Fluid Advection

However, precisely because velocity and space coordinates are independent variables, *in phase space the information travels in a much simpler way than in ordinary configuration space.*

In fact, the Boltzmann distribution $f(\vec{x}, \vec{v}; t)$ moves along straight lines defined by the molecular speed, namely

$$d\vec{x}_v = \vec{v}dt, \tag{1.31}$$

where $\vec{v}$ carries no dependence on space or time.

This is what we refer to as *molecular free-streaming*, as opposed fluid advection, which proceeds along the *material* lines in ordinary space (see Fig. 1.5).

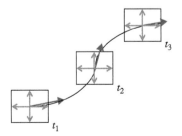

Figure 1.5 *Fluid trajectory (curved line) versus molecular streamlines, here represented by just four directions for simplicity. The four directions are independent of the spatial location of the fluid element.*

The latter are defined by

$$d\vec{x}_u = \vec{u}(\vec{x}; t)dt, \tag{1.32}$$

The inertial term $\vec{u} \cdot \nabla \vec{u}$ in the NSE is just the acceleration along a material line.

In general, the flow field $\vec{u}$ may be a fairly complicated function of space and time, in contrast to the simple straight molecular trajectories. In the kinetic picture, all this complexity is absorbed by the Boltzmann distribution $f(\vec{x}, \vec{v}; t)$, while the streaming of molecules keeps going along straight lines, no matter how complex the macroscopic field.

In slightly more technical language, hydrodynamic transport in ordinary space, i.e., *advection*, is nonlinear because the material lines are defined by the flow velocity itself, while molecular transport in phase space, i.e., *streaming*, is linear because the molecular trajectories do not depend on the transported quantity, namely $f(\vec{x}, \vec{v}; t)$.

1.3.2   Molecular Relaxation versus Momentum Diffusivity

The kinetic representation also has profound implications on the description of dissipative effects. In the Navier–Stokes picture, these are described by the divergence of a gradient, i.e., the Laplace operator. This is because, at a macroscopic level, dissipation is due to momentum diffusivity across the fluid and the equivalence between space and time has to be broken to describe irreversible behavior. Consider a fluid driven by a moving wall; the layers near the wall are set in motion by direct contact with the wall, and transmit these motions thanks to the diffusion of momentum in the direction perpendicular to the wall. If the momentum would not diffuse, the wall would move on its own with no effect on the fluid. Conversely, the fluid would move along a fixed wall with no friction (superfluid).

Kinetic theory, on the other hand, does not know about diffusion (an emergent phenomenon); it only knows about molecular collisions. These collisions act in such a way as to relax the Boltzmann distribution to a local equilibrium, on a time scale which is basically the average time traveled by a molecule before colliding with another molecule. The local equilibrium is a Gaussian (Maxwell–Boltzmann) distribution in velocity space, which depends parametrically on the local fluid density, velocity and temperature. Here, local means the same spatial position $\vec{x}$ and time t at which the Boltzmann distribution is evaluated.

By definition, fluid quantities show appreciable variations on space and time scales well above those associated with molecular relaxation. Such scale separation is quintessential to obtain hydrodynamics as the asymptotic limit of kinetic theory as the Knudsen number is sent to zero.

It can be shown that momentum diffusivity is precisely the macroscopic manifestation of molecular relaxation. However, since momentum diffusivity *emerges* from collisions, and since Boltzmann's collisions are completely local in space and time, they do not involve any space-time communication, such communication being in full charge of the streaming operator only. Nor do they demand any symmetry breaking between space and time, since relaxation to local equilibrium takes place through molecular collisions.

As a result, the kinetic formalism does not need any Laplacian operator to represent diffusion processes, which is again a significant simplification, both from the conceptual and computational points of view.

1.4 Summary

Summarizing, the NSEs of continuum fluid mechanics prove exceedingly difficult to solve, as they assemble two nightmares of computational physics: strong nonlinearity and complex geometry, within a fully three-dimensional, time-dependent formulation.

The kinetic picture trades three extra-velocity dimensions for linearity and locality: *linear streaming versus nonlinear advection* and *local relaxation versus non-local momentum diffusion*. More precisely, it disentangles non-locality and nonlinearity: nonlocality (streaming) is linear and nonlinearity (collisions) is local. It turns out that lattice versions of the Boltzmann-kinetic equation make a very attractive computational bargain of such disentanglement. This is the bottom line of the lattice Boltzmann story to be told in this book.

....

REFERENCES

1. L. Landau and E. Lifshitz, *Fluid Mechanics*, Pergamon Press, Oxford, 1987.
2. R. Peyret, *Computational Methods for Fluid Flow*, Springer Verlag, 1983.
3. U. Frisch, *Turbulence*, Cambridge Univ. Press, Cambridge, 1999.
4. L. Boltzmann, *Vorlesungen über Gastheorie*, in Lectures on Gas Theory, J.A. Barth Editor, unab. Dover, Leipzig, 1995.

....

EXERCISES

1. Prove the equivalence between eqns (1.1) and (1.19).
2. The ratio l_d/l_μ measures the separation between the smallest hydrodynamic and the microscopic scale. Does this ratio increase or decrease in the limit of infinite Reynolds number?
3. Derive the equations for incompressible flows from the general eqns (1.1). Hint: use the identities $\nabla \cdot (\rho \vec{u}) = \vec{u} \cdot \nabla \rho + \rho \nabla \cdot \vec{u}$ and $\nabla \cdot (\rho \vec{u}\vec{u}) = \vec{u} \cdot \nabla (\rho \vec{u}) + \rho \vec{u} \cdot \nabla \vec{u}$.

2

Boltzmann's Kinetic Theory

Kinetic theory is the branch of statistical physics dealing with the dynamics of non-equilibrium processes and their relaxation to thermodynamic equilibrium. Established by Ludwig Boltzmann (1844–1906) in 1872, his eponymous equation stands as its mathematical cornerstone. Originally developed in the framework of dilute gas systems, the Boltzmann equation has spread its wings across many areas of modern statistical physics, including electron transport in semiconductors, neutron transport, quantum liquids, to cite but a few. In this chapter, we shall provide a basic introduction to the Boltzmann equation in the context of classical statistical mechanics.

> *I am conscious of being only an individual struggling weakly against the stream of time. But it still remains in my power to contribute in such a way that, when the theory of gases is again revived, not too much will have to be rediscovered.*
>
> (L. Boltzmann)

2.1 Atomistic Dynamics

Let us consider a collection of N molecules moving in a box of volume V at temperature T and mutually interacting via a two-body intermolecular potential $V(\vec{r})$, $\vec{r}$ being the intermolecular separation between two generic molecules.[2]

If the linear size s of the molecules, basically the effective range of the short-range interaction potential, is much smaller than their *mean-interparticle separation* $d = (V/N)^{1/3}$,

---

[2] The symbol $V(\vec{r})$ denotes the interparticle potential, not to be confused with plain V, the volume of the system, and $\vec{V}$ the barycentric velocity of the two-body problem.

The Lattice Boltzmann Equation. Sauro Succi, Oxford University Press (2018).
© Sauro Succi. DOI: 10.1093/oso/9780199592357.001.0001

the molecules can, to a good approximation, be treated like *point-like structureless particles*.

To the extent where the De Broglie length $\lambda = \hbar/mv$ of these particles is much smaller than any other relevant length scale, their dynamics is governed by the classical Newton's equations:

$$
\begin{cases}
\dfrac{d\vec{x}_i}{dt} = \vec{v}_i, \\[2ex]
\dfrac{d\vec{v}_i}{dt} = \dfrac{\vec{F}_i}{m}, \quad i = 1, N
\end{cases}
\tag{2.1}
$$

where $\vec{x}_i$ is the position coordinate of the *i-th* particle, $\vec{v}_i$ its velocity and $\vec{F}_i$ is the force experienced by the *i*-th particle as a result of intermolecular interactions and possibly external fields (gravity, electric field, etc.).

Upon specifying initial and boundary conditions, equations (27.50) can in principle be solved in time, to yield a fully exhaustive knowledge of the state of the system, namely a set of $6N$ functions of time $\{\vec{x}_i(t), \vec{v}_i(t)\}$, $i = 1, N$.

This programme is totally unviable and, fortunately, needless as well. Unviability stems from two main reasons: first, N is generally of the order of the Avogadro number $N_{Av} \sim 10^{23}$, far too big for any foreseeable computer. Second, even if one could store it, tracking so much information for sufficiently long times would be utopia, since any tiny uncertainty on the initial conditions would blow up in the long run because of dynamical instability of phase space. By dynamical instability, we refer to the fact that any uncertainty δ_0 on the initial positions and/or momenta grows exponentially in time as $\delta(t) = \delta_0 e^{\lambda t}$. The coefficient λ, known as *Lyapunov exponent*, is a measure of the temporal horizon of deterministic behavior of the N-body system, in that at times greater than λ^{-1}, the growth of uncertainty is such to prevent any deterministic prediction of the state of the system. It is estimated that a centimeter cube of Argon in standard conditions (300 K, 1 *Atm*) produces as much as 10^{29} digits of information per second. This means that in order to keep an exact record of the state of the system over a $1s$ lifespan, we need a number with nothing less than 10^{29} digits. Fortunately, we manage to survive with less than that, reason being that *we are much larger than the molecules our body is made of* !

The physical observables we are interested in, say the fluid pressure, temperature, visible flow originate from a statistical average over a large number of individual molecular histories.

A rigorous definition of what is meant by statistical average is not trivial, but here we shall be content with the intuitive notion of spatial average over a thermodynamic volume, namely a region of space sufficiently small with respect to the global dimensions of the macroscopic domain, and yet large enough to contain a statistically meaningful sample of molecules.

Typical numbers help getting the picture. The density of air in standard conditions is about $n_L = 2.687 \; 10^{25}$ molecules/m^3 (known as as *Loschmidt number*). Hence, a centimeter cube of air contains about $2.7 \; 10^{19}$ molecules, corresponding to a statistical error of less than one part per billion.

2.2 Statistical Dynamics: Boltzmann and the BBGKY Hierarchy

Given the very huge numbers involved, it appears therefore wise to approach the collective behavior of the ensemble of molecules from a *statistical* point of view.

This can be done at various levels of complexity, but, for a start, we shall begin with the simplest one: the *single-particle kinetic* level.

The chief question of single-particle kinetic theory is:

What is the probability of finding a molecule around position $\vec{x}$ at time t with velocity $\vec{v}$?

Let $f(\vec{x}, \vec{v}, t)$ the probability density, more often simply denoted as distribution function.

The quantity $\Delta N = f \Delta \vec{x} \Delta \vec{v}$ represents the mean number of molecules in a finite volume $\Delta \vec{x} \, \Delta \vec{v}$, centered about $(\vec{x}, \vec{v})$ in the so-called single-particle *phase space*:

$$\Gamma_1 = \{\vec{z} \equiv (\vec{x}, \vec{v}); \vec{x}, \vec{v} \in R^3\}$$

Integration upon the velocity degrees of freedom delivers the number of particles per unit volume, i.e., the number density of the system at any given time t:

$$\int f(\vec{x}, \vec{v}; t) d\vec{v} = \frac{\Delta N}{\Delta V}$$

which recovers the continuum density $n(\vec{x}, t)$ in the limit $\Delta V \to 0$.

As a result, integration upon the entire phase space delivers the total number of molecules in the system at any given time t,

$$\int f(\vec{x}, \vec{v}); t) d\vec{x} d\vec{v} = N(t)$$

The distribution function $f(\vec{x}, \vec{v}; t)$ is the pivotal object of Boltzmann's kinetic theory.

In 1872, Ludwig Boltzmann (1844–1906) was able to derive an equation describing the evolution of $f(\vec{x}, \vec{v}; t)$ in terms of the underlying microdynamic interactions. This is the celebrated Boltzmann equation (BE), one of the greatest achievements of theoretical physics of the nineteenth century (1).

The BE represents the first quantitative effort to attack the grand-issue of why time goes "one-way only" on a macroscopic scale while the underlying microdynamics is apparently perfectly reversible.[3] In this book, we shall not be much concerned with fundamental issues, but rather keep the focus on the BE as a mathematical tool to investigate, analytically or numerically, the properties of fluid flows far from equilibrium.

---

[3] We shall stick to the common tenet that microscopic equations, either classical or quantum, are invariant under time reversal.

The kinetic equation for the one-body distribution function in the presence of an external force $\vec{F}(\vec{x})$ reads as follows (2):

$$\partial_t f + \vec{v} \cdot \partial_{\vec{x}} f + \vec{a} \cdot \partial_{\vec{v}} f = C_{12} \qquad (2.2)$$

where $\vec{a} = \vec{F}/m$ is the particle acceleration due to external and internal forces.

The left-hand side represents the streaming of the molecules along the trajectories associated with the force field $\vec{F}$ (straight lines if $\vec{F} = 0$) and C_{12} represents the effects of intermolecular (two-body) collisions taking molecules in/out the streaming trajectory.

Let us comment on the two sides separately.

Once it is accepted that the cloud of N molecules moves like a lump of fluid in phase space Γ_1, the streaming term reduces to a mere mirror of Newtonian mechanics.

To convince oneself, simply rewrite the streaming term as a Lagrangian derivative along the trajectory $\vec{x}(t)$,

$$\frac{df}{dt} \equiv \partial_t f + \frac{d\vec{x}}{dt} \cdot \partial_{\vec{x}} f + \frac{d\vec{v}}{dt} \cdot \partial_{\vec{v}} f$$

Using Newton's equations, $\frac{d\vec{x}}{dt} = \vec{v}$, $\frac{d\vec{v}}{dt} = \vec{F}/m$, this returns precisely the left-hand side of the Boltzmann equation. The streaming term carries the information contained in the distribution function untouched from place to place in phase-space.

Indeed, the solution of the collisionless Boltzmann equation $\frac{df}{dt} = 0$, with initial conditions $f(\vec{x}, \vec{v}, t = 0) = f_0(\vec{x}, \vec{v})$, is simply

$$f(\vec{x}, \vec{v}, t) = f_0[\vec{x}(t), \vec{v}(t)], \qquad (2.3)$$

where $\vec{x}(t)$ and $\vec{v}(t)$ is the solution of the Newton's equations with initial conditions $\vec{x}(t = 0) = \vec{x}$ and $\vec{v}(t = 0) = \vec{v}$. The key physical point is the following: the streaming term moves the distribution function in phase space with no loss of information, hence no loss of memory of the initial conditions: *reversible* motion.

2.3 The Born–Bogoliubov–Green–Kirkwood–Yvon (BBGKY) Hierarchy

The right-hand side of the BE, on the other hand, takes care of exchanging information across different trajectories, through intermolecular interactions.

The collision operator encodes two-body collisions, between, say molecule one, sitting at point $\vec{x}_1$ with speed $\vec{v}_1$ *and* molecule two, sitting at $\vec{x}_2$ with speed $\vec{v}_2$, both at time t.

Formally, this information is stored in the *two-body* distribution function

$$f_{12}(\vec{x}_1, \vec{v}_1, \vec{x}_2, \vec{v}_2; t),$$

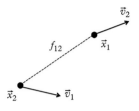

Figure 2.1 *Sketch of the two-body distribution function f_{12}. The two molecules are correlated to each other, so that if one moves the other must move too. In the one-body representation, each molecule moves independently, although it feels it the effects of the other molecules through short-range collisions described by the collision operator C_{12}.*

expressing the *joint* probability of finding molecule one around $\vec{x}_1$ with speed $\vec{v}_1$ *and* molecule two at $\vec{x}_2$ with speed $\vec{v}_2$, both at time t (see Fig. 2.1).

More precisely, the quantity

$$\Delta N_{12} = f_{12}\, \Delta\vec{x}_1 \Delta\vec{v}_1 \Delta\vec{x}_2 \Delta\vec{v}_2$$

gives the average number of *pairs* of molecules sitting at points $\vec{z}_1 \equiv (\vec{x}_1, \vec{v}_1)$ and $\vec{z}_2 \equiv (\vec{x}_2, \vec{v}_2)$ of phase space at time t.

Living as it does in a $(6 + 6 = 12)$-dimensional phase space (13 with time), it goes without saying that f_{12} is a very heavy-duty object to work with.

The one-body distribution is recovered by integrating over the second particle phase-space coordinates:

$$f_1(\vec{x}_1, \vec{v}_1; t) = \frac{2}{N-1} \int f_{12}(\vec{x}_1, \vec{v}_1, \vec{x}_2, \vec{v}_2; t) d\vec{x}_2 d\vec{v}_2 \tag{2.4}$$

where the factor 2 accounts for the fact that there are $N(N-1)/2$ symmetric pairs out of a pool of N particles.

Clearly, this projection from 13 to 7 dimensions erases a huge amount of information, the two-body correlations. The loss of this information prevents an exact reconstruction of f_{12} from $f_1 \equiv f(\vec{z}_1; t)$ and $f_2 \equiv f(\vec{z}_2; t)$, separately.

However, as we shall see shortly, educated guesses on the physical nature of the system under consideration, can (partially) make up for this fundamental limitation. In principle, it is not difficult to write down the dynamic equation for f_{12}, the only trouble being that this equation calls into play the three-body distribution function f_{123}, which in turn depends on f_{1234} and so on, down an endless line known as the BBGKY hierarchy, after Bogoliubov, Born, Green, Kirkwood and Yvon (4).

The physical origin of such open structure is that a N-body system can in principle host molecular collisions at all orders, from binary onward up to order N. If one could solve the BBGKY hierarchy, one would obtain a complete *statistical* knowledge of the full N-body problem described by the Newtonian equations for the N molecules. This is again utopia, only in statistical rather than dynamic vests!

Consequently, one must settle for less ambitious goals, i.e., approximate descriptions. The loss of information inevitably associated with such approximations is responsible for *irreversibility*, to be literally intended as our inability to reconstruct the initial conditions exactly (loss of memory).

Fortunately, powerful heuristics are available to guide the search for sensible approximations to the BBGKY hierarchy.

Indeed, in actual practice, the probability of a simultaneous interaction between, say, k molecules, decays very fast with k, approximately like $(s/d)^{3k}$, where

$$d = 1/n^{1/3}$$

is the mean-intermolecular separation and n is the number density, the two being related via $nd^3 = 1$, i.e., one particle on average in a cublet of volume d^3.

The ratio

$$\tilde{n} = \left(\frac{s}{d}\right)^3 \equiv ns^3 \tag{2.5}$$

sometimes called "granularity," provides a direct measure of the degree of diluteness of the system. Indeed, in a system at density n, each molecule inhabits a volume d^3, and $\tilde{n}$ is the fraction of that volume occupied by the molecule itself (here a cublet of side s).

From its very definition, it is clear that $\tilde{n}$ controls the strength of many-body interactions, which fade away as $\tilde{n} \to 0$. For instance, air at standard conditions features $\tilde{n} \sim 10^{-3}$, so that many-body interactions are largely negligible. Water in standard conditions, on the other hand, provides $\tilde{n} \sim 1$, which surely calls for careful consideration of many-body effects. Nevertheless, as we shall see in Chapter 3, the structure of the Navier–Stokes equations of continuum fluid dynamics is to a large extent independent of many-body effects. This is a great gift of mother nature, known as *Universality*.

2.4   Back to Boltzmann

The simple, yet basic, considerations previously suggested, set the stage for Boltzmann's clever way out of the BBGKY hierarchy.

To close equation (2.2), Boltzmann made a few stringent assumptions on the nature of the physical system: a *dilute* gas of *point-like, structureless* molecules interacting via a *short-range* two-body potential.

Under such conditions, intermolecular interactions can be described solely in terms of localized binary collisions, with molecules spending most of their lifespan on free trajectories (in the absence of external fields), merrily unaware of each other.

Within this picture, the collision term splits into Loss and Gain components:

$$C_{12} \equiv \mathcal{G} - \mathcal{L} = \int (f_{1'2'} - f_{12}) v_r \sigma(v_r, \vec{\Omega}) d\vec{\Omega} d\vec{v}_2 \tag{2.6}$$

corresponding to direct(inverse) collisions taking molecules out(in) the volume element $d\vec{v}_1 d\vec{v}_2$ respectively (see Fig. 2.2).

The right-hand side requires a number of detailed comments.

First, the shorthand f_{12} stands for $f_{12}(\vec{z}_1, \vec{z}_2; t)$.

In the above, v_r is the magnitude of the relative speed between particle 1 and particle 2 and $\vec{\Omega}$ denotes the solid angle associated with the scattering event (see Fig. 2.3). The

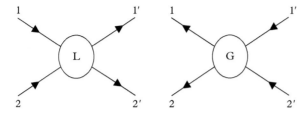

Figure 2.2 *Symbolic diagram of direct and inverse collisions-. Inverse collisions (Gain) place particles in state one, while direct collisions (Loss) take them away from it.*

symbol σ denotes the differential cross section, i.e. the effective area presented by a particle in the plane across its center and perpendicular to the relative velocity.

Likewise, $f_{1'2'}$ stands for $f_{12}(\vec{z}'_1, \vec{z}'_2; t)$, where prime indicates the molecular positions and velocities *after* a direct collision.

These two factors are purely statistical in nature. Note that *all* four spatial coordinates, both pre- and post-collisional, lie within a sphere of radius $s \ll d$, on account of the diluteness assumption. Therefore, in the Boltzmann limit $s/d \to 0$, they can be reconduced to the same spatial location $\vec{x} = \vec{x}_1 = \vec{x}'_1 = \vec{x}_2 = \vec{x}'_2$.

This is a major simplification of the Boltzmann equation, with far-reaching consequences on theoretical as well as computational aspects.

The pre- and post-collisional velocities $\vec{v}_1, \vec{v}_2$ and $\vec{v}'_1, \vec{v}'_2$ are related through the three basic Mass–Momentum–Energy conservation laws, that is

$$\begin{cases} m_1 + m_2 = m'_1 + m'_2 \\ m_1\vec{v}_1 + m_2\vec{v}_2 = m'_1\vec{v}'_1 + m'_2\vec{v}'_2 \\ m_1 v_1{}^2 + m_2 v_2{}^2 = m'_1 v'_1{}^2 + m'_2 v'_2{}^2. \end{cases} \tag{2.7}$$

Since mass can be assumed invariant across a collision, $m'_1 = m_1$ and $m'_2 = m_2$, the first equation is basically a statement of number conservation, $2 = 2$, two molecules before collisions, two molecules after.

The other two conservation laws, however, deliver a great deal of information, as we shall see in the sequel.

2.4.1 Two-Body Scattering

The two-body collision problem is best treated as the scattering of a single particle of reduced mass m_r impinging on a target particle of mass $M = m_1 + m_2$, sitting at the median position $\vec{X}$, with median velocity $\vec{V}$, defined as follows:

$$\begin{cases} \vec{X} = (m_1\vec{x}_1 + m_2\vec{x}_2)/M \\ \vec{V} = (m_1\vec{v}_1 + m_2\vec{v}_2)/M \end{cases} \tag{2.8}$$

The reduced mass is given by

$$m_r = \frac{m_1 m_2}{m_1 + m_2} \tag{2.9}$$

and it is seen to coincide with the lightest mass, say m_1, in the limit $m_2 \gg m_1$. For equal mass molecules, the case assumed hereafter, $m_r = m/2$ and $M = 2m$.

The two-body scattering problem is best treated in a frame with the origin located at $\vec{X}$.

Using a polar representation (r, θ) for the interparticle separation

$$\vec{r} = \vec{x}_1 - \vec{x}_2 \tag{2.10}$$

the total energy writes as

$$E = \frac{m_r v_r^2}{2} + \frac{\mathcal{J}^2}{2m_r r^2} + V(r) \tag{2.11}$$

where

$$\mathcal{J} \equiv m_r r^2 \dot{\theta} = m_r v_r b \tag{2.12}$$

is the angular momentum and $V(r)$ the interparticle potential.

In (2.12), b is the so-called impact parameter, i.e., the distance of the colliding molecule from the origin perpendicular to its relative velocity (see Fig. 2.2).

After noting that mass-momentum conservation yields

$$\vec{V}' = \vec{V}$$

it is readily appreciated that energy conservation implies that the relative velocity vector

$$\vec{v}_r = \vec{v}_2 - \vec{v}_1 \tag{2.13}$$

is conserved in magnitude.

As a result, the only effect of the collision is to rotate the relative velocity by an angle χ in the scattering plane defined by $\vec{r}$ and $\vec{v}_r$.

The kinematic identity $\vec{v}_i = \vec{V} - (m_j/M)\vec{v}_r$, $i = 1, 2$, $j = 2, 1$, delivers the following mapping between post- and pre-collisional velocities:

$$\begin{cases} \vec{v}'_1 = \vec{v}_1 - \frac{m_2}{M}\vec{v}_r \\ \vec{v}'_2 = \vec{v}_2 + \frac{m_1}{M}\vec{v}_r \end{cases} \tag{2.14}$$

It can be checked that this one-to-one mapping preserves the volume element in velocity space, i.e.,

$$|d\vec{v}'_1 d\vec{v}'_2| = |d\vec{v}_1 d\vec{v}_2| \tag{2.15}$$

a property which shall prove very useful in the sequel.

With the two-body kinematics in place, one can compute the number of molecules, dN, scattered around the solid angle $d\vec{\Omega} = \sin\chi\, d\chi\, d\alpha$, where α fixes the orientation of the scattering plane in three-dimensional space.

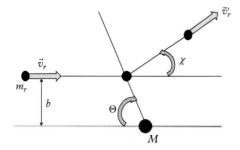

Figure 2.3 *Scattering angle associated with a binary collision. The collision takes place in the plane defined by the interparticle separation $\vec{x}_1 - \vec{x}_2$ and the relative speed $\vec{v}_r = \vec{v}_1 - \vec{v}_2$. The solid angle $\vec{\Omega}$ is defined by the scattering angle χ in the collisional plane and by the azimuthal angle ϕ around the collisional plane (not shown).*

The relation

$$dN = \sigma(v_r, \vec{\Omega})d\vec{\Omega}$$

defines the so-called *differential cross section σ*.

The next task is to compute the scattering angle $\chi = \chi(b, v_r)$ as a function of the impact parameter and the relative velocity (see Fig. 2.3).

To this purpose, let us consider all impinging particles sitting in the annulus of radius b and thickness db. The conservation of the number of these particles implies, namely:

$$2\pi b\,db = 2\pi\sigma(\chi, v_r)\sin\chi\,d\chi \tag{2.16}$$

$$\sigma(\chi, v_r) = \frac{b}{\sin\chi}\frac{db}{d\chi} \tag{2.17}$$

This reveals that the differential cross section is fixed by the functional relation $b = b(\chi, v_r)$, which in turn depends on the details of the scattering potential $V(r)$.

A quantity of major interest is the *cross section*:

$$\sigma(v_r) = \int \sigma(\chi, v_r)\sin\chi\,d\chi \tag{2.18}$$

and its integral version:

$$\Sigma(T) = 4\pi \int \sigma(v_r)f(v_r)v_r^2 dv_r = n\sigma_T \tag{2.19}$$

also known as total cross section. Note that σ_T, as defined above, is generally a function of the temperature T, so that one can write

$$\sigma_T = \kappa(T)s^2 \tag{2.20}$$

s being the size of the molecule, identified with the range of the potential. For short-range potentials one can assume further assume $\kappa(T) \sim O(1)$, i.e., the effective cross section does not differ drastically from the geometrical one.

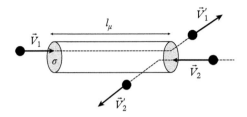

Figure 2.4 *Geometrical representation of the mean-free path. The molecule two travels a distance l_μ before colliding with molecule two. The associated cross section defines the collisional cylinder, whose volume is $V_{col} = \sigma l_\mu$. In the dilute gas limit $\tilde{n} \to 0$, the collisional cylinder collapses to a needle-like shape with $\sigma / l_\mu \to 0$.*

The total cross section defines the molecular mean-free path (see Fig. 2.4):

$$l_\mu = \frac{1}{\Sigma} \tag{2.21}$$

and the associated collisional timescale

$$\tau_\mu = \frac{l_\mu}{v_T} = \frac{1}{n \sigma_T v_T}. \tag{2.22}$$

where

$$v_T = \sqrt{\frac{k_B T}{m}}$$

is the thermal speed.

The mean-free path is the mean distance traveled by a molecule before colliding with another molecule and represents the pivotal lengthscale of kinetic theory and transport phenomena.

2.4.2   Spatial Ordering in Dilute Gases

Based on the definitions (2.19) and (2.21), one obtains

$$n \sigma_T l_\mu = 1$$

indicating that by construction, the so-called collisional cylinder of volume $\sigma_T l_\mu$ contains just a single colliding molecule.

Recalling the definition of the mean-interparticle distance, $nd^3 = 1$, the previous section yields

$$\frac{l_\mu}{d} = \frac{d^2}{\sigma_T}.$$

Based on the relation (2.20) with $\kappa \sim 1$, one further obtains

$$s \ll d \ll l_\mu$$

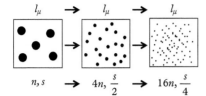

Figure 2.5 *Geometrical representation of the dilute gas limit in Boltzmann kinetic theory. By halving the size s, density quadruples and the total area ns², inversely proportional to the mean-free path, stays the same (at least this is the intention of the picture.).*

which is the typical scale ordering of dilute gases.

It is of interest to note that the proper meaning of "dilute gas" in the Boltzmann framework does *not* correspond at all to a gas in the ordinary sense, i.e., a fluid of vanishingly small density. Quite on the contrary, the proper limit is the one where density is formally sent to infinity! The point is that, at the same time, the size s is sent to zero, in such a way as to keep the product ns^2 constant. In other words, size goes to zero, density goes to infinity and the mean-free path is left constant. Putting all together, the mean-intermolecular distance scales like $d \sim n^{-1/3}$ while the molecular size scales like $s \sim n^{-1/2}$, so that the diluteness parameter $\tilde{n} = (s/d)^3$ scales like $n^{-1/2}$ and goes to zero in the limit $n \to \infty$, (see Fig. 2.5).

Summarizing, the dilute gas limit corresponds to the following limiting scenario:

$$s \to 0, \ n \to \infty, \ l_\mu = Const., \ \tilde{n} \to 0 \tag{2.23}$$

This further witnesses the crucial role of the mean-free path as the fundamental length scale of Boltzmann kinetic theory and shows that $\tilde{n}$ is the appropriate smallness parameter describing many-body effects in dense gases and liquids. We shall return to these matters in Chapter 7 devoted to the kinetic theory of dense fluids.

2.4.3 Two-Body Scattering Problem

To sort out the explicit dependence $\chi = \chi(b, v_r)$, one needs to solve the two-body scattering problem. To this aim, it proves expedient to move to polar coordinates in the scattering plane.

The equations of motion resulting from conservation of energy E and angular momentum $\mathcal{J}$ read as follows:

$$E = \frac{1}{2}m_r\dot{r}^2 + \frac{m_r^2 v_r^2 b^2}{r^2} + V(r) = \frac{m_r}{2}v_r^2 \tag{2.24}$$

$$\mathcal{J} = m_r r^2 \dot{\theta} = m_r v_r b \tag{2.25}$$

where the right-hand side corresponds to the limit $r \to \infty$.

Dividing the two, one obtains

$$\frac{dr}{d\theta} = \frac{r^2}{b}\left[1 - \frac{b^2}{r^2}\phi(r)\right]^{1/2} \tag{2.26}$$

where

$$\phi(r) = \frac{2V(r)}{m_r v_r^2} \tag{2.27}$$

is the ratio of potential to kinetic energy in the rest frame.

For purely repulsive potentials, there exists a minimum-approach distance, r_{min} defined by the condition $\frac{dr}{d\theta} = 0$.

Some algebra delivers the implicit relation:

$$\frac{r_{min}}{b} = \sqrt{\frac{m_r v_r^2/2}{E - V(r_{min})}} \tag{2.28}$$

Note that for repulsive potentials, $V(r) > 0$, $r_{min} > b$, while the opposite is true for attractive ones.

Integrating upon r from r_{min} to infinity, one obtains

$$\Theta \equiv \theta_{min} - \theta_\infty = b v_r \int_{r_{min}}^\infty \frac{dr}{r^2 \sqrt{1 - \phi(r)}} \tag{2.29}$$

which is known as the apse angle.

The scattering angle is finally derived as

$$\chi = \pi - 2\Theta. \tag{2.30}$$

This procedure shows that the dependence of the scattering angle on the potentials is generally pretty involved. As a general rule, however, small-impact parameters correspond to large scattering angles.

2.4.4   Distinguished Potentials

Once the atomistic potential is known, the procedure already outlined permits us to compute the scattering differential cross section σ, hence the collisional relaxation time and the mean-free path, starting from the atomistic potentials. This accomplishes the fundamental task of transferring information from the atomistic world of trajectories and intermolecular potentials to the kinetic world of statistical distributions, scattering cross sections and mean-free path.

Symbolically, the kinetic micro-meso bridge reads as follows:

$$V(r) \to \sigma(v_r) \leftrightarrow l_\mu. \tag{2.31}$$

Given that some specific potentials stand out for their importance, either from the mathematical point of view or for their applicability to realistic fluids, in the sequel we provide a cursory coverage of such potentials.

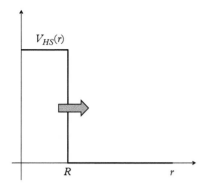

Figure 2.6 *The hard-sphere potential. For graphical purpose the potential step has a finite amplitude, leading nonetheless to an infinite force (arrow) at r = R. This is basically a solid wall, bouncing back the molecules the molecules which impinge on it.*

2.4.4.1   Hard Spheres

A particularly important and analytically solvable case is provided by the hard-sphere potential,

$$V_{HS}(r) = \begin{cases} \infty, & r \le R \\ 0, & otherwise \end{cases} \tag{2.32}$$

where R is the sphere radius (see Fig 2.6).

Detailed calculations yield $\sigma = \pi R^2$, i.e., the geometrical area of the particle cross section, as it should be.

Despite their simplicity hard-sphere potentials have played a major role in the kinetic theory of fluids and continue to provide valuable information for molecular dynamics simulations with hard-core repulsive interactions.

2.4.4.2   Lennard-Jones potential

Another potential which plays a prominent role in the physics of non-ideal fluids, is the so-called 12–6 Lennard-Jones potential, after the British physicist John Edward Lennard-Jones (1894–1954):

$$V_{LJ}(r) = 4\epsilon\,[(r/R)^{-12} - (r/R)^{-6}] \tag{2.33}$$

This potential consists of a hard-core repulsion (-12 branch), plus soft-core attraction (-6 branch) (see Fig. 2.7). The former stems for the strong repulsion between incipient overlap of electronic orbitals, when nuclei get seriously close together at distances around one third of nanometer and below. The latter is due to cohesive forces arising from screened multipole electrostatic interactions (Van der Waals interactions) and plays a defining role on the thermodynamic properties of the fluid.

The competition between short-range repulsion and long-range attraction leads to a minimum of depth $-\epsilon$ at a distance $r^* = 2^{1/6}R$, which fixes the typical scale of intermolecular separation in the fluid.

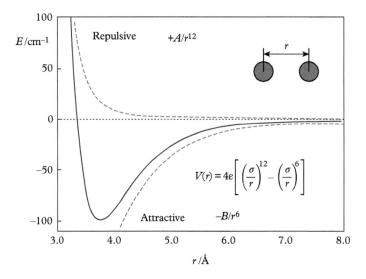

Figure 2.7 *The Lennard-Jones 6 - 12 potential. In the figure, σ represents the range of interaction, called R in the text, to avoid confusion with
the cross-section. From http://chemistry.stackexchange.com/questions/342
14/physical-significance-of-double-well-potential-in-quantum-bonding.*

The Lennard-Jones potential provides a microscopic basis for the celebrated van der
Waals equation of state of non-ideal fluids,

$$\left(p + \frac{a}{V^2}\right)(V - b) = Nk_B T \tag{2.34}$$

where N is the number of molecules in the volume V.

The attractive branch a/V^2 echoes the soft-core tail $(r/R)^{-6}$ and the covolume b is
related to the spatial scale, $b^{1/3}$, of the hard-core repulsion $(r/R)^{-12}$.

2.4.4.3 Maxwell molecules

A special case is provided by the so-called Maxwell molecules, characterized by a -4
power-law decay:

$$V_{MM}(r) \propto r^{-4}. \tag{2.35}$$

The calculations show that for such power-law potential $v_r \sigma(v_r) = Const.$, so that the
collision time scale is a constant, see eqn (2.22).

This constitutes a major simplification of the Boltzmann collision integral, whence
the special role of this potential in kinetic theory.

Even though Maxwell's molecules do not appear to have any realistic counterpart in the physical world, they provide nonetheless a very fruitful theoretical idealization for several mathematical developments in kinetic theory.

In particular, under appropriate simplifications, they permit us to obtain exact solutions of the Boltzmann equation.

Calculations in three spatial dimensions for inverse power-law potentials of the form

$$V(r) \sim 1/r^{\alpha} \tag{2.36}$$

show that

$$v_r \sigma(v_r) \sim v_r^{\frac{\alpha-4}{\alpha}}. \tag{2.37}$$

This highlights that Maxwell molecules, $\alpha = 4$, mark an qualitative borderline: for $\alpha < 4$, i.e., slower decay than for Maxwell molecules, the collision rate

$$\gamma = n v_r \sigma(v_r) \tag{2.38}$$

turns from an increasing to a decreasing function of the relative speed v_r, i.e., essentially the fluid temperature.

A moment's thought reveals that a collision frequency decreasing with the molecular speed implies that fast molecules experience less friction than the slow ones, which is clearly a portal to collective instability.

Indeed, this opens up *non-hydrodynamic* scenarios, whereby particles accelerated beyond a given critical speed by, say, a constant external field, do not experience a sufficient collisional drag to be drained back to the bulk distribution. As a result, no local equilibrium can be established and the system enters various sorts of unstable regimes, some of which are of great relevance to fusion and astrophysical plasmas and other states of matter typically governed by long-range microscopic interactions.

2.4.4.4 *Long-range potentials*

An important example of strongly non-hydrodynamic conditions is provided by long-range potentials, such as r^{-1} unscreened Coulomb electrostatics, or gravitation, formally corresponding to α.

For such potentials, the calculations provide a *divergent* cross section, due to the unbounded accumulation of many small-angle deflections (*grazing* collisions).

This is not surprising: the mean-free path is virtually zero, because owing to the infinitely long range of the acting force, the molecules are constantly interacting and the accumulation of very numerous small deflections leads to a logarithmic divergence of the cross section.

In practice, such infrared divergence is regulated by imposing a long-range cut off, typically via a so-called Debye screening, after the Dutch chemist Peter Debye (1884 – 1966):

$$\frac{1}{r} \rightarrow \frac{e^{-r/\lambda}}{r} \tag{2.39}$$

The Debye length, λ, marks the scale above which electrostatic interactions are screened out due to polarization effects, a condition typical of quasi-neutral plasmas, composed by a mixture of oppositely charged species, say ions and electrons.

The kinetic theory of such screened systems is described by a different collision operator, due to Landau and Balescu–Lenard, after the Hungarian–Belgian physicist Radu Balescu (1932–96) and the German, Philipp Lenard (one "n" only, not to be confused with Lennard-Jones!) (1862–1947) (5).

This takes the following form:

$$C_{LBL} = \partial_{\vec{v}} \cdot \int \vec{\vec{B}}(\vec{v}, \vec{v}') \cdot \left[\frac{\partial}{\partial \vec{v}} - \frac{\partial}{\partial \vec{v}'} \right] f(\vec{v}) f(\vec{v}') \ d\vec{v} d\vec{v}' \tag{2.40}$$

where $\vec{\vec{B}}(\vec{v}, \vec{v}')$ is a suitable-tensorial collision kernel.

This expression is obtained from the Boltzmann collision operator by expanding upon the velocity change, $\Delta \vec{v} = \vec{v}' - \vec{v}$, under the assumption of small deflections:

$$|\Delta \vec{v}| \ll v_T \tag{2.41}$$

as it is appropriate for soft-core grazing collisions.

The Balescu–Lenard collision operator belongs to the general class of Fokker–Planck kinetic equations, which we shall discuss in chapter 9. For the case of unscreened long-range interactions, say self-gravitating systems, the derivation of a suitable collision operator is still an open issue in modern statistical mechanics, with important implications in plasma physics, astrophysics, and cosmology.

2.4.5 Molecular Chaos (Stosszahlansatz)

Having discussed the details of the two-body scattering problem inherent to Boltzmann's collision operator, we next move on to consider the all-important *statistical* aspects of this operator.

In the first place, in order to derive a closed equation, one has to express the two-body distributions f'_{12} and f_{12}, in terms of the one-body ones f_1 and f_2.

The simplest such closure, which is precisely the one taken by Boltzmann reads as follows:

$$f_{12} = f_1 f_2 \equiv f(\vec{z}_1; t) f(\vec{z}_2; t) \tag{2.42}$$

and same for $f'_{12} \equiv f_{1'2'}$.

This closure is tantamount to assuming no correlations between molecules entering a collision (*molecular chaos or Stosszahlansatz*).

This assumption is fairly plausible for a dilute gas with short-range interactions, in which molecules spend most of their lifetime traveling in free space, only to meet occasionally for very short lived, in fact instantaneous, interactions.

Note that molecules are assumed to be correlated only *prior* to the collision, whereas after collision, they become strongly correlated on account of mass, momentum and energy conservation.

Within this picture, the probability for two molecules that met at time t, to meet again at some subsequent time $t + \tau$, with the *same* velocities $\vec{v}_1$ and $\vec{v}_2$, decays exponentially with τ.

More precisely, this probability scales like $e^{-\tau/\tau_{int}}$ where τ_{int} is the duration of a collisional event. Since in Boltzmann's theory $\tau_{int} \sim s/v_T$ (the thermal speed v_T is taken to be a typical particle speed and s a typical effective molecular diameter) is negligibly small, so is the (auto)correlation function at time τ.

The situation is obviously completely different in a liquid, where, due to the much higher density, the molecules are in constant interaction.

Violations of Boltzmann's molecular chaos can occur due to the onset of nonlinear correlations. A most notable example are the famous long-time tails, first detected by Alder and Wainwright (6), where molecular correlations exhibit anomalous persistence due to self-sustained vortices generated by the molecular motion itself.[4]

Summarizing, in view of the molecular chaos assumption, the Boltzmann equation takes the following form:

$$\partial_t f + \vec{v} \cdot \partial_{\vec{x}} f + \frac{\vec{F}^{ext}}{m} \cdot \partial_{\vec{v}} f = \int (f_{1'} f_{2'} - f_1 f_2) v_r \sigma (v_r, \vec{\Omega}) d\vec{\Omega} d\vec{v}_2 \tag{2.43}$$

The left-hand side is a mirror of reversible Newtonian single-particle dynamics, while the right-hand side describes intermolecular interactions, under the Stosszahlansatz approximation.

2.5   Local and Global Equilibria

Given that the collision operator naturally splits into a Gain minus Loss components, it is only natural to ask under what conditions would the two antagonists come to an exact balance.

This singles out a very special distribution function, characterizing the attainment of *local equilibrium*, a notion which proves central to the purpose of deriving hydrodynamic equations from Boltzmann's kinetic theory.

Mathematically, the local equilibrium is defined by the condition

$$C(f^e, f^e) \equiv \mathcal{G}(f^e, f^e) - \mathcal{L}(f^e, f^e) = 0 \tag{2.44}$$

where superscript "e" denotes "equilibrium."

---

[4] This is amazingly reminiscent of the mechanisms invoked by Aristoteles to explain the motion of arrows in air!

The identification of Gain and Loss follow straight from the expression of the collision operator:

$$\mathcal{G} \equiv \int f_{1'}f_{2'}v_r\sigma\left(v_r\right)d\vec{v}_2 \tag{2.45}$$

$$\mathcal{L} \equiv \int f_1 f_2 v_r\sigma\left(v_r\right)d\vec{v}_2. \tag{2.46}$$

This leads to the so-called *detailed balance* condition:

$$f_1'f_2' = f_1 f_2 \tag{2.47}$$

which holds *regardless of the details of the molecular interactions*.

This is a strong statement of universality: microscopic details affect the *rate* at which local equilibrium is reached, not the equilibrium itself, which depends only on conserved quantities.

Of course, detailed balance does by no means imply that molecules sit idle, but rather that any direct (inverse) collision is dynamically balanced by an inverse (direct) partner collision.

For instance, in a room at standard temperature, with no appreciable macroscopic flow, the typical molecule moves at the speed of sound, that is, about three times faster than a Ferrari!

For a fluid at rest, along any given spatial direction, there is, on average, another molecule doing exactly the same along the opposite direction, so that no net macroscopic flow results.

The detailed balance condition has far-reaching consequences on the shape of the equilibrium distribution in velocity space.

To appreciate this, let us first take the logarithm of the eqn (2.47), to obtain

$$log\,f_1' + log\,f_2' = log\,f_1 + log\,f_2 \tag{2.48}$$

This shows that the quantity $log\,f$ is an *additive collision invariant*, i.e., a microscopic additive property which does not change under the effect of collisions.

The immediate consequence is that, at thermodynamic equilibrium, $log\,f$ must be a function of the five collision invariants

$$\mathcal{I}(v) \equiv \{1, m\vec{v}, mv^2/2\}$$

associated with the conservation of number (mass), momentum and energy.

This yields (repeated Latin indices, denoting spatial directions, are summed upon):

$$log\,f^e(\vec{x}, \vec{v}; t) = A(\vec{x}; t) + B_a(\vec{x}; t)v_a + C(\vec{x}; t)\frac{v^2}{2} \tag{2.49}$$

where A, B_a, C are five Lagrangian multipliers, carrying the entire dependence on the space-time coordinates through the conjugate hydrodynamic fields

$$\mathcal{H} \equiv \rho(\vec{x}, t), \rho(\vec{x}, t) u_a(\vec{x}, t), \rho(\vec{x}, t) e(\vec{x}, t)$$

namely density of mass, momentum and energy.

These Lagrangian parameters can be computed by imposing conservation of mass-momentum energy:

$$\int f^e \left\{ m, m v_a \frac{m v^2}{2} \right\} d\vec{v} = \{\rho, \rho u_a, \rho e\} \tag{2.50}$$

where, $\rho = nm$ is the mass density, u_a, $a = x, y, z$, is the macroscopic flow speed and ρe is the energy density.[5]

Elementary quadrature of Gaussian integrals delivers the celebrated Maxwell–Boltzmann equilibrium distribution.

In D spatial dimensions, this reads as follows:

$$f^e = \frac{n}{(2\pi v_T^2)^{D/2}} e^{-c^2/2v_T^2} \tag{2.51}$$

where c is the magnitude of the *peculiar speed*

$$\vec{c} = \vec{v} - \vec{u} \tag{2.52}$$

namely the relative speed of the molecules with respect to the fluid, and

$$v_T = \sqrt{\frac{k_B T}{m}} \tag{2.53}$$

is the thermal speed associated with the fluid temperature T, k_B being the Boltzmann constant. With this definition, each direction carries $k_B T/2$ units of energy.

2.5.1 Local Equilibria and Equation of State

Local equilibria associate with perfect (or inviscid) fluids, i.e., fluids in which dissipative effects can be neglected.

At equilibrium, density, temperature and pressure are related through an equation of state:

$$p = p(\rho, T) \tag{2.54}$$

---

[5] The Latin subscript a denotes cartesian components of a vector, so that the notation v_a is an equivalent substitute for $\vec{v}$ and shall be used interchangeably throughout the text, see Appendix on Notation.

From a kinetic point of view, the temperature is defined as the variance of the equilibrium distribution:

$$\rho D \frac{k_B T}{2} = \int f(\vec{x}, \vec{v}; t) \frac{mc^2}{2} d\vec{v} \tag{2.55}$$

where c is the magnitude of the peculiar velocity and $\rho = nm$ is the mass density.

The definition shows that temperature is basically the variance of the kinetic distribution function, or, in different terms, the peculiar kinetic energy of the molecules with respect to the mean-flow motion (see Figs. 2.8 and 2.9).

Another quantity of chief interest for hydrodynamics is the momentum-flux tensor, often called pressure tensor for short:

$$P_{ab} = m \int f v_a v_b d\vec{v} \tag{2.56}$$

The definition indicates that the component P_{ab} of the pressure tensor represents the amount of momentum $m v_a$ along direction x_a, fluxing across the unit surface with normal oriented along direction x_b.

The ordinary pressure is given by the diagonal components of the pressure tensor, evaluated at zero-flow conditions $u_a = 0$.

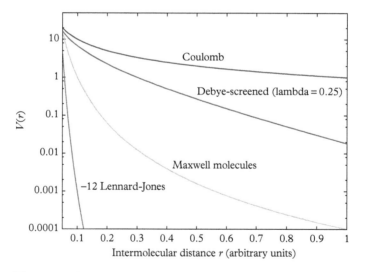

Figure 2.8 *From long-range to short-range potentials: top, repulsive bare Coulomb (1/r), Debye-screened Coulomb ($\frac{e^{-r/\lambda}}{r}$), Maxwell molecules ($1/r^4$) and the repulsive branch of the 6–12 Lennard-Jones. The prefactors have been adjusted to keep all four potentials on the same scale.*

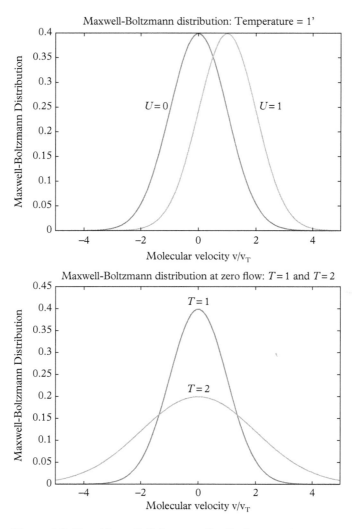

Figure 2.9 *Top: Maxwell–Boltzmann distribution at unit tempera-ture, $T = 1$, at rest, $U = 0$, and with flow, $U = 1$. Bottom: Maxwell–Boltzmann distribution at rest with $T = 1$ and $T = 2$.*

For an isotropic fluid at rest ($u_a = 0$), each component gives the same result, namely

$$p = P_{xx} = P_{yy} = P_{yy} \tag{2.57}$$

It is now instructive to compute the value of the macroscopic quantities corresponding to the local equilibrium distribution.

Elementary gaussian integration yields

$$\rho^e = \rho, \quad u_a^e = u_a, \quad T^e = T, \quad P_{ab}^e = \rho v_T^2 \delta_{ab} \tag{2.58}$$

This shows that local equilibrium only supports diagonal components of the pressure tensor, with a corresponding ideal equation of state

$$p = \rho v_T^2 = n k_B T$$

It is also of interest to note that, under such conditions, the thermal speed corresponds exactly to the sound speed, defined as the derivative of the pressure with respect to the density at a constant temperature, namely

$$c_s^2 = \frac{\partial p}{\partial \rho}\Big|_T. \tag{2.59}$$

Knowledge of the ratio of thermal to sound speed, $\theta \equiv \frac{v_T}{c_s}$, as a function of density and temperature is just another way of specifying the equation of state of the fluid, $\theta = 1$ denoting the ideal gas.

2.5.2   The Evershifting Battle

As shown in Section 2.5.1, local Maxwell equilibria are the result of a statistical balance between forward and backward collisions.

This balance between gain and loss terms annihilates the effect of the collision operator on the distribution function.

Quite remarkably, such balance holds true independently of whether or not the macroscopic fields exhibit a variation in space and/or time (whence the label "local"), as long as such variation occurs on scales longer than the mean-free path.

This property stems directly from the assumption that collisions take place in the limit $s \to 0$, or, more precisely, $s/l_\mu \to 0$.

A natural, and indeed often-asked question is:

Do local equilibria annihilate the effects of the streaming operator too?

A moment's thought reveals that this is *not* the case, unless the macroscopic fields are totally flat, i.e., constant in space and time, a condition which defines *global* equilibria, hence equilibrium thermodynamics.

To appreciate the point in a little more detail, let us compute the effect of the streaming operator on local equilibria.

A simple application of the chain rule yields

$$\frac{df^e}{dt} = \frac{\partial f^e}{\partial \rho}\frac{d\rho}{dt} + \frac{\partial f^e}{\partial \vec{u}} \cdot \frac{d\vec{u}}{dt} + \frac{\partial f^e}{\partial T}\frac{dT}{dt}. \tag{2.60}$$

By evaluating the partial derivatives of f^e with respect to ρ, $\vec{u}$ and T, simple algebra delivers

$$\frac{d\,\log f^e}{dt} = \frac{d\,\log\rho}{dt} + \vec{\xi} \cdot \frac{d\vec{u}}{dt} + \left(1 - \frac{\xi^2}{2}\right)\frac{d\,\log T}{dt} \tag{2.61}$$

where

$$\vec{\xi} \equiv \frac{\vec{v} - \vec{u}}{v_T} \tag{2.62}$$

is the peculiar speed in units of the thermal speed.

Expression (2.62) clearly shows that, by construction, local equilibria are not pre-served upon streaming, i.e., they can only be conserved if all macroscopic fields are constant in space and time, which, by definition means they are no longer local, but global ones.

Of course, this is all but a coincidence. The broken-invariance of local equilibria upon streaming, reflects a profound physical mechanism: *space-time inhomogeneity is the source of non-equilibrium*.

Differently restated, collisions act so as to achieve detailed balance, thereby couching the distribution function into the universal local Maxwell–Boltzmann distribution.

Streaming, on the other hand, works exactly in the opposite direction; it destroys the delicate (detailed) balance established by collisions, and revives non-equilibrium through inhomogeneity.

This is the famous "evershifting battle" between equilibrium and non-equilibrium, as evoked by Boltzmann, a battle which is not over until a featureless uniform macroscopic scenario is attained.

For those versed in philosophical aspects of science, the evershifting battle between equilibrium and non-equilibrium can be seen as a sort of metaphor of Life itself, which depends crucially on the ability to function far from equilibrium in local and temporary elusion of the Second Law ("life on borrowed time").

It is sometimes heard that the depth of any given equation is measured by the con-ceptual distance between its left- and right-hand side. If such criterion is anything to go by, there is little doubt that the Boltzmann equation scores very highly indeed.

Back to the ground. Broken invariance of local equilibria under streaming reflects the broken symmetries of classical mechanics. To this purpose, we note that Maxwellian equilibria inherit two basic symmetries of Newtonian mechanics, namely (7):

— *Space and time translational invariance*

$$\begin{cases} x'_a = x_a - \lambda_a \\ t' = t - \tau \end{cases} \tag{2.63}$$

where τ and λ_a are arbitrary constants.

The invariance under such transformation reflects the homogeneity of space and time. For the case where $\lambda_a = V_a \tau$, V_a being a constant velocity, (2.63) reflect invariance under Galilean transformations.

— *Rotational invariance*

$$x'_a = \sum_{b=1}^{D} R_{ab} x_b \tag{2.64}$$

where R_{ab} is a symmetric, unitary (norm-preserving) rotation matrix.

Rotational invariance, which applies to the particle velocities as well, is ensured by the fact that the peculiar speed $\vec{c}$ appears through its magnitude alone, so that any sense of preferential direction is erased.

These symmetries are built-in in continuum kinetic theory, but it would be a gross mistake to take them for granted also when space time and velocity are made discrete.

Actually, this is the leading theme of Discrete Kinetic Theory.

The hydrodynamic probe of rotational symmetry is the *momentum-flux tensor* P_{ab}, which plays a pivotal role in Discrete Kinetic Theory and most notably in Lattice Gas Cellular Automata and Lattice Boltzmann theories.

2.6   Summary

Summarizing, Boltzmann kinetic theory describes the dynamics of dilute gases in terms of a probability distribution function including, besides space and time. molecular velocities. The result is a complicated quadratic integro-differential equation describing the competition between free streaming and interparticle collisions. For sufficiently well-behaved (short-ranged) atomistic potentials, such competition ultimately ends up into a universal local equilibrium, which depends only on the local conserved fluid quantities, mass, momentum and energy. This local equilibrium plays a crucial role in the derivation of hydrodynamics from Boltzmann's kinetic theory.

REFERENCES

1. L. Boltzmann, *Lectures on Gas Theory*, University of California Press, California, 1964.
2. C. Cercignani, *Theory and Application of the Boltzmann Equation*, Elsevier, New York, 1975.
3. C. Cercignani, *Mathematical Methods in Kinetic Theory*, Plenum Press, New York, 1969.
4. K. Huang, *Statistical Mechanics*, Wiley, New York, 1987.
5. R. Balescu, *Equilibrium and non equilibrium statistical mechanics*, John Wiley and sons, New York, 1975.
6. B. Alder and J. Wainwright, Decay of the velocity autocorrelation function, *Phys. Rev.*, *A* 1, 18, 1970.
7. S. Goldstein, *Classical Mechanics*, Addison–Wesley, London, 1959.

....

EXERCISES

1. Prove the relation (2.15).
2. Prove the Maxwellian expression (2.51).
3. What fraction of molecules move faster than $2v_T$ in a local Maxwellian? And how many at $5v_T$?

3

Approach to Equilibrium, the *H*-Theorem and Irreversibility

Like most of the greatest equations in science, the Boltzmann equation is not only beautiful but also generous. Indeed, it delivers a great deal of information without imposing a detailed knowledge of its solutions. In fact, Boltzmann himself derived most, if not all, of his main results without ever showing that his equation did admit rigorous solutions. In this chapter, we shall illustrate one of the most profound contributions of Boltzmann, namely the famous *H*-theorem, providing the first quantitative bridge between the irreversible evolution of the macroscopic world and the reversible laws of the underlying microdynamics.

(We cannot stop the arrow of time, but perhaps we can try to flight with it)

It is amazing to observe how much information can be extracted from a great equation, without even knowing whether its solution(s) exist in the first place, but simply assuming they do. In fact, one might argue that the greatness of an equation is to a large extent measured precisely by its *generosity*, i.e., the amount of knowledge it delivers with no requirements of explicit knowledge of its solution. There is hardly any doubt that the Boltzmann equation is one of the greatest equations of all time; less known than the famous Einstein's $E = mc^2$, or Schrödinger's wave equation, its practical and philosophical implications are by no means any less. The Boltzmann equation tells us about profound and universal ideas, which concern all of us, such as the connection between the invisible micro and the tangible macro, a story intimately related to the nature of time and the way macrosystems evolve toward their equilibrium states, if any. The main import of the *H*-theorem is to place such deep questions onto a systematic mathematical basis.

The Lattice Boltzmann Equation. Sauro Succi, Oxford University Press (2018).
© Sauro Succi. DOI: 10.1093/oso/9780199592357.001.0001

3.1 Approach to Equilibrium: the Second Principle of Thermodynamics

Heat flows from hot to cold bodies, we all know this. The question is why? At skim value, one could simply observe that if it were otherwise, hot bodies would become increasingly hot, and cold bodies increasingly cold, so that no equilibrium could ever be attained. Thus, heat flow from hot to cold bodies is basically an instance of stability. Note that not all systems behave like this, the financial market being a much debated example in point these days. The way heat moves in space and time (Thermodynamics) was a major concern to nineteenth-century scientists, primarily Nicolas Sadi Carnot (1796–1832) in France and Rudolf Clausius in Germany (1822–88).

These scientists were eager to understand how to possibly optimize the "moving force of heat," i.e., the amazing ability of heat to convert into useful mechanical work (those were the heydays of the industrial revolution). In the course of such studies, they realized that while the heat exchanged between two bodies, name it δQ, depends on the specific process in point, say constant pressure, or constant volume, the *ratio* $\delta Q/T$ is process independent, i.e., depends only on the initial and final states. In the above, T denotes the temperature at which the heat exchange takes place. Such a ratio defines the change of entropy in the process, i.e., $dS = \delta Q/T$, S being the entropy of the system. The symbol dS stands for a true differential, one which depends only on the initial and final states, as opposed to δ, which denotes a process-dependent change. The inquisitive reader might have noticed that we are using a common temperature T for the hot and cold bodies, which formally contradicts the very notion of hot and cold in the first place. The point is that the heat exchange is supposed to take place under conditions where the difference between the two temperatures is much smaller than any of the two. Having clarified the point, the change of entropy in a process where a quantity δQ of heat flows from the hot body at temperature T_H to the cold body at temperature T_C is simply given by the sum of the two, namely

$$\Delta S = -\frac{\delta Q}{T_H} + \frac{\delta Q}{T_C} \tag{3.1}$$

where the minus sign indicates that the hot body loses entropy, while the cold one gains it. The very fact that, by definition, $T_H > T_C$ implies that

$$\Delta S \geq 0 \tag{3.2}$$

In other words, entropy can only increase in the process, and stays the same only once the two bodies come to a common temperature, for instance $(T_H + T_C)/2$ (see Fig. 3.1).

This defines the condition of thermal equilibrium. The inequality (3.2) is the mathematically expression of the Second Law of Thermodynamics, allegedly one of the most universal and inescapable constraints of the natural world, as we know it. Entropy is inexorably growing in time, a literal aging process which ends up in the bleaky "thermal death" picture. Viewed from this side, it is easy to see why entropy was immediately connected with the most enigmatic concept of all, Time, and colorfully named the "arrow of time" by Sir Arthur Eddington (1882–1944).

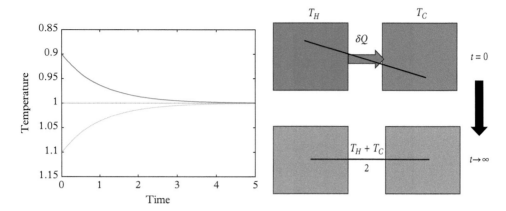

Figure 3.1 *Thermal equilibration: the hot body (upper curve) lowers its temperature by transferring heat to the cold body (lower curve). This process drives the system (hot body plus cold body) toward thermal equilibrium characterized by the same temperature for the two bodies. Time-reversal symmetry is clearly broken, the hallmark of irreversibility. The opposite process, spontaneous heat flow from cold to hot bodies, albeit not forbidden by Newton's laws, is not observed in the macroscopic world other than under the effect of thermal instabilities.*

It should be noted that the Second Principle makes no reference whatsoever to microscopic states of matter; as a matter of fact, in those days, the very existence of atoms was far from being accepted, a controversy which costed Boltzmann much anxiety and, according to some, might even have played a role in his tragic decision to take his life.

Leaving these fascinating matters to the historians of science, we next proceed to sketch the way that Boltzmann managed to lay down a mathematical bridge between the irreversible laws of Thermodynamics and the reversible equations of the underlying Newtonian mechanics.

3.2 Approach to Equilibrium, the *H*-Theorem

Thermal equilibria speak a language of dull uniformity, no room for heterogeneity. We have noted that at the level of *global* uniform equilibria, the fluid density ρ, the mean-flow speed $\vec{u}$ and temperature T must be constant throughout the physical domain. We have also noted that this restriction stems from the streaming operator, which has no effect once global uniformity is attained.

The collision operator, being local in space and time, does not place any such restriction on the spacetime dependence of the flow density, speed and temperature, with the only caveat that such dependence must occur on scales much longer than the molecular mean-free path. Whenever such slow spatial dependence is allowed in the local thermo-hydrodynamic fields, one speaks of *local* hydrodynamic equilibria, namely states which attain thermodynamic equilibrium only on a local scale. Thus, the system supports a variety of local equilibria, each with its own local values of density, velocity and temperature. On a longer timescale, global equilibration does then take place through the flux of mass, momentum and energy across the different regions of the system (transport phenomena).

This connects to the length scales l_μ, the particle *mean-free path*, the mean distance traveled by molecules between two subsequent collisions and l_M the typical scale of variation of macroscopic fields. Local equilibration takes places on a timescale $\tau_\mu \sim l_\mu/v_T$, whereas global equilibration requires much longer times, of the order of several $\tau_M = l_M/u_M$, u_M being a macroscopic velocity, or l_M^2/D, D being a typical diffusivity of the system. This identifies the *transport* regime in which hydrodynamic quantities diffuse and advect along macroscopic distances within the fluid domain. It is of course a regime of great practical relevance, since most real-life devices—to be of any use at all—must work away from thermodynamic equilibrium, and, sometimes, even very far from it.

Transport is naturally associated with the notion of *dissipation*, hence to the elusive concept of *irreversibility*, the subtle thread behind the Second Law of Thermodynamics. As hinted earlier on, one of the most profound contributions of Boltzmann to statistical mechanics rests with his discovery of a quantitative measure of irreversibility. We refer to the celebrated H-function and the attendant H-theorem.

Boltzmann showed that the following functional (H-function)

$$H(t) = -\int f(\vec{x}, \vec{v}; t) \log f(\vec{x}, \vec{v}; t) \ d\vec{v} \ d\vec{x} \tag{3.3}$$

is a monotonically increasing function of time, regardless of the underlying microscopic potential.

In equations

$$\frac{dH}{dt} \geq 0 \tag{3.4}$$

the equality sign holding at global equilibrium, when the evolutionary potential of the system is exhausted and the *entropy* of the system (basically H itself) is maximal.

The role of $H(t)$ as an evolutionary potential (the "time arrow") is adamant and its intellectual magic hardly escaped.

A full proof of the H-theorem can be found in most textbooks of kinetic theory. Therefore, for the mere sake of self-containedness, in the following we only sketch the main guidelines.

3.2.1 Sketch of the Proof of the *H*-Theorem

Let us begin by considering the force-free Boltzmann equation in the uniform case ($\nabla_x f_1 = 0$):

$$\partial_t f_1 = \int (f_1' f_2' - f_1 f_2) v_r \sigma \left(v_r, \vec{\Omega}\right) d\vec{\Omega} d\vec{v}_2 \tag{3.5}$$

where $f_1 \equiv f$ is the one-body distribution function.

Define a local H functional as

$$H(t) = -\int f_1 \log f_1 \, d\vec{v}_1 \tag{3.6}$$

where space dependence is removed because of the uniformity assumption.

Let us multiply both sides of eqn (3.5) by $1 + \log f_1$ and integrate upon $\vec{v}_1$. The left-hand side yields

$$\int (1 + \log f_1)\partial_t f_1 \, d\vec{v}_1 = \int \partial_t (f_1 \log f_1)\, d\vec{v}_1 \equiv -dH/dt$$

As a result, by equating the left- and right-hand side, we obtain

$$-\frac{dH}{dt} = \int (1 + \log f_1)(f_1' f_2' - f_1 f_2) v_r \sigma(v_r, \vec{\Omega}) d\vec{\Omega} d\vec{v}_1 d\vec{v}_2 \tag{3.7}$$

Since the previous expression does not change upon swapping the dummy subscripts $1 \leftrightarrow 2$, we can further write

$$-\frac{dH}{dt} = \int (1 + \log f_2)(f_2' f_1' - f_2 f_1) v_r \sigma(v_r, \vec{\Omega}) d\vec{\Omega} d\vec{v}_2 d\vec{v}_1 \tag{3.8}$$

By summing (3.7) and (3.8) and dividing by a factor 2:

$$-\frac{dH}{dt} = \frac{1}{2}\int [2 + \log(f_1 f_2)](f_2' f_1' - f_2 f_1) v_r \sigma(v_r, \vec{\Omega}) d\vec{\Omega} d\vec{v}_2 d\vec{v}_1 \tag{3.9}$$

By the same token, we can swap primed and unprimed quantities to obtain

$$-\frac{dH}{dt} = -\frac{1}{2}\int [2 + \log(f_1' f_2')](f_2' f_1' - f_2 f_1) v_r' \sigma(v_r', \vec{\Omega}) d\vec{\Omega} d\vec{v}'_2 d\vec{v}'_1 \tag{3.10}$$

Upon summing (3.10) and (3.9) and dividing again by 2, we finally obtain

$$-\frac{dH}{dt} = -\frac{1}{4}\int \log\left(\frac{f_1' f_2'}{f_1 f_2}\right)(f_2' f_1' - f_2 f_1) v_r \sigma(v_r, \vec{\Omega}) d\vec{\Omega} d\vec{v}_2 d\vec{v}_1 \tag{3.11}$$

where we have made use of the kinematic properties $d\vec{v}'_1 d\vec{v}'_2 = d\vec{v}_1 d\vec{v}_2$ and $v_r' = v_r$, whence $\sigma(v_r) = \sigma(v_r')$.

By calling $X \equiv f_1 f_2$ and $X' \equiv f_1' f_2'$, we next observe that the function $(X'-X)\log(X'/X)$ cannot be negative for any value of its arguments X and X' being both non-negative, since it is the product of two equally signed functions. In fact, this function attains its minimum, zero, at $X' = X$.

As a result, one finally concludes that

$$\frac{dH}{dt} \geq 0 \tag{3.12}$$

q.e.d. The identification of H with entropy S immediately delivers the second law of thermodynamics, namely

$$\frac{dS}{dt} \geq 0 \tag{3.13}$$

This proof relies upon a few ingenious tricks based on the symmetries of the Boltzmann equation, as combined with the kinematic properties of the two-body problem underlying the collision operator. Remarkably, it does *not* depend on the details of the molecular interactions, but only on the general conservation properties and symmetries of the Boltzmann equation.

The extension to the non-uniform case takes the form of a conservative equation for the local H-function $h(\vec{x}; t)$,

$$\partial_t h + \nabla \cdot \vec{\mathcal{J}}_h = S_h(\vec{x}; t) \tag{3.14}$$

where $\vec{\mathcal{J}}_h = -\int f \, log f \, \vec{v} d\vec{v}$ is the entropy flux and S_h is the entropy production due to the collisions.

Equation (3.14) shows that the local value $h(\vec{x}; t)$ has no definite sign, since the positive contribution of the collisions can always be outbalanced by the flux term. However, the second law is still recovered by integrating all over the spatial domain,

$$\frac{dH}{dt} = \int S_h(\vec{x}; t) d\vec{x} \geq 0 \tag{3.15}$$

The inequality (3.15) stems from the fact that the right-hand side is negative definite and the flux term does not contribute, since we have assumed zero flux at the boundaries of the spatial domain (isolated system).

3.2.2   *H*-theorem and Irreversibility

The H-theorem provides a formal basis for the emergence of macroscopic irreversibility: the streaming is reversible, i.e., invariant under time reversal, while collisions are not.

In mathematical terms, time reversal is defined by the *time-parity* (T) transformation:

$$t \rightarrow t' = -t \tag{3.16}$$

We can think of t and t' as of forward and backward times, respectively (see Fig. 3.2). Note that since only time is inverted, while space remains untouched, the velocities are also reversed under the T transformation:

$$\vec{v} \rightarrow \vec{v}' = -\vec{v} \tag{3.17}$$

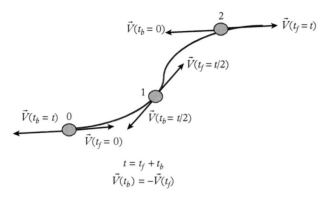

$$t = t_f + t_b$$
$$\vec{V}(t_b) = -\vec{V}(t_f)$$

Figure 3.2 *Geometrical representation of microscopic-time reversal. The time loop consists of a forward branch ($0 \rightarrow 1 \rightarrow 2$) and a backward one ($2 \rightarrow 1 \rightarrow 0$). Both forward and backward times lie in the range $[0, t]$ and each point along the trajectory carries both forward and backward time labels, with the constraint $t = t_f + t_b$. One might think of each trajectory as two-time sided. The time-parity transformation at $t_f = t \leftrightarrow t_b = 0$ implies the condition $\vec{V}(t_f) = -\vec{V}(t_b)$ along the entire trajectory. This is a formal statement of microscopic-time reversibility.*

The physical meaning of time-reversal transformation is quite transparent. By streaming a particle trajectory from $\vec{x}(0)$ at time $t = 0$ to $\vec{x}(t)$ at time t and then streaming back from t to $t - t = 0$, with inverted velocity $\vec{v}(t) \rightarrow -\vec{v}(t)$, the particle regains exactly its initial position $\vec{x}(0)$, with inverted velocity $-\vec{v}(0)$.

Exactly, here means literally exactly: no information is lost in the time loop: streaming is reversible, a mirror of Newtonian mechanics. Collisions, on the other hand, do break such symmetry, as they keep acting in the same direction, i.e., pull the local distribution function back to a local Maxwellian, regardless of whether the system is in the forward or backward branch of the time loop. The information lost in the loop is the one associated with the initial conditions.

The *H*-theorem provides an elegant and powerful formalization of the fundamental competition between reversible streaming (non-equilibrium) and irreversible collisions (equilibrium).

Yet, the *H*-theorem presents one of the most debated and controversial issues in the history of theoretical physics. We will not delve here into the details of the various paradoxes which were raised against the *H*-theorem as a bridge between micro- and macrodynamics. Beautiful accounts can be found in the literature (1; 2). Nor shall we comment on the fact that Boltzmann derived his theorem without demonstrating that his equation, a complicated integro-differential initial-value problem, does indeed have solutions in a rigorous mathematical sense.[6] While leaving rigor somehow behind, the H-theorem showed for the first time the way to the unification between two fundamental and previously disconnected domains of science: Mechanics and Thermodynamics.

Even though none of the original paradoxes raised against the H-theorem could stand the test of time, one should always keep in mind that Boltzmann's H-theorem was derived under very specific conditions, namely diluteness as discussed in Chapter 2. Such conditions rule out many important states of matter, such as dense fluids and liquids, not to mention more complex materials such as glasses, which exhibit anomalous (very long) relaxation to equilibrium. Even though the Boltzmann equation in its original form does not apply to the previous problems, suitable generalizations thereof (effective Boltzmann equations) are indeed capable of providing useful insights into the physics of complex states of matter out of equilibrium. This makes a very fascinating subject of modern non-equilibrium statistical mechanics, to which we shall turn considerable attention in the "Beyond Navier–Stokes" part of this book.

3.3 Collisionless Vlasov Equilibria

It should be mentioned that there exists a class of local equilibria which can be attained through purely reversible motion, i.e., in the total absence of collisional processes. These are called Vlasov equilibria, and play an important role in many contexts of statistical mechanics, such as astrophysics and plasma physics.

By definition, Vlasov equilibria annihilate the streaming operator:

$$\vec{v} \cdot \nabla_x f^{e,V} + \vec{a} \cdot \nabla_v f^{e,V} = 0 \tag{3.18}$$

Under the assumption that the force field derives from a potential, $\vec{F} \equiv m\vec{a} = -\nabla_x V(\vec{x})$, a simple quadrature yields the following separable solution:

$$f^{e,V}(\vec{x}, \vec{v}) = A n_0 e^{-\beta V(\vec{x})} e^{-\beta \frac{mv^2}{2}} \tag{3.19}$$

where $\beta = 1/k_B T$ and $A = (\beta/2\pi)^{D/2}$ is the normalization constant. As one can appreciate, Vlasov equilibria are a specific instance of the *canonical* distribution $Z^{-1}(\beta)e^{-\beta H}$, where $H = \frac{mv^2}{2} + V(\vec{x})$ is the Hamiltonian and Z the partition function.

Here, they are most conveniently regarded as global Maxwellian equilibria in velocity space, prefactored by a space-dependent density, the space dependence being dictated by the potential, $n(\vec{x}) = n_0 e^{-\beta V(\vec{x})}$.

These equilibria do not require any collisional relaxation and cannot exist in ordinary configuration and momentum space separately, but only as the result of a dynamic balance of the two in the double-dimensional phase space. The molecules flowing in the

---

[6] Rigorous proofs of the existence of solutions of the Boltzmann equation have made a rampant surge in modern mathematics, earning two Fields Medals, arguably the highest honor in mathematical research, to Pierre Louis Lions (1994) and Cedric Villani (2010).

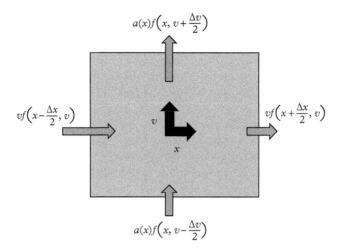

Figure 3.3 *Vlasov equilibria result from a flux balance in phase space. The flux gained(lost) along the spatial coordinate is lost(gained) along the velocity coordinate. This can only happen if the force derives from a potential.*

phase-space element centered about $(\vec{x}, \vec{v})$ via free-streaming in real space are exactly balanced by the molecules streaming away in velocity space under the effect of the external force (see Fig. 3.3). Note that such delicate balance can only occur if the force derives from a potential.

It should be noted that Vlasov equilibria are generally incompatible with local Boltzmann equilibria.

More precisely, they are compatible only for the velocities perpendicular to the flow velocity $\vec{u}$, as one can readily check by decomposing the molecular velocity into a longitudinal component $\vec{v}_l$, aligned with the local-flow velocity $\vec{u}$ and a transverse component $\vec{v}_t$, perpendicular to it. Upon such decomposition, one writes $(\vec{v} - \vec{u}) \cdot (\vec{v} - \vec{u}) = v_t^2 + u^2 + v_l^2 - 2v_l u$, which shows that the kinetic energy u^2 serves as a Vlasov potential for the transverse component of the equilibrium distribution function. Along the longitudinal direction, on the other hand, such splitting into a kinetic and potential energy is no longer possible. Vlasov equilibria play an important role in plasma physics and astrophysics where they offer the only mechanism to attain equilibria which would otherwise be impossible due to the paucity of collisions.

3.4   The Boltzmann Equation in Modern Mathematics

We shall soon focus mainly on the *practical* use of the Boltzmann equation, namely its capability to compute transport parameters, such as diffusivity, viscosity, and thermal conductivity, which characterize the approach to equilibrium on a macroscopic scale.

Before doing so, a few comments on more general issues related to the existence of solutions of the Boltzmann equations are in order.

The Boltzmann equation is not an easy piece of math: a nonlinear integro-differential equation living in 6 + 1 dimensions! No surprise that analytical solutions of such an equation do not abound in the literature, the few available being sort of precious flowers in the desert. Some of these precious flowers are listed in (3), as well as in (4).

In the sequel, we shall spend a few comments on a different sort of analytical work, the one which does not preoccupy itself so much with the identification of specific solutions of the Boltzmann equation, but rather with the general questions regarding its basic properties, such as existence, regularity and time-asymptotic convergence (5). These deep questions are the traditional hunting ground of pure mathematicians, with no implications of "impurity" for the applied ones (does such distinction make really sense?). Indeed the highest-caliber modern math has turned a close eye to the general properties of the Boltzmann's equation.

Forced by the author's lack of specific competence, here we only convey a few minimal ideas on this elegant and sophisticated topic, not because it would bear any direct connection to Lattice Boltzmann theory, but as a homage to its inherent beauty, in the hope of stimulating the math-minded reader to find perhaps some for the future.

The tool of the trade is *functional analysis*, i.e., the mathematics of objects inhabiting infinite-dimensional spaces. In such functional spaces, the distribution function f is most conveniently regarded as a vector (sometimes called *ray*) with an infinite number of components. This may sound like a very thin-air notion, but it is actually a most concrete and operational idea. Indeed, similarly to the way an ordinary vector in three-dimensional space is decomposed into three components along the x, y, z directions, a function in (suitable) infinite-dimensional spaces can be expanded as an infinite series:

$$f(v) = \sum_{n=1}^{\infty} f_n H_n(v) \tag{3.20}$$

In (3.20), each $H_n(v)$ represents a suitable "basis function" in Hilbert space, the analog of unit vectors in ordinary space, and f_n are the corresponding components of the "vector" $f(v)$ along the direction associated with $H_n(v)$. For a fully phase-space dependent $f(x, v; t)$, the basis functions remain the same and the space-time dependence is wholly picked up by the coefficients $f_n(x; t)$. It is as if at each spatial location x, one would attach a time-dependent infinite-dimensional vector.

In abstract notation, the Boltzmann equation reads as follows:

$$\partial_t f = \hat{B} f \tag{3.21}$$

where the Boltzmann evolution operator is formally given by

$$\hat{B} = \hat{C} - \hat{S} \tag{3.22}$$

In (3.22), $\hat{S}f \equiv \vec{v} \cdot \nabla_x f + \vec{a} \cdot \nabla_v f$ defines the streaming operator and $\hat{C}f \equiv C(f,f)$ stands for the collision operator.

The formal solution of (3.21) is

$$f_t = e^{\hat{B}t}f_0 \tag{3.23}$$

where f_0 is the initial condition and f_t is its image under the action of the Boltzmann *propagator*

$$\hat{P}_t = e^{\hat{B}t} \tag{3.24}$$

over a time span $[0, t]$.

Such propagator obeys the following commutative and additive relations, also known as *semi-group* properties:

$$\hat{P}_{t_1} \cdot \hat{P}_{t_2} = \hat{P}_{t_2} \cdot \hat{P}_{t_1} = \hat{P}_{t_1 + t_2} \tag{3.25}$$

for any $t_1, t_2 \geq 0$.

It also obeys $\hat{P}_0 = I$, I being the identity.

Note that t_1, t_2 are restricted to non-negative numbers. Indeed, the group is only *semi* because it does not obey the time-reversal property $\hat{P}_{-t}\hat{P}_t = I$, which would imply

$$\hat{P}_{-t} = \hat{P}_t^{-1} \tag{3.26}$$

This latter is a mathematical statement of reversibility; by rolling time backward, the image f_t maps exactly back into the starting point f_0, which implies the non-singularity of $\hat{P}_t$. For the case of irreversible evolution, the operator $\hat{P}_t$ cancels information, which cannot be retrieved by rolling time backward, whence its singularity. The "culprit" is, of course, the collision operator and the previous semigroup property is a formal echo of the *H*-theorem.

A major mathematical question concerns the main properties of the image f_t, say existence, finiteness, positivity, smoothness and so on, for any given class of initial conditions f_0.

As a first step in the rigorous study of such kind of questions, one has to define the proper functional space to work with. Such functional space is typically *metric*, i.e., equipped with a measure of the distance between its elements, for otherwise one could hardly address questions of convergence to any given limit.

In analogy with Euclidean space, a typical distance between two elements, say f and g, takes the form

$$d(f,g) = \left[\int_{R^3} |f(\vec{v}) - g(\vec{v})|^2 d\vec{v} \right]^{1/2} \tag{3.27}$$

This defines $||\phi||_2$, the norm of the displacement, $\phi \equiv f - g$, in Hilbert's space L_2 (the space of square-integrable functions).

In the case of kinetic theory, a more natural norm is Hilbert's L_1 defined by $||f||_1 = \int_{R^3} |f| d\vec{v}$. This space includes all distributions supporting a finite density.

More general Hilbert's spaces can be defined via the following norm:

$$||f||_s^p \equiv \int_{R^3} |f|^p (1 + v^2)^{s/2} d\vec{v} < \infty \tag{3.28}$$

This is statement of finiteness, not only for the density but also for higher-order quantities, such as the energy ($s = 2$).

To be noted that L_s^p gets narrower at increasing s.

Consider for instance the one-dimensional Lorentzian distribution:

$$f_L(v) = \frac{A}{1 + v^2}$$

where A is a normalization constant. Since $\int f_L(v) dv < \infty$ but $\int f_L v^2 dv$ is unbounded, f_L belongs to L_1^0 but not L_2^0 (see Fig. 3.4).

The parameter p plays a different role. For $p > 1$, raising p has the effect of enhancing the contrast between the maximum and minimum value attained by the function (sharpening). This emphasizes regions where f is substantial, typically the low-energy region with respect to the high-energy tails. The opposite is true for $p < 1$.

More complex spaces can be defined by involving not only the function f itself, but also its velocity gradients, providing a measure of *smoothness*:

$$H_s^p = \{ f : \int_{R^3} |\partial_v f|^p (1 + v^2)^{s/2} d\vec{v} < \infty \} \tag{3.29}$$

Of importance for entropic reasons is the so-called *LlogL* space

$$LlogL = \{ f : \int_{R^3} |f| d\vec{v} < \infty; \int_{R^3} |f| \log f d\vec{v} < \infty \} \tag{3.30}$$

The goal of an existence theorem is to show that an initial distribution which belongs to a given subspace with sufficient smoothness and regularity at time $t = 0$ remains within such subspace during its evolution under the effect of the Boltzmann equation.

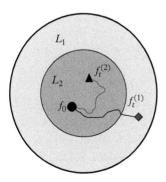

Figure 3.4 *Symbolic trajectory of the Boltzmann distribution in Hilbert space. Starting at time $t = 0$ in Hilbert space L_2, the distribution function f_0 may either remain within, $f_t^{(2)}$ ending in the triangle, or move outside it, $f_t^{(1)}$ ending in the diamond. Either way, in this example the distribution function remains within the larger Hilbert space L_1.*

Proving such property for short times and for small data, small departures of f_0 from a global Maxwellian is already a non-trivial task in itself. Proving the same for large data and long times, and particularly in the limit $t \to \infty$, is much more demanding. Much depends on the scattering potential, especially the use of regularized cutoffs, in both radial and angular directions.

Two important questions which remain (partially) unsolved are:

- *To what extent is smoothness preserved under Boltzmann evolution?*
- *How fast do solutions of the Boltzmann equation approach equilibrium?*

The first question, the regularity of the Boltzmann equation, is understood for small perturbations of equilibrium measures. There is a general framework of renormalized solutions, developed by R. DiPerna and P.L. Lions, but precise estimates of the solutions remain elusive. The second question, the decay to equilibrium for the Boltzmann equation, is historically related to the *Cercignanis's conjecture*, after the noted Italian mathematical physicist Carlo Cercignani (1939–2010), namely that, for a suitable class of potentials, entropy growth would obey the following inequality:

$$\dot{S} \geq K(S[f] - S[f_M]) \tag{3.31}$$

where K is a constant rate and f_M stands for a global Maxwellian. Clearly, such inequality implies exponential decay to global equilibrium, a very strong condition indeed. The conjecture was shown to be excessive by Cercignani himself, in joint work with Bobylev, and proven weakly wrong = nearly right, a few years later by Toscani and Villani.

In passing, we note that in an information-theoretic context the quantity $S[f] - S[f_M]$ represents the information lost in the process of converging to the global equilibrium.

For an accessible and yet scientifically informative account of these developments, see the recent book by Cedric Villani, Le Théorème Vivant, and Grasset (2012). For a rigorous mathematical discussion of the hydrodynamic limit of the Boltzmann equation, see the beautiful book by Laure Saint Raymond (6).

3.5   Summary

Summarizing, the Boltzmann equation provides a bridge between mechanics and thermodynamics, in the form of the celebrated H-theorem. Such a theorem expresses the inherent tendency of the system to loose memory of its initial conditions due to intermolecular interactions, a tendency which forms the microscopic basis for the one-sidedness of time evolution at a macroscopic level. Strictly speaking, the H-theorem is no theorem at all, since the existence of global, long-term solutions of the Boltzmann equation under general conditions still remains an open issue in the modern mathematics. This does not take a bit off the great conceptual value of the H-theorem as a paradigm of irreversibile behavior.

....

REFERENCES

1. C. Cercignani, "Ludwig Boltzmann: the man who trusted atoms", Oxford Univ. Press, Oxford, 1998.
2. G. Gallavotti, W. Reiter and J. Ynvgason editors, "Boltzmann legacy, ESI Series in Mathematical Physics, European Mathematical Society, Helsinki 2007.
3. S. Harris, "An introduction to the theory of the Boltzmann equation", Holt, Rinehart and Winston, NY, 1971.
4. C. Cercignani, "Exact solutions of the Boltzmann Equation, Modern Group Analysis: Advanced Analytical and Computational Methods" in *Mathematical Physics*, Springer, Berlin, 125–133, 1999.
5. C. Villani, "A review of mathematical topics in collisional kinetic theory", *Handbook of mathematical fluid dynamics*, Elsevier, Amsterdam, 2002.
6. L. Saint Raymond, "Hydrodynamic limit of the Boltzmann equation, Springer *Notes in Mathematics*, LNM, Berlin, 1971, 2009.

....

EXERCISES

1. Compute the entropy associated with a local Maxwellian at temperature T and velocity u along the x direction.
2. Compute the entropy associated with a global Maxwellian at rest and temperature T.
3. Show that the Euclidean distance fulfills the triangular inequality $d(x, y) \leq d(x, z) + d(z, x)$.

4

Transport Phenomena

In this chapter, we provide an elementary introduction to transport phenomena and discuss their intimate relation to non-equilibrium processes at the microscopic scale.

If you want to go fast, go alone. If you want to go far, go together.

(African saying)

4.1 Length Scales and Transport Phenomena

In Chapter 3, we discussed the notion of local and global equilibria and have shown that these equilibria represent the special forms taken by the distribution function once direct and inverse collisions come into balance.

In this chapter we focus on the dynamical mechanisms by which such equilibria are attained in time, namely transport phenomena.

Qualitatively, the approach to equilibrium is controlled by the timescales:

- $\tau_{\mathrm{m}} \sim s/v_T$: duration of a collisional event, the many-body scale.
- $\tau_\mu \sim l_\mu/v_T$: mean-flight time between collisions, single-body scale.
- $\tau_M \sim l_M/u_M, l_M^2/D$: macroscopic (advection and diffusion) timescales.

where we remind that s is the typical potential range (effective size of the molecule), l_μ the mean-free path and l_M is a characteristic macroscopic scale and u_M a typical macroscopic velocity (see Fig. 4.1).

The three time scales identify a corresponding hierarchy of dynamic relaxation regimes (see Fig. 4.2):

The Lattice Boltzmann Equation. Sauro Succi, Oxford University Press (2018).
© Sauro Succi. DOI: 10.1093/oso/9780199592357.001.0001

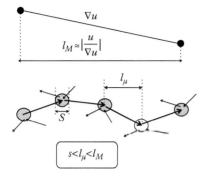

Figure 4.1 *The length scales involved in transport processes. For the sake of the example, the macroscopic variable is the fluid velocity. Many-body correlations die out within a length scale s, single-particle non-equilibrium lives on a scale l_μ and local equilibrium is attained on a scale l_M, taken several mean-free paths. Transport phenomena take places on scales l_M and above.*

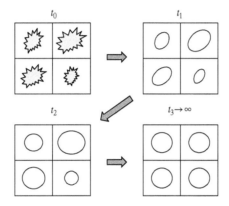

Figure 4.2 *Dynamical stages of the equilibration: f → local Maxwellian → global Maxwellian. The curves represent contours of a two-dimensional distribution in the plane $f(v_x, v_v; x, y)$ (four spatial cells for simplicity). At time $t = 0$ (top left) the contours are rough due to small-scale non-equilibrium fluctuations in the initial conditions. Such fluctuations are rapidly washed out by collisions, leaving only longer scale fluctuations driven by gradients of the hydrodynamic quantities, say shear (top right). Such gradients are also smoothed out by collisions on a longer time scale, leaving isotropic distributions, possibly at different temperatures in the four boxes (bottom left). This is of course a pedagogical simplification based on the assumption that temperature inhomogeneities leave longer than shear inhomogeneities. In the long term, all inhomogeneities are leveled off and the system attains a global uniform equilibrium (bottom right). The kingdom of the Second Principle of Thermodynamics.*

1. *Many-body regime* $(0 < t < \tau_m)$:

   Fast relaxation of the many-body to the single-particle distribution function: $f_{12\cdots N} \rightarrow f_1$ (the many-body distribution will be discussed in detail in chapter 7).

   This stage is particularly short for dilute systems, where multi-body collisions, sustaining many-body correlations, are exponentially rare. In the Boltzmann limit, it is literally zero.

2. *Kinetic regime* ($\tau_m < t < \tau_\mu$):

Relaxation to a local Maxwellian with density, flow speed and temperature varying smoothly on a mean-free path scale

$$f_1(\vec{v}, \vec{x}, t) \rightarrow f_M(\vec{\xi}(\vec{x}; t)) \tag{4.1}$$

where $\vec{\xi} \equiv \frac{\vec{v} - \vec{u}}{v_T}$ is the peculiar speed in units of the thermal speed $v_T \sim \sqrt{k_B T/m}$. Note that the distribution function f_M depends on space and time only through the macroscopic fields $\rho, \vec{u}, T$, which denotes the so-called *normal solutions*.

3. *Hydrodynamic regime* ($t_\mu < t < \tau_M$):

Slow drift of the *local* Maxwellian with spacetime-dependent flow speed and temperature to a *global* Maxwellian with constant speed and temperature:

$$f_M(\vec{\xi}(\vec{x}; t)) \rightarrow f_M(\vec{\xi}_0) \tag{4.2}$$

where the subscript 0 refers to constant values of density, flow speed and temperature.

Within Boltzmann theory, the first step is virtually erased since collisions are treated as instantaneous. The second step, local equilibration, is collision driven and attracts any initial distribution function toward a universal Maxwell–Boltzmann shape, whose space–time dependence is entirely carried by the hydrodynamic fields. Finally, the third step describes transport across the system on longer hydrodynamic scales of macroscopic interest.

4.2 Mean-Free Path (Again!)

We have noted before that the mean-free path, l_μ, stands out as the central length scale in kinetic theory. By definition, it represents the average distance travelled by a given molecule before colliding with another molecule (see Fig. 4.3). As mentioned in Chapter 3, the mean-free path is dictated by the underlying molecular interactions, more specifically by the cross section σ, according to

$$l_\mu = \frac{1}{n\sigma_T} \tag{4.3}$$

where n is the number density of the system and σ_T is the average cross section.

The term $v_r \sigma(v_r)$ in the Boltzmann collision integral is precisely the volume swept by the molecule per unit time. Therefore, the quantity

$$\gamma_c = n\langle v_r \sigma(v_r) \rangle = n v_T \sigma_T \tag{4.4}$$

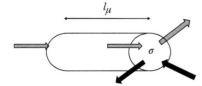

Figure 4.3 *The collisional cylinder, of area σ and height l_μ. In the Figure σ denotes the total cross section σ_T.*

provides an estimate of the average number of collisions per unit time (collision frequency). In (4.4), brackets denote averaging over the distribution function. Its reciprocal

$$\tau_\mu = \frac{1}{\gamma_c} \tag{4.5}$$

is the typical collision time, i.e., the average time lapse between two subsequent collisions.

For hard spheres, where σ is constant, the collision frequency scales like $v_r \sim T^{1/2}$, which is plausible, since thermal fluctuations enhance the probability for two molecules to get in near contact, i.e., within a distance D, D being the diameter of the sphere.

For Maxwell molecules $\sigma \sim 1/v_r$, and the collision frequency is a constant, independent of the temperature.

For Coulomb interactions $\sigma \sim v_r^{-4}$, which means that fast molecules collide less frequently than slow ones. This is a potentially unstable situation, since the friction experienced by molecules moving above a given threshold is no longer sufficient to stop their acceleration under the effect of, say, an external field. Such is the case for the so-called runaway electrons in plasmas, whose dynamics lies outside the hydrodynamic realm.

We have already noted that in a dilute gas, $\tilde{n} \equiv ns^3 \ll 1$, so that that $l_\mu > d$, where $d = n^{-1/3}$ is the mean intermolecular distance. This is reasonable, since the condition $l_\mu = d$ is realized only by molecules moving exactly along the direction connecting to their nearest molecule, which is in general not the case, since all directions in space are equally available, at least in an isotropic fluid.

As we shall see, this marks a key difference between continuum and lattice-kinetic theory.

The previous considerations identify the relevant hierarchy of length scales in the dilute gas limit, namely

$$s < d < l_\mu < l_M \tag{4.6}$$

The ratio s/l_μ defines the separation between the microscopic (atomistic or molecular) and the mesoscopic (kinetic) scales: many-body versus single-body physics. The ratio

$$Kn = \frac{l_\mu}{l_M} \tag{4.7}$$

also known as Knudsen number, defines the scale separation between the kinetic and the hydrodynamic scales. This number is crucial, as it controls the convergence of kinetic theory to hydrodynamics (see Fig. 4.4).

Some actual figures help putting this hierarchy in perspective. For air at standard conditions, $s \sim 3 \ 10^{-10}$, $d \sim 3 \ 10^{-9}$, $l_\mu \sim 10^{-7}$, all in meters, so that a clearcut scale-separation between the many-body, kinetic and hydrodynamic regime is indeed realized. For water, however, the corresponding values are approximately $s \sim 3 \ 10^{-10}$, $d \sim s$ and $l_\mu \sim d$, showing basically no scale-separation, as it is typical of the liquid state of matter.

As already observed, it is quite remarkable that the large-scale limit of both previous situations can be described by the same Navier–Stokes equations, only with different

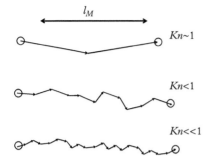

Figure 4.4 *Different Knudsen regimes: weakly coupled gas with Kn $\sim$ 1 (top), coupled fluid with Kn $<$ 1 (middle) strongly coupled fluid with Kn $\ll$ 1 (bottom).*

values of the transport coefficients. This universality, i.e., insensitivity to molecular individualism, stems directly from the conservation of mass, momentum and energy at a microscopic level.

4.3 Transport Parameters: Macroscopic Picture

According to the principles of Thermodynamics, a macroscopic system subject to spatial inhomogeneities supports a series of *transport phenomena*, whose role is to smooth out the inhomogenities, so as to drive the system toward the uniform state associated with thermodynamic equilibrium, in compliance with the Second Principle.

Such transport phenomena are vital to most natural and industrial phenomena involving fluid flows, Life itself, in the first place.

Transport phenomena may involve several physical quantities, but generally the most important ones concern the usual triad of microscopically conserved ones, namely mass-momentum-energy (MME). More precisely, the flux of these quantities, defined as the amount of MME transported across a given surface per unit time:

$$\Phi_Q = \frac{\Delta Q}{\Delta t} \tag{4.8}$$

where Φ_Q denotes the flux of the generic quantity Q across a given surface in a time lapse Δt. To compute these fluxes, we shall assume a simplified one-dimensional picture, whereby a generic molecule at position x in space, carries "on its back" the fluid quantity $Q(x)$, corresponding to the average value of the ensemble of molecules located around position x (for simplicity we shall drop the time dependence).

This value stays unchanged until the next collision occurs at $x + l_\mu$ (on average).

With reference to the planes located at $x^- \equiv x - l_\mu/2$, x and $x^+ \equiv x + l_\mu/2$, respectively, the net flux of Q crossing the plane at location x is given by the unbalance between the incoming fluxes from x^- and x^+, namely

$$\Phi_Q(x) = v_T \Delta y \Delta z [q(x^-) - q(x^+)] \tag{4.9}$$

where $q = nQ$ is the volumetric density of Q and we have taken v_T as a representative particle speed.

A Taylor expansion to first order of (4.9) expression delivers

$$\Phi_Q(x) = -v_T l_\mu \Delta y \Delta z \, \partial_x q(x) \tag{4.10}$$

The current, i.e., flux per unit area, is then given by

$$\mathcal{J}_Q(x) = -D \partial_x q(x) \tag{4.11}$$

where

$$D = v_T l_\mu \tag{4.12}$$

defines the so-called diffusion coefficient.

The expression (4.13) is commonly known as Fick's law: the current is linearly proportional to the gradient. It is often convenient to define a diffusion speed as

$$u_D = -D \frac{\partial_x q}{q} \tag{4.13}$$

which is clearly non-zero only as long as the system is non-uniform.

A number of considerations are in order.

First, we note that expression (4.13) holds only close to equilibrium, for otherwise, there would be no reason to truncate the Taylor expansion of $q(\vec{x})$ to first order, nor to assume v_T as a typical particle speed, because the local distribution function would be far from a local Maxwellian.

Thus, at least in principle, Fick's linear relation between fluxes and density gradients only applies near local equilibrium, where gradients are sufficiently small. This is precisely the realm of continuum hydrodynamics described by the Navier–Stokes equations.

Second, the minus sign in front of the gradient, indicates that a positive diffusivity drives the system toward uniformity by moving Q from Q-rich regions to Q-poor ones. This equalizing, "Robin Hood"-like, mechanism lies at the heart of thermodynamic stability. Indeed, negative diffusivities would promote unstable "Wall-Street"-like scenarios: rich, richer and poor, poorer.

In terms of spatial organization, segregation instead of mixing and resulting structure formation effects. Instabilities are a commonplace in natural systems *far* from equilibrium, where the current $\mathcal{J}$ receives dominant contributions from nonlinear and nonlocal terms beyond Fick's law. Such instabilities usually saturate because, beyond a given level of disparity between rich and poor, further growth is no longer sustainable.

Failing such self-control mechanism, the system may eventually transit from *evolutionary* to *revolutionary* changes (phase-transitions), leading to qualitatively new regimes, such as gas to liquid, liquid to solid and so on.

Third, for the sake of simplicity, we have illustrated the case of isotropic fluids, for which diffusivity is a simple scalar. For general systems, however, Fick's law acquires a tensorial character,

$$\mathcal{J}_x = D_{xx} \partial_x n + D_{xy} \partial_y n + D_{xz} \partial_z n$$

and similar for other directions.

A typical case in point are anisotropic fluids, with different diffusivity along different spatial directions, $D_{xx} \neq D_{yy} \neq D_{zz}$, like for instance in plasmas and magnetic fluids.

In the most general case, non-zero off-diagonal diffusivity may also occur, implying that gradients along a given direction can generate diffusive motion along a different one.

The tensorial structure of the diffusivity reflects the nature of the media hosting the motion of the molecules, and consequently it provides a very valuable source of information on the material properties of such media.

The previous consideration hold for any generic macroscopic property $q(\vec{x}, t) = \int f(\vec{x}, \vec{v}, t) Q_{micro}(\vec{v}) d\vec{v}$. In the following we shall discuss the explicit examples of mass, momentum and energy diffusivity, corresponding to the five microscopic invariants, $Q_{micro} = \{m, m\vec{v}, \frac{mv^2}{2}\}$, respectively.

4.3.1 Mass Diffusivity

Here we focus on mass diffusivity, hence we set $Q_{micro} = m$ and $q = nm = \rho$, the mass density. Consider a fluid at non-uniform density along, say, the x direction. As time unfolds, due to collisions, the molecules migrate from the high-density region to the low-density one.

The current density, i.e., the mass flow per unit time and area, is given by Fick's law:

$$\mathcal{J}_x = -D\partial_x\rho \tag{4.14}$$

where $D = v_T l_\mu$ is the mass diffusion coefficient and the minus sign indicates motion against the density gradient, from high to low density (see Fig. 4.5).

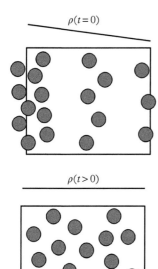

$\rho(t=0)$

$\rho(t>0)$

Figure 4.5 *Mass diffusivity: due to collisions, the molecules move from high- to low-density regions. The process comes to an end once uniform density is attained in the time-asymptotic limit.*

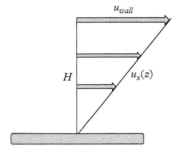

Figure 4.6 *Momentum diffusivity in Couette flow. The upper wall sets the near-wall fluid in motion and momentum diffuses from the upper layers to the underlying ones via intermolecular collisions. The non-slip constraint of zero velocity at the bottom layer results in a linear profile across the fluid. In the absence of momentum diffusivity, i.e., zero viscosity, the fluid would not move at all, due to lack of momentum transfer from the molecule bounced at the wall to the molecules inside.*

4.3.2 Momentum Diffusivity: Kinematic Viscosity

Consider a planar flow (Couette flow), driven by a force acting on the upper wall at a constant speed U_w, with a linear profile of the flow velocity:

$$u_x(z) = U_w z/H$$

where $z = 0$ and $z = H$ denote the lower- and upper-plane walls, respectively (see Fig. 4.6).

Momentum diffusivity is the basic mechanism by which the force exerted on the upper plate transmits across the fluid via molecular collisions.

More specifically, by the same argument used for the generic quantity $Q_{micro} = mv_x$, the flux of momentum (i.e., a force) along x, crossing a surface with normal along z, is given by ($\rho = nm$):

$$\Delta F_{xz} = \rho u_z \Delta x \Delta y [u_x(z^-) - u_x(z^+)] \tag{4.15}$$

The force along x transmitted per unit surface with normal along z is thus given by

$$P_{xz} = \frac{\Delta F_{xz}}{\Delta x \Delta y} = \rho u_z [u_x(z^-) - u_x(z^+)] \tag{4.16}$$

By Taylor expanding $u_x(z)$ to the first order, and replacing u_z with the typical thermal speed v_T, we obtain

$$P_{xz} = \rho v_T l_\mu \partial_z u_x \tag{4.17}$$

which is the force along x per unit area along z.

The transport coefficient measuring the momentum transport is called *dynamic-shear viscosity* and is given by

$$\mu = |\frac{P_{xz}}{\partial_z u_x}| = \rho v_T l_\mu \tag{4.18}$$

The same argument given here for transport of x momentum across the z direction readily extends to the general case of transport of momentum along the x_a direction across a surface with normal along x_b, $a, b = x, y, z$.

In general, the dynamic viscosity is a fourth-order tensor $\mu_{ab,cd}$, linking the *momentum-flux tensor* P_{ab}, to the *deformation tensor* $D_{cd} \equiv \nabla_c u_d$.

Note that the momentum-flux tensor has dimensions of energy per unit volume, while the deformation tensor is an inverse time. As a result, the dynamic viscosity is a measure of the rate of change of energy density due to dissipative effects.

4.3.3   Energy Diffusivity: Thermal Conductivity

By now, the reader needs no arm-twisting to accept that, by applying the same line of thinking to the kinetic energy, $Q_{micro} = mc^2/2$, c being the magnitude of the peculiar speed defined in chapter 2 one can derive a similar transport coefficient, known as *thermal conductivity* (see Fig. 4.7).

This is defined as the ratio of the energy flux to the temperature gradient, so that one can write this relation again in the form of Fick's law:

$$e_x = -\kappa \partial_x (n k_B T) \tag{4.19}$$

where e_x denotes the diffusive energy flux along direction x. The usual procedure yields

$$\kappa = v_T l_\mu \tag{4.20}$$

Since mass, momentum and energy are all transported the same way, i.e., "on the back" of single molecules, there is no reason why the diffusivity of mass, momentum and energy should be different. As we shall see, this is only true for ideal fluids, i.e., fluids where kinetic energy is the only form of active energy in the system.

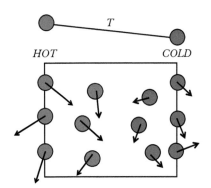

Figure 4.7 *Heat flows from hot to cold regions via collisions between fast and slow molecules.*

4.4 Transport Parameters: Kinetic Picture

The analysis of transport coefficients given in Section 4.3 makes use of basic kinetic quantities, the mean-free path and the thermal speed, without making any appeal to the formal apparatus of the Boltzmann equation. As a result, the mean-free path must be treated as an input parameter rather than as a quantity which can be derived from first principles.

In this section, we shall illustrate the kinetic foundations of transport phenomena and their intimate link with *non-equilibrium* effects at the microscopic level. From a kinetic-theory standpoint, transport processes arise as a direct consequence of the departures of the distribution function from local equilibrium. Leaving rigor behind, this basic fact can be quickly grasped through the following heuristic mathematical arguments.

Let us formally split the distribution function into a (local) equilibrium and non-equilibrium components, as follows:

$$f(\vec{x}, \vec{v}; t) = f^{eq}[M(\vec{x}; t), \vec{v}] + f^{neq}(\vec{x}, \vec{v}; t) \tag{4.21}$$

where M stands for the set of conserved-macroscopic fields.

Upon inserting this splitting into the Boltzmann's equation, in symbolic form, we obtain

$$\frac{df^{neq}}{dt} = C(f^{eq} + f^{neq}) - \frac{df^{eq}}{dt} \tag{4.22}$$

where d/dt is a shorthand for the streaming operator.

On the assumption $|f^{neq} \ll |f^{eq}|$, let us next expand the collision operator around f^{eq}, to obtain

$$\frac{df^{neq}}{dt} = \hat{L} f^{neq} - \frac{df^{eq}}{dt} \tag{4.23}$$

where $\hat{L}$ denotes the linearized collision operator:

$$\hat{L} \equiv \frac{\delta C}{\delta f}|_{f^{eq}} \tag{4.24}$$

The kernel of the linearized collision operator is given by the partial derivative

$$L(\vec{v}, \vec{v}') = \frac{\partial C(\vec{v})}{\partial f(\vec{v}')} \tag{4.25}$$

evaluated at $f = f^{eq}$.

This operator is central to linear transport theory, since its leading non-zero eigenvalues dictate the actual value of the transport coefficients of the fluid.

This immediately signals a major practical asset of the Boltzmann equation, namely the possibility of computing the transport coefficients from first-principles, i.e., from the underlying atomistic potentials, along the conceptual chain: from atomistic potentials to kinetic cross sections, to eigenvalues λ of the linearized collision operator:

$$V(r) \to \sigma \to \lambda \sim l_\mu$$

We already know that collisions always act in such a way as to erase the non-equilibrium component f^{neq}. Mathematically, this implies that all eigenvalues of $\hat{L}$ must be zero or negative. Zeros associate with the microscopic invariants which lie at the roots of inviscid hydrodynamics in the continuum limit. Negatives, on the other hand, control the dissipative structure of the emerging Navier–Stokes hydrodynamics.

Negative eigenvalues imply that (part of) the information contained in the initial distribution is irreversibly lost as the system evolves toward its local equilibrium. This loss of information is precisely the microscopic origin of dissipation.

The right-hand side of (4.23) also informs us that the non-equilibrium component does not die out completely, as long as the environment displays some form of heterogeneity.

This is because *the streaming operator does not annihilate the local equilibrium distribution,* but provides instead a source for the non-equilibrium component.

This is again a manifestation of Boltzmann's "Evershifting Battle," between equilibrium (collisions) and non-equilibrium (streaming), as already discussed.

Equation (4.23) further illustrates the point: the lifetime of non-equilibrium excitations, say τ_μ, is governed by the inverse eigenvalues of the linearized collision operator $\hat{L}$.

By pure dimensional arguments, the generic diffusivity D_Q of any physical quantity Q carried "on the back" of the molecules, is therefore expected to take the form

$$D_Q = l_{mu} v_T \tag{4.26}$$

This is the same expression derived by the macroscopic balance of the fluxes. However, the kinetic formulation permits, at least in principle, us to *compute* the mean-free path out from the eigenvalues of the linearized Boltzmann operator.

It also brings up a new conceptual angle to the discussion: *in a world where collisional relaxation could proceed instantaneously, i.e., infinite propagation speed of the interactions, there would be no dissipation.* This is the idealized world of perfect fluids described by the inviscid Euler equations.

Somehow counter-intuitively, perfect fluids correspond to infinitely strong interactions: molecules interact so strongly that they have no time to move any distance at all between two subsequent collisions! This stands in apparent contradiction with the fact that perfect fluids do not dissipate any momentum or energy.

The paradox is easily resolved by noting that, first, the collisions conserve MME and second, that the mean-free path defines the spatial horizon where dissipation takes place as an *emergent*-macroscopic effect. Zero mean-free path means the collapse of such region to a zero-sized volume, a literal and very subtle singularity of the Navier–Stokes equations. So subtle, in fact, that a rigorous understanding of the zero-viscosity limit of the Navier–Stokes equations still belongs to the list of the famous Clay's mathematical problems of the century (for the practically inclined, the solution is worth one million dollars . . .).

4.4.1   The Structure of Dissipation: Micro versus Macro

A final observation concerns the very different mathematical structure of the micro- and macroscopic faces of dissipation.

In the microscopic world, dissipation is inherently built-in within the H-theorem obeyed by the collision operator. Since collisions are zero ranged in the Boltzmann's picture, dissipation is completely local in space. As a result, space and time always come on the same footing, both first order, as reflected by the streaming operator.

At a macroscopic level, however, dissipation acts via diffusion, hence through second-order spatial derivatives. Such second-order spatial derivatives are indeed responsible for the breaking of spacetime reversal, $t \rightarrow -t$, $x \rightarrow -x$, the so-called TP (Time-Parity) invariance, hence introducing an arrow of time. As we shall detail in Chapter 5, such symmetry breaking is the direct result of the *enslaving* of non-conserved quantities, typically the momentum flux tensor, to their local equilibrium value.

Once again, the recurrent assumption of weak departure from local equilibrium.

4.5   Dimensionless Numbers

Mass, momentum and energy diffusion in fluids is characterized by two dimensionless numbers, the *Schmidt* number (momentum/mass)

$$Sc = \frac{\nu}{D} \tag{4.27}$$

and the *Prandtl* number (momentum/energy)

$$Pr = \frac{\nu}{\kappa} \tag{4.28}$$

As we have seen, for ideal gases $Sc = Pr = 1$, because mass, momentum and energy transport always takes places through the same mechanism, namely molecular diffusion, i.e., all quantities are carried "on the back" of molecules.

In liquids, potential energy plays a prominent role, and other mechanisms become available, which involve more orchestrated form of "collective" phenomena, typically leading to enhanced momentum and heat, versus mass transport, i.e., $Sc > 1$ and $Pr > 1$.

So much for diffusion.

As to advection, the relevant group for incompressible fluids is the *Reynolds* number:

$$Re = \frac{UL}{\nu} \tag{4.29}$$

where L and U are typical macroscopic length and fluid speed.

The analog quantities for mass and energy are readily derived via the Schmidt and Prandtl numbers. These are the *Peclet* number:

$$Pe = UL/D = Re\ Sc \tag{4.30}$$

and its thermal analog:

$$Pe_T = UL/\kappa = Re\ Pr \tag{4.31}$$

As discussed in this book, the Reynolds number measures advective versus diffusive transport. Indeed, *Re* can also be expressed as the ratio of the macroscopic fluid velocity U versus the diffusive one, defined as $u_D = \nu/L$. Based on the definition of the Knudsen number $Kn = l_\mu/L$ and recalling that $\nu \sim l_\mu v_T$, this also reads as $u_D = v_T\ Kn$. Since the Knudsen number is by definition very small ($Kn < 0.01$) for ordinary fluids, this shows that the diffusive velocity in a hydrodynamic flow is well below the mean velocity of single molecules.

This is yet another manifestation of molecular collisions, which prevent single molecules from following a directed trajectory over distances longer than the mean-free path, thus resulting in much slower collective motion than individual one.

This is also the reason why states of matter supporting more organized instances of collective motion, such as liquids, can host more efficient forms of transport.

4.6 Beyond Fick's Law: Non-Local and Nonlinear Transport

In this chapter we have kept the picture at its simplest, i.e., linear and local relation between the driving gradients and the resulting fluxes (Fickian transport).

Of course, things can get significantly more involved when the gradients are *not* small, and particularly when they are not small on the scale of the molecular mean-free path.

In this case, the gradient-flux relation takes the form of an integral expression

$$\mathcal{J}_a(x,t) = \int_0^t \int_x^{\prime} D_a b(x, x'; t, t') g_b(x', t') dx' dt'$$

where $D_{ab}(x, y)$ is the non-local diffusivity tensor, and g_b is a vector constructed with the density gradient and possibly higher powers thereof.

Sometimes, the flux-gradient relation is non-local and nonlinear at a time; for example capillary forces arising from surface tension effects in complex fluids interfaces, are typically expressed as the divergence of the Laplacian of the density and its gradient squared. We shall meet this kind of effects in the study of Lattice Boltzmann schemes for non-ideal fluids. Generally, non-local and non-linear flux-gradient constitutive relations point to the onset of physics beyond the hydrodynamic description.

4.7 Summary

Summarizing, the main take-home message of this chapter is that transport phenomena are intimately related to the departure of the Boltzmann distribution function from local equilibrium. Such departure is sustained by the system inhomogeneity against the equalizing action of intermolecular collisions. The magnitude of the transport coefficients scales with the lifetime of the non-equilibrium excitations, as well as with their thermal speed. In an idealized world with instantaneous collisions (zero mean-free path) and/or no thermal fluctuations (zero temperature), there would be no macroscopic dissipation and the chimera of *perpetuum mobile* would come true.

With the notable exception of exotic states of matter, such as superfluids, our ordinary world tells another story. The Boltzmann equation has an important role in this story, as its linearized version permits, in principle, to compute the transport coefficients from first principles, thereby illuminating the microscopc foundations of transport phenomena.

....

REFERENCES

1. Charles E. Hecht, "Statistical thermodynamics and kinetic theory", Dover unabridged, 1998.

....

EXERCISES

1. Estimate the value of the dimensionless density ns^3 in water and air.
2. Write down a non-local generalization of Fick's law.
3. Do repulsive interactions increase or decrease the speed of sound?

5

From Kinetic Theory to Navier–Stokes Hydrodynamics

In this chapter, we shall illustrate the derivation of the macroscopic fluid equations, starting from Boltzmann's kinetic theory. Two routes are presented, the heuristic derivation based on the enslaving of fast modes to slow ones, and the Hilbert–Chapman–Enskog procedure based on low-Knudsen number expansions (1; 2). Either ways, the assumption of weak departure from local equilibrium is crucial in recovering hydrodynamics as a large-scale limit of kinetic theory.

Παντα ρεῖ, "*Everything flows (No man steps in the same river twice, for it is not the same river and he's not the same man)*".

(Heraclitus)

5.1 From Kinetic Theory to Hydrodynamics: Heuristic Derivation

Kinetic theory lives in six-dimensional phase space, as opposed to the three-dimensional configuration space inhabited by fluids. As a result, in order to derive hydrodynamics from kinetic theory, the first step is to "project out" the information contained in the velocity space onto real-space macroscopic fields, also known as *kinetic moments*.

This generates an infinite sequence of partial differential equations, whose informed truncation leads to the Navier–Stokes equations of continuum hydrodynamics.

The heuristic derivation of hydrodynamic equations from Boltzmann's kinetic equation proceeds through two basic steps:

1. *Integration upon velocity space.*
2. *Closure, based on the assumption of weak departure from local equilibrium.*

The Lattice Boltzmann Equation. Sauro Succi, Oxford University Press (2018).
© Sauro Succi. DOI: 10.1093/oso/9780199592357.001.0001

The first step is purely mechanical, as it involves systematic integration upon the velocity space. The second step is an informed approximation based on the assumption of weak departure from local equilibrium.

The lowest order kinetic moments identify with well-known hydrodynamic quantities, such as density, flow speed, temperature and heat flux, at order zero to three, respectively.

Higher-order moments no longer enjoy such a direct physical interpretation, and in this book we shall often refer them as to *ghosts*, i.e., they must be there for structural consistency, but do not manage to emerge into the world of hydrodynamic observables. Eventually, however, they do take stage far from equilibrium.

Kinetic theory is equivalent to an infinite-dimensional field theory, since the number of kinetic moments of the distribution function $f(\vec{x}, \vec{v}; t)$ is virtually infinite. As a result, in order to obtain a close and finite formulation, such as hydrodynamics, some form of truncation is required. This is precisely the task accomplished in step 2 (see Fig. 5.1).

This qualitative discussion highlights that hydrodynamics is basically a low-dimensional asymptotic limit of the infinite-dimensional sequence of kinetic moments associated with the Boltzmann kinetic equation. The relevant smallness parameter is the Knudsen number, i.e., the scale separation between the mean-free path l_μ, and the typical macroscopic lengthscale l_M.

For the sake of simplicity, and with no loss of conceptual generality, we shall consider the force-free Boltzmann equation in single-relaxation Bhatnagar–Gross–Krook (BGK) form, namely (see Chapter 8):

$$\partial_t f + v_a \partial_a f = \omega(f^e - f) \tag{5.1}$$

where f^e is the local equilibrium and ω is a suitable relaxation frequency, assumed to be constant for simplicity. As usual, Latin indices run over spatial dimension and repetition stands for summation.

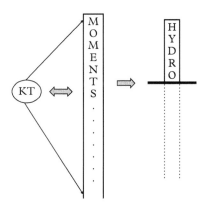

Figure 5.1 *The logical path from kinetic theory to hydrodynamics. The first step is a mere change of representation: the kinetic equation is equivalent to an infinite set of kinetic moments, which are exposed by suitable integration in velocity space (projection step, left). Such infinite hierarchy is then truncated to a finite set of low-order hydrodynamics moments based on the assumption of weak departure from local equilibrium (right). The latter responds to the principle of enslaving of high-order moments (lower lying in the figure, since they do not emerge to the macroscale) in the limit of small Knudsen numbers.*

5.1.1 Moment Balance Equations

Let us consider a generic microscopic quantity, $\phi = \phi(\vec{v})$, function of velocity space alone and define the corresponding macroscopic field as follows:

$$\Phi(\vec{x}, t) = \int \phi(\vec{v}) f(\vec{x}, \vec{v}; t) d\vec{v} \tag{5.2}$$

The macroscopic field is the average of $\phi(\vec{v})$ taken over the actual distribution function $f(\vec{x}, \vec{v}; t)$. More precisely, such an average is formally defined as

$$\langle \phi \rangle (\vec{x}; t) = \frac{\int f(\vec{x}, \vec{v}; t) \phi(\vec{v}) d\vec{v}}{\int f(\vec{x}, \vec{v}; t) d\vec{v}} \tag{5.3}$$

so that one writes $\Phi = n \langle \phi \rangle$, $n = n(\vec{x}; t)$ being the number density of the fluid. Essentially, the expression (5.2) probes in the portion of velocity space where the product $f\phi$ is most substantial and transfers this information to the macroscopic field $\Phi(\vec{x}; t)$. The latter inherits the space and time dependence from the distribution function.

Upon multiplying the BGK equation by $\phi(\vec{v})$ on both sides, and integrating upon the velocity variable, we obtain

$$\partial_t \Phi + \partial_a \Phi_a = \omega (\Phi^e - \Phi) \tag{5.4}$$

where Φ_a the current density of ϕ is defined as

$$\Phi_a(\vec{x}; t) = \int \phi(\vec{v}) v_a f(\vec{v}, \vec{x}; t) d\vec{v} \tag{5.5}$$

If ϕ is a collisional invariant, i.e., $\phi = \{1, v_a, v^2/2\}$, by definition $\Phi^e = \Phi$, so that the right-hand side is automatically zero and the field Φ obeys a standard continuity equation.

Otherwise, $\Phi(\vec{x}; t)$ relaxes to its local equilibrium $\Phi^e(\vec{x}; t)$ on a timescale $\tau = 1/\omega$.

Let us now specialize $\phi(\vec{v})$ to the four collisional invariants $\{1, v_x, v_y, v_z\}$.

For $\phi \equiv 1$, we obtain

$$\partial_t \rho + \partial_a \mathcal{J}_a = 0 \tag{5.6}$$

where we have defined the fluid density

$$\rho(\vec{x}; t) = m \int f(\vec{v}, \vec{x}; t) d\vec{v} \tag{5.7}$$

and fluid current density

$$\mathcal{J}_a(\vec{x}; t) = m \int v_a f(\vec{v}, \vec{x}; t) d\vec{v} \equiv \rho u_a \tag{5.8}$$

Note that (5.8) is also a definition of the flow-speed $u_a \equiv \mathcal{J}_a/\rho$ (the spacetime dependence is omitted for notational simplicity).

Next, we identify $\phi \equiv v_a$, with $a = x, y, z$, and integrating again upon velocity space we obtain

$$\partial_t \mathcal{J}_a + \partial_b P_{ab} = 0 \tag{5.9}$$

where we have defined

$$P_{ab}(\vec{x}; t) = m \int v_a v_b f(\vec{v}, \vec{x}; t) d\vec{v} \tag{5.10}$$

as the momentum-flux tensor.

So far, collisions have taken no part in the game, they are soon to make their entry, though.

Next, we move one step higher and identify $\phi \equiv v_a v_b$, where Latin indices run over the $D(D + 1)/2$ components of the symmetric dyadic in D spatial dimensions: $v_x v_x, v_x v_y \ldots$ and so on.

Integration upon velocity space delivers

$$\partial_t P_{ab} + \partial_c Q_{abc} = \omega(P^e_{ab} - P_{ab}) \tag{5.11}$$

where the triple tensor (flux of momentum-flux) is defined as follows:

$$Q_{abc}(\vec{x}; t) = m \int v_a v_b v_c f(\vec{v}, \vec{x}; t) d\vec{v} \tag{5.12}$$

Such triple-indexed field is the tensorial version of the energy (heat) flux in ordinary thermo-hydrodynamics.

While the left-hand side is just a tensorial escalation of the previous equations, the right-hand side is not, as it now shows a non-zero collisional contribution. This is because, the dyadic-microscopic tensor $v_a v_b$ is *not* a collision invariant, only its trace, $m Tr(v_a v_b) = mv^2$, is (twice the kinetic energy).

The attentive reader will notice that this mechanical projection procedure generates an endless ladder of tensors of ascending order: density changes are driven by the divergence of the current density, whose time change is in turn driven by the divergence of the moment-flux tensor, whose time change is in turn driven by the divergence of the momentum-flux-flux, and so on up along an endless ladder of increasing dimensionality as one moves "downwards" toward the microscopic world (see Fig. 5.2).

Why is such hierarchy open ended? And, under what conditions may we hope to close it in a sensible way?

The "culprit" of open endedness is *molecular freedom*, i.e., the fact that, at non-zero temperature, molecules have a finite spread around the mean-flow velocity u_a, the peculiar speed, $c_a = v_a - u_a$ defined earlier in this book.

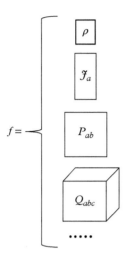

$f =$

Figure 5.2 *The distribution function contains an infinite hierarchy of kinetic moments represented by tensors of ascending rank.*

The point is immediately appreciated by considering a kinetic moment of order p (the tensorial nature being now ignored for simplicity):

$$M_p(\vec{x}; t) \equiv n < v^p > = \int f(\vec{x}, \vec{v}; t) v^p d\vec{v} \tag{5.13}$$

The time evolution of M_p is driven by the divergence of $M_{a,p+1}$, because of the v_a term in the streaming operator of the Boltzmann (BGK) equation.

Upon splitting $v_a = u_a + c_a$, according to the definition of peculiar speed, we obtain

$$M_{a,p+1} \equiv < v_a v^p > = < (u_a + c_a) v^p > = u_a M_p + < c_a v^p > \tag{5.14}$$

where we have used the definition $u_a = < v_a >$.

We see that the $(p + 1)$-th moment is the sum of the product of the moment of order 1 (the flow speed u_a) times M_p, plus the kinetic current

$$\tilde{M}_{a,p+1} \equiv < c_a v^p >$$

The former is the usual convective transport along the material fluid lines, while the latter is a purely kinetic contribution which vanishes at zero temperature, $T = 0$, a condition in which $c = 0$ by definition.

Since $T = 0$ is ruled out by the third Principle of Thermodynamics, the kinetic advection term does not vanish and the hierarchy remains open ended.

The problem would dissolve, at *any* temperature, by taking the average with the local equilibrium distribution f^e, rather than with the actual distribution f because then the kinetic current could be expressed in terms of the hydrodynamic density, momentum and temperature. Thus, the kinetic current remains unknown, solely because of non-equilibrium effects.

This provides a powerful heuristic to break the endless hierarchy. Indeed, sufficiently close to local equilibrium, non-equilibrium terms are expected to become weaker and weaker as the order p is increased, so that higher-order kinetic moments can be replaced by the corresponding equilibrium values.

This is known as *enslaving closure*, because non-equilibrium is enslaved to equilibrium quantities, i.e., density, velocity and temperature, on a timescale $\tau = 1/\omega$. The enslaving principle is a powerful heuristics which applies in all realms of physics where scale-separation arguments permit us to split the dynamic field into a "slow" and "fast" component.

It is quite fortunate that continuum (isothermal) fluid dynamics can be obtained from kinetic theory by closing the hierarchy just at the third floor of the ladder, $p = 0, 1, 2$, as we shall detail in the sequel.

5.1.2 Closures: the Enslaving Principle

The equation for the momentum-flux tensor can be closed by assuming that such tensor is enslaved to its local equilibrium value, namely

$$|\partial_t P_{ab}| \ll \omega |P^e_{ab} - P_{ab}| \tag{5.15}$$

This means that P_{ab} decays to its equilibrium value, P^e_{ab}, on a very short timescale $\tau = 1/\omega$, so that on timescales of order τ, the time derivative (5.15) can safely be neglected.

The next additional assumption is that on the same timescales, the momentum-flux-flux triple tensor can be replaced by its equilibrium value, $Q_{abc} \sim Q^e_{abc}$.

This is tantamount to neglecting non-equilibrium contributions to the kinetic currents discussed earlier on.

Under such combined assumptions, equation (5.11) can be solved in closed form, to deliver

$$P_{ab} = P^e_{ab} - \tau \partial_c Q^e_{abc} \tag{5.16}$$

By inserting this expression back into the momentum equation (5.9), we finally obtain

$$\partial_t \mathcal{J}_a + \partial_b P^e_{ab} = \tau \partial_b \partial_c Q^e_{abc} \tag{5.17}$$

The question now comes: are these *exactly* the Navier–Stokes equations of fluid dynamics? Answer: nearly, but not yet!

A direct comparison with the actual Navier–Stokes equations shows that an exact match requires the following identities to hold:

$$\begin{cases} P^e_{ab} = \rho(u_a u_b + v^2_T \delta_{ab}) \\ \tau \partial_b \partial_c Q^e_{abc} = \partial_b \sigma_{ab} \end{cases} \tag{5.18}$$

where σ_{ab} is the dissipative tensor of the Navier–Stokes equations (see Chapter I). Clearly, these expressions provide a list of constraints to be matched by the local equilibria.

At this point, it proves convenient to list the moments up to order three in compact notation.

Contribution from the equilibrium distribution:

$$< 1 >_e = 1 \tag{5.19}$$

$$< v_a >_e = u_a \tag{5.20}$$

$$< v_a v_b >_e = u_a u_b + v_T^2 \delta_{ab} \tag{5.21}$$

$$< v_a v_b v_c >_e = u_a u_b u_c + v_T^2 (u_a \delta_{bc} + u_b \delta_{ac} + u_c \delta_{ab}) \tag{5.22}$$

Contribution from the non-equilibrium distribution:

$$< 1 >_{ne} = 0 \tag{5.23}$$

$$< v_a >_{ne} = 0 \tag{5.24}$$

$$< v_a v_b >_{ne} = \theta_{ab} \tag{5.25}$$

$$< v_a v_b v_c >_{ne} = u_a \theta_{bc} + u_b \theta_{ac} + u_c \theta_{ab} + \theta_{abc} \tag{5.26}$$

where we have set $\theta_{ab} \equiv< c_a c_b >_{ne}$ and $\theta_{abc} \equiv< c_a c_b c_c >_{ne}$.
Using the above expressions, we obtain

$$P_{ab}^e = \rho(u_a u_b + v_T^2 \delta_{ab}) \tag{5.27}$$

$$Q_{abc}^e = \rho v_T^2 (u_a \delta_{bc} + u_b \delta_{ac} + u_c \delta_{ab}) \tag{5.28}$$

The first is immediately identified with the conservative component of the Navier-Stokes tensor, while the second gives the following dissipative tensor:

$$\sigma_{ab} = \mu[\partial_a u_b + \partial_b u_a + (\partial_c u_c)\delta_{ab}] \tag{5.29}$$

corresponding to a shear viscosity $\mu = \rho v_T^2 \tau$ and a bulk viscosity $\zeta = \frac{5}{3}\mu$.
The final result is:

$$\partial_t(\rho u_a) + \partial_b(\rho u_a u_b) = -\partial_a p + \partial_b \sigma_{ab} \tag{5.30}$$

i.e. the Navier-Stokes equations for a fluid with an ideal equation of state

$$p = \rho v_T^2 \tag{5.31}$$

and kinematic shear and bulk viscositities

$$\nu = v_T^2 \tau, \quad \nu_{bulk} = \frac{5}{3}\nu \tag{5.32}$$

Mission accomplished (sweeping the non-zero bulk viscosity under the carpet for now).

5.2   The Hilbert expansion

In the previous section, we have provided a heuristic derivation of Navier-Stokes hydrodynamics from Boltzmann's kinetic theory, by projecting upon velocity space and closing the resulting hierarchy by an argument of weak departure from local equilibrium and the associated enslaving of kinetic moments to their equilibrium values. The procedure makes full sense, but remains silent on the actual convergence properties of the closure (see Fig. 5.3).

The question of a systematic derivation of hydrodynamics from kinetic theory dates back to the historical work of David Hilbert (1862–1943) and later, Sydney Chapman (1888–1970) and David Enskog (1884–1947) (1; 4).

The first step in the Hilbert–Chapman–Enskog procedure is to recast the Boltzmann equation in dimensionless form.

Scaling space and time with characteristic macroscopic-scales τ_M and l_M, we obtain

$$\epsilon \partial f = C[f,f] \tag{5.33}$$

where ϵ is the Knudsen number and ∂ is a shorthand for the streaming operator.

Expression (5.33) naturally invites the following (Hilbert) expansion in powers of the smallness parameter:

$$f = \sum_{n=0}^{\infty} \epsilon^n f_n = f_0 + \epsilon f_1 + \ldots \epsilon^n f_n + \ldots \tag{5.34}$$

where all arguments have been suppressed for simplicity.

Each term expresses a higher-order departure from the local equilibrium f_0, due to non-equilibrium effects. On the assumption that $f_1 \ldots f_n$ are all of the same order, each successive term should be about ϵ time smaller than the previous one, leading to a convergent series in the limit $\epsilon \to 0$.

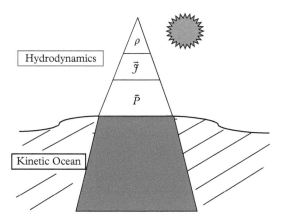

Figure 5.3 *A pictorial view of the kinetic iceberg, with emergent hydrodynamics floating in the kinetic ocean. Note that the representation is upside down, the lowest orders of the hierarchy being the emergent ones. One might also represent this as an ascending ladder, with hydrodynamics at the ground levels.*

Inserting (5.34) into (5.33) and ordering powers term by term, we obtain the following hierarchy of equations:

$$C[f_0, f_0] = 0 \tag{5.35}$$

$$\partial f_0 = \hat{L} f_1 \tag{5.36}$$

$$\partial f_1 = \hat{L} f_2 + S_1 \tag{5.37}$$

$$\cdots \tag{5.38}$$

$$\partial f_n = \hat{L} f_{n+1} + S_n \tag{5.39}$$

where

$$\hat{L} f_n \equiv 2C[f_0, f_n] \tag{5.40}$$

defines Boltzmann's linearized collision operator $\hat{L}$.

Note that the source terms $S_1 \ldots S_n$ stem from the nonlinear reminder of the collision operator, namely $S_2 = C[f_0, f_1]$, $S_3 = 3C[f_1, f_2]$ and so on.

The hierarchy exhibits a number of remarkable properties.

To begin with, the first equation confirms that the zeroth-order solution can be taken in the form of a local Maxwellian, i.e., a *normal* solution, in which the space-time dependence is fully in charge of the hydrodynamic fields.

The second consideration is that, owing to the sequential nature of the hierarchy, each successive term f_{n+1} can be obtained by solving a *linear, integral, non-homogeneous* equation of the form:

$$\hat{L} f_{n+1} = \partial f_n - S_n \tag{5.41}$$

Since the right-hand side depends only on n, it can be evaluated explicitly from the previous iteration, notwithstanding the nonlinearity of the collisional source S_n. The expression (5.41) informs us that nonequilibrium fluctuations at any order $n > 0$ are fueled by both inhomogeneity and nonlinearity at the previous level.

Given the sequential nature of the hierarchy, all terms, at any order, can ultimately be traced back to level zero, the local equilibrium!

This is a nontrivial property, as it indicates that the local equilibrium contains *all* the information it takes to reconstruct any solution of the Boltzmann equation, on the tacit assumption, though, that the Hilbert series is convergent. As one may expect, this is not generally true away from the limit $\epsilon \to 0$.

To perform the actual task, one has to solve the sequence of integral equations (5.41).

Interestingly, this is always the *same* equation, at each iteration, the only change from level to level being the source term.

Equations like (5.41) are well-known faces in applied math; they fall within the class of so-called Fredholm integral equations of second kind, whose formal solution reads

$$f_{n+1} = \hat{L}^{-1}(\partial f_n - S_n) \tag{5.42}$$

where $\hat{L}^{-1}$ is the inverse of $\hat{L}$.

Expression (5.42) is purely formal, since the inverse operator $\hat{L}^{-1}$ is readily recognized to be necessarily singular. The reason is mass-momentum-energy conservation, which forces the first five eigenvalues of $\hat{L}$ to zero.

This property can be checked by direct inspection of the explicit expression of the linearized collision operator (for full details see Cercignani's book):

$$\hat{L}h = \int v_r \sigma\,(v_r) f_0(w)[h(\vec{v'}) + h(\vec{w'}) - h(\vec{v}) - h(\vec{w})]d\vec{w} \qquad (5.43)$$

where direct collisions take $(\vec{v}, \vec{w})$ into $(\vec{v'}, \vec{w'})$ and vice versa for inverse ones.

In (5.43), the notation is as follows:

$$f = f_0(1 + h) \qquad (5.44)$$

explicitly indicating that the linear collision operator acts on the relative non-equilibrium distribution:

$$h = \frac{f - f_0}{f_0}$$

Based on (5.43), it is immediately seen that $\hat{L}$ annihilates the five collisional invariants $I_k \equiv \{1, v_a, v^2\}$, $k = 0, 4$.

Due to the singularity of $\hat{L}$, the Fredholm equation (5.41) is subject to solvability constraints, stating that the right-hand side should be orthogonal to the five collision invariants, namely

$$\int I_k(\vec{v})\,(\partial f_n - S_n)d\vec{v} = 0, \quad k = 0, 4 \qquad (5.45)$$

Given that both the full and the linearized Boltzmann collision operators annihilate the five invariants, so does the nonlinear source term S_n too, at any order.

As a result, the solvability conditions reduce to

$$\int I_k(\vec{v})\partial f_n d\vec{v} = 0, \quad k = 0, 4 \qquad (5.46)$$

By unrolling the definition of $\partial = \partial_t + \vec{v} \cdot \nabla$, conditions in (5.46) can be cast in the form of a series of generalized continuity equations:

$$\partial_t \rho_{n,k} + \nabla \cdot \vec{\mathfrak{J}}_{n,k} = 0 \qquad (5.47)$$

where

$$\rho_{n,k} = \int f_n I_k(\vec{v})d\vec{v} \qquad (5.48)$$

$$\vec{\mathfrak{J}}_{n,k} = \int \vec{v} f_n I_k(\vec{v})d\vec{v} \qquad (5.49)$$

are the generalized density and currents at order n, respectively.

Boundary layer

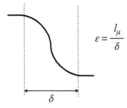

$$\epsilon = \frac{l_\mu}{\delta}$$

Figure 5.4 *Sketch of a boundary layer of width δ. The curve represents the spatial profile of any macroscopic quantity, say fluid density of velocity. The Hilbert expansion looses validity whenever the width of the boundary layer becomes comparable to the molecular mean-free path, i.e., $\epsilon \sim 1$.*

One can then proceed to solve the previous hierarchy. At order zero, one recovers local equilibria associated with inviscid Euler hydrodynamics. At the next level, f_1 inherits the derivatives in space and time of f_0, which yield dissipative corrections to Euler inviscid hydrodynamics. However, such corrections are generally not sufficient to recover Navier–Stokes hydrodynamics, the reason being that the Hilbert expansion is unable to capture steep features of the flow, such as boundary layers (see Fig. 5.4).

Since the detailed computations are straightforward but rather lengthy, the detail-thirsty reader is best directed to the excellent textbooks available in the literature.

We conclude this section by noting that the Hilbert expansion (5.34) is especially valuable for establishing that general solutions of the Boltzmann equation can be expressed solely in terms of the conserved hydrodynamic fields (and their space-time derivatives at all orders), the so-called *normal* solutions.

As noted before, however, the convergence of the Hilbert expansion is conditioned to the smallness of the expansion parameter.

From the practical point of view, this means that the Hilbert expansion fails to capture solutions with steep gradients, where $\epsilon \sim O(1)$, as it is typical of shocks, as well as boundary and initial layers.

This is where Chapman–Enskog makes its glorious entry.

5.3 Beyond Hilbert: the Chapman–Enskog Multiscale Expansion

The Hilbert expansion appears ill-equipped to catch on with "sneaky" solutions, flows whose relevant features are segregated within increasingly narrower regions, down to the order of the molecular mean-free path. This is typical of non-linear hydrodynamic interactions, and notably of the inertial forces discussed in Chapter I, which are very effective in transferring energy from large to small scales.

An ingenious and effective way to cope, at least in part, with this problem is to extend the non-equilibrium Knudsen expansion from the distribution function f alone, to space time itself.

The basic idea of this *multiscale* analysis is to represent space– and time–variables in terms of a hierarchy of slow and fast scales, such that each variable is $O(1)$ at its own relevant scale. In a nutshell, tiny regions are "stretched out," so as to reveal their features on a larger and more treatable scale.[7] This is a great help not only for the theoretical analysis, but also for the development of stable numerical methods.

The timescale hierarchy discussed previously suggests the following multiscale representation of spacetime variables (vector notation relaxed for simplicity):

$$x = \sum_{n=0}^{\infty} \epsilon^{-n} x_n, \quad t = \sum_{n=0}^{\infty} \epsilon^{-n} t_n \tag{5.50}$$

The physical meaning of such an expansion is best appreciated through a simple example. Consider the case of a second-order expansion with $t_0 = t_1 = t_2 = 1$. The "absolute" time is then $t = 1 + \epsilon^{-1} + \epsilon^{-2}$. Phenomena on a timescale $0 < t < 1$ are appropriately described by $0 < t_0 < 1$ at order zero. The next-order phenomena take place in the time interval $1 < t < 1 + 1/\epsilon \gg 1$, and are well captured in terms of the natural scale $0 < t_1 < 1$, rather than t_0. The same argument applies to t_2 versus t_1 on a still longer timescale $1 + 1/\epsilon < t < 1 + 1/\epsilon + 1/\epsilon^2$.

The multiscale expansion (5.50) induces a corresponding representation of the differential operators:

$$\partial_x = \sum_{n=0}^{\infty} \epsilon^n \partial_{x_n}, \quad \partial_t = \sum_{n=0}^{\infty} \epsilon^n \partial_{t_n} \tag{5.51}$$

By expanding both space and time, along with the distribution functions, the Boltzmann equation is again "dissected" in powers of ϵ, including separate contributions from the streaming term. The inclusion of such contributions permits us to capture the Navier–Stokes equations and deal with flows with relatively sharp features.

It should be noted that the nature of the emergent hydrodynamic equations depends on the specific way that space and time are multiscaled. In particular, the relevant scaling for Navier–Stokes is

$$\begin{cases} x = x_1/\epsilon \\ t = t_1/\epsilon + t_2/\epsilon^2 \end{cases} \tag{5.52}$$

so that diffusive phenomena can emerge at second order, with the ratio x^2/t kept constant as the smallness parameters is sent to zero.

As we shall see later in this book, the Chapman–Enskog expansion plays a crucial role in the theory of the Lattice Boltzmann equation.

---

[7] A daily life example of multiscale device is the common watch, with separate arrows for seconds, minutes and hours. When timing a 100 m run, the minutes and hours are blank. The relevant arrow is seconds, and below. Thus, the common watch is basically a three-scale device with "Knudsen" number $\epsilon = 1/60$.

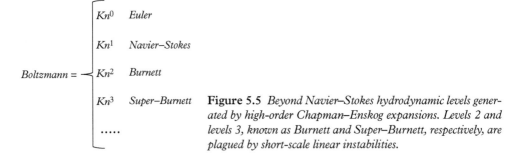

$$Boltzmann = \begin{cases} Kn^0 & Euler \\ Kn^1 & Navier\text{–}Stokes \\ Kn^2 & Burnett \\ Kn^3 & Super\text{–}Burnett \\ \;\;..... \end{cases}$$

Figure 5.5 *Beyond Navier–Stokes hydrodynamic levels generated by high-order Chapman–Enskog expansions. Levels 2 and levels 3, known as Burnett and Super–Burnett, respectively, are plagued by short-scale linear instabilities.*

5.3.1 Higher-Order Hydrodynamics

The Chapman–Enskog procedure leads to the inviscid Euler equations to zeroth order in the Knudsen expansion and to dissipative Navier–Stokes hydrodynamics at second order. A natural question arises: having prepared the full machinery at any arbitrary order in the ϵ expansion, what does one gain by escalading the hierarchy beyond the Navier–Stokes level?

The answer is rather disappointing: most of the time, not much. Higher-order closures, known as Burnett–Knudsen and super-Burnett equations (see Fig. 5.5), corresponding to second- and third-order closures in the Knudsen number, are not only clumsy, but often even plagued by basic linear small-scale instabilities, as shown in a seminal paper by Bobylev (5). This does not come as a big surprise to the math-inclined reader. A glance at the starting equation (5.33) shows that this is a singular perturbation problem, because the small parameter upfronts the highest order differential operator (6) (it is just first order, but still the highest since collisions are local). Singularly perturbed problems are notorious for being only *asymptotically* convergent, an elegant byword for failure to deliver increased accuracy in return for inclusion of additional extra terms.[8] Even though regularization and exact resummation schemes have been developed in the recent years (7; 8), the stability of higher-order Chapman–Enskog expansions remains a rather thorny issue, to which we shall return shortly in Chapter 6.

5.4 Summary

Summarizing, Navier–Stokes hydrodynamics emerges from kinetic theory in the limit of small departures from local equilibrium, namely low Knudsen numbers. Such limit can be derived heuristically by assuming enslaving of non-conserved modes to the conserved ones, or via systematic expansions in the Knudsen number. The latter however are only asymptotically convergent, which means that higher-order expansion beyond the level

---

[8] Asymptotic convergence is pretty common in series expansions of ordinary functions. Take for instance the well-known Taylor series $\frac{1}{1-x} = 1 + x + x^2 + \ldots x^n$, which takes the value zero for $x \to \infty$, even though each of its terms, except the first, are infinite in this limit.

of Navier–Stokes does not lead to any substantial improvement. To the contrary, the resulting equations are often plagued by basic instabilities. Hence the Navier–Stokes equations appear to offer a very general and robust mathematical framework for the study of the physics of fluids.

....

REFERENCES

1. S. Chapman, T.G. Cowling, "The Mathematical Theory of Non-Uniform Gases", 3rd ed., Cambridge Univ. Press, Cambridge, 1990.
2. C. Cercignani, *Theory and Application of the Boltzmann Equation*, Elsevier, New York, 1975.
3. C. Cercignani, *Mathematical Methods in Kinetic Theory*, Plenum Press, New York, 1969.
4. S. Harris, *An introduction to the theory of the Boltzmann equation*, Holt, Rinehart and Winston, NY, 1971.
5. A.V. Bobylev, *Sov. Phys. Dokl.* 27, **29**, 19832.
6. L. Saint Raymond, *Hydrodynamic limits of the Boltzmann equation*, Springer Verlag (2009).
7. H. Struchrup and M. Torrillon, H theorem, Regularization, and Boundary Conditions for Linearized 13 Moment Equations, *Phys. Rev. Lett.*, 99 (1), **014502**, 2007.
8. A. Gorban and I. Karlin, Hilbert's 6th Problem: Exact and Approximate Hydrodynamic Manifolds for Kinetic Equations, *Bulletin of the Am. Math. Soc.*, 51 (2), **187**, 2017.

....

EXERCISES

1. Compute the Knudsen number for air at standard conditions, by taking $L = 1$ mm as the shortest hydrodynamic scale.
2. Prove that the local Maxwell distributions reduces to a Dirac's delta in the limit of zero temperature.
3. Compute the equilibrium value of the first three central moments $\langle c^p \rangle$, $p = 0, 1, 2$.

6

Generalized Hydrodynamics Beyond Navier–Stokes

In this chapter, we discuss derivations of generalized hydrodynamics beyond the realm of the Navier–Stokes description, with special reference to Grad's thirteen-moment formulation.

Παντα ρεῖ, καὶουδεν μενεῖ,
Everything flows, nothing stands still.

(Heraclitus, still him!)

6.1   Grad's Thirteen-Moment Theory

The work of Chapman and Enskog opened a long period, lasting about three decades, in which most of the activity in kinetic theory was directed to the computation of the transport coefficients for different types of intermolecular potentials. Seeking the solution of the full Boltzmann equation itself was not much in focus, mostly on account of its daunting complexity. This situation took a sharp turn in 1949, with the publication of Harold Grad's thesis (1).

Following a mathematical procedure pretty similar to the one used in quantum mechanics, the American mathematical physicist Harold Grad (1923–86) proceeded to provide approximate solutions to the Boltzmann equation by a systematic expansion of the distribution function onto a suitable set of basis functions in velocity space.

In cartesian coordinates, and taking the thermal speed $v_T = 1$ for simplicity, such basis functions identify with the tensor Hermite polynomials, $H_k(\vec{v})$, defined as follows:

$$H_k(\vec{v}) = (-1)^k \, w(v)^{-1} \, (\nabla_{\vec{v}})^k \, w(v) \tag{6.1}$$

where $v = |\vec{v}|$ and

$$w(v) = (2\pi)^{-3/2} e^{-v^2/2} \tag{6.2}$$

The Lattice Boltzmann Equation. Sauro Succi, Oxford University Press (2018).
© Sauro Succi. DOI: 10.1093/oso/9780199592357.001.0001

is the weight function entering the orthonormality condition

$$\int H_k(\vec{v}) w(v) H_l(\vec{v}) d\vec{v} = \delta_{kl} \tag{6.3}$$

Here, we just list the first five (Latin subscripts denoting as usual the three cartesian components x, y, z):

$$H_0 = 1 \tag{6.4}$$
$$H_{1,a} = v_a \tag{6.5}$$
$$H_{2,ab} = v_a v_b - \delta_{ab} \tag{6.6}$$
$$H_{3,abc} = v_a v_b v_c - P_3[v_a \delta_{bc}] \tag{6.7}$$
$$H_{4,abcd} = v_a v_b v_c v_d - P_6[v_a v_b \delta_{cd}] + P_3[\delta_{ab}\delta_{cd}] \tag{6.8}$$

Here $P_3[v_a \delta_{bc}]$ denotes the set of three permutations $P_3[v_a \delta_{bc}] \equiv v_a \delta_{bc} + v_b \delta_{ac} + v_c \delta_{ab}$ and same goes for the three permutations of $\delta_{ab}\delta_{cd}$ and the six permutations of $v_a v_b \delta_{cd}$.

Note the one-to-one link of the first four with the collisional invariants $\{1, v_a\}$, while the trace of the fifth one relates to kinetic energy through the condition $Tr(H_{2,ab}) = v^2 - 3$.

Grad's expansion reads as follows:

$$f(\vec{x}, \vec{\xi}; t) = \rho w(c) \sum_{k=0}^{\infty} a_k(\vec{x}; t) H_k(\vec{\xi}) \tag{6.9}$$

where $\vec{\xi} = (\vec{v} - \vec{u})/v_T$ is the peculiar velocity introduced previously in this book, rescaled with the thermal speed, so as to make it dimensionless. As a result, the weight

$$w(\xi) = \frac{e^{-c^2/2}}{(2\pi v_T^2)^{3/2}}$$

identifies with the *local* Maxwell distribution and the ratio $g = f/w$ describes the departure from such local equilibrium. Note that the thermal speed has been explicitly reinstated for dimensional clarity.

In the previous expression, the "coefficient" a_k is the kinetic moment of order k with respect to the k-th Hermite basis, defined as

$$a_k(\vec{x}; t) = (f, H_k) \equiv \int f(\vec{x}, \vec{\xi}; t) H_k(\vec{\xi}) d\vec{\xi} \tag{6.10}$$

where $(.,.)$ is a shorthand for scalar product in Hilbert's space. Note that a_k are *central* moments, i.e., defined in terms of the peculiar velocity $\vec{c}$, as opposed to the absolute moments discussed in Chapter 5. A main advantage of the central moments is to secure Galilean invariance term by term, at any order of the expansion.

The main point of Grad's method is quickly recognized: by including an increasing number of terms, the Hermite expansion (6.9) *systematically* "fills up" the whole

Table 6.1 *Grad's thirteen moment hierarchy.*

Order	Hermite	Macrofield	Number
0	H_0	ρ	1
1	$H_{1,a}$	$j_a = 0$	3
2	$H_{2,ab}$	p_{ab}	6
3	$H_{3,a}$	q_a	3

Hilbert space including regions which are eventually inaccessible to both Hilbert and Chapman–Enskog's expansion. This is guaranteed by the completeness of the Hermite's functional basis, i.e., the fact that any function belonging to the appropriate Hilbert space can be represented by a (finite or infinite) superposition of Hermite polynomials.

As a practical example, Grad's expansion proves able to generate non-equilibrium fluxes depending on both density and temperature gradients, which cannot be represented via normal solutions and derivatives thereof.

To turn such strength into an actual asset, two basic questions must be addressed.

First, how many terms should be retained in the expansion in order to describe flow regimes at non-negligible Knudsen numbers?

Second, under what conditions does the expansion converge?

In other words, under what conditions does the inclusion of additional terms lead to a more accurate description of non-equilibrium effects?

Grad's choice for the first question was to proceed up to the third level of the Hermite expansion, namely $\{\rho, j_a, p_{ab}, q_a\}$, (see Table 6.1) for a total of $1 + 3 + 6 + 3 = 13$, whence the nomer "Grad's thirteen moments method." In Table 6.1 $H_{3,a} \equiv H_{3,abb}$ i.e., the contraction of the third other Hermite tensor over the second and third indices. It should now be clear why Hermite polynomials hold a special place in kinetic theory; the microscopic invariants $\{1, v_a, v^2/2\}$ coincide precisely with the first three (five in $D = 3$) Hermite polynomials (the energy must be shifted by a constant).

Moreover, it can be shown that the Hermite polynomials are the eigenfunctions associated with Maxwellian molecules interacting via the $V(r) \sim r^{-4}$ potential. Since they form a complete basis in Hilbert space, i.e., the space of L_1 integrable functions, i.e., such that $\int f(\vec{v})d\vec{v} < \infty$,[9] Hermite functions can represent any well-behaved distribution function, regardless of the specific form of the underlying atomistic potential, whence their special place in the development of mathematical methods for kinetic theory, Grad's method in the first place. As we shall see, lattice transcriptions of the Hermite polynomials play a important role in Lattice Boltzmann theory as well.

Grad's truncation is clearly informed by the concrete physical meaning of the thirteen moments. The first is the density, the second is the current, here equal to zero because the expansion is based on central moments, the third is the non-equilibrium component

---

[9] The spectral theory of linear operators is a classical subject covered by most textbooks in quantum mechanics. It entails some basic notions of functional analysis, which are normally available to physicists and mathematicians, somewhat less to engineers. Here, we confine our discussion to the essential physics, without delving into any mathematical sophistication.

of the momentum flux and the fourth is the non-equilibrium component of the heat flux. Beyond these thirteen, a direct macroscopic interpretation is no longer available.[10]

By itself, however, this offers little guarantee for the second pending question. In particular, one problem with such *truncation* is immediately apparent: while the original distribution f is non-negative definite by construction, any finite-order truncation of the Hermite expansion, being based on global non-positive-definite polynomials, offers no guarantee to preserve non-negativeness.

Thus, abrupt truncations represent a serious threaten to the *microscopic realizability*. This is the price to pay for the highly valuable orthonormality of Hermite polynomials: global polynomials cannot be made orthogonal unless they change sign across velocity space!

Because of these and other technical reasons, mostly associated with boundary conditions, the formal beauty and elegance of Grad's method does not fully carry on to a correspondingly powerful computational tool. Yet, the conceptual import of Grad's method and its major bearing on the theory of the Lattice Boltzmann equation are such to warrant some further digression on its mathematical structure.

6.2   Grad's Approach in (Some) More Detail

The Chapman–Enskog procedure is entirely based on the so-called *normal* solutions, i.e., local equilibria, which depend on space and time only through the hydrodynamically conserved quantities. Non-equilibrium processes are thereby described via perturbative contributions, by taking spacetime derivatives of the local equilibria and expanding in powers of the Knudsen number. Grad's approach presents an alternative route, no longer centered around the explicit splitting between equilibrium and non-equilibrium, but rather on a systematic representation of the *full*-distribution function, in terms of Hermite basis functions.

In Chapter 5 we have derived the hydrodynamic equations by integrating the Boltzmann equation over successive powers of the molecular velocity using a BGK model collision operator for simplicity.

The same goal can be achieved in a more systematic and general way by projecting the full Boltzmann equation upon the Hermite polynomials.

To this end, let us consider the force-free Boltzmann equation with a general collision operator, that is

$$\partial_t f + \vec{v} \cdot \nabla f = C(f,f) \tag{6.11}$$

---

[10] In the theory of probability, the third-order moment is related to the so-called "skewness," defined as $s \equiv \frac{<c^3>}{<c^2>^{3/2}}$, while the fourth one is known as "flatness" or "kurtosis", a Greek word for curved or arching, and is defined as $\kappa \equiv \frac{<c^4>}{<c^2>^2}$. The skewness measures the unbalance between above and below average events, while the kurtosis measures the impact of "heavy" tails in the distribution. For a Gaussian distribution, $s = 0$ and $\kappa = 3$.

Upon inserting the expansion (6.9) into (6.11) and taking into account the orthonormality of the Hermite polynomials, Grad obtained the following set of nonlinear differential equations for the thirteen moments:

$$\partial_t \rho + \partial_a(\rho u_a) = 0 \tag{6.12}$$

$$\partial_t(\rho u_a) + \partial_b(\rho u_a u_b + p_{ab}) = 0 \tag{6.13}$$

$$\partial_t p_{ab} + \partial_c(u_a p_{bc} + p_{abc}) + \pi_{ab} = C_{ab} \tag{6.14}$$

$$\partial_t q_a + \partial_b(u_b q_a + q_{ab}) + \pi_a = C_a \tag{6.15}$$

where

$$q_{ab} = \frac{5}{2}\frac{p^2}{\rho}\delta_{ab} + \frac{7}{2}\hat{p}_{ab} \tag{6.16}$$

and

$$p_{abc} = \frac{2}{5}(q_a \delta_{bc} + q_b \delta_{ac} + q_c \delta_{ab}) \tag{6.17}$$

and we have used the shorthands:

$$\pi_a = p_{abc}\partial_c u_b + q_b \partial_b u_a - \frac{p_{ca}}{\rho}\partial_b p_{cb} - \frac{Tr(p)}{\rho}\partial_b p_{ab} \tag{6.18}$$

and

$$\pi_{ab} = p_{ca}\partial_c u_b + p_{cb}\partial_a u_c \tag{6.19}$$

In the previous expressions, $\hat{p}_{ab}$ denotes the traceless (deviatoric) version of p_{ab}.

This set of equations does not follow straight from Hermite projection, but requires some further algebraic rearrangements which are described in Grad's original work (a very terse yet rather dense read for the non-math inclined). For a very clear and pedagogical exposition see also the book by Kremer (2).

The computation of the collision terms C_a and C_{ab} is also a bit involved, but finally provides a very transparent expression, namely

$$C_{ab} = -\frac{p_{ab}}{\tau_c} \tag{6.20}$$

$$C_a = -\frac{2}{3}\frac{q_a}{\tau_c} \tag{6.21}$$

where τ_c is the collisional relaxation time.[11]

---

[11] This can be computed in terms of the following integrals $C^{l,m} = \int_0^\infty \int_0^\infty e^{-x^2/2} x^l [1 - cos^m \chi(y)] \, y \, dy \, dx$, where $\chi(y)$ is the scattering angle as a function of the impact parameter y and x is a dimensionless velocity. See Chapter 5 on the kinematics of two-body collisions.

The relations (6.12) close the Grad's hierarchy for the thirteen hydrodynamic fields:

$$\mathcal{G}_{13} = \{\rho, u_a, p_{ab}, q_a\} \tag{6.22}$$

A few remarks are in order.

First, we observe that Grad's thirteen-moment equations represent a superset of the Navier–Stokes equations, in which neither an equation of state nor constitutive relations are required, simply because the momentum flux and heat flux obey their own evolution equations.

Second, this superset is *hyperbolic*, i.e., space and time derivatives always appear on the same footing, both first order. This dispenses with the enslaving principle already invoked and reflects into a basic physical distinction between Grad's and Navier–Stokes equations: in the former information travels at *finite speed*, as opposed to the infinite propagation speed inherent to diffusion processes, such as dissipation in the Navier–Stokes equations.

Both previous items stand out as major forward steps on the way to extracting the physical content of the Boltzmann equation beyond the Navier–Stokes description.

Indeed, Grad's equations have met with some practical success in describing flow phenomena with fast transients and sharp features, beyond the capabilities of the Navier–Stokes equations. However, by and large, the list of these success stories remains somehow below the expectations spawned by the formal beauty and elegance of the Grad's procedure.

As mentioned before, the culprit is basically the lack of positive definiteness of the Grad's representation.

Finally, it should be noted that, although formally elegant, Grad's equations do not look particularly simple, especially in view of the numerous nonlinear coupling terms with no counterpart in the Navier–Stokes description. Such terms are potentially liable to instabilities like in the case of Burnett and super-Burnett equations mentioned in Chapter 5.

For all the formal beauty of the Hermite expansion, the abrupt truncation to thirteen moments turns out to be a rather crude form of closure. Significant progress achieved in the recent years thanks to various forms of regularization of Grad's approach (3). Yet, the stability and convergence of truncated Hermite expansions of the Boltzmann equation remains a difficult subject.

As we shall see in the second part of this book, the Lattice Boltzmann theory bears many conceptual links with Grad's theory, with the twist of significant benefits due to the discrete nature of phase space inherent to the lattice formulation.

6.3   R13: Regularized Grad-13 Expansion

It is fair to mention that efforts to amend the idyosyncrasises of Grad's 13 moment approach, as well as of higher-order hydrodynamics have led to formulations which are free from the linear instabilities which plague these approaches. The idea is to augment the Grad-13 equations with additional terms corresponding to educated (read stable) approximations of higher-order moments which are set to zero in the standard Grad's approach. The insertion of such additional terms is shown to stabilize the thirteen Grad's equations without requiring any extra kinetic moment, whence the nomer R13. The cure does not appear to be universal and the equations still look less than simple, but the reader should not walk away from this chapter under the impression that instability is the last word on this (thick) subject, because it is not.

6.4 Linearized Collision Operator

The Hermite polynomials are tightly related to the linearized-Boltzmann collision operator. Since this operator plays a central role in kinetic and in Lattice Boltzmann theory alike, in the sequel we provide some further information on its main properties. From a mathematical viewpoint, the linear operator $\hat{L}$ maps the non-equilibrium distribution, f^{ne}, into the collisional rate of change, defined as (see Fig. 6.1)

$$g(\vec{x}, \vec{v}; t) \equiv \int L(\vec{v}, \vec{v}') f^{ne}(\vec{x}, \vec{v}'; t) d\vec{v}' \tag{6.23}$$

$L(\vec{v}, \vec{v}')$ being the *kernel* of the operator. Like any other linear operator, $\hat{L}$ is fully characterized by its *spectrum*, $\Sigma(\hat{L})$, namely the set of *eigenvalues* λ_k, $k = 1, \infty$ and the corresponding eigenfunctions $\{\phi_k(v)\}$, defined by the condition

$$\int L(\vec{v}, \vec{v}') \phi_k(\vec{v}') d\vec{v}' = \lambda_k \phi_k(\vec{v}) \tag{6.24}$$

Geometrically, the effect of $\hat{L}$ on ϕ_k is to multiply ϕ_k by a factor λ_k, without mixing it with any other eigenfunction. The kernel of $\hat{L}$ can then be expressed in spectral form as a double expansion over its eigenfunctions:

$$L(\vec{v}, \vec{v}') = \sum_k \phi_k(\vec{v}) \lambda_k \phi_l(\vec{v}') \tag{6.25}$$

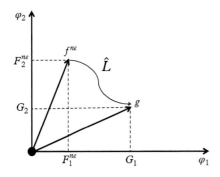

Figure 6.1 *The action of the linearized operator $\hat{L}$ in a fictitious two-dimensional "Mickey Mouse" Hilbert's space. In this cartoon $\lambda_1 > 1$ and $\lambda_2 < 1$.*

The details of the spectrum $\Sigma(\hat{L})$ are dictated by the underlying microscopic potential, and the study of its properties makes an important subject of mathematical kinetic theory.

For simplicity, in (6.23) we have assumed that all the eigenvalues are discrete and real, which can indeed be shown to be the case if the *kernel* is real and symmetric, namely

$$L(\vec{v}, \vec{v}') = L(\vec{v}', \vec{v})$$

This is a specific instance of a more general property known as hermiticity involving the symmetric conjugate of the complex operator, $L^{\star}(\vec{v}', \vec{v}) = L(\vec{v}, \vec{v}')$.

In general, however, the spectrum needs not be discrete. For generic potentials giving rise to a velocity dependence of the collision-frequency $\gamma = nv\sigma(v)$, the spectrum contains a continuum component associated with the eigenvalues fulfilling the condition

$$\gamma(v)\phi(v; \lambda) = \lambda\phi(v; \lambda) \tag{6.26}$$

The corresponding eigenfuctions are singular, i.e., Dirac's deltas centered about the velocity-solving eqn (6.26), namely

$$\phi(v; \lambda) = \delta[\gamma(v) - \lambda] \tag{6.27}$$

For instance, for power-law potentials decaying faster than r^{2-2D} in D spatial dimensions (Maxwell molecules) and with an angular cut-off, the collision frequency can be shown to be a monotonically increasing function of the velocity. This means that fast molecules undergo more collisions than slow ones, which surely helps the stability of the fluid. In this case, the continuum spectrum lies below (more negative) the discrete one and does not compromise the exponential decay to equilibrium. On the other hand, if the velocity decay is slower than for Maxwell molecules, then $\gamma(v)$ is a decreasing function of the velocity, in which case the fast movers experience *less* friction than the slow ones setting the stage for potential instabilities. In this case the continuum spectrum lies above (less negative) the discrete one, and should it reach up to $\lambda = 0$, exponential decay is no longer realized.

We shall not dig any deeper into these important matters, and restrict our attention to the case of a fully discrete spectrum.

If the set of eigenfunctions $\{\phi_k\}$ is complete, any well-behaved distribution function can be expanded as follows:

$$f(\vec{x}, \vec{v}; t) = w(v) \sum_k F_k(\vec{x}; t)\phi_k(\vec{v}) \tag{6.28}$$

where the coefficients of the expansion, $F_k(\vec{x}; t)$, are the kinetic moments associated with the eigenfunctions ϕ_k. In Chapter 5 we referred to M_k as to kinetic moments associated with (tensor) powers of the velocity. While linearly related to F_k they are less practical for high-order representations, because they do not stem from an orthogonal set of eigenfunctions. On the assumption that these eigenfunctions constitute an ortho-normal set, namely

$$(\phi_k, \phi_l) \equiv \int \phi_k(\vec{v}) w(v) \phi_l d\vec{v} = \delta_{kl} \tag{6.29}$$

the kinetic moments are simply given by

$$F_k = (f, \phi_k)$$

By inserting (6.28) into the right-hand side of (6.11) and taking into account the definition (6.1), we readily obtain

$$g(\vec{x}, \vec{v}; t) = \sum_k F_k^{ne}(\vec{x}; t) \int L(\vec{v}, \vec{v}') \phi_k(\vec{v}') d\vec{v}' = \sum_k \lambda_k F_k^{ne}(\vec{x}; t) \phi_k(\vec{v}) \tag{6.30}$$

In other words, the collision function $g(\vec{x}, \vec{v}; t)$ has just the same expansion as f^{ne}, only with kinetic moments prefactored by the corresponding eigenvalue, namely

$$G_k(\vec{x}; t) = \lambda_k F_k^{ne}(\vec{x}; t) \tag{6.31}$$

Each operator defines its own set of eigenfunctions, and the Hermite polynomials provide the most natural and convenient basis for the case of a cartesian representation of velocity space. For other representations, say spherical or cylindrical, different polynomials should be used (see (4)). Based on the material presented in Chapter 5, it is clear that the spectrum of $\hat{L}$ is entirely dictated by the underlying microscopic potential. This spectrum contains all of the near-equilibrium physics in a most compact and elegant mathematical form. Hence, the connection

$$V(r) \to \sigma(v) \to \Sigma(\hat{L}) \tag{6.32}$$

provides the mathematical bridge between the atomistic world and Boltzmann's kinetic theory.

Working out the specific form of this connection is not an easy task, at least not for a generic potential. However, as already mentioned, short-ranged, repulsive potentials lead to discrete and stable spectra and in this book we shall not need anything more complicated than this.

6.4.1 The CSF Spectral Decomposition

Based on what we have learned in Section 6.4, it proves convenient to group the eigenvalues/vectors into three main families:

1. *Conserved* (C): $\lambda = 0$,
2. *Slow* (S): $\Lambda < \lambda < 0$,
3. *Fast* (S): $\lambda < \Lambda < 0$

where $\Lambda < 0$ denotes a conventional spectral borderline between Slow and Fast modes.

This grouping induces a corresponding spectral decomposition of the distribution function:

$$f = f^C + f^S + f^F \tag{6.33}$$

where the three components are defined as follows:

$$f^\alpha = \sum_{\lambda_k \in \Sigma^\alpha} F_k \phi_k, \quad \alpha = C, S, F \tag{6.34}$$

In Figure 6.2

$$\Sigma^C = \{\lambda_k = 0\}, \ \Sigma^S = \{\Lambda < \lambda_k < 0\}, \ \Sigma^F = \{\lambda_k < \Lambda < 0\} \tag{6.35}$$

are the three components of the spectrum.

As usual, the kinetic moments carry the full spacetime dependence, while the eigenfunctions depend only on the velocity.

The number of modes in each of the three C-S-F sectors is easily counted. The C sector contains the five conserved modes: density (scalar), current (vector), energy (scalar). The S sector contains the quasi-conserved transport modes, momentum flux tensor and heat flux vector, for a total of $5 + 3 = 8$. The F sector collects all $\infty - 13$ remaining ones, the complement to Grad in Hilbert's space. The reader will surely notice that the S and F sectors can be identified with what we have previously called Transport and Ghost modes.

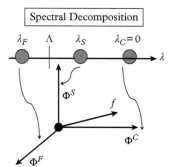

Figure 6.2 *Sketch of the spectral decomposition of the Boltzmann distribution. Boltzmann's distribution f lives is an infinite-dimensional functional space, named after Hilbert. The symbols $\Phi^{C,S,F}$ stand for "unit vectors" of the C, S, F subspaces of Hilbert's space, respectively, spanned by the corresponding basis functions ϕ_k. Thus, the C space has five dimensions, while S and F subspaces have as many dimensions as the number of corresponding eigenvalues. These three subspaces are orthogonal by construction.*

The spectral decomposition applies to the linear operator as well, as is apparent from the expression (6.25). By definition, we have

$$\hat{L}^C \equiv 0$$

since the sum that defines this operator runs over zero eigenvalues only. Despite L^C being identically zero, the corresponding f^C is not, signaling that the operator is a singular one. Mathematically, f^C is said to belong to the *null-space* of $\hat{L}$, i.e. $\hat{L}f^C = 0$.

By the same token,

$$L^S(\vec{v}, \vec{v}') = \sum_{\lambda_k \in \Sigma^S} \phi_k(\vec{v}) \lambda_k \phi_k(\vec{v}') \tag{6.36}$$

and an analog expression holds for L^F, with the summation running on the fast eigenvalues only.

It is readily recognized that these operators project the actual distribution function f onto its Slow and Fast components, respectively, i.e., $f^S = \hat{L}^S f$, and similarly for the F component.

This is why they are appropriately called *projectors*.

The projector formalism offers an elegant and powerful tool to deal with a variety of problems in non-equilibrium statistical mechanics at large, not only kinetic theory. Here, we simply invite the reader to appreciate the analogy with ordinary vector calculus: the Boltzmann distribution lives in an infinite dimensional space, known as Hilbert's space, spanned by the set of the eigenfunctions of the linear operator $\hat{L}$. With all due mathematical precautions, the spectral representation and the ensuing projector formalism permit us to handle such infinite-dimensional space much like an ordinary vector space.

It is of some interest to compare the CSF spectral representation

$$f = f^C + f^S + f^F \tag{6.37}$$

with the Chapman–Enskog treatment based on the double splitting equilibrium versus non-equilibrium.

$$f = f^e + f^{ne} \tag{6.38}$$

How do these two splittings, equilibrium-nonequilibrium, versus CSF connect to each other?

The C modes are conserved; hence, by definition, they do not contribute to non-equilibrium. The S and F modes, on the other hand, contribute to both f^e and f^{ne}, but their equilibrium values are entirely dictated by the C modes, the enslaving principle invoked in Chapter 5. The S moments are weakly enslaved, in that they relax slowly, hence have a dynamics of their own before decaying to equilibrium. The F moments, on the other hand, are strongly enslaved, i.e., they converge rapidly to their equilibrium value, and such value should be negligible as compared to the corresponding contribution of the C and S modes.

The equilibrium moments F_k^e can be computed exactly at any order k, thanks to another beautiful property of the Hermite polynomials: the local Maxwellian is their *generating function*.

In one dimension for simplicity:

$$e^{-\frac{1}{2}\left(\frac{v-u}{v_T}\right)^2} = e^{-\frac{v^2}{2v_T^2}} \sum_{k=0}^{\infty} H_k\left(\frac{v}{v_T}\right)\left(\frac{u}{v_T}\right)^k \tag{6.39}$$

This uniquely identifies the equilibrium moment at any order:

$$F_k^e = \left(\frac{u}{v_T}\right)^n \tag{6.40}$$

In other words, the equilibrium kinetic moment of order k is simply proportional to the k-th power of the Mach number, the proportionality constant being the ratio of fluid to thermal speed. Thus, high-order equilibrium moments are vanishingly small for low-Mach number flows. Clearly, in $D > 1$ this power is a k-th-order tensor (see Fig. 6.3).

Incidentally, the (6.40) expansion informs us that any finite-order truncation of the local maxwellian would fail to preserve Galilean invariance for an arbitrary flow velocity u. This simple observation bears major implications on the Lattice Boltzmann theory. Also to be noted that temperature only enters as a scale factor for the molecular velocity, reflecting the scale invariance of the local equilibria under space and time dilatations which leave the velocity unchanged.

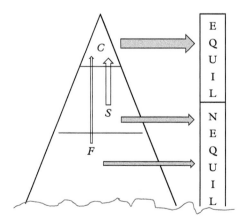

Figure 6.3 *Cartoon of the CSF versus equil–non-equil representations of generalized hydrodynamics. The S and F modes contribute to the equilibrium only through the intermediate of the C modes (enslaving vertical arrows in the sketch). Note the emergence of higher-order modes, which were left below the kinetic sea-level in the Navier–Stokes description.*

6.5 Summary: Grad versus Chapman–Enskog

In closing this chapter, it is appropriate to summarize the main differences between the Chapman–Enskog and Grad's approaches.

In the former, the non-equilibrium component of the distribution function is entirely generated from the equilibrium one, based on the Knudsen number as an expansion parameter:

$$f = f_0 + \epsilon f_1 + \epsilon^2 f_2 + \ldots \ldots$$

where f_0 is the local equilibrium and higher-order terms f_k, $k > 0$ are "boot-strapped" from the local equilibrium by taking spacetime derivatives of f_0 to all orders in the Knudsen number. The bootstrap procedure involves the inclusion of spacetime gradients of increasing order, leading to a non-local formulation. This has major intuitive appeal but a few inconveniences too.

First, the expansion is only asymptotically convergent, there is no guarantee that by increasing the order of the expansion one would obtain a more accurate description of non-equilibrium phenomena beyond the Navier–Stokes description. As we have seen in Chapter 5, the corresponding equations, Burnett and super-Burnett, are often plagued by basic instabilities and remedies for such inconveniences are not easily found. Second, although convergent, this expansion may still not be complete, i.e., there is no guarantee that it can reach up to every "corner" of the Hilbert space, particularly the ones inhabited by flows with boundary layers and the like.

In Grad's approach, the notion of a *perturbative* expansion in the Knudsen smallness parameter is abandoned in favor of a systematic *non-perturbative* expansion onto the appropriate functional basis: the Hermite functions. This offers three immediate advantages: first, it preserves the hyperbolic nature of the Boltzmann equation order by order. Second, it is formally guaranteed to "fill up" the Hilbert space, thanks to the completeness of the Hermite basis. Third, it remains local.

In Grad's expansion, each kinetic moment contributes an equilibrium and non-equilibrium component $F_k = F_k^e + F_k^{ne}$, where the former can be given the exact and compact form, eqn (6.40). By suitable rearrangements, a set of *hyperbolic* nonlinear partial differential equations can be obtained for the F_k's, with no approximation formally required.

However, this set is inevitably open, and some form of closure is still needed. Instead of smallness of the Knudsen number, in Grad's representation such closure comes in the form of a *truncation* of the Hermite expansion to the first thirteen F_k's, the ones carrying a direct physical meaning. While preserving the hyperbolic nature of the Boltzmann equation, such truncation undermines its realizability, i.e., the non-negative definiteness of the distribution function, thereby opening again a loophole into the building, in the form of instabilities similar to the ones plaguing the Burnett and Super-Burnett equations. This, together with the difficulty of setting up proper boundary conditions for the thirteen macroscopic fields, severely impairs the practical impact of Grad's approach.

Remedies have been developed in the past decades, through ingenious forms of regularization, but the subject remains a rather difficult one.

....

REFERENCES

1. H. Grad, On the Kinetic Theory of Rarefied gases, *Commun. Pure Appl. Math.* 2(4), **331**, 1949.
2. G. Kremer, *Moment Methods. An Introduction to the Boltzmann Equation and Transport Processes in Gases*, Springer, Berlin Heidelberg, 2010.
3. H. Struchtrup and M. Torrillon, Regularization of Grad's 13 Moment Equations: Derivation and Linear Analysis, *Phys. Fluids*, 15(9), **2668**, 2003 and *Phys. Rev. Lett.*, 99, **014502**, 2007.
4. S. Harris, *An introduction to the theory of the Boltzmann equation*, Holt, Rinehart and Winston, NY, 1971.

....

EXERCISES

1. Generate the first three tensor Hermite polynomials based on their definition, eqn (6.1).
2. Prove the identity $\sum_k \phi_k(v)\phi_k(v') = \delta(v - v')$, where ϕ_k is a complete set of orthonormal-basis functions.
3. The time decay associated with a continuum spectrum with spectral density $n(\lambda)$ in the strip $\lambda_1 < \lambda < \lambda_2 < 0$ is given by $\int_{\lambda_1}^{\lambda_2} n(\lambda)e^{\lambda t}d\lambda$. Compute the time decay associated with a spectral density $n(\lambda) \propto \lambda^{-p}$, with $p > 0$.

7

Kinetic Theory of Dense Fluids

In this chapter we present the basic elements of the kinetic theory of non-ideal fluids, to which both kinetic and potential energy contribute on comparable footing. Non-ideal fluids lie at the heart of many complex fluid-dynamic applications, such as those involving multiphase and multicomponent flows and make for one of the major streams of modern Lattice Boltzmann research.

If you want to learn what high-density fluids are, the metro of Rome is a great teacher ...

(this author).

7.1 Cautionary Remark

This chapter presents a degree of abstraction which may not come by handy to the reader with limited interest to the formal theory of classical many-body systems. This reader can safely skip the math and retain the basic bottomline. He/she may just as well skip this chapter altogether, but in this author's opinion, this is likely to come with a toll on the full appreciation of Lattice Boltzmann theory for non-ideal fluids, in fact one of the most successful offsprings of Lattice Boltzmann theory.

7.2 Finite-Density Extension of the Boltzmann Equation

As discussed in Chapter 2, Boltzmann derived his equation on the heuristic assumption of a dilute gas, whose dynamics is dominated by two-body instantaneous collisions. This is, in principle, a stringent limitation, both on conceptual and practical grounds, for it rules out many fluids of great practical importance, dense fluids and liquids in the first place.

Indeed, owing to the paramount role of dense fluids and liquids, extensions of the Boltzmann's equations to non-dilute gases have made the object of intense research since the early days of kinetic theory.

The Lattice Boltzmann Equation. Sauro Succi, Oxford University Press (2018).
© Sauro Succi. DOI: 10.1093/oso/9780199592357.001.0001

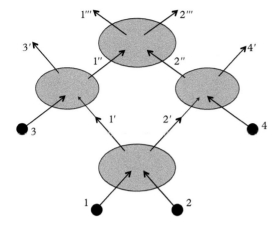

Figure 7.1 *Example of a reducible four-body collision described by the $C_4(f,f,f,f)$ operator. Note that owing to the intermediate collisions $(2',4) \to (2'',4')$ and $(3,1') \to (3',1'')$ molecules 1 and 2 can collide again shortly after the initial collision $(1,2) \to (1',2')$. Such recollision events, which manifestly violate Boltzmann's molecular chaos assumption, are responsible for the divergence of finite density expansions of the collision operator.*

The prime practical goal of these extensions is to describe finite-density, many-body corrections to the equation of state and transport coefficients. To this purpose, the most natural move is to explore suitable expansions in the granularity parameter:

$$\tilde{n} = ns^3 \tag{7.1}$$

where, as usual, n is the number density and s the effective size of the molecules.[12]

One is naturally led to formulate kinetic equations with collisions operators of increasing order, say

$$C = \tilde{n}^2 C_2(f,f) + \tilde{n}^3 C_3(f,f,f) + \tilde{n}^4 C_4(f,f,f,f) + \ldots \tag{7.2}$$

where C_k is in charge of k-th body collisions (see Fig. 7.1).

Consistently with such a finite density expansion, non-ideal equations of state are sought in the form

$$\tilde{p} \equiv \frac{P}{nk_BT} = 1 + b_1(T)\tilde{n} + b_2(T)\tilde{n}^2 + \ldots \tag{7.3}$$

where the *virial* coefficients $b_k(T)$ incorporate the effects of k-body collisions.

While reasonably successful in predicting finite-density corrections to the equation of state of non-ideal fluids, such efforts have met with limited success in capturing reliable extensions of the transport coefficients (1).

Intuitively, one of the reasons is that the transport mechanisms in dense fluids are *qualitatively* different from those in ideal or weakly non-ideal gases.

Think for instance of momentum diffusivity, characterizing the shear viscosity. In a gas of point-like particles, $s \ll d$, or $\tilde{n} \ll 1$, momentum can be transported across any given surface only on condition that the molecule itself would cross that surface. At

---

[12] This is also known as *reduced density*, and often denoted as $\tilde{n}$ in the molecular dynamics literature.

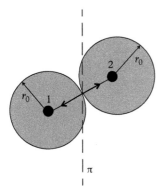

Figure 7.2 *Sketch of the non-local momentum transfer between two finite-size particles of radius r_0. The particles can exchange momentum across the plane Π even if their centermass points remain on separate sides of the plane.*

high density, however, particles come much closer to each other, to the point that their mean-interdistance, d, becomes comparable with their range of interaction, $r_0 \sim s$ for short-range interactions. At this point, potential energy takes central stage in controlling the mass, momentum and energy exchanges within the fluid. In particular, for finite-size particles, $s > 0$, momentum can be exchanged across a surface even though the center-mass of neither interacting particles does cross the surface itself. This is a qualitatively new transport mechanism which simply does not exist in weakly non-ideal gases (see Fig. 7.2).

Owing to such qualitative difference, power expansions of the mean-free path of the form

$$l_\mu = \frac{1}{n\sigma} \left(\frac{1}{c_1 \tilde{n} + c_2 \tilde{n}^2 + \dots} \right) \tag{7.4}$$

have met with mixed results in describing transport phenomena in dense fluids.

The reason, as usual, is that the non-perturbative regime where $\tilde{n} \sim 1$ would require an all-term resummation of the entire series. In this respect, much can be learned by a statistical formulation of the many-body problem, as developed by several authors in the 50s, primarily Nikolay Bogoliubov (1909–22) in Russia, Herbert Green (1920–99) and John Kirkwood (1907–59) in the USA and Jacques Yvon (1903–79) in France, whence the BBGKY acronym (2; 3).

7.3 The BBGKY Hierarchy

The BBGKY theory provides an elegant route to derive the Boltzmann equation from the general framework of many-body non-equilibrium classical statistical mechanics.

Since this matter is covered in many (advanced) textbooks, here we only sketch the baseline arguments. The reader with little penchant for formal developments of statistical mechanics can skip this section without any serious damage for the subsequent reading, except perhaps for a narrower appreciation of the lattice Boltzmann formulation of non-ideal fluids. For those willing to go through this chapter, the good news is that a great deal of the physics of dense fluids can be gleaned by stopping at just the second level of the full N-body BBGKY hierarchy. Nevertheless, some level of abstraction must be digested.

7.3.1 Statistical Mechanics: The Liouville Equation

Let us begin by considering again Newton's equation for a collection of N classical point-like particles, interacting through a two-body potential V:

$$\begin{cases} \frac{d\vec{x}_i}{dt} = \vec{v}_i \\ \frac{d\vec{p}_i}{dt} = \vec{F}_i = -\partial_{\vec{x}_i} \sum_{j>i=1}^{N} V(\vec{x}_i, \vec{x}_j), \ i = 1, N \end{cases} \tag{7.5}$$

where $\vec{p}_i = m\vec{v}_i$ are the particle momenta.

For the purpose of formulating the BBGKY equations, it proves convenient to recast the Newton's equation in the elegant Hamiltonian formalism, namely (4):

$$\begin{cases} \frac{d\vec{q}_i}{dt} = \frac{\partial H}{\partial \vec{p}_i} \\ \frac{d\vec{p}_i}{dt} = -\frac{\partial H}{\partial \vec{q}_i} \end{cases} \tag{7.6}$$

where $\{\vec{q}_i, \vec{p}_i\}$, $i = 1, N$, are the generalized positions and momenta of the N particles and

$$H(\vec{q}_1 \dots \vec{p}_N) = \sum_{i=1}^{N} \frac{p_i^2}{2m} + \sum_{i=1}^{N} \sum_{j>i=1}^{N} V(\vec{q}_i, \vec{q}_j) \tag{7.7}$$

is the N-body *Hamiltonian* of the system, on the assumption of pairwise interactions.

The dynamical state of the N-body system at each given time, t, can be represented in a compact form as a single point moving around in the $6N$-dimensional *phase space* Γ_N

$$\zeta(t) \equiv \{\vec{\zeta}_1(t) \dots \vec{\zeta}_N(t)\} \tag{7.8}$$

where we have used the shorthand

$$\vec{\zeta}_i = \{\vec{q}_i, \vec{p}_i\} \tag{7.9}$$

for the point in the six-dimensional single-particle phase space.

In cartesian coordinates:

$$\zeta \equiv \{x_1, y_1, z_1, p_{x,1}, p_{y,1}, p_{z,1} \dots x_N, y_N, z_N, p_{x,N}, p_{y,N}, p_{z,N}\}$$

for a total of $6N$ variables.

As time unfolds, for each given initial condition $\vec{\zeta}(t = 0) = \vec{\zeta}_0$, the point $\zeta(t; \zeta_0)$ traces a smooth trajectory in the $6N$-dimensional phase space Γ_N.

The same information can be encoded in the so-called N-body *Klimontovich distribution* defined as follows:

$$\delta_{1...N}(\zeta;t) = \delta[\zeta - \zeta(t)] \tag{7.10}$$

which, by definition, is zero everywhere except for $\zeta(t) = \zeta$, i.e., when the trajectory lands exactly on top of the point ζ in $6N$-dimensional space Γ_N.

In explicit position-momentum coordinate form:

$$\delta_{1...N}(\vec{q}_1, \dots \vec{p}_N; t) = \prod_{i=1}^{N} \delta[\vec{q}_i - \vec{q}_i(t)]\delta[\vec{p}_i - \vec{p}_i(t)] \tag{7.11}$$

Clearly, the Klimontovich distribution contains the same information as the full set of Newton–Hamilton equations. However, it proves more convenient for subsequent formal manipulations, to be described shortly.

Let us now consider a generic phase-space physical observable $Q(\zeta)$. The observed value of $\bar{Q}(t)$ in a given experiment is defined through a time-average over an observational time τ.

Formally,

$$\bar{Q}(t;\tau) = \frac{1}{\tau} \int_{t-\tau/2}^{t+\tau/2} Q(\zeta(t'))dt' \tag{7.12}$$

where $\zeta(t')$ is the generic point along the phase-space trajectory traced in the time interval $t-\tau/2 < t' < t + \tau/2$.

Since the observational time τ is supposedly much longer than the time-scale variation of $Q(\zeta(t))$, the experimentally observed value of Q at statistical equilibrium can be defined as the limit of (7.12) for $\tau \to \infty$

$$\bar{Q}(t) = \lim_{\tau \to \infty} Q(t;\tau) \tag{7.13}$$

Note that, by omitting the dependence on the initial conditions, we are implicitly assuming that sampling over the entire trajectory is equivalent to averaging over the ensemble of initial conditions, a property called *ergodicity*, to which we shall return shortly (see Fig. 7.3).

Let us now consider a "cloud" of initial conditions, $\zeta^\alpha(t = 0) = \zeta_0^\alpha, \alpha = 1, N_\alpha$, with the property that each different ζ_0^α maps to the *same* value of the macroscopic observable $\bar{Q}$, say the total energy of the system.

Such cloud is called *ensemble*, a subset of the full phase space Γ_N. For instance, if Q stands for kinetic energy, the ensemble would live on the hypersurface of Γ_N defined by the equation

$$E_{kin} = \sum_{i=1}^{N} \frac{mv_i^2}{2} = Const \tag{7.14}$$

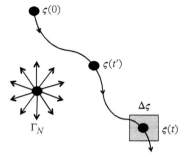

Figure 7.3 *Trajectory $\zeta(t)$ in phase space Γ_N. The N-body distribution $f_N(\zeta; t)$ receives contributions from all trajectories $\zeta(t)$ which lie within the phase-space element $\Delta\zeta$ at time t. The contribution is proportional to the time Δt spent by the trajectory within $\Delta\zeta$.*

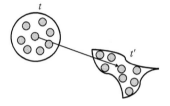

Figure 7.4 *The cloud of points evolving in 6N-dimensional phase-space Γ_N, here a simple plane for graphical purpose. The shape is very different but the area is the same (at least this is the intention in the Figure ...).*

As time unfolds under the effect of the Newton–Hamilton equations, the initial cloud moves and morphs into a different shape, like a fluid leaving in the hyperdimensional phase space Γ_N (see Fig. 7.4).

Let

$$\Delta\mathcal{N} \equiv f_{1...N}(\zeta)\,\Delta\zeta$$

represent the average number of points within a volumlet $\Delta\zeta = \Delta\vec{x}_1 \Delta\vec{v}_1 \ldots \Delta\vec{x}_N \Delta\vec{v}_N$ centered around ζ in Γ_N. The quantity $f_{1...N}$ is the N-body distribution function, i.e., the N-dimensional extension of Boltzmann's one-body distribution function f_1.

In explicit terms, $f_{1...N}\Delta\vec{\zeta}$ represents the average number of N-body configurations associated with particle 1 around position $\vec{x}_1$ with velocity $\vec{v}_1$, *and* particle 2 around position $\vec{x}_2$ with velocity $\vec{v}_2$ *and* ... particle N around position $\vec{x}_N$ with velocity $\vec{v}_N$, all at the same time t.

Hence, in terms of the Klimontovich distribution introduced in Figure 7.4

$$f_{1...N}(\zeta; t) = \,<\delta_{1...N}>(\zeta; t)$$

where brackets denote ensemble averaging.

The corresponding probability is given by the normalized N-body distribution:

$$p_{1...N} = \frac{f_{1...N}}{N!}$$

where the denominator counts the number of ways of selecting N distinguishable particles out of a pool of N, in exclusive sequence.

In other words, there are N ways of selecting the first particle, $N-1$ to select the second and so on, up to the last particle, for which there is just one only option left. Clearly, this holds for classical, hence distinguishable, particles.

7.3.2   Liouville Theorem

Having defined the N-body distribution function, we next proceed to consider the motion of the $6N$-dimensional cloud over a small time interval, from t to $t' = t + \Delta t$.

By definition, all points belonging to the cloud C at time t are to be found in the deformed cloud C' at t', that is

$$f_N(\zeta)|\Delta\zeta| = f_N(\zeta')|\Delta\zeta'| \tag{7.15}$$

where f_N is a shorthand for $f_{1...N}$ and $|\Delta\zeta|$ denotes the volume of the phase-space element $\Delta\zeta_N$ (in a more sophisticated mathematical language, the volume is called "measure").

Differently restated,

$$f_N(\zeta') = f_N(\zeta) \left|\frac{d\zeta}{d\zeta'_N}\right| \equiv |\mathcal{J}| f_N(\zeta) \tag{7.16}$$

where $\mathcal{J}$ is the determinant of the Jacobian of the transformation $\zeta \to \zeta'$.

At this point, it proves expedient to write Newton's equations as a first-order flow in phase space:

$$\frac{d\zeta}{dt} = g(\zeta) \tag{7.17}$$

where g is a $6N$-dimensional vector in Γ_N.

It is left as an exercise to the reader to show that under the condition

$$div_{Ng} \equiv \sum_{i=1}^{N} \partial_{\zeta_i} g_i = 0$$

the determinant of the Jacobian is unity for any Δt.

In other words, the $6N$-dimensional phase-space fluid is incompressible and the volume of the cloud remains unchanged, no matter how complex its shape. The same brave reader can also check that this is indeed the case for Newtonian equations, as long as the forces depend only on the spatial coordinates.

As a result, we obtain

$$f_N(\zeta') = f_N(\zeta) \tag{7.18}$$

This is an important result, also known as *Liouville's theorem*, after the French mathematician Joseph Liouville (1809–82), stating that the N-body distribution is invariant under the underlying microscopic dynamics.

The Liouville theorem is readily turned into a differential equation by taking the limit $\Delta t \to 0$. In this limit, the Jacobian still tends to the unit value, with no need of invoking incompressibility.

The result is the N-body Liouville equation, namely

$$\partial_t f_N + \mathcal{L}_N f_N = 0 \tag{7.19}$$

where

$$\mathcal{L}_N \equiv \sum_{i=1}^{N} g_i \partial_{\zeta_i} \tag{7.20}$$

is the N-body Liouville operator, actually a streaming operator in Γ_N.

In explicit Newtonian form:

$$\mathcal{L}_N = \sum_{i=1}^{N} \left(\vec{v}_i \cdot \partial_{\vec{x}_i} + \frac{\vec{F}_i}{m} \cdot \partial_{\vec{v}_i} \right) \tag{7.21}$$

For the case of two-body interactions, the forces are given by

$$\vec{F}_i = \sum_{j>i} \vec{F}_{ij} \equiv -\sum_{j>i} \frac{\partial V_{ij}}{\partial \vec{x}_i} \tag{7.22}$$

More general cases also fit within the same formalism.

7.3.3 Taking Averages

The Liouville equation strikes for its formal beauty and elegance.

But, how about its practical use? In other words, what did we gain as compared to solving the Newton equations?

As it stands, the answer is a pretty close proxy to nothing.

In fact, solving a partial differential equation in $6N$ dimensions, with Avogadro-like values of N, is a *worst* nightmare than solving the $6N$ Newton equations themselves!

So, what is the point of the Liouville equation?

As we shall see in a moment, the point is quite crucial indeed: the Liouville formulation permits us to define macroscopic observables through (computable) statistical averages.

To this purpose, let us return a moment on the process of taking averages. We start with a phase-space function of the microscopic observables, $Q = Q(\zeta)$, and define macroscopic averages via either time or ensemble averaging.

In the latter case, one first turns the phase-space dependence into a time dependence, i.e., $Q(t) \equiv Q(\zeta(t))$, or, in a more systematic form, via a phase-space convolution with the Klimontovich distribution:

$$Q(t) = \int_{\Gamma_N} \delta_N[\zeta - \zeta(t)]Q(\zeta)d\zeta$$

Note that this convolution implies no average as yet, since the Klimontovich distribution contains the very same information of the instantaneous N-body configuration. Ensemble averaging of the previous expression delivers

$$<Q>(t) = \int_{\Gamma_N} <\delta_N[\zeta - \zeta(t)]> Q(\zeta)d\zeta = \int_{\Gamma_N} f_N[\zeta;t]Q(\zeta)d\zeta \qquad (7.23)$$

This expression shows that f_N permits us to formally compute macroscopic averages in the form of an integral over $6N$-dimensional phase space Γ_N.

Before facing the task of computing such a hyper-dimensional integral, a more fundamental question arises in the first place: *does the phase-space average, $<Q>(t)$, coincide with the time-trajectory average $\bar{Q}(t)$?*

This is by no means a given.

In fact, the equality

$$\bar{Q}(t) = <Q>(t) \qquad (7.24)$$

identifies a specific class of dynamical systems known as *Ergodic* systems (5).

In a nutshell, a system is ergodic if its trajectories visit every region phase space uniformly, so that the time spent in a given subregion of phase space, hence the probability of finding the system in that region, is directly proportional to the measure of the volume of the region itself.

Under ergodicity, time averages and ensemble averages are clearly equivalent.

Ergodicity is *not* the rule for dynamical systems: at the time of writing, a rigorous proof of ergodicity has been given only for a handful of dynamical systems, the most popular one being perhaps the so-called Sinai's billiard (a rigid disc undergoing specular reflections on the solid walls of a confining box).

Notwithstanding this state of affairs, ergodicity is often assumed in actual practice, for it relieves from the unwieldy task of computing trajectories over very long periods of time, or taking expensive averages over large sets of initial conditions.

Here comes the high point of the phase-space approach: most physical quantities result from *averaging over a very low-dimensional subset of phase space*. Hence, the hyper-dimensional phase-space integral (7.23) can be performed in much lower dimensions, in many instances just $N = 2$.

A few examples illustrate the point.

Consider the total kinetic energy of the system, eqn (7.14).

According to (7.23), its average value is

$$E_{kin}(t) = \frac{1}{2} \int \sum_{i=1}^{N} m v_i^2 p_N(\zeta) \, d\zeta, \tag{7.25}$$

where $p_N(\zeta) \equiv p_N(\zeta_1 \ldots \zeta_N)$. Since the integrand $Q = \frac{m}{2} \sum_i v_i^2$ depends only on the single-particle velocity, the integration over the space coordinates simply delivers

$$E_{kin}(t) = \frac{m}{2} \int \sum_{i=1}^{N} v_i^2 p_N(\vec{v}_1 \ldots \vec{v}_N) d\vec{v}_1 \ldots d\vec{v}_N \tag{7.26}$$

where

$$p_N(\vec{v}_1 \ldots \vec{v}_N) \equiv \int p_N(\vec{\zeta}_N) d\vec{x}_1 \ldots d\vec{x}_N$$

The left integrand is simply a sum of single-particle velocity terms $\vec{v}_i$, and therefore the integration over velocity space reduces to

$$E_{kin}(t) = \frac{m}{2} \int \sum_{i=1}^{N} v_i^2 p_1(\vec{v}_i) d\vec{v}_i = \frac{m}{2} N \int v_1^2 p_1(\vec{v}_1) d\vec{v}_1 \tag{7.27}$$

where we have defined the one-body probability distribution as

$$p_1(\vec{v}_1) = \int p(\vec{v}_1 \ldots \vec{v}_N) d\vec{v}_2 \ldots d\vec{v}_N \tag{7.28}$$

Clearly, p_1 is a close relative of the Boltzmann distribution.

In other words, the calculation of the total energy $E_{kin}(t)$ just requires the one-body distribution function in velocity space!

In hindsight, this is obvious, since the total kinetic energy is the sum of N replicas of the same one-particle quantity.

By the same token, the average potential energy is given by an integral over the two-body spatial distribution

$$E_{pot}(t) = \int \sum_{i=1}^{N} \sum_{j>i=1}^{N} p_{12}(\vec{x}_i, \vec{x}_j) V(\vec{x}_i, \vec{x}_j) d\vec{x}_i d\vec{x}_j = \frac{N(N-1)}{2} \int p_{12}(\vec{x}_1, \vec{x}_2) V(\vec{x}_1, \vec{x}_2) d\vec{x}_1 d\vec{x}_2 \tag{7.29}$$

where the two-body spatial distribution is defined as

$$p_{12}(\vec{x}_1, \vec{x}_2) = \int p_N(\vec{x}_1, \vec{v}_1 \ldots \vec{x}_N, \vec{v}_N) d\vec{x}_3 \ldots d\vec{x}_N d\vec{v}_1 \ldots d\vec{v}_N \tag{7.30}$$

These examples clearly highlight the strategic role played by low-dimensional projections of the full N-body distributions, also called *reduced* or *marginal* distribution functions.

The BBGKY hierarchy to be presented in Section 7.3.4 provides a systematic framework to derive dynamical equations for these reduced distribution functions at any order.

7.3.4 Many-Body Equilibria of the Liouville Equation

The Liouville equation admits an important class of steady-state solutions (equilibria) defined by the condition

$$\mathcal{L}_N f_N^{eq} = 0$$

It is readily seen that these solutions take the general form

$$f_N^{eq}(p, q) = f[H(p, q)] \tag{7.31}$$

where f is any function of the energy $E = H(p, q)$ and we used the shorthand

$$q = \{\vec{q}_1 \ldots \vec{q}_N\}, \ p = \{\vec{p}_1 \ldots \vec{p}_N\}$$

Such solutions are readily checked to descend from the Hamilton equations (7.6). This is best seen by writing the Liouville equation in Hamiltonian form, namely:

$$\partial_t f_N + \{H, f_N\} = 0$$

where the Poisson brackets stand for

$$\{H, f_N\} = \sum_i \frac{\partial H}{\partial \vec{p}_i} \frac{\partial f_N}{\partial \vec{q}_i} + \frac{\partial H}{\partial \vec{q}_i} \frac{\partial f_N}{\partial \vec{p}_i}$$

This clearly shows that any f_N function of the hamiltonian (energy) alone (see 7.31) annihilates the Poisson brackets hence it realizes equilibrium. Of particular relevance to equilibrium statistical mechanics is the so-called *canonical distribution*:

$$p_N^{eq}(p, q) = Z^{-1} e^{-\beta H(p, q)} \tag{7.32}$$

where $\beta = 1/k_B T$ is the inverse temperature and $Z = Z(\beta)$ is the *partition function* of the system, i.e.,

$$Z(\beta) = \int_{\Gamma_N} e^{-\beta H(p, q)} \, dp dq \tag{7.33}$$

with

$$dp = \prod_{i=1}^{N} dp_{i,x} dp_{i,y} dp_{i,z}, \ \ dq = \prod_{i=1}^{N} dq_{i,x} dq_{i,y} dq_{i,z}$$

in cartesian coordinates, $q_{i,x} = x_i$, $q_{i,y} = y_i$, $q_{i,z} = z_i$, for simplicity (for general coordinates one should use the appropriate measure $d\mu(p,q)$).

This distribution describes many-body systems in contact with a so-called *heat-bath* at temperature T and represents a cornerstone of equilibrium statistical mechanics.

The canonical distribution can in principle be used to compute equilibrium averages of any phase-space quantity $Q(\zeta)$, if only one could perform the $6N$-dimensional integral (7.33).

With a few notable exceptions, for instance quadratic potentials leading to direct product of one-dimensional gaussian distributions, such integral can be computed only approximately, typically by Monte Carlo techniques.

This is a major mainstream of classical statistical mechanics.

7.4   Reduced Liouville Equations and the BBGKY Hierarchy

We have already noticed that the Liouville equation is by no means any easier to solve than the full set of Newton's equations. In fact, the opposite is true: it is basically a computationally untractable object!

Nevertheless, it is of great conceptual and practical value, for it permits us to develop systematic approximations to the classical many-body problem.

The first observation is that, as we did in Section 7.3, one can lower the rank of the many-body problem, from N to any $M < N$, by integrating over the excluded phase-space coordinates $d\vec{\zeta}_{M+1} \ldots d\vec{\zeta}_N$.

This defines the so-called *reduced* M-*body distribution*:

$$f_{1\ldots M}(\vec{\zeta}_1 \ldots \vec{\zeta}_M; t) = \frac{(N-M)!}{N!} \int f_{12\ldots N}(\vec{\zeta}_1 \ldots \vec{\zeta}_N) d\vec{\zeta}_{M+1} \ldots d\vec{\zeta}_N \tag{7.34}$$

a process sometimes called "marginalization".

Geometrically speaking, this corresponds to projecting the full $6N$-dimensional space Γ_N onto the lower $6M$-dimensional space Γ_M.

This projection permits us to derive the dynamic equation for the M-body marginal distribution in a systematic way.

Piece by piece integrations yield the following results.

The time derivative is straightforward, by definition, and leaving aside constant prefactors:

$$\int \partial_t f_N \ d\vec{\zeta}_{M+1} \ldots d\vec{\zeta}_N = \partial_t f_M \tag{7.35}$$

where we have set $f_M \equiv f_{12\ldots M}$ for notational simplicity.

Spatial derivatives are nearly as straightforward.
With the shorthand $D\zeta_M \equiv d\zeta_{M+1} \dots d\zeta_N$, results are as follows.
For $i = 1, M$:

$$\int \vec{v}_i \cdot \partial_{\tilde{x}_i} f_N \, D\zeta_M = \vec{v}_i \cdot \partial_{\tilde{x}_i} f_M \tag{7.36}$$

again just by definition of f_M.

For $i = M + 1 \dots N$, and imposing the boundary condition $f_N = 0$ at the spatial boundary, we obtain

$$\int \vec{v}_i \cdot \partial_{\tilde{x}_i} f_N D\zeta_M = 0 \tag{7.37}$$

Velocity derivatives are a bit more elaborate, due to the two-body forces.
We have

$$\int \vec{F}_i \cdot \partial_{\vec{v}_i} f_N D\vec{\zeta}_M = \sum_{j \neq i} \int \vec{F}_{ij} \cdot \partial_{\vec{v}_i} f_N D\vec{\zeta}_M \tag{7.38}$$

For $i = M+1 \dots N$, this is zero in view of the boundary condition $f_N = 0$ at the boundary of the velocity domain.

For $i = 1, \dots M$, and $j = 1, M \neq i$, the result is

$$\sum_{i=1}^{M} \sum_{j \neq i=1}^{M} \vec{F}_{ij} \cdot \partial_{\vec{v}_i} f_M$$

because $\vec{F}_{ij}$ does not interfere with the integration over the phase-space volume $D\vec{\zeta}_M$.

For $j > M$, on the other hand, $\vec{F}_{ij}$ couples to f_N.

By symmetry, all j's can be collected within a single contribution at $j = M + 1$, so that, by integrating over $d\vec{\zeta}_{M+2} \dots d\vec{\zeta}_N$, we obtain

$$\sum_{i=1}^{M} \int \vec{F}_{i,M+1} f_{M+1} \, d\vec{\zeta}_{M+1}$$

The final result for the reduced M-body distribution is

$$\partial_t f_M + \mathcal{L}_M f_M = \sum_{i=1}^{M} \int \hat{\vec{F}}_{i,M+1} \cdot \partial_{\vec{v}_i} f_{M+1} \, d\vec{\zeta}_{M+1} \tag{7.39}$$

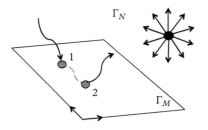

Figure 7.5 *A smooth trajectory in full phase space Γ_N landing into the lower-dimensional phase space Γ_M, where it spends some time (dashed curve between points 1 and 2) before taking-off again into Γ_M. Here, for simplicity (M = 2 and N = 10).*

where the reduced M-body Liouvillean is defined as follows:

$$\mathcal{L}_M = \sum_{i=1}^{M} \left(\vec{v}_i \cdot \partial_{\vec{x}_i} + \sum_{j \neq i} \frac{\hat{F}_{ij}}{m} \right) \tag{7.40}$$

and $\hat{F}_{ij}$ is the following two-body force operator:

$$\hat{F}_{ij} = \vec{F}_{ij} \partial_{\vec{v}_i} + \vec{F}_{ji} \partial_{\vec{v}_j} = \vec{F}_{ij} \cdot (\partial_{\vec{v}_i} - \partial_{\vec{v}_j}) \tag{7.41}$$

where use has been made of Newton's third law:

$$\vec{F}_{ij} + \vec{F}_{ji} = 0$$

This also rewrites as

$$\hat{F}_{ij} = \vec{F}_{ij} \cdot \frac{\partial}{\partial \vec{v}_{ij}} \tag{7.42}$$

where $\vec{v}_{ij} = \vec{v}_i - \vec{v}_j$ is the relative velocity between the two particles.

Full details of the calculations can be found in the beautiful book by Ishihara (6).

It is important to observe that the right-hand side is now non-zero, indicating that, unlike the full N-body distribution, the marginals f_M are subject to an effective collisional process.

This is natural, because a smooth N body trajectory goes in and out of the M-dimensional space Γ_M, thus leading to "jumps" of the projected trajectories (see Fig. 7.5). It is this sudden jump "in and out" that give rise to a non-zero left-hand side in the eqn (7.39).

The beauty of the reduced Liouville equations rests with their generality: no assumption has been made on the nature of the interactions, other than stipulating that they come in two-body form.

The second major strength of this procedure is that it can be iterated down the lowest (Boltzmann) level, namely $N = 1$, in a fairly systematic way.

7.5 Two-Body Closures

As mentioned before, for systems interacting through two-body potentials, the BBGKY hierarchy can be truncated at the level of the two-body distribution f_{12}, which in turns requires a closure statement on the three-body distribution f_{123}.

The two-body Liouville equation reads explicitly as follows:

$$(\partial_t + \vec{v}_1 \cdot \partial_{\vec{x}_1} + \vec{v}_2 \cdot \partial_{\vec{x}_2} + \vec{F}_{12} \cdot \partial_{\vec{v}_1} + \vec{F}_{21} \cdot \partial_{\vec{v}_2}) f_{12} = \int (\hat{F}_{13} + \hat{F}_{23}) f_{123} d\vec{\zeta}_3 \qquad (7.43)$$

The corresponding one-body equation is

$$(\partial_t + \vec{v}_1 \cdot \partial_{\vec{x}_1}) f_1 = \int \hat{F}_{12} f_{12} \, d\vec{\zeta}_2 \qquad (7.44)$$

By closing for f_{123}, one can, in principle, solve (7.43) for f_{12}. However, this task is hardly practical because eqn (7.43) lives in 12-dimensional phase space plus time. On can try lower-hanging fruits, i.e., solve (7.44) for f_1, upon closing for f_{12}. This is still demanding, but no more complicated than solving the Boltzmann equation, since both equations live in six dimensions plus time.

Indeed, being the lowest step in the BBGKY ladder, equation (7.44) is expected to map into the Boltzmann equation, although the precise nature of this mapping remains open at this point.

Further insight can be gained by elaborating a bit on the right-hand side of the eqn (7.43) A simple integration by parts shows that the term proportional to $\partial_{\vec{v}_2}$ makes no contribution, as long as the two-body force does not depend on the velocity coordinates. The other piece can be written as an effective one-body term $\vec{F}_1^{eff} \cdot \partial_{\vec{v}_1} f_1$, where the effective one-body force is defined as follows:

$$\vec{F}_1^{eff} = \frac{\int \vec{F}(\vec{x}_1, \vec{x}_2) f_{12} d\vec{\zeta}_2}{\int f_{12} d\vec{\zeta}_2} \qquad (7.45)$$

The central object of two-body theories is the two-body correlation function, also known as *radial correlation function* (RCF), which is defined as follows:

$$g(\vec{\zeta}_1, \vec{\zeta}_2; t) \equiv \frac{f_{12}(\vec{\zeta}_1, \vec{\zeta}_2; t)}{f_1(\vec{\zeta}_1; t) f_2(\vec{\zeta}_2; t)} \qquad (7.46)$$

This is more than a formal rewrite, because, as we shall see, the RCF lends itself to major and physically sensible simplifications (7).

In particular, since relaxation in velocity space usually occurs much faster than in configuration space, the velocity and time dependence of the RCF can often be neglected, thus bringing the two-body problem down to six spatial dimensions, plus time.

If the fluid is structurally *homogeneous*, this dependence further reduces to the intermolecular distance $\vec{r} = \vec{x}_2 - \vec{x}_1$, a three-dimensional vector.

If, besides being homogeneous, the fluid is also *isotropic*, the RCF depends only on the magnitude $r = |\vec{r}|$, a simple scalar.

In this case, the integral

$$N_{12}(r_0) = \int_0^{r_0} g(r)4\pi r^2 dr \tag{7.47}$$

measures the number of molecules in a sphere of radius r_0, pairing with a reference molecule placed in the origin at $r = 0$.

The RCF encodes the structural correlations of the fluid at distances comparable with the interaction range of the two-body potential, typically nanometers and below (see Fig. 7.6). Beyond such distances, the ideal gas (no spatial correlation) condition is attained, namely

$$g(r) = 1 \tag{7.48}$$

Inserting the expression (7.46) into (7.43), and integrating by parts, we obtain the following compact expression for the effective one-body force:

$$\vec{F}_1^{eff}(\vec{x}_1) = \partial_{\vec{x}_1} \int V(\vec{x}_1, \vec{x}_2)g(\vec{x}_1, \vec{x}_2)n(\vec{x}_2)d\vec{x}_2 - \int V(\vec{x}_1, \vec{x}_2)n(\vec{x}_2)\partial_{\vec{x}_1}g(\vec{x}_1, \vec{x}_2)d\vec{x}_2 \tag{7.49}$$

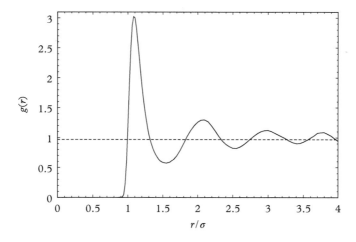

Figure 7.6 *The radial distribution function of a Lennard–Jones fluid. Due to the hard-core repulsion the probability to find the partner molecule closer than the interaction range is basically, but not literally, zero. The decaying oscillations represent the alternating sequence of "voids" and dense regions which characterize the structure (spatial memory) of the system, as dictated by the combination of short-range repulsion and mid-range attraction.*

By introducing the barycentric coordinates

$$\vec{x} = \frac{\vec{x}_1 + \vec{x}_2}{2}, \ \vec{r} = \frac{\vec{x}_1 - \vec{x}_2}{2},$$

and assuming a central potential $V(\vec{x}_1, \vec{x}_2) = V(\vec{r})$, the previous expression simplifies to

$$\vec{F}^{\text{eff}}(\vec{x}) = -\partial_{\vec{x}} \int V(\vec{r}) g(\vec{x}, \vec{x} + \vec{r}) n(\vec{x} + \vec{r}) d\vec{r} - \int V(\vec{r}) n(\vec{x} + \vec{r}) \partial_{\vec{x}} g(\vec{x}, \vec{x} + \vec{r}) d\vec{r} \quad (7.50)$$

This expression shows that for the case of a structurally homogeneous fluid, i.e.,

$$g(\vec{x}, \vec{x} + \vec{r}) = g(\vec{r}) \quad (7.51)$$

the second term in equation (7.50) vanishes, so that the effective force can be expressed as the gradient of the following effective potential:

$$V^{\text{eff}}(\vec{x}) = \int V(\vec{r}) g(\vec{r}) n(\vec{x} + \vec{r}) d\vec{r} \quad (7.52)$$

However, in the general case of a inhomogeneous fluid, say near interphase boundaries or confining walls, the effective force receives a contribution from the heterogeneity of the RCF, and it is no longer of gradient type.

By very definition, exact knowledge of the radial correlation function is tantamount to the full solution of the two-body problem. This is available only in exceptional cases, such as hard-sphere interactions, but in general, an educated guess (closure) is required.

Since for an ideal gas, $g = 1$, a plausible guess for a weakly inhomogeneous fluid, sometimes called *imperfect gas*, is given by

$$g(\vec{r}) = Z^{-1} e^{-\beta V(\vec{r})} \quad (7.53)$$

where Z^{-1} is a normalization constant (the inverse-partition function with one single degree of freedom) and $\beta = 1/k_B T$ is again the inverse temperature. However, for strongly interacting fluids, with $\beta V \sim 1$, this guess is no longer adequate.

In general, the RCF depends on the local-fluid density and temperature, so that one looks for density expansions of the form

$$g(\vec{x}, \vec{r}) = e^{-\beta V(\vec{r})} \sum_{k=0}^{\infty} g_k(\vec{r}) \tilde{n}^k(\vec{x}) \quad (7.54)$$

where $\tilde{n}(\vec{x}) = n(\vec{x}) s^3$ is the local diluteness parameter.

We shall elaborate a bit further on such an expression shortly.

7.5.1 Short and Mid-Range Interactions

Once an educated guess for the RCF is available, the expression (7.50) permits us to evaluate the effective force acting upon the one-body distribution function.

For this purpose, it proves expedient to split the integration in the radial coordinate r into a *short-range* ($r < s$) and *mid-range* ($r > s$) contributions, respectively:

$$F_{short}^{eff}(\vec{x}) = -\partial_{\vec{x}} \int_{r<s} V(\vec{r})g(\vec{r})n(\vec{x}+\vec{r})d\vec{r} \tag{7.55}$$

and

$$F_{mid}^{eff}(\vec{x}) = -\partial_{\vec{x}} \int_{r>s} V(\vec{r})g(\vec{r})n(\vec{x}+\vec{r})d\vec{r} \tag{7.56}$$

The "short-" and "mid-" range forces, sometimes also called *hard and soft core* for the reasons described in the next paragraph, behave in a qualitatively different way.

A prototypical example is provided by the Lennard–Jones potential discussed earlier in this book, which combines a short-range repulsion $V_{short} \propto (r/s)^{-12}$ with a mid-range attraction $V_{mid} \propto -(r/s)^{-6}$. The term "mid-range" indicates that the attractive tail is still sizeable at distances above the interaction range s, while the repulsive hard-core is not. Yet, mid-range forces decay must faster than long-range ones, such as Coulomb electostatics. Indeed both components of the LJ potential decay faster than for Maxwell molecules, which formally qualifies both of them as short range in the language used previously (see Fig. 7.7).

As we have learned earlier in this book, short-range interactions result in large scattering angles and generally lead to abrupt changes in the particle trajectory. It is intuitively clear that these short-range collisions are precisely the ones forming Boltzmann's collision operator. For the same reason, the mid-range interactions give rise to small

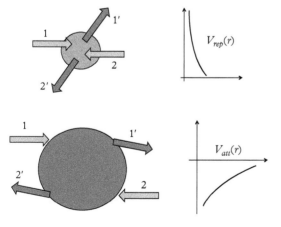

Figure 7.7 *Scattering from a short-range repulsive (top) and mid-range attractive potential.*

scattering angles, hence smooth changes in the trajectory, sometimes also called "grazing collisions." It is intuitively clear that this type of interactions behave like an effective "mean-field," resulting from the interaction of a given molecule with all other partner molecules.

As a result, the soft-core interactions can often be treated like *self-consistent* forces leading to smooth (differentiable) trajectories, whence their place on the left-hand side of the kinetic (Boltzmann–Vlasov) equation.

Based on the previous ideas, the effective one-body kinetic equation accounting for two-body interactions can be cast in the following form:

$$[\partial_t + \vec{v}_1 \cdot \partial_{\vec{x}_1} + \vec{F}^{eff}_{mid} \cdot \partial_{\vec{v}_1}]f_1 = -\vec{F}^{eff}_{short} \cdot \partial_{\vec{v}_1}f_1 \tag{7.57}$$

where the right-hand side is purely formal, since the action of short-range forces does not lead to differentiable trajectories.

A detailed proof that the right-hand side can be reconduced to the form of a Boltzmann collision operator can be found in (6).

Here, we simply sketch the basic steps.

The term on the right-hand side writes as

$$\int (\vec{v}_r \cdot \nabla_r f_{12}) \, d\vec{r} \, d\vec{v}_2,$$

where $\vec{r} = \vec{x}_2 - \vec{x}_1$ is the interparticle separation and $\vec{v}_r = \vec{v}_2 - \vec{v}_1$ is the relative velocity.

By performing the integration over $d\vec{r}$ in a cylindrical frame with the axis z aligned with the relative velocity, and assuming molecular chaos $f_{12} = f_1 f_2$, we obtain

$$\int \partial_z (f_1 f_2) \, b \, db dz d\chi \, d\vec{v}_2$$

where (b, χ) are the radius and polar angle in the plane perpendicular to the axis.

As shown in Chapter 2, $bdb \equiv \sigma(g, \chi)d\vec{\Omega}$, σ being the differential cross section. The integration over the axial coordinate delivers the statistical factor $f_1' f_2' - f_1 f_2$ in the Boltzmann equation, under the assumption that the mean-free path be much larger than the interaction range, so that the post-collisional distributions (primed) can be replaced by their asymptotic values at $z \to \infty$.

Such an assumption is readily recognized as Boltzmann's definition of a dilute gas.

It should be noted that this procedure only holds in the case of a homogeneous, spatially uniform fluid. Otherwise, the spatial dependence of f_1 and f_2 should also be taken into account, let alone separability.

All of this is reassuring, as it shows that the BBGKY treatment recovers Boltzmann's heuristic derivation in the special case where the distribution function shows no appreciable variation on the scale of the interaction range.

No new physics perhaps, but the derivation from the formal apparatus of many-body statistical mechanics provides Boltzmann's heuristic derivation with a more solid mathematical and physical background.

Moreover, it also opens the way to a number of strategies to deal with non-diluteness, as we shall briefly discuss in the sequel.

7.6 Theories of Correlation Functions

The kinetic theory of dense fluids is not an easy piece: the mathematics is elegant but thick . . . . The full solution of many-body Liouville equations is not viable even at the last but one level of the hierarchy, the two-body distribution function, which lives on a (12 + 1)-dimensional space. On the other hand, much of the many-body physics governed by pair potentials can be reconduced to the knowledge of the two-body RCF, $g(\vec{r})$, which explains the considerable focus placed in the 60s on the derivation of closure equations for this quantity.

The advantage of such perspective is that the RCF, in principle a function of 13 variables too, $g = g(\vec{\zeta}_1, \vec{\zeta}_2; t)$ lends itself to major and yet physically sensible simplifications.

In the first place, the dependence on $\vec{v}_1$, $\vec{v}_2$ can safely be neglected on timescales of hydrodynamic interest. In addition, as we have discussed before, for homogeneous fluids, the spatial dependence reduces to the interparticle distance $\vec{r} = \vec{x}_1 - \vec{x}_2$, and for isotropic ones, the vector $\vec{r}$ can be replaced by its magnitude, so that we end up with a one-dimensional problem!

The prototypical equation for two-body correlations due to Ornstein and Zernike (OZ) reads as follows (indices $1, 2, 3$ are shorthands for $\vec{\zeta}_1, \vec{\zeta}_2, \vec{\zeta}_3$):

$$h_{12} = c_{12} + n_1 \int c_{13} h_{23} \, d3 \tag{7.58}$$

where

$$h_{12} \equiv g_{12} - 1 \tag{7.59}$$

is the *total correlation function* (TCF). This goes to zero in the ideal gas limit, whereas c_{12} is the *direct correlation function* (DCF), since it accounts for direct two-particle correlations, while h_{12} collects the whole series of indirect contributions of the form $\int c_{13} c_{34} c_{42} \, d3 d4$ and higher orders.

In combinatorial terms, the end-points 1 and 2, representing h_{12} in diagrammatic form, can be connected through a virtually infinite number of different paths, each involving n intermediate points. Depending on the topology of the connecting path, some points may not need to be traversed in order to connect 1 and 2, while some others just cannot be skipped. The latter are called "nodes."

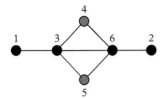

Figure 7.8 *Graph corresponding to fourth-order contributions to the total correlation function h_{12}. Points 3 and 6 are nodes, while 4 and 5 are not.*

For instance, in the diagram depicted in Fig. 7.8, 3 and 6 are nodes, while 4 and 5 are not.

The OZ equation can be regarded as a formal definition of the TCF, c_{12}. Indeed, by formally inverting the relation (7.58), we obtain ($\star$ denoting convolution and omitting indices for simplicity)

$$h = (1 - n \star c)^{-1} c = c + nc \star c + n^2 c \star c \star c \dots .$$

In spite of its purely symbolic form, this structure exposes the summation to all orders beneath the OZ equation. It is therefore clear that, once the information about c_{12} is available, the OZ represents an integral equation for h_{12}, which can be solved, typically by iterative techniques.

Such information makes the object of different types of closures.

The simplest closure is the imperfect-gas approximation:

$$g_{12} = e^{-\beta V_{12}} \tag{7.60}$$

or simply $y(r) = 1$ in terms of the so-called *cavity function* $y \equiv g e^{\beta V}$.

Within this picture, the probability of finding a second particle around the reference one, exhibits a single peak corresponding to the minimum of the interaction potential. In other words, the CRF is one-to-one related to the two-body potential, and carries no additional many-body effects.

These effects generally reflect into an explicit dependence of the CRF on the fluid density, which motivates power-law expansions of the form

$$g(\vec{r}; n) = e^{-\beta V(\vec{r})} \sum_{k=0}^{\infty} g_k(\vec{r}) n^k \tag{7.61}$$

It is customary to express the series on the right-hand side in a compact form as $e^{\beta W(\vec{r}; n)}$, so that

$$g(r; n) = e^{-\beta(V(r) - W(r;n))}$$

This identifies $W(\vec{r}; n)$ as an *effective pseudo-potential*, where "pseudo" refers to the explicit dependence on the fluid density $n = n(\vec{x})$ which betrays its non-atomistic nature (atomistic potentials depend explicitly on the intermolecular distance).

Such pseudo-potential makes then the target of the various closures of the OZ equation. A prominent example in point is the *Hypernetted–Chain* (HCN) approximation, which reads as follows:

$$c_{12}^{HCN} = h_{12} - \ln g_{12} - \beta V_{12} \qquad (7.62)$$

or, in equivalent terms, $c_{12} = h_{12} - \ln y_{12}$.

This approximation amounts to neglecting contributions of all diagrams without nodes, in algebraic terms $W = W_n$, where subscript n reminds that only diagrams with nodes are included.

By the very definition of nodes previously given, it appears intuitively clear that nodeless diagrams should contribute less to the TCF than node-carrying ones.

Another popular closure is the one due to *Percus–Yevick* (PY), which reads as follows:

$$c_{12}^{PY} = g_{12}(1 - e^{\beta V_{12}}) \qquad (7.63)$$

The algebraic approximation here is $W = e^{W_n} - 1$.

It has been found by several authors that the PY and HNC closures are somehow complementary, in that the former works well for hard-core interactions, while the latter yields satisfactory results mostly for soft-core potentials. An appealing feature of the PY theory is that it can be solved analytically for hard-spheres. Other types of closures combining the merits of HCN and PY can be found in the specialized literature.

These closures lie beyond the scope of the present work, and consequently here we simply remark that, unlike the two-body Liouville equation, the OZ closures are computationally demanding, but not unviable. However, they all depend on some (clever) form of low-density approximation, which prevents their systematic applicability to dense fluids, particularly at liquid state densities $\tilde{n} \sim 1$.

For this reason, alternatives routes have been explored. Among others, particularly noteworthy is Enskog's heuristic generalization of the Boltzmann equation, which we now describe in some detail.

7.7　The Enskog Equation

Despite considerable efforts, the problem of generalizing the Boltzmann equation to the case of dense fluids has not yet received a full and systematic solution. The bottom-up Liouville approach described earlier on is formally rigorous, but its practical applicability ultimately depends on some form of low density closure.

A useful, top-down, heuristic approach was developed by Enskog, back in 1922, who postulated a generalized version of the Boltzmann collision operator coping with some of the major limitations of the original dilute-gas form.

More precisely, Enskog's relinquished the notion of localized collisions and also took into account the short-scale spatial dependence of the Boltzmann distribution function.

These two ingredients permit us to account for non-local momentum exchanges due to potential energy interactions. It is to be noted that the ideal gas dynamics, where only kinetic energy is included takes into account only the motion of the centermass of the particles. This is adequate for pointlike particles, but as soon as particles acquire a non-zero size, it is clear that momentum can be exchanged across a plane, even though the centermass of the two interacting particles remain on the two opposite sites of the plane (see Figure 7.2).

To acknowledge this fundamental aspect, Enskog abandoned Boltzmann's Stosszahlansatz for pre-collisional particles, i.e., no correlation for particles entering a collision, and assumed instead that they are correlated, as if they were at equilibrium.

As a result, the loss term (direct collisions), proportional to f_{12}, in the Boltzmann equation is replaced by

$$\mathcal{L}_E = f_1 f_2^< g_{12}^<$$

where $f_1 \equiv f(\vec{x}_1, \vec{v}_1; t)$ as usual and we have set

$$f_2^< \equiv f(\vec{x}_1 - 2\vec{r}_0, \vec{v}_2; t) \tag{7.64}$$

and

$$g_{12}^< \equiv g_{12}^e(\vec{x} - \vec{r}_0) \tag{7.65}$$

to denote the spatially back-shifted distributions.

In (7.65) $\vec{r}_0$ is the vector joining the centermass of particles 1 and 2 and g_{12}^e is the equilibrium RDF for hard-spheres, namely (7)

$$g_{12}^e(\vec{x}, \vec{r}) = 1 + K\frac{4\pi r_0^3}{3}\rho(\vec{x})\frac{\vec{r}}{r_0} + O(r^3) \tag{7.66}$$

where $\vec{r}$ is the interparticle separation, $\vec{x}$ is the barycentric coordinate and $K = 5/2$.
Similarly, the gain term (inverse collisions) becomes

$$\mathcal{G}_E = f_1' f_2'^> g_{12}^>$$

where $f_1' \equiv f(\vec{x}_1, \vec{v}_1'; t), f_2'^> \equiv f(\vec{x}_1 + 2\vec{r}_0, \vec{v}_2; t)$ and $g_{12}^> \equiv g_{12}^e(\vec{x} + \vec{r}_0)$.
All factored in, the Enskog collision operator reads as follows:

$$C_E[f,f] = \int v_r \sigma(v_r, \vec{\Omega}) [f_1' f_2'^> g_{12}^> - f_1 f_2^< g_{12}^<] \, d\vec{v}_r d\vec{\Omega} \tag{7.67}$$

This is the desired non-local generalization of Boltzmann's collision operator, to which it reduces in the limit $r_0 \to 0$.

It is instructive to expand Enskog's collision term in powers of r_0, to obtain

$$C_E[f,f] = \int v_r \sigma(v_r, \vec{\Omega}) \{[f_1' f_2' - f_1 f_2] + \vec{r}_0 \cdot [f_1' \nabla_x (f_2' g_{12}) - f_1 \nabla_x (f_2 g_{12})]\} \, d\vec{v}_r d\vec{\Omega} \quad (7.68)$$

It is readily appreciated that, in the limit of homogeneous ($\nabla_x g_{12} \to 0$) and uniform ($\nabla_x f \to 0$) fluids, or, more precisely,

$$Kn_0 \equiv r_0 \nabla_x \ll 1 \quad (7.69)$$

the bracketed term on the right-hand side vanishes, so that the standard Boltzmann operator is recovered.

To be noted that (7.69) plays the role of an inner-Knudsen number, the size of the interacting molecule taking the place of the mean-free path, l_μ.

For a dilute gas $r_0 \ll l_\mu$ and the inner-Knudsen number is of the order of 10^{-3} or below, hence completely negligible. For liquids, however, especially in the presence of strong inhomogeneities (interfaces, nano-confinement), Kn_0 can surge to sizeable values, of order 1.

The Chapman–Enskog equation adds further complexity to the Boltzmann equation. Besides its obvious non-local nature, *streaming and collisions are no longer two separate processes*, as they overlap on a spatial scale of the order of r_0's.

Despite its heuristic nature and mathematical complexity, the Enskog equation has played and continues to play a very valuable role in the theoretical and computational study of dense and non-uniform fluids.

7.8 Entry Computers: Molecular Dynamics

The BBGKY equations display a sort of platonic beauty; they can accommodate literally any many-body configuration; those that occur with high probability, as well as those that do not stand any realistic chance of ever being realized!

Since the latter are overwhelming, the BBGKY equations are basically unviable as a practical computational tool (we refer of course only to the head-on solution of the BBGKY equations as hyper-dimensional PDE's, not to the two-body closures previously discussed).

However, with the advent of electronic computers in the 50s, kinetic theory took a major turn toward what is known today as *Computer Simulation*, and most notably for the case in point, molecular dynamics (MD).

The idea is to integrate Newton's equations of motion, but for a much smaller, in fact *outrageously* smaller, number of molecules. Pioneering efforts by Alder and Wainwright, in the mid 50s provided remarkably rich physics with as little as hundreds of particles (8)!

These heroic efforts pale in front of the multi-billion particle simulations affordable on present-day supercomputers, but their ingenuity still stands tall.

Multi-billion simulations[13], in turn pale in front of the Avogadro numbers of molecules in a centimeter cube of water!

Molecular dynamics is sometimes referred to as to "brute force," an expression which does not meet much empathy with this author. In the first place performing efficient MD simulations requires a great deal of ingenuity and technical skills, let alone a very solid knowledge of statistical mechanics, to tell "computer hallucinations" apart from genuine physics.

Second, and most importantly, like all direct simulation methods, MD interrogates Nature on very basic questions, such as *how many molecules does it take to establish collective behavior out of individualistic one?*

This is quintessential statistical physics, with very practical implications for computer simulations.

Indeed, computational physics is the art of making up with the enormous gap between the degrees of freedom available to mother nature versus those affordable by our computers.

Thus, the predictive capabilities of computer simulation depend to a large extent on our ability to draw benefit from nature's "generosity," i.e., its degree of redundancy and insensitivity to molecular details. In a word, *Universality*, as we have invoked it many times in this book.

For instance, we know that a mere micron cube of water contains of the order of 30 billion molecules. However, the pioneering simulations by Alder and Wainwright, which marked the dawning of molecular dynamics were able to show evidence of highly non-trivial collective behavior, such as violations of Boltzmann's molecular chaos and phase-transitions with hard spheres, using just a few hundreds computational molecules! This shows that direct simulations of the Newton equations (that's what MD is) are ideally placed to probe and draw benefit from the aforementioned redundancy, read generosity, of many-body classical systems: the molecules just go where interactions drive them, without wasting much information and computation on "empty" regions of phase space, the computational doom of the BBGKY hierarchy.

Needless to say, this scouting property does not come for free: MD simulations tick at extremely short timesteps, of the order of femtoseconds, and below. Hence they have very short lags in time; current supercomputers can hardly cover time spans in excess of hundreds of microseconds. A similar problem, although much less stringent, goes for space. Thus, while many phenomena are becoming accessible to MD simulations, especially in the realm of material science of microbiology, many others remain out of reach for want of computer power.

---

[13] The current MD world record stands at 4.125 trillions (10^{12}) particles, achieved by researchers at Technical University of Munich, Germany. The longest ones, with less than hundred thousand particles, reach up to 1 ms on the special-purpose supercomputer Anton, at DE Shaw Research.

These considerations explain why, like for the Navier–Stokes equations, there is a continued need for *accelerating* and *up-scaling* MD simulations. This represents a vigorous and fast-growing sector of modern computational statistical mechanics known as *Multiscale Modelling*.

The Lattice Gas Cellular Automaton method, the ancestor of Lattice Boltzmann, inscribes precisely into this line of thinking: a form of supra-molecular dynamics which strives to discard irrelevant molecular details and focus computational power on emergent hydrodynamic phenomena. And so does Lattice Boltzmann.

7.9 Summary

Summarizing, the task of extending the Boltzmann equation to non-dilute gas regimes can be formally accomplished by formulating a full many-body kinetic theory, the Liouville equation and the ensuing BBGKY hierarchy. However, such formulation meets with a number of practical problems, primarily the ultra-dimensionality of the many-body phase space.

Fortunately, much can be learned on the physics of dense fluids by truncating the BBGKY hierarchy at the level of two-body correlations. Yet, such truncations cannot capture the full picture when the diluteness parameter is of order unity, as it is the case for liquids. To this purpose, empirical extensions of the Boltzmann equation, due to Enskog, provide a valuable source of information.

Computer simulation via molecular dynamics can provide information at virtually any value of the diluteness parameter, but it is limited to small regions of space and time. This explains why the study of dense fluids, liquids and complex states of flowing matter in general, still takes a central stage in modern statistical physics.

....

REFERENCES

1. E. Cohen, 50 Years of Kinetic-Theory, Physica A, 191, 229, 1993.
2. V. Bogolubov, "Problems of a dynamical theory in statistical physics", in *Studies in Statistical Mechanics*, J. de Boer and G. Uhlenbeck eds, Vol. I, p. 11, North–Holland, Amsterdam, 1962.
3. K. Huang, *Statistical Mechanics*, Wiley, New York, 1987.
4. S. Goldstein, *Classical Mechanics*, Addison–Wesley, London, 1959.
5. T. Bohr, M. Jensen, G. Paladin and A. Vulpiani, *Dynamical Systems Approach to Turbulence*, Cambridge Nonlinear Science Series, Cambridge, 1998.
6. A. Ishihara, *Statistical Physics*, Academic Press, 1971
7. V.I. Kalikmanov, *Statistical Physics of Fluids*, Springer Verlag, 2001.
8. B. Alder and J. Wainwright, "Decay of the velocity autocorrelation function", *Phys. Rev. A* **1**, 18, 1970.

....

EXERCISES

1. Starting from the Hamiltonian equations of motion (7.7) prove that the jacobian of the transformation $\vec{\zeta} \rightarrow \vec{\zeta}'$ in (7.16) takes a unit value in the limit where the time interval goes to zero.
2. Prove that the steady-state collisionless Boltzmann equation for a system of particles in a potential $V(x)$ is solved by $f = f(E)$, where $E = mv^2/2 + V(x)$.
3. What is the value of $g(r)$ at $r = 0$ for hard-sphere potentials?

8

Model Boltzmann Equations

In this chapter, we discuss simplified models of the Boltzmann equation aimed at reducing its mathematical complexity, while still retaining the most salient physical features.

All models are wrong, but some are useful . . .

(G. Box)

8.1 On Models and Theory

According to George Box's maxim above, all models are wrong, which reflects the notion that, by definition, a model is a wrong approximation to the right Theory.

Yet, as recognized by the second part of the sentence, this Cindirella role to Theory is sometimes useful indeed.

In this author's view, models may even have an edge to theory, in that, just by the very fact of being approximations, they stand higher chances of being useful without being right.

In this Chapter we shall discuss a model which has proved extremely useful indeed for the study of the kinetic theory of fluids.

8.2 Bhatnagar–Gross–Krook Model Equation

As we have observed many times in this book, the Boltzmann equation is all but an easy equation to solve. The situation surely improves by moving to its linearized version, but even then, a lot of painstaking labor is usually involved in deriving specific solutions for the problem at hand.

In order to ease this state of affairs, in the mid-fifties, stylized models of the Boltzmann equations were formulated, with the main intent of providing a facilitated access to the

The Lattice Boltzmann Equation. Sauro Succi, Oxford University Press (2018).
© Sauro Succi. DOI: 10.1093/oso/9780199592357.001.0001

main qualitative aspects of the actual solutions of the Boltzmann equation, without facing head-on its mathematical complexity. As is always the case with models, the art is not to throw away the baby with the tub water. The baby, in this case, is the basic properties of the Boltzmann collision operator, namely:

1. *Local equilibrium as a zero-point of the collision operator:*

$$C(f^e, f^e) = 0 \qquad (8.1)$$

2. *Mass-Momentum-Energy (MME) conservation:*

$$\int C(f, f)\{1, \vec{v}, v^2\} d\vec{v} = 0 \qquad (8.2)$$

3. *H-theorem:*

$$\int C(f, f) h(f) d\vec{v} \leq 0 \qquad (8.3)$$

where $h(f)$ is any convex function, not necessarily $\log f$.

The simplest collision operator fulfilling these properties was proposed back in 1954 by Prabhu Bhatnagar (1912–76), Eugene Gross (1926–91) and Max Krook (1913–85), and independently by Pierre Welander (1925–96) (1).

For some unknown reason, the most common acronym, BGK, retains only the first three.

Be that as it may, the BGK or BGKW model takes the following simple form (1):

$$C_{BGK}(f) = -\frac{f - f^e}{\tau} \qquad (8.4)$$

where f^e is the local equilibrium parametrized by the local conserved quantities, density ρ, speed $\vec{u}$ and temperature T, while $\tau > 0$ is a typical timescale associated with collisional relaxation to the local equilibrium.

The first property, (8.1), is fulfilled by construction, and so is the second, as long as the local equilibrium is designed to carry the same MME of the actual distribution. Compliance with the H-theorem is less straightforward, and we shall gloss over this point here, except for observing that a positive τ guarantees that f^e is a stable attractor of f. Indeed, if $f > f^e$, $C_{BGK} < 0$ and vice versa, so that in either case, f is brought back to f^e.

This identifies the negative of the Euclidean distance,

$$S = -\int (f - f^e)^2 d\vec{v},$$

as a suitable local entropy.

Direct comparison with the Boltzmann collision operator shows that the BGK representation implies that particles at the reference velocity $\vec{v}$ are gained at a rate

$$\mathcal{G} = \omega f^e \tag{8.5}$$

and lost at a rate

$$\mathcal{L} = \omega f \tag{8.6}$$

where $\omega \equiv 1/\tau$.

It is worth noting that the loss term is exactly the same as in the original Boltzmann operator, where the collision frequency ω is generally a function of the molecular speed, (except for the case of Maxwell molecules).

On the other hand, the Boltzmann gain term can always be formally written as $\mathcal{G} = \omega f^+$, with f^+ a complicated functional of f.

By *simultaneously* setting

$$f^+ = f^e \tag{8.7}$$

and

$$\tau = const \tag{8.8}$$

BGK stipulates that: i) f is not too far from local equilibrium, and ii) close to equilibrium, the whole spectrum of eigenvalues collapses to a single, constant value.

It should be appreciated that, albeit related, the two conditions are independent of each other. Indeed, closeness to local equilibrium does not generally imply that the relaxation spectrum collapses to a single, constant, value.

By taking $f^+ = f$ and τ a proper model-functional of f, one could still describe far-off equilibrium physics, a point to which we shall return later in this book.

For the moment, let us proceed by the standard route, that is by taking stock that the BGK assumptions entail a number of limitations.

First, momentum and energy diffuse at the same rate, reflecting in a unit Prandtl and Schmidt numbers, which is only true for an ideal gas. More generally, all non-conserved moments relax at the same rate, which is unrealistic even for an ideal gas.

Nevertheless, such limitations are most often overcompensated by the gains in mathematical simplicity. The major simplification is that the quadratic global coupling in velocity space is channeled into a functional dependence on the conserved hydrodynamic variables, serving as *order parameters* for the particle distribution function.

At a first glance, it might look like the nonlinearity has been wiped out in the process, because no *explicit* particle–particle quadratic coupling appears anymore. However, this is obviously not the case; in fact, formally, BGK is even *more* nonlinear than the Boltzmann equation itself, because the local equilibrium depends exponentially on the fluid speed and inverse temperature, both being linear functionals of the particle distribution.

Notwithstanding this hidden exponential nonlinearity, the BGK equation proves very useful for analytical manipulations and sometimes it even permits us to obtain exact solutions.

As we shall see, the BGK equation plays a pivotal role in the theory of the Lattice Boltzmann equation.

8.3   Generalized BGK Models

We have mentioned that the major physical limitation of the BGK operator is the fact that mass, momentum and energy all diffuse at the same rate. In addition, even in the case of isothermal flows, BGK does not allow separate values of shear and bulk viscosities.

These limitations can be lifted by suitable generalizations of the basic BGK model, some of which we briefly illustrate in the sequel.

8.3.1   Multi-Relaxation BGK

The most natural generalization is to move from a single-relaxation time to a multi-relaxation time (MRT) kernel of the form

$$C_{MTR} = \int \Omega(\vec{v}, \vec{v}') f^{ne}(\vec{v}') \, d\vec{v}'' \tag{8.9}$$

where $f^{ne} \equiv f - f^e$, as usual.

As discussed in a previous Chapter, the collision kernel can be expressed as a weighted sum over its spectral eigenvectors/values:

$$\Omega(\vec{v}, \vec{v}') = \sum_{k=0}^{N} A^k(\vec{v}) \omega_k A^k(\vec{v}'') \tag{8.10}$$

where the first five eigenvalues are set to zero, so as to encode mass, momentum and energy conservation, whereas the subsequent eigenvalues describe the (inverse) relaxation rate of other quantities of physical interest, such as the momentum flux (viscosity) and the energy flux (conductivity), plus of course a sequence of higher-order eigenvalues associated with the tower of kinetic moments on top of hydrodynamics (or rather underneath it, in the iceberg analogy already described).

In continuum kinetic theory, emphasis is placed on deriving the spectral properties of the linearized collision operator, starting from the microscopic interactions.

In Lattice Boltzmann theory, it proves often expedient to turn this strategy upside down, i.e., *design* the collision operator "top-down," in such a way as to recover the desired hydrodynamic equations, rather than deriving it from first-principles.

Generalized BGK models coping with the most obvious limitations of BGK have been developed since long in continuum kinetic theory (see Harris's book). According to Harris, such generalizations are at risk of self-defection, since once deprived of its simplicity, the BGK model may just lose its scope altogether.

On the other hand, *lattice* kinetic theory seems to tell a slightly different story. The MTR collision operator has gained a very visible role in Lattice Boltzmann theory, because, besides broadening the physical applicability to non unit Prandtl number, it may also provide better numerical stability. As we shall see, the Lattice Boltzmann method was in fact born in matrix (hence inherently multi-time) form, and only later refined and renamed in what goes now by the name of MTR.

8.3.2 Multi-Stage BGK

Another useful variant of the BGK operator, coping with the non-unit Prandtl problem, is the multi-stage version, first proposed by Shakov and revived in the lattice context by a number of authors (2).

The idea goes as follows. The Boltzmann collision operator can be regarded as a non-linear mapping from the actual distribution f to the corresponding local equilibrium f^e. This mapping can be pictured as a trajectory in the infinite-dimensional kinetic space (the Hilbert space discussed already), taking f at time $t = 0$ to f^e at $t \to \infty$.

In practice, as long as exponential relaxation applies, the local equilibrium is attained in a matter of a few collision times.

This dynamic relaxation process can be formalized as a sequence of N steps, a discrete path in kinetic space, starting at f and ending at f^{eq}. Formally,

$$f \equiv f_0 \to f_1 \to f_2 \ldots f_\infty \equiv f^e \tag{8.11}$$

In (8.11), the f_k's represent a series of intermediate quasi-equilibria, connecting the initial and final states, f and f^e.

One may then design these quasi-equilibria in such a way that on the k-th segment, the k-th order kinetic moment M_k, would relax to its equilibrium value, while the corresponding non-equilibrium component would die out on timescale τ_k.

For simplicity, let us consider just a two-step chain, represented by the following two-time BGK collision operator (see Fig. 8.1):

$$C_{BGK2} = -\omega_1(f - f_1) - \omega_2(f_1 - f^e) \tag{8.12}$$

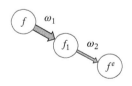

Figure 8.1 *Schematics of a two-stage BGK collision operator. The initial state f decays fast (thick arrow) to the quasi-equilibrium f_1, which in turns decays more slowly (thin arrow) into the local equilibrium f^e.*

Figure 8.2 *Schematics of the one-stage BGK "renormalized" collision operator, equivalent to the two-stage relaxation previously portrayed.*

This can also be recast in standard single-relaxation form

$$C_{BGK2} = -\omega_1 (f - \tilde{f}^e) \tag{8.13}$$

with a "renormalized" equilibrium:

$$\tilde{f}^e = \frac{\omega_2}{\omega_1} f^e + \left(1 - \frac{\omega_2}{\omega_1}\right) f_1 \tag{8.14}$$

See Fig. 8.2.

Clearly, in the limit $\omega_1 \to \omega_2 = \omega$, the renormalized equilibrium recovers the original one, i.e., $\tilde{f}^e = f^e$.

By choosing $f_1 = f^e(T)$, i.e., the local equilibrium f^e evaluated at the *current* temperature T instead of the average one T_0, one secures that $f_1 \to f^e$ in the limit $T \to T_0$. It is readily checked that in such limit, one recovers again a single relaxation scheme, with local equilibrium $\tilde{f}^e = f^e$.

The idea is then to choose $\omega_1 \gg \omega_2$, so as to describe a two-stage relaxation whereby the local distribution is first attracted to the quasi-equilibrium f_1 on a short timescale ω_1^{-1}, and subsequently relax f_1 to the isothermal local equilibrium $f^e(T_0)$ on a longer timescale ω_2^{-1}.

By allowing some form of "artificial" energy exchange at non-unit Prandtl number, this hierarchical organization of the collisional process has been shown to achieve better numerical stability for the simulation of isothermal flows.

8.4   Discrete Velocity Models

The Boltzmann collision operator involves an integration in velocity space which can rarely be performed analytically, a precious exception being the case of Maxwell molecules.

As a result, it is natural to replace it with discrete versions, especially with computer implementations in mind. In principle, for a "blind" discretization, one would expect that accurate representations of the Boltzmann integral would require a very fine subdivision of velocity space in thin cells, much smaller in size than the typical thermal speed v_T:

$$\Delta v \ll v_T \tag{8.15}$$

This is a steep road for numerics, since a fine discretization of velocity space and configuration space at the same time rapidly leads to unmanageable demands of computer resources: a 32-grid point discretization of single-particle phase space gives $32^6 \sim 2 \, 10^9$ grid points overall.

Fortunately, for hydrodynamic purposes, much can be learned by pretty economic velocity-space discretizations, with

$$\Delta v \sim v_T$$

See Fig. 8.3. Provided the basic symmetries and dynamical constraints of the original collision operator, the three basic properties of the collision operator are preserved.

Ultimately, this is possible *because phase space is by no means symmetric with respect to its space and velocity dimensions: equilibration in velocity space is usually much faster than in configuration space, owing to microscopic conservation laws.*

Metaphorically it is a bit like the Pacific versus the Atlantic Ocean ...

Let us next consider a sequence of discrete velocities $\vec{v}_i$, $i = 1, N_v$. For the case of $N_v = 4$ velocities in two spatial dimensions:

$$v_{ix} = c \, cos(\phi_i), \quad v_{iy} = c \, sin(\phi_i) \tag{8.16}$$

where c is the maximum speed in town and the angles $\phi_i = (i-1)\frac{\pi}{2}$, $i = 1, 4$, span the four nearest neighbors, while the magnitude is fixed at the maximum speed c.

The Boltzmann integral is turned into a discrete sum, comprising all direct collisions $(i,j) \rightarrow (k,l)$, the Loss term, and the inverse ones $(k,l) \rightarrow (i,j)$, the Gain term. In equations:

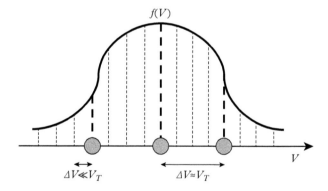

Figure 8.3 *Fine-grained and coarse-grained (filled circles) discretization of the distribution function $f(v)$. If the coarse-grain mesh points are chosen properly, a number of low-order kinetic moments (integrals of the distribution) can be computed* exactly. *This is the principle of Gaussian quadrature, to be described in detail in chapter 15.*

$$C_i = \sum_{j=1}^{N_v} (A_{kl;ij} f_k f_l - A_{ij;kl} f_i f_j) \equiv \mathcal{G}_i - \mathcal{L}_i \tag{8.17}$$

Note that the index i is left free, while j runs. Once the source state (i,j) is given, the destination state (k,l) is uniquely determined by the conservation laws, so that j is left as the only runner.

One can then hope that just a few discrete speeds with the right symmetry may be able to retain the essential of the true Boltzmann collision operator, namely the existence of a unique attractor (local equilibrium) and an H-theorem securing that any initial condition would converge to such an attractor.

The attentive reader might have noticed a crucial difference between discrete velocity models and BGK-like ones.

In the former, collisions do take place explicitly, if only on a discretized space. In the latter, they are only implicitly accounted for through the *specification* of the local equilibrium at the outset. Since collisions are explicitly realized, it is no surprise that besides the three properties (8.1,8.2,8.3), a further basic requirement needs to be imposed.

This is the so-called *detailed balance*, namely

$$p_{ij} A_{ij,kl} = p_{kl} A_{kl,ij} \tag{8.18}$$

where p_{ij} and p_{kl} are the probabilities to realize (i,j) and (k,l) as a pre(post)-collisional states, respectively.

The condition

$$A_{ij,kl} = A_{kl,ij} \tag{8.19}$$

is also called *microreversibility*, as it implies a one-to-one reciprocity between direct and inverse collisions (see Figs. 8.4 and 8.5).

As combined with detailed balance, it implies that at equilibrium

$$p_{ij} = p_{kl} \tag{8.20}$$

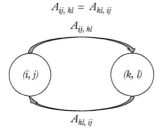

Figure 8.4 *Schematics of the microreversibility condition. Loosely speaking, it means that the two states are equally accessible, no "shut doors behind" in the transition from one to the other.*

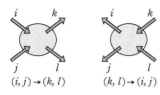

Figure 8.5 *Schematics of a direct collision $(i, j) \rightarrow (k, l)$ and inverse ones $(k, l) \rightarrow (i, j)$. The latter is obtained by simply inverting the arrows.*

As a result, under molecular chaos, the discrete-equilibrium distribution must obey the relation

$$f_i^e f_j^e = f_k^e f_l^e \tag{8.21}$$

This show that, like in the continuum case, $log\, f^e$ is a collisional invariant, which sets the stage for Boltzmann's H-theorem in discrete-velocity space.

A weaker condition, still sufficient to prove the H-theorem, is the so-called *semi-detailed balance* condition:

$$\sum_{ij} p_{ij} A_{ij,kl} = 1, \forall k, l \tag{8.22}$$

This states that, upon summing over all possible source states (i, j), any destination state (k, l) is reached with probability 1.

With these basic constraints in place, the discrete velocity model is well poised to capture many qualitative, and sometimes even quantitative, features of the true Boltzmann equation. Maybe still a cartoon of a real gas of molecules, but a very valuable and informative one: no "wallpaper syndrome" (beautiful pictures with no physical content).

Many Boltzmann model equations have been developed since the mid-fifties, starting with pioneering work of T. Carleman, M. Kac, J. Broadwell, then H. Cabannes, R. Gatignol, H. Cornille, N. Bellomo, G. Toscani and others.

The main aim of such discrete-velocity models is to simplify the solution of the Boltzmann equation, either numerically or analytically, and possibly come up with exact solutions, typically in the form of nonlinear waves far from equilibrium.

Hence, no special emphasis on the hydrodynamic regime.

Thus, while holding an important place in kinetic theory, including the role of conceptual precursors of Lattice Boltzmann methods, their impact on hydrodynamics has remained comparatively mild.

8.5   BGK Is Not (Necessarily) a Dilute-Gas Approximation

Before closing this brief survey on model Boltzmann equations, it is worth spending a few more comments on the broad import of the BGK approximation, beyond its historical realm of the kinetic theory of rarefied gases, and also to dispel some rather widespread misconceptions.

The BGK model is often regarded as a very crude approximation of Boltzmann physics, and, in many respects, it really is. Yet, it carries a much subtler insight than one might anticipate at first glance. After all, if it didn't, the original BGK paper surely would not score the many thousands citations it does, sixty years down the line.

This was presciently recognized by Wannier, who in his wonderful book writes:

> It is widely supposed that the relaxation time approximation is a crude device having no theoretical basis. This is not so. The Boltzmann operator is very naturally analyzed in terms of its relaxation rate spectrum, that is, in terms of particular solutions which decrease exponentially in time. (G. Wannier, *Statistical Physics*, p. 404.)

Wannier goes on to analyze the nature of the relaxation approximation in a most lucid and insightful way, which is strongly recommended to any reader, particularly those with a penchant for hasty dismissal of BGK, not to mention its lattice versions.

A rather common catch (based on many emails I have received), sometimes even fueled by boldly titled review papers is that BGK should apply only to dilute gases, the reason being that it derives from an approximation of the true Boltzmann equation, which is itself valid only for dilute gases.

In fact, things are a bit subtler than this.

The main point is that BGK is based on two fairly general assumptions, namely:

- i) *There exists a local attractor to the collisional dynamics (local equilibrium)*
- ii) *In the vicinity of this local equilibrium, the collisional process can be described in terms of a single relaxation timescale τ fixing the (linear) transport parameters of the fluid.*

Assumption i) is true for any statistical system supporting microscopic invariants, while assumption ii) is also fairly plausible, at least for systems with short-range interactions.

Neither of the two is necessarily restricted to dilute gases!

On the one hand, local Maxwellian equilibria hold true for liquid and gases alike, they do not depend on the assumption of diluteness. On the other hand, by promoting the relaxation time to a self-consistent space-time dependent field $\tau \rightarrow \tau(\vec{x}; t)$, the BGK formalism can effectively describe systems far from local equilibrium.

For instance, such extensions play a crucial role for the application of the Lattice Boltzmann method beyond the realm of Navier–Stokes hydrodynamics.

Indeed, although kinetic theory was originally meant to describe weakly-interacting (dilute) systems, modern developments in non-equilibrium statistical mechanics have shown that suitably generalized kinetic equations can prove very effective in describing the dynamics of *collective* degrees of freedom, best known as *quasi-particles* in the language of modern statistical mechanics.

The import of the BGK representation of quasi-particle dynamics is further accrued by considering the wide-open scenario of modern kinetic theory, in which kinetic-like equations are nowadays used across a wide spectrum of disciplines besides physics and chemistry, such as population dynamics, cellular communication and traffic flows in urban environments, finance and socio-dynamics, to name but a few (4).

8.6 Summary

Summarizing, many model versions of the Boltzmann equations have been developed in the past sixty years to cope with the mathematical complexity of the original Boltzmann equations. The main distinctive features of these model equations are a drastic, but not simplistic, simplification of the collision operator, often associated with economic discretization of velocity space. As we shall see, Lattice Boltzmann theory falls directly into the footsteps of these model equations, although with a rather unique focus on hydrodynamic applications.

....

REFERENCES

1. P. Bhatnagar, E. P. Gross, and M. Krook, A Model for Collision Processes in Gases .1. Small Amplitude Processes in Charged and Neutral One-Component Systems, *Phys. Rev.*, **94**(3), 511, 1954.
2. E.M. Shakov, Generalization of the Krook Relaxation Kinetic Equation, *Fluid Dyn.*, 3(5), 95–96, 1968.
3. G. Wannier, "Statistical Physics", Dover Publications, 1987.
4. R. Liboff, "Kinetic Theory: Classical, Quantum, and Relativistic Descriptions", John Wiley and sons, 1998.

....

EXERCISES

1. Write down the general solution of the BGK equation in terms of an integral along the single-particle trajectory.
2. Prove that in a small time interval $\Delta t \ll \tau$, this solution reduces to $f(x, v; t + \Delta t) = (1 - \omega \Delta t) f(x, v; t) + \omega \Delta t f^e(x, v; t)$. where $\omega = 1/\tau$.
3. Discuss the stability of the general solution in terms of τ.

9

Stochastic Particle Dynamics

As we have seen in the previous chapters, in dense fluids and liquids molecules are in constant interaction, hence they do not fit into the Boltzmann' picture of a clearcut separation between free-streaming and collisional interactions. Since the interactions are soft and do not involve large scattering angles, an effective way of describing dense fluids is to formulate stochastic models of particle motion, as pioneered by Einstein's theory of Brownian motion and later extended by Paul Langevin. Besides its practical value for the study of the kinetic theory of dense fluids, Brownian motion bears a central place in the historical development of kinetic theory. Among others, it provided conclusive evidence in favor of the atomistic theory of matter. In this chapter, we present the basic notions of stochastic particle dynamics, and point out its connection with other important kinetic equations, primarily the Fokker–Planck equation, which bear a complementary role to the Boltzmann equation in the kinetic theory of dense fluids.

Get your kicks on Route 66,

(Bing Crosby and the Andrews Sisters)

9.1 Langevin Dynamics

Molecules in dense fluids and liquids constantly interact, their dynamics receiving comparable contributions from kinetic and potential energy. As a result Boltzmann's clearcut splitting between streaming and instantaneous-localized collisions is no longer adequate to describe the dynamics of dense fluids. We have previously discussed potential extensions of the Boltzmann equation to the case of dense fluids and the serious difficulties they are faced with. In view of such difficulties, it appears reasonable to look for alternative approaches, betting more on physical intuition than on a hard-core tackling of the mathematical physics of many-body collisions. Opening a route which has proven exceedingly beneficial in the physics of many-body systems up to now, the coarse-graining Langevin approach replaces details of the many-body interactions with three

The Lattice Boltzmann Equation. Sauro Succi, Oxford University Press (2018).
© Sauro Succi. DOI: 10.1093/oso/9780199592357.001.0001

qualitatively new-*emergent* mechanisms with no counterpart in Newtonian mechanics: *Noise, Dissipation and Memory*. In this chapter we shall focus mostly on the first two.

The physical intuition behind this move goes as follows.

We have learned that the soft-core interactions in dense fluids give rise to small deflections of the molecular trajectories. As a result, it appears appropriate to represent the interaction of each molecule with the surrounding ones (the *bath* in physics parlance) as a series of frequent, small-amplitude random kicks. Since the kicks are uncorrelated, on average they balance each other, thereby preventing molecules from keeping a directed motion. Instead, the molecules move along "drunk-walker" paths, known as Brownian motion, after the botanist Robert Brown (1773–1858), who first investigated the erratic motion of pollen particles suspended in water.[14] The first quantitative description of Brownian motion is due to Einstein (1879–1955), in one of his epoch-making 1905 trittico. Since Einstein's treatment made use of simplifying assumptions, here we discuss the general formulation, due to Paul Langevin (1872–1946).

Mathematically, the Langevin dynamics reads as follows (1):

$$\begin{cases} \dfrac{d\vec{x}}{dt} = \vec{v} \\[2ex] \dfrac{d\vec{v}}{dt} = \vec{a}(\vec{x}) - \gamma\vec{v} + \vec{\xi} \end{cases} \tag{9.1}$$

subject, as usual, to suitable initial and boundary conditions. As anticipated, we assume no memory, i.e., the friction acts instantly on the actual fluid velocity, while memory effects would imply a convolution over the past values of the velocity as well, see Fig. 9.1.

The Langevin equation bears a close resemblance to Newtonian dynamics, with two very substantial departures expressed by the second and third terms on the right-hand side of the momentum equation.

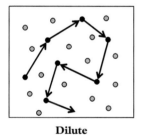

Dilute

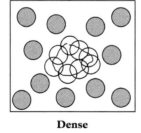

Dense

Figure 9.1 *Typical trajectory of a tagged molecule in a dilute gas and liquid-like fluid, respectively.*

---

[14] Apparently, the Dutch-born biologist Jan IngenHousz observed and described Brownian motion in 1785, earlier than Brown himself! This is yet another manifestation of *Stygler's law of eponymy*: no discovery is named after the real discoverer!

The first term is the Newtonian acceleration due to, say, an external force field. The second term, with no counterpart in Newtonian mechanics is the systematic drag experienced by any given molecule as a result of the frictional interaction with the surrounding bath. Being proportional to the molecular speed, it is immediately appreciated that this frictional force is *non-conservative*, i.e., it leads to a steady decrease of the kinetic energy of the molecule, the energy being lost on the surrounding fluid, which acts as a dissipative bath, so long as $\gamma > 0$.

For a spherical particle of radius R and mass M the friction coefficient (inverse time) is given by the Stokes relation:

$$\gamma = 6\pi \frac{\eta R}{M} \tag{9.2}$$

where η is the dynamic viscosity of the fluid solvent assumed at rest.

It is readily seen that for small particles, the mass in the denominator makes for very large values of γ.

Typical values for colloidal particles are $1 < R < 10^3 \, nm$, which combined with $\eta = 10^{-3} \, kg \, m^2/s$ for water gives γ^{-1} of the order of picoseconds (10^{-12} s) for $R = 1 \, nm$. For simplicity we have assumed colloids with the same density as water. It is of some interest to recast the expression (9.2) in the language of particle collisions, as we did with the Boltzmann equation.

By writing $\eta = nmv = nml_\mu v_T$ (see Chapter 4), we have

$$\gamma = 6\pi \frac{m}{M} n v_T R l_\mu \tag{9.3}$$

where m is the mass of solvent molecules and l_μ their mean-free path.

This also rewrites as

$$\gamma \equiv \tau^{-1} = 6 \frac{m}{M} \frac{l_\mu}{R} \gamma_{HS}$$

where $\gamma_{HS} = n v_T \pi R^2$ is the collision rate for hard spheres. Thus, for single-species dense fluids, with $m = M$ and $l_\mu \sim R$, the collision frequency is pretty close to the one of hard spheres.

After this short digression, let us come back to the stochastic term in the Langevin equations. This term, also a new entry with respect to Newtonian dynamics, describes the effects of random collisions with the surrounding solvent molecules. These are, of course, the same collisions as before, but the stochastic term $\vec{\xi}$, also referred to as noise takes charge only of their random component (the "kicks").

The noise is typically taken as a totally uncorrelated (white), namely

$$\begin{cases} \langle \vec{\xi}(t) \rangle = 0 \\ \langle \vec{\xi}(t)\vec{\xi}(t') \rangle = D_v \hat{I} \delta(t - t') \end{cases} \tag{9.4}$$

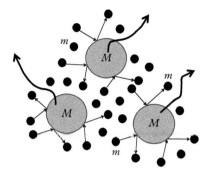

Figure 9.2 *Colloidal particles of mass M floating in a solvent bath of molecules with mass m << M. The colloidal particles perform a Brownian motion under the effect of erratic collisions with small-solvent particles.*

where $\hat{I}$ stands for the unit tensor, see Fig. 9.2.

Since drag and noise are two faces of the same medal, it comes as no surprise that they relate one to one through the so-called *Fluctuation–Dissipation* expression:

$$D_v = \gamma v_T^2 \qquad (9.5)$$

where v_T is the thermal speed of the fluid, as usual, while D_v is the diffusion coefficient in *velocity* space. Summarizing, we can interpret the motion of the Langevin particle as the result of three concurrent forces, *Conservative, Dissipative and Random*, respectively.

Namely,

$$\vec{F}_C = m\vec{a}(\vec{x}), \quad \vec{F}_D = -m\gamma\vec{v}, \quad \vec{F}_R = m\vec{\xi} \qquad (9.6)$$

The conservative force describes the coupling with the external environment, the dissipative force reflects the systematic friction resulting from the collective interaction with the surrounding molecules, and the random force reflects the stochastic component of the same collective interaction.

An exact (non-formal) solution of the Langevin equation is generally unavailable even in one spatial dimension. However, a great deal of information can be gleaned without actually solving it (a hallmark of great equations in science ...).

We shall return to this later. For the moment, let us proceed with a discussion of Brownian motion as a special case of the Langevin dynamics.

9.2 Brownian Motion

Let us consider the case where the mass of the molecule is sufficiently small to neglect its inertial effects versus frictional damping, namely

$$|\frac{d\vec{v}}{dt}| \ll |\gamma\vec{v}|$$

Under such reasonable *overdamping* assumption, the second Langevin's equation reduces to

$$\vec{v} = \frac{\xi}{\gamma} \qquad (9.7)$$

where we have considered the case of no external forces, i.e., $\vec{a} = 0$.

Replacing (9.7) into the first Langevin equation, we obtain

$$\frac{d\vec{x}}{dt} = \frac{\xi}{\gamma} \qquad (9.8)$$

This is the equation of Brownian motion, as it was used by Einstein in his 1905 celebrated paper.

Note that while the trajectory $\vec{x}(t)$ is continuous, its time derivative is not. In other words, the Brownian trajectory is continuous but not differentiable, a consequence of having sent the particle mass to zero, allowing for infinite curvature ("kicks").

In mathematics, Brownian motion is also known as a *Wiener process*.

It is of pedagogical interest to discuss Brownian motion in some further mathematical detail. To this purpose, let us now consider the one-dimensional case for simplicity. Upon integrating (9.8) over a small interval $[t, t + dt]$, we obtain

$$x(t + dt) = x(t) + \frac{1}{\gamma} \int_t^{t+dt} \xi(t')dt' \qquad (9.9)$$

This describes the trajectory of a single Brownian particle.

Since the random term under the integral is given by a series of uncorrelated kicks, the trajectory is erratic. In particular, since, on average, the kicks balance each other, the ultimate result is that, on average, the molecule moves a lot without achieving any net displacement; no directed motion.

Mathematically, this is seen by taking an ensemble average of (9.9).

$$\langle x \rangle (t + dt) = \langle x \rangle (t) + \frac{1}{\gamma} \int_t^{t+dt} \langle \xi(t') \rangle dt' \qquad (9.10)$$

Since the random force has zero average, see (9.4), one obtains

$$\langle x \rangle (t + dt) = \langle x \rangle (t) \qquad (9.11)$$

i.e., on average, the particle does not move.

Much ado about nothing?

Not really, as it is readily recognized by inspecting the variance of the displacements.

By taking the product of both sides of (9.8) with x, integrating in time, and taking ensemble average, we obtain

$$\sigma(t + dt) = \sigma(t) + \frac{1}{\gamma} \int_t^{t+dt} \langle x(t')\xi(t')\rangle dt' \tag{9.12}$$

where we have set $\sigma = \langle x^2\rangle/2$.

Substituting $x(t') = x(t) + \int_t^{t+t'} \xi(t'')dt''$, and recalling that $\langle x\rangle = 0$ at any time, we are left with

$$\sigma(t + dt) = \sigma(t) + \frac{1}{\gamma^2} \int_t^{t+dt} \int_t^{t+t'} \langle \xi(t')\xi(t'')\rangle dt' dt'' \tag{9.13}$$

Using the second relation (9.4), we finally obtain

$$\sigma(t + dt) = \sigma(t) + \frac{v_T^2}{\gamma} dt \tag{9.14}$$

This is a most distinctive property of Brownian motion: net motion is zero but the walkers "spread out" in space over a region of size proportional to the square root of time. Take a collection of Brownian walkers starting from the same position, say $x = 0$ at time $t = 0$, and let them evolve each under different noise realizations: the result is a "cloud" centered about zero (within statistical accuracy) and radius growing linearly in time, according to the relation (9.14). The ratio

$$D = \lim_{t\to\infty} \frac{\langle x^2\rangle}{t} = \frac{v_T^2}{\gamma} \tag{9.15}$$

defines the diffusion coefficient of the Brownian process in real space.

For typical values $v_T \sim 10^3$ (m/s) and $\gamma \sim 10^{12}$ 1/s, one has $D \sim 10^{-6}\, m^2/s$. This means a root-mean-square displacement of about $1\, mm$ a second, namely $1\, cm$ in 100 seconds, and $1\, m$ in 10^6 s, nearly two weeks. The upshot is: diffusion is a really slow process at a macroscopic scale. This is only natural for particles which collide frequently and randomly, see Fig. 9.3 and Fig. 9.4.

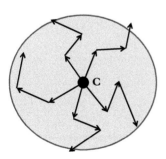

Figure 9.3 *A very rough Brownian motion: just five, three-step realizations emanating from the central point C. The average displacement over the five realizations is approximately zero, while the average-radial distance squared $< r^2 > = < x^2 + y^2 >$ is approximately 2Dt, t = 3 being the number of steps.*

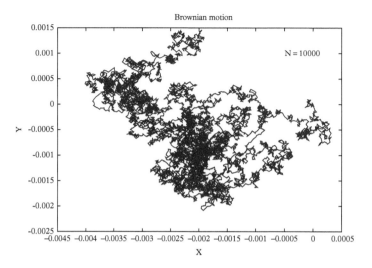

Figure 9.4 *A single Brownian motion realization with* 10,000 *steps.*

9.2.1   Link to the Diffusion Equation

It can be shown that the time evolution of the statistical distribution of the Brownian walkers is described by the well-known diffusion equation:

$$\partial_t C = D\partial_x^2 C \tag{9.16}$$

where $C \equiv C(x; t)$ is the density of walkers around the position x at time t.

A corresponding probability is readily defined as

$$p(x; t) = \frac{C(x; t)}{\int_{-\infty}^{+\infty} C(x; t)\, dx} \tag{9.17}$$

where we have assumed a spatially unbounded domain. Since the diffusion equation is linear, the probability $p(x; t)$ obeys the same diffusion equation, which can therefore be regarded as a kinetic equation in its own right.

The analytical solution in an unbounded space, with initial condition $p(x, 0) = \delta(x)$, reads

$$p(x; t) = \frac{1}{\sqrt{2\pi Dt}}\, e^{-\frac{x^2}{2Dt}} \tag{9.18}$$

From this expression, one readily observes that the average position does not change in time, i.e.,

$$\langle x \rangle (t) = \int x p(x; t)\, dx = 0 \tag{9.19}$$

The variance, on the other hand, reads as follows:

$$\langle x^2 \rangle(t) = \int x^2 p(x; t)\,dx = 2Dt \tag{9.20}$$

which coincides with the expression (9.14).

These simple expressions illustrate the one-to-one correspondence between stochastic dynamics and kinetic equations.

This correspondence goes beyond the simple diffusion process described here and leads to a more general kinetic equation, known as Fokker–Planck equation (FPE).

The basic idea is that to each stochastic trajectory $x(t)$, one can associate a probability distribution function $p(x; t)$ (PDF), defined as the probability density of finding the position $x(t)$ in the interval $x - \Delta x/2 < x(t) < x + \Delta x/2$ at time t.

On a (very) heuristic mode, and with reference to a statistical steady–state process, one writes

$$p(x)\Delta x = \frac{\Delta t}{T} \tag{9.21}$$

where Δt is the cumulative time spent by the stochastic trajectory in the time interval, T being the total-time lapse.

With due apologies to the mathematician, one takes an infinitesimal time interval dt and recasts the equation as follows:

$$p(x; t) = \frac{dt/dx}{T} = \frac{1}{T\dot{x}(t)} \tag{9.22}$$

This expression shows that the PDF associated with the stochastic trajectory $x(t)$ receives substantial contributions from the regions where the time derivative is small. This is natural, since these are the regions where the velocity is small hence the residency time is high, virtually infinite if the velocity is strictly zero.

For instance, the reader can readily check that the PDF associated with the periodic motion $x(t) = sin(t)$ is given by $p(x) = \frac{1}{\sqrt{1-x^2}}$. This is singular at $x = \mp 1$, precisely where $\dot{x} = cos(t) = 0$.

Before proceeding further, a few general comments on the duality between trajectories versus probability distributions are in order.

9.3 A Note on Coarse-Graining: Boltzmann versus Langevin

In this book, we have already met with two equivalent, yet very different faces of the diffusion coefficient and transport coefficients in general.

First, a macroscopic definition, the ratio between the mass current generated by a density gradient and the gradient itself (Fickian transport). Second, a microscopic interpretation, as the product of the mean-free path times the thermal speed.

Here, we encounter yet another microscopic definition: the asymptotic ratio of the square distance traveled by a Brownian walker in a given time, over the time span itself. The interesting point is that the two microscopic definitions arise from very different microscopic frameworks: Boltzmann kinetic theory on one side and Brownian dynamics on the other.

These stand for two distinct and equally important avenues to non-equilibrium statistical physics: probability distributions (Boltzmann's avenue) versus stochastic trajectories (Langevin's avenue).

It is important to realize that these two approaches correspond to very different ways of *coarse graining* the atomistic dynamics, as given by Newton's equations.

In Boltzmann's approach, this implies a radical change of mathematical language, from trajectories to phase-space PDF's, through the ergodicity assumption. The Langevin approach is less radical and somehow more intuitive: still trajectories, although *stochastic* ones, due to the inclusion of two new ingredients with no counterpart in Newtonian dynamics, i.e., *noise and dissipation (and eventually memory as well)*.

It only adds to the beauty and richness of kinetic theory that such a dual representation, kinetic equations versus stochastic trajectories proves capable of describing a very broad spectrum of statistical systems far from equilibrium, ranging from dilute (Boltzmann) to dense fluid systems (Langevin).

We shall return on this point shortly, when discussing one of the most important kinetic equations of all, i.e., the Fokker-Planck equation. Before doing so, let us spend some further considerations on the Langevin equation.

9.4   More on Langevin Dynamics

Having discussed Brownian motion as a special case of Langevin's dynamics, we now proceed to illustrate some qualitative properties of the Langevin equation, which make it one of the tools of choice of non-equilibrium statistical mechanics, with many applications even beyond the realm of physical sciences.

9.4.1   The Kramers Problem and First Passage Time

Of particular interest, especially for modern nanotechnology is the case of the motion of colloidal particles in a conservative force field deriving from a potential.

In one spatial dimension for simplicity:

$$F_C(x) = -\frac{dV}{dx} \tag{9.23}$$

For further sake of simplicity, let us for the moment leave noise apart, i.e., study the behavior of the system at zero temperature, i.e.,

$$\frac{d^2x}{dt^2} = -\gamma \frac{dx}{dt} + F_c(x)/m \tag{9.24}$$

Under the drive of the conservative force $F_C(x)$, the particle rolls down the potential landscape $V(x)$ and dissipates energy in the process, due to the friction with the surroundings.

Such dissipation eventually leads to a time-asymptotic rest point, a *static equilibrium* characterized by zero force and zero velocity, namely

$$\ddot{x} = \dot{x} = 0 \tag{9.25}$$

Such double condition can only be met at a stable zero point of the force field, i.e., a minimum of the potential, defined by the condition

$$\frac{dV}{dx}|_{x^*} = 0 \tag{9.26}$$

as combined with

$$\frac{d^2 V}{dx^2}|_{x^*} > 0 \tag{9.27}$$

for the sake of stability.

Depending on the shape of the potential, this equation may have a unique solution, multiple solutions, or no solution at all.

Let us begin by considering the simplest instance of single solution, the harmonic potential

$$V_H(x) = \frac{m}{2}\Omega^2 x^2 \tag{9.28}$$

where m is the mass of the particle and Ω a typical oscillation frequency of the harmonic oscillator. This confining potential is very relevant to the so-called "optical tweezers," i.e., the trapping of colloidal particles by laser beams. Clearly, there is a single stable zero point at $x^* = 0$.

In chemical kinetics it is known as the *Kramers problem*, a paradigm for the computation of escape rates of Brownian particles over potential barriers of various sorts.

For the case of the harmonic oscillator, the equations of motion take the well-known form of a damped harmonic oscillator, i.e.

$$\frac{d^2 x}{dt^2} + \gamma \frac{dx}{dt} + \Omega^2 x = 0 \tag{9.29}$$

with initial conditions $x(t_0) = x_0$ and $v(t_0) = v_0$.

In the limit of zero damping, $\gamma \to 0$, the particle exhibits harmonic oscillations around $x = 0$, whose amplitude is dictated by the initial conditions. In the opposite limit of pure damping, $\Omega \to 0$, the particle is monotonically attracted to the time asymptotic position $x = 0$ with no oscillation. In the general case, the solution shows a blend of damping and oscillations depending on the relative strength of the two.

Of particular interest to soft matter and biological applications is the so-called the *over-damped* regime, defined by the condition:

$$\frac{\gamma}{2} > \Omega$$

In this case, the exact solution reads as

$$x(t) = e^{-\frac{\gamma t}{2}} [x_0 \cosh(\omega t) + \frac{v_0 + \frac{\gamma}{2} x_0}{\omega} \sinh(\omega t)] \tag{9.30}$$

where

$$\omega^2 = \frac{\gamma^2}{4} - \Omega^2$$

In other words, starting from the initial position x_0, the particle is monotonically attracted to its time-asymptotic rest position $x(t \to \infty) \equiv x^* = 0$. Note that $x(t \to \infty) = 0$, regardless of the initial conditions, since $x^* = 0$ is a unique stable attractor of the dynamics.

In the *underdamped* regime, $\frac{\gamma}{2} < \Omega$, the solution exhibits damped oscillations toward the asymptotic position $x^* = 0$. In the critical case $\gamma/2 = \Omega$, the solution decays to $x^* = 0$ like $[x_0 + (v_0 + \gamma x_0)t] \, e^{-\gamma t}$. Finally, in the overdamped regime, the solution is a combination of two exponential decays, with rates $\gamma/2 \pm \Omega$.

For generic potentials, with multiple minima, the situation is obviously more involved, the final attractor x^* depending on the initial position and velocity.

Upon rolling down the potential landscape, the particle eventually gets near the closest zero-force point. The equation of motion can be linearized around such a point by setting $y = x - x^*$, so that the analytical solution (9.30) then applies to the local displacement $y(t)$, with the oscillation frequency given by the local curvature of the potential, i.e., $\Omega(x) \equiv \sqrt{d^2 V/dx^2}$ at $x = x^*$.

In other words, the damping term secures that a zero-motion position is always attained; which one exactly depends on the initial conditions and on the strength of the damping.

With strong damping the particle is likely to end up at the closest stable attractor, while when damping is small, the chances to travel away from the closest minimum are enhanced.

So much for the zero temperature dynamics.

How does the stochastic noise affect this picture?

In the first place, noise keeps the system "alive," with random excursions of the order of $\delta x = v_T/\gamma$ around the attractor x^*. The rate at which a particle at x^* jumps to reach another minimum, separated from x^* by a potential barrier ΔV, can be shown to scale like

$$k \propto \frac{\Omega_m \Omega_M}{\gamma} \, e^{-\frac{\Delta V}{k_B T}} \tag{9.31}$$

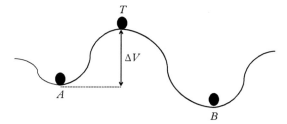

Figure 9.5 *Sketch the exit-time Kramer's problem. To transit from A to B, the particle must go over the top T of the potential barrier. The barrier for the A → B process is* $\Delta V_{AB} = E_T - E_A$, *which is smaller than the barrier* $\Delta V_{BA} = E_T - E_B$ *for the inverse transition B → A. Hence the former is more likely than the latter.*

where Ω_m and Ω_M are the second-order derivatives of the potential evaluated at the positions x_m and x_M, where $V(x)$ attains its minimum (valley) and maximum (crest), respectively, see Fig. 9.5.

More precisely, since k is an attempt rate (inverse time), the formula (9.31) delivers the first-passage time (FPT), namely the time it takes, on average, to cross the barrier separating the valley from the crest, that is:

$$\tau = 1/k \propto \frac{\gamma}{\Omega_m \Omega_M} e^{\frac{\Delta V}{k_B T}} \qquad (9.32)$$

This formula, also known in chemistry as Arrhenius law, after the Swedish chemist Svante Arrhenius (1859–1927), plays a crucial role in describing *thermally activated* chemical reactions, in which a reactant species R must overcome an energy barrier ΔE in order to react and transform to product species P.

The Arrhenius formula highlights the very steep, non-analytic dependence of the FPT on the temperature.

A few numbers help conveying the idea: given that $k_B T \sim 1/40 \; eV$ and taking $\omega_m \omega_M / \gamma \sim 1$ picosecond, the FPT associated with an energy barrier of $1 eV$ in standard conditions, i.e., $T = 300$ Kelvin degrees, is about $2 \; 10^5$ seconds, i.e., about two and half days. At $T = 400$ Kelvin, the FPT reduces to just about 10 seconds, and to less than half a millisecond at $T = 600$ Kelvin!

To go with the ancient Romans, "Ignis mutat res" (fire changes things) fits well the situation!

Hence, the key role played by temperature is to allow the system to explore its phase space; a macroscopic manifestation of molecular individualism.

Because of this property, multi-dimensional Langevin equations are much used to sample high-dimensional potential landscapes of a broad variety of complex non-equilibrium systems, especially, but not exclusively, for biological applications.

9.4.2 Driven Systems

Next, let us consider the case of a potential $V(x)$ that does not admit any stable minimum. The typical case is a charged particle moving under the effect of a constant electric field, E_0:

$$a(x) = a_0 = \frac{qE_0}{m} \qquad (9.33)$$

where q is the electric charge of the particle and m its mass.

In the long time limit, the particle keeps moving down the potential slope at a constant drift velocity

$$v_E = \frac{qE_0}{m\gamma} \tag{9.34}$$

This is basically Ohm's law.

At finite temperature, the particle experiences random fluctuations of the velocity, on top of the systematic drift. The average distance traveled after a time t is given by

$$< x(t) > = v_E t \tag{9.35}$$

with a variance

$$\sigma^2(t) \equiv < x^2(t) > - v_E^2 t^2 = \frac{v_T^2}{\gamma} t \tag{9.36}$$

The root-mean square grows like the square root of time, so that the noise/signal ratio scales like $\sigma / < x > \propto t^{-1/2}$, hence vanishing in the time asymptotic limit.

The relation (9.36) defines the diffusion coefficient as

$$D = lim_{t\to\infty} \frac{\sigma^2(t)}{t} = \frac{v_T^2}{\gamma} \tag{9.37}$$

This, combined with the expression (9.2), delivers the celebrated Einstein–Stokes relation:

$$D = \frac{k_B T}{6\pi \eta R} \tag{9.38}$$

Thus, the diffusivity of the Brownian particle is directly proportional to the temperature and inversely proportional and the dynamic viscosity of the solvent. The inverse dependence on its radius highlights the increasing role of thermal fluctuations on the dynamics of small particles.

9.5 Beyond Brownian Motion

These relations pertain to the original Langevin picture: Gaussian fluctuations (uncorrelated noise) and a friction scaling linearly on the particle velocity (Stokes friction). Neither assumption needs to hold in general. Indeed, modern statistical physics, especially in the context of biological applications is littered with fascinating examples wherein either of the two assumptions is violated, if not both.

Here, for the sake of completeness, we just convey the flavor of this active sector of modern non-equilibrium statistical mechanics.

9.5.1 Correlated Noise and Memory Effects

The assumption of uncorrelated noise made so far implies that the random kicks experienced by the Brownian walker bear no memory of the previous ones. This is generally a plausible assumption whenever the environment does not develop any organized long-term response to the passage of a given particle, so that the next passing particle finds the very same environment as its forerunner: the environment is a *passive* media with no memory.

In many applications, especially but not exclusively in biology, this picture no longer holds; the environment bears a finite-time memory of the particle passage, and through such memory, the motion of the particles becomes correlated.

The media is then said to be *active*, for it is no longer a passive backdrop of material motion.

Mathematically, correlated noise is expressed as follows:

$$\langle \xi(t)\xi(t')\rangle = D_v K(t, t') \tag{9.39}$$

where $K(t, t')$ is the *memory kernel* and D_v is the diffusion coefficient in velocity space.

The friction term in Langevin equation with colored noise develops an integral structure of the form

$$F_D(t) = -\gamma M \int_0^t K(t - t')v(t')dt' \tag{9.40}$$

For a stationary (time-homogeneous) process, the memory kernel depends only on the time lag $t - t'$, i.e., $K(t, t') = K(t - t')$.

A proto-typical expression for the memory kernel is

$$K(t - t') = ke^{-k(t-t')} \tag{9.41}$$

sometimes called *colored* noise, as opposed to *white* noise, characterized by a total lack of correlation.

Colored random noise keeps memory of the past over a time-span $1/k$. Clearly, the zero memory limit $k \to \infty$ recovers the case of delta-correlated white noise.

In the language of probability theory, colored noise is also called a *non-Markovian* process, i.e., a process in which the future does not depend only on the present state of affairs, but also on a sequence of past events. In other words, *history* enters the game.

A fascinating instance of a non-Markovian process is particle motion in *active media*, i.e., materials which modify their structure in response to particle motion. As previously mentioned, under such conditions, particles traveling across the media feel the effects of their predecessors and get thereby correlated to each other.

In some instances they may also exhibit "smart" behavior, i.e., probe the changes of the environment and exploit them to perform functional tasks. This is typical of biological matter; for instance there are evidences that the intracellular motion of biological molecules responds to these criteria, and so does the motion of cells in tissues (2).

9.5.2   Nonlinear Friction

The assumption of linear friction derives from Stokes law and is an accurate description of a colloidal particle moving *slowly* through the solvent.

If the particle is not slow, or the media modifies upon exchanging energy with the traveling particle, such linearity no longer holds. The same happens, although for different reasons, in the case of long-range interactions, like in plasmas, where memory is directly related to non-locality.

A velocity-dependent friction coefficient can give rise to a host of fascinating phenomena, which we shall briefly comment upon in the following.

9.5.3   Runaway Electrons in Plasmas

The runaway of plasma electrons is an important phenomenon in astrophysics and thermonuclear plasmas. A useful model, due to Russian physicist A. Vedenov leads to this collision frequency for the electrons in a neutral plasma:

$$\gamma(v) = \frac{\gamma_0}{[1 + (v/v_T)^2]^{3/2}} \tag{9.42}$$

At low speed, $v \ll v_T$, this reduces to the standard picture with constant friction frequency γ_0. At high speed, however, the collision frequency decays like v^{-3}, which means a friction force vanishing like v^{-2}. As a result, at any given value of the electric field, E, there corresponds a critical speed above which friction is no longer able to balance the momentum input by the electric field. Such critical value is readily computed by equating the friction and the electric forces, i.e., $qE = -m\gamma(v)v$, yielding

$$v_c = v_T \left(\frac{\gamma_0 v_T}{qE/m} \right)^{1/2} \tag{9.43}$$

This implies that for sufficiently strong electric fields, $E > E_c = \frac{\gamma_0 m v_T}{q}$, the critical speed becomes comparable to the thermal speed, so that a substantial fraction of electrons transits to the so-called "runaway" regime characterized by a progressive and untamed acceleration, see Fig. 9.6.

In the time asymptotic limit, this process depletes the bulk of the distribution function, all electrons accumulating around the speed of light. It is clear that the runaway regime lies outside the realm of hydrodynamics and requires a fully kinetic treatment.

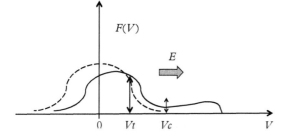

Figure 9.6 *Cartoon of a runaway probability-distribution function. The bulk is depleted to populate the supercritical region $v > v_c$.*

9.5.4 Active Matter

Another context in which nonlinear friction plays a major role is transport phenomena in complex and adaptive dynamical systems, and *active matter* in general.

In a broad sense, active matter is made of "smart particles," which are endowed with internal degrees of capable of probing the surroundings and organize a consequential response accordingly. Typically, smart particles prove capable of taking up energy from the environment storing it within some internal degree of freedom (*depot*) and eventually convert it into kinetic theory, so as to sustain their own motion (self-propulsion, chemotaxis being important examples in point).

Since these smart particles show some form of intelligence and planning ability, they are often called *agents*.

A simple yet representative example of active particles capable of making good use of the resources of the environment is now briefly described.

Consider a Langevin friction of the form

$$\gamma(v) = \gamma_0 - de(t) \tag{9.44}$$

where $e(t)$ is the energy stored in the environment (depot), a function of the particle speed, and d is a coupling coefficient.

Expression (9.44) implies that, by storing a sufficient amount of energy in the depot, the particle can turn its friction from positive to negative, thereby activating the capability of *extracting* energy from the environment instead of loosing it, the so-called *pumping regime*.

A typical rate equation for the depot reads as follows:

$$\dot{e} = q - (c + dv^2)e(t) \tag{9.45}$$

where q is the energy flux to the depot, c its friction coefficient and d is the rate of conversion from internal to kinetic energy, see Fig. 9.7.

Expression (9.45) informs us that the depot level decreases at increasing particle speed, so that fast particles are progressively excluded from the perks of the pumping regime, since they are less and less likely to need them . . . .

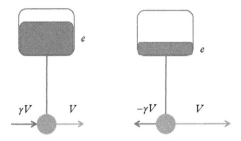

Figure 9.7 *An active particle in the pumping regime (left) and the standard-dissipative regime (right). Note that in the pumping regime, the depot is high and friction is cooperating (aligned) with the velocity. In the standard case (right), the opposite is true.*

At steady state, this gives the following effective friction coefficients:

$$\gamma(v) = \gamma_0 - \frac{qd}{c + dv^2} \qquad (9.46)$$

The crossover condition $\gamma(v) = 0$ identifies a critical speed

$$v_c^2 = q/\gamma_0 - c/d \qquad (9.47)$$

fixing the border line between the needy "slow" and the non-needy "fast" particles.

It is therefore clear that, depending on the specific values of these parameters, namely $qd/c > \gamma_0$, the system transits from a low-speed active state, in which it extracts energy from the environment, to a high-speed passive state, where it dissipates energy to the environment, as in the standard Langevin picture.

In other words, the environment acts in a equalizing "Robin-Hood" mode: it feeds low-energy particles with the energy gained from the high-energy ones, a sort of social toll system.

The study of transport phenomena in active media is a fast-growing front of modern statistical physics, with many applications in biology, quantum materials, as well as finance and social systems alike (3).

9.6 Back to Phase Space: The Fokker–Planck Equation

We have already mentioned the beautiful duality between the particle trajectories and phase-space distributions in the study of non-equilibrium fluids. We have also shown that the diffusion equation can be regarded as the kinetic equation associated with Brownian motion, see Fig. 9.7.

How about Langevin dynamics?

The kinetic equation describing the statistical dynamics of Langevin particles is known as Fokker–Planck, after the Dutch physicist Adriaan Fokker (1887–1972) and Max Planck (1858–1947), (yes, the quantum Planck!) (4).

Let us proceed with a heuristic derivation of the Fokker–Planck equation.

Based on the dynamical equations (9.1), the one-body kinetic distribution obeys the following Liouville equation (vector notation relaxed for simplicity):

$$\partial_t f + v \partial_x + [a(x) - \gamma v + \xi] \partial_v f = 0 \tag{9.48}$$

Note that the right-hand side is zero because in the Langevin dynamics there are no explicit binary encounters between molecules, other than those described by friction and noise.

It is immediately recognized that expression (9.48) is mathematically ill-posed because, as we have observed earlier on, stochastic trajectories are continuous but not differentiable. This means that the term $\xi \partial_v$ is ill-defined since the incremental ratio $\Delta v / \Delta t$ diverges in the limit $\Delta t \to 0$. A convergent quantity is provided instead by the diffusive ratio $\Delta v^2 / \Delta t$, whose limit is the diffusion coefficient in *velocity* space, i.e., $D_v = \gamma v_T^2$.

Formally, this difficulty is circumvented by passing to the finite-difference formulation of the second Langevin equation, namely[15]

$$\Delta v = (a - \gamma v) \Delta t + \xi(t) \Delta t^{1/2} \tag{9.49}$$

where the 1/2 exponent in the noise term reflects the diffusive nature of the process.

This stochastic difference equation is also known as *Ornstein-Uhlenbeck* process.

Likewise, the finite-time Liouville equation reads then

$$f(x + \Delta x, v + \Delta v, t + \Delta t) - f(x, v; t) = 0 \tag{9.50}$$

By Taylor expanding to first order in space and time, and *second* order in velocity, we obtain a drift-diffusion equation

$$\partial_t f + v \partial_x + a(x) \partial_v f = \gamma \partial_v (v f + v_T^2 \partial_v f) \tag{9.51}$$

This is precisely the desired Fokker–Planck equation (FPE).

In (9.51), we have made use of the fluctuation-dissipation relation:

$$\frac{\langle \Delta v^2 \rangle}{\Delta t} = \gamma v_T^2 \tag{9.52}$$

---

[15] Indeed, (9.49) is the formally correct way of writing down the Langevin equation since the noise term ξ is not differentiable. The eq. (9.1) should then read

$$d\vec{x} = \vec{v} dt$$
$$d\vec{v} = (\vec{a}(\vec{x}) - \gamma \vec{v}) dt + \vec{\xi} dt^{1/2}$$

With this notation, the second equation in (9.4) would in turn read as: $\langle \vec{\xi}(t) \vec{\xi}(t') \rangle = v_T^2 \hat{I} \delta(t - t')$. In the text we have preferred to stick to the "incorrect" but simpler notation of ordinary differential equations. This does no harm as long as the noise term is integrated in time, which is indeed the case here.

as well as of the drift relation

$$\frac{\langle \Delta v \rangle}{\Delta t} = -\gamma < v >$$

(9.53)

The FPE is paramount to a broad variety of dynamical phenomena in natural and also socio-economical sciences.

For the case of fluids, it is commonly regarded as appropriate for liquids, and more generally, for many-body systems with soft-core interactions, i.e., those where kinetic and potentially energy always compete moreless on the same footing. In this respect, the FPE appears to stand in a complementary camp as compared to the Boltzmann equation. The formal distance between the two is however shorter than it might seem at first sight.

For one thing, it should be mentioned that the FPE can be derived from the Boltzmann equation for a mixture of heavy and light fluids, in the limit of vanishing mass ratio $m/M \to 0$. As a result, the FPE appropriately describes the kinetics of colloidal particles.

From a mathematical standpoint, the FPE can also be described as a multi-relaxation time BGK equation of the sort previously discussed. The reader may check (see Risken's book) that the scattering kernel associated with the FPE collision operator can be expressed as a double expansion in Hermite polynomials, with a characteristic spectrum of eigenvalues, $\omega_k = k\omega_0$, for the case of a harmonic potential. Each potential gives rise to its own spectrum of eigenvalues.

9.6.1   Fokker–Planck Equation for Nonlinear Processes

The FPE is by no means restricted to Langevin equations with linear friction. For more general processes, the FPE collision operator can be cast in the general form:

$$\partial_t f + v \partial_x f + a(x) \partial_v f = \partial_v [R(v)f + D\partial_v f]$$

(9.54)

where $R(v)$ is a generic velocity-dependent friction and $D = \gamma v_T^2$ is the diffusion coefficient in velocity space. The right-hand side comes in the form of the divergence of a flux, or current, in velocity space:

$$\mathcal{J}(v) \equiv R(v)f + D\partial_v f$$

(9.55)

The divergence of this flux vanishes at equilibrium, and assuming that so does the flux itself, the following equilibrium distribution is obtained:

$$f^e(v) = Z^{-1} e^{-\Phi(v)}$$

(9.56)

where Z is a normalization constant (inverse-partition function) and we have set

$$\Phi(v) = \int_{-\infty}^{v} \frac{R(v')}{D} \, dv'$$

(9.57)

with boundary condition $\Phi(-\infty) = 0$.

By setting $R(v) = \gamma_0 v \phi(v)$, where $\phi(v)$ is a measure of nonlinearity, and recalling that $D = \gamma_0 v_T^2$, we finally obtain

$$\Phi(v) = \frac{1}{v_T^2} \int_{-\infty}^{v} v \phi(v) dv \qquad (9.58)$$

Clearly, the linear case $\phi(v) = 1$ recovers standard Maxwell–Boltzmann equilibria, as for the case of Boltzmann collisions, even though the microscopic dynamics is markedly different (soft versus hard collisions).

The function $\Phi(v)$ is a close relative of the Rayleigh Dissipation function of classical mechanics, associated with the dissipative force $F_D = -\partial_v \Phi(v)$.

The FPE in general form (9.54) is of great importance in chemical kinetics, where it is used for the computation of the escape rates from potential barriers, the Kramer's problem discussed earlier on in this chapter. It is also extremely useful for the simulation of rare events, which are typically associated with "fat-tail" distributions decaying to zero much slower than the Gaussian, typically power laws or stretched exponentials. Unlike the Maxwell–Boltzmann equilibria, such fat-tail local equilibria support only a finite number of convergent kinetic moments, another typical hallmark of non-hydrodynamic behavior.

9.7 Summary

Summarizing, in this chapter we have taken a bird's eye view of a very broad territory of kinetic theory which is not directly associated with the Boltzmann equation.

We have learned that phase-space distributions and stochastic trajectories are two equivalent (dual) representations of the same phenomenon. This speaks for the beauty and richness of kinetic theory, but also raises a natural question, as to which one of the two should be used, and when.

In general, phase-space distributions are useful because they dispense from the onerous task of taking ensemble-averages over many stochastic realizations in order to achieve acceptable statistics. On the other hand, owing to the large dimensionality of phase space, the numerical solution of kinetic equations by standard numerical methods is extremely demanding. This why kinetic equations in full six-dimensional phase space, including Boltzmann's, are typically solved by particle methods such as Direct Simulation on Monte Carlo.

As we shall see, the Lattice Boltzmann method can be regarded as an economic and efficient way of dealing with the discretization of the velocity space, whenever the physics does not generate large departures from local equilibrium.

....

REFERENCES

1. R. Zwanzig, "Non equilibriums statistical mechanics", Oxford University Press, Oxford 2001.
2. F. Schweitzer, "Brownian agents and active particles in," *Springer Series in Synergetics*, Springer Verlag, Heidelberg, 2003.
3. N. Bellomo, "Modeling Complex Living Systems: A Kinetic Theory and Stochastic Game Approach", Birkhauser, 2008.
4. H. Risken, "The Fokker–Planck equation", *Springer Series in Synergetics*, 3rd printing, Springer Verlag, Heidelberg, 1996.

....

EXERCISES

1. Compute the damping coefficient γ for a particle of $1\ nm$ radius and a mass of 10^3 a.m.u., suspended in water at room temperature.
2. Show that the cross section of a colloidal particle of mass M and radius R in a solvent of molecular mass m is given by $\sigma = 6\pi \frac{m}{M} l_\mu R$, where l_μ is the mean-free path of the solvent.
3. Compute the dissipative potential $\Phi(v)$ for the case of the Vedenov collision operator.

10

Numerical Methods for the Kinetic Theory of Fluids

In this chapter we provide a bird's eye view of the main numerical-particle methods used in the kinetic theory of fluids, the main purpose being locating lattice Boltzmann in the broader context of computational kinetic theory.

Computers are incredibly fast, accurate and stupid, humans are incredibly slow, inaccurate and brilliant. Together they are powerful beyond imagination.

(A. Einstein)

10.1   Introduction

The leading numerical methods for dense and rarefied fluids are *Molecular Dynamics* (MD) and *Direct Simulation Monte Carlo* (DSMC), respectively. These methods date of the mid 50s and 60s, respectively, and, ever since, they have undergone a series of impressive developments and refinements which have turned them in major tools of investigation, discovery and design.

However, they are both very demanding on computational grounds, which motivates a ceaseless demand for new and improved variants aimed at enhancing their computational efficiency without loosing physical fidelity and vice versa, enhance their physical fidelity without compromising computational viability.

The tradeoff between computational efficiency and physical accuracy is a hallmark of computational physics and such tradeoff pretty much depends on the questions to be asked for the given problem at hand.[16] For the case of Lattice Boltzmann, like for any *mesoscale* method in general, the relevant questions are rarely detail driven, but rather

---

[16] We kindly remind the reader that simulations are supposed to answer a question! For the aberrations of using simulations to answer questions that nobody ever asked, see Daan Frenkel's "Simulations: the dark side," a quintessential Frenkelian masterpiece of interlaced humor and depth (1).

The Lattice Boltzmann Equation. Sauro Succi, Oxford University Press (2018).
© Sauro Succi. DOI: 10.1093/oso/9780199592357.001.0001

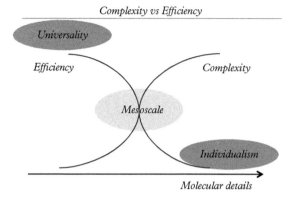

Figure 10.1 *Computational Complexity scales with the degree of molecular detail (Individualism) while Efficiency follows the opposite trend. Mesoscale methods try to strike an optimal balance between the two conflicting trends by relinquishing as much molecular details as possible without compromising the essential physics resulting from a mix of Universality and Individualism.*

the result of a problem-dependent blend of Universality and Individualism, a regime we shall denote later in this book as *Weakly Broken Universality* (see Fig. 10.1).

For the sake of conceptual completeness, in this short chapter we provide a very cursory view of a few major computational methods for both microscale and mesoscale fluids.

10.2   Molecular Dynamics

Molecular dynamics arises from the idea of *imitating* Newtons's law on electronic computers.

More precisely, the fluid is described by Newton's equation of motion

$$
\begin{cases}
\dfrac{d\vec{r}_i}{dt} = \vec{v}_i, \\[2mm]
\dfrac{d\vec{v}_i}{dt} = \dfrac{\vec{F}_i}{m}, \quad i = 1, N
\end{cases}
\tag{10.1}
$$

The "only" difference with mother nature is that instead of Avogadro's number of molecules, one works with a much smaller number of computational particles, just a few hundreds in the pioneering days of computer simulations, now in the order of trillions on most powerful present-day computers.

As a result, "all one has to do" in MD is to integrate a large number of ordinary differential equations long enough in time to be able to appreciate the desired physical effects to an acceptable level of statistical accuracy.

Many time-marching techniques have been devised to advance Newton's equations in time, which are described in excellent textbooks (2; 3).

Here we report one of the simplest and most popular, i.e., the Velocity–Verlet (VV) scheme.

The VV scheme consists of three basic steps:

1. *Advance positions*:

$$\vec{r}_i(t + \Delta t) = \vec{r}_i(t) + \vec{v}_i(t)\Delta t + \vec{a}_i\frac{\Delta t^2}{2}$$

2. *Compute forces (accelerations)* based on the new positions:

$$\vec{a}_i(t + \Delta t/2) = \frac{\vec{a}_i(t) + \dot{\vec{a}}_i(t + \Delta t)}{2}$$

where

$$\vec{a}_i(t) \equiv \vec{a}_i(\vec{r}(t)) = \sum_{j>i} \frac{\vec{F}(\vec{r}_i(t), \vec{r}_j(t))}{m}$$

3. *Advance velocities*:

$$\vec{v}_i(t + \Delta t) = \vec{v}_i(t) + \vec{a}_i\left(t + \frac{\Delta t}{2}\right)\Delta t$$

This simple scheme can be shown to be second-order accurate toward global errors associated with the entire lifespan of the simulation most importantly it is, *symplectic*, i.e., it preserves the measure $|drdv|$ in phase space (Liouville theorem), a property which is key to perform very long-time integration without incurring destructive round-off errors.

The timestep is usually a small fraction of the typical interaction time, typically: $\Delta t \sim 10^{-3}r_0/c_s$, where r_0 is the range of the interaction potential, and c_s is the sound speed. With $r_0 \sim 0.3\,nm$ and $c_s \sim 1000\,m/s$, the resulting timestep is of the order of $\Delta t \sim 10^{-15}\,s$, or below.

The most expensive part of the VV scheme is the calculation of the forces, in principle a N^2 computational task. However, for short-range potentials, only a limited number of neighbors must be included in the force summation, typically within a cutoff distance of the order of $r_c \sim (3 \div 5)r_0$. Since there is about one molecule in the interaction sphere at $\tilde{n} \sim 1$, this means about $3^3 \div 5^3 \sim 50 \div 100$ interacting neighbors. These can be computed with $O(N)$ complexity by using linked lists, i.e., the sequence of interacting pairs for each molecule at each timestep (see Fig. 10.2).

With $N \sim 10^9$ particles and $r_0 \sim 0.3\,nm$, MD can simulate regions of the order of a tenth of a micron cube.

Even though this falls short of describing macroscopic fluids, it provides nonetheless a number of insightful clues on many fundamental problems in the kinetic theory of fluids, such as the famous violation of Boltzmann's molecular chaos assumption, as first detected in the pioneering work of Alder and Wainwright.

However, the major challenge in MD is not space but rather time. By ticking at a pace of femtoseconds, a million timestep simulation takes no farther than one nanosecond, which is generally way too short to observe significant meso/macroscale physics. This

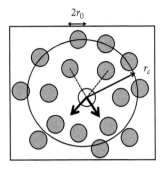

Figure 10.2 *Computing the forces (thick arrows) in MD. The given molecule (open circle) interacts with all molecules whose centers lie within the interaction sphere of radius r_c. For simplicity, only two pair interactions are shown.*

motivates a very active research activity aimed at accelerate MD algorithms, so as to bridge the daunting gap between physical time and simulated one.

10.3   Direct Simulation Monte Carlo

For dilute gases, the stochastic particle analogue of the Boltzmann equation is provided by the DSMC dynamics, first pioneered by Graeme Bird in the early 60s as a numerical method to solve the Boltzmann equation (4).

In the following we shall outline just the main ideas behind DSMC, leaving full details to the vast and excellent original literature.

We have commented several times in this book on the complexity of solving the Boltzmann equation, owing to its integral, nonlinear and high-dimensional nature.

The latter is possibly the most constraining feature from the numerical point of view. With, say, a modest resolution of 64 grid points for each of the six dimensions of phase space, one generates $64^6 \sim 64 \; 10^9$ discrete variables, close to the limit of the most powerful current-day supercomputers. The problem would be considerably alleviated if, instead of discretizing phase space uniformly, one could represent the Boltzmann distribution only by a statistical sample of particles. This is the basic strategy of the DSMC method.

What Bird achieved was a Monte Carlo formulation of the Boltzmann statistical dynamics, whence the namer DSMC. DSMC is thoroughly described in a vast literature, Bird's book in the first place and therefore here we shall only provide the basic ideas.

Let us consider a 2 + 2 phase space for simplicity, i.e., a two-dimensional space with planar velocities (Flatland). Let us partition the two-dimensional space into a sequence of square cells, of size Δ comparable with the interaction range, r_0, of the intermolecular potential.

For a system at density n, such cell contains on average $N = n\Delta^2$ particles, out of which one can form $N(N-1)/2$ colliding pairs.

The DSMC algorithm proceeds through the following steps (see Fig. 10.3):

1. *Choose the timestep*:
   Choose a collisional timestep $\Delta t < \tau_c$, where $\tau_c = 1/(n\sigma v_T)$ is the typical collision time.

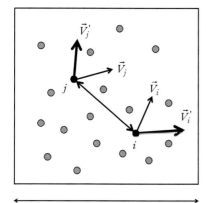

Figure 10.3 *Performing a collision in the DSMC scheme. Note that the cell-size Δ must be significantly smaller than the mean-free path l_μ for matter of accuracy. This sets a limitation on the timestep $\Delta t < \Delta/v_T < \tau_\mu$, v_T being the thermal speed.*

$\Delta < l_\mu$

2. *Select collision pairs*:
   For each cell, select a pair of particles (i,j) at random out of the $N(N-1)/2$ available ones.

3. *Compute the collision probability*:
   The collision probability of the selected pair (i,j) in the time-lapse Δt is given by

$$p(i,j) = n\sigma(i,j)v(i,j)\Delta t,$$

where $v(i,j)$ is the magnitude of the relative velocity of the pair and $\sigma(i,j)$ is the corresponding cross section.

4. *Perform the collision*:
   Draw a random number r, uniformly distributed in $[0,1]$.
   If $r < p$: Perform the collision;
   Else: select another pair and try again.
   Collision:
   The velocities $\vec{v}(i)$ and $\vec{v}(j)$ are turned into the corresponding post-collisional values $\vec{v}'(i)$ and $\vec{v}'(j)$ based on the kinematics of two-body collisions described in Chapter 2.

Steps 2–4 are repeated till completion of all $N(N-1)/2$ collisions and the time counter is then updated from t to $t + \Delta t$.

Once the collision procedure is completed, particles fly freely to the new positions defined by the post-collisional velocities, i.e.,

$$\vec{r}_i(t + \Delta t) = \vec{r}_i(t) + \vec{v}'_i \Delta t \tag{10.2}$$

This is a very poor-man version of DSMC, a major method which has undergone progressive developments and refinements over five decades of successful applications since Bird's original 1963 paper. We wish to note that the idea of a stochastic dynamics underneath the Boltzmann equation is far from being trivial, as witnessed by the long time it took before Bird's algorithm could be recognized and accepted as a realistic microdynamic stochastic realization of Boltzmann's equation.

Bird's method is conceptually transparent and well suited to handle strongly non-equilibrium flows in complex geometries, thanks to the simple structure of the particle trajectories.

Its Achille's heel, however, is the statistical efficiency of the collision procedure, which may face severe issues of efficiency whenever particles interacting too weakly to provide substantial collision probabilities, thereby leading to frequent rejection of the Monte Carlo moves. In addition, like MD, DSMC is affected by significant statistical noise, which require extensive ensemble averaging to extract smooth signals from the particle simulations.

Both liabilities have been clearly recognized since the inception of DSMC and many remedies proposed and implemented in the form of variance reduction techniques. Notwithstanding major progress, DSMC still remains a pretty intensive computational tool.

10.4   Mesoscale Particle Methods

Both MD and DSMC can be regarded as microscopic methods, in that they provide a physically realistic description of the basic physics contained in the Newton's and Boltzmann's equations, respectively. As we have commented, this entails stringent limitations in terms of computational demand of the corresponding algorithms.

In order to gain computational ground, it is natural to look for simplified versions of the MD or DSMC dynamics based on *meso-particles* representing a collection of a large number of microscopic molecules. Given that the ultimate target is hydrodynamics, the dynamics of such mesoparticles can be designed in such a way to sieve out the unnecessary and time-consuming microscopic details.

As usual, the coarse-graining recipe is far from being unique, and in the following we present two popular mesoparticle strategies, Multi-Particle Collisions, and Dissipative Particle Dynamics, the mesoscale proxies of DSMC and MD, respectively.

10.5   Multi-Particle Collisions

As discussed in Section 10.4, the main liability of DSMC is the computational efficiency of the Monte Carlo collisions.

The basic observation that much of this efficiency is lost in mimicking collisions to a degree of fidelity which is *not* required if hydrodynamics is the main purpose. As a result, one seeks to replace DSMC collisions with a simplified version which would retain the main ingredients to achieve hydrodynamic behavior at macroscopic scales.

The main idea of the Multi-Particle Collision method (MPC) is to replace collisions based on the actual scattering rates with stochastic rotations of the relative velocity, based on the kinematics of the two-body problem described in Chapter 2 (5).

More precisely, each particle in the grid cell with velocity $\vec{V}_i$ takes the post-collisional velocity:

$$\vec{V}_i' = \vec{V} + \hat{\Omega} \cdot (\vec{V}_i - \vec{V}) \tag{10.3}$$

where $\hat{\Omega}$ is a one-parameter (in two spatial dimensions) rotation matrix.

In explicit cartesian notation:

$$\begin{cases} V'_{i,x} = V_x + cos(\pm\chi)\ V_{i,x} + sin(\pm\chi)\ V_{i,y} \\ V'_{i,y} = V_y - sin(\pm\chi)\ V_{i,x} + cos(\pm\chi)\ V_{i,y},\ i = 1, N \end{cases} \qquad (10.4)$$

where $\vec{V} = \frac{1}{N}\sum_{i=1}^{N} \vec{V}_i$ is the centermass velocity in the cell and the rotation angle takes the value $\pm\chi$ with probability 1/2.

The obvious advantage of MPC over DSMC is that collisions are deterministic, hence zero chance of rejection. In addition, every particle in the cell undergoes the same collision *simultaneously*, rather than sequentially as in DSMC.

The obvious disadvantage is that the random choice of the collision angle does not reflect the actual physics of intermolecular collisions, as encoded in the cross section. The rationale is that the details of such physics are largely immaterial to the fluid dynamic level. In particular, the rotation angle, together with the number of particles in the cell, $N = n\Delta^2$, provides the transport parameters of the MPC fluid.

For instance, the kinematic viscosity reads as follows:

$$\nu_{MPC} = v_T^2 \Delta t\, f_{MPC}(N, \chi)$$

where f_{MPC} is a (non-trivial) dimensionless function of N and χ reflecting the artificial mesoscale nature of the method.

Typical values in the applications are $N \sim 10 \div 50$ and $\chi \sim 130^o$. As for any particle method, the ratio of the mean-free path to the linear cell size, $\lambda \equiv v_T \Delta t / \Delta$ plays an important role on the accuracy and efficiency of the method.

In particular, it is found that a good match between predicted values of the transport coefficients and those observed in the simulations is obtained for $\lambda \sim 1$. In other words, the MPC particles should free-stream over distances comparable to the cell size, so as to undergo uncorrelated collisions with particles in a different cell at each timestep, see Fig. 10.4.

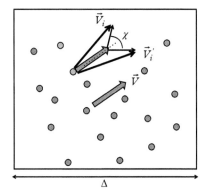

Figure 10.4 *Performing a MPC collision in two space and velocity dimensions. The relative velocity of each particle in the cell undergoes a rotation by the same angle χ.*

Like in DSMC, the cell size Δ is significantly larger than the range of molecular interactions.

MPC appears to have found an interesting niche of applications for low-Reynolds flows, particularly in the presence of suspended bodies.

10.6   Dissipative Particle Dynamics

Dissipative Particle Dynamics (DPD) can be regarded as a many-body extension of the Langevin equation with momentum conservation (6).

As usual, the idea is to formulate the dynamics of mesoscale particles representing a collection of a large number of molecules (See Fig. 10.5). Each particle is assigned a mass, position and velocity according to a direct coarse-graining in space:

$$\begin{cases} M_I = \sum_{i=1}^{N} m_i W(\vec{x}_i - \vec{X}_I) \\ M_I \vec{X}_I = \sum_{i=1}^{N} m_i \vec{x}_i W(\vec{x}_i - \vec{X}_I) \\ M_I \vec{V}_I = \sum_{i=1}^{N} m_i \vec{v}_i W(\vec{x}_i - \vec{X}_I) \end{cases} \qquad (10.5)$$

where $W(\vec{x}, \vec{X}_I)$ is a compact-shape function centered about the mesoparticle position $\vec{X}_I$ and zero above a given distance h controlling the size of the mesoparticles. Note that the capital index I runs from 1 to the number of DPD particles, $N_{DPD} << N$.

The mesoparticles are postulated to obey Newton-like equations of motion of the form

$$\begin{cases} \dfrac{d\vec{X}_I}{dt} = \vec{V}_I, \\ \dfrac{d\vec{V}_I}{dt} = \dfrac{\vec{F}_I^C + \vec{F}_I^D + \vec{F}_I^R}{M_I}, \quad I = 1, N_{DPD} \end{cases} \qquad (10.6)$$

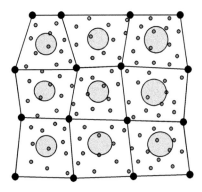

Figure 10.5 *The geometric construction of DPD particles (large circles in gray) using two-dimensional polygons as shape functions. More elaborate schemes, based on Voronoi tessellation, are typically used in current DPD simulations.*

The DPD forces consist of three components:
Conservative forces:

$$\vec{F}_I^C = -\nabla_I \sum_{J>I} V(R_{IJ}) \tag{10.7}$$

In (10.7) $V(R)$ is typically a short-range radial potential which depends only on the relative distance $R_{IJ} = |\vec{X}_I - \vec{X}_J|$. Typical choices are piecewise polynomials within a cut-off distance R_c and zero above it.
Dissipative forces:

$$\vec{F}_I^D = -\gamma \sum_{J>I} W^D(R_{IJ}) \frac{\vec{R}_{IJ} \cdot (\vec{V}_I - \vec{V}_J)}{R_{IJ}} \tag{10.8}$$

where $W^D(R)$ is a weight function vanishing above R_c.
Random forces:

$$\vec{F}_I^R = -W^R(R_{IJ})\xi_{IJ}\vec{R}_{IJ} \tag{10.9}$$

where ξ_{IJ} is a random variable with zero mean and variance σ and $W^R(R)$ is another weight function vanishing above R_c.

The DPD equations bear an obvious similarity to molecular dynamics, with two essential twists: Dissipation and Noise.

Because of this, the time integration of the DPD equations raises specific issues which make a very technical and important sector of computational statistical physics, with applications to many other fields where stochastic modeling is key, finance being one of the most popular (or maybe unpopular ...) examples in point.

Far from entering these highly technical matters, here we just report a popular choice of stochastic integrator consisting of a modified version of the velocity Verlet (VV).

This reads as follows:

$$\vec{X}_I(t + \Delta t) = \vec{X}_I(t) + \vec{V}_I(t)\Delta t + \frac{\Delta t^2}{2}\vec{A}_I(t) \tag{10.10}$$

$$\vec{V}_I^*(t + \Delta t) = \vec{V}_I(t) + \theta \Delta t \vec{A}_I(t) \tag{10.11}$$

$$\vec{A}_I(t + \Delta t) \equiv \vec{A}_I\left[X(t + \Delta t, \vec{V}_I^*(t + \Delta t)\right] \tag{10.12}$$

$$\vec{V}_I(t + \Delta t) = \vec{V}_I(t) + \frac{\Delta t}{2}\left(\vec{A}_I(t) + \vec{A}_I(t + \Delta t)\right) \tag{10.13}$$

where $\vec{A}_I \equiv \vec{F}_I/M_I$ is the mesoparticle acceleration.

In (10.11) $0 < \theta < 1$ is a parameter accounting for the velocity dependence of the dissipative force, which would reduce to the standard VV value $\theta = 1/2$ for the case of purely conservative forces.

Special attention must be paid to the random force, which must secure compliance with the Fluctuation–Dissipation theorem, namely

$$< A_I^R A_J^R > = \frac{\gamma v_T^2}{\Delta t} \, \delta_{IJ} \qquad (10.14)$$

This shows that the stochastic forcing, $A^R \Delta t$, scales like the square root of the timestep; hence, it vanishes much more slowly than the deterministic one in the limit $\Delta t \to 0$. This signals the slow convergence of the time integrator at increasing temperature, a Monte Carlo signature in time.

One the main DPD assets is the natural incorporation of thermal fluctuations, leading to a thermally consistent statistical thermodynamics, along with natural adaptivity to sharp flow features, typically colloidal interfaces in fluid solvents.

Among the computational downsides must be reckoned the complexity of Voronoi tessellations in three spatial dimensions.

The DPD method has nevertheless gained significant popularity for the computational study of low-Reynolds complex flows, such as colloids and polymer flows.

10.7   Grid Methods for the Boltzmann Equation

As mentioned earlier on in this chapter solving the Boltzmann equations on a phase-space grid is very daunting, mostly on account of the six-dimensional nature of the problem. However, with the advances of computer technology and the concurrent progress of numerical methods, it is nowadays possible to tackle full-phase-space simulations of the full Boltzmann equation on grids of moderate resolution, such as 64^6. Moreover, a plethora of advanced techniques, such as locally adaptive phase-space grid refinement, fast summation of the discrete Boltzmann collision operator and others, will certainly allow further significant advances in the next decade (7).

In this book, however, we shall often be concerned with the solution of the Boltzmann equation under near-hydrodynamic conditions, i.e., not too far from local equilibrium.

To this end, it should be observed that velocity space and configuration space are by no means equivalent, the former being generally much smoother owing to the existence of microscopic invariants.

As a result, there is plenty of scope for methods whereby velocity space could and should be discretized much more economically than configuration space. This is the spirit of Discrete Velocity Methods, whereby the Boltzmann collision operator is handled by direct discretization on a limited set of discrete speeds, often just the six nearest neighbors or simple extensions thereof. We shall touch briefly at these precursors of the Lattice Boltzmann method in the sequel of this book.

10.8   Lattice Particle Methods

The four-particle methods described above differ considerably in nature and scope: DSMC and MPC are both stochastic and *collision-driven*, albeit the latter cuts the collisions to their bone. MD and DPD are *force-driven*, but the former is deterministic and microscopic while the latter is stochastic and mesoscopic. Notwithstanding these crucial differences, they all share a common feature: particles are *grid free*, i.e., they move freely in space, the grid serving only as a bookkeeping device to locate and monitor the evolution of Eulerian fluid observables.

This is in line with the behavior or real molecules in continuum spacetime, but incurs the cost of identifying and locating the interacting neighbors of each given molecule at every timestep.

Such penalty can be obviated by forcing the dynamics of the particles, propagation and collisions, to take place entirely on the nodes of a regular lattice (*grid-bound* particles).

The nature of this *lattice particles* and the constraints imposed by the lattice on their dynamics, may vary for the different lattice methods. For Lattice Gas Cellular Automata (LGCA), particles are reduced to mere "ontological" statements of presence/absence, which can be coded with a single bit, thus leading to a maximally stylized *boolean* lattice dynamics.

For Lattice Boltzmann (LB), they are discrete-velocity versions of the Boltzmann distribution, whereby both configuration and velocity degrees of freedom are lattice-based.

Richard Feynman (1918–1988) once said that a good piece of theory should admit at least three independent derivations. Lattice Boltzmann seems indeed to be compliant with Feynman's test, as illustrated in Fig 10.6.

As usual, lattice particles come with their pros and cons, but it is fair to say that the idea has proven extremely productive for the computational study of fluid flows over nearly three decades.

For LGCA see Chapter 11, for LB, see the rest of this entire book!

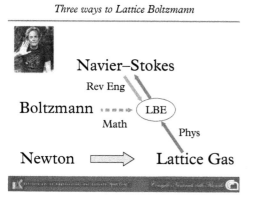

Figure 10.6 *Three ways to Lattice Boltzmann. The bottom-up (physicist) way, the top-down (reverse engineering) way and the side-up (mathematician) way. Side-up means that LBE belongs to the same mesoscopic level as the Boltzmann equation, but lives closer to the macroscopic level because of the drastic coarsening of phase-space.*

10.9 Summary

Summarizing, the kinetic theory of fluids counts on a broad and rich array of computational techniques covering both particle and phase-space grid methods.

The mainstreamers are Molecular Dynamics and DSMC, for liquids and gases, respectively, which are both based on grid-free (particles) degrees of freedom.

Grid methods are extremely demanding, but major progress in computer technology and computational techniques show promise of reaching out to moderate resolutions of the full-Boltzmann equation in the next decade.

Near the hydrodynamic regime, the broken symmetry between configuration space and velocity space spawns substantial opportunities for mesoscale methods relinquishing much of the computational complexity of MD and DSMC, while retaining the essential macroscale physics.

This is precisely the framework Lattice Boltzmann inscribes to.

REFERENCES

1. D. Frenkel, "Simulations: The dark side", *Eur. Phys. J. Plus* 128, **10**, 2013.
2. D. Frenkel and B. Smit, "Understanding molecular simulation: from algorithms to applications," Academic Press, New York, 1996, reprinted 2002.
3. D. Rapaport, "The art of molecular dynamics simulation," Cambridge University Press, Cambridge, 2004.
4. G. Bird, "Molecular Gas Dynamics and the Direct Simulation of Gas Flows," Claredon, Oxford, 1994.
5. A. Malavanets and R. Kapral, Mesoscopic Model for Solvent Dynamics, *J. of Chem. Phys*, 110, 8605–13 1999.
6. P. Hoggerbruegge and J. Koelman, "Simulating microscopic hydrodynamic phenomena with dissipative particle dynamics". *Europhysics Letters*, 19, **155**, 1992.
7. R. Gatignol, H. "Struchtrup editors, Proceedings of the 31th Int. Symp. on Rarefied Gas Dynamics Symposium," Victoria, 2016.

EXERCISE

1. Simulate a Poiseuille flow using two methods of your choice out of the four discussed in this chapter.

Part II

Lattice Kinetic Theory

According to Richard Feynman, any good piece of theory should be derived in at least three independent ways.

The LB method complies with the Feynman criterion, as it can indeed be derived at least along the three following routes: a) Bottom-up; the one-body Boltzmann kinetic equation stemming from the many-body Lattice Gas Cellular Automata, a boolean analogue of Newtonian mechanics; b) Top-down; The lattice kinetic equation with minimal symmetries required to recover the Navier-Stokes equations of continuum fluid mechanics in the macroscopic limit; c) Side-up; a suitable phase-spacetime discretization of the corresponding kinetic equation in continuum phase-space. The three routes are not strictly equivalent and each them comes with its own ups and downs; a) is the historical route and derives LB from the general framework of classical statistical mechanics. Intellectually the richest, it also entails a number of unnecessary limitations which severely hamper its viability for practical hydrodynamic simulations. In b), LB is designed top-down, rather than derived bottom-up, in such a way as to recover the Navier-Stokes equations in the continuum limit, with minimal requirements on the underlying microscopic dynamics. It achieves dramatic computational advantages out of a top-down reverse engineering of the Navier-Stokes equations. However, it also leaves an open end towards the second principle, with ensuing issues of numerical stability in the limit of very low viscosity.

Finally, route c) is the most direct, as it minimizes foreknowledge of statistical mechanics. It offers the benefits of methodological systematicity, although at the price of working in a conceptually narrower context. In addition, not all LBE's can be reached this way.

Be as it may, all three routes contribute to the richness and versatility of the LB method. Following this author's didactical experience and also to do justice to the historical development of the subject, the second Part of this book starts by taking the scenic route a): lattice gas cellular automata.

11

Lattice Gas-Cellular Automata

In this chapter we present the ancestor of the Lattice Boltzmann, the Boolean formulation of hydrodynamics known as lattice Gas-Cellular Automata.

Digital arithmetic is to scientific computing as collisions of molecules are to fluid mechanics: crucial to the implementation, but irrelevant to most scientific questions.

L.N. Trefethen, "Maxims about numerical mathematics, computer science and life,"
SIAM News, Jan/Feb, 1998.

11.1 Boolean Hydrodynamics

In 1986, Uriel Frisch, Brosl Hasslacher and Yves Pomeau sent big waves across the fluid dynamics community: a simple cellular automaton, obeying nothing but conservation laws at a microscopic level, was able to reproduce the complexity of real fluid flows (1). This discovery spurred great excitement in the fluid-dynamics community. The prospects were tantalizing: a round-off free, intrinsically parallel computational paradigm for fluid flows. However, a few serious problems were quickly recognized and addressed with great intensity in the following years.

The LBE developed in the wake of the Lattice Gas Cellular Automata (LGCA) method and was generated precisely in response to its initial drawbacks (2). Shortly after its inception, LBE evolved into a self-standing research subject which could be discussed independently with no need to refer to LGCA at all. This presentation route, if viable on practical grounds, would not do any justice to the chronological development of the subject, nor would it benefit the pedagogical side of the matter. Since we attach major value at both of these items, this chapter is devoted to a presentation of the basic ideas behind LGCA. Thorough accounts of the subject can be found in the very nice monographs available in the literature (3).

The Lattice Boltzmann Equation. Sauro Succi, Oxford University Press (2018).
© Sauro Succi. DOI: 10.1093/oso/9780199592357.001.0001

11.2   Fluids in Gridland: The Frisch–Hasslacher–Pomeau Automaton

Let us begin by considering a regular lattice with hexagonal symmetry such that each lattice site is surrounded by six neighbors identified by six connecting vectors $\vec{c}_i \equiv c_{ia}$, $i = 1, \ldots, 6$, the index $a = 1, 2$ running over the spatial dimensions (see Fig. 11.1). Each lattice site hosts up to six particles with the following prescriptions:

- All particles have the same mass $m = 1$
- Particles can move only along one of the six directions defined by the discrete displacements c_{ia}, see Fig. 11.1.
- In a time cycle (made unit for convenience) the particles hop to the nearest neighbor pointed by the corresponding discrete vector c_{ia}. Longer and shorter jumps are both forbidden, which means that all lattice particles have the same energy.
- No two particles sitting on the same site can move along the same direction c_{ia} (*exclusion principle*).

These prescriptions identify a very stylized-gas analog, whose dynamics represents a sort of cartoon of real-molecule Newtonian dynamics. In a real gas, molecules move along any direction (*isotropy*) whereas here they are confined to a six-barred cage.

Also, real molecules can move at virtually any (subluminal) speed, whereas here only six monochromatic beams are allowed. Finally, the exclusion principle sounds a bit weird for a system of classical particles whose momentum and position can be specified simultaneously with arbitrary accuracy.

Amazingly, this apparently poor cartoon of molecular dynamics has all it takes to simulate realistic hydrodynamics!

With the prescription kit given, the state of the system at each lattice site is unambiguously specified in terms of a plain *yes/no* option, telling whether or not a particle sits on the given site. That is all one needs to know.

This dichotomic, *tertium non datur*, situation is readily coded with a single binary digit (bit) per site and direction so that the entire state of the lattice gas is specified by $6N$ bits,

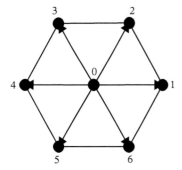

Figure 11.1 *The FHP hexagonal lattice.*

N being the number of lattice sites. Borrowing the parlance of statistical mechanics, we introduce an *occupation number* n_i, such that

$$n_i(\vec{x}, t) = \begin{cases} 0 & \text{particle absence at site } \vec{x} \text{ and time } t, \\ 1 & \text{particle presence at site } \vec{x} \text{ and time } t \end{cases} \quad (11.1)$$

The collection of occupation numbers $n_i(\vec{x}, t)$ over the entire lattice defines a $6N$-dimensional time-dependent Boolean field whose evolution takes place in a Boolean phase space consisting of 2^{6N} discrete states. This Boolean field takes the intriguing name of *Cellular Automaton* (CA) to emphasize the idea that not only space and time, but also the dependent variables (matter) take on discrete (Boolean) values.

The fine-grain microdynamics of this Boolean field can *not* be expected to reproduce the true molecular dynamics to any reasonable degree of microscopic accuracy. However, as it is well known since Gibbs (Josiah Gibbs, 1839–1903), many different microscopic systems can give rise to the same macroscopic state, and it can therefore be hoped that the macroscopic dynamics of the Lattice Boolean field would replicate real-life hydrodynamic motion even if its microdynamics does not.

We shall dub elementary Boolean excitations with such a property simply as *fluons* to denote "quanta" of flowing matter.

11.3 Fluons in Action: LGCA Microdynamic Evolution

So much for the definitions. Let us now prescribe the evolution rules of our CA. Since we aim at hydrodynamics, we should cater for two basic mechanisms:

- *Free streaming*
- *Collisions*

Free streaming consists of simple-particle transfers from site to site according to the set of discrete speeds $\vec{c}_i \equiv c_{ia}$. Thus, a particle sitting at site $\vec{x}$ at time t with speed $\vec{c}_i$ moves to site $\vec{x}_i \equiv \vec{x} + \vec{c}_i$ at time $t + 1$.

In equations:

$$n_i(\vec{x} + \vec{c}_i, t + 1) = n_i(\vec{x}, t) \quad (11.2)$$

This defines the discrete free-streaming operator Δ_i as

$$\Delta_i n_i \equiv n_i(\vec{x} + \vec{c}_i, t + 1) - n_i(\vec{x}, t) \quad (11.3)$$

This equation is a direct transcription of the Boltzmann free-streaming operator $\partial \equiv \partial_t + v_a \partial_a$ to a discrete lattice in which spacetime are discretized according to the synchronous "light-cone" rule:

$$\Delta \vec{x}_i = \vec{c}_i \Delta t \quad (11.4)$$

The relation between the discrete and continuum streaming operators reads as follows:

$$\Delta_i = e^{\partial_i} - 1 \tag{11.5}$$

where $\partial_i \equiv \partial_t + c_{ia}\partial_a$ is the generator of spacetime translations along the i-th direction (sum over repeated indices is implied, see Fig. 11.2 and Fig. 11.3).

Upon formally expanding the exponential, one notes that the discrete streaming operator Δ_i contains an infinite series of differential ones.

Manifestly, the magnitude c plays the role of the "light speed" in the discrete world, in that no signal can propagate faster than c in the lattice.

Once on the same site, particles interact and reshuffle their momenta so as to exchange mass and momentum among the different directions allowed by the lattice (see Fig. 11.4).

This mimics the real-life collisions taking place in a real gas, with the crude restriction that all pre- and post-collisional momenta are forced to "live" on the lattice. As compared to continuum kinetic theory, the LGCA introduces a very radical cut of degrees of freedom in momentum space: just one speed magnitude (all discrete speeds share the same magnitude $c = 1$, hence the same energy) and only six different propagation angles. Not bad for an originally set of ∞^6 degrees of freedom!

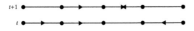

Figure 11.2 *Free streaming in a discrete one-dimensional lattice.*

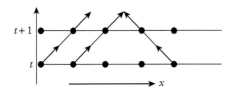

Figure 11.3 *Free streaming along the light cones of x–t spacetime.*

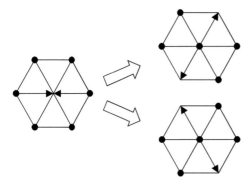

Figure 11.4 *A FHP collision with two equivalent outcomes.*

Spacetime is also discretized (see eqn (11.4)), but this is common to all dynamical systems destined to computer simulation. At this stage, it is still hard to believe that such a stylized system would display all the complexities of fluid phenomena. And yet it does!

The reader acquainted with modern statistical mechanics surely smells the sweet scent of *universality*: for all its simplicity, the FHP automaton belongs to the same class of universality of a real fluid. The name of this magic is *Symmetry and Conservation*.

Let us dig a bit deeper into this matter.

Consider the FHP collision depicted in Fig. 11.4.

Albeit stylized, this collision shares two crucial features with a real molecular collision:

- It conserves particle number (2 before, 2 after)
- It conserves total momentum (0 before, 0 after)

These properties are a *"conditio sine qua non"* to achieve hydrodynamic behavior once a sufficiently large group of particles is considered; this is what sets a fluid apart from a "wild bunch of particles." It is a necessary but not sufficient condition, though. One may ask why not make things even simpler and consider for instance just a four-state automaton like the one depicted in Fig. 11.5.

This is the so-called Hardy–Pomeau–de Pazzis (HPP) cellular automaton predating FHP by more than ten years (7). Like FHP, the HPP automaton can also secure conservation laws. However, it fails to achieve a further basic symmetry of the Navier–Stokes equations, namely *rotational invariance*.

Making abstraction of dissipative terms for simplicity, the hydrodynamic probe of isotropy is the momentum-flux tensor $P_{ab} \equiv \rho u_a u_b + p \delta_{ab}$, which has been and discussed in Part I of this book.

For a 2D Navier–Stokes fluid, this reads

$$P_{xx} = p + \rho u^2, \tag{11.6}$$

$$P_{xy} = P_{yx} = \rho uv, \tag{11.7}$$

$$P_{yy} = p + \rho v^2, \tag{11.8}$$

where p is the fluid pressure and u, v are the Cartesian components of the flow field. This is an isotropic tensor because its components are invariant under arbitrary rotations of the reference frame.

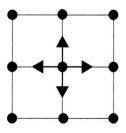

Figure 11.5 *The square grid of the HPP automaton.*

With HPP, we would obtain instead

$$P_{xx} = p + \rho(u^2 - v^2), \tag{11.9}$$

$$P_{xy} = P_{yx} = 0, \tag{11.10}$$

$$P_{yy} = p + \rho(v^2 - u^2) \tag{11.11}$$

Besides missing isotropy, this "Squareland" tensor has little to do with real fluids altogether.[17] Thus, the four HPP particles do not qualify for the status of *fluons*, namely, elementary excitations whose collective dynamics reproduces macroscopic hydrodynamics, while the six FHP particles do.

It is important to realize that this lack of isotropy cannot be cured by going to finer spacetime resolutions, no matter how much finer. In mathematical language, this is because the continuous group of rotations $SO(1)$ is compact. In two dimensions, the group of discrete rotations Z_6 (rotations by multiple of $2\pi/6$) can substitute for the continuous group $SO(1)$ while Z_4 cannot. More precisely, the lattice must generally provide enough symmetry to ensure the following tensorial identity (spatial indices are summed upon):

$$\left[\sum_{i=1}^{b} c_{ia}c_{ib} \left(c_{ic}c_{id} - \frac{c^2}{D}\delta_{cd} \right) \right] u_c u_d = u_a u_b \tag{11.12}$$

for any choice of the dyadic $u_a u_b$. Here D is the space dimensionality and b is the number of discrete speeds (not to be confused with the subscript b denoting spatial indexing, $a, b = x, y, z$). Also, note that the sum over discrete velocities is explicit, whereas repeated latin indices are summed upon.

This is a very stringent condition that weeds out most discrete lattices. As it stands, the constraint (11.12) comes a bit out of the blue. Intuitively, however, the point is that as opposed to, say, a scalar field, fourth-order tensor fields are demanding probes of the lattice since they sense more details of the spacetime "fabric" (tensors are the geometrical encoding of extended objects, hence they sense multiple directions at a time). As a result, it is more difficult to cheat on them, i.e., have them the same for a square and a circle simply does not work!

This is not surprising, after all. More surprising, and pleasing as well, is that a simple hexagon does serve the purpose.

Even after this blend of definitions of isotropy and hand-waving arguments, the inquiring reader may still ask why fourth-order tensors and not some other, say second or sixth?

To make the picture truly compelling, a little plunge into the mathematics of discrete kinetic theory cannot be helped (9). A partial justification, if not an explanation will nonetheless be given later in this chapter. Let us for the time being suspend criticism and accept for a fact that a hexagon is "good enough" to replace a circle, whereas a square is not.

Now, back to the collision operator.

---

[17] As personally experienced by this author, it produces beautiful *square* vortices (sic!).

Symbolically, its effect on the occupation numbers is a change from n_i to n'_i on the same site

$$n'_i - n_i = C_i \left(\underline{n} \right) \tag{11.13}$$

where $\underline{n} \equiv [n_1, n_2, \ldots, n_b]$ denotes the set of occupation numbers at a given lattice site.

To formalize the expression of C_i it proves expedient to label the Boolean phase space via a bit-string $\underline{s} = [s_1, s_2, \ldots, s_b]$ spanning the set of all possible (2^b) states at a given lattice site. For instance, numbering discrete speeds 1–6 counterclockwise starting from rightward propagation, $c_{1x} = 1$, $c_{1y} = 0$, the pre- and post-collisional states associated with a collision $(1, 4) \rightarrow (2, 5)$, read $\underline{s} = [100100]$ and $\underline{s}' = [010010]$, respectively.

It is natural to define a *transition matrix* $A(\underline{s}, \underline{s}')$ flagging all permissible collisions from source state $\underline{s}$ to destination state $\underline{s}'$, $\underline{s} \rightarrow \underline{s}'$ as follows:

$$A \left(\underline{s}, \underline{s}' \right) = \begin{cases} 1 & \text{collision allowed,} \\ 0 & \text{collision forbidden} \end{cases} \tag{11.14}$$

"Allowed" here means compliant with conservation laws (see Fig. 11.6).

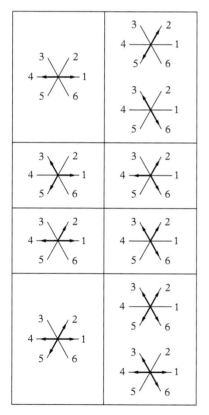

Figure 11.6 *Table of all possible FHP collisions. (Courtesy of L.S. Luo.)*

The transition matrix obeys the *semi-detailed balance* condition:

$$\sum_{\underline{s}} A\left(\underline{s}, \underline{s}'\right) = 1 \qquad (11.15)$$

for any $\underline{s}'$.

This means that any destination state can be reached from at least one source state within the phase space of the automaton.

This condition does not imply a one-to-one equivalence between the source and destination states, as it is the case of *detailed balance*:

$$A\left(\underline{s}, \underline{s}'\right) = A\left(\underline{s}', \underline{s}\right) \qquad (11.16)$$

The latter ensures microreversibility, i.e., one can come and go from source to destination and vice versa with the same probability, a property not shared by the weaker condition of semi-detailed balance.

Indeed, it is easily shown that a given pre-collisional FHP input state can land into more than one (actually, two) post-collisional output states, both compliant with conservation laws; head-on collisions can equally well rotate particle pairs $\pi/6$ left or right. Consequently, unlike HPP, the time evolution of the FHP automaton is no longer deterministic. In practice, the resulting lack of chiral invariance is readily accommodated by choosing either collision with equal probabilities.

Next, let us define the probability to have $\underline{s}$ as input state with occupation number $\underline{n}$:

$$P\left(\underline{s}, \underline{n}\right) = \prod_{i=1}^{b} n_i^{s_i} \, \bar{n}_i^{\bar{s}_i} \qquad (11.17)$$

Here the overbar denotes complement to one, i.e., $\bar{n}_i \equiv 1 - n_i$ as it befits to fermionic degrees of freedom.

Let us clarify with an example. The probability of occupying state $\underline{s} = [100100]$ as an input state is given by $P[100100] = n_1 \bar{n}_2 \bar{n}_3 n_4 \bar{n}_5 \bar{n}_6$.

Manifestly, this quantity is always zero, except when a particle with speed $\vec{c}_1$ and a particle with speed $\vec{c}_4$ are sitting simultaneously on the node. In passing, we note that "particle absence" is to all intents tantamount to "hole presence," echoing the particle–hole symmetry of fermionic matter.

With these preparations, the collision operator can be formally recast in the traditional gain minus loss form:

$$C_i = \sum_{(\underline{s}, \underline{s}')} \left(s_i' - s_i\right) P\left(\underline{s}, \underline{n}\right) A\left(\underline{s}, \underline{s}'\right) \qquad (11.18)$$

The reader is encouraged to check that C_i is a "trit," namely a three-state variable taking values -1 (annihilation), 0 (no action), $+1$ (generation) (see Fig. 11.7).

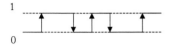

1

0

Figure 11.7 *The railway analogy of Boolean computing. Only* $0 \to 1,\ 1 \to 0$ *transitions can occur, leaving the system on either upper or lower rails. No drifts allowed.*

The trit associated with the head-on collision depicted in Fig. 11.4 is easily computed: $C_i = [-1, 1, 0, -1, +1, 0\rangle$. It is also readily checked that the sum over all discrete speeds yields identically zero, in fulfilment of the requirement of mass conservation. This is formally traced back to the identity:

$$n_i = \sum_{\underline{s}, \underline{s}'} s_i P\left(\underline{s}, \underline{n}\right) A\left(\underline{s}, \underline{s}'\right) \tag{11.19}$$

which is easily checked by direct computation.

Owing to its definition, the collision operator C_i obeys the remarkable and crucial property of preserving the Boolean nature of the occupation numbers. This can be verified directly by noting that the Boolean-breaking combinations, $n_i = 0, C_i = -1$ or $n_i = 1, C_i = 1$, can never occur. They are automatically ruled out by the fact that if $n_i = 0$, the collision operator cannot subtract particles to the input state, and conversely, if $n_i = 1$, it cannot add them to the pre-collisional state.

This closely evokes the well-known properties of fermionic annihilation/generation operators as described by the language of second quantization in quantum statistical mechanics (see chapter 33).

To sum up, the final LGCA update rule reads as follows:

$$\Delta_i n_i = C_i, \tag{11.20}$$

or, in equivalent form

$$n_i\left(\vec{x} + \vec{c}_i, t + 1\right) = n_i'\left(\vec{x}, t\right), \tag{11.21}$$

where all quantities have been defined previously.

The expressions (11.20) and (11.21) represent the microdynamic equation for the Boolean lattice gas, the analog of the N-body Liouville equation for real molecules.

As already pointed out, this formulation evokes closely the second quantization formalism, and it casts fluid dynamics into a many-body language probably more familiar to condensed matter and quantum field theorists than fluid dynamicists. Yet, it is a very useful language for computer implementations (11).

This equation constitutes the starting point of a lattice BBGKY hierarchy, ending up with the Navier–Stokes equations. At each level, one formulates a lattice counterpart of the various approximations pertaining to the four levels of the hierarchy (see Fig. 11.8).

The remarkable point is that, notwithstanding the drastic reduction of microscopic degrees of freedom, the lattice Navier–Stokes equation can be made basically to coincide

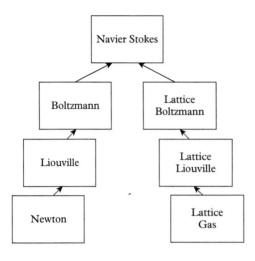

Figure 11.8 *The BBGKY and lattice BBGKY hierarchies. The lack of microscopic detail of the lattice branch becomes less and less important as one walks upward along the hierarchy.*

with its continuum counterpart. In the LGCA jargon, the lattice is "erased" from the macroscopic dynamics.

The guiding lights of this clever procedure are the fundamental conservation laws of classical mechanics (9).

In Boolean terms:

- *Mass and momentum conservation*:

$$\sum_i C_i = 0 \tag{11.22}$$

$$\sum_i c_{ia} C_i = 0 \tag{11.23}$$

- *Angular-momentum conservation (rotational invariance)*:

$$\sum_i c_{ia} c_{ib} c_{ic} c_{id} = \frac{bc^4}{D(D+2)}\left(\delta_{ac}\delta_{bd} + \delta_{ad}\delta_{bc} - \frac{2}{D}\delta_{ab}\delta_{cd}\right) \tag{11.24}$$

As it stands, the microdynamic eqn (11.20) represents the dynamics of a $6N$-body system of hopping Boolean particles with local mass and momentum conservation, a very rich and interesting model of non-equilibrium statistical mechanics. That the large-scale dynamics of this model does indeed reproduce the fluid dynamic equations still remains to be proven. We shall only sketch the main ideas directing the reader fond of full details to the excellent monographs available in the literature (3).

11.4   From LGCA to Navier–Stokes

The standard *bottom-up* procedure[18] taking from many-body particle dynamics (Newton–Hamilton) all the way up to continuum fluid-like equations proceeds through the three formal steps familiar from classical statistical mechanics, which we repeat here for the sake of concreteness:

1. *From Newton–Hamilton to Liouville*

   Upon ergodicity assumption[19] the deterministic Newton–Hamilton equations of motion are replaced by the probabilistic Liouville equation for the N-body distribution function $f_N(\vec{x}_1, \vec{v}_1, \ldots, \vec{x}_N, \vec{v}_N; t)$ expressing the probability of finding particle 1 around $\vec{x}_1$ with speed $\vec{v}_1$, *and* particle 2 around $\vec{x}_2$ with speed $\vec{v}_2$, *and* particle N around $\vec{x}_N$ with speed $\vec{v}_N$, at the same time t.

   Symbolically: $z(t) \rightarrow f_N(z, t)$, where $z \equiv [\vec{x}_1, \ldots, \vec{v}_N]$ is the state vector in $6N$-dimensional phase space.

2. *From Liouville to Boltzmann*

   Upon diluteness assumption, high-order distribution functions are expressed in terms of the low-order ones by integration over many-body phase-space coordinates. This procedure ends up at the lowest level (Boltzmann) dealing with the single-body distribution function. The hierarchy is not closed because the distribution function at a given level depends on the upper-lying one. To close the hierarchy, a truncation in the diluteness parameter $(s/d)^3$, s being the effective molecular diameter and d the mean-intermolecular distance, is used.

   Symbolically: $f_N(z, t) \rightarrow f_1(\vec{z}_1, t)$.

3. *From Boltzmann to Navier–Stokes*

   Upon integration over momentum space degrees of freedom, a set of partial differential equations for the spacetime dependent moments of the Boltzmann distribution function is obtained.

   Upon retaining a finite, say $m = 1, \ldots, M$, number of moments, the set of moment equations takes the form of an M-dimensional *generalized continuity equation*

   $$\partial_t \rho_m + \partial_a \mathcal{J}_{am} = C_m \qquad (11.25)$$

---

[18] By "bottom" we imply the most fundamental, microscopic level, the idea being that upper levels "emerge" to the status of macroscopic observables from coarse-graining of the lower levels, see the iceberg analogy in chapter 5.

[19] We remind that, loosely speaking, ergodicity means that the probability of visiting a given region of phase space, i.e., the time spent by the system in that region, is proportional to the volume of the region. This permits us to replace the phase space averages with time averages.

where

$$\rho_m(\vec{x}; t) = \int f(\vec{x}, \vec{v}; t) \phi_m(\vec{v}) \, d\vec{v} \qquad (11.26)$$

$$\mathcal{J}_{am}(\vec{x}; t) = \int f(\vec{x}, \vec{v}; t) v_a \phi_m(\vec{v}) \, d\vec{v} \qquad (11.27)$$

are the generalized densities and currents, respectively, and

$$C_m(\vec{x}; t) = \int C \left[f(\vec{x}, \vec{v}; t) \right] \phi_m(\vec{v}) \, d\vec{v} \qquad (11.28)$$

represents the effect of intermolecular interactions.

Finally, $\{\phi_m\}$ is a complete set of basis functions in momentum space, typically Hermite polynomials for cartesian coordinates.

The most relevant hydrodynamic moments are

Density: $\qquad\qquad\qquad \rho = m \int f \, d\vec{v},$

Velocity: $\qquad\qquad\quad\; u_a = m \int f v_a \, d\vec{v}/\rho$

Momentum flux tensor: $P_{ab} = m \int f v_a v_b \, d\vec{v}$

The system (11.25) is not closed either since the time derivative of a given generalized density involves the divergence of the corresponding generalized flux lying one step higher in the hierarchy. An approximation (closure) is required to close the hierarchy.

This is achieved by splitting up the one-particle distribution into a local equilibrium and non-equilibrium components:

$$f = f^e + f^{ne},$$

with the assumption that $f^{ne} \sim O(Kn)f^e$, Kn being the Knudsen number (the *Chapman–Enskog procedure* discussed in Part I).

To zeroth order $Kn \to 0$, the power expansion in Knudsen number of the Boltzmann equation delivers the Euler equations of inviscid flows, whereas at the next order $O(Kn)$ the dissipative Navier–Stokes equations are obtained.

Exactly the same steps are involved in the process of deriving the lattice Navier–Stokes equations from the lattice BBGKY, with the notable caveat that lattice discreteness needs to be handled with great care since continuum symmetries (Galilean invariance, roto-translations, parity) are always at risk of being broken by the lattice discreteness.

The most dangerous effects bear upon the discrete lattice equilibria, to which we turn our attention next.

11.4.1 Discrete Local Equilibria

Since LGCA microdynamics obeys an exclusion principle, lattice equilibria are expressed by a Fermi–Dirac distribution

$$f_i^{\text{eq}} = \frac{\rho/b}{1 + e^{\Phi_i}}, \tag{11.29}$$

where Φ_i is a linear combination of the collisional invariants, mass and momentum. Restricting our attention to isothermal ideal fluids, where energy and mass are the same, we obtain

$$\Phi_i = A + B u_i, \tag{11.30}$$

where we have set

$$u_i = \frac{c_{ia} u_a}{c_s^2} \tag{11.31}$$

and A and B are free Lagrangian parameters, to be adjusted in order to secure mass and momentum conservation.

As usual, the goal is to express the Lagrange multipliers A, B as a function of fluid density ρ and momentum u_a, by using the conservation relations (11.22) and (11.23). This task is easy in the continuum, much less so in a discrete momentum space. Let us see why.

By inserting the Fermi–Dirac expression into the conservation relations (11.22) and (11.23), we obtain two nonlinear equations for the Lagrangian parameters A, B as a function of the fluid density ρ and velocity u_a. Actually, ρ can always be scaled out since it is just a multiplicative prefactor (we are considering incompressible fluids). The dependence on u_a is non-trivial, in the sense that it is generally *not* possible to come up with a closed, analytical expression of the form $A = A(u_a), B = B(u_a)$, valid for *any* value of the fluid velocity u_a.

The point is by no means a mere technicality!

What we would like to obtain is something in the form $\Phi = A + B_a(c_{ia} - u_a)$ to ensure Galilean invariance. The exponential form (11.29) stems from the minimization of the Boltzmann entropy

$$H = -\int \tilde{f} \ln \tilde{f} \, d\vec{v} \tag{11.32}$$

where $\tilde{f} \equiv f/(1 - f)$ for fermions, which is strictly related to Boltzmann's Stosszahlansatz assumption. The subsequent possibility to express the Lagrangian parameters in a closed form is in turn strictly tied up to the continuity of momentum space.

In a discrete momentum world, the exponential function plays no special role, if it did, the plain replacement $v_a = c_{ia}$ would work! Hence, we may expect it to be replaced by some other functional form, perhaps some suitable polynomials.

In the absence of a closed analytical solution for local equilibria, it appears quite natural to seek approximate solutions obtained by expanding the exponential in powers of the flow field, or more precisely, of the Mach number $M = u/c_s$. Conservation constraints can then be imposed order by order in the perturbation series and since the ultimate target, the Navier–Stokes equations exhibit a quadratic nonlinearity (the convective term $\partial_b u_a u_b$), it is legitimate to expect that a second-order expansion should fill the bill.

This is indeed the case, in modulo there are some anomalies that will be commented on shortly.

The specific expression of LGCA equilibria is

$$f_i^{\text{eq}} = \frac{\rho}{b}(1 + u_i + Gq_i), \tag{11.33}$$

where

$$q_i = \frac{Q_{iab}u_a u_b}{2c_s^4} \tag{11.34}$$

and

$$Q_{iab} = c_{ia}c_{ib} - c_s^2 \delta_{ab} \tag{11.35}$$

is the projector along the i-th discrete direction.

Here

$$c_s = \frac{c}{\sqrt{D}} \tag{11.36}$$

is the lattice sound speed.

The breaking of Galilean invariance is signaled by the anomalous prefactor:

$$G(\rho) = \frac{1-2d}{1-d} \neq 1 \tag{11.37}$$

where

$$d = \frac{\rho}{b} \tag{11.38}$$

is the *reduced-fluid density*, i.e., the density per link.

The isotropy constraint (11.12) unveils now part of his mysterious nature: plugging discrete quadratic equilibria (11.33) into the definition of the (equilibrium) momentum flux tensor (see Chapter 1), $P_{ab} = \int f^e v_a v_b \, d\vec{v}$, reveals that (11.12) is just the condition to match the tensorial equality:

$$\int f^e v_a v_b \, d\vec{v} = \rho \left(u_a u_b + c_s^2 \delta_{ab}\right) \tag{11.39}$$

discussed in Part 1.

Let us now take a short break from theory and spend a few comments on the practical use of lattice-gas cellular automata as a tool for computer simulation.

11.5 Practical Implementation

The main computational assets of the LGCA approach to fluid dynamics are:

- *Exact computing (round-off freedom)*
- *Virtually unlimited parallelism*

The Boolean nature of the LGCA update rule implies that the corresponding algorithm can be implemented in pure Boolean logic, without ever needing floating-point computing. This is very remarkable, since it offers a chance to sidestep a number of headaches associated with the floating-point representation of real numbers, primarily round-off errors. Among others, the Boolean representation eases out the infamous problem of numerical drifts plaguing long-time floating-point simulations, both in fluid dynamics and fundamental studies in statistical mechanics.

To see this in more detail, let us consider the head-on collision of Fig. 11.4. The collision is encoded by the following logic statement:

If there is a particle in state 1 AND a hole in state 2 AND a hole in state 3 AND a particle in state 4 AND a hole in state 5 AND a hole in state 6, then the collision occurs. Otherwise nothing happens.

All we need to put this plain logical statement into practice are the elementary Boolean operations AND (logical exclusion), OR (logical inclusion), NOT (negation) and XOR (exclusive OR) (see Table 11.1). The effect of these operations is usually represented by the so-called "truth table" (1 = TRUE and 0 = FALSE) which we report below for simplicity.

The statement "the collision occurs" corresponds to setting up a collision mask flagging collisional configurations $\mathcal{M} = 1$ and $\mathcal{M} = 0$ all others:

$$\mathcal{M} = n_1.\text{AND}.\bar{n}_2.\text{AND}.\bar{n}_3.\text{AND}.n_4.\text{AND}.\bar{n}_5.\text{AND}.\bar{n}_6 \tag{11.40}$$

where the overbar means negation.

Table 11.1 *Truth table of the four basic Boolean operators.*

A	B	A.AND.B	A.OR.B	A.XOR.B	.NOT.A
0	0	0	0	0	1
0	1	0	1	1	
1	0	0	1	1	0
1	1	1	1	0	

Once the collision mask $\mathcal{M}$ is set up, the post-collisional state

$$n_i' = n_i + C_i \tag{11.41}$$

is obtained by simply "XORing" the pre-collisional state with the collisional mask.

$$n_i' = \mathcal{M}.\text{XOR}.n_i \tag{11.42}$$

This simple procedure is applied simultaneously in lock-step mode to all lattice sites in the typical "blindfold" fashion so dear to vector computers. More importantly, each site is updated independently of all others, thus making the scheme ideal for parallel processing. Amenability to massively parallel processing is probably worth a few additional words of comment.

Roughly speaking, parallel computing works on the time-honored Roman principle *Divide et Impera* (divide and conquer). To solve a large problem, first break it into small parts, then solve each part independently, and finally glue all the parts together to produce the global solution. By doing so, a collection of, say, P processors would ideally solve a given problem at a fraction $1/P$ of the cost on a single computer.

Apparently, nothing could be easier, but actual practice often tells another story because many sources of inefficiency, both practical and conceptual, may hamper the practical realization of this general principle.

Among others, a chief quantity to consider is the communication/computation ratio. Generally speaking, a numerical algorithm is good for parallel computing whenever the amount of communication between the different processors needed to coordinate their interaction can be made negligible in comparison with the in-processor computations (see Appendix C). This is quite intuitive, because if communication is scanty, all processors can proceed (almost) independently without spending much time in exchanging information (the cell-phone syndrome).

LGCA are fairly well positioned in this respect since the collision step, by far the most time-consuming operation, is completely local and requires no communication at all among the processors. This low communicativity makes LGCA outstanding candidates to massively parallel processing.

The elegance and the practical appeal of round-off free-parallel computing justify the blue-sky scenarios envisaged in the late 80s, when LGCA held promise of simulating turbulence on parallel machines at unprecedented resolutions.

However, a closer look at the method revealed a number of flaws that ultimately ground it almost to a halt for fully developed turbulence (at least in the academic environment) in the early 90s. Basically, it was realized that LGCA are plagued by a number of anomalies, namely broken symmetries which cannot be restored even in the limit of zero-lattice spacing. Ingenious cures were found, but still the final picture did not live up to the (real-high) initial expectations, at least for the simulation of fully developed turbulent flows.

These anomalies surface neatly only once the task of deriving quantitative hydrodynamics out of the LGCA Boolean dynamics is undertaken.[20] This is a serious technical task, which proceeds along the lines of standard statistical mechanics, with the additional burden of "slalomizing" through the technical catches littered by spacetime–momentum discreteness. In this context, it is only fair to mention the very thorough and detailed account of the basic theory of cellular automata fluids by Steven Wolfram (yes, the Mathematica Wolfram), which appeared only a few months after the seminal FHP paper (10).

11.6   Lattice Gas Diseases and How to Cure Them

The formal derivation of the "Navier–Stokes look-alike" equations from LGCA microdynamics highlights two basic diseases related to the lattice discreteness, namely:

A1. *Lack of Galilean invariance*
A2. *Anomalous velocity dependence of the fluid pressure*

Additional drawbacks coming along with phase-space discretization are:

A3. *Statistical noise*
A4. *High viscosity (low Reynolds number)*
A5. *Exponential complexity of the collision rules*
A6. *Spurious invariants*

Anomalies A1, A2 trace back to the fact that a finite number of speeds does not allow a *continuum* family of (Fermi–Dirac) equilibria parameterized by the flow speed $\vec{u}$. As a result, hydrodynamic equilibria can only be defined via a perturbative expansion in the flow field, and this expansion cannot match exactly the form of the Navier–Stokes inertial and pressure tensors. It delivers instead the following:

$$\sum_i f_i^e c_{ia} c_{ib} = \rho \left[g u_a u_b + c_s^2 (1 - gM^2) \, \delta_{ab} \right] \tag{11.43}$$

where

$$g = \frac{D}{D+2} \frac{1-2d}{1-d} \tag{11.44}$$

is the Galilean breaking factor.

---

[20] As opposed to the "wallpaper" prejudice looming on LGCA in the early days. According to this prejudice, CA are very good at pointing out colorful (wallpaper) analogies, much less so at nailing down quantitative hard numbers. It is only fair to say that this kind of criticism did not strike home with LGCA. Shortly after the seminal FHP paper, Frisch, d'Humières, Lallemand, Hasslacher, Pomeau and Rivet hastened to produce a thorough quantitative description of the way the large-scale LGCA dynamics yields real-life hydrodynamic equations (9).

The quadratic term in this expression is responsible for the advection term $u\nabla u$ in the Navier–Stokes equation, whereas the remaining term is associated with fluid pressure. Both terms are anomalous, and in a somewhat devious way. To make advection Galilean invariant we need $g = 1$. First, there is *no* value of the fluid density for which this can occur. Second, the value $g = 1$ just maximizes the anomalous contribution gM^2 to the fluid pressure. This spells true frustration.

Admirably enough, pioneers in the field responded quite promptly to this disheartening state of affairs.

It was rapidly recognized that for quasi-incompressible flows (low-Mach number $u/c_s \ll 1$), the pressure anomaly becomes negligibly small, the density is virtually a constant, and so is the Galilean breaking factor g. Under these conditions, a plain rescaling of time $t \rightarrow gt$ and viscosity $\nu \rightarrow \nu/g$ makes it possible to reobtain anomaly free Navier–Stokes equations!

11.6.1   Statistical Noise

Like, and in fact more than any particle method, LGCA are exposed to a substantial amount of statistical fluctuations.

This problem was quickly recognized by Orszag and Yakhot (12) and shortly later demonstrated by actual numerical simulations (13) against pseudospectral methods for the case of two-dimensional turbulence. The problem was that, like any reasonable N-body Boolean system, the automaton was still crunching a lot of needless many-body details. Very noticeable efforts, mainly in Nice, Paris and Shell Research brought the second and third generation FHPs to the point of competitiveness with conventional methods. Besides clever theory, computer hardware has some say too. Comparing LCGA and floating-point methods on so-called general purpose computers is rather unfair, simply because what we call a general purpose computer is de facto a special purpose computer for floating-point calculations! Think of the following nightmare: computational fluid dynamics with floating-point operations in software … !).

Why not do the same for LGCA, namely build a LGCA-chip performing, say, a collision (in fact, very many) in just a single clock cycle? On such a computer, LGCA would probably win hands down over "traditional" methods.

Some prototypes in this breed were indeed built at MIT and at the Ecole Normale in Paris and lately envisaged also at EXA Corporation. Discussing why these prototypes did not make it into a fully fledged LGCA machine would be too long a story, that certainly has to do with the mind-boggling escalation of computer power afforded by "off the shelf" commercial floating-point processors. But, again, that is another story.

Before moving on to the next item, we wish to mention that there is a positive side to noise, too. As shown by Grosfils *et al.* (14) the statistical noise inherent to LGCA dynamics has many *quantitative* features in common with true noise in actual thermodynamic systems. This puts LGCA in a privileged position to address problems related to modern statistical (micro)hydrodynamics.

11.6.2   Low Reynolds Number

The maximum Reynolds number achievable by a LGCA simulation is controlled by the minimum mean-free path one can reach in the lattice, namely the maximum number of collisions per unit time that the automaton is able to support. More precisely, it is not just the number of collisions which matters, but also their quality, namely the amount of momentum they are able to transfer across the different directions of the discrete lattice, i.e., the change of momentum flux.

Phase-space discreteness plays a two-faced role here. On the one side, the limited number of discrete speeds combined with compliance to mass–momentum conservation places severe restrictions on the number of possible collisions. For instance, even in FHP III, a low-viscosity variant of FHP with rest particles (7 states per site), only 22 out of 128 configurations are liable to collisions, which means that only about twenty percent of phase space is collisionally active. This is to be contrasted with continuum kinetic theory in which virtually all of phase space is collisionally active. This lowly twenty percent efficiency is partially rescued by the streaming step. Here lattice discreteness plays in our hands because the discrete speeds force particles to meet at lattice sites with no chance for "near miss" events (a sort of "highway" effect).

One may thus hope to achieve good collisionality, namely a mean-free path smaller than the lattice spacing a. Indeed, that is exactly what happens.

The figure of merit of LGCA collisionality is the dimensionless Reynolds number R^*, defined as

$$R^* = g \frac{c_s a}{\nu} \qquad (11.45)$$

With this definition, the Reynolds number becomes

$$Re = R^* M N \qquad (11.46)$$

where N is the number of lattice site per linear dimension ($L = Na$) and M is the Mach number.

By recalling that fluid viscosity ν is basically the mean-free path l_μ times the particle speed c, we obtain $R^* = ga/l_\mu$. This definition shows that R^* is essentially the number of mean-free paths in a lattice spacing a. Since the computational power required to simulate a three-dimensional flow of size L scales roughly like L^4 (three spatial dimensions times the temporal span of the evolution), it is clear that raising R^* by a factor of two buys out a factor 8 in memory occupation and 16 in computer time. This explains the intense efforts spent on the optimization of this parameter in the late 80s.

The earliest FHP model featured $R^* = 0.4$, which means a collision every two and a half streaming hops. This also means that hydrodynamics cannot be attained at the scale of a single lattice spacing, and consequently that heavy coarse-graining is needed to extract sensible hydrodynamics out of the noisy LGCA signal. Once again, we are steered in the direction of trading spatial degrees of freedom for emergence of coherent signals. This contrasts with conventional Computational Fluid Dynamics (CFD) where

Lattice Fluid

Real Fluid   **Figure 11.9** *True and LGCA mean-free path.*

the problem simply does not exist and the lattice pitch, a, can be freely tuned in the close vicinity of the smallest hydrodynamic scale. Since, by definition, hydrodynamics works in the regime of low Knudsen numbers $Kn \ll 1$, it follows that $l_\mu/a = Kn \ll 1$.

Therefore, conventional computational fluid dynamics works at $R^* \sim 1/Kn \sim 100$ or more. In other words, the situation is just the opposite as for early LGCA: many mean-free paths in a lattice spacing. The impact on the Reynolds number is obvious; consider just simple figures: a 1000 cubed grid with a flow at $Mach = 0.1$ and $R^* = 0.2$, yields a lowly $Re = 0.1 \times 0.2 \times 1000 = 20$, at least two orders of magnitude below conventional CFD, see Fig. 11.9.

On the other hand, since the computational work grows with the 4-th power of R^*, it is clear that even minor improvements on this parameter bring a huge pay-off in terms of computational efficiency. This explains the intense research efforts spent in the late 80s to enhance LGCA collisionality. This activity yielded a remarkable top value of $R^* = 13$; that is, almost the hardly sought for two orders of magnitude.

The gap was basically bridged even on "general purpose" machines: a very remarkable achievement indeed.

The large number of speeds (24) required by the algorithm, however, made it rather impractical even on most powerful computers of those days. This takes us to the next issue, namely exponential complexity of the collision rule.

11.6.3   Exponential Complexity

A weird and intriguing feature of LGCA is that there is no three-dimensional crystal ensuring the isotropy of fourth-order tensors, as required to recover isotropy of the Navier–Stokes equations. For a (short) while, this seemed yet another unforgiving "no go" to three-dimensional LGCA hydrodynamics. A clever way out of this problem was found by d'Humières, Lallemand and Frisch in 1989 (15). These authors realized that a suitable lattice can be found, provided one is willing to add an extra space dimension to the LGCA world. In particular, they showed that the *four*-dimensional Face Centered HyperCube (FCHC) has the correct properties. The FCHC consists of all neighbors of a given site (center) generated by the speeds $c_i = [\pm 1, \pm 1, 0, 0]$ and permutations thereof. This yields 24 speeds, all sharing the same magnitude $c_i^2 = 2$.

The three-dimensional projection of FCHC is shown in Fig. 11.10, from which we see that the 24 speeds are generated as follows:

- *6 nearest neighbors (nn) (connecting the center C to the centers of the six faces of the cube).*

- *12 next-to-nearest neighbors (nnn) connecting C to the centers of the edges of the cube.*

- *The remaining six speeds are accounted for by counting a factor of two (degeneracy) to the nn connections as they possess a nonzero speed along the fourth dimension.*

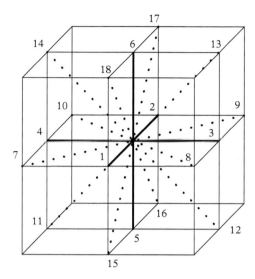

Figure 11.10 *The face centered hypercubic lattice. Thick solid lines represent degenerate directions along the fourth dimension ($c_4 = \pm 1$).*

Thus, *nn* links can be thought as of carrying two particles instead of one.

The use of extra dimensions to remove anomalies is not a germane practice in theoretical physics; the famous ten/eleven dimensions of superstring theories also come from a quest for cancellations of unwanted anomalies.

But let us come back to the point. The FCHC restores the missing symmetry, namely isotropy, but not for nothing. In fact, the number of Boolean operations needed to exhaustively represent the collisional states is a $O(1)$ fraction of the single site phase space of the automaton. As a result, a *b*-bit automaton generates a computational complexity of order $O(2^b)$.

In 2D $b = 6, 7$ and the collision operator is nearly a matter of "back of the envelope" calculation. In 3D, however, $b = 24$, and the situation is much less favorable. Some actual numbers help in putting the picture straight.

With $b = 24$, we must make allowance for $2^{24} = 16, 128$ millions Boolean operations. Even assuming a closed, analytical, Boolean expression for such a "monster" operator could be found, its practical value would be defeated by the huge amount of computer time required to process it.

Because of these daunting difficulties, researchers turned to the use of pre-hard-wired look-up tables storing the post-collisional state $\underline{s}'$ as a function of the pre-collisional one $\underline{s}$. Interestingly enough, the compilation of these table look-ups required a computer program on its own taking a few CRAY-2 hours Central Processing Unit (CPU) time! As of today, these figures don't look that scary anymore, but we are talking mid 80s, a long while ago.

Once the table is computed, the simulation code needs "only" to read-off the post-collisional state $\underline{s}'$ corresponding to each pre-collisional state $\underline{s}$. The conceptual simplicity of this strategy should not deceive the reader. From a practical point of view reading off the output state given the input one is a potentially costly operation. First, a matter of size: the mapping $A(\underline{s}, \underline{s}')$ generates a table look-up consisting of 24×2^{24} bits, namely 48 Mbytes (1 byte = 8 bits) of storage. What is worse, this huge piece of storage is to be accessed almost randomly because there is no way of storing the state $\underline{s}(\vec{x})$ at site $\vec{x}$ and those of its neighbors $\underline{s}(\vec{x} + \vec{c}_i)$ in contiguous memory locations.

As a result, reading off the table look-up may imply relatively long and expensive memory searches.

Pioneering simulations with FCHC–LGCA were performed by Rivet, Frisch, and Henon in Nice, who used a highly tuned version for the CRAY-2 machine to study transitional flows past a flat plate (16). By then, R^* was raised by more than a factor of ten with respect to the original FHP, $R^*_{\text{FCHC}} \sim 7$. Subsequently, P. Rem and J. Somers, then at Shell Research devised ingenious strategies to work with reduced look-up tables (64 Kbytes instead of 48 Mbytes), by sacrificing approximately a factor of five in R^*.

11.6.4 Spurious Invariants

In the absence of external sources/sinks continuum fluids conserve mass, momentum and energy: nothing less, nothing more either. Discrete fluids, as a result of leaving in a cage, may exhibit additional conserved quantities that draw their existence solely from the lattice discreteness. Being pure artifacts, they are currently referred to as *spurious invariants*. The effect of these spurious invariants is to shrink the accessible portion of phase space (lack of ergodicity) with undesirable effects on the coarse-graining procedures leading to hydrodynamic equations. Systematic procedures to detect these invariants have been devised by a number of authors.

Although spurious invariants may be regarded as subtler pathologies as compared to the anomalies discussed in this chapter, they must be watched very carefully in the simulations, especially in the vicinity of generic boundaries non-aligned with the lattice grid lines, where spurious modes are more likely to be excited and propagate inside the fluid region. Fortunately, no serious damage seems to occur for simple boundaries aligned with the flow.

11.7 Summary

In spite of the remarkable progress achieved in the late 80s, leading to major improvements over the initial state of affairs, the LGCA algorithms remained nonetheless rather heavy and stiff with respect to changes in the collision rules. On the other hand, these inconveniences were not compensated by any dramatic advantage in terms of accessible Reynolds numbers. As a result, the interest in LGCA as a tool for high Reynolds flow simulations leveled off in the early 90s. This was the time the lattice Boltzmann method took off.

As of 2017, it is fair to add that the figures that let LGCA down back then look now much less dreadful. Even though competitiveness with floating-point CFD for high-Reynolds flows is still ruled out, maybe the role of LGCA for molecular hydrodynamics is now worth being revisited.

....

REFERENCES

1. U. Frisch, B. Hasslacher and Y. Pomeau, "Lattice gas automata for the Navier–Stokes equations", *Phys. Rev. Lett.*, **56**, 1505, 1986.
2. G. McNamara and G. Zanetti, Use of the Boltzmann equation to simulate lattice gas automata, *Phys. Rev. Lett.*, **61**, 2332, 1988.
3. D. Rothman and S. Zaleski, *Lattice Gas Cellular Automata: Simple Models of Complex Hydrodynamics*, Cambridge University Press, 1997.
4. B. Chopard and M. Droz, *Cellular Automata Modeling of Physical Systems*, Cambridge University Press, Cambridge, 1998.
5. J.P. Rivet and J.P. Boon, *Lattice Gas Hydrodynamics*, Cambridge University Press, 2000.
6. D.W. Gladrow, *Models Lattice Gas Cellular Automata and Lattice Boltzmann*, "Springer Notes in Mathematic", 1725, 2000.
7. J. Hardy, O. de Pazzis and Y. Pomeau, "Molecular dynamics of a lattice gas: transport properties and time correlation functions", *Phys. Rev. A.*, **13**, 1949, 1973.
8. J. Hardy, O. de Pazzis and Y. Pomeau, "Time evolution of a two-dimensional model system I: invariant states and time correlation functions", *J. Math. Phys.*, **14**, 1746, 1976.
9. U. Frisch, D. d'Humières, B. Hasslacher, P. Lallemand, Y. Pomeau and J.P. Rivet, "Lattice gas hydrodynamics in two and three dimensions", *Complex Systems* **1**, 649, 1987, reprinted in *Lattice Gas Methods for Partial Differential Equations*, p. 11, Addison–Wesley, Reading, Massachusetts, 1989.
10. S. Wolfram, "Cellular Automaton Fluids 1: Basic Theory", *J. Stat. Phys.*, **45**, 471 1986.
11. R. Feynman, *Feynman lectures on computation*, edited by T. Hey and R.W. Allen, Westview publications, 1996.
12. S. Orszag and V. Yakhot, "Reynolds number scaling of cellular automaton hydrodynamics", *Phys. Rev. Lett.*, **56**, 1691, 1986.
13. S. Succi, P. Santangelo and R. Benzi, "High resolution lattice gas simulation of two-dimensional lattice gas turbulence", *Phys. Rev. Lett.*, **60**, 2738, 1988.
14. P. Grosfils, J.P. Boon and P. Lallemand, *Phys. Rev. Lett.*, **68**, 1077, 1992.
15. D. d'Humières, P. Lallemand and U. Frisch, "Lattice gas models for 3D hydrodynamics", *Europhys. Lett.*, **2**, 291, 1986.
16. J.P. Rivet, M. Henon, U. Frisch and D. d'Humières, "Simulating fully three-dimensional external flow by lattice gas methods", *Europhys. Lett.*, **7**, 231, 1988.

....

EXERCISES

1. Construct the collision mask for a triple collision [101010] going into [010101].
2. Estimate the signal-to-noise ratio in a FHP block with 16×16 Boolean sites per fluid variable.
3. Suppose to simulate a 3D flow at $Re = 10^4$ on a 1000^3 grid. What R^* would you plan?

12

Lattice Boltzmann Models with Underlying Boolean Microdynamics

In this chapter we shall take a walk into the Jurassics of LBE, namely the earliest Lattice Boltzmann model that grew up out in response to the main drawbacks of the underlying LGCA.

Stars do not make noise

(J. Stephens)

12.1 Nonlinear LBE

The earliest LBE was first proposed by G. McNamara and G. Zanetti in 1988, with the explicit intent of sidestepping the statistical noise problem plaguing its LGCA ancestor (1). The basic idea is simple: just replace the Boolean-occupation numbers n_i with the corresponding ensemble-averaged populations

$$f_i = \langle n_i \rangle, \tag{12.1}$$

where brackets stand for suitable ensemble averaging. The change in perspective is exactly the same as in Continuum Kinetic Theory (CKT); instead of tracking single-Boolean molecules, we content ourselves with the time history of a collective population representing a "cloud" of microscopic degrees of freedom.

To implement this program, it is formally convenient to split up the average and fluctuating part of the Boolean occupation numbers as follows:

$$n_i = f_i + g_i, \tag{12.2}$$

where the fluctuation g_i averages to zero by definition ($\langle g_i \rangle = 0$).

The Lattice Boltzmann Equation. Sauro Succi, Oxford University Press (2018).
© Sauro Succi. DOI: 10.1093/oso/9780199592357.001.0001

By inserting (12.2) into the microdynamic Boolean eqns (11.21) given previously, we obtain

$$\Delta_i f_i = C_i\left(f\right) + G_i, \tag{12.3}$$

where f stands for full set of discrete populations and G_i collects all contributions from interparticle correlations. It is important to remark that the left-hand side of eqn (12.3) is a direct transcription of the corresponding Boolean counterpart, with the plain replacement

$$n_i \rightarrow f_i \tag{12.4}$$

This reflects to the linearity of the streaming operator. This is not the case for the multibody collision operator which, being a b-th order polynomial, introduces particle–particle correlations up to order b. These are collected in the term G_i, see Fig. 12.1.

A simple example for the 4-state HPP automaton yields

$$\langle n_1 \bar{n}_2 n_3 \bar{n}_4 \rangle = f_1 \bar{f}_2 f_3 \bar{f}_4 + \mathcal{P}\left(f_1 \bar{f}_2 G_{34}\right) + G_{1234} \tag{12.5}$$

where $\mathcal{P}$ denotes all possible permutations and $\bar{f}_i = 1 - f_i$ are the hole populations. We note that the correlation chain of the lattice BBGKY hierarchy and the attendant Mayer cluster expansions are significantly more cumbersome than is usual in kinetic theory because of the multibody nature of the lattice collision operator. Nonetheless, the fact that the kinetic space is finitely dimensional permits interesting simplifications.

In line with CKT, one makes the "Stosszahlansatz" assumption of no correlations between particles entering a collision, namely

$$G_i = 0 \tag{12.6}$$

By weeding out all multiparticle correlations, we are left with a closed, nonlinear, finite difference equation for the one particle distribution f_i:

$$\Delta_i f_i = C_i(f_1, \ldots, f_b), \quad i = 1, b \tag{12.7}$$

This is the earliest Lattice Boltzmann Equation (LBE).

In order to underline the nonlinear nature of the collision operator, we shall hereafter refer to it as *nonlinear LBE*.

It is interesting to point out that nonlinear LBE is a *déjà vu* from LGCA theory, since it was derived as an intermediate stopover in the journey from LGCA to Navier–Stokes. However, in the LGCA treatment LBE was regarded just as an intermediate

Figure 12.1 *Cluster diagram $f_{12} = f_1 f_2 + g_{12}$. The springy line denotes correlations.*

conceptual step, with no hint at the possibility that it might constitute a valuable tool for fluid dynamics computations on its own.

McNamara and Zanetti's merit was precisely to point out a new way to interpret and use a known equation! In a way, they reinvented the LBE by looking at it from a different angle.

Having assessed that the nonlinear LBE is a direct transcription of LGCA micro-dynamics with the plain replacement $n_i \rightarrow f_i$, the trade-off is clear: noise is erased because f_i is *by definition* an averaged, smooth quantity. That is why LBE is sometimes also called a *pre-averaged* LGCA.[21]

On the other hand, the price to pay for this smoothness is twofold: on the physical side, the physics of many-body correlations ("non-Boltzmann effects") is lost. This means that fundamental issues such as the breakdown of molecular chaos, long time tails, mode–mode coupling and related points go off limits.

Second, and more practical, round-off freedom falls apart since, being real numbers, the f_i are no longer amenable to exact Boolean algebra. The ground-breaking idea of "exact" computing fades away.

Even if the nonlinear LBE is less revolutionary in scope than LGCA, we are still left with several reasons for excitement.

First, the LBE can be used, *easily* and profitably on any type of computer.

Second, it still inherits many of the best LGA features, such as exact streaming.

Third, it can accommodate additional mesoscopic physics not easily accessible to continuum-based models.

Thus, leaving the fundamental world of many-body systems does not mean landing in a culturally flat landscape either.

McNamara and Zanetti demonstrated the nonlinear LBE by means of a shear-wave decay numerical experiment with the six-state FHP scheme. By proving that the fluid viscosity is in an excellent accord with the theoretical expectation given by LGCA theory, they showed that nonlinear LBE does indeed correctly capture (linear) fluid-transport phenomena. However, they did not go further and perform any large-scale flow simulations to test nonlinear LBE for more complex flows.

At this stage it should be noted that, apart from statistical noise, all other "dark sides" of LGCA are still left with the nonlinear LBE.

In particular, the unviability to 3D computations is just as bad as for LGCA, due to the overwhelming quest for compute power (floating-point instead of Boolean) raised by the collision operator.

This is precisely the problem taken up and solved by the next member of the LBE family, introduced by Higuera and Jimenez (HJ) basically at same time. Before going into the details of the HJ theory, a few more comments on nonlinear LBE are in order.

As already noticed, nonlinear LBE does away with statistical noise but still retains a multibody collision operator, which is where it draws its complexity reduction from. This is in a marked contrast with the Boltzmann equation in CKT, which deals with two-body

---

[21] To the best of this author's knowledge, this witty definition is due to Daan Frenkel.

collisions since higher-order collisions are ruled by the combined assumption of low-density and short-range interactions, the typical scenario of rarefied gas dynamics.

Such an assumption is *not* made here: the interaction, although short ranged, does involve multiple particles, a situation evoking a solid-state-like dense medium rather than a rarefied gas. This is a genuine lattice effect: since they can only move along the links of the lattice, particles are forced to meet at the lattice sites even at low density.

The question is: what are these higher-order collisions for? Do we really need them for the purpose of describing large-scale hydrodynamics?

A minute's thought reveals the answer: yes, we do need them *to make the mean-free path small*, actually much smaller than purely binary collisions would allow.

Thus multibody collisions help to cushion the detrimental effect of discrete momentum space (the cage) on the mean-free path, namely momentum diffusivity. Of course we can always trade efficiency (R^* as defined in Chapter 11) for complexity; in other words, we can reduce the complexity of the collision operator by compensating the raise of diffusivity by a larger number of sites according to the relation

$$Re = R^* \, M \, N \tag{12.8}$$

where N is the number of sites per direction and M is the Mach number.

This is pure Very Large Size Integration (VLSI) line of thought: do something simple, the simplest non-simplistic, on each site and replicate it on as many sites as possible.

However, since the computational work goes like the fourth power of R^*, we know that there is a wide scope for making R^* as large as possible without stressing the size of the spatial grid.

Another LGCA asset retained by the nonlinear collision operator is the existence of an H-theorem, with the attendant nonlinear stability of the numerical scheme.

In other words, the fact of sticking to the underlying microdynamics implies not only the same mean-free path, but also the same physical behavior *vis-à-vis* the second principle of thermodynamics. All this, of course, modulo the numerical drifts possibly introduced in the numerical scheme by round-off errors.

12.1.1   Lattice Quantum Fluids

Before quitting nonlinear LBE, we wish to point out a formal analogy with the quantum Boltzmann equation for quasiparticles in quantum liquids, i.e., elementary fermionic excitations whose de Broglie wavelength $\lambda_B = \hbar/p$ is much smaller than the mean-free path. The quantum nature surfaces only via the explicit presence of holes $\bar{f}_i \equiv 1 - f_i$ in the collision operator, as dictated by the exclusion principle. This is again a legacy of LGCA, and, as we shall see shortly, it has no reason to survive in a framework where the final goal is only classical hydrodynamics. It is natural to ask whether other quantum (Bose–Einstein) or classical (Maxwell–Boltzmann) or maybe fancier fractional statistics might find some use here.

In fact, one can easily generalize the collision operator by defining a conjugate hole function $\tilde{f}_i$ as follows:

$$\tilde{f}_i = (1 + \gamma f_i) \tag{12.9}$$

where $\gamma = -1, 0, 1$ corresponds to fermionic, classical and bosonic degrees of freedom, respectively.

The various statistics reflect into the expression of the Galilean breakdown factor, which takes the form

$$g \sim \frac{b + 2(\gamma + 1)\rho}{b + (\gamma + 1)\rho} \tag{12.10}$$

This expression favors classical statistics ($\gamma = 0$) for two reasons. First, in this case $g = 1/2$ independently of ρ, and therefore it can be safely rescaled out at any density.

As a further bonus, Maxwell–Boltzmann statistics halves the complexity of the collision operator since the conjugate populations become all unit constants. Summarizing, once classical hydrodynamics is the mandate, there is no reason to retain quantum statistics. That said, it would be interesting to learn whether fermionic/bosonic LBEs could be used as numerical tools for quantum fluids. Some interesting efforts along these lines have indeed been deployed in the last decade.

12.2   The Quasilinear LBE

The problem of the unviability of nonlinear LBE to three-dimensional computations did not last long; indeed it was circumvented by HJ basically at the same time as the McNamara–Zanetti paper.[22]

The spirit is always the same; since the final aim is macroscopic hydrodynamics, one tries to spot the unnecessary degrees of freedom and work out systematic procedures to get rid of them.

One of the crucial steps in the Chapman–Enskog treatment, leading from kinetic theory to hydrodynamics, is the low-Knudsen, small-mean-free path assumption. In discrete language this reads

$$f_i = f_i^e + f_i^{ne} \tag{12.11}$$

where superscripts e and ne stand for equilibrium and non-equilibrium, respectively. As usual, the ne component is of order $O(Kn)$ with respect to the equilibrium one, Kn being the Knudsen number of the flow.

---

[22] Historical note: the HJ paper was delayed by a rejection from *Physical Review Letters*.

Being aware that, at some point, the discrete speeds force a low Mach-number expansion into the lattice Chapman–Enskog treatment, HJ resolved to perform this expansion right away on top of (12.11). This yields:

$$f_i = f_i^{e0} + f_i^{e1} + f_i^{e2} + f_i^{ne} + O\left(kM^2\right) \qquad (12.12)$$

where superscripts 0, 1, 2 refer to the order of the Mach-number expansion. With specific reference to the equilibria introduced in Chapter 11, these perturbations read as follows:

$$f_i^{e0} = \frac{\rho_0}{b} \qquad (12.13)$$

$$f_i^{e1} = \frac{\rho_0}{b} \frac{c_{ia} u_a}{c_s^2} \qquad (12.14)$$

$$f_i^{e2} = \frac{\rho_0}{b} \frac{Q_{iab} u_a u_b}{2c_s^4} \qquad (12.15)$$

where ρ_0 is the uniform, global, density.

Next, expand the collision operator around *global* equilibria f_i^{e0} as follows:

$$C_i\left(f\right) = C_i^0 + C_{ij}^0 \phi_j + \frac{1}{2} C_{ijk}^0 \phi_j \phi_k, \qquad (12.16)$$

where $\phi_i = f_i - f_i^{e0}$, $C_{ij} = \partial C_i / \partial f_j$, $C_{ijk} = \partial^2 C_i / \partial f_j \partial f_k$ and superscript 0 denotes evaluation at the global equilibrium $f_i = f_i^{e0}$.

Recalling that $C_i^0 = 0$ because global equilibria annihilate the collision operator and discarding terms higher than $O(kM^2)$, one obtains

$$C_i(f) = C_{ij}^0 f_j^1 + \frac{1}{2} C_{ijk}^0 f_j^1 f_k^1 + C_{ij}^0 f_j^{ne} \qquad (12.17)$$

Next let us specialize (12.17) to the case of a *local* equilibrium $f_i = f_i^e$, namely $f_i^{ne} \equiv 0$, to obtain

$$C_{ij}^0 f_j^1 + \frac{1}{2} C_{ijk}^0 f_j^1 f_k^1 = 0 \qquad (12.18)$$

This permits us to express the Hessian term as a function of the Jacobian, which is mostly welcome since it leaves us with the simple quasilinear expression:

$$C_i = C_{ij}^0 f_j^{ne} \equiv C_{ij}^0 \left(f_j - f_j^e\right) \qquad (12.19)$$

This finally leads to the *quasilinear LBE* in the form proposed by Higuera and Jimenez:

$$\Delta_i f_i = A_{ij} \left(f_j - f_j^e \right) \tag{12.20}$$

The notation $A_{ij} \equiv C_{ij}^0$ emphasizes that the link with the transition matrix $A(\underline{s}, \underline{s}')$ governs the underlying LGCA microdynamics.

Several remarks are in order.

First, we note that (12.20) looks like a linear (quasilinear) equation, but it is definitely not: since it describes the nonlinear Navier–Stokes dynamics it simply cannot be linear!

The (quadratic) nonlinearity is hidden in the local-equilibrium term f_i^e via the quadratic dependence on the flow-field $\rho u_a = \sum_i f_i c_{ia}$. This is a typical *mean-field* coupling where the slow-hydrodynamic degrees of freedom couple back into the kinetic dynamics by dictating the local form the non-equilibrium distribution function must relax to.

Second, we note that the scattering matrix A_{ij} is still one-to-one related to the underlying LGCA dynamics since it is the second-order derivative (Hessian) of the nonlinear collision operator C_i, as derived from LGCA Boolean dynamics.

The distinctive feature here is that since this Hessian is evaluated at the global uniform equilibrium f_i^{e0}, *the scattering matrix C_{ij}^0 can be computed analytically from the LGCA transition-matrix $A(\underline{s}, \underline{s}')$ once and for all.*

In equations, it can be shown that the exact relation is

$$A_{ij} = \sum_{\underline{s}\,\underline{s}'} \left(s_i' - s_i \right) A \left(\underline{s}, \underline{s}' \right) \left(s_j' - s_j \right) d_0^p \, (1 - d_0)^q, \tag{12.21}$$

where $p = \sum_i^b s_i$ is the number of particles in state $\underline{s}$ and $q = b - p + 1$ is the complementary number of holes.

Once the transition matrix $A(\underline{s}, \underline{s}')$ is prescribed, this explicit formula allows precomputation of A_{ij} once and for all at the outset. The resulting reduction of complexity is dramatic: C_i entails $O(2^b)$ operations whereas the scattering matrix only requires $O(b^2)$. Everybody agrees that flipping base and exponent when b is large ($b = 24$) makes a great deal of a difference!

Actually, it is precisely this nice property that paved the way to fully three-dimensional LBE simulations.

Having highlighted the most important advantages associated with quasilinear LBE, we next move on to discuss its basic properties.

12.3 The Scattering Matrix A_{ij}

A few important properties of the scattering matrix are easily identified.

The matrix is:

- *Cyclic and isotropic: A_{ij} depends on $|i - j|$ only*

- *Symmetric: just a consequence of isotropy*
- *Negative definite.*

Isotropy and symmetry stem from the fact that the scattering between populations f_i and f_j only depends on $\mu_{ij} = |\vec{c}_i \cdot \vec{c}_j/c^2|$, the absolute value of the cosine of the scattering angle between directions i and j (see Fig. 12.2). This reduces the number of independent matrix elements from $O(b^2)$ to $O(b)$ only. Owing to the periodicity of the cosine, it is easily seen that the matrix A is cyclic; that is, it repeats itself in cycles of a few elements. This is a very important property, as it permits analysis of its spectrum in full depth *on purely analytical means.*

As an example, consider the 6×6 scattering matrix of the FHP model. There are only four possible scattering angles, $\theta = [0, \pi/6, \pi/3, \pi]$, with which we associate the four independent matrix elements a, b, c, d. With this notation, the FHP scattering matrix is explicitly written down as follows:

$$A_{ij} = \begin{pmatrix} a & b & c & d & b & c \\ c & a & b & c & d & b \\ b & c & a & b & c & d \\ d & b & c & a & b & c \\ c & d & b & c & a & b \\ b & c & d & b & c & a \end{pmatrix} \tag{12.22}$$

Negative definiteness hinges on the spectral properties of A_{ij}. First, we focus on the all-important mass and momentum conservation principles.

The requirement of mass and momentum conservation translates into the following sum rules:

$$\sum_i A_{ij} = 0, \quad j = 1, \dots, b \tag{12.23}$$

$$\sum_i c_{ia} A_{ij} = 0, \quad j = 1, \dots, b, \quad a = 1, \dots, D = 2 \tag{12.24}$$

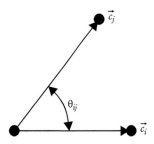

Figure 12.2 *The scattering angle θ_{ij}.*

where use has been made of the symmetry $A_{ij} = A_{ji}$. These simply mean that $D + 1$ eigenvalues are zero, and describe the collisional invariant subspace $\mathcal{H}$ of the b-dimensional kinetic space $\mathcal{K}$ spanned by the discrete populations f_i, $i = 1, \ldots, b$.

Next come the leading nonzero eigenvalues associated with quasihydrodynamic, transport, modes. It can be checked by direct computation that these eigenvalues are associated with the kinetic projector $Q_{iab} = c_{ia}c_{ib} - c_s^2 \delta_{ab}$ and yield the expression of the kinematic viscosity

$$\nu = \frac{D}{D+2} c_s^2 \left(-\frac{1}{\lambda} - \frac{1}{2} \right) \tag{12.25}$$

well known from the theory of lattice gas-cellular automata.

Finally, the purely kinetic modes have eigenvalues that must lie in the strip $[-2, 0]$ for reasons of numerical stability. Once it is proved that all eigenvalues lie in this strip, the negative-definiteness property immediately follows

$$\sum_{ij} V_i A_{ij} V_j = \sum_{k=1}^{b} \lambda_k \|E_k\|^2 < 0, \tag{12.26}$$

where V_i is a generic vector in kinetic space and $\|E_k\|$ is the norm of the k-th eigenvector of the scattering matrix.

Conservation laws guarantee compliance with the first principle of thermodynamics, whereas negative definiteness ensures fulfillment of the second principle.

It is worth noting that the sum rules (12.23–24) enforce mass and momentum conservation even if the local equilibrium does *not* carry exactly the same amount of mass and momentum of the actual distribution f_i. Actually, (12.23) implies that *any* weak departure from local equilibrium is reabsorbed in a time lapse of the inverse-nonzero eigenvalues of the scattering matrix.

Viewed from this angle, the quasilinear LBE is most conveniently regarded as a *multiple-time relaxation* method. As such, in principle, it can generate other sorts of PDEs besides Navier–Stokes, depending on the form of the non-equilibrium departure as a function of the slow fields $\rho(x, t)$ and $u_a(x, t)$.

The quasilinear LBE wipes out the three-dimensional barrier, but *how about the high-diffusivity/low-Reynolds limitation?*

Equation (12.25) stands firm to say that this limitation is still with us.

More precisely the momentum diffusivity of the quasilinear LBE gas is still given by eqn (12.25), where λ is the leading nonzero eigenvalue of the scattering matrix. At a first glance one might come to the (incorrect) conclusion that since A_{ij} has a simple two-body structure, the associated mean-free path should be higher than for nonlinear LBE.

Fortunately, this is not true; we should not let ourselves be fooled by the two-body notation. Here, "effect of population i on population j" does *not* mean two-body collisions, but *all* collisions, including many-body ones, involving populations f_i and f_j!

The much coveted reduction of degrees of freedom *does not result from weeding out multibody collisions* but rather from clustering their effects into the factor $d^p(1-d)^q$ in eqn (12.21). This factor only depends on the number of particles in a given configuration, regardless of the specific configuration itself.

Clearly this clustering is made possible only by the fact of evaluating the collision operator at the global uniform equilibrium, because only then all directions collapse into a single value $f_i^e = \rho_0/b$. This is the key to the magic which allows pre-computation of the scattering matrix once and for all, independently of the flow configuration.

As usual, no free lunch, this magic comes at a price; specifically nonlinear stability (*H*-theorem) must be surrendered. For the simple (linear) case of a uniform equilibrium, $f_i^e = \rho_0/b$, quasilinear LBE has a straightforward *H*-theorem

$$-\frac{dH}{dt} = -\sum_{ij} \phi_i A_{ij} \phi_j > 0, \qquad (12.27)$$

where

$$H = \sum_i \phi_i^2 \qquad (12.28)$$

and $\phi_i \equiv f_i - f_i^{e0}$ is the departure from global equilibrium.

As already noted, the *H*-theorem is reconduced to the negative definiteness of the scattering matrix A_{ij}. Under general-flow conditions, and notably when the quadratic term f_i^{e2} is included, general statements on the existence of a *H*-theorem become a fairly non-trivial matter.

This means that quasilinear LBE is potentially exposed to nonlinear numerical instabilities, as typically observed whenever the local flow field u_a and/or its spatial gradients become "too large" for the hosting grid.

Again, one more nice property of the LGCA ancestor falling apart, or better said, hard to keep standing.

Is the gain worth the pain?

Surely it is, for otherwise three-dimensional simulations would be totally out of reach. Nonetheless, the lesson is that there is still room for improvement, as we shall see in Chapter 13.

12.4   Numerical Experiments

The first numerical simulations using a two-dimensional quasilinear LBE were performed by Higuera (3), Succi-Higuera-Benzi (4) and Higuera-Succi (5) for the case of moderate Reynolds number flow past a cylinder.

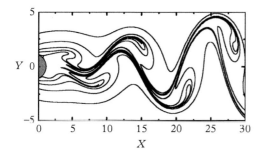

Figure 12.3 *Streaklines of a moderate Reynolds flow past a cylinder. From Higuera and Succi,* Europhys. Lett. *8, 517, 1989.*

These simulations compare successfully with existing literature at a competitive cost in terms of computer resources.[23] Soon after, a number of three-dimensional simulations of flow in porous media were performed (6), which, although still restricted to moderately low Reynolds numbers, proved the viability of the method versus existing floating-point techniques, see Fig. 12.3.

Somewhat ironically, the Boolean lattice gas paradigm had generated a floating-point spin-off!

12.5 Summary

Summarizing, the HJ matrix version of the nonlinear Lattice Boltzmann equation achieves a dramatic boost of efficiency. However, since the collision matrix was still one-to-one related to the underlying LGCA microdynamics, the scheme remained confined to low Reynolds numbers, hence still non competitive in the general arena of computational fluid dynamics.

....

REFERENCES

1. G. McNamara and G. Zanetti, "Use of the Boltzmann equation to simulate lattice gas automata", *Phys. Rev. Lett.*, **61**, 2332, 1988.
2. F. Higuera and J. Jimenez, "Boltzmann approach to lattice gas simulations", *Europhys. Lett.*, **9**, 663, 1989.

[23] Practical hint: note that, by symmetry arguments, C_{ij}^0 annihilates both ϕ_i^1 and ϕ_i^2, so that (12.19) further simplifies to

$$C_i = C_{ij}^0 f_j^{ne} \equiv C_{ij}^0 \left(f_j - f_i^{e2} \right) \tag{12.29}$$

This reduces the number of floating operations needed to encode the collision operator.

Second practical hint: in computational practice it may prove convenient to work with reduced populations $\tilde{f}_i \equiv (f_i - f_i^{e0})/f_i^{e0}$, in such a way that full machine precision can be directed to departures from uniform equilibria, rather than to uniform equilibria themselves!

3. F. Higuera, "Lattice gas simulation based on the Boltzmann equation", in *Discrete Kinetic Theory, Lattice Gas Dynamics and Foundations of Hydrodynamics*, R. Monaco ed., p. 162, World Scientific, Singapore, 1989.
4. S. Succi, R. Benzi, and F. Higuera Lattice Gas and Boltzmann simulations of homogeneous and inhomogeneous hydrodynamics, p. 329 World Scientic, Singapore, 1989.
5. F. Higuera and S. Succi, "Simulating the flow around a circular cylinder with a lattice Boltzmann equation", *Europhys. Lett.* **8**, 517, 1989. As a personal recollection: the paper was accepted on the spot by the then EPL Editor, a rather unique experience in my entire career...
6. S. Succi, E. Foti and F. Higuera, "Three-dimensional flows in complex geometries with the lattice Boltzmann method", *Europhys. Lett.*, **10**(5), 433, 1989.

....

EXERCISES

1. Write down the explicit form of the nonlinear LBE for the FHP scheme.
2. Do the same with the quasilinear LBE.
3. Compute the eigenvalues of the quasilinear matrix of the FHP scheme.

13

Lattice Boltzmann Models without Underlying Boolean Microdynamics

In Chapter 12 we have learned how to circumvent two major stumbling blocks of the LGCA approach: statistical noise and exponential complexity of the collision rule. Yet, the ensuing LB still remains confined to low Reynolds flows, due to the low collisionality of the underlying LGCA rules.

The high-viscosity barrier was broken just a few months later, when it was realized how to devise LB models top-down, i.e., based on the macroscopic hydrodynamic target, rather than bottom-up, from underlying microdynamics.

Most importantly, besides breaking the low-Reynolds barrier, the top-down approach has proven very influential for many subsequent developments of the LB method to this day.

For some players luck itself is an art.
The Color of Money

13.1 LBE with Enhanced Collisions

Now that two major drawbacks of LGCA models, statistical noise and practical inviability to three-dimensional simulations, have been lifted, we can turn our attention to the third major limitation of LGCA: high-momentum diffusivity, also known as low-Reynolds flows. This further step was taken in short sequence by Higuera, Succi and Benzi (1), just a few months after quasilinear LBE was successfully demonstrated by the numerical simulations described in Chapter 12.

The main observation is that the high-diffusivity constraint is again a needless limitation inherited from LGCA via the one-to-one correspondence between the transition matrix $A(\underline{s}, \underline{s}')$ and the scattering matrix A_{ij}. In other words, neither nonlinear LBE nor quasilinear LBE are allowed to support a single collision that the subjacent LGCA would

The Lattice Boltzmann Equation. Sauro Succi, Oxford University Press (2018).
© Sauro Succi. DOI: 10.1093/oso/9780199592357.001.0001

not permit. This tyrannic dependence is encoded in the one-to-one relation between the transition matrix $A(\underline{s}, \underline{s}')$ and the scattering matrix A_{ij} given in Chapter 12. On the other hand, since quasilinear LBE is known to yield the Navier–Stokes equations, why not take quasilinear LBE itself as a *self-standing* mathematical model of fluid behavior, regardless of its microscopic-Boolean origin?

This stance of "independence" is perfectly legal provided one secures that the scattering matrix (a structured set of free parameters at this stage) is chosen with the correct symmetries ensuring the Navier–Stokes equation as a macroscopic limit. The right properties of the scattering matrix are easily guessed from what we learned with quasilinear LBE: it is simply the array of conservation properties listed in Chapter 12, securing compliance with the first and second principles of thermodynamics:

- *Conservativeness (sum-rules)*
- *Isotropy and symmetry*
- *Negative definiteness*

The basic idea is that *the spectral properties of A_{ij} controlling the transport coefficients are no longer dictated by any underlying microscopic theory, but they are handled as free parameters of the theory instead.* Consequently, they can be freely tuned in such a way as to achieve minimum viscosity, hence maximum Reynolds numbers. This idea has proved very fruitful for many subsequent developments of LBE methods.

Technically, the name of the game is simply "spectral decomposition": once the spectrum of eigenvalues and corresponding set of eigenvectors is known, any matrix A can be reconstructed via the spectral decomposition

$$A_{ij} = \sum_{k=1}^{b} \lambda_k P_{ij}^{(k)} \tag{13.1}$$

where $P_{ij}^{(k)}$ is the projector along the k-th kinetic eigenvector E_i^k.

We are in fairly good shape, because in a discrete-velocity space the linear eigenvalue problem associated with the spectrum of the scattering matrix can be solved exactly! This conveys a concrete feel for what "minimal" kinetic theory means and the fringe benefits it brings on this table.

Standard analysis shows that if the set of eigenvectors $\underline{E} = [E_i^1, \ldots, E_i^b]$ is orthonormal, the projectors P_{ij}^k are expressed by the dyadic matrix $E_i^k E_j^k$, whose matrix element (i,j) is given by the product of the i-th and j-th element of the k-th column eigenvector. Once the formal constraint (brackets denote sum over Boolean states with weight $d^p(1-d)^{b-p+1}$ as defined in Chapter 12)

$$A_{ij} = \left\langle \left(s_i' - s_i \right) A \left(\underline{s}, \underline{s}' \right) \left(s_j' - s_j \right) \right\rangle \tag{13.2}$$

is relaxed, it is a rather straightforward task to minimize the viscosity, since the relation between the leading nonzero eigenvalue and the momentum diffusivity is still given by

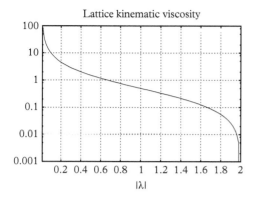

Lattice kinematic viscosity

Figure 13.1 *Lattice viscosity as a function of $|\lambda|$ (the prefactor $c_s^2 D/(D+2)$ made unit). Note that the viscosity plummets in the vicinity of $|\lambda| = 2$. Note that the curve is dual around $|\lambda| = 1$, i.e. by letting $\epsilon = |\lambda| - 1$, and $\tilde{\nu} \equiv \frac{D+2}{Dc_s^2} \nu$, one has*

$$\tilde{\nu}(-\epsilon) = \frac{1}{\nu(\epsilon)}.$$

the basic relation obtained by the Chapman–Enskog analysis of LGCA microdynamics, namely (see Fig. 13.1)

$$\nu = c_s^2 \frac{D}{D+2} \left(-\frac{1}{\lambda} - \frac{1}{2} \right) \tag{13.3}$$

This expression prompts a greedy question: can we achieve *zero* viscosity by simply choosing $\lambda = -2$?

Should this be the case, we would have a real ground-breaking tool at our fingertips, the dream of every turbulence researcher!

Unfortunately, the answer is a very down-to-earth "no," as we shall detail shortly.

Before doing so, let us discuss the spectrum of the quasilinear LBE collision operator in some detail.

As anticipated, the spectral analysis of the scattering collision operator can be taken to the very end on purely analytical means (2).

For the sake of conciseness, we shall refer to the FCHC lattice, not because this is the simplest case, but because it represented the workhorse of the early-day LGCA and LBE simulations.

Our task is to analyze the 24×24 matrix A_{ij} generated by the FCHC discrete speeds. Many properties of the spectrum $\sigma[A]$ are known beforehand.

First, direct inspection of the cosines $\cos\theta_{ij}$ between directions $\vec{c}_i$ and $\vec{c}_j$ shows that there are only *five* independent scattering channels/elements denoted by a, b, c, d, e and corresponding to $\theta = 0, \pi/3, \pi/2, 2\pi/3, \pi$, respectively (don't forget we are in $D = 4$).

This means that the spectrum consists of only five distinct eigenvalues

$$\sigma(A) = \{\lambda_k, \ k = 1, \ldots, 5\} \tag{13.4}$$

each with multiplicity μ_k, such that $\sum_{k=1}^{5} \mu_k = 24$.

These eigenvalues are real since the matrix is symmetric, and in addition they must lie within the strip $[-2, 0]$ for reasons of numerical stability.

Mass and momentum conservation enforce the following equalities:

$$a + 8b + 6c + 8d + e = 0 \tag{13.5}$$

$$a + 4b - 4c - e = 0 \tag{13.6}$$

which leave us with only *three* distinct nonzero eigenvalues.

This completes the reconstruction of the conservative part of the spectrum. Now comes the dissipative/transport component. At this stage we turn our attention to the eigenvectors $\underline{E}^k$. Four eigenvectors are immediately spotted from mass and momentum conservation

$$E_i^0 = 1_i \tag{13.7}$$

$$E_i^a = c_{ia}, \quad a = 1, \dots, 4 \tag{13.8}$$

The remaining ones can be constructed by the well-known Gram–Schmidt procedure. A quicker alternative is to recognize (3) the *circulant* property of the scattering matrix, which allows us to write down the analytical expression of the spectrum. We shall however proceed through a different route because we feel it conveys more physical insight into the problem.

For the sake of clarity, we report here the explicit computation. We seek a set of eigenvectors orthogonal to $\underline{E}^0$, $\underline{E}^1$. A sensible guess is $E_i^{ab} = c_{ia}c_{ib} - K\delta_{ab}$, where the scalar K is to be computed by imposing orthogonality to the previous ones, namely

$$\sum_i (c_{ia}c_{ib} - K\delta_{ab}) 1_i = 0 \tag{13.9}$$

$$\sum_i (c_{ia}c_{ib} - K\delta_{ab}) c_{ic} = 0 \tag{13.10}$$

The second of these equations is automatically satisfied by the parity of the lattice (all odd tensors vanish identically by parity), whereas the first one yields

$$K_{ab} = \sum_i c_{ia}c_{ib} = \frac{bc^2}{D}\delta_{ab} \tag{13.11}$$

This delivers a set of $5 \times 4/2 - 1 = 9$ distinct eigenvectors

$$E_i^k = c_{ia}c_{ib} - c_s^2\delta_{ab}, \quad k = 6, \dots, 14 \tag{13.12}$$

associated with eigenvalue $\lambda = \lambda_1$, $\mu_1 = 9$. In (13.12), $c_s^2 = c^2/D$ is the sound speed, as usual.

On the right-hand side we recognize the kinetic projector $Q_{iab} = c_{ia}c_{ib} - c_s^2\delta_{ab}$.[24]

---

[24] There is a manifest link with tensor Hermite polynomials: see for instance Ref. (4), p. 205.

One can proceed further with the diagonalization procedure and derive the remaining eigenvalues and eigenvectors

$$E_i^{15} = \chi_i', \; E_i^{16} = \chi_i'', \tag{13.13}$$

$$\lambda = \lambda_2, \mu_2 = 2 \tag{13.14}$$

and

$$E_i^k = \chi_i' c_{ia}, \; E_i^k = \chi_i'' c_{ia}, k = 17, \ldots, 24 \tag{13.15}$$

$$\lambda = \lambda_3, \; \mu_3 = 8, \tag{13.16}$$

where the weights χ_i', χ_i'' will be discussed shortly.

The final relation between the matrix elements and the nonzero eigenvalues is

$$\lambda_1 = a - 2c + d \tag{13.17}$$

$$\lambda_2 = \frac{3}{2} (a - d) \tag{13.18}$$

$$\lambda_3 = \frac{3}{2} (a + 6c + d) \tag{13.19}$$

with multiplicities $\mu_1 = 9$, $\mu_2 = 2$, $\mu_3 = 8$, respectively.

The spectral analysis of the scattering matrix is summarized in the following eigenvalue–multiplicity–eigenvector–conjugate fields synoptic table:

λ,	μ,	E_i,	M_k,
0,	5,	$[1_i, c_{ia}]$,	$[\rho, \mathcal{J}_a]$,
λ_1,	9,	$[Q_{iab}]$,	$[\hat{P}_{ab}]$,
λ_2,	2,	$[\chi_i', \chi_i'']$,	$[\rho', \rho'']$,
λ_3,	8,	$[\chi' c_{ia}, \chi'' c_{ia}]$,	$[\mathcal{J}_a', \mathcal{J}_a'']$.

This completes the spectral analysis of the FCHC scattering matrix A_{ij}.

It is now instructive to examine the physical meaning of the conjugate macroscopic fields $M_k(x, t) = \sum_i f_i(x, t) E_i^k$.

13.2 Hydrodynamic and Ghost Fields

In line with the material previously presented, we shall define conjugate macroscopic fields as the projection of the distribution function f_i over the set of eigenvectors $\underline{E}^k$ in momentum space (underline stands for vector in momentum space $\underline{E}^k \equiv E_i^k$). For simplicity, we shall refer to the three-dimensional projection, so that we shall deal with a 18×18 matrix. The remaining six quenched populations are reabsorbed into a set of effective weights:

$$\underline{w} = [(2, 2, 2, 2, 2, 2) \mid (1, 1, 1, 1) \mid (1, 1, 1, 1) \mid (1, 1, 1, 1)] \tag{13.20}$$

where the bars group the following sublattices:

- Six direct links along x, y, z, respectively
- Four cross links along xy, xz, yz

With this definition:

$$M_k(x, t) = \sum_{i=1}^{18} w_i E_i^k f_i. \tag{13.21}$$

[25]The hydrodynamic sector of the spectrum is well known, and generates mass and current densities associated with conserved quantities and null eigenvalues:

$$\rho \equiv M_0(x, t) = \sum_{i=1}^{18} w_i f_i \tag{13.22}$$

$$\mathcal{J}_a \equiv M_a(x, t) = \sum_{i=1}^{18} w_i f_i c_{ia}, \quad a = 1, 2, 3 \tag{13.23}$$

The transport sector of the spectrum defines the slowest non-equilibrium decaying quantities. The corresponding macrofield is the deviatoric part of the momentum flux tensor, namely

$$\hat{P}_{ab} \equiv M_{ab}(x, t) = \sum_{i=1}^{18} w_i f_i Q_{iab}, \quad a, b = 1, 2, 3 \tag{13.24}$$

with the relation

$$P_{ab} \equiv \sum_{i=1}^{18} w_i f_i c_{ia} c_{ib} = \hat{P}_{ab} + p \delta_{ab} \tag{13.25}$$

where $p = \rho c_s^2$ is the fluid pressure.

Being symmetric and traceless, the tensor $\hat{P}_{ab}$ generates $3 + 3 + 2 + 1 = 9$ independent fields associated with the leading nonzero eigenvalue of the matrix A_{ij}.

Next comes the "fast," mesoscopic component of the spectrum; that is, the modes that do not manage to make it from the microscopic to the macroscopic level. They were dubbed *ghost fields* to stress the idea that they live "behind the scenes," but cannot be erased from the dynamics for pain of disrupting essential symmetries.

---

[25] This definition, with integer weights w_i explicitly exposed in front of the populations f_i reflects the early days, notation. Nowadays, the weights are typically given in fractional form and incorporated within the populations.

By sheer definition:

$$\rho' = \sum_{i=1}^{18} \chi'_i f_i \tag{13.26}$$

$$\rho'' = \sum_{i=1}^{18} \chi''_i f_i \tag{13.27}$$

where

$$\underline{\chi'} = [(1, 1, 1, 1, -2, -2) \mid (-2, -2, -2, -2) \mid (1, 1, 1, 1) \mid (1, 1, 1, 1)], \tag{13.28}$$
$$\underline{\chi''} = [(1, 1, -1, -1, 0, 0) \mid (0, 0, 0, 0) \mid (1, 1, 1, 1) \mid (1, 1, 1, 1)] \tag{13.29}$$

From these definitions, we see that the ghost-densities ρ' and ρ'' correspond to a pair of staggered densities: $\rho' = \rho^x + \rho^y - 2\rho^z - 2\rho^{xy} + \rho^{xz} + \rho^{yz}$ and $\rho'' = \rho^x - \rho^y + \rho^{xz} + \rho^{yz}$, where the suffixes label the four distinct sublattices introduced previously.

Being differences rather than sums, the staggered densities carry higher-order short-lived information, corresponding to derivatives in velocity space.

The remaining ghost fields are easily identified as ghost currents

$$\mathcal{J}'_a = \sum_{i=1}^{18} \chi'_i f_i c_{ia} \tag{13.30}$$

$$\mathcal{J}''_a = \sum_{i=1}^{18} \chi''_i f_i c_{ia} \tag{13.31}$$

These relations show that the ghost fields can also be given a transparent physical interpretation.

The full set of equations for the 24 moments of the FCHC–LBE scheme can be written in full splendor as follows:

$$\partial_t \rho + \partial_a \mathcal{J}_a = 0 \tag{13.32}$$

$$\partial_t \mathcal{J}_a + \partial_a P_{ab} = 0 \tag{13.33}$$

$$\partial_t \hat{P}_{ab} + \partial_c R_{abc} = \lambda_1 \left(\hat{P}_{ab} - \hat{P}^e_{ab} \right) \tag{13.34}$$

$$\partial_t \rho' + \partial_a \mathcal{J}'_a = \lambda_2 \rho' \tag{13.35}$$

$$\partial_t \rho'' + \partial_a \mathcal{J}''_a = \lambda_2 \rho'' \tag{13.36}$$

$$\partial_t \mathcal{J}'_a + \partial_b P'_{ab} = \lambda_3 \mathcal{J}'_a, \tag{13.37}$$

$$\partial_t \mathcal{J}''_a + \partial_b P''_{ab} = \lambda_3 \mathcal{J}''_a \tag{13.38}$$

The hydrodynamic and ghost sectors of the theory make contact through the triple tensor:

$$R_{abc} = \sum_{i=1}^{18} w_i f_i Q_{iab} c_{ic} \tag{13.39}$$

Direct substitution of the hydrodynamic component

$$f_i^H \equiv \rho \left[1 + \frac{c_{ia} u_a}{c_s^2} + \frac{Q_{iab} \hat{P}_{ab}}{2 c_s^4} \right] \tag{13.40}$$

into the expression of R_{abc} shows that the divergence of this triple tensor reorganizes sensibly into an isotropic form corresponding to the Navier–Stokes stress tensor:

$$H_{ab} \equiv \partial_c R_{abc}^H = \partial_a \mathcal{J}_b + \partial_b \mathcal{J}_a - \frac{2}{D} (\partial_c \mathcal{J}_c) \, \delta_{ab} \tag{13.41}$$

The same is *not* true for ghost contributions because, owing to the different weights χ_i', χ_i'', the corresponding ghost tensor $G_{ab} \equiv \partial_c R_{abc}^G$ does not fulfill the isotropy requirements.

It should be appreciated that we have made use of the freedom on the choice of the hydrodynamic component of f_i in order to fine-tune the coefficients of expansion (13.40) in such a way as to achieve the desired anomaly-free result.

The system of partial differential equations (13.32)–(13.38) has a rich and formally elegant structure, which invites some daring analogies.

13.3   Field-Theoretical Analogies

Disclaimer: this section is admittedly pretty abstract: readers with no penchant for field theory may safely skip this section.

The 24-moment FCHC equations exhibit a beautiful symmetry between hydro and ghost fields: they turn one into another under the dual transformation

$$w_i \to \tilde{w}_i \tag{13.42}$$

where w_i are the weights associated with the degeneracies and $\tilde{w}_i = [w_i \chi_i, w_i \chi_i'']$ are the corresponding weights for the ghost fields.

These weights can also be interpreted as elementary masses of the particles propagating along the various directions of the lattice. Note that, in tune with their physical meaning, the hydrodynamic (normalized) weights are all positive and sum to one, whereas some of the ghost weights are forcibly negative because they must sum to zero.

The physical content of the hydrodynamic-and ghost-sectors of the theory is very different.

Consistent with this difference, the duality is broken by the nonzero mass (eigenvalues) of the ghost fields, which goes with the short-lived nature of these excitations. In the numerical analyst's language, hydro-fields are smoothed-out (integral) variables in momentum space, whereas the ghost fields represent the short-scale, low-amplitude detail of the distribution function.

The question is: do we really want/need these ghosts fields, or should we rather try to erase them from the dynamics?

Let us handle this question in two separate parts.

First and foremost, why do ghost fields appear at all?

The reason is that *they represent the excess of information from kinetic space required to formulate the Navier–Stokes equations in a fully hyperbolic form*. This short-scale information, if unable to emerge from the meso to the macroscopic scale, is necessary to secure the basic local symmetries which guarantee correct hydrodynamics.

A formal analogy with the ghost fields of quantum field theory is tempting: here come the arguments, (5):

- *Quantum Field Theory*
  i) Ghost fields are required to cancel unwanted degrees of freedom arising from the covariant treatment of massless gauge vector fields (photons, gluons).
  ii) Their task is to restore unitarity. They live only inside Feynman loops, and must disappear from the ultimate physical picture.
- *Lattice Hydrodynamics*
  i) Ghost fields are required to ensure the basic *local* spacetime symmetries of classical physics within a covariant formulation of hydrodynamics.
  ii) Their task is to restore Galilean and rotational invariance. They are short-lived and do not surface to the macroscopic level.

In light of this analogy, the condition (13.41) plays a role similar to the Ward identity in quantum-field theory, namely anomalies cancellation. In our case, the anomaly is lack of Galilean and rotational invariance, whereas quantum-field theory involves more sophisticated symmetries, typically related to internal degrees of freedom not (necessarily) associated with spacetime transformations. However, if one regards the particle momentum as an internal degree of freedom, the one instructing the particle "where to go next," the difference gets hazier.

Summarizing, duality permits us to write the ghost dynamics in conservative form, like hydrodynamics, thus explicitly acknowledging the fact that ghost and hydro fields are just two manifestations of the same underlying (kinetic) theory. In this author's opinion, this represents a beautiful and elegant result. To date, attempts to draw concrete benefits from such result have not led to any conclusive evidence. But, who knows for the future?

13.3.1 Extra-Dimensions

Let us spend a few final comments on the present hydrodynamic derivation for the FCHC–LBE model. First of all, the 24-moment equations do correspond to the Navier–Stokes dynamics of a *four-dimensional fluid* (4 +1 in field theory language).

Actually, the fluids we are interested in only need three, so that it is natural to ask what to do with the extra spatial dimension.[26]

Normally, the fourth dimension is quenched by imposing a zero current along the fourth dimension, $\mathcal{J}_4 = 0$. This introduces a degeneracy in the populations with $c_{i4} = 0$, which collapses into three doublets, thus leaving only eighteen independent populations out of the original twenty-four. The corresponding numerical simulations are then performed by shrinking the fourth "Kaluza–Klein" dimension into a single layer across which periodicity is imposed so as to erase any dependence on x_4, see Figs 13.2 and 13.3.

On the other hand, if this degeneracy is lifted by allowing $\mathcal{J}_4 \neq 0$, and the dimensional curling of the fourth dimension is still retained, the resulting scheme corresponds to a three-dimensional fluid carrying along the current $\mathcal{J}_4$ as a passive scalar.

This is formally understood as the limit $\partial_4 \to 0$, namely no observable dependence on the quenched dimension.

As a result, the 4D FCHC scheme with 18 discrete speeds and 24 populations describes a three-dimensional fluid plus a passive scalar, for free.

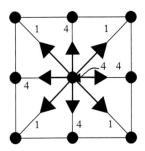

Figure 13.2 *The 2D projection of FCHC and the corresponding weights.*

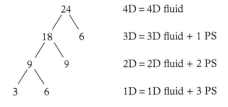

4D = 4D fluid	
3D = 3D fluid + 1 PS	
2D = 2D fluid + 2 PS	
1D = 1D fluid + 3 PS	

Figure 13.3 *The hierarchical tree of the FCHC lattice. The rightmost figure counts the number of degenerate populations resulting from the four-dimensional projection.*

[26] Again, another amazing analogy with field theory, namely the famous five-dimensional Kaluza–Klein attempt to unify gravitation and electromagnetism. By now, we know that even if the Kaluza–Klein theory fails to achieve the goal, the idea of pursuing unification in higher-dimensional spaces has proved overly productive in modern developments of quantum-field theory and quantum gravity.

By the same token, a 2D FCHC scheme, with nine discrete speeds and 24 populations, describes a two-dimensional fluid, plus *two* passive scalars and a 1D FCHC with three speeds describes a 1D fluid plus three passive scalars.

Such a peculiar property can be leveraged to study a number of fluid-transport problems, such as fast-chemistry combustion, magneto-hydrodynamics and thermal flows.

13.3.2   Ghost-Busters

Since ghosts carry no hydrodynamic value, it is natural to ask how to decouple them from dynamics of the fluid system.

A simple, if approximate, recipe is readily available, just take

$$\lambda_2 = \lambda_3 = -1 \tag{13.43}$$

so as to damp out ghost excitations in a single timestep.

The effect of this move is readily appreciated from the homogeneous version of eqn (13.32–38):

$$\rho'(t+1) - \rho'(t) = -\rho'(t) \tag{13.44}$$

This yields $\rho'(t+1) = 0$, which means that the ghost density at time $t+1$ is erased from the dynamics. In the non-homogeneous case, however, ghosts are revived at each timestep by the divergence of the ghost currents, which in turn are excited by the ghost momentum flux tensor:

$$\rho'(t+1) - \rho'(t) = -\rho'(t) + \partial_a \mathcal{J}'_a(t) \tag{13.45}$$

It easily seen that the same recipe guarantees that these terms cannot live longer than a single timestep, so that the prescription (13.43) gives rise to a sort of low-amplitude sawtooth oscillations, whose amplitude should remain small unless pathological initial or boundary conditions are met (remember the staggered invariants discussed earlier in this book).

With the basics of ghost dynamics grasped out, we can move one step further and ask ourselves whether we can make any good use of them, for instance put them at work to "improve" hydrodynamic behavior.

Here "improve hydrodynamic behavior" essentially stands for sub-grid-scale modeling, namely generate an effective viscosity corresponding to the effect of small, unresolved scales on the resolved ones. Such attempts have been made in the past but, to the best of the author's knowledge, they have not been demonstrated numerically. Since these models are rather technical, we direct the interested reader to the original reference and take as a final stance that ghosts should be kept at a minimum. Of course, the brave reader is encouraged to challenge this conservative statement and find his own way to a smart use of ghost fields. In fact, the so-called Multi-Relaxation Time version of LB, to be discussed in Chapter 14, goes somehow in this direction, and so do recent developments of the entropic method to be discussed in chapter 21.

13.4 The Route to Navier–Stokes: Enslaving Assumption

The set of eqns (13.32)–(13.38) is a 24-moment *hyperbolic superset of the Navier–Stokes equations*. It contains *exactly* the same amount of information as the LBE, only recast (projected) onto a different basis set. As for any discrete kinetic model, closure is automatically provided by the finite dimensionality of the kinetic space $\mathcal{K}$. To recover the Navier–Stokes dynamics we must necessarily break the hyperbolic character of this system and cast it into a dissipative form where time and space derivatives are no longer in balance.

The route is *enslaving assumption* on the slow, non-conserved modes, in our case the (traceless) momentum-flux tensor $\hat{P}_{ab}$. By enslaving, we mean exactly the same as we did in continuum kinetic theory, namely that the momentum-flux tensor relaxes to its local-equilibrium value on a timescale much shorter than any typical hydrodynamic scale. Mathematically, this is tantamount to dropping the time derivative in the evolution equation for the momentum-flux tensor, eqn (13.34), and obtaining a simple-algebraic solution

$$\hat{P}_{ab} \simeq \hat{P}_{ab}^{e} + \frac{H_{ab}}{\lambda_1} \tag{13.46}$$

where ghosts have been (partly) phased out through the condition (13.43).

By plugging (13.46) into (13.32), we obtain

$$\partial_t \mathcal{J}_a + \partial_b \left(\hat{P}_{ab}^{e} + \frac{H_{ab}}{\lambda_1} \right) = 0 \tag{13.47}$$

All we need at this stage is simply to recall that the equilibrium (traceless) momentum-flux tensor reads as

$$\hat{P}_{ab}^{e} = \rho u_a u_b \tag{13.48}$$

and realize that eqn (13.47) is indeed the Navier–Stokes equation with a kinematic viscosity $\nu \sim -1/\lambda_1$.

This is of course a very cheap treatment of the hydrodynamic limit, in the sense that the subtleties of the multiscale Chapman–Enskog approach have been swept under the carpet. We recall that in the multiscale approach not only the dependent variables, f_i but also spacetime variables are expanded in multiple scales driven by a smallness parameter, typically the Knudsen number. This implies that different equations are obtained at different orderings of this smallness parameter.

From LGCA theory, it is known that being cavalier about multiscale considerations does not spoil the qualitative structure of the macroscopic equations, but it does introduce quantitative errors in the transport coefficients. In particular, it is known that the Taylor expansion of the finite difference streaming operator $\Delta_i f_i \equiv f_i(\vec{x} + c_{ia}, t + 1) -$

$f_i(\vec{x}, t)$ to second order in the lattice spacing brings about a factor $-1/2$ in front of the second-order spatial derivative.

It is fortunate that such an purely numerical effect can be likened to a *physically negative* viscosity known as *propagation viscosity*.

The final quantitative expression therefore reads

$$v = -c_s^2 \frac{D}{D+2} \left(\frac{1}{\lambda_1} + \frac{1}{2} \right)$$

(13.49)

which matches exactly the corresponding LGCA expression.

This brings us back to the issue of zero viscosity.

13.5 The Mirage of Zero Viscosity

One of the crucial results of the spectral analysis is the expression relating the fluid viscosity to the leading nonzero eigenvalue of the scattering matrix A_{ij}. This equation was known long before LBE, but the change in perspective makes the whole difference. Here the eigenvalue λ_1 is no longer imposed by the underlying microdynamics, but becomes a free handle that can be fine-tuned to minimize fluid viscosity. In principle, zero viscosity could be achieved by selecting $\lambda_1 = -2$, which corresponds to the collisional diffusivity coming in perfect balance with the negative propagation viscosity $v_p = -1/2$.

As usual, when two quantities become serious about balancing each other, new effects pop out that call for full attention.

This is the thorny issue of nonlinear stability, namely the effect of large macroscopic gradients on the numerical stability of LBE once the viscosity becomes really small, i.e., so small that the grid-Reynolds number $Re_\Delta \equiv u\Delta x/v > 1$.

This is strictly related to the so-called *sub-grid* scale regime, whose distinctive feature is the potential existence of dynamic scales shorter than the lattice pitch,

$$l_d < \Delta x$$

(13.50)

where l_d is the smallest-hydrodynamic scale (the Kolmogorov scale in a turbulent flow, see chapter 20).

This important issue is examined apart in chapter 16. What we wish to point out here is that LBE actually bridges the gap with conventional CFD, in that *the minimal fluid viscosity is dictated only by the mesh resolution and not by intrinsic limitations of the LGCA collisionality (lattice mean-free path)*.

In other words, the issue of "low collisionality" is lifted, and one can go as low in viscosity as allowed by the lattice.[27]

---

[27] For a while it was even hoped that one could go *below* the grid ("sub-grid modeling" in CFD jargon), but this exciting perspective was abandoned as it was unsupported by numerical evidence. This line of thinking has been revived by recent developments of the entropic method, see chapter 21.

In equations

$$l_\mu < \Delta x < l_d \tag{13.51}$$

This contrasts with the LGCA ordering

$$\Delta x < l_\mu < l_d \tag{13.52}$$

in which hydrodynamic behavior cannot be attained on the scale of a single lattice spacing, thus requiring much larger grids at a given Reynolds.

It is natural to ask why this is so given the fact that the structure of the lattice, and in particular the number of discrete speeds, has not been touched at all.

The point is the following: by freely adjusting the scattering matrix A_{ij} *and* prescribing the discrete equilibrium f_i^e at the outset, we implicitly take into account *all* possible collisions involving the discrete speeds $\vec{c}_i$ and $\vec{c}_j$, even though only a (very) limited set is retained in the actual evolution scheme. We thus have a sort of set of virtual collisions with a continuous set of speeds even if, in the end, only a fraction of these discrete speeds keeps being tracked all along. In other terms it is an effective form of coarse-graining in velocity space.

Once this picture is realized, the issue of Galilean invariance dissolves, because the local equilibrium can be chosen at will (within hydrodynamic constraints) without being tied down to specific statistical distributions of continuum physics.

In particular, the following expression

$$f_i^e = \frac{\rho}{b} \left(1 + \frac{c_{ia} u_a}{c_s^2} + \frac{Q_{iab} u_a u_b}{2 c_s^4} \right), \tag{13.53}$$

which has already been used in this chapter to obtain the correct form of the Navier–Stokes stress tensor, is manifestly compliant with all hydrodynamic constraints and Galilean invariant as well.

In conclusion, the self-standing LBE contains two dynamical "knobs":

1. *The scattering matrix A_{ij}*
2. *The discrete local equilibrium f_i^e*

By tuning the associated set of parameters, one can flexibly generate a whole class of nonlinear partial hyperbolic equations including of course Navier–Stokes. Even though little has been changed in the formal LBE apparatus, this standpoint dissolves two issues in one swoop, low collisionality and Galilean invariance, thus making full contact with standard computational fluid dynamics.

Moreover, by capitalizing on the freedom to play with the scattering matrix and the local equilibria, the present LBE paves the way for a host of generalized hydrodynamic applications, such as magnetohydrodynamics, thermal convection and many other situations involving *field–fluid* interactions. Even though there was no need for such a restriction, the LB with enhanced collisions, such was named the self-standing LBE in those days continued to be used in a form where Galilean invariance had to be recovered via rescaling of the LGCA non galilean factor $g(\rho)$.[28]

A systematic procedure to dispose of both galilean non-invariance and pressure-anomalies affecting LGCA versions was developed in a subsequent influential paper by H. Chen, S. Chen and W. Matthaeus (7).

13.6   Early Numerical Simulations

The LBE scheme with enhanced collisions was demonstrated shortly after quasilinear LBE for a 2D transitional flow past a thin plate at moderate Reynolds number (1). Subsequently, high-Reynolds 2D LBE turbulence was tested against spectral methods (6). These simulations proved that LBE with enhanced collisions performs competitively against spectral methods even on moderate resolutions (512^2). In terms of the quality-factor R^*, these simulations showed that one can safely reach as high as $R^* \sim 30$ by choosing $\lambda_1 \sim -1.9$, corresponding to a lattice-viscosity $\nu \sim 0.01$, basically the smallest value compatible with the given lattice resolution.

The calculation is simple: take $U = 0.1$, $N = 1000$ lattice sites and $\nu = 0.01$, all in lattice units, the resulting Reynolds is $Re = 10^4$, more or less the state of the art in the direct-numerical simulation of fluid turbulence at the time.

Caveat: we hasten to emphasize that having broken the umbilical cord with underlying LGCA, *the top-down LB is no longer guaranteed to be microscopically realizable*. At a macroscopic level, this reflects in the potential crumbling of the H-theorem and ensuing stability issues. Indeed, having swept the *H*-theorem under the carpet (at no point have we discussed compliance with the *H*-theorem in the nonlinear flow regime), it is not obvious a priori that the top-down LBE can be taken to the lowest viscosities allowed by the grid resolution without incurring numerical instabilities.

The fact that the top-down LB could be taken close to lowest viscosities allowed by the grid resolution is once again a lucky break. Indeed, the theoretical justification and understanding for such smiley behavior became apparent only nearly years later, with the development of the entropic method.

Besides turbulence, LBE with enhanced collisions was used for the earliest calculations of three-dimensional single and multiphase flows in porous media.

Notwithstanding these early sign of success, its prime time as a tool of choice for LBE simulations did not last long: just the few years it took to realize that the LBE formalism could be simplified still further. This takes us to the last "release" of the Jurassic-LBE family, the lattice Bhatnagar–Gross–Krook version to be discussed in Chapter 14.

---

[28] This goes to the full discredit of the present author: junior author, at that time, M. Vergassola repeatedly pointed out that there was no need to retain a non-unit $g(\rho)$, to whom this author consistently replied that this was an unimportant detail . . . .

13.7   Summary

Summarizing, the LB with enhanced collisions first promoted the idea of designing LB top-down, i.e., based on sole symmetry requirements to recover macroscopic hydrodynamics, instead of deriving it bottom-up from the underlying lattice-gas microdynamics. This permitted us to lift all limitations on the Reynolds number, other than those imposed by the grid resolution, thus placing LB for the first time on the map of computational fluid dynamics. More importantly, the top-down strategy proved very influential for most subsequent developments of the LB theory.

....

REFERENCES

1. F. Higuera, S. Succi and R. Benzi, "Lattice gas with enhanced collisions", *Europhys. Lett.*, **9**, 345, 1989.
2. R. Benzi, S. Succi and M. Vergassola, "The lattice Boltzmann equation: theory and applications", *Phys. Rep.*, **222**(3), 145, 1992.
3. S. Wolfram, "Cellular automaton fluids I: basic theory", *J. Stat. Phys.*, **45**, 471, 1986.
4. R. Liboff, *Kinetic Theory, Classical, Quantum and Relativistic Description*, Wiley, New York, 1998.
5. I. Aitchinson and A. Hey, *Gauge Theories in Particle Physics*, Adam Hilger, Bristol, 1982.
6. R. Benzi and S. Succi, "Two-dimensional turbulence with the lattice Boltzmann equation", *J. Phys. A*, **23**, L1, 1990.
7. H. Chen, S. Chen and W. Matthaeus, Recovery of the Navier-Stokes equations using a lattice-gas Boltzmann method, *Phys. Rev. A*, **45**, R5349, 1992.

....

EXERCISES

1. Verify by direct computation that the FCHC only admits five distinct scattering angles (don't forget the fourth dimension!).
2. How many ghost fields does one have in 3D-FCHC?
3. Compute the weights of the 2D and 1D projection of the FCHC scheme.

14

Lattice Relaxation Schemes

In Chapter 13, we have shown that the complexity of the LBE collision operator can be cut down dramatically by formulating discrete versions with prescribed local equilibria. In this chapter, we take the process one step further by presenting a minimal formulation whereby the collision matrix is reduced to the identity upfronted by a single relaxation parameter fixing the viscosity of the lattice fluid. The idea is patterned after the celebrated Bhatnagar–Gross–Krook (BGK) model Boltzmann introduced in continuum kinetic theory as early as 1954. In the second part of the chapter, we shall also describe the comeback of the early LBE in optimized multi-relaxation form, as well as few recent variants thereof.

If it doesn't look easy it is that we have not tried hard enough yet.

(F. Astaire)

14.1 Single-Relaxation Time

The main lesson taught by the self-standing LBE of Chapter 13 is that the scattering matrix and local equilibria can be regarded as free parameters of the theory, to be tuned to our best purposes within the limits set by the conservation laws and numerical stability. Within this picture, the viscosity of the LB fluid is controlled by a single parameter, namely the leading-nonzero eigenvalue of the scattering matrix A_{ij}. The remaining eigenvalues are then chosen in such a way as to minimize the interference of non-hydrodynamic modes (ghosts) with the dynamics of macroscopic observables. This spawns a very natural question: since transport is related to a single-nonzero eigenvalue, why not simplify things further by choosing a one-parameter only scattering matrix? The point raised almost simultaneously by a number of authors (1) is indeed well taken. Such a formal change is tantamount to choosing the scattering matrix in a diagonal form

$$A_{ij} = -\omega\delta_{ij} \tag{14.1}$$

The Lattice Boltzmann Equation. Sauro Succi, Oxford University Press (2018).
© Sauro Succi. DOI: 10.1093/oso/9780199592357.001.0001

where the parameter $\omega > 0$ is the inverse-relaxation time to local equilibrium, a constant here and throughout.

This means moving from a multi-time to a single-relaxation time scheme in which *all* modes relax on the same timescale $\tau = 1/\omega$.

Upon such a change, the diagonal LBE reads as follows:

$$\Delta_i f_i = -\omega \ (f_i - f_i^e) \tag{14.2}$$

Given the direct link with the time-honored Bhatnagar–Gross–Krook model Boltzmann equations in continuum kinetic theory (4), the eqn (14.2) has appropriately been named *Lattice BGK, LBGK* for short.

It is difficult to imagine something simpler than this equation to recover the Navier–Stokes equations.[29]

Still a few points call for some clarification.

As already noted, the diagonalization of the collision operator implies that all modes decay at the same rate, which is unphysical.

Leaving aside ghosts for a while, let us consider the hydrodynamic fields.

The fix is simple: since the conservation laws can no longer be encoded at the level of the scattering matrix, no way a diagonal matrix would fulfill the sum-rules obeyed by the matrix A_{ij}. They must necessarily be enforced on the local equilibrium.

As a result, at variance with quasilinear LBE, we shall explicitly require local equilibria to carry the same density and momentum as the actual-distribution function:

$$\sum_i f_i^e = \sum_i f_i = \rho \tag{14.3}$$

$$\sum_i f_i^e c_{ia} = \sum_i f_i c_{ia} = \rho u_a \tag{14.4}$$

Note that these equalities need *not* be obeyed by the quasilinear LBE equilibria because sum rules of the scattering matrix take automatically care of conservation. This is why quasilinear LBE local equilibria can be reduced to the second-order term in the Mach expansion, upfronted by the global-density ρ_0 instead of the local one ρ.

The first task with LBGK is therefore to derive explicit equilibria fulfilling (14.3) and (14.4).

14.2   LBGK Equilibria

In facing the task of defining LBGK equilibria, we shall aim at a greater generality than previously done with LBE. In particular we shall cater for the possibility of *multi-energy* levels, which means lifting the constraint of the same magnitude for *all* discrete speeds.

---

[29] Personal note: indeed, the question "are you REALLY solving the Navier–Stokes equations"? kept coming time and again for quite a long while in the early LB days.

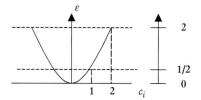

Figure 14.1 *Energy–momentum relation with three energy levels, $0, 1/2, 4/2$.*

By introducing two distinct "quantum numbers," $j = 0, \ldots, j_E$ labeling the energy level and $i = 1, \ldots, b_j$ labeling the momentum eigenstate within energy level j, we can write

$$\epsilon_j = \frac{1}{2} c_{ji}^2, \quad j = 0, \ldots, b_E, \quad i = 1, \ldots, b_j \tag{14.5}$$

for a lattice with $b_E + 1$ energy levels.

With these preparations, a generic family of LBGK equilibria can be expressed as the low Mach expansion of a multi-energy Maxwellian, see Fig. 14.1. This reads

$$f_I^e = \rho w_I \left(1 + \frac{c_{Ia} u_a}{c_s^2} + \frac{Q_{Iab} u_a u_b}{2 c_s^4} \right) \tag{14.6}$$

where I is a shorthand for (i, j) and the weights w_I and the constant c_s (sound speed) depend on the specific choice of the discrete speeds c_{Ia}.

Mass and momentum conservation, as well as isotropy, impose the following equalities:

$$\sum_I w_I = 1 \tag{14.7}$$

$$\sum_I w_I c_{Ia} = 0 \tag{14.8}$$

$$\sum_I w_I c_{Ia} c_{Ib} = c_s^2 \delta_{ab} \tag{14.9}$$

$$\sum_I w_I c_{Ia} Q_{Ibc} = 0 \tag{14.10}$$

$$\sum_I w_I c_{Ia} c_{Ib} c_{Ic} c_{Id} = c_s^4 (\delta_{ab}\delta_{cd} + \delta_{ac}\delta_{bd} + \delta_{ad}\delta_{bc}) \tag{14.11}$$

where $Q_{Iab} = c_{Ia} c_{Ib} - c_s^2 \delta_{ab}$ is the second-order kinetic projector introduced in chapter 13.

These equations are generally under-constrained, i.e., they admit several solutions because, due to the extra freedom provided by the multi-energy lattice, there are more discrete speeds than constraints.

Optimal use of this extra freedom is key to the success of the LBGK approach.

D1Q3   D1Q5   **Figure 14.2** *The D1Q3 and D1Q5 lattices.*

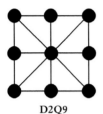

D2Q9   **Figure 14.3** *The D2Q19 lattice.*

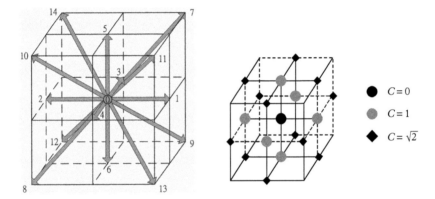

Figure 14.4 *The D3Q19 (left) and D3Q15 (right) lattices.*

Qian *et al.* (1992) provide a whole family of solutions dubbed DnQm for m speed model in n dimensions.

Popular examples are D1Q3, D1Q5 and D2Q9, whose diagrams are sketched in Figs 14.2 and 14.3, together with the synoptic table of their main parameters.

An important asset of the multi-energy schemes is that *there is no problem in going to three dimensions*. In fact there are a number of viable three-dimensional LBGKs, such as the D3Q15 and D3Q19 depicted in Fig. 14.4.

The D3Q15 is the crystal made up by the center of the cell, the six face-centers (nearest neighbor, *nn*) and the eight vertices (next-nearest-neighbor, *nnn*). Note that there is a spectral gap, namely no particles with speed $\sqrt{2}$.

D3Q19 is a close relative of the three-dimensional projection of FCHC, with different weights though (see Table 14.1).

Table 14.1 *Main parameters of some DnQm BGK lattices.*

	c_s^2	Energy	Weight
D1Q3	1/3	0	4/6
		1/2	1/6
D1Q5	1	0	6/12
		1/2	2/12
		2	1/12
D2Q9	1/3	0	16/36
		1/2	4/36
		1	1/36
D3Q15	1/3	0	16/72
		1/2	8/72
		3/2	1/72
D3Q19	1/3	0	12/36
		1/2	2/36
		1	1/36
D3Q27	1/3	0	64/216
		1/2	16/216
		1	4/216
		3/2	1/216

Note that D2Q9 and D3Q27 are the direct product of D1Q3 in two and three spatial directions, respectively. The weights of all three lattices obey the scaling relation

$$w_j = w_0 \left(\frac{1}{4} \right)^j \tag{14.12}$$

formally corresponding to an inverse temperature $e^{-\beta/2} = 1/4$, i.e., $T = 1/(4 ln 2) \sim 1/2.8$, not far from the sound-speed square $c_s^2 = 1/3$.

Also to be noted that D3Q15 and D3Q19 result from D3Q27 (shown in Fig. 14.5) upon pruning the appropriate-energy subshells. Remarkably, such pruning does not compromise isotropy.

The reader with some background in numerical analysis will probably notice a similarity between the LBGK weights and the numerical stencils, sometimes called "computational molecules," used in finite difference and finite element calculations. This analogy is real, and will become apparent in the Chapter 15 devoted to the Gauss–Hermite derivation of LBE.

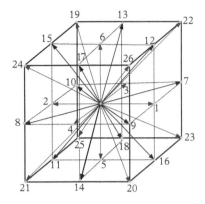

Figure 14.5 *The D3Q27 mother lattice: the tensor product of D1Q3 in three dimensions.*

14.3 LBGK versus Early LBE

The Lattice BGK seemingly represents the ultimate version of LBE in terms of simplicity, which explains why, despite substantial criticism, the vast majority of LBE applications still make use of it.

However, a few words of comment are in order.

First, the gain in efficiency over LBE is less dramatic than what one would expect from the diagonal form of the collision operator. This is because LBE does not need the full equilibrium distribution but only the second-order term in the Mach number expansion. In addition, LBE is more robust as it tolerates weak density fluctuations in the equilibria since it can work with a constant, uniform density ρ_0, instead of the local value ρ.

The main drawback of BGK is that all modes decay at the same time-rate ω^{-1}, which is clearly unphysical.

Depending on initial and boundary conditions, this may also imply unphysical short-wave oscillations which degrade numerical stability in the low-viscous regimes.

Another LBE bonus not shared by LBGK is the possibility of seamless inclusion of passive scalars arising from the dimensional projection of the FCHC lattice in dimensions $D < 4$ (see Chapter 13). Such a bonus is not available with LBGK where, in principle, each additional passive scalar calls for a separate set of at least $2D$ populations to implement a diffusion equation in D spatial dimensions. This represents a burden for applications involving many chemical species.

Finally, a single relaxation time implies that mass, momentum and energy transfer, all take place at the same rate, which is appropriate only for ideal gases.

This latter restriction can be partially cured by using a two-time relaxation operators (5) of the form

$$C_i = -\omega(f_i - f_i^e) - \omega^*(f_{i*} - f_{i*}^e) \tag{14.13}$$

where i^* denotes the parity-conjugate of i ($\vec{c}_i + \vec{c}_{i*} = 0$). By tuning the two relaxation parameters, one can change the momentum and heat diffusivity ν and χ independently, thereby achieving a non-unit Prandtl number ν/χ. The dynamic range of realizable Prandtls remains relatively narrow, though. More recently, significant benefits of the two-time formulation have been pointed out by Ginzburg and collaborators (6).

More sophisticated multi-BGK versions, based upon a sequence of *BGK* equilibria and associated relaxation rates have been proposed in the recent past to improve the stability of the LBGK scheme (7).

Notwithstanding these improvements, and its unquestionable top simplicity in the LBE family, the LBGK has made the object of substantial criticism in the last decade. In particular, a number of authors have argued that much is to be gained by going back to the original LBE in matrix form, with a suitably optimized choice of the relaxation frequencies.

Given the popularity gained by this line of thinking in the last decade, in Section 14.4 we convey the basic ideas behind this approach.

14.4   Multiple–Relaxation Time (MRT)

In the early 2000s the scattering matrix LBE approach has been revisited and optimized in view of increased stability and accuracy. These developments have gained significant popularity in the last decade and got a name on their own, MRT, for Multi-Relaxation Time.[30]

The starting point is to lift the main limitation of LBGK, namely the fact that all non-conserved modes decay at the same rate. The idea is to devise spectral representations of the scattering matrix which, besides allowing non-unit Prandtl numbers, would also prove capable of enhancing (linear) numerical stability. The practical implementation of the MRT schemes proceeds via a dual real–momentum space representation of the discrete populations, say f_i and M_k, which diagonalize the streaming and collision operators, respectively (the former being the standard representation).

First, one defines a set of orthogonal eigenvectors E_i^k of the scattering matrix A_{ij}, such that, by definition

$$\sum_{j=0}^{b} A_{ij} E_j^k = \omega_k E_i^k, \ \ k = 0, b \tag{14.14}$$

where the eigenvalues $\{\omega_k\}$, $k = 0, b$ run over a discrete set of $b + 1$, generally distinct values.

The discrete distribution is then be projected upon such discrete basis, according to

$$f_i = \sum_{k=0}^{b} T_{ik} M_k \tag{14.15}$$

---

[30] In passing, it should be noted that the LBE was actually *born* in matrix form (Higuera–Jimenez, 1989, Higuera–Succi–Benzi, 1989).

where the transformation matrix is defined as

$$T_{ik} \equiv E_i^k \tag{14.16}$$

i.e., the matrix whose k-th column is the k-th eigenvector of the scattering-matrix A_{ij}.

By construction, T_{ik} takes the kinetic moments M_k into the discrete distributions f_i, and its inverse $T_{ki} = T_{ik}^{-1}$ does just the same job in reverse, i.e.,

$$M_k = \sum_{i=0}^{b} T_{ki} f_i \tag{14.17}$$

where $\sum_{k=0}^{b} T_{ik} T_{kj} = \delta_{ij}$.

The action of the scattering matrix on the discrete distribution is written as

$$\sum_{j=0}^{b} A_{ij} f_j = \sum_{j=0}^{b} \sum_{k=0}^{b} A_{ij} T_{jk} M_k$$

Upon using the orthogonality condition $\sum_{j=0}^{b} A_{ij} T_{jk} = \omega_k T_{ik}$, this delivers

$$\sum_{j=0}^{b} A_{ij} f_j = \sum_{k=0}^{b} \omega_k T_{ik} M_k \tag{14.18}$$

As a result, the MRT collision operator takes the form

$$C_i = \sum_{k=0}^{b} \omega_k T_{ik} (M_k^e - M_k) \tag{14.19}$$

and the overall MRT scheme reads as follows:

$$f_i(\vec{x} + \vec{c}_i; t + 1) - f_i(\vec{x}, t) = \sum_{k=0}^{b} \omega_k T_{ik} (M_k^e - M_k) \tag{14.20}$$

where the moments on the right-hand side are computed via (14.17).

The result is an elegant-numerical scheme, whereby streaming takes place in the standard discrete-velocity basis, whereas collisions are performed in the moment basis, according to (14.19).

This very compact form reduces to the standard LBGK in the limit where all eigenvalues ω_k collapse to the same value. Hence, once the transformation matrix is available, the computation of the collision operator reduces to a set of b matrix-vector scalar products, b^2 complexity at most.

It is reported that a careful optimization of matrix-product operations keeps the computational overhead only marginally above the one associated with "plain" LBGK, of the order of $10 \div 20$ percent for D2Q9 and about 40 percent for D3Q15. It is further argued that this is more than compensated by the gains in numerical stability, as we shall briefly discuss in the sequel.

14.4.1　Constructing the MRT Matrix

Let us now see how to construct the transformation matrix in actual practice.

Following continuum kinetic theory, the eigenvectors are most conveniently associated with three orthogonal subspaces: *Conserved, Transport, Ghost* (CTG) representation. These are the discrete analogs of the Conserved–Slow–Fast partition discussed in Part I.

The C-sector is spanned by the kinetic eigenevectors associated with mass and momentum conservation:

- $A_i^{(0)} = \{1_i\}, A_i^{(1,2,3)} = \{c_{ix}, c_{iy}, c_{iz}\}$

The corresponding eigenvalues are zero by construction.

The T-sector is spanned by the eigenvectors associated with the momentum flux tensor:

- $A_i^{(4 \div 9)} = \{Q_{ixx}, Q_{ixy}, Q_{ixz}, Q_{iyy}, Q_{iyz}, Q_{izz}\}$

for a total of ten. These are the discrete analogs of the zeroth-, first- and second-order tensor Hermite polynomials. The corresponding eigenvalues are non-zero and control the shear and bulk viscosities, not necessarily the same along the different directions (anisotropic fluids).

The remaining eigenvectors in the G sector, five for D3Q15, nine for D3Q19 and so on, are subject to some deliberation.

In principle, the simplest route is to construct them by Gram–Schmidt orthogonalization paying due attention to the linear dependencies due to lattice discreteness. For instance, all discrete lattices living in the first Brillouin cell, with $c_{ia} = \{0, \pm 1\}$, suffer the following degeneracies:

$$c_{ia}^3 = c_{ia}, \quad c_{ia}^4 = c_{ia}^2, \quad a = x, y, z \tag{14.21}$$

As a result, only some of the higher-order Hermite polynomials are linearly independent in the lattice. This gives rise to some degree of arbitrariness in constructing the G sector.

Some explicit expressions, not necessarily following the CTG decomposition given, can be found in the original literature (8; 9).

14.4.2　Enhanced MRT Stability

The main advantage of MRT is that by a suitable choice of the different eigenvalues, one can fine-tune the *linear* properties of the numerical scheme, namely *dispersion, dissipation, isotropy, Galilean invariance and ultimately linear stability.*

In a very informative paper, Lallemand and Luo performed a detailed linear stability analysis of the MRT scheme for the D2Q9 lattice at *finite wavenumber* and *non-zero, constant-flow speed* (9).

In particular, the authors analyze the discrete-dispersion relation associated with the MRT scheme as a function of the wave-vector $\vec{k} = (k_x, k_y)$ and the (constant) fluid speed, $\vec{U} = (U_x, U_y)$.

Formally, the discrete dispersion relation reads as follows:

$$\Omega_l = D_l(\vec{k}\Delta x, \omega_0\Delta t \ldots \omega_8\Delta t; \vec{U}), \; l = 0, 8 \qquad (14.22)$$

where ω_l are the free-relaxation parameters of the MRT matrix and Ω_l are the resulting propagation frequencies. The two sets coincide only in the continuum limit $\vec{k}\Delta x \to 0$, $\omega_l\Delta t \to 0$. At any finite resolution, there are departures which depend on the flow velocity $\vec{U}$.

Since no fluid simulation takes place at infinite resolution and/or zero speed, departures from the continuum dispersion relation are clearly relevant to the practical operation of the LB scheme.

The authors derive analytical expressions for the shear and bulk viscosities and show that proper tuning of two out of the six free eigenvalues (three are frozen to zero by conservation) permits us to optimize the numerical stability of the MRT scheme, especially in the strongly over-relaxed regime, $\omega \to 2$, which is the one relevant to high-Reynolds simulations.

Similar benefits have been reported in connection with boundary conditions, where MRT is credited for a better control of the artificial dependence of the effective boundary location on the relaxation frequency ω, which is known to affect the LBGK scheme. The same analysis has been extended to the three-dimensional D3Q19 lattice in a subsequent paper by D. Humieres *et al.* (10).

The case for MRT has been pursued very assertively and, in this author's opinion, often beyond its merits.

Indeed, it should be mentioned that, for all its unquestionable virtues, MRT is not immune from criticism either.

First, the MRT stability analysis is by construction restricted to the linear regime. Indeed, while linear instability dominates the early stage of the evolution, the subsequent stages are often characterized by the onset of nonlinear instabilities which, by definition escape linear stability analysis. As a result, there is no a-priori guarantee that a set of relaxation parameters carefully tuned to optimize linear stability, would also provide optimal operation in the nonlinear regime. And even less likely that they would do so *universally*, i.e., across different types of flows and boundary conditions.

Such limitations were clearly recognized by the original authors (9). It should also be pointed out, while it is true that MRT has been showing superiority over LBGK for a number of flows, counter-examples are available too (11). This is not to diminish the merits of MRT, but simply to caution the reader that unconditional superiority of MRT versus LBGK, as it is frequently assumed in the recent literature (based on this author's experience as a referee), should by no means be taken for granted.

Since the MRT versus LBGK has developed into a rather opinionated debate, it may be worth to observe that, from a distance, the question does not appear really worth much controversy at all.

Given that MRT is a superset of LBGK, it is indeed quite plain that it must be able to deliver *at least* the same performance in terms of stability and accuracy, if not necessarily efficiency. For any given flow, a parameter set is most likely to be found, such that MRT would outperform LBGK in terms of stability and accuracy.

However, to the best of this author's knowledge, systematic and universal criteria to identify such optimal set in the *nonlinear* regime and under general flow conditions, remain to be found. Failing such general criteria, the judgement about the superiority of MRT versus LBGK should not be taken off any "expert's shelf," but rather formed by direct and critical experimentation on the specific problem at hand.

14.5   Further Relaxation Variants

In the recent years, a number of relaxation variants have been developed, which improve on the main limitations discussed. Among those already mentioned before in this chapter, we recall here the so-called Regularized-BGK scheme (RLBGK) and the Cascaded LBE (CLBE).

14.5.1   Regularized LBGK

The basic idea of the so-called "regularized" LB (12; 13) is to protect the transport modes from unwanted interference of the higher-order ghost modes. To this purpose, a simple and effective strategy consists of filtering out the ghost modes from the discrete distributions after completion of the streaming step, i.e., prior to next collision.

In equations:

Streaming:

$$f_i(\vec{x}; t) = f_i(\vec{x} - \vec{c}_i \Delta t; t - \Delta t) \qquad (14.23)$$

Collision:

$$f_i(\vec{x}; t) = f_i'(\vec{x}; t) \equiv (1 - \Omega_{ij}) f_j^H(\vec{x}; t) + \Omega_{ij} f_j^{eq}(\vec{x}; t) \qquad (14.24)$$

In (14.24), f^H is the hydrodynamic component of the discrete distribution, i.e., the one including density, current and momentum-flux tensor (the CT sector), but no ghost contributions. It is worth pointing out that the local equilibrium f^{eq} in eq. (14.24) stays unchanged under filtering because, by construction, it receives no contribution from the ghost modes. Note that we have taken the matrix form for generality, but the idea clearly applies to LBGK as well.

This "ghostbuster" regularization proves beneficial for the numerical stability of the scheme, at a very moderate extra-computational cost and programming effort. Further benefits appear to emerge for flows beyond the hydrodynamic regime, in connection with higher-order lattices (see chapter 29). For full details see the original papers (12; 13). As a result, the Regularized LBGK shows promise to improving on LBGK in a very simple and effective way.

14.5.2   Cascaded LBE

The theory of the Lattice Boltzmann method discussed so far is the Galilean invariant only up to second order in the Mach number. Even though this is sufficient for athermal and nearly incompressible flows, it has been pointed out that lack of Galilean invariance at higher orders provides a liability toward numerical stabilities in the low viscous regime.

To cope with this liability, Geier *et al.* (14) propose to formulate the LB equation no longer in terms of the *absolute* kinetic moments, $M_k = \int f(v)v^k dv$, but rather in terms of the *central* ones, defined as

$$\mu_k = \int f(v)(v-u)^k dv \qquad (14.25)$$

where vector notation has been relaxed for simplicity.

The advantage of the central moment representation is that, by design, a consistent match between continuum and discrete central moments secures Galilean invariance at *all* orders.

In actual practice, however, it still is more convenient to work with absolute moments, and consequently a transformation rule between the two sets is required. This is provided by the standard-binomial expression

$$\langle (v-u)^k \rangle = \sum_{l=0}^{k} \frac{k!}{l!(k-l)!} (-1)^{k-l} \langle v^l \rangle u^{k-l} \qquad (14.26)$$

where brackets stand for integration in velocity space. This binomial transform permits us to compute central moments from absolute ones. The most important feature of this transformation is that a central moment of order k depends only on absolute moments up to order k and not on higher orders. As a result, the collision operator based on central moments can be constructed by starting from the lowest order moments first and proceed to the highest ones in ascending sequence. The cascade is an algorithmic trick that spares new elementary collisions at each scattering step. For the rather laborious derivation and algorithmic details the reader is pointed to the original publication (14). Swifter derivations are also available in the more recent literature on the subject.

14.6 Summary

Summarizing, the LB was born in matrix form, and quickly superseded by the simpler single-relaxation-time-BGK version, LBGK.[31]

LBGK tops simplicity, but at a price of restrictions in physical realizability, numerical accuracy and stability. Such weaknesses can be partially cured by turning back to the LB in matrix form, with proper fine-tuning of the relaxation times. This fine-tuned version, now known Multi-Relaxation Time (MRT) has shown better performance as compared to LBGK for a number of flows. Such superiority, however is neither unconditional nor universal, hence it should not be taken for granted a-priori, but rather investigated case-by-case for each given flow under consideration. Other variants, such as Regularized and Cascaded LBGK, may offer further options to improve on LBGK, although more validation work is needed before a solid pronunciation can be made.

....

REFERENCES

1. S. Chen, H. Chen, D. Martinez and W. Matthaeus, "Lattice Boltzmann model for simulation of magnetohydrodynamics", *Phys. Rev. Lett.*, **67**, 3776, 1991.
2. J.M. Koelman, "A simple lattice Boltzmann scheme for Navier–Stokes fluid flow", *Europhys. Lett.*, **15**, 603, 1991.
3. Y. Qian, D. d'Humières and P. Lallemand, "Lattice BGK models for the Navier–Stokes equation", *Europhys. Lett.*, **17**, 479, 1992.
4. P. Bhatnagar, E. Gross and M. Krook, "A model for collisional processes in gases I: small amplitude processes in charged and neutral one-component system", *Phys. Rev.*, **94**, 511, 1954.
5. Y. Chen, H. Ohashi and M. Akiyama, "Thermal lattice Bhatnagar–Gross–Krook model without nonlinear deviations in macrodynamic equations", *Phys. Rev. E,* **50**, 2776, 1994.
6. I. Ginzburg, F. Verhaeghe and D. D'Humieres, "Two-relaxation-time Lattice Boltzmann scheme: About parametrization, velocity, pressure and mixed boundary conditions", *Comm. in Comp. Phys.*, **3**, 427, 2008.
7. S. Ansumali *et al.*, "Quasi-equilibrium lattice Boltzmann method", *Europ. Phys. J. B*, **56**, 135, 2007.
8. D. d'Humières, "Generalized lattice Boltzmann equations, Rarefied Gas Dynamics: Theory and Simulation", *Prog. Astronaut. Aeronaut.*, **159**, 450, AIAA Washington, 1992.

---

[31] It is interesting to note that several authors in the recent literature regard LBGK as the starting point of the LBE family, with MRT coming as a subsequent development. I hope this book makes it clear that LB was actually *born* in matrix form three years before LBGK.

9. P. Lallemand and L.S. Luo, "Theory of the lattice Boltzmann method: Dispersion, dissipation, isotropy, Galilean invariance, and stability", *Phys. Rev. E*, **61**, 6546, 2000.

10. D. d'Humieres, I. Ginzburg, M. Krafczyk *et al.* "Multiple-relaxation-time lattice Boltzmann models in three dimensions", *Phil. Trans. Roy. Soc. A*, **360**, 437, 2002.

11. R.K. Freitas *et al.*, "Analysis of Lattice-Boltzmann methods for internal flows", *Computers and Fluids*, **47**, 115, 2011.

12. J. Latt and B. Chopard, "Lattice Boltzmann method with regularized pre-collision distribution functions", *Mathematics and Computers in Simulation*, **72**, 165, 2006.

13. H. Chen, R. Zhang, I. Staroselsky and M. Jhon, "Recovery of full rotational invariance in lattice Boltzmann formulations for high Knudsen number flows", *Physica A*, **362**, 125, 2006.

14. M. Geier, A. Greiner and J. Korvink, "Cascaded digital lattice Boltzmann automata for high Reynolds number flow", *Phys. Rev. E*, **73**, 066705, 2006.

···

EXERCISES

1. Verify that the D2Q9 equilibria fulfill the constraints (14.7)–(14.11).
2. For what value of ν (in lattice units) does one enter the over-relaxation regime?
3. Imagine to assign a different relaxation rate to each discrete speed ω_i. What do you expect to gain in terms of generality and numerical stability?

15

The Hermite–Gauss Route to LBE

In this chapter we shall describe the side-up approach to Lattice Boltzmann, namely the formal derivation from the continuum Boltzmann-(BGK) equation via Hermite projection and subsequent evaluation of the kinetic moments via Gauss–Hermite quadrature.

> *I turn away with fright and horror from the lamentable evil of functions which do not have derivatives.*
>
> (Charles Hermite)

15.1 LBE from Continuum-Kinetic Theory: The Hermite–Gauss Approach

Notwithstanding its physics-inspired roots, at the end of the day, the LBGK takes the form of a set of finite-difference equations. As a result, it is natural to wonder how such finite-difference formulation relates to the discretization of the continuum BGK equation, namely

$$\partial_t f + v_a \partial_a f = -\omega \ (f - f^e) \tag{15.1}$$

To develop this formal connection, let us begin by expanding the distribution function in Hermite polynomials (in one space and velocity dimensions for simplicity):

$$f(x, v; t) = w(v) \sum_{k=0}^{\infty} M_k(x; t) H_k(v) \tag{15.2}$$

where $w(v)$ is the standard Gaussian-weight function and the moments are given as usual by the scalar product in Hilbert space:

$$M_k(x; t) = \int_{-\infty}^{+\infty} f(x, v; t) H_k(v) dv \tag{15.3}$$

The Lattice Boltzmann Equation. Sauro Succi, Oxford University Press (2018).
© Sauro Succi. DOI: 10.1093/oso/9780199592357.001.0001

By multiplying (15.1) by $H_l(v)$, integrating upon the velocity space and using the orthonormality condition $\int H_k(v)w(v)H_l(v)\,dv = \delta_{kl}$, we obtain a chain of evolution equations for the kinetic moments:

$$\partial_t M_l + \partial_x \mathcal{J}_l = \omega(\rho_l^e - \rho_l) \tag{15.4}$$

where the current $\mathcal{J}_l$ is given by

$$\mathcal{J}_l(x; t) = \int_{-\infty}^{+\infty} vf(x, v; t)H_l(v)\,dv \tag{15.5}$$

Let us next evaluate the velocity integrals by Gauss-Hermite numerical quadrature, namely

$$M_l(x; t) = \sum_{i=0}^{b} w_i p(x, v_i; t)H_l(v_i) \tag{15.6}$$

and

$$\mathcal{J}_l(x; t) = \sum_{i=0}^{b} w_i v_i p(x, v_i; t)H_l(v_i) \tag{15.7}$$

where we have defined the polynomial distribution

$$p(x, v; t) \equiv \frac{f(x, v; t)}{w(v)} \tag{15.8}$$

We hasten to note that the weights of the Gauss–Hermite quadrature, w_i do *not* identify with $w(v_i)$, the absolute Maxwellian evaluated at the Gaussian node $v = v_i$.

It is well known that Gauss–Hermite quadrature with $(b + 1)$ nodes, such as (15.6) integrates *exactly* polynomials $p(v)$ up to order $2(b + 1) - 1 = 2b + 1$. For instance, a simple D1Q3 lattice reproduces moments up to order five.

Being exact, Gauss–Hermite quadrature is at a vantage point within the LB theory. Because it permits us to choose large spacings in velocity space, of the order of the sound(thermal) speed, namely

$$\Delta v \sim v_T \tag{15.9}$$

This stands in contrast with finite-difference practice, say trapezoidal or Simpson numerical integration, which usually demands much smaller velocity spacings

$$\Delta v \ll v_T \tag{15.10}$$

to obtain reasonable accuracy.

From a slightly different angle, one may also interpret the Gauss–Hermite quadrature as an optimal *sampling* of velocity space, or, better still, an *exact* sampling of the bulk of the distribution function, the one contributing most to the lowest-order kinetic moments (frequent events). Capturing higher-order moments, beyond hydrodynamics (rare events), requires an increasing number of nodes and weights.

By inserting (15.6) and (15.7) into (15.4) and collecting common prefactors in front of H_i^k, we obtain

$$\partial_t f_i + \partial_x (v_i f_i) = -\omega \ (f_i - f_i^e) \tag{15.11}$$

where we have set

$$f_i \equiv w_i p_i = \frac{w_i}{w(v_i)} f(x, v_i; t) \tag{15.12}$$

By marching each population in time with an explicit Euler-forward scheme along the characteristics $\Delta x_i = v_i \Delta t$, and integrating explicitly in time the collision term, we obtain

$$f_i \ (x + v_i \Delta t, t + \Delta t) - f_i \ (x, t) = \omega \Delta t \ (f_i^e - f_i) \tag{15.13}$$

which is precisely LBGK.

This shows that Hermite expansion of the continuum BGK, as combined with Gauss–Hermite quadrature, gives rise to the LBGK scheme (1; 2). The same reasoning can be applied to higher dimensions by expanding on tensor Hermite polynomials (see Part I).

This procedure encapsulates the LBGK formalism within the general box of computational kinetic theory. This an elegant result, which shows that the discrete speeds, so far chosen based on pure symmetry requirements, can be identified with the nodes of the Gauss-Hermite quadrature, namely the zeros of the Hermite polynomials.

For this reason, the originary LBE and LBGK schemes were somehow dismissed as "primitive" numerical approximations of the continuum BGK, a rather debatable assessment in this author's opinion.

To begin with, it should be noted that not every LBE scheme falls under the Hermite umbrella; among those confined to the first Brillouin cell, only D1Q3 and its tensor products D2Q9 and D3Q27.

Second, lattices beyond the first Brillouin cell involve non-integer nodes which do not fit a space-filling lattice, unless they are complemented by some interpolation procedure filling the gap between abutting cells.

As a result, the relation between LBE (LBGK) schemes and Gauss–Hermite quadrature is by no means a one-to-one connection, nor an inclusive one. In fact, the early LBGK schemes represent a *smart "mutation,"* capable of reproducing hydrodynamics with a lesser number of discrete nodes than demanded by the standard Gauss–Hermite formulation.

As we shall see shortly, this is because *isotropy is less demanding than separability.*

15.2   Multi-Dimensions, Higher-Order Lattices and Separability

The extension of the Gauss–Hermite approach to higher dimensions is straight-forward, based on tensor Hermite polynomials, along the way opened by Grad. In the lattice the recipe is very simple (for low-order schemes confined to the first Brillouin cell): just take the direct product of the one-dimensional nodes and weights.

For instance starting with D1Q3, i.e., the three nodes D1Q3 = $\{-1, 0, 1\}$, one generates D2Q9 by direct product of two three-discrete-velocity stencils along x and y respectively, i.e., in "chemical notation," $D_2Q_9 = \{-1, 0, 1\}^2$. By the same token, the three-dimensional D3Q27 lattice is obtained as $D_3Q_{27} = \{-1, 0, 1\}^3$. In fact, the discrete velocity stencils can be regarded as "computational molecules," a notation which is often used in numerical analysis and computational physics.

This simple procedure entirely owes to the *separability* of the Maxwell distribution along the $x - y - z$ velocity directions, namely

$$e^{-(v_x^2 + v_y^2 + v_z^2)} = e^{-v_x^2}\, e^{-v_y^2}\, e^{-v_z^2} \tag{15.14}$$

Interestingly, this was noted by J.C. Maxwell (1831–1879) *before* Boltzmann developed his own equation.

Separability sends a very neat message, namely that *not all* LBE schemes can be derived from straight Gauss–Hermite quadrature.

For instance, $D3Q15$ and $D3Q19$ are *not* separable, and yet they fulfill the isotropy requirements for hydrodynamic applications. To understand why, it proves expedient to decompose D3Q27 in four topological subsets based on the magnitude of the corresponding discrete velocities (*energy shells*), namely:

1. One cell center (C), ($c_0 = 0$)
2. Six face centers (F), ($|c_i| = 1$)
3. Twelve edge centers (E), ($|c_i| = \sqrt{2}$)
4. Eight vertices (V), ($|c_i| = \sqrt{3}$)

Thus, again in "chemical" notation,

$$D_3Q_{27} = C_1 \oplus F_6 \oplus E_{12} \oplus V_8 \tag{15.15}$$

By depriving $D3Q27$ of the eight vertices, we obtain

$$D_3Q_{19} = C_1 \oplus F_6 \oplus E_{12} \tag{15.16}$$

Similarly, by stripping off the 12 face centers

$$D_3Q_{15} = C_1 \oplus F_6 \oplus V_8 \tag{15.17}$$

Finally, by eliminating both vertices and face centers

$$D_3Q_{13} = C_1 \oplus E_{12} \tag{15.18}$$

Thus the early LBE lattices are sublattices of D3Q27, deprived by some of the four topological classes, an operation technically known as *pruning*.

This shows why the original LBE was not that "primitive" at all.

A fully fledged discussion of Gauss–Hermite quadrature for LBE schemes within and *beyond* hydrodynamics, and not restricted to direct product quadratures, is given in a very informative paper by X. Shan (3).

The authors define the general polytope $E_{D,n}^d$ as the set of discrete velocities in D-dimensional space achieving algebraic accuracy of order n, with d discrete velocities, see Fig. 15.1. For instance, in this language, $D2Q9 = E_{2,5}^9$, eventually rescaling the velocities with the corresponding sound speed. The glorious FHP is hereby pigeonholed into the rather engaging symbol $E_{2,5}^7$, but such are the rules in crystallography ....

The same analysis shows a number of cute results, for instance that $D3Q13 = E_{3,5}^{13}$ is the smallest degree-five quadrature scheme known in three dimensions. Shan *et al.* go on to discuss higher-order lattices not restricted to the first Brillouin cell, such as $E_{2,7}^{12}$, $E_{2,7}^{17}$, $E_{3,7}^{39}$ and so on.

A full list of higher-order stencils up to sixteenth order (yes, sixteenth!) in two dimensions is given in (4).

In the last decade, higher-order lattices have proved instrumental in taking LBE beyond the hydrodynamic limit, as well as to formulate LB models of non-ideal fluids with multi-range interactions and also relativistic LB equations.

These remarkable developments shall be discussed in the sequel to this book.

Here, we wish to reiterate that only some of these higher-order lattices are *congruent*, i.e., can tile Euclidean space leaving no hole in between. For instance, the $E_{2,7}^{16}$ consisting of four speeds $(0, \pm c_-)$, $(\pm c_-, 0)$, another four $(0, \pm c_+)$, $(\pm c_+, 0)$, and finally eight diagonals $(\pm c_-, \pm c_+)$, with $c_\pm^2 = 3 \pm \sqrt{6}$ is clearly non space-filling, as the reader can readily check by direct visualization.

Higher order lattices

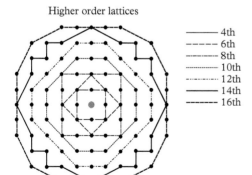

——— 4th
----- 6th
—·—·— 8th
·········· 10th
—··—··— 12th
——— 14th
------- 16th

Figure 15.1 *Higher-order two-dimensional lattices with isotropy up to order 16. Reprinted with permission from M. Sbragaglia et al., Phys. Rev. E, 75, 026702, 2007. Copyright 2007 by the American Physical Society.*

Incongruent lattices can still be used, provided they are supplemented with an interpolation step placing off-lattice particles back to the lattice. Unfortunately, such interpolation spoils one of the most precious properties of genuine LBE schemes, namely *exact* streaming.

15.3   Non-Cartesian Quadrature Schemes

The Gauss–Hermite procedure described thus far is strictly Cartesian. One may then wonder whether it could be extended to more general geometries, say cylinders, spheres and similar. Indeed, such geometries do possess analog quadratures known as Gauss–Laguerre, for energy, and Gauss–Legendre for angular variables.

LBE schemes in spherical coordinates have indeed been developed in the recent past (5). It is perhaps of interest to note that such schemes bear a relation to a general class of numerical quadrature in spherical coordinates named after the Russian mathematician V.I. Lebedev (6).

Lebedev quadratures apply to spherical integrals of the form

$$I = \int_0^\infty \int_0^\pi \int_0^{2\pi} f(v,\theta,\phi) sin(\theta) v^2 \, d\theta \, d\phi \, dv \tag{15.19}$$

and the corresponding quadrature reads as follows:

$$I = \sum_{i=1}^{N_v} W_i \sum_{j=1}^{N_i} W_j f(v_i,\theta_j,\phi_j) \tag{15.20}$$

where j runs over the discrete angles at a given energy shell $v = v_i$.

It is plausible to expect that the LBE discrete velocities would correspond to a shell-by-shell pruning of the Lebedev quadratures. To the best of this author's knowledge, such kind of connection remains to be explored in depth. Spherical quadratures have also been used for relativistic LBE formulations to be discussed in the final part of the book (see Fig. 15.2).

15.3.1   Personal Afterthoughts

When the Hermite pathway to LBE first appeared, this author's reaction was kind of sowattish: so what? Of course such a correspondence must be there! Besides, I also found the dismissal of previous LBGK as "primitive numerics" rather far-fetched, especially in the face of the fact that, at the time, the Hermite approach had not unveiled any previously unknown scheme.

As to the claims of rigor, I must admit that I understand them as little now as I did back then, unless rigor is taken as a byword for "systematic," which actually it is not.

However, fifteen years down the line, I believe it is only fair to acknowledge that systematicity proved a very helpful ace for the development of the subject.

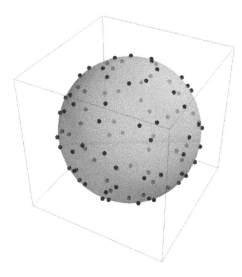

Figure 15.2 *High order numerical quadrature in spherical coordinates. Usually a few points would suffice. However, if one insists in constraining the quadrature points to a cubic lattice, as it is the case for this figure, then their number increases substantially. Reprinted with permissions from M. Mendoza et al.,* Phys. Rev. D 87, 065027, 2013. *Copyright 2013 by the American Physical Society.*

When faced with a complex-flow problem, in which hydrodynamic equations are either shaky or maybe non-existent at all, it really helps to proceed along the formal steps outlined by the Gauss–Hermite procedure. This proved to be the case in many instances, complex flows, entropic-LBE and relativistic-LBE schemes as well.

15.4 Relation to Discrete-Velocity Models

This chapter concludes our introduction to the three basic pathways to LBE. Before moving on to the illustration of how this theory can be put to practice, a few remarks are in order.

In particular, we wish to mention the relation between LBE and Discrete-Velocity Models (DVM) of the Boltzmann equation.

First and foremost, one should never forget that DVMs were invented long before LBE. Actually, they date back to Kac's and Carleman's work in the late 50s and several other authors in the 60s.

How does LBE relate to this work?

Surely DVM and LBE are close relatives in mathematical terms, since they are both based on grid-bound particles moving along a set of discrete speeds chosen by symmetry argument rather than standard finite-differencing of velocity space. However, as far as we can tell, they depart significantly as to their practical aim and purpose. DVM is genuinely concerned with kinetic theory, the main aim being to develop stylized-Boltzmann equations potentially amenable to exact analytical solutions or even to theorem demonstration and easier to solve numerically (7). Computer simulation is certainly in focus, but with no particular obsession on hydrodynamics.

LBE is less and more ambitious at the same time.

Less ambitious, because the idea of a credible description of kinetic phenomena was not pursued at all in the early days. As we shall see later in this book, this is much less true today than it was in 2001, since LBE has developed some genuinely "kinetic" dimension in the last decade. By and large, however, the major body of LB work still remains within hydrodynamics.

It was, and still is, more ambitious, because it challenges the Lion King in his own den, aiming as it does at capturing the hydrodynamic phenomena traditionally described by Navier–Stokes equations of continuum mechanics, for which extremely powerful numerical methods have been developed since the early days of computer simulation.

15.5   Summary: Three Ways to Lattice Boltzmann

We have already mentioned that, according to Richard Feynman, a good piece of theory should be accessible through at least three independent routes.

At this point, the three routes to LB have indeed been exposed, as symbolically represented in Fig. 15.3.

The first branch (*Phys.*) is what we may call the "statistical physics" approach: *derive* LB bottom-up, from its microscopic lattice foundations, i.e., Lattice Gas Cellular Automata. This route is very rich and scenic, but it also carries along a number of needless limitations, as discussed in Chapter 11.

The second route (*Eng*), that we may call "reverse engineering" or "inverse kinetic theory," is epitomized by the top-down LBE with enhanced collisions described previously in this book. It provides dramatic gains in computational efficiency and, more importantly, it first put forward the idea of top-down lattice-kinetic *design* based on macroscopic target equations. This idea proved very productive and influential for many subsequent developments of the field, also beyond hydrodynamics, where the macroscopic target is not limited to the Navier–Stokes equations.

However, the top-down approach comes at the price of potential loss of touch with the second principle. Quite remarkably, entropic LB schemes to be discussed later in this book, fix this potential flaw.

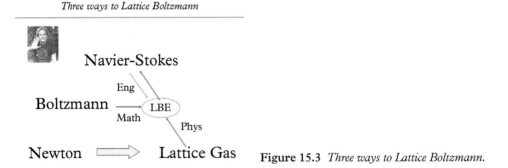

Figure 15.3 *Three ways to Lattice Boltzmann.*

Finally, the side-up branch (*math*): LBE from BE, the "mathematician way," states the citizenship of LBE as a discrete version of the corresponding continuum kinetic equation.

It was this author's opinion that the relevance of this route had been slightly over-emphasized, and, at its nadir, sometimes even taken as an excuse to short-circuit the trailblazing developments previously described, i.e., the way LB developed in actual historical fact. In hindsight, however, this third branch has proved very valuable: higher-order thermal, non-ideal and entropic, as well as relativistic LB schemes, all have drawn significant benefit from this systematic route connecting lattice kinetic theory with its continuum counterpart. In summary, all three routes make their own original contribution to the richness and flexibility of the subject.

REFERENCES

1. X. He and L.S. Luo, "*A priori* derivation of the lattice Boltzmann equation", *Phys. Rev. E*, **55**, 6333, 1997.
2. X. Shan and X. He, "Discretization of the velocity space in the solution of the Boltzmann equation", *Phys. Rev. Lett.*, **80**, 65, 1998.
3. X. Shan, X.F. Yuan and H. Chen, "Kinetic theory representation of hydrodynamics a way beyond the Navier–Stokes equation", *J. Fluid Mech.*, **550**, 413, 2006.
4. M. Sbragaglia, R. Benzi, L. Biferale, S. Succi, K. Sugiyama, and F. Toschi, "Generalized lattice Boltzmann method with multirange pseudopotential", *Phys. Rev. E*, **75**, 026702, 2007.
5. V.E. Ambrus and V. Sofonea, "High-order thermal lattice Boltzmann models derived by means of Gauss quadrature in the spherical coordinate system", *Phys. Rev. E*, **86**, 016708, 2012.
6. V.I. Lebedev, "Quadratures on a sphere", *Zh. Vchisl. Mat. Mat. Fiz.*, **16**, 293, 1976, (see also Wikipedia).
7. J. Broadwell, "Study of the rarefied shear flow by the discrete velocity method", *J. Fluid Mech.*, **19**, 401, 1964.
8. R. Gatignol, *Theorie Cinetique des Gaz a Repartition Discrete de Vitesses*, Lecture Notes in Physics 36, Springer–Verlag, Berlin, 1975.
9. H. Cabannes, "The discrete models of the Boltzmann equation", *Transp. Theory and Stat. Phys.*, **16**, 809, 1987.
10. C. Cercignani, *Mathematical Methods in Kinetic Theory*, Plenum Press, New York, 1969.
11. R. Monaco and L. Preziosi, *Fluid Dynamic Applications of the Discrete Boltzmann Equation*, Series on Advances in Mathematics for Applied Sciences 3, World Scientific, Singapore, 1991.

....

EXERCISES

1. Compute the integrals $I_{2n} = \int_{-\infty}^{+\infty} e^{-x^2/2} x^{2n} dx$, with $n = 0, 1, 2, 3$ using a three-node Gauss–Hermite quadrature and analyse the corresponding error.
2. The same exercise in two dimensions with the integrand $(x^2 + y^2)^n$, using a 3×3 Gauss–Hermite quadrature.
3. Same exercise in three dimensions with the integrand $(x^2 + y^2 + z^2)^n$, using a $3 \times 3 \times 3$ Gauss–Hermite quadrature.

16

LBE in the Framework of Computational-Fluid Dynamics

In this chapter, we shall outline the main properties of LB as a numerical scheme within the general framework of computational-fluid dynamics (CFD). The matter has witnessed significant developments in the last decade, and even though the bottomline picture of LB as a very effective numerical scheme stands intact, a number of assessments made in my previous book need some revision. Since the matter is fairly technical, only general notions shall be discussed leaving in-depth details to the original literature.

> *Before you make a calculation, make sure you know the result.*
>
> (J. Wheeler)

> *The purpose of computing is insight, not numbers.*
>
> (R. Hamming, L.N. Trefethen)

16.1 LBE and CFD

The earliest LBE's were brainchildren of statistical mechanics rather than numerical analysis: this is the "humus" from which they drew their main assets and liabilities as well. Indeed, the LB method took off the day it was realized that, besides a number of beautiful properties, statistical mechanics was also placing needless restrictions on lattice gas fluids as particle-based solvers of the hydrodynamic equations. The most poignant examples are the anomalies discussed in Part II, lack of Galilean invariance, spurious invariants, low Mach numbers and others. Most of these flaws could be lifted by cutting the umbilical cord with lattice gas cellular automata and regarding LBE as a self-standing kinetic model of the Navier–Stokes equations, the top-down approach described in the previous chapters. This "information removal" procedure takes LBE within the territory of numerical computing and specifically computational fluid dynamics.

The Lattice Boltzmann Equation. Sauro Succi, Oxford University Press (2018).
© Sauro Succi. DOI: 10.1093/oso/9780199592357.001.0001

To begin with, let us characterize the distinctive features of LBE in the language of numerical analysis (1).

The main properties of any numerical scheme can be classified as follows (the order corresponds to presentation convenience):

Causality, Accuracy, Stability, Consistency, Efficiency, Flexibility.[32]

With respect to these properties, LBE can be classified as a hyperbolic, *finite-compressibility*, superset of the Navier–Stokes equations, with the following highlights:

- *Causality*: fully local in space and time
- *Accuracy*: order in space and in time (the latter via a locally implicit time integration)
- *Stability*: conditional linear and nonlinear stability, pointwise conservativeness, to machine round-off
- *Consistency*: conditioned to low Mach numbers
- *Efficiency*: good efficiency on serial computers, outstanding one on parallel computers (no free-lunch though, hard work must be invested for complex geometries!)
- *Flexibility*: easy handling of grossly irregular boundary conditions. Limited suitability to non-uniform grids (in its original version) and adaptive timesteps. Handy inclusion of additional physics beyond fluid-dynamics

16.1.1 Causality

Numerical schemes for time-dependent equations fall within two broad categories depending on the way the time variable is discretized, namely:

- *Explicit*
- *Implicit*

In explicit schemes the current state of a given variable is computed in terms of a (usually small for the case of partial differential equations) number of neighbors at a preceding time. Thus, the current state is uniquely and explicitly specified in terms of the past state of the neighborhood. For a simple one-dimensional equation, a typical-explicit scheme reads as follows:

$$f(x_l, t_n) = \sum_{l-p \le m \le l+p, k < n} T_{lm,nk} f(x_m, t_k) \tag{16.1}$$

---

[32] Flexibility is not really a canonical category of numerical analysis! It reflects however the daily experience that simple schemes often tend to outlive more sophisticated ones, just because of better ease of use. This is not rigorous, yet important nonetheless to actual computational practice.

where the coefficients $T_{lm,kn}$ are the spacetime connectors between the actual spacetime location ($x_l = l\Delta x$, $t_n = n\Delta t$) and its neighborhood. Note that causality implies $t_n > t_k$, whereas purely spatial connections are generally two-sided, $l - p \leq m \leq l + p$, for a neighborhood of $2p + 1$ spatial sites, see Fig. 16.1.

In implicit schemes the current state of a given variable depends not only on the past but also on the *current* state of its neighborhood, the last term on the right-hand side of the expression

$$f(x_l, t_n) = \sum_{l-p\leq m\leq l+p, k<n} T_{lm, nk}f(x_m, t_k) + \sum_{l-p\leq m\leq l+p} T_{lm, nn}f(x_m, t_n) \tag{16.2}$$

Simultaneity introduces the non-physical notion of "action at distance" thus breaking the aforementioned causal link. The immediate operational effect is a *loss of locality*: in order to compute the actual value at (x_l, t_n), we need to solve the linear system (16.2) involving *all* space locations at time t_n. In fact, even if the spatial connection is local (that is, the matrix $T_{lm,nk}$ is sparse) with only a few non-zero entries around the diagonal, its inverse is generally full, which means that all space locations are simultaneously coupled. As a result to advance in time, we must solve a linear system at each timestep, which makes implicit schemes significantly more demanding on a per timestep basis. The reward is larger timesteps without breaking the stability thresholds (beware of accuracy though . . . , see Fig. 16.1).

Standard LBE belongs to the family of *explicit* time-marching schemes. To back up this statement, let us consider the differential form of the streaming operator:

$$\partial f \equiv (\partial_t + v_a \partial_a) f \tag{16.3}$$

and discretize it according to a first-order Lagrangian scheme along the characteristic $dx_a = v_a\, dt$, $a = x, y, z$.

The result is

$$\Delta_a f \equiv f(\vec{x} + v_a\, \Delta t, v_a, t + \Delta t) - f(\vec{x}, v_a, t) \tag{16.4}$$

Now specialize $v_a = c_{ia}$, $i = 1, \ldots, b$, to the set of discrete speeds and what we obtain is precisely the left-hand side of the Lattice Boltzmann equation with $\Delta t = 1$. The explicit

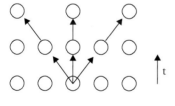

Figure 16.1 *Space–time diagram of an explicit method based on nearest-neighbor connections in one space dimension.*

nature of the discretization scheme, as well as its *apparently* first-order accuracy in time are clearly exposed.

However, on the streaming side, what appears to be a first-order time marching is in fact *exact*, because the discrete velocities are constant. On the other hand, the collision operator in the standard LB is also treated explicitly, namely

$$-\omega \int_t^{t+\Delta t} [f_i(x(t'), t') - f_i^{eq}(x(t'), t')] dt' = -\omega [f_i(x, t) - f_i^{eq}(x, t)] \Delta t \tag{16.5}$$

Since (16.5) is only first-order accurate, one concludes that the time accuracy is first order only. Since space and time are linked one to one, one would also conclude that space accuracy is also first order only. This is in principle true for each component f_i separately, but since hydrodynamic quantities are generated via weighted sums of all discrete populations, whereby each component is matched by a mirror partner moving along the opposite spatial direction, second order in space may result despite first-order accuracy in time.

Second-order accuracy in time can be regained, *without loosing the explicit LBGK structure*, by adopting a locally implicit method to be discussed later.

This amounts to working with the following time-advanced distribution function:

$$\bar{f} \equiv f + \frac{\omega \Delta t}{2}(f - f^e) \tag{16.6}$$

This brings LBGK back to second order in both space and time, as we shall detail in chapter 22.

It should be pointed out that the *kinetic* accuracy of LBGK, i.e., toward the continuum BGK, versus the *hydrodynamic* one, i.e., toward the weakly-compressible Navier–Stokes equations are two separate issues. The reader interested in the details of these matters is kindly directed to the thorough treatment by Junk *et al.*, as well as Ohwada and Asinari *et al.* (2; 3).

It should also be pointed out that the order of accuracy is by no means the whole story in practical applications: all simulations work at finite size, which means that pre-factors may matter a lot! LBE scores well in this respect, because streaming is exact and the collision operator, although only second order, is conservative to machine round-off.

This is ultimately the reason why LB, a "lowly" second-order scheme, can compete with higher-order accuracy schemes, including spectral methods.

16.1.2 Accuracy

Accuracy refers to the errors introduced by replacing differential operators with finite difference ones. If this error is zero for polynomials up to degree p, we say the discretization has p-th order accuracy.

For instance, a one-sided forward finite difference of the form

$$\Delta^> f \equiv f(x + \Delta x) - f(x)$$

reproduces exactly the derivative of a linear function, but not a quadratic (or higher) one, which means that accuracy is only first order.

It can easily be checked that a centered finite difference such as

$$\Delta^0 f \equiv \frac{f(x + \Delta x) - f(x - \Delta x)}{2}$$

is second-order accurate.

To discuss the spatial accuracy of LBE it is expedient to consider a parity-conjugate pair of discrete speeds $\vec{c}_i$ and $\vec{c}_{i*} = -\vec{c}_i$.

Now introduce the sum of the streaming operators $\Delta_i^+ \equiv \Delta_i + \Delta_{i*}$ associated with free motion along the i-th direction.

This reads

$$\Delta_i^+ f = \left(f(\vec{x} + \Delta \vec{r}_i) - 2f(\vec{x}) + f(\vec{x} - \Delta \vec{r}_i) \right) / 2$$

where we have set $\Delta \vec{r}_i = \vec{c}_i \Delta t$.

It is immediately checked that, upon dividing by Δr_i^2, this is a centered discretization of the second-order spatial derivative along the i-the direction, which is known to be second-order accurate.

Now consider the difference $\Delta_i^- \equiv \Delta_i - \Delta_{i*}$.

This reads as

$$\Delta_i^- f = \frac{f(\vec{x} + \Delta \vec{r}_i) - f(\vec{x} - \Delta \vec{r}_i)}{2}$$

which is again a second-order accurate discretization of the first-order directional derivative along the i-th direction.

Since it is precisely these sums and differences of streaming operators that ultimately enter the LBE discretization of the Navier–Stokes equations, the expressions tell a story of second-order accuracy.

On the other hand, the collision operator is handled by explicit time marching, which makes it only first-order accurate, unless one uses the locally implicit time marching to be discussed later.

Second-order spatial accuracy is an expected result, since the existence of a mirror conjugate for every discrete speed $\vec{c}_i$ implies a centered finite differencing scheme. Second-order accuracy applies to the bulk of the computational domain, and might eventually degrade to first order at the boundaries. For instance, the bounce-back procedure used to impose no-slip boundaries is apparently only first order because it breaks the space-centered nature of the streaming operator.

Ways to restore second-order accuracy such as mid-grid placement of the boundary shall be described in chapter 17 devoted to boundary conditions.

Before closing this section, we would like to mention that LBE methods are often empirically observed to provide higher quality results than second-order finite difference schemes. This is definitely the case for the turbulence computations discussed in chapter 20, where LBE is shown to yield results of quasi-spectral (asymptotically *exponential*) accuracy. This may be tentatively attributed to the *exact conservativeness* (to machine round off) of the LBE scheme: the streaming operator is exactly conservative (no floating-point operation), while the collision is conservative to machine round off. Such properties are likely to make the prefactor in front of the quadratic error very small indeed.[33]

16.2 Stability

Stability is a key property of any numerical scheme.

For explicit schemes, the basic notion is that the lattice is a discrete world which can only support signals with a finite-propagation speed. Once this is recognized, the necessary criterion for stability is simply that *physical information should not travel faster than the fastest speed supported by the lattice*. This is the physical content of the well-known Courant–Friedrichs–Lewy (CFL) conditions.

For a simple one-dimensional convective equation of the form

$$\partial_t f + u \partial_x f = 0, \tag{16.7}$$

the CFL condition reads

$$C_A = \frac{u \Delta t}{\Delta x} < 1 \tag{16.8}$$

where C_A is the advective Courant number of the scheme.

The physical interpretation of this inequality is that the physical speed U should not exceed the highest speed supported by the discrete grid $c_l \equiv \Delta x / \Delta t$, on pain of generating "tachyonic" instabilities, i.e., signals propagating "faster-than-light," c_l being the lattice-light speed.

A similar condition applies to the diffusion equation, namely

$$C_D = \frac{D \Delta t}{\Delta x^2} < 1/2 \tag{16.9}$$

stating that the physical diffusivity D cannot exceed the lattice one $D_l \equiv \Delta x^2 / \Delta t$.

In d spatial dimensions the CFL conditions become more restrictive, i.e.

$$C_A < 1/d, \quad C_D < 1/2\,d \tag{16.10}$$

It is worth noting that the diffusive CFL is much more constraining than the advective one; indeed the maximum diffusive timestep scales like the square of the lattice spacing, while the advective one scales only linearly, see Fig. 16.2.

---

[33] The author is grateful to Dr. Hudong Chen for bringing this point up, and to Dr. Irina Ginzburg and D. d'Humières for further elucidations.

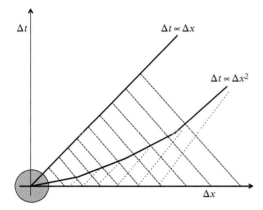

Figure 16.2 *CFL conditions for advection and diffusion processes. The latter sets a very stringent-quadratic constraint on the timestep. Note that in LB units C_A is the fluid-speed u_{lb} and C_D is the kinematic diffusivity v_{lb}.*

This means that doubling the resolution forces a four times smaller timestep. In lattice units, $\Delta x = \Delta t = 1$, the CFL conditions (16.8–9) read simply

$$u_{lb} < 1, \quad v_{lb} < 1/2 \tag{16.11}$$

where v refers to the momentum diffusivity and subscript *lb* denotes the values in LB units. Whenever advection and diffusion act simultaneously, the two constraints sum up together, leading to a more restrictive cumulative constraint.

Typical values in LB simulations are $u_{lb} \sim 0.1$ and $v_{lb} \sim 0.01$, indicating that the CFL condition, although not directly relevant to LB since diffusion is not described by second-order derivatives, is nevertheless usually well fulfilled.

The CFL condition alone is no guarantee of *practical* stability, since numerical debris due to imperfect conservations may build-up systematically and ruin the calculation. Before entering this discussion, which touches upon the thorny issue of nonlinear stability, let us address the sweeter cake first, linear stability.

Linear Stability

Let us consider the *linearized*-BGK equation recast in the following form:

$$\Delta_i f_i = -\Omega \Delta t \, (\delta_{ij} - L_{ij}) f_j \tag{16.12}$$

where L_{ij} is the matrix mapping the actual distribution f_i into the *linear* local equilibrium, $f_i^e = w_i \rho (1 + \frac{\vec{u} \cdot \vec{c}_i}{c_s^2})$. Here by Ω we denote the LBGK relaxation frequency leaving the symbol ω for the generic frequency in the Fourier analysis.

Little algebra shows that

$$L_{ij} = w_i \left(1 + \frac{\vec{c}_i \cdot \vec{c}_j}{c_s^2} \right) \tag{16.13}$$

A Fourier transform of the eqn (16.12) delivers the following dispersion relation:

$$Det\{(z_i - 1 + \Omega \Delta t)\delta_{ij} + \Omega \Delta t L_{ij}\} = 0 \tag{16.14}$$

where we have set

$$z_i = e^{j(\vec{k} \cdot \vec{c}_i - \omega)},$$

and j denotes the imaginary unit, to avoid confusion with the discrete velocity index i.

For the idealized case of propagation in the vacuum ($\rho = 0$), $f_i^e = 0$, hence $L_{ij} = 0$. Equation (16.14) then delivers a set of independent solutions in the form:

$$z_i = 1 - \Omega \Delta t \qquad (16.15)$$

Upon splitting the running frequency ω into real and imaginary parts, $\omega = \omega_r + i\gamma$, eqn (16.15) delivers two equalities:

$$\begin{cases} e^{-\gamma \Delta t} \cos \theta_i = 1 - \Omega \Delta t \\ e^{-\gamma \Delta t} \sin \theta_i = 0 \end{cases} \qquad (16.16)$$

where $\theta_i \equiv \vec{k} \cdot \vec{c}_i - \omega_r$ is the phase of the Fourier modes propagating along the i-th direction.

The latter equation delivers $\theta_i = 0$, namely $\omega_r = \vec{k} \cdot \vec{c}_i$, which is precisely the continuum-dispersion relation of a free wave propagating along the i-the direction. Note that such free-propagation speed is the same for all directions, as it should be.

This informs us that *the lattice free-streaming operator preserves spatial coherence*, as it does not introduce any distortion of the spatial profile (*dispersion*).

This nice property is strictly related to the light-cone discretization.

By squaring both eqns (16.16) and (16.16), and summing them up, we obtain

$$e^{-2\gamma \Delta t} = (1 - \Omega \Delta t)^2 \qquad (16.17)$$

This yields

$$\gamma = -\frac{log(1 - \Omega \Delta t)}{\Delta t} \sim \Omega + O(\Omega^2 \Delta t^2) \qquad (16.18)$$

Linear stability implies $\gamma < 0$, which in turns requires $|1 - \Omega \Delta t| < 1$, namely

$$0 < \Omega \Delta t < 2 \qquad (16.19)$$

Based on the expression of the fluid viscosity in lattice units:

$$\nu_{lb} = \left(\frac{1}{\Omega_{lb}} - \frac{1}{2} \right) \qquad (16.20)$$

where $\Omega_{lb} = \Omega \Delta t$, we recognize that (16.19) is precisely the condition for the fluid viscosity to be positive, apart from the factor $-1/2$ coming from second-order terms in the Taylor expansion of the streaming operator. This matches the intuitive notion of negative viscosity as a byword for physical instability.

Both upper and lower bounds are, so to say, "virtual."

The limit $\Omega \Delta t \rightarrow 0$ (*non-interacting gas, formally infinite viscosity*) breaks the enslaving assumption that begets the convergence of a hyperbolic equation like LBE to an advective-diffusive equation like the Navier–Stokes equation.

The other extreme, $\Omega \Delta t \rightarrow 2$, (*strongly interacting, zero viscosity*) cannot be reached either, because it gives rise to sub-grid scales shorter than the lattice spacing, which the scheme is not able to dissipate, thereby leading to ultraviolet instabilities.

However, as we shall see, it can be approached *very* closely, a property which is key to the simulation of high-Reynolds flows.

Note in fact that by choosing $\Omega_{lb} = 2 - \epsilon$, with $\epsilon \ll 1$, the viscosity is given by $\nu_{lb} = c_s^2 \epsilon/(4 - 2\epsilon) \sim c_s^2 \epsilon/4$, while the timestep remains $\Delta t = 1$, all in lattice units.

Thus, *LB provides the opportunity of simulating high-Reynolds flows without collapsing the timestep*. Realizing this opportunity in actual practice, however depends on the nonlinear stability of the scheme.

Before delving into these issues, it is important to remark that this simplified stability analysis is confined to the idealized condition of free propagation *in-vacuo*.

Once a non-zero background flow is considered, and even more so, once such flow is allowed to change in space and time, as it is certainly the case for practical flows, the detailed analysis presented in (4) shows that even linear stability may go in jeopardy at sufficiently high speed and/or wavenumbers.

Nonlinear stability

An exact theoretical analysis of nonlinear stability of LBGK schemes is generally unviable, for it would amount to solving the LBE itself!

However, a number of general guiding criteria can still be formulated, which prove fairly useful in assessing the nonlinear stability of LBE scheme (5).

One of these is the concept of *Conservativeness*, namely the ability of a numerical scheme to ensure that physically conserved quantities remain *exactly* conserved in the lattice.

The actual value of the conserved quantity might not be exactly the same as in continuum, but as long as it does not change in time, one is at least guaranteed that the numerical trajectory remains adhered to an invariant (hyper)surface in phase space and consequently it cannot run astray. A property which is very important for very long simulations.

LGCA are ideally placed to guarantee this (remember the excitement of "exact computing" in the early LGCA days). Less so for LBE, because in any floating-point computation, conservativeness can only be guaranteed up to machine round-off, at best. Still, as compared to many floating-point techniques, LBE offers some valuable assets.

The streaming operator is perfectly conservative, since it implies a plain transfer from one site to another of the same distribution function, *no floating-point* involved: still Boolean! The collision operator is also conservative, to machine round-off owing to conservation laws explicitly encoded in the scattering matrix or in the local equilibria, or both. This makes LBE an "exactly" conserving numerical scheme.

This does not automatically protect against numerical blow-ups, but it surely benefits the numerical robustness of the method.

This takes us to a more detailed discussion of nonlinear stability and its connections with consistency requirements.[34]

16.2.1   Consistency

As we shall see in chapter 20, devoted to turbulent flows, the nonlinear regime of the Navier–Stokes equations implies energy transfer from large to small scales, hence the excitation of short-scale motion in the flow. Such short scales are the most exposed to numerical artifacts due to the lattice discreteness. In particular, if dissipation fails to compensate the energy flux from large scales, small scales can grow unstable.

A nonlinearly stable numerical scheme should be able to tame such instabilities, without corrupting the basic physics of large-scale motion and possibly short-scale as well. Not an easy task.

A typical manifestation of short-scale under-representation is the so-called Gibbs phenomenon; that is, spurious oscillations (7) that threaten the *positivity* of the distribution function:

$$f \geq 0 \tag{16.21}$$

This consistency requirement is sometimes called a *realizability* constraint, in the sense that negative values are unphysical, i.e., not realizable in the physical world, as we know it. (For the sake of completeness, the full-realizability constraint should read as $0 \leq f \leq 1$, but the troublemaker is typically the lower end of the bound.)

Realizability is neither a sufficient nor a necessary condition for nonlinear stability, but it is generally agreed that negative f switch a red lamp on the computation even in the absence of numerical blow-up.

To analyze the realizability condition in the nonlinear regime, let us recast LBGK in the form of a pseudo-ordinary differential equation as follows:

$$\partial_t f = -\omega f + S \tag{16.22}$$

where the non-equilibrium source S is given by

$$S = \omega f^e - v_a \partial_a f \tag{16.23}$$

---

[34] Before doing so, let us take an exception. When we talk about LBE, we refer to the quasilinear versions with prescribed local equilibria, not the full nonlinear McNamara–Zanetti version. This latter has an *H*-theorem securing positivity even in the nonlinear regime and, consequently, most likely also a better stability. Unfortunately, as we have seen, it is not practical for three-dimensional high-Reynolds computations and therefore it will not be discussed any further.

If S is slow compared to f, the formal solution of eqn (16.22) reads

$$f = f_0 e^{-\omega t} + \left(1 - e^{-\omega t}\right) \frac{S}{\omega} \tag{16.24}$$

If the equivalent system is linearly stable ($\omega > 0$ in the continuum, $0 < \omega < 2$ in the lattice), this solution relaxes to a gradient-corrected equilibrium

$$\hat{f}^e = (1 - \tau v_a \partial_a) f^e \tag{16.25}$$

in a time-lapse $\tau = 1/\omega$ [35].

Let us now consider smooth equilibria, such that $|\tau v_a \partial_a f^e| \ll |f^e|$ (small Knudsen number, $Kn \ll 1$). For these smooth equilibria, the positivity constraint translates into an equivalent, but much more controllable, positivity constraint on the local equilibrium:

$$f^e > 0 \tag{16.26}$$

This requirement is met by construction in continuum kinetic theory, but in the discrete world it cannot be taken for granted. For low Mach, non-thermal flows, the condition (16.26) is met for sufficiently subsonic speeds, and the situation is fine.

However, discrete-lattice equilibria are *not* positively definite for all values of the flow speed. For the sake of concreteness, let us refer to the simple D1Q3 BGK equilibria:

$$f_1^e = \frac{\rho}{6} \left(1 + 3u + 3u^2\right) \tag{16.27}$$

$$f_2^e = \frac{\rho}{6} \left(1 - 3u + 3u^2\right) \tag{16.28}$$

$$f_0^e = \frac{4\rho}{6} \left(1 - \frac{3}{2} u^2\right) \tag{16.29}$$

where all speeds are measured in units of the sound-speed $c_s = 1/\sqrt{3}$.

Consider a region of the flow where $u > 0$. The "most endangered" species is f_0, which goes negative for $u > \sqrt{2/3}$ (see Fig. 16.3). This is significantly above the weakly compressible regime and therefore it does not constitute a serious problem.

Consider now the left moving distribution f_2: the local equilibrium attains a minimum value

$$f_2^e = \frac{\rho}{6} 1/4 = \frac{\rho}{24} \tag{16.30}$$

at $u = 1/2$.

---

[35] Based on the identity $e^{v_a \tau \partial_a} f(x_a) = f(x_a + v_a \tau)$, we see that the renormalized equilibrium is basically a spatial shift of the original one, i.e., $\hat{f}^e(x_a) \sim f^e(x_a - v_a \tau)$.

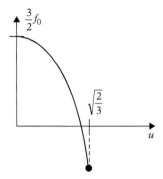

Figure 16.3 *Endangered species going negative under the combined effect of high local-flow speed and flow gradients.*

Based on eq. (16.25), the positivity constraint for the gradient-corrected equilibrium reads as:

$$3|\tau(\partial_x u + 2u\partial_x u)| \le 1/4 \qquad (16.31)$$

This shows that once velocity gradients are sufficiently strong to violate such constraint, instabilities would result, even though the bare equilibrium f_2^e remains positive.

This shows that high local speeds (high Mach number) transport) *and* high local gradients (high local-Knudsen number) conspire to threaten realizability.

Even though it has been recognized by several authors that realizability is neither necessary nor sufficient for numerical stability, it still remains a highly desirable feature to preserve.

For this purpose, several empirical recipes have been devised, a popular one being the so-called *fix-up*, which consists in altering the local-relaxation frequency $\omega \to \omega^*$ in such a way to force the post-collisional distribution to be non-negative, namely

$$f' = f - \omega^*(f - f^e) \ge 0 \qquad (16.32)$$

This yields

$$\omega^* \le |\frac{f}{f - f^e}| = 1 + |\frac{f^e}{f^{ne}}| \qquad (16.33)$$

This expression shows that for weak non-equilibrium, $|f^{ne}| < |f^e|$, the linear bound $\omega < 2$ is sufficient to secure nonlinear stability as well.

It should be noticed that fixup is virtually guaranteed to violate the conservation laws, but hopefully just by tiny tolerable amounts.

The opposite situation of strong non-equilibrium, $|f^{ne}| > |f^e|$, however, implies $\omega^* < 2$. This is where nonlinear stability becomes more stringent than the linear one and new criteria are needed. Positivity is always a physically sound option. However, as we shall see, more powerful and systematic criteria can be derived by imposing local compliance with the second principle, i.e., by enforcing the local H-theorem. The corresponding LB variant, known as entropic-Lattice Boltzmann (ELB), shall be discussed later.

16.2.2 Efficiency

By mere counting of degrees of freedom, LBE is by definition *more* than a Navier–Stokes solver. Indeed, together with density, flow and pressure, it also tracks the momentum-flux tensor and a bag of additional fields which we called ghosts. This makes LBE memory and CPU intensive as compared to most explicit methods.

The extra information brings however several significant rewards.

First, Poisson-freedom, i.e., no need to solve the Poisson equation for the pressure field since the method is weakly compressible. The price is small timesteps to track sound waves, but by and large, this proves to be a good bargain, especially with parallel computers in mind.

Second, the ghost fields can eventually put at service of numerical stability, for instance via the so-called multi-relaxation version of LBE.

Third, no second-order derivatives are required to describe diffusion: diffusion emerges from enslaving of the momentum-flux tensor to local equilibrium.

This is an important property, since it circumvents the diffusive CFL constraint eqn (16.9), the most restrictive one in standard explicit finite difference methods.

The previous properties make of LBE an attractive-explicit finite-difference solver.

Here, some words of cautions are in order.

As shown in (2), the correct scaling to assess systematic convergence of LBE to the *incompressible* Navier–Stokes equations is not the convective one:

$$\Delta x = c\Delta t = O(\epsilon) \tag{16.34}$$

but rather the diffusive one:

$$\Delta x^2 = \nu \Delta t = O(\epsilon^2) \tag{16.35}$$

In other words, under convective scaling, and the ensuing Chapman–Enskog analysis, one must accept a non-zero *finite-compressibility* error even at the finest resolutions. This is apparent by casting the convective Courant number in LB units:

$$C_A \equiv \frac{u\Delta t}{\Delta x} = u_{lb} \tag{16.36}$$

where subscript "lb" stands for LB units, i.e., $\Delta x_{lb} = \Delta t_{lb} = 1$.

Given the expression of the Reynolds number in LB units,

$$Re = \frac{u_{lb}N}{\nu_{lb}} \tag{16.37}$$

it is clear that by doubling the resolution at a given Reynolds number in the *convective scaling* scenario implies the following protocol:

$$N \to 2N, \ \nu_{lb} \to 2\nu_{lb}, \ u_{lb} \to u_{lb} \tag{16.38}$$

From this, it is clearly seen that, even in the continuum limit $N \to \infty$, u_{lb} stays unchanged and so do the compressibility errors.

On the other hand, in the *diffusive scaling* protocol

$$N \to 2N, \; v_{lb} \to v_{lb}, \; u_{lb} \to u_{lb}/2 \tag{16.39}$$

so that compressibility errors decay quadratically at increasing grid resolution.

It is therefore clear that, although more costly (the timespan of the simulation scales inversely with u_{lb}), the proper framework to perform convergence studies to the hydrodynamic limit is the diffusive one.

Since in the authors words "LBE is intended to solve the *incompressible* Navier–Stokes equations ...," phenomena occurring on the short timescale, typically sound waves are "mere numerically artifacts."

Beware though: while formally impeccable, such a statement may often turn out to be over-restrictive from the practical standpoint.

For one, while it is *formally* true that LBE was intended to solve the incompressible Navier–Stokes equations, it is *practically* no less true that small compressibility errors are often well worth being tolerated in exchange for a less restrictive constraint on the timestep.

A simple example illustrates the point. Given that compressibility errors usually scale like u_{lb}^2, a convergence study in convective scaling with $u_{lb} = 0.01$, would be unaffected by such errors until the error tolerance reaches down to 10^{-4}. Although unsuitable for benchmark purposes, this is a tolerance one may well live with, in applications affected by many other sources of uncertainty.

It is often the case in physics that (small) errors are merrily tolerated as long as they come along with substantial computational gains (these are sometimes called *soft constraints*).

The previous statements regarding "numerical artifacts" assumes that zero is the only answer one should take ("infinitely hard constraint").

In this author's opinion, there are little apologies to be offered for deciding to live with an efficient "soft-constrained" system, the weakly compressible Navier–Stokes equations, rather than a much more demanding "hard-constrained" one, the strictly incompressible Navier–Stokes equations.

Of course, this should not be taken as a cheap excuse to tolerate errors in our numerics.

Nevertheless, the worst danger are not errors per-se, but tolerating them without knowing that they exist! This is why rigorous warnings are very healthy and helpful fence-setters. The safety distance from the fence, however, is problem-dependent, hence ultimately the user's choice and responsibility.

16.2.3 Parallel Computing

Lattice Boltzmann is valued for its excellent amenability to parallel computing. In the following section, we expand the point in some detail.

As described in the corresponding Appendix, an efficient parallel scheme should feature the following highlights:

- *High fraction of parallel content (Amdahl's law)*
- *Low Communication to Computation Ratio (CCR)*
- *Even distribution of the computational workload across computational spacetime (load balance)*

It is readily checked that LBE scores well on all three items.

A high fraction of parallel content is guaranteed by the local nature of the method: most of the CPU time of typical LBE applications is taken by the collision step, literally a communication-free process with ideal parallel efficiency. This is a precious inheritance of LGCA.

Low CCR is secured again by the dominance of collision versus streaming steps of the algorithm, typical value being CCR $\sim O(10)$. A word of caution is in order, though: LBE is computationally slightly more intensive than standard explicit finite difference method, because there are more populations than hydrodynamic fields to keep track of. Typical CCRs for a plain explicit finite difference code is $O(1)$, which explains why these algorithms do not always easily achieve very high parallel efficiency.

Of course, there is little point in heralding a high parallel efficiency obtained by raising the computational load instead of reducing the communication costs!

The fact remains that modern computer architectures place a high reward on the policy of "*data reusability*," namely once data has been read-off from memory to the local buffer (high-speed buffer, or simply "cache" in computer jargon) performing two floating-point operations on such data costs less than twice a single operation, just because data transfer from/to memory to/from CPU is an expensive operation.

To be sure, this is rapidly becoming *the* limiting factor in modern computing, LB is no exception to the rule.

Bottom line: the computational overload of LBE does not really reflect into a corresponding toll on execution time as long as data traffic can be kept at a minimum, so as to comply with the reusability principle. This turns out to be a highly non-trivial task, especially in large scale applications. In this respect, parallel computers provide extra help, for they reduce the amount of data each single processor has to deal with. Even so, the issue of data traffic from CPU to memory and back is possibly the most crucial one in the design of efficient LB algorithms on large-scale parallel machines including recent Graphic Processor Units (GPU).

Finally, optimal-load balance is secured by the uniform nature of the computation: the same operation repeated over and over again on all sites. This is a property shared by all CFD schemes working with regular data structures. Of course, not all applications conform to such single-instruction-multiple-data (SIMD) paradigm; in particular complex geometries often require more sophisticated data structures, and again, LB is no exception to the rule.

Once these issues are properly addressed, which takes a great deal of programming ingenuity and hard work, LB invariably pays off with excellent parallel performance across virtually any parallel architecture. Parallel efficiency, as combined with user-friendliness is one of the major assets of LBE.

16.2.4   Flexibility

Having assessed that LBE corresponds to a second-order integration along a set of constant characteristic speeds, we can proceed to examine further consequences of this peculiar feature. The main point is that space and time are *not* treated independently, but they are coupled instead via the light-cone condition:

$$d\vec{x} - \vec{c}_i dt = 0 \qquad (16.40)$$

This light-cone discretization reflects the covariant, hyperbolic nature of LBE, in which space and time are treated on the same footing. This confers to LBE its character of *finite-hyperbolicity* approximation (recall the enslaving approximation of Part I). Within this picture, the molecules are *grid-bound*, i.e., they move in lock-step mode from one lattice site to another, never to stop in between: they cannot live other than on lattice sites.

This property, a strict legacy of LGCA, yields important formal and practical advantages. For instance, the fact that the characteristics are constant trivializes the Riemann problem. It also makes the scheme very simple in terms of data structure organization, as it permits the discretization of the streaming operator without any floating-point operation. By necessity, however, it implies limitations as well, most notably the inability to apply plain LBE to the more general case of non-uniform and body-fitted grids.

Researchers in the field have come up with several recipes to do away with this constraint, mostly based on mergers of LB with finite difference and finite volume methods.

A much heralded virtue of LB methods is the ability to accommodate grossly irregular boundary conditions, such as those encountered in flows through porous media. This emphasis is certainly well deserved, for it is true that accounting for whatever irregular distribution of solid obstacles within the flow is *conceptually* straightforward. Credit for this goes directly to the dual "fluid–particle" nature of the LB scheme: move "swarms" of particles along simple particle trajectories.

Slightly underemphasized, perhaps is the fact that the actual implementation of these boundary conditions may be somewhat painful, due to the relatively high number of populations, often demanding a separate treatment for different propagation directions (outward/inward movers, corners being the typical headache of the hasty programmer).

With this practical caveat in mind, the fact remains that flows through porous media, and generalizations thereof, prove one of the most successful applications of LBE methods. A few comparative studies report that LBE superiority versus finite element methods surfaces more and more neatly as the "randomness" of the media is increased (8).

Physical flexibility and user-friendliness

Ease of use and a gentle learning curve is another major practical asset of LB.

This is again due to the simple stream-collide paradigm, on a uniform-spacetime grid.

Explicit finite-difference methods are "vanilla CFD," and as such, they were (and possibly still are) regarded as just too expensive; too many timesteps, too many grid points, to compete with well rugged implicit solvers, moving in large timesteps.

The dramatic growth of computing power and the attendant drop of computing costs is changing all that: quoting J. Boris (9),

> "we are reaching the point of diminishing returns in terms of trading off computational cost for accuracy."

And Boris goes further,

> "It is now more effective to increase the number of grid points to improve spatial resolution and hence accuracy than to seek greater accuracy through high-order algorithms."

Bottomline: low order, fast-explicit solvers are often a very appealing choice, especially with parallel computers at hand.

Besides computational efficiency, ease of use brings other important bonuses to the table. A major one is the ability to embody beyond hydro-physics via external-effective potentials and/or generalized equilibria. We have already seen that the whole nonlinear physics of the Navier–Stokes equations translates into the few more lines required to code quadratic terms in the local equilibrium!

The good news is that the same idea extends to generalized hydrodynamics: *substantial new physics can be added within a few lines of extra code!*

Whenever additional meso/macroscopic physics can be molded into an effective force of the form $F(x, t; M(x, t), \nabla M)$, where M denotes any macro/mesoscopic field (typically density), this physics can also be encoded within a few lines into the very same LBE harness in charge of hydrodynamics, i.e., the local equilibria. The power of discrete kinetic theory in action! This is the leading theme of the "Beyond" part of this book, to be described.

16.3   Link to Fully Lagrangian Schemes

As already mentioned, there are mixed opinions as to whether the physical roots of LB methods should be regarded as an asset rather than a liability. Work on the former line has been mentioned, when it was shown that LB methods can be viewed as specific instances of discretization schemes for the continuum BGK equation using Hermite expansions and numerical Gaussian quadrature for the actual evaluation of moments of the distribution function. The conclusion drawn by the authors was that "by realizing that LBE models are merely simple and rather primitive finite difference representations of the discrete BGK equation, we can employ more sophisticated numerical techniques in solving these equations with better efficiency, stability and flexibility."

In a early and still remarkable paper, Ancona advocated a rather opposite point of view (10), namely that,

"the lack of separation of physics and numerics by conventional kinetic theory approach is a major drawback when the purpose is solving partial differential equations."

To substantiate this statement, the author presents an in-depth analysis of LBE schemes and points out their relation to a wider class of so-called Fully Lagrangian (FL) methods for partial differential equations. He shows that LB methods are a subclass of FL methods characterized by the following distinctive properties:

- *Pointwise conservation of physical quantities*
- *Perturbative expansion of the Lagrangian variables*
- *Use of discretization errors to represent physical effects*

It is instructive to analyze these points in a little more detail. First let us clarify what we mean by fully Lagrangian methods. The basic idea is to transform Eulerian fields, such as density, flow, and pressure as a function of x and t into a set of "moving fields," the Lagrangian variables in point.

Of course, the way to achieve this is by no means unique. For instance, in the simple case of a 1D fluid, density ρ and speed u would be turned into a pair of right- and left-propagating fields $R(x, t)$, $L(x, t)$ obeying the following relations:

$$\rho = \frac{R + L}{2} \tag{16.41}$$

$$\rho u = c \frac{R - L}{2} \tag{16.42}$$

where c is the magnitude of the common speed of the Lagrangian movers ("particles"). By replacing (16.41–42) into the continuity and Navier–Stokes equations in 1D, one readily obtains the evolution equations for the Lagrangian movers:

$$\partial_+ R = S \tag{16.43}$$

$$\partial_- L = -S \tag{16.44}$$

where $\partial_\pm \equiv \partial_t \pm c\partial_x$ are the left/right Lagrangian derivatives and

$$S = 2c\partial_x \frac{RL}{R + L} + \frac{\partial_x P}{2c} \tag{16.45}$$

is the source term, mediating the interaction between the two Lagrangian fields. Finally, $P = -p + 2\mu\partial_x u$ is the one-dimensional stress tensor.

It is immediately seen that the Lagrangian formulation discloses a beautiful duality which was left hidden in the Eulerian version. The transformation

$$c \to -c, \ (R, L) \to (L, R) \tag{16.46}$$

leaves the eqns (16.43) and (16.44) invariant (in the limit $\mu \to 0$). It is not hard to see that LB falls within this kind of formulation. For this purpose all we need is to march in time with a first-order Euler-forward scheme and discretize space with a so-called *upwind* technique, namely $\partial_x R = R(x + \Delta x) - R(x)$ and $\partial_x L = L(x - \Delta x) - L(x)$.

This shows that LB belongs to the class of FL solvers.

In particular coming to the first point of the list, it is readily shown that not all FL formulations lead to *pointwise* conservation of mass and momentum, on a node-by-node basis. LBE does, by construction, which surely marks a point in its favor.

The second point, perturbative treatment of the Lagrangian variables is also easily illustrated. The source term S is a nonlinear (quadratic, since we only have two fields) function of the Lagrangian fields. How to discretize this term is again a matter of choice, each method spawning its own numerical scheme. There is a clear link to hyperbolic methods with sources (11) and kinetic-flux splitting methods in gas dynamics (14). The distinctive LBE feature is that these nonlinear source terms are expanded *perturbatively* around local equilibria, thus making the scheme fully explicit.

This is a physically inspired move, which buys out a major computational simplification. Ancona points out that the replacement of the nonlinear-source S with the BGK relaxation term $\omega(R - R^e)$ with a constant ω makes the method only *conditionally consistent*. Recovery of the original continuum PDEs limit is no longer guaranteed for any generic value of ρ and u, but becomes conditioned to the small-Mach number assumption. This is basically the positivity/realizability issue discussed previously in this chapter.

Helas, a weakness of LBE.

The final point using discretization errors to model physical effects refers to the idea of retaining second-order terms in the Taylor expansion of the discrete free-streaming operator in powers of the differential one $\partial_i \equiv \partial_t + c_{ia}\partial_a$: $\Delta_i \sim \partial_i + \partial_i^2/2$.

As we learned earlier on, this leads to the so-called propagation viscosity, $\nu_p = -c_s^2/2$ (in lattice units) which adds to the collisional one to produce the effective viscosity $\nu = \nu_c - \nu_p$. This is no coincidence and responds to the physical arguments behind the Chapman–Enskog procedure which takes LBE into Navier–Stokes.

Always in the spirit of locating LBE within the framework of numerical analysis, Ancona shows that the LBE discretization of the 1D Navier–Stokes equation, better known as Burger's equation is *exactly* equivalent to the Dufort–Frankel scheme.[36] Indeed, the

---

[36] For a 1D diffusion equation $\partial_t f - D\partial_x^2 f = 0$, the Dufort–Frankel scheme reads as follows:

$$f(i, n + 1) - f(i, n - 1) = \frac{2D\Delta t}{\Delta x^2} \left[f(i + 1, n) - (f(i, n - 1) + f(i, n + 1)) + f(i - 1, n) \right]$$

Dufort–Frankel (DF) scheme shares many general features with LBE, explicitness, second-order accuracy in space, unconditional stability (for diffusion equations) and conditional consistency.

On the other hand, the author himself recognizes that this identity between LBE and DF does not apply generically, which means that LBE is not just a matter of "reinventing the wheel," mostly on account of the way isotropy is achieved in higher dimension (see also the remark on separability versus isotropy in Chapter 15).

In fact, very informative work from the numerical analysis/applied mathematics community, has highlighted that LBE is a kind of *multi-difference* method in which propagation along different directions (nearest-neighbor, next-to-nearest neighbor) are subject to a distinct type of finite differencing (15). It is probably this multi-differential nature which makes of LBE more than yet another finite-difference scheme!

Both Ancona's and He–Luo's analyses show that, while it is true that LBE can be derived by combining ideas drawn from other numerical methods (how could it possibly be otherwise?), it is also true that this combination does not follow plainly from any single other existing method.

On the other hand, heralding the physics, as opposed to numerics, oriented character of LBE as a source of superiority may be no less misleading.

The most obvious danger are "computer hallucinations" or, in a more professional wording, "uncontrolled approximations," i.e., intuitive procedures which offer no guarantee of error decay at increasing resolution. As discussed earlier warning against uncontrolled approximations should be taken seriously, but not as diktat: we all work at finite resolution, and at finite resolution awareness of errors is often more important than (asymptotic) absence of errors altogether. A critical, yet balanced, view is probably the healthiest attitude toward a fair appreciation of the real power and limits of the LB method.

16.4   Summary

Summarizing, the LBE is a weakly compressible, hyperbolic superset of the Navier–Stokes equations, with the following highlights:

- Nonlinearity and non-locality are disentangled
- Transport is linear and *exact* (exact streaming)
- Nonlinearity is local (local equilibrium) and conservative to machine roundoff
- Pressure is local in space and time (finite-compressibility)
- Diffusion emerges from relaxation, no Laplace operators
- Easy handling of grossly irregular-boundary conditions (information travels along straight lines)
- Excellent amenability to parallel computing
- Excellent flexibility toward inclusion of beyond-hydro physics

From a strict numerical point of view, LBE is second order in space and time, conditionally stable in both linear and nonlinear regimes. Thanks to the negative propagation viscosity it can attain very low viscosities, say $\nu \sim O(\epsilon)$, with timestep of order $O(1)$. This property is key to simulate high-Reynolds flows.

On the dark side of the LB moon:

- The (standard) scheme lives in a uniform spacetime lattice: for many engineering applications in complex geometries, this may still be perceived as a limitation. Ways to overcome this limitation shall be discussed later

- The LB scheme is natural-born dynamic: if steady state is the only target, implicit CFD solvers may be more advantageous

- Thermal flows; incorporating energy exchanges within the LB formalism requires larger sets of discrete velocities affording higher-order isotropy. This leads to major implications on numerical stability, an item that will also be covered later in the book

Twenty-five years down the line, it is fair to say that, by and large, the upsides of LB largely overcome the downsides. Yet, much room for improvement still remains.

· ·

REFERENCES

1. B. Elton, D. Levermore and G. Rodrigue, "Convergence of convective–diffusive lattice Boltzmann methods", *SIAM J. Num. Anal.*, 32(5), 1327, 1995.
2. M. Junk, A. Klar and L.S. Luo, "Asymptotic analysis of the lattice Boltzmann equation", *J. Comp. Phys.*, 210, 676, 2005.
3. T. Ohwada and P. Asinari, "Artificial compressibility method revisited: Asymptotic numerical method for incompressible Navier-Stokes equations", *J. Comp. Phys.*, 229, 1698, 2010
4. L. S. Luo and P. Lallemand, "Theory of the lattice Boltzmann method: dispersion, dissipation, isotropy, Galilean invariance and stability", *Phys. Rev. E,* 61(6), 6546, 2000.
5. J. Sterling and S. Chen, "Stability analysis of lattice Boltzmann methods", *J. Comp. Phys.* 123, 196, 1996.
6. P. Pavlo, G. Vahala, L. Vahala and M. Soe, "Linear stability analysis of thermo-lattice Boltzmann models", *J. Comp. Phys.*, 139, 79, 1998.
7. D. Potter, *Computational Physics*, Wiley, London, 1977.
8. D. Kandhai, D. Vidal, A. Hoekstra, H. Hoefsloot, P. Iedema and P. Sloot, "A comparison between lattice Boltzmann and finite element simulations of fluid flow in static mixer reactors", *Int. J. Mod. Phys. C,* 9(8), 1123, 1998.

9. J. Boris, "New directions in computational fluid dynamics", *Ann. Rev. Fluid Mech.*, **21**, 695, 1987.

10. M. Ancona, "Fully Lagrangian and lattice Boltzmann methods for solving systems of conservation equations", *J. Comp. Phys.*, **115**, 107, 1994.

11. F. Coquel and B. Perthame, "Relaxation of energy and approximate Riemann solvers for general pressure laws in fluid dynamic equations", *SIAM J. Num. Anal.*, **35**(6), 2223, 1998.

12. E. Godlewski and P. Raviart, *Numerical Approximation of Hyperbolic Systems of Conservation Laws*, Springer–Verlag, Berlin, 1996.

13. D. Aregba-Driollet and R. Natalini, "Discrete kinetic schemes for multidimensional systems of conservation laws", *SIAM J. Num. Anal.*, **37**(6), 1973, 2000.

14. K. Xu, "A new class of gas–kinetic relaxation schemes for the compressible Euler equations", *J. Stat. Phys.*, **81**(1/2), 147, 1995;

15. M. Junk, "Discretization for the incompressible Navier–Stokes equations based on the lattice Boltzmann method", *SIAM J. Sci. Comp.*, **22**(1), 1, 2000.

EXERCISES

1. Find the CFL condition for your preferred finite-difference 1D diffusion equation.
2. Write the finite-difference scheme in relaxation form.
3. Show that compressibility effects scale like the square of the Mach number.

Part III

Fluid Dynamics Applications

Li Wenzi always thinks three times before he acts. Twice is enough.
(Confucius)

Having covered the basic theoretical ground of the LB method, we now move on to the discussion of some selected fluid dynamic applications, namely:

1. Flows at moderate Reynolds number
2. Low Reynolds flows in porous media
3. Turbulent flows

This list is far from exhaustive of the wide spectrum of LBE applications developed to date, but only responds to criteria of simplicity and space economy.

By now, the portfolio of fluid dynamic LB applications has grown exponentially (a glance at the Science Citation Index shows that this is—literally—the case). As a result, any attempt to provide a comprehensive overview of the current state of affairs within reasonable space limits is beyond the knowledge and energy of the present author.

For a comprehensive view of the various LB applications, the web can certainly offer valuable help.

Since no application can be discussed without specification of boundary conditions, we begin this part with a chapter devoted to this subject. Indeed, boundary conditions deserve a word apart, as they have made the object of intense developments in the last decade. By their very nature, these developments tend to be highly technical and detail-driven, which makes them hardly amenable to pedagogical purposes. As a result, an account of (some of) these developments is provided in connection with the applications they refer to, leaving the general description not far from where it was in the previous book.

The elementary warm-up computer program formally available on my website is now explicitly provided in the final Appendix, along with a few explanatory comments on the various routines. It is this author's teaching experience that reading through this little program (in old but still good Fortran) and then writing their own in their language of choice proves extremely beneficial to absorb the essence of the method, and walk out with a working knowledge of the matter. For a much more exhaustive illustration, see the website and particularly the open-source Lattice Boltzmann software Palabos. (www.palabos.org).

17

Boundary Conditions

The actual dynamics of fluid flows is highly dependent on the surrounding environment, whose influence is mathematically described through the prescription of suitable boundary conditions. Boundary conditions play a crucial role, as they select solutions which are compatible with external constraints. Accounting for these constraints may be comparatively simple for idealized geometries but for general ones it represents a delicate (and sometimes nerve-probing!) task. In fact, the treatment of the boundary conditions often makes the difference in the quality of fluid dynamic simulations. In this chapter we illustrate the most common ways to impose boundary conditions to LB flows. The subject is very technical and has grown considerably for the last decade, which means that this chapter can only serve as a guiding introduction to the vast and still growing original literature.

The only place where success comes before work, is the dictionary

(V. Lombardi)

17.1 Initial Conditions

We begin with boundary conditions in time, also known as Initial Conditions.

Initial conditions may significantly affect the transient dynamics of complex flows, i.e., the way such flows attain their steady state, if any. A very popular practice in LB simulations is to initialize the discrete populations to the local equilibrium corresponding to a prescribed value of the hydrodynamic fields, i.e.,

$$f_i(\vec{x}; t = 0) = f_i^e[\rho_0(\vec{x}), \vec{u}_0(\vec{x})] \tag{17.1}$$

where

$$\rho_0(\vec{x}) \equiv \rho(\vec{x}; t = 0), \quad \vec{u}_0(\vec{x}) \equiv \vec{u}(\vec{x}; t = 0) \tag{17.2}$$

The Lattice Boltzmann Equation. Sauro Succi, Oxford University Press (2018).
© Sauro Succi. DOI: 10.1093/oso/9780199592357.001.0001

are the chosen values of the initial density and velocity. Most often, the density is fixed to a constant value throughout, while the velocity field may be chosen as close as possible to its steady state, or a best guess thereof, in order to minimize the transient time. For high viscous flows, departures from the equilibrium initial conditions decay rapidly to zero. However, as the viscosity is decreased, transients become considerably longer, and departures from the expression (17.1) not compatible with the quasi-incompressibility of the fluid may disrupt the simulation. A number of improvements, taking into account density and flow inhomogeneities, are discussed in the useful book by Guo and Shu (1) and Krueger et al. (2).

17.2   General Formulation of LBE Boundary Conditions

Consider a fluid flowing in a bounded domain Ω confined by a surrounding boundary $\partial\Omega$. Generally speaking, the problem of formulating boundary conditions within the LBE formalism consists in finding an appropriate relation expressing the unknown *incoming* (wall to fluid) populations $f_i^<$ as a function of the known *outgoing* (fluid to wall) ones $f_i^>$.

Outgoing populations at a boundary-site $\vec{x}$ are defined by the condition

$$\vec{c}_i \cdot \vec{n}(\vec{x}) > 0 \qquad (17.3)$$

where $\vec{n}$ is the outward normal at the boundary location $\vec{x}$, where outward means from the fluid to the boundary.

Incoming populations are defined by the opposite sign of the inequality. In mathematical terms, this relationship translates into a linear-integral equation

$$f_i^<(\vec{x}) = \sum_y \sum_j B_{ij}(\vec{x} - \vec{y}) f_j^>(\vec{y}) \qquad (17.4)$$

where the kernel $B_{ij}(\vec{x} - \vec{y})$ of the boundary operator generally extends over a finite range of values $\vec{y}$ inside the fluid domain. This boundary operator reflects the interaction between the fluid molecules and the molecules sitting at the boundary. Consistent with this molecular picture, boundary conditions may be regarded as special (sometimes even simpler) collisions between fluid and boundary molecules.

Physical fidelity can make the boundary kernel quite complicated, which is generally not the idea with LB. As usual, we look instead for expressions minimizing the mathematical burden without compromising the essential physics.

In particular, we shall require minimal kernels fulfilling the desired constraints on the macroscopic variables (density, speed, temperature and possibly the associated fluxes as well) at the boundary sites $\vec{x}$. This may lead to a mathematically under-determined problem, more unknowns than constraints *This opens up an appealing opportunity to accommodate more interface physics into the formulation of the boundary conditions, but it also calls for some caution to guard against mathematical ill-posedness.*

A major element controlling the complexity of the boundary problem is whether or not the collection of boundary points lies on a surface aligned with the grid. The latter case is significantly more complex and requires extra care. In this book we shall focus mostly on geometrically simple applications where the former situation applies. Advanced applications of the method capable of dealing with much more complicated situations lie beyond the scope of this book.

17.3   Survey of Various Boundary Conditions

We shall distinguish two basic classes of boundary conditions:

- *Grid-aligned*
- *Grid-cutting*

By "grid-aligned" we imply that *the physical boundary is aligned with the grid coordinates.* Grid-aligned boundaries need not be smooth surfaces, such as straight lines or planes, for instance they could be staircase "Legoland" approximates of fairly complex surfaces. The distinctive mark is that they *do not cut through mesh cells*, (see Fig. 17.1).

Grid-cutting, on the contrary, can take virtually any shape and consequently they can provide a more accurate description of complex geometries, see Fig. 17.2. They are typically more difficult to implement, as they generally require interpolations, but the pain is often worth the gain for practical applications, and a true necessity for some flows, such as low-viscous flows over slaneted bodies.

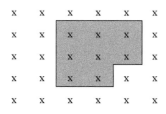

Figure 17.1 *Grid-aligned boundaries staircasing through the lattice. Note that the boundary, while running between lattice sites, still remains aligned with the grid lines.*

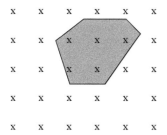

Figure 17.2 *Grid misaligned boundaries cutting through the lattice.*

We shall consider the following classes of boundary conditions:

- *Periodic*
- *No-slip*
- *Free-slip*
- *Frictional slip*
- *Sliding walls*
- *Open inlet/outlets*

Boundary conditions for grid-aligned boundaries can be formulated via elementary geometrical operations, such as bounce-back and similar. Grid-cutting boundaries require more sophisticated techniques, such as:

- Interpolation
- Extrapolation
- Multireflection
- Surface elements (*Surfels*) dynamics

All of these boundary conditions apply to the velocity field. For many open flows of engineering interest, it is more appropriate to deal with mixed pressure and velocity boundary conditions. Since the native LBE is inherently a weakly compressible method, pressure boundaries are somewhat sensitive and will therefore be considered separately.

17.3.1   Periodic Boundary Conditions

Periodic boundary conditions are one the simplest instances of boundary conditions. They are typically intended to isolate bulk phenomena from the actual boundaries of the real physical system and consequently they are adequate for physical phenomena where surface effects play a negligible role. A typical example is homogeneous isotropic-incompressible turbulence to be discussed in chapter 20.

The practical implementation of periodic boundaries goes as follows: to fix these ideas, let us refer to the nine-speed D2Q9 scheme numbered 0–8 in counterclockwise direction starting with $\vec{c}_1 = (1, 0)$ (see Figure 17.3).[37]

Consider a square box consisting of $N \times M$ lattice sites; the discrete fluid is then represented by nine matrices $f_i(l, m)$, $l = 1, \ldots, N$, $m = 1, \ldots, M$, $i = 0, \ldots, 8$.

Let us now introduce four extra layers (buffers) of sites N, S, W, E (standing for North, South, West, East), see Fig. 17.4:

$$N = [i = 0, \ldots, N + 1; j = M + 1], \quad W = [i = 0; j = 0, \ldots, M + 1],$$
$$S = [i = 0, \ldots, N + 1; j = 0], \qquad E = [i = N + 1; j = 0, \ldots, M + 1].$$

---

[37] The counterclockwise ordering is more natural for illustration purposes. In coding practice, it is more efficient in terms of memory access to order the discrete speeds within each energy shell separately, namely $i = 0$ for rest particles, $i = 1, 4$ for speed 1 particles and $i = 5, 8$ for speed $\sqrt{2}$ particles.

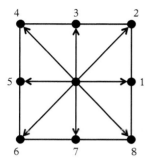

Figure 17.3 *Counterclockwise numbering of the D2Q9 lattice.*

nw	n	n	n	n	n	n	n	n	n	n	ne	6
w	f	f	f	f	f	f	f	f	f	f	e	5
w	f	f	f	f	f	f	f	f	f	f	e	4
w	f	f	f	f	f	f	f	f	f	f	e	3
w	f	f	f	f	f	f	f	f	f	f	e	2
w	f	f	f	f	f	f	f	f	f	f	e	1
sw	s	s	s	s	s	s	s	s	s	s	se	0
0		1	2	3	4	5	6	7	8	9	10	11

Figure 17.4 *Numbering a 10×5 periodic lattice with buffers. f: fluid, w: west boundary, e: east boundary, n: north boundary, s: south boundary.*

Note that four corners are shared and may require separate treatment.

Periodicity along x it is imposed by simply filling the W buffer with the populations of the rightmost column of the physical domain:

$$f_{in}(W) = f_{out}(E) \tag{17.5}$$

$$f_{in}(E) = f_{out}(W) \tag{17.6}$$

The subscripts "in" and "out" denote inward and outward populations, respectively. With the speed numbering adopted here, the map is

```
{in ,W} = {1,2,8}
{out,W} = {4,5,6}
{in ,E} = {out,W}
{out,E} = {in, W}
```

The four corners require special treatment:

```
f_{in}(NW) = f_{out}(SE)
f_{in}(SW) = f_{out}(NE)
f_{in}(NE) = f_{out}(SW)
f_{in}(SE) = f_{out}(NW)
```

Once the buffers are filled by the boundary procedure, everything is in place for the streaming step to move particles into the proper location at the next timestep. A typical code sequence looks as follows:

1. *Boundary*
2. *Move*
3. *Collide*

Without buffers, periodicity is encoded directly in the Move routine.

17.3.1.1   *Technical aside*

The buffer policy described here is just one out of several possible strategies. One could of course dispense with buffers and copy variables directly in the physical domain. The crudest way to do this is to place "IF" statements within the "DO" loops. This solution is quickly programmed but certainly not elegant, and possibly inefficient on most computers, since conditional statements usually break the pipeline execution, thus resulting in some CPU overhead. Another possibility is to treat the borderline fluid nodes with separate loops. This is a bit more laborious, but avoids conditional statements, see Figs 17.4 and 17.5.

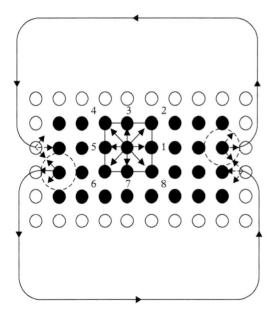

Figure 17.5 *Periodic boundaries. Particles leaving from the right fill in the left buffers and vice versa.*

17.3.2   No-Slip Boundary Conditions

The next simplest boundary condition is the so-called "no-slip" situation, namely zero fluid velocity at a given solid surface. This is physically appropriate whenever the solid wall has a sufficient rugosity to prevent any net fluid motion relative to the wall. Again, we shall assume the solid surface to be aligned with the grid, like for instance the north/south walls in a two-dimensional rectangular channel.

Here we distinguish two types of implementations:

- *On-grid*
- *Mid-grid*

The on-grid condition means that the physical boundary lies exactly on a grid line, whereas the mid-grid case refers to the situation where the boundary lies in between two grid lines, still aligned though, see Figs 17.6 and 17.7.

The on-grid situation is easy; just reverse all populations sitting on a boundary node:

$$f_{in}(N) = f_{out}(N) \tag{17.7}$$

$$f_{in}(S) = f_{out}(S) \tag{17.8}$$

where N and S denote the North and South rows of the domain, respectively.

In terms of the boundary kernel (17.4), the bounce-back kernel (on the top wall) is the 3×3 matrix:

$$\begin{bmatrix} f_6 \\ f_7 \\ f_8 \end{bmatrix} = \begin{pmatrix} 1 & 0 & 0 \\ 0 & 1 & 0 \\ 0 & 0 & 1 \end{pmatrix} \begin{bmatrix} f_2 \\ f_3 \\ f_4 \end{bmatrix} \tag{17.9}$$

all populations being evaluated at site (x, y).

This complete reflection guarantees that both tangential and normal to the wall fluid speed components, identically vanish.

Figure 17.6 *On-grid bounce-back, no-slip boundary condition. The solid versus dashed arrows stand for incoming and outgoing particles, respectively.*

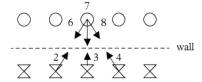

wall

Figure 17.7 *Mid-grid bounce-back, no-slip boundary condition. Populations 6, 7, 8 on the buffer receive populations 2, 3, 4, respectively, from the mirror conjugate sites in the fluid.*

0 1 2 3 4 5 6 7

n n n **N** n n n n   **Figure 17.8** *The north site* n = 3, *(marked N), interacts with the fluid-*
- - - - - - - - - - - - - - - - - -   *sites 2, 3, 4 (marked F) in the lower lying fluid row. The wall (– – –) is in*
f f **F** **F** **F** f f f   *between.*

For an North boundary, the two components of the current density are given by:

$$\mathcal{J}_x = (f_1 + f_2 + f_8) - (f_4 + f_5 + f_6) = f_1 - f_5 = 0 \tag{17.10}$$
$$\mathcal{J}_y = (f_2 + f_3 + f_4) - (f_6 + f_7 + f_8) = 0 \tag{17.11}$$

The first equality is ensured by initializing $f_1 = f_5$ at $t = 0$, these values not being altered by the dynamics at any subsequent timestep. The second equality is guaranteed by the complete reflection (17.7) and (17.8). All this is very simple, and yet sufficient to set up concrete LBE simulations of fluid flows in simple bounded domains, see Fig. 17.8.

The on-grid bounce-back is generally credited first-order accuracy only because of the one-sided character of the streaming operator at the boundary. Mid-line reflection buys second-order accuracy at the price of a modest complication:

$$f_{in}(N) = f_{out}(NF) \tag{17.12}$$
$$f_{in}(S) = f_{out}(SF) \tag{17.13}$$

where N and S denote a generic site on the top (North) and bottom (South) walls, respectively, whereas NF, SF is the *set* of fluid sites connected to N, S, respectively (*domain of influence*), somehow the range of the boundary potential discussed previously. Since particles meet at mid-time $t + 1/2$ between the N, S and NF, SF rows, this implies a fully centered spacetime scheme with resulting second-order accuracy.[38]

The explicit scheme reads

$$\begin{bmatrix} f_6\,(x,y) \\ f_7\,(x,y) \\ f_8\,(x,y) \end{bmatrix} = \begin{pmatrix} 1 & 0 & 0 \\ 0 & 1 & 0 \\ 0 & 0 & 1 \end{pmatrix} \begin{bmatrix} f_2\,(x-1,y-1) \\ f_3\,(x,y-1) \\ f_4\,(x+1,y-1) \end{bmatrix} \tag{17.14}$$

where $(x,y) \equiv N$ is the generic site on the North wall. The South wall is obviously handled in the same way.

We remark that this weakly non-local formulation requires some care in handling the corners.

For instance in the leftmost site $x = 1$, the left neighbor falls within the inlet buffer, and must therefore be drawn therefrom.

---

[38] The reader should be cautioned that according to some authors this distinction between mid-grid and on-grid accuracy is not very meaningful, since on-grid can also be made second-order accurate. The only difference between the two methods is the time it takes to attain a zero speed at the wall.

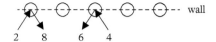

Figure 17.9 *On-site specular reflection. Population 2 goes into 8 and 4 into 6, so that the tangential momentum is conserved.*

17.3.3 Free-Slip Boundary Conditions

Free-slip boundary conditions apply to the case of smooth boundaries with negligible friction exerted upon the flowing gas or liquid.

In this case, the tangential motion of the fluid flow on the wall is free and no momentum is to be exchanged with the wall along the tangential component.

The simplest on-grid implementation consists of an elastic scattering on the wall, namely, (see Fig. 17.9):

$$f_{in}(N) = f_{out}(N) \tag{17.15}$$

where N is again the North-wall layer.

In terms of our cherished boundary matrix,

$$\begin{bmatrix} f_6 \\ f_7 \\ f_8 \end{bmatrix} = \begin{pmatrix} 0 & 0 & 1 \\ 0 & 1 & 0 \\ 1 & 0 & 0 \end{pmatrix} \begin{bmatrix} f_2 \\ f_3 \\ f_4 \end{bmatrix} \tag{17.16}$$

all evaluated locally at (x, y).

This implies no tangential momentum transfer to the wall, as required for a free-slip fluid motion. The condition $f_3 = f_7$ implies no normal-to-wall speed. This is easily checked by simple algebra:

$$\delta \mathcal{J}_x \equiv \mathcal{J}'_x - \mathcal{J}_x = (f_8 - f_2) - (f_6 - f_4) = 0 \tag{17.17}$$

where δ refers to the change due to the boundary collision and prime means after boundary collision.

A similar relation applies to the mid-grid version corresponding to the transformation

$$\begin{bmatrix} f_6(x, y) \\ f_7(x, y) \\ f_8(x, y) \end{bmatrix} = \begin{pmatrix} 0 & 0 & 1 \\ 0 & 1 & 0 \\ 1 & 0 & 0 \end{pmatrix} \begin{bmatrix} f_2(x-1, y-1) \\ f_3(x, y-1) \\ f_4(x+1, y-1) \end{bmatrix} \tag{17.18}$$

Again, the bottom wall is handled with symmetrical arguments and the corners require special care, see Fig. 17.10.

17.3.4 Frictional Slip

Often in applications involving liquid–gas flows, liquid droplets impinge on the solid wall and are only partially reflected back into the fluid, a certain fraction of molecules forming a film that moves along the wall subject to frictional drag.

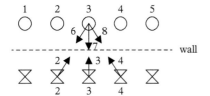

Figure 17.10 *Region of influence of the mid-site specular reflection. The top site n = 3, marked N, interacts with the fluid sites 2, 3, 4 (left, mid, right) in the low-lying row. The wall (- - -) lies in-between.*

This corresponds to no motion normal to the wall ($\mathcal{J}_y = 0$) and damped tangential motion along it. The lattice analog of this boundary condition is a blend of no-slip bounce-backs and free-slip reflections.

In terms of the boundary matrix B_{ij}, for the simple on-grid implementation:

$$
\begin{bmatrix} f_6 \\ f_7 \\ f_8 \end{bmatrix} = \begin{pmatrix} r & 0 & s \\ 0 & 1 & 0 \\ s & 0 & r \end{pmatrix} \begin{bmatrix} f_2 \\ f_3 \\ f_4 \end{bmatrix}
\tag{17.19}
$$

where again all populations are evaluated at site (x, y).

In the above, r and $s = 1 - r$ are the proportions of bounce-back and specular reflections, respectively. The special choices $(r = 1, s = 0)$ and $(r = 0, s = 1)$ recover pure no-slip and free-slip, respectively. Note that by removing the condition $r + s = 1$ absorbing/emitting boundaries can also be modeled.

If bounce-back is present in any proportion, $r > 0$, frictional drag results at the wall that ultimately puts the fluid at rest. In fact, any $r > 0$ does indeed produce a steady-state zero speed at the wall; however, the time it takes the fluid to attain this steady state grows exponentially with $(1 - r)$ and consequently, with r sufficiently small, the wall velocity, which is constantly replenished by the populations coming from the fluid region, never goes to zero.

This is easily seen by computing the momentum transfer to the wall

$$
\delta \mathcal{J}_x = (f_8 - f_2) - (f_6 - f_4) = -2r\,(f_2 - f_4)
\tag{17.20}
$$

$$
\delta \mathcal{J}_y = (f_2 + f_3 + f_4) - (f_6 + f_7 + f_8) = 0
\tag{17.21}
$$

From this expression it is clear that any positive value of r results in a $\delta \mathcal{J}_x$ opposite to $\mathcal{J}_x$, a net frictional effect which produces exponential decay of the fluid momentum at the wall. Such an exponential decay has indeed be detected by Lavallée *et al.* (3) in early lattice gas and Lattice Boltzmann simulations of straight channels. These authors also show that the value of the coefficient r affects the effective location of the boundary where the fluid speed vanishes. This is not surprising, since in a truly microscopic picture the boundary is always diffuse and its effective location depends on the interaction potential between the wall atoms and the fluid molecules. This is a very active research topic in nano-engineering, microhydrodynamics and related disciplines dealing with systems with large surface-to-volume effects. The computational tool of choice is molecular dynamics or Direct-Simulation Monte Carlo for rarefied systems. Both LGCA and LBE can provide a valuable complement to these *ab-initio* simulations, by giving access to larger scales once the expressions of r and s are given as a function of the wall parameters (temperature, rugosity, and so on), see also (4).

17.3.5 Sliding Walls

Another important generalization of the no-slip boundary is the prescription of a given tangential speed at the solid boundary (wall):

$$u(W) = u_w \qquad (17.22)$$

where W denotes a generic site on the sliding wall. Mathematically, this is known as Dirichlet-boundary condition. A proto-typical example is the shear-flow driven by a sliding wall (planar Couette flow), or the lid-driven cavity.

Several recipes are available to impose the constraint (17.22).

17.3.6 Wall Equilibria

The simplest one is to set the wall sites at the their equilibrium value corresponding to u_w and a density taken from, say, the nearest fluid neighbor. This choice is appealing as it does not require any information of geometrical complexity of the boundary.

However, it suffers from an obvious shortcoming: it neglects the non-equilibrium components precisely there where they are usually most substantial, i.e., near solid boundaries, where the flow gradients typically attain their highest values.

Sometimes the actual error is not so large and the method performs reasonably well, at a minimum programming effort (5). Therefore, even though neglecting gradients at solid boundaries is nothing to be recommended, the resort to simplified local treatments does not necessarily require any deep apologies, depending on the specific problem at hand.

17.3.7 The Inamuro Method

Inamuro *et al.* (6) developed an elegant method to impose the Dirichlet boundary condition $u(W) = u_w$.

The idea is to draw the reinjected populations from a local equilibrium with the same normal and different tangential speed at the wall:

$$u = u_w + u' \qquad (17.23)$$
$$v = v_w \qquad (17.24)$$

where u', the so-called counterslip speed, is adjusted so as to achieve mass flux conservation on the wall: $\rho' u = \rho_w u_w$.

By treating the wall density ρ_w and the fictitious wall-equilibrium density ρ' as unknowns, one obtains the following set of equations:

$$\sum_i f_i \{1, c_{ix}, c_{iy}\} = \{\rho_w, \rho_w u_w, \rho_w v_w\} \tag{17.25}$$

In the above equations, the outgoing populations $f_{2,3,4}$ from the north wall are provided by the bulk dynamics, whereas the incoming ones $f_{6,7,8}$ are drawn from the aforementioned equilibria.

This leaves us with three equations for the three unknowns ρ_w, ρ', u' which are readily solved to deliver the desired solution.

Inamuro *et al.* tested this scheme for a 2D Poiseuille flow for a series of values of the relaxation parameter τ, see Fig. 17.11. Their results show that the no-slip condition is accurately reproduced for values of $0.5 < \tau < 20$ (see Fig. 17.12), whereas standard bounce-back would yield significant errors already beyond $\tau = 2$.

This is in line with physical expectations, since τ goes with the molecular mean-free path. Besides its simplicity and physical soundness, the Inamuro method shines for its grid and dimensional independence. In fact, in D dimensions one can dispose of the $D+1$

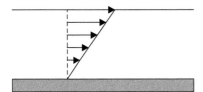

Figure 17.11 *Planar Couette flow. The upper wall moves tangentially to the flow and drags the underlying fluid.*

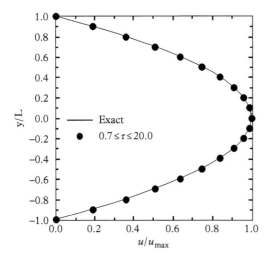

Figure 17.12 *Calculated velocity profile for Poiseuille flow as a function of the relaxation parameter τ. (After (6), courtesy of O. Inamuro.)*

unknowns, wall density, equilibrium density and $D-1$ tangential speed components by means of $D+1$ constraints, namely D velocity components $u_a = u_{wa}$ and one from equating the wall density to the fluid one $\rho = \rho_w$.

Other strategies have been proposed (7; 8) which focus directly on the unknown populations rather than expressing them via adjustable local equilibria.

These strategies are more general but tend to be involved because they must cope with an inevitable mismatch between the number of unknowns and the number of macroscopic constraints. This mismatch is usually closed by heuristic rules involving a series of extra steps, including extrapolation and mass redistributions among the populations. As repeatedly stated, this extra freedom is a blame and a blessing at a time.

17.3.8   A Word of Caution: Analytical Results

By analytically solving the LBGK equation for some simple flows, such as two-dimensional Couette and Poiseuille flows, He *et al.* provide a very detailed and illuminating discussion of the artificial slip effects associated with various types of boundary conditions (9).

For instance, these authors show that the on-grid bounce back gives the following artificial slip flow (in the D2Q9 lattice):

$$\frac{U_s}{U_c} = \frac{2}{3}[8(\tau - 1/2)(\tau - 3/4) - 3N]/N^2 \qquad (17.26)$$

where U_c is the centerline speed, N the channel height in lattice units and τ the LBGK relaxation parameter.

This expression shows that, for any value of τ reasonably above $1/2$, the on-site bounce-back gives rise to terms of the order $1/N$, hence it is only first-order accurate. On the other hand, with half-way bounce back the error in the bulk is shown to scale like $\frac{4\tau(4\tau-5)+3}{(N-1)^2}$. Hence the scheme is second-order accurate and with moderate prefactors since the numerator remains below 3.25 in the range $0.5 < \tau \le 1.25$. This is true everywhere, except in the bounce-back rows, which are still left with a first-order, yet wall-confined, correction.

One can rephrase the situation by saying that with LBGK, the actual position of the wall, the location where the flow speed is exactly zero depends on the relaxation time, hence on the fluid viscosity, which is clearly unphysical. This situation can be improved and possibly even completely cured by resorting to multi-relaxation LBE. In fact, a simple two-time relaxation is often sufficient (10).

LBGK has been intensely criticized for such shortcoming. This criticism is well taken, except for neglecting that for macroscopic flows, where $1/N \ll 1$, first-order errors may eventually be tolerated without any major harm. On the other hand, for flows where the fluid is locally constricted in small regions with $N \sim 10$ or less, such as flows in porous media or in micro-channels, tolerance criteria should be significantly tightened.

17.4   Moving Boundaries

The more general case of a solid boundary moving along both tangential and *normal* directions implies a major leap of complexity.

This is because the normal motion of the wall induces *compressed* fluid motion, whereby the fluid volume changes in time. A typical example is piston motion in a combustion engine.

This implies that fluid sites must be erased and regenerated in the course of time depending on the actual position of the moving boundary. It is not hard to see that this calls for significant extra book keeping, which becomes pretty demanding whenever the wall motion proceeds along a generic direction not aligned with the grid coordinates.

This significantly more complex problem has been tackled by Ladd for the simulation of colloidal suspensions and, more generally, flows with suspended objects. Ladd's method will be described in the sequel to this book, together with a short survey of the main variants developed in the last decade.

17.5   Open Boundaries

Many flows of practical and theoretical interest involve *open* boundaries such as fluid inlets and outlets. Such open boundary conditions, especially the outlet are notorious for requiring special care, due to the onset of compressibility effects and ensuing generation of spurious sound waves propagating from the outlet back into the fluid.

For these open flows, it is common to assign a given density ρ_{in} and velocity profile $u_{\text{in}}(y)$ at the inlet, while at the outlet either a given pressure P_{out} or a no-flux condition normal to the wall, $\partial_n \vec{u} = 0$, is imposed.

Let us consider the inlet boundary condition first.

A prescribed inlet flow is easily implemented by constantly refilling the buffer with the equilibrium population corresponding to the desired values of density and flow speed:

$$f_{\text{in}}(y) = f^e \left[\rho_{\text{in}}, u_{\text{in}}(y) \right] \tag{17.27}$$

The outlet is more delicate.

Naively, one might think of imposing the zero-gradient condition by simply copying the last fluid column into the buffer, namely column 6 into column 7 in Fig. 17.13. Unfortunately, this does not necessarily work, at least, not unless the outlet is placed far enough downstream to allow the flow to settle down to a one-dimensional zero-gradient profile. If this condition is not met, trains of backward propagating disturbances are triggered, which may invade the domain and jeopardize the simulation.

This situation exposes the LB quasi-incompressibility as a potential liability.

A practical way of dealing with these disturbances is to apply the so-called "*porous plug*" boundary condition. The idea is that particles reaching the outlet should slow down, as if they were entering a porous media (the plug). To implement such conditions,

```
0  1  2  3  4  5  6  7
n  n  n  n  n  n  n  n    north wall
i  f  f  f  f  f  f  o
i  f  f  f  f  f  f  o
i  f  f  f  f  f  f  o
i  f  f  f  f  f  f  o
s  s  s  s  s  s  s  s    south wall
0  1  2  3  4  5  6  7
```

Figure 17.13 *A* 6 × 4 *grid with inlet (i) and outlet (o) sites.*

the outlet populations are reflected with a probability $0 < r < 1$, which is adjusted in such a way as to ensure global mass conservation throughout the fluid.

With reference to the D2Q9 lattice with plain counterclockwise numbering, the outlet populations obey the relation

$$f_{4,5,6}^{out} = r\, f_{1,2,8}^{out} \tag{17.28}$$

As a result, the outgoing half-current at the outlet is given by

$$\mathcal{J}^{out} = (f_1 + f_2 + f_8)(1 - r) \tag{17.29}$$

and consequently, global-mass conservation, $\mathcal{J}_{out} = \rho u_{in}$, implies

$$p \equiv 1 - r = \frac{\rho u_{in}}{f_1 + f_2 + f_8} \tag{17.30}$$

where p is the probability of being transmitted, e.g., non-reflected, at the outlet. As usual, it is instructive to replace the outgoing populations with their equilibrium values. By taking the same density at inlet and outlet, and neglecting quadratic contributions, one obtains

$$p = \frac{u_{in}}{\frac{1}{6} + \frac{u}{2}} \tag{17.31}$$

From this expression, it is seen that the condition $r < 1$ ($p > 0$) sets an upper bound to the inlet speed, namely $u_{in} < 1/6 + u/2$. Such bound is readily understood by realizing that the equilibrium outgoing half-current

$$\mathcal{J}^{out,e} = f_1^e + f_2^e + f_8^e = \rho(1/6 + u/2)$$

is larger than the current itself, $\mathcal{J} = \rho u$, as long as $u < 1/3$, which is typically the case in LB simulations. Indeed, under the very same condition, the equilibrium incoming half-current

$$\mathcal{J}^{in,e} = -(f_4^e + f_5^e + f_6^e) = -\rho(1/6 - u/2)$$

is negative. Hence, there is scope for acting on such half-current via the reflection coefficient, so as to achieve the desired mass outflow.

Outlet absorbing boundary conditions for compressible LB flows have been developed in the recent years, see for instance (11).

17.6 Exactly Incompressible LBE Schemes

As noted, many problems of engineering interest involve open outlets. At such outlets, the most natural and convenient boundary conditions consist in specifying the value of the pressure.

Quasi-incompressible isothermal LBE schemes do not easily deal with pressure (density) boundary conditions, simply because density and pressure are tied up together via the ideal gas equation of state, $p = \rho c_s^2$.

Since c_s is $O(1)$ in lattice units, any pressure change is necessarily associated with a comparable change in density, which may easily break the quasi-incompressibility constraint on the LB flow.

In conventional CFD, this problem is usually side-stepped by allowing an artificially high sound speed $c_s \to \infty$, so that sizeable pressure drops δP can be sustained by small density changes $\delta \rho = \delta P / c_s^2$.

The conventional way to impose pressure drops in open LB flows consists of replacing the "true" pressure gradient with a corresponding volume force $\vec{F}$ providing exactly the same momentum input to the flow. Such replacement works fine as long as there is no need to know the detailed spacetime distribution of the pressure field. Whenever this information cannot be passed out, exactly incompressible models are called for.

Ideally, these models should be tailored to the requirement of strict incompressibility of the flow-field div $\vec{u} = 0$, which corresponds to an infinite sound speed.

Before mentioning how these strictly incompressible models work (12), it is worth mentioning a simple strategy (6) which apparently yields acceptable results within the simple class of quasi-incompressible (as opposed to strictly incompressible) LBE models.

The idea is simply to impose a density drop between inlet and outlet by augmenting the outgoing populations by a constant term:

$$f_{\text{in}} = f_{\text{out}} + C \tag{17.32}$$

The constant C is computed by using the equilibrium values of the distribution function and imposing a prescribed value for the inlet density ρ_{in}. By doing so, the flow proves capable of sustaining minor density drops, yet sufficient to produce the pressure gradients needed by many incompressible applications.

As repeatedly stated, standard LBE methods are weakly compressible approximations of the incompressible Navier–Stokes equations. The continuity equation, rewritten in Lagrangian form

$$\text{div}\,\vec{u} = -\frac{D \ln \rho}{Dt}, \tag{17.33}$$

shows that the LBE steady states ($\partial_t = 0$) are affected by a local compressibility

$$\text{div}\,\vec{u} = -\vec{u} \cdot \nabla(\ln \rho), \tag{17.34}$$

This error is of order Ma^2, hence negligible for many practical purposes, but still annoying, as it hinders the application of LBE to flows with significant pressure changes $\Delta p/p > Ma^2 \sim 0.1$. It is therefore highly desirable, especially for open flows, to extend LBE so as to ensure exact incompressibility at steady state.

An effective way to achieve this is to use the current density $\vec{J}$ to represent the flow speed $\vec{u}$. Then, the LBE steady state automatically delivers a solenoidal flow $\text{div}\,\vec{u} = 0$.

The trick is to modify standard LBE equilibria so as to fulfill the following equalities:

$$\sum_i f_i^e = \rho, \quad \sum_i f_i^e c_{ia} = u_a \tag{17.35}$$

With specific reference to the D2Q9 model, the sought set of equilibria is readily found (8):

$$f_0^e = \frac{4}{9}\left(\rho - \frac{3}{2}u^2\right) \tag{17.36}$$

$$f_{i1}^e = \frac{1}{9}\left(\rho + 3u_i + \frac{9}{2}u_i^2 - \frac{3}{2}u^2\right)$$

$$f_{i2}^e = \frac{1}{36}\left(\rho + 3u_i + \frac{9}{2}u_i^2 - \frac{3}{2}u^2\right)$$

where $u_i = c_{ia}u_a$ and the subscripts 1, 2 denote the nearest neighbors and diagonal links, respectively. The crucial ingredient is that in order to match the second eqn (17.35), at variance with standard LBGK, *the density ρ is no longer prefactoring the polynomials in u_i*. As shown in He-Luo, this is tantamount to neglecting density changes $\delta\rho/\rho$ in front of the velocity-dependent terms in the local equilibrium, where the local density is replaced by a constant average value ρ_0 (fixed to 1 in 17.36). The continuity equation takes then the form

$$\frac{1}{c_s^2}\partial_t p + \text{div}\,\vec{u} = 0 \tag{17.37}$$

whose steady-state limit recovers the condition of exact incompressibility.

Zou and He (8) used *D2Q9i* (*i* stands for incompressible) for various two-dimensional flows and demonstrate a dramatic drop of the compressibility error, typically from about 10^{-4} to 10^{-9}.

17.6.1 Remark

The compressibility error is computed by taking the departure of the numerical solution from the analytical one. For more complex flows, in which the analytical solution is not known, the local compressibility error can be computed as $E = |div\vec{u}/\nabla\vec{u}|$, i.e., the ratio of the (inverse) time scale of longitudinal and transversal deformation. Structures for which $E \ll 1$ are largely unaffected by compressibility effects. To the best of this author's knowledge, such quantity has been investigated only recently in (13).

Coming back to pressure boundary conditions, the good news is that one is free to accept a space-dependent density, hence pressure, without hindering incompressibility, simply because the compressibility errors have been "swallowed" by the new definition of u_a. The flip side of the medal is that the trick only works at steady state, since there is no way to reabsorb the time derivative of the fluid density. In this respect, these exactly incompressible LBGK models are best thought of as *false-transient, iterative schemes*, similar to those widely in use for steady-state CFD computations (14).

17.6.2   Zou-He Non-Equilibrium Bounce-Back

A treatment that has met with increasing popularity in the last decade is the one due to Zhou and He, which we now describe in some detail.

Let us refer to a D2Q9 scheme for a planar flow, with on-grid top and bottom walls, and pressure-velocity boundary conditions at the inlet and outlet, see Fig. 17.14.

At the bottom wall, the outward movers, f_4, f_7, f_8 are known after streaming, while the inward movers, f_2, f_5, f_6 are unknown. In this, the discrete velocity indices run from 1–4 for speed 1 and 5–8 for speed $\sqrt{2}$, both counterclockwise. As usual, subscript 0 denotes rest particles.

The three inward populations obey the three hydrodynamic conditions:

$$f_2 + f_5 + f_6 = \rho - (f_0 + f_1 + f_3 + f_4 + f_7 + f_8) \tag{17.38}$$

$$f_5 - f_6 = \rho u_x - (f_1 - f_3 - f_7 + f_8) \tag{17.39}$$

$$f_2 + f_5 + f_6 = \rho u_y + (f_4 + f_7 + f_8) \tag{17.40}$$

Note that subscripts x and y stand here for the velocity components along the horizontal and vertical axes, respectively.

The first and the third combine into the consistency condition for the wall density:

$$\rho = \frac{1}{1 - u_y}[f_0 + f_1 + f_3 + 2(f_4 + f_7 + f_8)] \tag{17.41}$$

still leaving one of the three inward populations unknown. To close the system, Zhou and He invoke bounce-back of the non-equilibrium populations normal to the wall, namely

$$f_2 - f_2^e = f_4 - f_4^e \tag{17.42}$$

This delivers

$$f_2 = f_4 + \frac{2}{3}\rho u_y \tag{17.43}$$

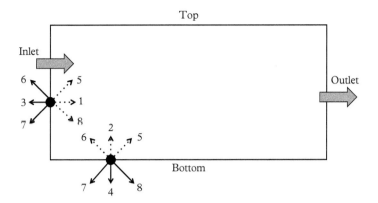

Figure 17.14 *Geometrical set up for the Zhou–He boundary conditions. The dotted arrows indicate the unknown populations. Note that the discrete speeds are numbered by energy shells.*

Combining this with (17.38–40), one finally obtains

$$f_5 = f_7 - \frac{f_1 - f_3}{2} + \frac{1}{2}\rho u_x + \frac{1}{6}\rho u_y \tag{17.44}$$

$$f_6 = f_8 + \frac{f_1 - f_3}{2} - \frac{1}{2}\rho u_x + \frac{1}{6}\rho u_y \tag{17.45}$$

which solves the problem.

The same idea is applied to the inlet section. Let us assume that on the inlet boundary the density $\rho = \rho_{in}$ is imposed, as well as $u_y = 0$. The unknowns are the incoming populations f_1, f_5, f_8 and the streamwise velocity u_x. Applying the same procedure discussed for the bottom wall, i.e., bouncing back the non-equilibrium distribution normal to the inlet

$$f_1 - f_1^e = f_3 - f_3^e \tag{17.46}$$

one ends up with the consistency condition

$$\rho_{in} u_x = \rho_{in} - [f_0 + f_2 + f_4 + 2(f_3 + f_6 + f_7)] \tag{17.47}$$

and the explicit expression for the inward movers

$$f_1 = f_3 + \frac{2}{3}\rho_{in} u_x \qquad f_5 = f_7 - \frac{f_2 - f_4}{2} + \frac{1}{6}\rho_{in} u_x \tag{17.48}$$

$$f_8 = f_6 + \frac{f_2 - f_4}{2} + \frac{1}{6}\rho_{in} u_x \tag{17.49}$$

which completes the inlet boundary.

An analog procedure can be applied at the outlet, with the specification of $\rho = \rho_{out}$.

As noted by the authors, nodes at the intersection of two perpendicular boundaries call for a special treatment. Consider for instance the inlet-bottom node. After streaming, f_1, f_2, f_5, f_6, f_8 need to be determined.

Using bounce-back for the populations normal to the walls, i.e.,

$$f_1 = f_3, \quad f_2 = f_4 \tag{17.50}$$

the currents along x and y give

$$f_5 - f_6 + f_8 = f_7 + f_3 - f_1 \tag{17.51}$$
$$f_5 + f_6 - f_8 = f_7 + f_4 - f_2 \tag{17.52}$$

namely

$$f_5 = f_7 \tag{17.53}$$

$$f_6 = f_8 = \frac{1}{2}(\rho - f_1 + f_2 - f_3 - f_4 - f_5 - f_7) \tag{17.54}$$

We are still missing one relation to fix ρ. However, since density at the inlet is assumed to be constant, one usually takes $\rho = \rho_{in}$.

17.6.2.1 *Remark*

Although the authors make no comment regarding the physical meaning of the non-equilibrium bounce-back rule, it is reasonable to speculate that this rule is equivalent to imposing continuity of the component of the stress tensor across the wall.

This is readily seen, by writing the non-equilibrium distribution in general form as

$$f_i^{ne} = w_i Q_{iab} P_{ab}^{ne} + h.o.t. \tag{17.55}$$

where all symbols have been previously defined, and h.o.t. denotes higher-order terms.

To first order in the Knudsen number, the non-equilibrium momentum flux tensor is proportional to the stress tensor $\mu(\partial_a u_b + \partial_b u_a)$, whence the assertion. This indicates that the Zhou-He non-equilibrium bounce-back rule amounts to a statement of continuity not only on mass and momentum, but also on momentum flux. It is natural to expect that this would benefit robustness and accuracy.

17.7 Curved Boundaries

The examples illustrated thus far dealt with straight boundaries, globally aligned with the grid lines. The case of curved boundaries can still be dealt with *locally* grid-aligned techniques (staircasing) or by more general grid-cutting methods. In the sequel we shall provide a brief account of both.

s s s s s

n n n n n n wall   **Figure 17.15** *f: fluid, s: solid, n: north wall. The populations are numbered*

f f f f f f   *according to Fig. 17.3.*

17.7.1 Staircased Boundaries

The quickest approach to curved boundaries is to "staircase" the cutting boundary by replacing it with a zig-zagging contour lying entirely on the grid (a sort of "lattice polymer"). This has the great merit of simplicity and honors the heralded assets of LGCA and LBE methods: easy handling of complex boundary conditions, see Fig. 17.15.

However, it also introduces an artificial rugosity of order $1/L$, L being a typical linear size of the obstacle in lattice units, which may deteriorate the accuracy of the near-wall flow computation when L becomes of the order of ten lattice sites or less.

Staircased boundaries are commonly used to simulate flows in porous media and for the sake of concreteness they shall be described in more detail in the chapter 19 specifically devoted to this application.

17.7.2 Extrapolation Schemes

An elegant and simple method to handle boundary conditions in a systematic rather than heuristic way was proposed by Chen and Martinez (15). The method places the wall on boundary nodes and *lets them undergo the same collisional step as the fluid nodes*, with the important proviso that the equilibrium population is explicitly tuned on the desired wall speed u_w. In this picture, the wall is nothing but a portion of fluid with a prescribed equilibrium. To complete the streaming step, boundary nodes need to import information from solid nodes lying inside the solid wall. This information is obtained by a straightforward second-order interpolation. For the north-east, north and north-west moving populations at a generic north-boundary node $(x, y + 1)$:

$$f_2 (x, y + 1) = 2f_2 (x - 1, y) - f_2 (x - 2, y - 1) \tag{17.56}$$

$$f_3 (x, y + 1) = 2f_3 (x, y) - f_3 (x, y - 1) \tag{17.57}$$

$$f_4 (x, y + 1) = 2f_4 (x + 1, y) - f_4 (x + 2, y - 1) \tag{17.58}$$

This scheme preserves second-order accuracy, does not depend on the specific lattice topology, nor on its dimensionality and it is easy to implement. The authors demonstrate its validity for the case of Poiseuille flow, Couette flow, lid-driven square cavity and others. They also point out potential generalizations to include von Neumann-type boundary conditions involving the fluid velocity and its derivative along the normal to the wall.

17.7.3 Multi-Reflection

It is well known from continuum kinetic theory that the presence of a wall produces a thin layer of the order of the mean-free path (Knudsen layer) whose steady state differs from the one attained in the bulk region (Kramers problem). On a discrete speed lattice, the Kramers problem can be solved analytically, as shown by d'Humières and co-workers (16; 17). On a six-state hexagonal lattice the main results are as follows:

- *Parallel/perpendicular orientation*
  For parallel and perpendicular boundaries, anisotropic Knudsen layers are completely suppressed by bounce-back or reflecting boundary conditions. However, the effective location of the wall is slightly displaced off-lattice with respect to the geometrical position.
- *General orientation*
  For a general orientation (irrational with respect to the lattice coordinate lines) there is an infinity of spurious modes that disappear only once the slope of the boundary becomes rational.

These spurious modes must be watched carefully in applications dealing with flows with irregular and jagged boundaries. For a general treatment of boundaries with an arbitrary inclination to the lattice grid lines, using the so-called multi-reflection scheme, see Ginzburg and d'Humieres (2003) (18). In this approach borrowing verbatim from the authors

> the boundary condition is written as a closure relation between an unknown population entering the fluid and some others known from the fluid dynamics; the populations are then replaced in the closure relation by their second-order approximations; finally a Taylor expansion of the result at the boundary node gives a second-order estimate of the perturbation of the kinetic solution by the boundary condition.

Formally, the multi-reflection boundary condition reads as follows:

$$f_i(\vec{r}_b; t + 1) = \sum_{m=-2}^{2} k_m f_i(\vec{r}_b + m\vec{c}_i; t + 1) w_i t_p^* \mathcal{J}_{iw} + t_p F_{\bar{i}}^{pc} \tag{17.59}$$

In (17.59), $\bar{i}$ denotes the mirror conjugate to i, k's are adjustable coefficients, t_p^*, $p \equiv c_i^2$ are related to the lattice weights and w_i are additional parameters to set up the Dirichlet boundary condition, $\mathcal{J}_{iw} = \vec{\mathcal{J}}_w \cdot \vec{c}_i$, where subscript "$w$" means wall.

Finally, the term $F_{\bar{i}}^{pc}$ is a further correction ensuring the Dirichlet condition to second order.

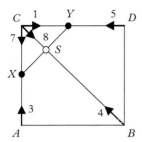

Figure 17.16 *A surfel* $S = \overline{XY}$ *joining west and north cell boundaries.*

The formalism is relatively dense and the interested reader is referred to the original work. Here, we simply note that the multi-reflection approach implies a multi-point relation between lattice nodes, hence it can deliver up to third-order accuracy. This makes it particularly interesting for low-resolution simulations and, more generally, whenever Knudsen layers have a significant impact on the flow, such as in microflows or flows with moving boundaries.

17.7.4 The Surfel Method

Another powerful method to deal with arbitrary boundaries is based on the notion of *surfels*, that is surface elements.[39] Surfels are elementary portions of the boundary cutting through a single volume cell, called *voxel*, and must viewed as *independent computational elements with a dynamics on their own*. They act as *flux scattering elements* receiving incoming fluxes from the flow and reinstituting a corresponding set of fluxes, consistent with the basic conservation laws at the surface (19).

Let us clarify with an example. Consider a 2D cell of the D2Q9 lattice identified by the four vertices (A, B, C, D). The cell is cut through by a surfel (segment) joining points X and Y on sides AC and CD. The mass entering the surfel per unit time is given by (see Fig. 17.16).

$$M_{\text{in}} = a_7 f_7 \, (C) \, \overline{AX} + a_8 f_8 \, (C) \, \overline{SB} + a_1 f_1 \, (C) \, \overline{YD} \tag{17.60}$$

where $a_i = c_{ia} n_a$, n_a being the outward normal of the surfel. In order to ensure mass conservation (non-absorbing boundary), the surfel must reinstitute an equal amount of mass M^{out}.

This is given by

$$M_{\text{out}} = a_3 f_3 \, (A) \, \overline{XC} + a_4 f_4 \, (B) \, \overline{SC} + a_5 f_5 \, (D) \, \overline{YC} \tag{17.61}$$

where the meaning of the notation is again apparent from Fig. 17.16.

---

[39] TM EXA Corporation.

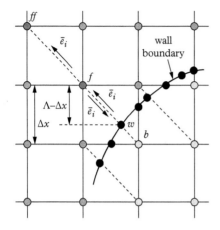

Figure 17.17 *Interpolation at a curved boundary cutting through the mesh. The symbol b denotes a boundary node, w the wall site and f, ff the nearest and next-to-nearest fluid sites, respectively.* Reprinted from Journal of Computational Physics, O. Filippova et al, *J. Comp. Phys*, 147, 219, 1998, copyright 1998, with permission from Elsevier.

By the same token, one writes down the momentum fluxes along x and y directions:

$$\mathcal{J}_x^{\text{in}} = f_1\,(C)\,a_1\,\overline{YD} + f_8\,(C)\,a_8\overline{SB} \tag{17.62}$$

$$\mathcal{J}_x^{\text{out}} = f_4\,(B)\,a_4\overline{CS} + f_5\,(D)\,a_5\,\overline{YD} \tag{17.63}$$

$$\mathcal{J}_y^{\text{in}} = f_7\,(C)\,a_7\overline{AX} + f_8\,(C)\,a_8\overline{SB} \tag{17.64}$$

$$\mathcal{J}_y^{\text{out}} = f_3\,(A)\,a_3\overline{CX} + f_4\,(B)\,a_4\overline{CS} \tag{17.65}$$

The three conditions

$$M^{\text{out}} = M^{\text{in}}, \quad \mathcal{J}_x^{\text{out}} = \mathcal{J}_x^{\text{in}}, \quad \mathcal{J}_y^{\text{out}} = \mathcal{J}_y^{\text{in}} \tag{17.66}$$

provide a 3×3 system for the 3 unknowns $f_3(A), f_4(B), f_5(D)$.

This procedure applies to generic shapes of the voxels but it requires a certain degree of sophistication to work in full generality. Since surfel dynamics is naturally expressed in terms of incoming–outcoming fluxes crossing the surface of each cell, the surfel method deploys its full power when formulated in the language of finite volumes (19). Successful finite-difference analogs are also reported in the literature (20; 21; 15), see Fig. 17.17.

17.7.5 Bouzidi–Lallemand Bounce-Back Rule

As we have previously seen, in link-based methods for curved boundaries only a fraction of the fluid population is bounced back, and such fraction is a function of the geometrical parameters of the boundary. Such parameters might either represent the solid fraction in the cell surrounding a lattice node, or other geometrical quantities (22; 21). These methods interpolate from fluid nodes adjacent to the boundary, as depicted in Fig. 17.18, (23).

The quantity q is defined as the distance of the boundary location B from the closest fluid site F, in units of the lattice spacing $\Delta x_i = c_i \Delta t$.

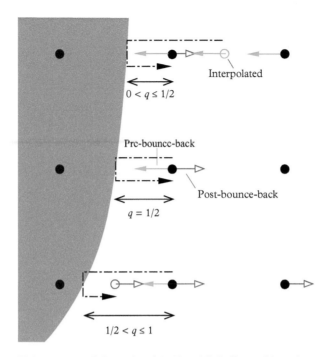

Figure 17.18 *Schematics of the Bouzidi–Lallemand boundary condition, for the three cases 0 < q < 1=2 (top), q = 1=2 (middle) and 1=2 < q < 1 (bottom). From Aidun-Clausen, Annu. Rev. Fluid Mech., 42, 439, 2010. In this picture, the solid site S is the square within the solid, the fluid site F is the square to its left and the boundary site B is the intersection of the link (S − F) with the solid line. The virtual site V is the open circle.*

For the case of Fig. 17.18:

$$x(B) = q\Delta x \tag{17.67}$$

where we have made the conventional choice

$$x(F) = 0 \tag{17.68}$$

and

$$x(S) = \Delta x \tag{17.69}$$

In (17.69), S denotes the nearest neighbor of F within the solid.

With these notations, the bounce-back rule for the standard case $q = 1/2$ reads as follows:

$$f_i(S; t) = f'_{i*}(F; t) + \frac{2w_i \rho \vec{u}(B) \cdot \vec{c}_i}{c_s^2} \tag{17.70}$$

$$f_i(F; t + 1) = f_i(S; t) \tag{17.71}$$

In (17.71) i^* denotes the mirror mate of i.

This corresponds to a trajectory of the fluid population leaving the site F at time t traveling a time-lapse $\Delta t/2$ toward the boundary being instantly bounced back exactly at time $t + \Delta t/2$, and traveling backward another $\Delta t/2$ stretch in time, thus landing with reverse velocity exactly on the same site F at time $t + \Delta t$.

When the boundary is no longer sitting at the mid-link location, the same picture applies, with a minor, yet powerful generalization.

One defines a virtual site at location

$$x(V) = (2q - 1)\Delta x \tag{17.72}$$

as the starting point for the fluid population to be bounced back at the boundary location $x(B) = q\Delta x$. Since $q \neq 1/2$, the population at the virtual site must be defined through interpolation with the neighboring sites, namely $[x - \Delta x, x]$ for the case $q < 1/2$ and $[x, x + \Delta x]$ for the case $q > 1/2$. With reference to Figure 17.18, the S site is the square within the solid, F the intermediate square left to S within the fluid, B is the intersection of the link $(S - F)$ with the solid curve and, finally, V is the location of the circle.

The particle trajectory starting at site V at time t travels a fraction $(1 - q)\Delta t$ forward toward the boundary, and a fraction $q\Delta t$ backward, after bouncing back at $t = q\Delta t$ to land exactly on the F node at time $t + \Delta t$. For instance, with $q = 1/4$ the virtual node sits at $x(V) = -\frac{\Delta x}{2}$ and the V population travels 3/4 timestep units rightward (toward the wall) and 1/4 units leftward after reflection. Conversely, with $q = 3/4$, the V node sits at $x(V) = \frac{\Delta x}{2}$, and travels 1/4 timestep units forward and 3/4 unit backward.

Summarizing, the Bouzidi–Lallemand modified bounce-back rule reads simply as follows:

$$f_i(F; t + 1) = f_{i*}(V; t) + \frac{w_i \rho \vec{u}(B) \cdot \vec{c}_i}{c_s^2} \tag{17.73}$$

where

$$f_{i*}(V; t) = (1 - 2q)f_{i*}(x - \Delta x) + 2qf_{i*}(x); \quad q < 1/2 \tag{17.74}$$

$$f_{i*}(V; t) = 2(1 - q)f_{i*}(x) + (2q - 1)f_{i*}(x + \Delta x); \quad q > 1/2 \tag{17.75}$$

The Bouzidi–Lallemand boundary condition is flexible and efficient. Moreover, it applies to both static and generic moving boundaries, wherein the coefficient q becomes a function of space and time. Indeed, this boundary condition has become very popular for LB simulations with moving boundaries.

It should be reminded that its use in conjunction with the BGK single-relaxation time operator is subject to the usual pathologies, namely the wall location depends on the actual value of the relaxation parameter τ. Such corrections, largely negligible for macroscopic flows, cannot be passed out in the case of nearby particle encounters. For such situations, one can switch to the multi-relaxation-time collision operator, as recommended in (24), or develop local subgrid strategies (25). For further material and applications to engineering flows, see the recent book by Gu and Shu (1) and also the even more recent one by Krueger et al. (2).

17.8 Summary

Summarizing, boundary conditions represent an asset of the LB method because they can be formulated in terms of simple mechanical relations for particles moving along straight trajectories. This is why the ease of handling boundary conditions in complex geometries is justly heralded as one of the main highlights of the LB method. However, it should be borne in mind that, whenever the boundaries are not aligned with the grid, their practical implementation requires nonetheless a significant amount of labor. This said, the treatment of non-aligned and curved boundaries has witnessed a major progress in the last decade.

···

REFERENCES

1. Z. Guo and Y. Shu, "Lattice Boltzmann Method and its Applications in Engineering", *Springer Series in Computational Fluid Dynamics,*World Scientific, Singapore, 2013.
2. T. Krüger, H. Kusumaatmaja, A. Kuzmin, O. Shardt, G. Silva, E.M. Viggen, *The Lattice Boltzmann Method. Principles and Practice*, Springer, Berlin, 2017.
3. P. Lavallée, J.P. Boon and A. Noullez, "Boundaries in lattice gas flows", *Physica D*, 47, 233, 1991.
4. S. Succi, "Mesoscopic modeling of slip motion at fluid-solid interfaces with heterogeneous catalysis", *Phys. Rev. Lett.*, **89**, 064502, 2002.
5. A. Mohammad and S. Succi, "A note on equilibrium boundary conditions in lattice Boltzmann fluid dynamic simulations", *The Europ. Phys. Journal Special Topics*, **171**, 213, 2009.
6. O. Inamuro, M. Yoshino and F. Ogino, "A non-slip boundary condition for lattice Boltzmann simulations", *Phys. Fluids*, 7(12), 2928, 1995.
7. R. Maier, R. Bernard and D. Grunau, "Boundary conditions for the lattice Boltzmann method", *Phys. Fluids*, 8, 1788, 1996.

8. Q. Zou and X. He, "On pressure and velocity boundary conditions for the lattice Boltzmann BGK model", *Phys. Fluids*, **9**, 1591, 1997.

9. X. He, L.S. Luo and M. Dembo, "Analytic solutions of simple flows and analysis of nonslip boundary conditions for the lattice Boltzmann BGK model", *J. of Stat. Phys.*, 87 115, 1997.

10. I. Ginzbourg and P.M. Adler, "Boundary flow condition analysis for the three-dimensional lattice Boltzmann model", *J. Phys. II France*, **4**, 191, 1994.

11. E, Vergnault, O. Malaspinas, P. Sagaut, "Lattice Boltzmann simulations of impedance tube flows", *J. Comp. Phys.*, **231**, 7335, 2012.

12. X. He and L.S. Luo, "Lattice Boltzmann model for the incompressible Navier–Stokes equations", *J. Stat. Phys.*, **88**(3/4), 927, 1997.

13. A. Montessori *et al.*, "Regularized lattice Bhatnagar-Gross-Krook model for two-and three-dimensional cavity flow simulations", *Phys. Rev. E* **89**, 053317, 2014.

14. S. Patankar, *Numerical Heat Transfer and Fluid Flow*, McGraw–Hill, New York, 1980.

15. S. Chen, D. Martinez and R. Mei, "On boundary conditions in lattice Boltzmann methods", *Phys. Fluids*, **8**, 2527, 1996.

16. R. Cornubert, D. d'Humières and D. Levermore, "A Knudsen layer theory for lattice gases", *Physica D*, **47**, 241, 1991.

17. I. Ginzbourg and D. d'Humières, "Multireflection boundary conditions for lattice Boltzmann models", *Phys. Rev.*, *E* **68**, 066614, 2003.

18. I. Ginzbourg and D. d'Humières, "Local second-order boundary methods for lattice Boltzmann models", *J. Stat. Phys.*, **84**(5/6), 927, 1996.

19. H. Chen, "Volumetric formulation of the lattice Boltzmann method for fluid dynamics: basic concept", *Phys. Rev. E,* **58**(3), 3955, 1998.

20. O. Filippova and D. Haenel, "Lattice Boltzmann simulation of gas–particle flows in filters", *Comput. Fluids* **26**(7), 697, 1997.

21. R. Mei, L. S. Luo and W. Shyy, "An accurate curved boundary treatment in the lattice Boltzmann method", *J. Comp. Phys.* **155**, 307, 1999.

22. O. Filippova and D. Haenel, "Grid refinement for Lattice BGK models", *J. Comp. Phys.* **147**, 219, 1998.

23. M. Bouzidi, M. Firdaouss and P. Lallemand, "Momentum transfer of a Boltzmann-lattice fluid with boundaries", *Phys. Fluids* **13**, 3452, 2001.

24. P. Lallemand and L.S. Luo, "Lattice Boltzmann method for moving boundaries", *J. Comp. Phys.* **184**, 406, 2003.

25. B. Chun and AJC Ladd, "Interpolated boundary condition for lattice Boltzmann simulations of flows in narrow gaps", *Phys. Rev. E*, **75**, 066705, 2007.

....

EXERCISES

1. A particle hits a solid wall at an incident angle θ and it is reinjected in the fluid at an angle θ'. Compute the particle-to-wall momentum transfer as a function of

the reinjection angle. Which values of θ and θ' maximize the exchange of normal momentum?

2. Write down the explicit Inamuro solution for sliding walls and analyze the solution as a function of the wall speed u_w. Discuss the admissible range of wall speeds.

3. Compute the explicit solution of the 3×3 system (17.25) and discuss its positivity domain.

18

Flows at Moderate Reynolds Numbers

In this chapter we present the application of LBE to flows at moderate Reynolds numbers, typically hundreds to thousands. This is an important area of theoretical- and applied-fluid mechanics, one that relates, for instance, to the onset of nonlinear instabilities and their effects on the transport properties of the unsteady flow con- figuration. The regime of Reynolds numbers at which these instabilities take place is usually not very high, of the order of thousands, hence basically within reach of present day computer capabilities. Nonetheless, following the full evolution of these transitional flows requires very long-time integrations with short timesteps, which command substantial computational power. Therefore, efficient numerical methods are in great demand. Also of major interest are steady-state or pulsatile flows at moderate Reynolds numbers in complex geometries, such as they occur, for instance, in hemodynamic applications. The application of LBE to such flows will also briefly be mentioned.

To many, total abstinence is easier than perfect moderation.

(Augustine of Hyppo)

18.1 Moderate Reynolds Flows in Simple Geometry

Let us consider a rectangular channel of length L and height H. The fluid enters the channel from the left side (inlet) and flows steadily down the channel under the effect of a constant pressure drop $\Delta P = P_i - P_o$ between the inlet and outlet section. At the top and bottom boundaries the flow speed vanishes (no-slip boundary condition) as a result of the drag exerted by the rigid walls. This is how dissipation balances the momentum input from the pressure gradient, see Fig. 18.1. If the Reynolds number is small enough, typically below 2000 in a cylindrical pipe, occasional fluctuations are unable to break the one-dimensional symmetry of the flow, so that the nonlinear terms of the Navier– Stokes equations vanish identically, hence have no effect on the evolution of the flow.

The Lattice Boltzmann Equation. Sauro Succi, Oxford University Press (2018).
© Sauro Succi. DOI: 10.1093/oso/9780199592357.001.0001

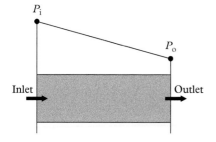

Figure 18.1 *Sketch of channel duct. The fluid enters from the left (inlet) and proceeds rightward to the outlet under the effect of the pressure gradient. Top and bottom walls are solid.*

Under these conditions, the steady-state Navier–Stokes equations reduce to a simple momentum balance between dissipation and pressure gradients (Stokes equation):

$$\rho \nu \partial_{yy} u = -\partial_x P \tag{18.1}$$

Note that the one-directional flow configuration $u = u(y)$, $v = 0$, automatically fulfills the incompressibility condition:

$$\partial_x u + \partial_y v = 0 \tag{18.2}$$

Within LBE, this condition is fulfilled only approximately, typically to order Mach squared, because, as pointed out several times in this book, LBE is a *weakly compressible* approximation of the NSE. We shall return on this point in more detail later.

On the assumption that the pressure gradient is constant throughout the flow, say $g = (P_i - P_o)/\rho L = \text{constant}$, the linear-Stokes equation is easily solved to yield

$$u(y) = U_c \left(1 - \tilde{y}^2\right) \tag{18.3}$$

where

$$U_c = \frac{gH^2}{8\nu} \tag{18.4}$$

is the centerline speed and $\tilde{y} \equiv \frac{2y}{H}$, $-\frac{H}{2} < y < \frac{H}{2}$ is the scaled crossflow coordinate.

The corresponding stress tensor reduces to its transverse component only, namely

$$\tau_{xy} = \frac{\rho \nu U_c}{H} \tilde{y} \tag{18.5}$$

which attains its extrema at the wall ($\tilde{y} = \pm 1$) and vanishes in the centerline of the flow, $\tilde{y} = 0$, see Fig. 18.2.

This solution is stable to transverse perturbations for Reynolds numbers below a critical value Re_c. Beyond this value, dissipation is no longer able to damp spontaneous

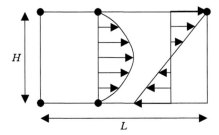

Figure 18.2 *Sketch of the Poiseuille flow and the corresponding stress tensor.*

fluctuations and instabilities develop, which drive the transition to turbulence.[40] The Poiseuille flow test is often used to calibrate the flow viscosity, see Fig. 18.2. In practice, one selects a given viscosity via the theoretical expression, $\nu = c_s^2(1/\omega - 1/2)$, and verifies that the maximal speed attained by the flow agrees with the expression (18.4). In this way, the LBE code serves as a "numerical viscosimeter" (1).

18.2   LBE Implementation

Let us now present in some more detail how to set up a LBE simulation in practice. For the sake of concreteness let us refer to the usual nine-speed 2D lattice, numbered counterclockwise starting from the East.

The first problem we face in the LBE implementation is the representation of the pressure drop ΔP. This bears on a rather delicate theoretical issue that was left silent in the theoretical presentation. As we have already learned, the LBE model applies to quasi-incompressible flows because of the low-Mach expansion that lies behind the whole LBE theory presented so far.

We have also learned that since there is no potential energy between the LBE "molecules," the LBE fluid obeys an ideal gas equation of state $P = \rho c_s^2$. Since the flow is isothermal, it is clear that a regime of no density variation as required by incompressibility implies no pressure variations as well. So, the conundrum is:

Where does the pressure gradient come from?

The answer is that the pressure gradient $\partial_a P$ in LBE is entirely in charge of pressure fluctuations acting on top of a uniform background P_0, namely

$$\partial_a P = \partial_a P_0 + \partial_a p' \sim \partial_a p' \tag{18.6}$$

where the fluctuations $p'/P_0 = \rho'/\rho_0$ are of the order of the Mach number squared. In terms of kinetic theory, pressure gradients are entirely in charge of the non-equilibrium

---

[40] Practical hint: in numerical practice, it is often hard to trigger these instabilities even well beyond the critical threshold. A useful remedy is to generate artificial perturbations of the initial conditions, such as appropriate combinations of sines and cosines. In planning these perturbations, one should not forget compliance with the divergence-freedom condition.

diagonal component of the momentum-flux tensor P_{ab}^{ne}. As a result, it is difficult to sustain a substantial and systematic pressure gradient within the LBE flow. This is the price to pay for being dispensed from solving the pressure field for each instantaneous flow configuration, as it is done in exactly incompressible methods.

The non-equilibrium kinetic fluctuations may break divergence-freedom, but owing to the quadratic dependence on the flow speed, this violation can normally be tolerated, provided the Mach number is sufficiently small, say below 0.1.

The lack of exact incompressibility is not unique to LBE, but a common feature of all weakly compressible approximations of the Navier–Stokes equations. What is unique to LBE, and makes life a bit more difficult here is that the LBE sound speed is not so easily disposable as in conventional CFD, but it is frozen to values $c_s \sim O(1)$ in lattice units.

As a result, at variance with conventional-computational fluid dynamics, it is not possible to send the sound speed virtually to infinity, so as to support significant pressure drops with only minor attendant density changes $\delta\rho = \delta P/c_s^2$, as it is done in *artificial compressibility* methods (2). In a way, one might say that being LB a *natural* weak-compressibility method, it must live with a finite sound speed.

It is worth recalling that since the sound speed (squared) is the pressure change per unit change of density, very small violations of the incompressibility conditions can be achieved even in the presence of significant pressure changes. According to the continuity equation rewritten in the form

$$\partial_a u_a = -u_a \partial_a \ln \rho \tag{18.7}$$

we see that departures from divergence freedom are correspondingly small. The standard way out of this difficulty is to *mimic* the pressure gradient by means of an *equivalent-volume force*, say F_a designed in such a way as to produce the same momentum input to the flow as the actual pressure gradient.

Formally, this is tantamount to invoking an external force term, $F_a(x)$, in the Boltzmann equation. The effect of this force on the generic-macroscopic quantity $\Phi(x,t) \equiv \int f(x,v,t)\phi(v)\,dv$ is given by

$$\frac{\delta\Phi(x,t)}{\delta t} = \langle F_a \frac{\partial f}{\partial v_a}, \phi(v)\rangle = -F_a(x)\langle f, \frac{\partial\Phi}{\partial v_a}\rangle \tag{18.8}$$

where $\phi(v)$ is any function of the velocity coordinates and $\langle\cdot,\cdot\rangle$ denotes scalar product in velocity space.

It is therefore possible to taylor the external force in such a way as to impose the required conditions on the macroscopic variation $\delta\Phi$.

In a Poiseuille flow this is easy: all we need is to bias the collision rule so as to enhance the populations moving along the pressure gradient and place a corresponding penalty on counterstreamers. Let Δf_i be the bias per population at a given site, the modification reads as follows:

$$f_i' = f_i + \Delta f_i \tag{18.9}$$

where it is understood that f_i refers to the state *after* streaming and collision.

For convenience, let us define a set of allied variables $g_i \equiv \Delta f_i / \Delta t$, the population change rates due to the external force. Let us further posit the following expression:

$$g_i = w_i \rho \frac{g c_{ix}}{c_s^2} \qquad (18.10)$$

where $\vec{g} \equiv (g, 0)$ is the acceleration (fictitious gravity) mimicking the effect of the pressure gradient along the x direction. It is readily checked that the expression obeys the following equations:

$$
\begin{cases}
\dfrac{\delta \rho}{\delta t} \equiv \sum_i g_i = 0 \\[2ex]
\dfrac{\delta J_x}{\delta t} \equiv \sum_i g_i c_{ix} = \rho g \\[2ex]
\dfrac{\delta J_y}{\delta t} \equiv \sum_i g_i c_{iy} = 0
\end{cases}
\qquad (18.11)
$$

As a result, the position $\rho g = \nabla P$, as combined with the equation (18.4), delivers

$$g = 8 U_c \frac{\nu}{H^2} \qquad (18.12)$$

This solves the problem of injecting the desired amount of momentum input within the flow.

On practical grounds, the change in the computer program is really trivial, just a few extra lines to add the source term g_i to each population after (or within) the collision routine.

A few remarks are in order.

First, it should be appreciated that the procedure outlined here is fairly general and, with some due caution, it extends to more complex situations described by space, time and also state-dependent forces. *This feature proves exceedingly useful to enrich the basic LBE schemes with additional physics and it has been used for a variety of generalized fluid dynamic applications such as: magnetohydrodynamics, thermal convection, granular flows, suspensions, flows with chemical reactions, flows with phase transitions, to name but a few.* A selected collection of these applications will be presented in part V, dealing with flows beyond the Navier-Stokes description.

The procedure is general, but one must nonetheless make sure that the source term does not spoil stability. With specific reference to the case in point, one must guarantee that g is sufficiently small to prevent counterstreaming populations from going negative (or co-streaming ones exceeding the unit value).

Let us analyze this issue in a more detail. The stability condition on the external forcing terms amounts to requiring that the absolute change of the current density in a single timestep be a small fraction of the current itself.

In equations:

$$\left|\frac{\dot{\mathcal{J}}\Delta t}{\mathcal{J}}\right| \ll 1 \tag{18.13}$$

or, in view of the expression (18.12), $8\nu U_c/H^2 \ll 1$.

It is easily checked that for most parameter values of interest, this is largely fulfilled: by taking $\nu < 0.1$, $U < 0.1$ and $H \sim 100$, the left-hand side is of order 10^{-5} or less.

Note that the group ν/H^2 is the inverse momentum-diffusion time τ_ν and consequently the inequality (18.13) is basically the familiar stability condition $\Delta t < \tau_\nu$, typical of explicit schemes, with a lattice and geometry dependent prefactor.

It is therefore apparent that pressure drop across the flow should account for a small fraction of the reference fluid pressure

$$\left|\frac{\Delta P}{P}\right| \ll 1$$

Breaking such condition is an excellent recipe for failure, even for the simplest flows, such as Poiseuille.

18.3 Boundary Conditions

Coming back to the Poiseuille flow simulation, we still have to impose boundary conditions. To make life easy, we select periodic conditions at the inlet/outlet and no-slip on the top and bottom rigid walls. As per the discussion in Chapter 17, periodic boundary conditions are implemented by simply reinjecting at the inlet the populations leaving the outlet.

With reference to the Sketch 18.1 and Fig. 18.3

$$f_{1,2,8} \text{ (I)} = f_{1,2,8} \text{ (E)} \tag{18.14}$$
$$f_{4,5,6} \text{ (O)} = f_{4,5,6} \text{ (W)} \tag{18.15}$$

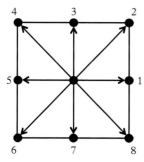

Figure 18.3 *The 9 velocity 2D lattice, again.*

where I and O denote the Inlet and Outlet columns and W and E the leftmost and rightmost (West and East) fluid columns, respectively. No-slip boundary conditions are implemented by bouncing-back the populations leaving the walls:

$$f_{6,7,8}\,(N) = f_{2,3,4}\,(N-1) \tag{18.16}$$

$$f_{2,3,4}\,(S) = f_{6,7,8}\,(S+1) \tag{18.17}$$

where N and S denote the North and South rows and N − 1 and S + 1 the immediate lower-lying and upper-lying fluid rows, respectively.[41] This naive implementation is less than perfect in terms of accuracy, yet good enough for the present purposes.

A map of a channel flow with 10×5 fluid sites, each with its identity (i = inlet, o = outlet, n = north, s = south, f = fluid) is given in the following sketch.

```
6    n   n   n   n   n   n   n   n   n   n   n   n
5    i   f   f   f   f   f   f   f   f   f   f   o
4    i   f   f   f   f   f   f   f   f   f   f   o
3    i   f   f   f   f   f   f   f   f   f   f   o
2    i   f   f   f   f   f   f   f   f   f   f   o
1    i   f   f   f   f   f   f   f   f   f   f   o
0    s   s   s   s   s   s   s   s   s   s   s   s

     0   1   2   3   4   5   6   7   8   9  10  11
```

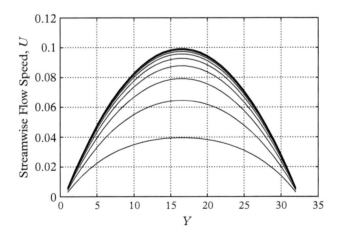

Figure 18.4 *Time development of a LBE-simulated Poiseuille flow.*

---

[41] Practical remark: it is good practice to assign each node of the lattice its own specific identity card, say Fluid, Boundary, Inlet, Outlet and so on (see sketch). This notation, if redundant for simple geometries like the one discussed here, becomes instrumental to an efficient management of the data structures appropriate for more complex geometries.

A simple LB Poiseuille flow is represented in Fig 18.4, which refers to the following parameters: $L = 128$, $H = 32$, $\nu = 0.1$, $U_0 = 0.1$. This has been produced with the sample computer program given in the Appendix, which the reader is kindly encouraged to try it out on their own.

18.4   Flows Past Obstacles

The Poiseuille flow is an easy test because only the linear components of the Navier–Stokes equations take part to the dynamics of the system, the nonlinear ones being consistently left dormant by the symmetries of the problem.

This symmetry spontaneously breaks down for Reynolds numbers roughly above the critical value marking the transition to chaos (loss of temporal coherence) and subsequently full turbulence (loss of both temporal and spatial coherence). The loss of symmetry, and the resulting instabilities can be triggered at much lower Reynolds in the presence of solid obstacles. This type of turbulence is often sought after in practical engineering devices aimed at optimal heat and mass transfer, such as heat exchangers or catalytic converters, simply because turbulence means enhanced heat and mass transfer. As an example of such a situation, we shall discuss the flow in a plane channel containing a periodic array of identical obstacles along the streamwise direction.

As anticipated previously, the idea is to promote flow instabilities at low Reynolds numbers, about one order of magnitude below the critical Reynolds of a free channel flow. This is achieved by exciting large-scale instabilities known as Tollmien–Schlichting waves triggered by the shear-layer instabilities past the thin plate (3).

This flow is obviously much richer than the laminar Poiseuille flow since all nonlinear terms are now in action. Their effect is well visible in terms of coherent structures (vortices) ejected in the wake of the obstacle. From the point of view of the LBE implementation, all proceeds exactly as in the case of the Poiseuille flow, with the minor addition of an internal boundary, the solid obstacle, thin plate. This is easily accommodated in the LB code by adopting the same technique used for the North–South walls of the channel.

This is a nice feature of LBE: *the fully nonlinear regime does not require any extra coding.* For the sake of concreteness we shall again refer to the D2Q9 lattice. Assuming for simplicity the thin plate consisting of a vertical shell spanning h sites in height ($y_s < y < y_n$) and only one lattice unit in width ($x_c < x < x_c + 1$), the non-slip internal boundary condition reads as follows (see Fig. 18.5):

- West-side story:

$$f_5(x_c, y) = f_1(x_c, y), y = y_s, \ldots, y_n \tag{18.18}$$

$$f_6(x_c, y) = f_2(x_c, y), y = y_s, \ldots, y_n \tag{18.19}$$

$$f_4(x_c, y) = f_8(x_c, y), y = y_s, \ldots, y_n \tag{18.20}$$

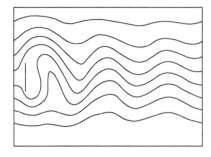

Figure 18.5 *Streamlines of a moderate Reynolds flow past a thin plate (after (4)).*

- East-side story:

$$f_1\,(x_c + 1, y) = f_5\,(x_c + 1, y)\,, y = y_s, \ldots, y_n \tag{18.21}$$

$$f_8\,(x_c + 1, y) = f_4\,(x_c + 1, y)\,, y = y_s, \ldots, y_n \tag{18.22}$$

$$f_2\,(x_c + 1, y) = f_6\,(x_c + 1, y)\,, y = y_s, \ldots, y_n \tag{18.23}$$

Corners require special care, as witnessed by the different bounds of the y intervals. Since we consider a periodic array of obstacles, the inlet and outlet boundary conditions are cyclic and can be implemented exactly as in the case of the Poiseuille flow.

A specific example of LB simulations for this type of flow was first presented in (5) for a series of plates 1 : 10 of the channel height H centered and laid down a distance $3.3\,H$ apart along the streamwise direction.

The stability of the flow was inspected by placing a series of probes at $x_n = x_c + L/n$, $n = 8, 4, 2$ and $y = y_c$, and monitoring the transversal component of the flow speed as a function of time $v = v(t)$. Under stable conditions this signal is flat, only to take an oscillatory waveform $v = A\sin(\omega t)$ for Re above a critical value Re_c.

As the Reynolds number is increased, a second harmonic develops, then a third one and so on until the flow becomes turbulent. The typical dependence of the harmonic amplitudes on the square root of the excess Reynolds number $Re - Re_c$ predicted by the Landau–Ginzburg theory is qualitatively visible in Fig. 18.6.

Extensive early work on time-dependent LBE flows past obstacles can be found in L.S. Luo's thesis work (6), see Fig. 18.8.

18.5 Hemodynamics

Another important category of flows at moderate Reynolds number deals with relatively slow flows in complex geometries, such as for instance the human body, see Fig. 18.6. Here, one is interested in both steady-state and pulsatile solutions, the goal being to provide an accurate description of the spacetime distribution of the blood flow inside the human body, the arterial and brain systems being prime targets in the field.

Indeed, it is by now increasingly accepted that many serious pathologies can be traced to anomalous blood circulation patterns near geometrical irregularities, such as stenoses and aneurisms. Computational Hemodynamics is a highly developed discipline,

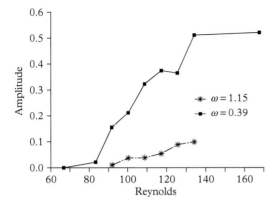

Figure 18.6 *Amplitude of the bifurcated modes of a moderate Reynolds flow past a thin plate (after (5)), as a function of the Reynolds number.*

which relies mostly on finite-volume and finite-element solutions of the Navier–Stokes equations to deal with the major complexity of real-life physiological geometries.

In the last decade, the LB has known major developments in this direction. Several groups have developed hemodynamic LB solvers which can handle the geometric complexity of real-life physiology, such as the human arterial system, as reconstructed from medical imaging techniques. These geometries are typically "painted" and staircased out of regular uniform lattices using however more sophisticated data structures than plain "ijk" cartesian arrays, so as to store only the active (fluid+boundary) nodes in the domain. This is no luxury, but a necessity instead, since the actual fraction of the effective-fluid volume within its embedding box can be easily 1 : 100.

An effective practice is to pre-process the geometry by generating lists of "fluid" and "boundary" sites, and access these lists at run time making use of a connectivity matrix, specifying the list of neighbors of each lattice site. This is obviously more involved and memory-consuming than plain cartesian "ijk" structures, but it permits us to handle geometries of real-life complexity, such as the one shown in Fig. 18.7.

Hemodynamics a topic of major scientific and societal impact. Here, we just wish to point out the major benefits of the LB technique for this kind of flows, namely geometrical flexibility, as combined with outstanding amenability to parallel computing.

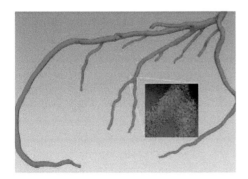

Figure 18.7 *An LB simulation of the coronary-arterial system. The inset shows the flow of red-blood cells represented as finite-size bodies moving within the fluid plasma. Reprinted from* Computer Physics Communications, *184, M. Bernaschi et al, 'Petaflop hydrokinetic simulations of complex flows on massive GPU clusters' 329–341, copyright 2013, with permission from Elsevier.*

None of the two comes for free in real-life applications; the motto "simple models for complex fluids" appears no longer appropriate.

However, once the proper hard work and technical skills are put in place, the result is truly outstanding (8; 9; 10; 11). For instance a recent multiscale LB simulator has been capable of running a one-billion site simulations, with two hundred million floating bodies (red-blood-cells lookalikes) on a two-million core Graphic Processing Unit (GPU) machine, delivering about 0.7 Petaflop/s (1 Petaflop/s = 10^{15} floating point operation per second), with a parallel efficiency of about 90%. To the best of our knowledge, LB is the only method which has proven such kind of parallel scalability on anatomically realistic geometries (11).

The rest can be debated, staircased boundaries versus body-fitted coordinates, Poisson-freedom, versus artificial compressibility methods, accuracy of boundary conditions (especially at the outlet). But the parallel scalability of LB sticks out of these discussions.

18.6 More on the Pressure Field: Poisson Freedom

Before closing this chapter, some further remarks on the pressure gradient implementation are in order. Earlier on, we introduced the notion of "functional equivalence" between the pressure gradient and a volume force designed in such a way as to supply the same momentum input. For the Poiseuille flow, such an equivalence is error-free, since the force was gauged on the exact solution! In general, this equivalence is only an approximate one, since there is no way one can possibly compute a force strictly equivalent to an unknown field!

Still, the procedure keeps making operational sense. The reason is that the pressure field in an incompressible flow *does not obey a prognostic equation* (12) but serves instead as a kinematic constraint enforcing the divergence-freedom of the flow. This constraint acts instantaneously because in an incompressible fluid sound waves propagate at a virtually infinite speed.

In mathematical terms, this is readily checked by taking the divergence of the Navier–Stokes equations and combining it with the solenoidal condition $div\vec{u} = 0$. This yields

$$\Delta P = -\nabla \cdot [(\vec{u} \cdot \nabla)\,\vec{u}]$$ (18.24)

This equation shows that, once the velocity field is known, the pressure field follows instantaneously from the solution of the elliptic Poisson equation. In terms of our equivalence procedure, this means that we are entitled to replace the term ∇P with an equivalent volume force, *as long as this force does not break the incompressibility condition*. This is a necessary but not sufficient condition, though. To be quantitatively consistent, the volume force should match exactly the gradient of the pressure field everywhere within the flow. To this end, one should solve the Poisson problem at each timestep, something we would rather steer clear of, since this is a very costly operation, taking a major fraction of the whole CPU time in exactly incompressible computational fluid dynamics codes.

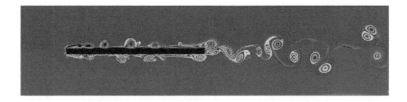

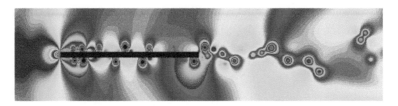

Figure 18.8 *Vorticity and pressure patterns of a von Karman street past a plane slab. (Courtesy of L.S. Luo.)*

The operational shortcut is to give up the idea of matching ∇P pointwise, achieving nonetheless the same global Reynolds number. To that purpose, all we need is to adjust the volume force so as to attain the desired Reynolds based on some average flow speed. This average flow speed is not known beforehand, but it can be adjusted in a few cut-and-try experiments. This is far from elegant, but often more efficient than solving the Poisson problem at each timestep.

Just to clarify things let us refer to the transitional flow past a plate discussed previously.

We know beforehand that by calibrating the volume force on a Poiseuille flow according to eqn (18.12), we would obtain an effective Reynolds number smaller than desired, simply because the internal plate contributes a lot of additional dissipation as compared to a Poiseuille flow.

Let ν_{eff} be the effective viscosity of the flow accounting for the additional dissipation and call $E = \nu_{\text{eff}}/\nu$ the enhancement factor over the bare molecular value. We do not know this value exactly, but very often we have some educated guess for it, i.e., for the coefficent E. What we can do, then is to enhance the volume force by a factor E and achieve a Reynolds in the vicinity of the desired one. This permits us to compute integral figures such as drag and lift coefficients of solid bodies of assorted shapes.

A practical way of doing this is to define a control volume, say an enboxing rectangle, and compute the momentum change per unit volume across the box as a surface integral of the momentum-flux tensor P_{ab}:

$$F_a = \int_{S_{\text{ext}}} P_{ab} \, dS_b + \int_{S_{\text{int}}} P_{ab} \, dS_b \tag{18.25}$$

where $P_{ab} = \rho u_a u_b + P\delta_{ab} - \sigma_{ab}$ is the Navier–Stokes momentum-flux tensor and "ext/int" label the external boundaries of the control volume (the perimeter of the box in our

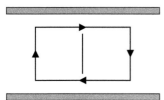

Figure 18.9 *Sketch of a control volume around a thin plate.*

example) and the internal ones (the surface of the body, a straight segment in our case), see Fig. 18.9.

A nice feature of the LB method is that *the momentum-flux tensor is available locally* from its very definition: $P_{ab}(x;t) = \sum_i f_i(x;t)c_{ia}c_{ib}$.

Consequently, the computation of the surface integrals does not involve any explicit space derivative, since these are implicitly contained in the non-equilibrium component of the momentum-flux tensor.

The convective term $\rho u_a u_b$ contributes nothing to the external surface integral if the control volume is large enough to justify the assumption that the flow is one-directional along all four sides of the box. On the top and bottom surfaces, the integral is identically zero because so is the vertical component, $v = 0$. At the inlet/outlet surfaces the integrals are nonzero but they cancel each other on account of the continuity equation $u_{\text{in}} = u_{\text{out}}$. The integral on the body surface is also identically zero by virtue of the no-slip boundary condition $u = v = 0$.

Once the equivalent pressure drop $\Delta P = F_x/H$ across the control volume is computed, one obtains the desired value of the drag coefficient $C_D = 2\Delta P/(\rho U^2)$.

The strength of this procedure is that all "universal" quantities, the ones controlled by the Reynolds number alone, can be computed even if the external force does not solve the Poisson constraint pointwise.

Coming back to general issues, it is clear that the volume–force procedure is ill-suited if one is keen on the precise value of the Reynolds number, but works quite well if the scope is rather a parametric scan over a range of Reynolds numbers. Since the latter scenario is at least as relevant as the former, the volume–force approach appears justified in spite of its empirical character.

However, this leaves a loose end for those engineering studies where the internal distribution of the pressure field is needed in detail. Even more so, if one considers that very often engineering practice requires the specification of pressure boundary conditions.

The weakly compressible LBE is ill-positioned to meet these needs, and must be turned either into an *exactly* incompressible method or into a thermodynamically consistent compressible method, depending on the application.

18.7   Summary

In conclusion, it appears as though LBE is a useful tool for the efficient computation of isothermal, incompressible, transitional flows, as well as for steady-state flows in complex geometries. Much of this efficiency stems from doing away with the Poisson problem for the pressure field and being able to deal with complex geometries in a very efficient way, thanks to the usual property: information moves along straight lines.

The flip side is the inability to reconstruct the internal structure of the pressure field if the incompressibility condition has to be strictly, and not just up to fluctuations of the order of Ma^2. To this purpose, better boundary conditions, possibly coupled with exactly incompressible LB solvers, represent an appealing option. In this respect, the Zou-He boundary condition described in Chapter 17, appears to have gained significant popularity in the last decade.

For a detailed coverage of LB engineering simulations see the very informative books by Guo and Shu (13) and Krugg et al (14).

....

REFERENCES

1. L. Kadanoff, G. McNamara and G. Zanetti, "A Poiseuille viscosimeter for lattice gas automata", *Complex Systems*, 1, 791, 1987, reprinted in *Lattice Gas Models for PDEs*, G. Doolen ed., p. 383, Addison–Wesley, Menlo Park, 1989.
2. A. Chorin, "A numerical method for solving incompressible viscous flows problems", *J. Comp. Phys.*, 2, 12, 1967.
3. H. Tennekes and J. Lumley, *A First Course in Turbulence*, MIT Press, 1972.
4. S. Succi, R. Benzi, F. Higuera, "The Lattice Boltzmann Equation: a new tool for computational fluid dynamics", *Physica D*, 47, 219, (1991).
5. S. Succi, R. Benzi, E. Foti, F. Higuera and F. Szelenyi, "Lattice Boltzmann computing on the IBM 3090/VF", *Cellular Automata and Modelling of Complex Physical Systems*, P. Manneville *et al.* eds, p. 178, Springer–Verlag, Berlin, 1989.
6. L.S. Luo, *Lattice Gas Automata and Lattice Boltzmann Equations for Two-Dimensional Hydrodynamics*, Ph.D. Thesis, Georgia Institute of Technology, 1993.
7. S. Melchionna, M. Bernaschi, S. Succi, E. Kaxiras, F.J. Rybicki, D. Mitsouras, A.U. Coskun and C.L. Feldman, "Hydrokinetic approach to large-scale cardiovascular blood flow", *Comp. Phys. Comm.*, 181, 462 (2010);
8. M.D. Mazzeo, P.V. Coveney, Heme L.B. "A high performance parallel lattice-Boltzmann code for large scale fluid flow in complex geometries", *Comp. Phys. Comm.*, 178, 894, 2008.
9. D.J.W Evans, P.V. Lawford, J. Gunn, D. Walker, D.R. Hose, R.H. Smallwood, B. Chopard, M. Krafczyk, J. Bernsdorf and A. Hoekstra, "The application of multiscale modelling to the process of development and prevention of stenosis in a stented coronary artery", *Phil. Trans. R. Soc. A*, 28, 366, 3343, 2008.
10. J. Latt, B. Chopard, O. Malaspinas, M. Deville, A. Michler Straight velocity boundaries in the lattice Boltzmann method, *Phys. Rev. E*, 77 23, 056703, 2008.

11. M. Bernaschi, M. Bisson, M. Fatica, S. Melchionna, S. Succi "Petaflop hydrokinetic simulations of complex flows on massive GPU clusters", *Comp. Phys. Comm.*, 184, 329 2013.
12. S. Orszag, *Lectures on the Statistical Theory of Fluid Turbulence*, MIT University Press, Boston, 1976.
13. Z. Guo and Y. Shu, "Lattice Boltzmann Method and its Applications in Engineering", Springer Series in Computational Fluid Dynamics, World Scientific, Singapore, 2013.
14. T. Krüger, H. Kusumaatmaja, A. Kuzmin, O. Shardt, G. Silva, E.M. Viggen, The Lattice Boltzmann Method. Principles and Practice, Springer, Berlin, 2017.

....

EXERCISES

1. Set up the thin plate calculation using the sample code provided in this book and run it for various Reynolds numbers between $10 \div 100$. Monitor the velocity field in a series of stations along the centerline $y = H/2$. At what value of Re do you start seeing unsteady flow?
2. For the previous simulations, monitor the local-density fluctuations. Can you predict the order of magnitude?
3. Write down the dynamic LB equation for the pressure field. Is it equivalent to the continuity equation, as combined with the equation of state? If not, why?

19

LBE Flows in Disordered Media

In this chapter we discuss the use of Lattice Boltzmann method for simulation of low-Reynolds flows in highly irregular geometries, such as porous materials and disordered systems. As of today, this represents one of the most consolidated areas of application of the lattice Boltzmann techniques.

There's a crack in everything, that's how the light gets in.

(L. Cohen)

19.1 Introduction

The study of transport phenomena in disordered media is a subject of wide interdisciplinary concern, with many applications in fluid mechanics, condensed matter, life and environmental sciences as well. Flows through grossly irregular (porous) media is a specific fluid mechanical application of great practical value in applied science and engineering. It is arguably also one of the applications of choice of the LBE methods. The dual field–particle character of LBE shines brightly here: the particle-like nature of LBE (populations move along straight particle trajectories) permits a transparent treatment of grossly irregular geometries in terms of elementary mechanical events, such as mirror and bounce-back reflections. On the other hand, the field-like nature of LBE, with populations carrying smooth hydrodynamic information, permits us to achieve fluid-dynamic behavior also in tiny interstitial regions of the flow which might not be easily accessible to the macroscopic approach. These assets were quickly recognized by researchers in the field, and still make LBE (and eventually LGCA) an excellent numerical tool for flows in porous media.

19.2 Flows Through Porous Media

Flow through porous media is a typical *multiscale* phenomenon encompassing several orders of magnitude in scale and involving three and possibly even four basic levels of description, which we conventionally classify as follows:

The Lattice Boltzmann Equation. Sauro Succi, Oxford University Press (2018).
© Sauro Succi. DOI: 10.1093/oso/9780199592357.001.0001

- *Microscopic* (molecular level, say $nm - \mu m$)
- *Mesoscopic* (single-pore level, say $\mu m - mm$)
- *Macroscopic* (many pores, sample-size level, say $mm - m$)
- *Megascopic* (field-size level, say m and above)

An all-embracing approach is clearly out of the question, and consequently specific techniques have been developed for each of these levels.

At the mega/macroscopic level the internal structure of the porous media is ignored and all microscopic knowledge is lumped into coarse-grained quantities, such as the volume of solid/fluid, via so-called *homogeneization* procedures.

A typical output of this analysis is the *permeability*, a property measuring the propensity of the porous media to let fluid flow across, as a function of the aforementioned average quantities. The megascopic level is usually described by a heterogeneous networked assembly of macroscopic units, with locally varying transport properties.

The chief relation at the macroscopic level is Darcy's law:

$$Q = -\frac{\kappa}{\mu} \frac{\Delta P}{L} \tag{19.1}$$

where $Q = UA$ is the volumetric flow rate across the surface of area A, driven by a pressure drop ΔP across the length L. Here μ is the dynamic viscosity of the fluid and κ, the sought macroscopic transport property, is the so-called *permeability* of the porous media. In general κ is a tensorial function of the spatial location $\vec{x}$, $\kappa_{ab}(\vec{x})$ (heterogeneous media), but in the following we shall restrict our analysis to the simple homogeneous isotropic case.

The mesoscopic methods come mainly from statistical mechanics, say network and percolation theory (1). At this scale, the porous media is represented as a network of micropores traversed by the fluid flow, see Fig. 19.1. The permeability of each single micropore/micropipe takes the form

$$\kappa (h) = Ch^2 \tag{19.2}$$

where h is the pore size (width of the micropipe) and C is a numerical constant depending on the specific geometry of the microchannel. For instance, as shown in Chapter 18, the Poiseuille flow between parallel plates yields $C = 1/8$. Assuming the pore network obeys a probabilistic distribution $P(h)$, the effective permeability is obtained as a weighted average over this distribution

$$\kappa = \int_{h_{\min}}^{h_{\max}} P(h) \, k(h) \, dh \tag{19.3}$$

where $h_{\min}$ and $h_{\max}$ are the smallest- and largest-size of individual pores, respectively. Note that on the assumption that the prefactor C in (19.2) does not change much with

$$Q_{ij} = K_{ij}(P_i - P_j) \quad Q_{ik} = K_{ik}(P_i - P_k)$$
$$Q_{jl} = K_{jl}(P_j - P_l) \quad Q_{kl} = K_{kl}(P_k - P_l)$$

$$P_i > P_k > P_j > P_l$$

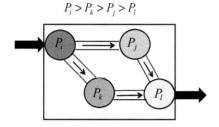

Figure 19.1 *A toy network made of four nodes with pressures P_i, P_k, P_j, P_l. The flow between any two (connected) nodes is proportional to the corresponding pressure unbalance, times a parameter K associated with the permeability of the connecting channel, see formulas on top. In this figure, the pressure decreases from right to left, namely $P_i > P_k > P_j > P_l$, so that the fluid flows from i to j, from i to k, from j to l and from k to l.*

the pore morphology, the expression (19.3) equates the permeability with the variance of the probability distribution $P(h)$.

Of course, at this level, an educated guess for the probability distribution $P(h)$ is needed. Typically, this is taken in the form of a power law reflecting an assumption of scale-freedom of the underlying porous media. This approach is elegant and well positioned to borrow most of the powerful tools of modern-statistical mechanics, such as percolation theory renormalization group methods and others (2). Nevertheless, none of these powerful methods can account for the full details of what happens inside the micropipes. In line with percolation theory, the typical network model would place solid/-void sites on a lattice (not to be confused with lattice gas-cellular automata, or Lattice Boltzmann!) and give a probability p that a given pipe be traversed by the fluid. The factor p responds to morphological constraints, but is totally unaware of hydrodynamic details inside the pipe itself. What happens inside the channel is taken for granted, in the sense that a fluid link is associated with a microscopic permeability which is to be taken as an input parameter of the model. On the other hand, if (as is often the case) the network topology is complex and tortuous, the microscopic flow inside the channel may depart significantly from simple Poiseuille flow and so might the micropermeability $k(h)$.

In other words, eqn (19.3) consists of two components, the distribution function $P(h)$, which carries the *morphological* information, and the micropermeability $\kappa(h)$ which carries the *hydrodynamic* one.

The network approach focuses on the former and takes the latter for granted. This is certainly wise, but not foolproof. In fact, as the network complexity increases, the two parts, morphological and hydrodynamical, can no longer be treated independently. Microflows inside tortuous meanders may give rise to additional losses of momentum (*micro turbulence*) which might sum up coherently to produce appreciable variations of the coefficient in front of Darcy's law, and sometimes even yield nonlinear deviations from Darcy's law, see Fig. 19.2.

This is precisely the framework in which LGCA and LBE may prove valuable sources of *genuinely microhydrodynamic information*. By simulating the Navier–Stokes equations in actual microgeometries, they can tell us whether microhydrodynamic details are irrelevant to Darcy's law or if they produce significant deviations instead.

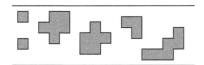

Figure 19.2 *Sketch of a disordered media.*

19.3   LBE Flows Through Porous Media

Let us now discuss how to set up a LBE simulation of porous media. We shall consider a square sample of porous material (say a rock) and put across it a low Reynolds two-dimensional fluid flow driven by an external pressure gradient $G = \Delta P/L$,[42] L being the linear size of the sample.

For the sake of simplicity the internal solid is represented by a collection of squares. This is of course only a sketch of a real porous media, but one that fits well our present purposes. In general, full detailed knowledge of the porous structure can be encoded into a porosity function $p(\vec{x})$, a Boolean array flagging the fluid/solid status of the lattice site $\vec{x}$:

$$p(\vec{x}) = \begin{cases} 1 & \text{(fluid)} \\ 0 & \text{(solid)} \end{cases} \tag{19.4}$$

In the jargon of computer graphics this is known as a "bitmap"; that is, a map of bits (binary digits). From knowledge of this bitmap, one can derive all statistical indicators, such as for instance fluid-fluid spatial autocorrelation

$$A(\vec{r}) = \frac{1}{V} \sum_{\vec{x}} p(\vec{x} + \vec{r}) \, p(\vec{x}) \tag{19.5}$$

where V is the volume of the domain.

Of particular interest for practical applications is the *void fraction* ϕ, defined as the void (fluid) versus the total volume of the sample:

$$\phi = \frac{V_f}{V_s + V_f} \tag{19.6}$$

and the *solid fraction*, its complement to one: $\phi_s \equiv 1 - \phi$. It is easily checked that, by definition, $\phi = A(0)$.

Void fraction and *porosity* will be used interchangeably in the sequel.

In nature these values change greatly from material to material. A few actual figures, from (3), are reported in Table 19.1

The porosity is only a zeroth-order indicator of the porous geometry. Other useful quantities are the minimum, maximum and average grain size, whereby grain we mean a cluster of solid sites (squares in our simple example) (2). Another important property

---

[42] We assume an incompressible fluid whose density can be safely taken as $\rho = 1$.

Table 19.1 *Typical porosity of some representative real materials.*

Material	Porosity (%)
Sandstone	10–20
Clay	45–55
Gravel	30–40
Soils	50–60
Sand	30–40

of the porous structure is the *tortuosity*, roughly speaking, a measure of the average curvature and torsion of the micropipes. To make life easy, here we shall consider a *monodisperse* media where all grains (squares) have the same size h, with their centers laid down randomly across the sample (random media).

The chief theoretical (and practical) question is how much fluid one can push across a sample of section A by applying a given pressure drop ΔP across the longitudinal size L of the sample. As we said, this is measured by the *permeability* κ of the rock sample.

Generally, the permeability is sought in the form of a functional (constitutive) relation with the *porosity* ϕ:

$$\kappa = \kappa\,(\phi) \tag{19.7}$$

This is of course only a simplification that does not take into account other important morphological factors, but it is nonetheless very useful for many practical purposes. The correlation $\kappa = \kappa(\phi)$ is either measured experimentally or, alternatively, it is derived analytically by means of statistical models based on network representations of the porous geometry (1).

The range of validity of these models is typically associated to very low or very high solid fractions ϕ_s, intermediate, non-asymptotic regions being hardly accessible to analytical treatment. This leaves wide scope for the application of direct numerical simulation.

19.4   Setting Up the LBE Simulation

Let us next discuss how to set up a LB flow simulation through a two-dimensional porous media. As for all simulations, one has first to decide about initial and boundary conditions. The latter typically are set to a uniform flow:

$$u(x,y) = U \quad v(x,y) = 0$$

The initial populations can then be taken in the form of local equilibria corresponding to the velocity field.

Next come the boundary conditions.

External boundaries can be dealt with as for solid bodies, the main practical issue being the LBE implementation of internal solid-boundary conditions, the porous media. An effective strategy consists of tagging all boundary-solid sites with a special flag and then associating to each of these nodes a list with the discrete speeds involved in the boundary procedure. Once this topological information, which is built once and for all at the outset, is available, the boundary conditions are implemented via a straightforward sweep over the boundary nodes. A possible pseudo-code illustrating the idea is given here:[43]

```
=======================================================================
For b = 1,NB{               /* sweep thru all boundary nodes       */
  xs = ISOLID[b]
  ys = JSOLID[b]

  For k=1,NF[b]{            /* scan thru all active links connecting the
                               given boundary node with a fluid site */
    xf=IFLUID[b][k]
    yf=JFLUID[b][k]
    is=speed[b][k]         /* set of discrete speeds emanating from
                              the solid boundary site b              */
    if=Mirror[is]          /* set of mirror conjugate discrete
                              fluid speeds                           */
    fpop[is][xs][ys] = fpop[if][xf][yf]    /* bounce-back */
  }
}
=======================================================================
```

The data structure is as follows:

- NB: number of boundary solid sites
- NF(b): number of active links at site b (discrete speeds) connecting the boundary node b to fluid sites
- ISOLID[b], JSOLID[b]: indirect address of the x, y Cartesian coordinates of the solid boundary site
- IFLUID[b][k], JFLUID[b][k]: indirect address of the x, y Cartesian coordinates of the k-th fluid site interacting with the boundary site along the link identified by the mirror conjugate discrete-speed indices if, is
- is=speed[b][k]: actual index of the discrete speed emanating from solid site site b along the k-th entry of the set of active links
- if: actual index of the discrete speed emanating from the fluid site interacting with b along the link (is, if)
- fpop[i][x][y]: i-th population at Cartesian site x, y (see Table 19.2)

---

[43] *Cave canem!* This pseudo-code is only meant to convey the general ideas in the simplest possible way. As such, it has not been tested as a real piece of code, a task which is left to the brave reader.

Table 19.2 *The set of active links at location xs = 4, ys = 3.*

k	site	xf	yf
1	nw	xs-1	ys+1
2	w	xs-1	ys
3	sw	xs-1	ys-1
4	s	xs	ys-1
5	se	xs+1	ys-1

For the sake of concreteness let us illustrate the idea with an example:

At the South-West corner of the solid block ($xs = 4$, $ys = 3$), denoted by B, we have (as usual, counterclockwise numbering starting from East is assumed):

This means that only the North-West, West, South-West, South and South-East fluid neighbors are involved in the population exchanges needed to set the desired boundary condition, typically $u = v = 0$ at the solid site.

An elementary way to impose this boundary condition is simply to bounce-back the populations along the direction they were coming from (standard counterclockwise speed numbering is assumed, starting from eastward propagation).

```
fpop(is,xs,ys)=fpop(if,xf,yf)
```

By scanning the full list of nodes, one can store the set of discrete directions pointing to fluid sites and indices of the corresponding fluid sites (see Table 19.3). This is a bit of a cumbersome pre-processing operation, but fortunately one that needs be performed only once, see Fig. 19.3.

All we need to do at run time is to read off this information from the list and process only *active* sites, namely fluid sites and solid-boundary sites. This data structure is quite general and permits us to address fairly irregular and disordered geometries.

Table 19.3 *Discrete velocities involved in the bounce-back collision with different neighbor sites.*

neighbor site	is	if
nw	4	8
w	5	1
sw	6	2
s	7	3
se	8	4

```
7    f   f   f   f   f   f   f

6    f   f   f   b   b   b   f

5    f   f   f   b   s   b   f

4    f   f   f   b   s   b   f

3    f   f   f   B   b   b   f

2    f   f   f   f   f   f   f

1    f   f   f   f   f   f   f

     1   2   3   4   5   6   7
```

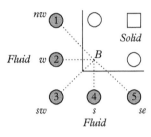

Figure 19.3 *Sketch: numbering of the fluid (f), solid boundary (b) and dead-solid (s) sites. Figure: the five-fluid sites connected to the solid boundary site B.*

```
s   s   s   s   f   f   f   s   s   s

s   s   s   s   f   f   f   s   s   s

s   s   s   s   s   f   s   s   s   s

f   f   f   f   f   f   S   f   f   f

s   s   s   s   s   s   s   s   s   s
```

Figure 19.4 *A pathological porous media with no fluid channel connecting the inlet to the outlet sections.*

Apart from the issue of indirect addressing (which could be dispensed with by using more flexible data structures in C language), it is also fairly efficient.[44] Care needs to be exercised to guard against topological defects, bottlenecks or dead ends, such as the one depicted in Fig. 19.4.

It is clear that this geometry does not allow *any* net flow across the sample simply because there is no fluid path (a path composed by fluid sites only) connecting the inlet and outlet sections of the sample. One such path materializes just by removing a single-solid site, the one labeled S in the previous picture.

Even then, one has to be careful because the communication channel between fluid regions is only one-lattice-site wide, which makes the whole issue of a hydrodynamic

---

[44] This data structure is a bit spendthrift in computer storage. Reserving a memory location also for dead-solid sites (internal solid sites) is a waste of memory, especially at high solid fractions. To cope with this waste, one could think of including only active sites, namely fluid sites and solid boundaries. This stores only the information strictly needed. The resulting data structure is more cumbersome, however, since now *all* sites, even the internal fluids need topological data (the list of interacting neighbors). Thus, part of the memory savings must be surrendered anyway.

representation highly questionable unless a very small-mean-free path is realized. Differently restated, *small pores may not be amenable to hydrodynamic treatment due to high local Knudsen numbers and the attendant free-slip effects.*

It is therefore important to identify the smallest channel width supporting fluid-like behavior. This width marks the gap in scale between the microscopic and meso-scopic levels, which is why the shorter the better. Early computational studies showed that four lattice sites is basically the minimal channel width supporting Poiseuille-like behavior within a single channel. This is a good figure because it shows that the LBE fluid approach can, in principle be taken down almost to the level of a single lattice spacing.

Caveat: This issue has been revisited by a number of authors, see for instance Luo *et al.* (4), who presented a detailed analysis of the inaccuracies associated with the use of single-relaxation time LBGK in small pores, in conjunction with various types of boundary conditions. Use of the multiple-time relaxation with fine-tuned parameters is suggested to provide substantial benefits. Among others, the authors highlight a significant suppression of the spurious dependence of the permeability on the viscosity, thanks to a tight control of the actual location of the solid boundaries.

However, such conclusions are pretty questionable, since the Knudsen dependence presented as a deficiency of the LBGK approach is in fact a genuinely physical effect, see for instance (5). Therefore, as long as the Knudsen number is kept within the proper range, one should not hasten to discard the use of LBGK for flows in porous media.

With this caveat in mind, let us illustrate the point by means of concrete numbers.

Consider water flowing in a cubic rock sample one millimeter in linear size ($L = 10^{-3}$ m) and assume a monodisperse media with $h = 10^{-5}$ meters pore size traversed by an average flow speed $U = 10^{-3}$ m/s.

Taking for simplicity $\nu = 10^{-6}$ m^2/s as a kinematic viscosity of water yields a global Reynolds $Re_L = UL/\nu = 1$, and a local Reynolds $Re_h = 10^{-2}$.

Assuming an average molecular speed $v = 10^3$ m/s, the resulting mean free path in water at room temperature is about $l_\mu \sim v/\nu = 10^{-9}$ m, which means a local Knudsen number $Kn_h = l_\mu/h = 10^{-4}$.

No question that in the physical-world hydrodynamics applies to this microchannel flow.

What about the lattice flow?

Assuming just one lattice spacing for the smallest pore (in fact, we need at least four, but let us look at the order of magnitude alone for the sake of simplicity), we need $N = L/h = 1000$ lattice sites per linear size. This is close to the best one can do with the present computer technology in three dimensions.

The resulting lattice pitch is $\Delta x = 10^{-6}$ m, namely $h/10$. Assuming a sound speed $c_s \sim 10^3$ m/s the corresponding lattice timestep is given by $\Delta t \sim \Delta x/c_s \sim 10^{-9}$ s.

The lattice speed going with this timestep is given by $u_{lb} \equiv U/c_s \sim 10^{-6}$ and the lattice viscosity $\nu_{lb} = \nu \, \Delta t/\Delta x^2 = 10^{-3}$.

Summarizing, in order to keep full consistency with the physical units the lattice simulation should proceed with the following parameters (subscript *lb* means lattice units):

- $N = 1000$ grid points/linear size
- $\Delta x_{lb} = 1$, $\Delta x = 10^{-6}$, m
- $\Delta t_{lb} = 1$, $\Delta t = 10^{-9}$, s
- $u_{lb} = 10^{-6}$, $u = 10^{-3}$, m/s
- $v_{lb} = 10^{-3}$, $v = 10^{-6}$, m^2/s

All is well, except that in order to simulate, say, one physical second, we need one billion timesteps! In order to guarantee correct hydrodynamics (local Knudsen 10^{-4}), LBE is forced to tick with very tiny timesteps, with the consequent heavy loss of efficiency.

How to get around such an awkward state of affairs?

First, let us analyse the reasons of such an inefficiency. This traces back to a lattice sound speed that cannot be made higher than $O(1)$. *The result is a very small timestep because the numerical scheme is bound to track sound waves.* This is again the issue of artificial compressibility discussed in Chapter 18. In a strictly incompressible scheme, sound waves propagate at infinite speed and the numerics can be instructed to not even "think" of tracking them, just let them go! The price is that the pressure field has to be computed via a costly elliptic Poisson problem. The reward is large timestep.

Since we wish to do away with the Poisson problem, a viable strategy is to *make the lattice-Mach much higher than the real one* by raising the lattice-flow speed u and the lattice viscosity v both at the same time so as to *keep the same Reynolds number.*

Given the von Karman relation:

$$Re = \frac{Ma}{Kn} \tag{19.8}$$

it is clear that raising the Mach number at a given Reynolds number means increasing the Knudsen number as well. This is the stage where hydrodynamics is endangered. In our specific example, since the Knudsen number is about 10^{-4} we may hope to be able to raise it by say, two or three orders of magnitude and still observe hydrodynamic behavior at $Kn = 0.01 - 0.1$. A similar argument holds for the Mach-number. This artificial "inflationary" procedure draws on the Reynolds-invariant scale transformation:

$$\begin{cases} Ma \to \lambda\, Ma < 0.1 \\ Kn \to \lambda\, Kn < 0.1 \\ Re \to Re \end{cases} \tag{19.9}$$

where $\lambda >> 1$ is the artificial-inflation factor.

On a ski-oriented analogy, we could say that real molecules perform a special slalom, whereas lattice molecules go for a giant one!

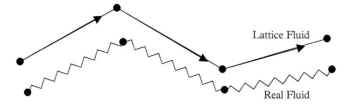

Figure 19.5 *Tight slalom: real fluid. Coarse slalom: lattice fluid.*

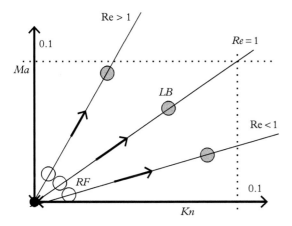

Figure 19.6 *LB fluids (LB, gray circles) and real fluids (RF, open circles) work at the same Reynolds number, but the Knudsen and Mach number of the LB fluid are artificially "inflated" to gain computational efficiency.*

The strategy is illustrated in Fig. 19.5 and Fig. 19.6.

What we are taking here is a bold step, namely assuming that the hydrodynamics settles down in *just a few* mean-free paths. This assumption, which is supported by several molecular-dynamics simulations of simple fluids is *key* to the whole lattice approach. *Should hydrodynamics unforgivingly require hundreds of mean free paths to set in, the lattice approach would be plagued by unacceptable inefficiency.* This said, it is clear that the low-Reynolds regime, $Re < 1$, poses a problem of efficiency, since the Mach number is forced to be proportionally lower than the Knudsen number: at say $Re = 10^{-2}$ and $Kn = 0.1$, the von Karman relation delivers $Ma = 10^{-3}$, a limit which cannot be overcome on pain of leaving the hydrodynamic regime.

Early numerical experiments with the FCHC–LBE scheme were performed to assess up to what Knudsen number lattice hydrodynamics is tenable (6). These tests consisted in running a Poiseuille flow in a rectangular channel of decreasing width and monitoring at what width h the simulation starts to depart from the hydrodynamic relation

$$Q \propto \frac{Gh^2}{\nu} \qquad (19.10)$$

These experiments indicated visible departures from hydrodynamics for Knudsen numbers well above 0.01, see Fig. 19.7 and 19.8.

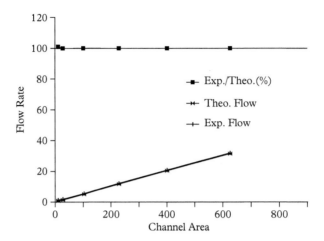

Figure 19.7 *Theoretical and numerical-flow rates as a function of the channel area for $R^* = 7.57$, $u = 0.05$ (after (6)). The linear dependence of the flow rate on the channel area indicates compliance with the Darcy's law.*

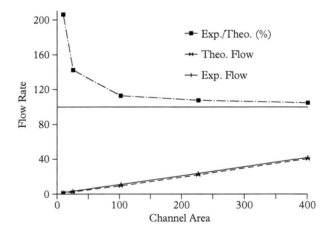

Figure 19.8 *Theoretical and numerical flow rates as a function of the channel area. Significant departures are visible for narrow channels (after (6)).*

This places obvious constraints on the granularity of the porous media. In order to observe hydrodynamic behavior, the smallest pore size should cover at least four to five lattice units: with $R^* = 10$ namely $l_\mu = 0.1\,\Delta x$, $h \sim 5\,\Delta x$, whence $Kn_h = 1/50$. To achieve a granularity $h/L \sim 0.01$, the computational box should consequently cater for 500 lattice sites, which is nowadays easily affordable on most computers.

19.5   Deposition Algorithm

An important ingredient of LBE simulations of porous flow is the preparation of the porous geometry.

For simple geometries such as the one described here this task is simple:

1. First define a coarsening length Λ and a corresponding coarse-grained lattice made up of $(L/\Lambda)^3$ boxes of side Λ, with $\Lambda \gg h$, h being the size of solid feature (a cubelet, either solid or fluid).
2. At a random location $\vec{x}$ within each box B place the center of a solid cubelet.
3. For each box B flip a random number r between 0 and 1. If $r < q$ place the solid cubelet, otherwise move to the next box. Here $0 < q < 1$ is a user-specified parameter which controls the desired solid fraction.

The following procedure grows a monodisperse-porous media with an average solid volume

$$V_s \sim \left(\frac{L}{\Lambda}\right)^3 \times h^3 \times q \tag{19.11}$$

corresponding to an average solid fraction

$$\phi_s = \frac{V_s}{L^3} = \left(\frac{h}{\Lambda}\right)^3 \times q \tag{19.12}$$

This relation fixes the free parameter q in terms of the desired solid fraction. This is just a "vanilla" version of a deposition algorithm. Much more sophisticated procedures designed in such a way as to conform to realistic morphologies can be found in the specialized literature (2; 7; 8).

19.6   Early-days Numerical Simulations

The first LB simulation of porous media was performed over twenty-five years ago on a low resolution 32^3 cubic lattice (9).

These simulations permitted assessment of the validity of Darcy's law and also provided a reasonable estimate of the permeability as a function of the porosity. These early investigations were subsequently refined by Cancelliere *et al.* (10), who doubled the resolution by a factor of two along each direction (64^3). The enhanced resolution allowed a better representation of the microgeometry, namely a random collection of overlapping spheres (see Fig. 19.9).

The spheres were represented in "Legoland" format; that is, as a sequence of cubelets four lattice units in size. Since the centers of the spheres were placed at random

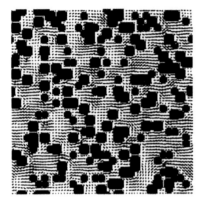

Figure 19.9 *Cross section of a three-dimensional LBE flow in a porous media (after (10)).*

on a 64/4 coarse grid, some degree of overlapping between the spheres was obtained for all but the most diluted porous media. This was desirable since the prime focus of this study was porous media at high solid fraction, typically beyond 0.8. In spite of the rough representation of the spherical surface, the simulations provided a fairly good match with analytical results at both low and high solid fractions. They also showed that the intermediate porosity regime inaccessible to analytical computations connects rather smoothly the low and high solid fraction-asymptotic regions. A number of authors, primarily D. Rothman and co-workers at MIT, and G. Doolen, S. Chen and co-workers at Los Alamos extended these simulations to the case of complex-multiphase flows (7; 11), still today one of the applications of choice for both LGCA and LBE simulations.

Generally speaking, there is a wide consensus that LBE is well suited for simulating microhydrodynamic flows in porous media (12; 13) (see Fig. 19.10 and 19.11).

19.7 Synthetic Matter and Multiscale Modeling

Besides the issue of numerical resolution, the simple numerical example presented in this chapter highlights a general feature of LBE simulations, namely the fact that the *focus is more on dimensionless numbers, such as Reynolds, Mach, Knudsen, than on specific values of the physical parameters, say the sound speed or the fluid viscosity.* Thus, with LBE

Figure 19.10 *Flow across a fibrous-porous media representing a sample of paper (13). Reprinted with permission from A. Koponen et al, Phys. Rev. Lett., 80, 716, (1998). Copyright 1998 by the American Physical Society.*

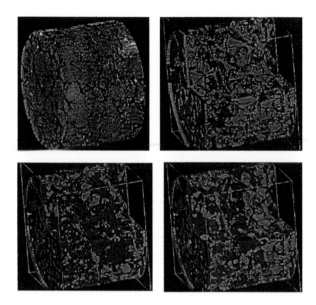

Figure 19.11 *A sandstone sample used in multiphase LBE simulations through porous media (after (7)).*

one should often be prepared to work with *synthetic matter* rather than real materials. The assumption is that real and synthetic matter belong, loosely speaking, to the same class of universality.

The LGCA and LB methods are indeed only gas "analogs" (14), in that they certainly can *not* be used to compute the actual viscosity of air, water or whatever.

However, it is precisely the flip-side of this medal that buys them the capability to describe meso and macroscopic scales which are off-limits for genuinely microscopic methods. Some physics is lost along the way, for sure, but hopefully not the one associated with universal features governed by dimensionless numbers rather than specific physical properties.

This is why LBE (and to some extent LGCA as well) qualify as valuable candidates for *multiscale* problems involving flows in disordered media.

19.8   Some Recent Developments

Flows in porous media have consolidated into a LB mainstream for the last decade. Disordered media of increasing complexity, such as gels, fibers, intracellular environments, eventually including homogeneous- and heterogeneous-chemical reactions have been simulated by several authors. Such models are typically combined with actual X-ray macro- and microtomographic or NMR measurements of real-life samples yielding an increasingly more accurate description of the actual microgeometry of the disordered/porous media.

These studies leverage the symbiotic growth of computing power and technical upgrades of the method, especially in connection with multiscale simulations. Indeed, the modeling of multiscale-heterogeneous media, whereby the local transport coefficients, such as the permeability, change from place to place, and from scale to scale, still poses an outstanding computational challenge. By necessity, in this case, the LB must be coupled to some form of effective media representation, whereby several groups of nodes, say macronodes for convenience are treated like an "effective" porous media, with a given permeability and an ensuing Darcy-like constitutive relation attached to it.

More specifically, each macronode is equipped with a resistance tensor

$$R_{ab}(\vec{x}) = \nu \kappa_{ab}^{-1}(\vec{x}) \tag{19.13}$$

where κ_{ab} is the permeability tensor attached to the macro-node centered in $\vec{x}$. The effective resistance tensor turns into an effective drag force:

$$F_a^{drag} = R_{ab} u_b \tag{19.14}$$

acting on the macronode. The matrix R_{ab} takes value 0 at fluid sites, and goes to infinity at solid wall, and takes intermediate values at porous macronodes. For a unified LB treatment of the various type of macronodes, regardless of the scale they refer to, see Kang *et al.*, (15). More recently, other types of so-called "gray node" LB models have been introduced, which draw their name from the fact that the macronodes result from a collection of 0 and 1 pixels (see Fig. 19.12).

At each gray node, only a fraction ϕ of the LB populations is bounced back, while the complementary fraction $1 - \phi$ goes through, see for instance Zhu and Ma, (16). Many variants, which differ in the details of the definition of the macronodes, as well as of their evolution rules have been proposed in the last decade. The reader keen on specifics of these methods is kindly referred to the current literature.

The modeling of flows in multiscale, heterogenous porous media represents and active front of modern LB research. By and large, however, at least at a pedagogical level, the fundamentals of LB methods for porous flows stay moreless where they stood at the time of the previous book. For novel technical developments and applications the reader is best directed to the original literature, as well as to the books by Guo and Shu (17) and Krueger et al. (18).

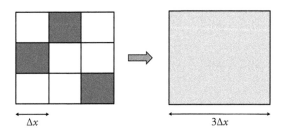

Figure 19.12 *A gray macronode obtained by collecting a 3×3 block of fluid (white) and solid (black) node. To a first approximation, particles entering the gray node have probability $p = 6/9$ (number of white/black cells) to go across it and $q = 3/9$ to be reflected.*

19.9 Summary

Summarizing, flows in porous media represent one of the most suitable applications for LB techniques. They are particularly useful to compute effective transport coefficients, such as the permeability based on actual microgeometric details. The efficiency of the LB simulations would be poor, due to the small timestep imposed by the low-values of the Mach number. Such shortcoming can partially be tamed by artificially inflating the value of the Mach and Knudsen numbers within the bounds compatible with quasi-incompressible hydrodynamics, see Fig. 19.6.

REFERENCES

1. D. Stauffer and A. Aharony, *Introduction to Percolation Theory*, Taylor & Francis, London, 1992.
2. M. Sahimi, "Flow phenomena in rocks: from continuum models to fractals, percolation, cellular automata and simulated annealing," *Rev. Mod. Phys.*, **65**(4), 1393, 1993.
3. J. Bear, *Dynamics of Fluids in Porous Media*, Elsevier, New York, 1972.
4. C. Pan, L.S. Luo and C.T. Miller, "An evaluation of lattice Boltzmann schemes for porous medium flow simulation", *Computers and Fluids*, 35 898, 2006
5. A. S. Ziarani, R. Aguilera, "Knudsen's permeability correction for tight porous media, Transport in Porous Media", **91**, 239 2012.
6. S. Succi, E. Foti and M. Gramignani, "Flow through geometrically irregular media with lattice gas automata", *Meccanica*, **25**, 253, 1990. See also (6; 9).
7. S. Chen, K. Diemer, G. Doolen, K. Eggert, C. Fu, S. Gutman and B. Travis, "Lattice gas models for non-ideal fluids", *Physica D*, **47**, 97, 1991.
8. M. Pilotti, Generation of realistic porous media by grains sedimentation, *Transp. Porous Media*, **33**, 257, 1998.
9. S. Succi, E. Foti and F. Higuera, "Three-dimensional flows in complex geometries with the lattice Boltzmann method", *Europhys. Lett.* **10**(5), 433, 1989.
10. A. Cancelliere, C. Chang, E. Foti, D. Rothman and S. Succi, "Permeability of a three-dimensional random media: comparison of simulation with theory", *Phys. Fluids, A* **2**, 2085, 1990.
11. A. Gunstensen, D. Rothman, S. Zaleski and G. Zanetti, "Lattice Boltzmann model of immiscible fluids", *Phys. Rev. A*, **43**, 4320, 1991.
12. P. Adler, *Porous Media*, Butterworth–Heinemann, London, 1992.
13. A. Koponen, D. Kandhai, E. Hellen, M. Alava, A. Hoekstra, M. Kataja, K. Niskasen and P. Sloot, "Permeability of three-dimensional random fiber webs", *Phys. Rev. Lett.*, **80**(4), 716, 1998.
14. G. Bird, "A contemporary implementation of the direct simulation Monte Carlo method", in *Microscopic Simulations of Complex Flows*, NATO ASI Series B: Vol. 292, 239, 19.

15. Q. Kang, D. Zhang and S. Chen, "Unified lattice Boltzmann method for flow in multiscale porous media", *Phys. Rev. E*, **66**, 056307, 2002.
16. J. Zhu and J. Ma, "An improved gray lattice Boltzmann model for simulating fluid flow in multi-scale porous media", *Adv. in Water Res.* **56**, 61, 2013.
17. Z. Guo and C. Shu, "Lattice Boltzmann Method and its Applications in Engineering, Springer Series in Advances in Computational Fluid Dynamics", vol. 3, 2013.
18. T. Krueger *et al.*, "Lattice Boltzmann Method and its Applications in Engineering, Springer Series in Advances in Computational Fluid Dynamics", vol. 3, 2013.

....

EXERCISES

1. Discuss the physical meaning of the autocorrelation function $A(r)$? *Hint: the product $p(x)p(y)$ is always zero unless both x and y fall on fluid sites.*
2. What value of R^* is needed to achieve hydrodynamic behavior (Darcy's law) on a $h = 4$ sites wide micropore?
3. Based on pseudo-code provided in this book, write up your own porous media simulation into the Fortran code provided in the Appendix. Enjoy the simulation!

20

Lattice Boltzmann for Turbulent Flows

In this chapter, we shall present the main ideas behind the application of LB methods to the simulation of turbulent flows. We restrict our attention to the case of direct numerical simulation, in which all scales of motion are retained within the grid resolution. Turbulence modeling, in which the effect of unresolved scales on the resolved ones is taken into account by various forms of modeling', will be treated subsequently.

Knocking on Heaven's doors ...

(Bob Dylan)

20.1   Fluid Turbulence

Turbulence is the peculiar state of flowing matter, typical of gases and liquids characterized by the loss of coherence of the flow velocity both in time and space. This is the result of the simultaneous nonlinear interaction of a broad spectrum of scales of motion, both in space and time. Such simultaneous interaction gives rise to a host of morphodynamical complexity which makes the mid/long-term behavior of turbulent flows very hard to predict, weather forecasting possibly offering the most popular example in point.

The importance of turbulence, from both theoretical and practical points of view cannot be overstated. Besides the intellectual challenge associated with the issue of predicting the dynamics of spatially extended non-linear systems, the practical relevance of turbulence is even more compelling as one thinks of the pervasive presence of fluids across most natural and industrial endeavors: air and blood flow in our body, gas flows in car engines, geophysical and cosmological flows, to name but a few. Even though the basic equations of motion of fluid turbulence, the Navier–Stokes equations, have been known for nearly two centuries, the problem of predicting the behavior of turbulent flows, even only in a statistical sense is still open to this day.

The Lattice Boltzmann Equation. Sauro Succi, Oxford University Press (2018).
© Sauro Succi. DOI: 10.1093/oso/9780199592357.001.0001

In the last few decades, numerical simulation has played a leading role in advancing the frontier of knowledge of this over-resilient problem (often referred to as the last unsolved problem of classical physics).

To appraise the potential and limitations of the numerical approach to fluid turbulence, it is instructive to revisit some basic facts about the physics of turbulent flows.

The degree of turbulence of a given flow is commonly expressed in terms of a single dimensionless parameter, the Reynolds number, defined as

$$Re = \frac{UL}{\nu} \tag{20.1}$$

where U is a typical macroscopic-flow speed, L the corresponding spatial scale and ν is the kinematic molecular viscosity of the fluid. The Reynolds number measures the relative strength of advective over dissipative phenomena in a fluid flow: $Re \sim u\nabla u/(\nu\Delta u)$. Given the fact that many fluids feature a viscosity around $10^{-6} \div 10^{-5}\,\text{m}^2/\text{s}$ and many flows of practical interest work at speeds above $U = 1\,\text{m/s}$ within devices sized around and above $L = 1\,\text{m}$, it is readily checked that $Re = 10^6$ is commonplace in real life applications. In other words, nonlinear inertia far exceeds dissipation, a hallmark of macroscopic-fluid flow.

The basic physics of turbulence is largely dictated by the way energy is transferred across scales of motion.

To date, turbulence energetics is best understood in terms of an *energy cascade* from large scales ($l \sim L$) where energy is fed into the system, down to small scales where dissipation takes central stage (see Fig. 20.1).

This cascade is driven by the nonlinear mode–mode coupling in momentum space associated with the advective term of the Navier–Stokes equations, $\vec{u} \cdot \nabla\vec{u}$. In real space, this is the familiar steepening of ocean waves as they approach the shore.

The smallest scale reached by the energy cascade is known as the Kolmogorov length, l_k, and marks the point where dissipation takes over advection and organized fluid motion dissolves into that form of incoherent molecular motion that we call heat.

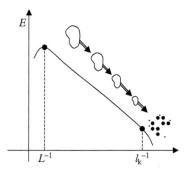

Figure 20.1 *Energy cascade in a turbulent flow: the dam analogy with energy flowing downhill from large to small scales.*

According to the celebrated Kolmogorov (1941) scaling theory (1), the Kolmogorov length can be estimated as

$$l_k \sim \frac{L}{Re^{3/4}} \tag{20.2}$$

The estimate (20.2) follows straight from Kolmogorov's assumption of a *constant-energy flux across all scales of motion*.

Let

$$\epsilon(l) \sim \frac{\delta u^2(l)}{\tau(l)}$$

be the rate of change of the kinetic energy associated with a typical eddy of size l (mass=1 for simplicity). By taking $\tau(l) \sim l/\delta u(l)$, we obtain

$$\epsilon(l) \sim \frac{\delta u^3(l)}{l}$$

Kolmogorov's assumption of scale invariance then implies

$$\delta u(l) = U \, (l/L)^{1/3}$$

where U is the macroscopic velocity at the integral scale L. Note that this relation implies that the velocity-gradient $\delta u(l)/l$ goes to infinity in the limit $l \to 0$ indicating that the flow configuration is singular, i.e., non-differentiable.

By definition, the Kolmogorov (or dissipative) length is such that dissipation and inertia come to an exact balance, hence:

$$\delta u(l_k) l_k / v = 1$$

namely, by writing $\delta u(l_k) = U(l_k/L)^{1/3}$,

$$U \, (l_k/L)^{1/3} (l_k/L) = v/L$$

whence the expression (20.2).

The corresponding-energy spectrum $E(k) \equiv u^2(k)$, $u(k)$ being the Fourier transform of $u(l)$, is readily seen to scale like

$$E(k) = const. \, k^{-5/3} \tag{20.3}$$

the famous Kolmogorov $-5/3$ spectrum.

The qualitative picture emerging from this analysis is fairly captivating: a turbulent flow in a cubic box of size L, at a given Reynolds number Re is represented by a collection of $N_k \sim (L/l_k)^3$ Kolmogorov eddies.

By identifying each Kolmogorov eddy with an independent degree of freedom, a "quantum of turbulence" (sounds close like a good title for a Bond movie . . .), we conclude that the number of degrees of freedom involved in a turbulent flow at Reynolds number Re is given by

$$N_{\text{dof}} \sim Re^{9/4} \tag{20.4}$$

According to this estimate, even a standard flow with $Re = 10^6$ features more than 10^{13} degrees of freedom, enough to saturate the most powerful present-day computers!

This sets the current bar of Direct-Numerical Simulation (DNS) of turbulent flows, manifestly one falling short of meeting the needs raised by many real life applications.

The message comes down quite plainly: *computers alone won't do!*

Of course, this does not mean that computer simulation is useless. Quite the contrary, it plays a pivotal role as a complement and sometimes even an alternative to experimental studies.[45]

Yet, the message is that sheer increase of raw-computer power must be accompanied by a corresponding advance in computational methods.

Turbulent flows are very sensitive to the dimensionality of the space they live in. For instance, three-dimensional fluids support finite dissipation even in the (singular) limit of zero viscosity, while two-dimensional ones do not. This is rather intuitive, since three-dimensional space offers much more morpho-dynamical freedom than two or one-dimensional ones, hence the flow can go correspondingly "wilder."

Therefore, we shall begin our discussion with the earliest LB simulations of two-dimensional turbulence (2).

20.1.1 Two-Dimensional Turbulence

As noted previously, two-dimensional turbulence differs considerably from three-dimensional turbulence. In particular, two-dimensional turbulence supports an infinite number of (so-called Casimir) invariants which can only exist in "Flatland."

These read as follows:

$$\Omega_{2p} = \int_V |\omega|^{2p} \, V, \quad p = 1, 2, \ldots \tag{20.5}$$

where

$$\vec{\omega} = \nabla \times \vec{u} \tag{20.6}$$

is the *vorticity* of the fluid occupying a region of volume V.

---

[45] As pointed out by P. Moin and K. Mahesh, "DNS need not attain real-life Reynolds numbers to be useful in the study of real life applications." (P. Moin and K. Mahesh, Direct numerical simulation: a tool in turbulence research, *Ann. Rev. Fluid Mech.*, 30, 539, 1998.)

By taking the curl of the Navier–Stokes equations, one obtains

$$D_t \vec{\omega} = \nu \Delta \vec{\omega} + \vec{\omega} \cdot \nabla \vec{u} \tag{20.7}$$

where D_t is the material derivative. It is easily seen that the second term on the right-hand side acts as a source/sink of vorticity. It is also readily checked that this term is identically zero in two dimensions because the vorticity of a flow confined to a plane is orthogonal to the plane itself.

As a result, in the inviscid limit $\nu \to 0$, vorticity is a conserved quantity (topological invariant), and so are all its powers. The fact that vorticity, and particularly *enstrophy* Ω_2, is conserved has a profound impact on the scaling laws of 2D turbulence. It can be shown that the enstrophy cascade leads to a fast decaying (k^{-3}) energy spectrum, as opposed to the much slower $k^{-5/3}$ fall-off of 3D turbulence. This regularity derives from the existence of long-lived metastable states, vortices, that manage to escape dissipation for quite long times. The dynamics of these long-lived vortices has been studied in depth by various groups and reveals a number of fascinating aspects whose description goes however beyond the scope of this book. For a very elegant account on turbulence theory see Frisch's book (3).

20.1.2 Turbulence and Kinetic Scales

Since the Kolmogorov length goes to zero in the limit of an infinite Reynolds number, one might wonder whether fully turbulent incompressible flows run against a lack of separation between hydrodynamic and kinetic scales similar to the one encountered in the propagation of shock fronts in gas dynamics.

It is easy to show that, at least within the Kolmogorov theory, just the opposite is true: *the higher the Reynolds number, the better is the separation between kinetic and hydrodynamic scales.* This follows from the Karman relation, namely

$$Re = \frac{Ma}{Kn} \tag{20.8}$$

By writing the mean-free path as $l_\mu = \nu/c_s$ and expressing the viscosity as $\nu = UL/Re$, we obtain $l_\mu/L = Ma/Re$. Based on (20.2), we conclude that the hydrodynamic to kinetic scale separation is given by

$$\frac{l_k}{l_\mu} \sim \frac{Re^{1/4}}{Ma}$$

This shows that, at any finite-Mach number, the kinetic and hydrodynamic scales separate away in the limit of infinite-Reynolds number.

This matches the intuitive notion that kinetic relaxation of molecular trajectories around a given fluid eddy proceeds faster for the smallest eddies than for the largest ones. This also means that *incompressible-fluid turbulence does not require a detailed description of the kinetic-relaxation mechanisms* and provides further motivation for the use of mesoscale methods, such as LB.

20.2 Early LBE Simulations of Two-Dimensional Turbulence

Simulating a homogeneous incompressible turbulent flow with LBE does not require any additional techniques besides those we have already familiarized with previously.

Since we are looking at a tiny internal squarelet of flow, the effects of real boundaries must be minimized by considering periodic-boundary conditions.

The second ingredients are initial conditions.

A customary choice is to start with a turbulent initial flow whose energy spectrum decays with some power law, say $E(k) \sim k^{\alpha}$, while the phase is made random.

On a regular grid (with j, l, m, n integers): $x_j = j\Delta x$, $y_l = l\Delta y$, and a reciprocal grid $k_{x,m} = m\Delta k_x$ $k_{y,n} = n\Delta k_y$, with $\Delta x \Delta k_x = \Delta y \Delta k_y = 2\pi$, the following procedure applies:

1. *Prescribe the initial velocity field*:

$$u_a(x_j, y_l, 0) = \text{IFT}[\hat{u}_a(m, n)].$$

where IFT denotes the inverse-Fourier transform:

$$u_a(j, l) \equiv u_a(x_j, y_l) = \sum_{m,n} e^{i(k_{x,m}x_j + k_{y,n}y_l)} \hat{u}_a(m, n).$$

2. *Set the initial populations*:
   The initial populations are usually defined by the local equilibria associated with $u_a(x, y, t = 0)$:

$$f_i(x, y, 0) = f_i^e[u_a(x, y, t = 0)].$$

Next, one decides between freely decaying or forced turbulence. Free decay does not require any additional ingredient. In the case of forced turbulence, one can still start from the initial configuration described previously, but the turbulence must be sustained by a large-scale forcing of the form

$$\begin{cases} F_x(y; k_f) = F_0 \sin(k_f y) \\ F_y = 0. \end{cases} \tag{20.9}$$

This forcing is a Dirac delta in k space and corresponds to monochromatic energy input at scale $l_f = 2\pi k_f/L$. Eventually a broad-band excitation can be used over a whole range of wavenumbers.

The nominal-Reynolds number is estimated as we did for the Poiseuille flow; that is, by ignoring nonlinear terms and solving the Stokes equation.

The sinusoidal forcing yields a sinusoidal velocity profile of amplitude

$$U_0 = \frac{F_0}{\nu k_f^2} \tag{20.10}$$

Due to nonlinear effects, the flow develops its own bootstrap viscosity so that the observed velocity amplitude is typically only a fraction of U_0 in Equation (20.10). This empirical fraction is then used to upscale the forcing term accordingly. Also, it should be realized that in homogeneous isotropic turbulence the role of U_0 is played by the rms speed $u_{rms} = \sqrt{\langle u^2 \rangle}$.

A correct understanding of the LBE units is key to the interpretation of the simulation results. Consider a reference 1024^2 simulation, and compare this with a pseudospectral method, the most popular technique for this type of flows (see Appendix) (4).

In the LBE code, space is normalized in units of the lattice pitch, $L_{LBE} = 1024$, speed $u_{LBE} = u_{rms} \equiv \sqrt{\langle u^2 \rangle}$, whereas for the pseudospectral codes the common normalization is $L_{PS} = 2\pi$ (the size of the periodic box) and $u_{PS} = 1$, so that the ratio $t_{PS} = L_{PS}/2\pi u_{PS}$, defining the typical large-eddy lifetime, is made one in pseudospectral units. This means that the LBE and pseudospectral clocks are related as follows: $t_{LBE}/t_{PS} = (1024/2\pi)(1/u_{rms})$. Since $u_{rms} \sim .1$ (we cannot exceed the lattice sound speed), we obtain $t_{LBE}/t_{PS} \sim 8000$.

In other words, it takes about eight thousand LBE timesteps to cover a large-eddy turnover time $L_{PS}/2\pi u_{PS}$, i.e., the typical lifetime of a large-scale, energy containing eddy.[46]

Thus, a typical simulation spanning, say, ten spectral units required a few tens of thousands of LBE timesteps. Since a typical LB code involves $O(100)$ floating-point operations/site/step, a single timesteps sweeping out a 1024^2 lattice took approximately one second on a mid-range workstation of those days (say 100 Mflops/s). As a result, a complete LBE simulation took about 40 000 seconds, namely about half a day. The earliest LB simulations of homogeneous isotropic incompressible turbulence were performed on a 2D-decay experiment (2), the main goal being to assess the fidelity of the (then) newborn method for moderately high Reynolds flows, as well as its computational efficiency against state-of-the-art pseudospectral methods.

The specific experiment consisted of a 512^2 simulation using LBE with enhanced collisions with $\nu \sim 0.05$ and $Re \sim 1000$.

A typical LBE vorticity field is shown in Fig. 20.2.

This figure highlighted the presence of various vortices at all scales below the forcing length $L/4$. The global indicators of turbulence dynamics, the total energy $E(t)$ and the

---

[46] Practical warning: since the LB clock is configuration-dependent via the rms speed, it cannot help but meandering in time by some few percent. This is something to bear in mind when making detailed comparisons between the two methods: the two clocks drift slightly apart, a (fake) relativistic effect!

Figure 20.2 *Vorticity map of a* 512^2 *LBE turbulent-flow simulation (after (2)).*

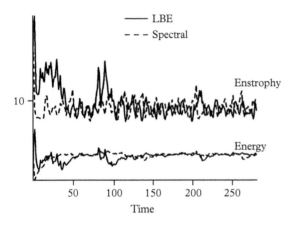

Figure 20.3 *Total energy* $E(t)$ *and enstrophy* $\Omega(t)$ *for the case of Fig. 20.2. Also shown are the same quantities for a corresponding spectral calculation (from (2)).*

total enstrophy $\Omega(t)$ provided a good agreement with pseudospectral calculations (see Fig. 20.3).

Despite this nice global agreement, the vorticity map betrays the absence of truly small scale structures as compared to the corresponding pseudospectral calculations, see Fig. 20.3. This sensation is made quantitative by inspecting the energy spectrum as shown in Fig. 20.4.

From this spectrum one notes that the inertial k^{-3} range was satisfactorily represented, see Fig. 20.4. However, the high k region presented a tiny but persistent energy tail at $k > 128$. Such a phenomenon is also visible in pseudospectral calculations, as witnessed by the energy spectra of LB and pseudospectral simulations at $L = 128^2$ resolution.

A *tentative* explanation goes as follows. High resolution pseudo-spectral simulations do not generally use plain Laplacian dissipation, but rather the so-called *hyperviscous* dissipators of the form $\nu_p \Delta^p$, with p some integer greater than one. The task of these hyper-Laplacians is to sharpen the dissipative boundary, namely keep dissipation at a minimum below the dissipative-Kolmogorov wavenumber $k_d = 2\pi/l_k$ and sharply raise it above it: in k space: $\nu(k/k_d)^{2p}$, see Fig. 20.5.

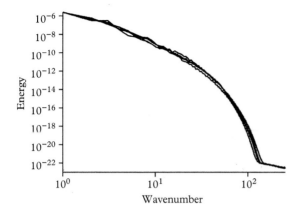

Figure 20.4 *Energy spectrum $E(k)$ for the case of Fig. 20.2 (from (2)).*

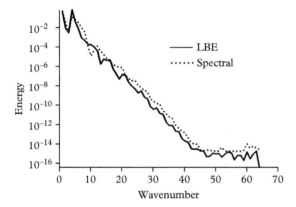

Figure 20.5 *Comparison of LBE and spectral energy at 128×128 grid resolution. Both spectra show underdamped tails, a sign that the grid could probably host a higher-Reynolds number (from (2)).*

The net result is that large structures are less affected by dissipation, whereas small ones are sharply dissipated as they cross the threshold-wavenumber k_d.

Indeed pseudo-spectral methods, especially in 2D, make heavy use of hyperviscous dissipation with fairly high exponents, $p = 8, \ldots, 16$. This produces a twofold effect: first, the inertial range extends deeper down in the vicinity of l_k; second, energetic bursts are more likely to be absorbed, since they meet with a sharper dissipation barrier. While in the spectral formalism the implementation of hyperviscosity is fairly straightforward, just a power law, in the real space LBE formalism the task is more cumbersome, since it requires smoothing averages over remote sites (p sites apart) which compromise LBE locality. We may therefore conclude that, as far as optimization of grid resolution is concerned, LBE lags behind the spectral method (no shame, the spectral method has asymptotic exponential accuracy!).

On the other hand, many researchers would contend that hyperviscosity is just a numerical trick. On such a premise, LBE could well be held as good as pseudo-spectral methods even on serial computers. This is, however, highly technical matter which is better left for the specialized literature.

20.2.1   Sub-Grid Scales and Numerical Stability

Let us now turn another practical aspect of hyperviscosity, namely the ability to protect against numerical instabilities caused by occasional burst of turbulence (*intermittency*). We have already mentioned that LBE was poised to a less ambitious task than LGCA *vis-à-vis* of turbulence. Still, at its inception, there was some hope (at least as far as this author is concerned) that LBE would support a sort of built-in sub-grid scale modeling, namely the ability to describe the effect of unresolved scales ($l < \Delta$ in lattice units) on the resolved ones (5).

Such a property is readily tested by making the viscosity ν (via the eigenvalue λ) so small as to send the dissipative Kolmogorov scale l_k below the lattice spacing, i.e., $l_k < \Delta$ in lattice units.

This means that the flow is left with residual energy to dissipate even below the smallest resolved scale. Usually, this is a source of instability, and the question is whether the numerical scheme can be instructed to dissipate this energy input in a physically plausible form, i.e., not only to prevent unstable blow-up (too little dissipation) but also excessive damping of the resolved structures (too much dissipation).

Of course, this is a rather tall order . . . .

In principle, LBE can still yield hydrodynamic behavior at subgrid scales because the nominal mean-free path can be made a tiny fraction of the lattice spacing. Nevertheless, at these sub-grid scales it is the Taylor expansion of the streaming operator that becomes questionable, because the expansion parameter $c\Delta t$ is by definition $O(1)$ at the scale of the lattice spacing.

In other words, at the scale of a single lattice site, LB is no longer solving the Navier–Stokes equations, but Navier–Stokes plus HOT (higher-order terms) in the *lattice-Knudsen number* $\Delta x/l_k$. In analogy with continuum kinetic theory, we shall dub Navier–Stokes + HOT as a *lattice Burnett* equation. Lattice Burnett aspects have been investigated in depth for the last decade, and some authors have argued strongly against near-grid LB as a fatally doomed regime because of uncontrolled high Knudsen effects.

In the early days, when asymptotic analysis of the subgrid regime was not available yet, numerical experimentation proved the most direct probe of sub-grid behavior. Several experiments in this direction demonstrated that once l_k goes below about two lattice units, destructive instabilities may develop. This is the problem of nonlinear stability which has been addressed before in this book.

Linear instabilities have been argued to contaminate the near and sub-grid physics *before* nonlinearities come into play. The conclusion is that, *with standard LBGK*, ν cannot be brought below a lower nonlinear threshold $\nu_{\min}$ such that

$$Re < Re_{\max} \sim \left(\frac{L}{l_{\text{lbe}}}\right)^{1/\alpha} \qquad (20.11)$$

where L is the linear size of the box in lattice units, α the scaling exponent of the spectrum, namely $\alpha = 1/2$ in 2D and $\alpha = 3/4$ in 3D and l_{lbe} some empirical LBE scale typically around 2 lattice units. More sophisticated estimates based on the strength of the gradients are given in Qian and Orszag (6). The previous expression yields $Re_{max} \sim 10^4$ for $L = 150$ and $Re_{max} \sim 10^6$ for $L = 1500$ in two dimensions.

These figures significantly overestimate the current practice of LBE simulations. In fact, the estimate (20.11) does not take into account numerical prefactors, which cannot be neglected when it comes to pin down hard numbers. These factors, and to a weaker extent the scaling exponents as well are sensitive to the phenomenon of *intermittency*, the propensity of turbulent flows to generate occasional bursts of velocity fluctuations at scale l well in excess of the Kolmogorov estimate $\delta u(l) = (\epsilon l)^{1/3}$, where ϵ is the average dissipation rate in the fluid.

These bursts generate high-local-grid Reynolds numbers

$$Re(l_k) = u(l_k)l_k/\nu > 1$$

which may trigger fatal numerical instabilities because of insufficient short-scale dissipation.

Let us consider a specific example. Suppose we aim at simulating $Re = 10^4$ on a 150^2 square grid. Since the Mach number must be small, a plausible set of parameters is $U = 0.1$ and $\nu = 0.001$. The condition for nonlinear stability reads, in lattice units $(l = \Delta x = 1)$:

$$l\delta u(l) < \nu \qquad (20.12)$$

This means that any velocity fluctuation larger than 10^{-3}, that is one percent of the u_{rms}, between any two neighboring sites of the lattice is a potential trigger of numerical instabilities. Often, but not always, these instabilities materialize in the form of negative populations, which rapidly propagate around the system due to the action of the streaming operator. Under such conditions, stability is endangered by strong non-equilibrium effects triggered by large-local gradients. Of course, a ten times finer lattice would tolerate ten times higher fluctuations, thus ensuring a roughly tenfold better stability.

In the 2001 version I wrote:

> The "bare" LBE scheme has no means of protection against these short-scale fluctuations, probably because at the single lattice spacing, LBE is simulating some sort of Burnett rather than Navier–Stokes equations. Finding a discrete kinetic formulation ensuring stable sub-grid scale behavior would be a major, so far unachieved, breakthrough.

As of today, this statement has been obsoleted by the development of several "post-modern" LBE's, such as entropic schemes, Multi-Relaxation Time and cascaded versions discussed earlier in chapter 14.

The entropic LB enhances the local viscosity "on demand," so as to pass the stability condition (20.12) in a pretty elegant and effective way. The procedure is informed by a powerful physical principle, completely self-driven and parameter-free and it has led to a number of impressive simulations in the recent years.

According to MRT supporters, similar results can be obtained by fine-tuning the matrix eigenvalues, especially in connection with the suppression of spurious pressure fluctuations.

As to supporters of the cascaded version, a minority at the time of this writing, most of the blame for the instability is placed on the lack of Galilean invariance at the grid level. Since preservation of Galilean invariance is the prime target of the cascaded version, the claim is that this is the best way to go. The matter is highly specialized and rather hotly debated among the specialists.

Coming back to LB versus spectral, more details and refined comparative analyses of LBE versus pseudospectral methods in both two and three dimensions can be found in the original papers (7; 9), as well as in subsequent work. For a handy summary see also the books by Guo-Shu and Krueger et al., and references therein.

These authors converge basically to the same conclusions:

* *Excellent quantitative agreement on global observables (total energy, enstrophy as a function of time).*
* *Fairly good agreement on energy spectra, minor discrepancies on long-time fine-grain morphological features such as the specific location of coherent vortices at late stages of the evolution.*
* *Finally, comparable efficiency on serial computers.*

Let us leave this 2D turbulence section with a few considerations on *computational efficiency*. We have already mentioned that LBE takes about $O(100)$ floating-point operations per site and timestep.

In modern LB parlance, this translates in 10 *MLUP/s*, i.e., ten-million lattice updates per second, on a computer delivering one Gflops/s. This means that, on such a computer, one would update ten-million sites per elapsed (wall clock) second.

For the current-grid sizes, say $N \sim 10^3$ grid points, this yields a *serial* performance pretty close to the one delivered by pseudospectral methods.

In view of what we learned in Chapter 19, namely that LBE does not need to solve the Poisson problem for the pressure field, the reader might wonder why would LB not outperform pseudo-spectral methods?

The point is that spectral methods are even better than LBE in handling the Poisson problem since the Laplace operator reduces to a simple algebraic factor $\Delta = -k^2$!

The next ace of the pseudo-spectral method is the underlying Fast Fourier Transform (FFT) algorithm which brings the originally N^2 complexity of the Fourier transformed convective term, down to a mildly super-linear $N \log_2 N$ dependence. For sizes of actual interest $N = O(10^3 \div 10^4)$ the operation count of LBE and pseudospectral is quite comparable so that a similar performance results on *serial computers*.

The pseudospectral method shines for the computation of flows with *periodic* boundary conditions. On the other hand, as soon as periodicity is lost, as it is the case for most geometries of engineering interest, this nice property falls apart, and so does much of the pseudo-spectral power.

LBE, on the other hand, has no problem in accommodating generic boundary conditions for complex geometries Thus, homogeneous-incompressible turbulence in a periodic box is possibly the *least*-favorable test for LBE against pseudospectral methods. The fact that even this test results basically in a tie, must therefore be regarded as an encouraging result for the use of LBE in simulating incompressible-turbulent flows.

20.3   Three-Dimensional-Incompressible Turbulence

The results discussed in the previous section indicate that LBE performs competitively with state-of-the-art pseudo-spectral methods for the case of two-dimensional homogeneous-incompressible turbulence. The same conclusions, possibly reinforced extend to the three-dimensional case.

In the sequel, we discuss briefly the LB simulation of three-dimensional incompressible turbulence starting with the earliest simulations of three-dimensional channel flow.

20.3.1   Three-Dimensional Channel Flow

From the methodological point of view, 3D LB simulations proceed along the same line as two-dimensional ones: cyclic-boundary conditions at inlet/outlet, no-slip boundary conditions on lateral walls and a volume force surrogating a pressure gradient.

The idea was to explore the dynamics of coherent structures, typically vorticity "blobs" ejected from the walls, and understand their effects on the statistical properties of the turbulent flow, primarily the way energy is transported from/to the bulk to/from the wall. This information is of great practical import as it dictates the effective drag exerted by the walls, hence the amount of energy required to push the fluid across the channel. A typical flow configuration showing the presence of these structures is reported in Fig. 20.6.

These simulations were performed on APE, a special-purpose Italian computer designed to perform numerical investigations of lattice quantum chromodynamics, the theory of strongly interacting-elementary particles. APE was a Single-Instruction-Multiple-Data (SIMD) computer (10) featuring up to 2 GB of central memory and at 25 Gflop peak-processing speed. This computer was extremely well suited to any kind of "crystal computation," i.e., computation on regular domains (regular lattices) and simple data structures. Much like quantum chromodynamics, lattice hydrodynamics has all that. And in fact excellent parallel performance was obtained by running the three-dimensional channel-flow LB version on a series of configurations.

This picture reports isosurfaces of the magnitude of the mainstream flow $|u|(x, y, z) =$ const. Coherent patches sprouting off the walls are clearly visible in Fig. 20.6. Just "colorful fluid dynamics"? Not really.

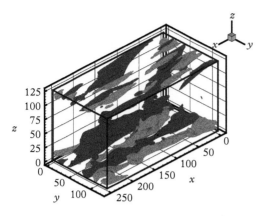

Figure 20.6 *Coherent structures in a 3D LBE channel-flow simulation. Isosurfaces of stream-wise flow at speed magnitude. The turbulent activity in the near-wall region is easily visible. (Courtesy of G. Amati.)*

After a series of validation tests, the parallel LBE code spawned a variety of new (back then) physical results relating to the scaling properties of incompressible turbulent flows. Among others, a new form of scaling known as extended self-similarity (12), which extends beyond the range of validity of Kolmogorov's theory.

20.3.2 Three-Dimensional Turbulence: Parallel Performance

A feature highlighted by the third dimension is the excellent LBE amenability to parallel computing, *provided proper care is taken to optimize memory access*. We have repeatedly stated that one of the major assets of LBE methods is their amenability to parallel computers, see Fig. 20.7. By now, such excellent amenability has been confirmed by several groups around the world and across virtually the entire spectrum of parallel architectures.

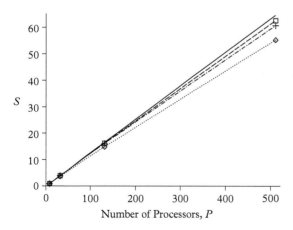

Figure 20.7 *Parallel speed-up of a 3D-parallel-LBE code on the "APE" SIMD computer as a function of the number of processors. The dotted, dash-dotted and dashed lines correspond to 25, 118 and 295 Mbytes of memory, respectively, and the solid line is the theoretical-upper-limit $S = P$) (from (11))*

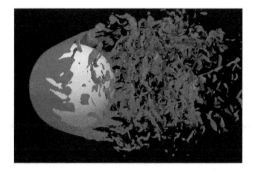

Figure 20.8 *Isocontours of the vorticity field past a sphere from a D3Q27 LB simulation with a special trick to achieve double resolution along each direction corresponding to an effective grid with about 10-billion grid points. Courtesy of S. Ansumali (14).*

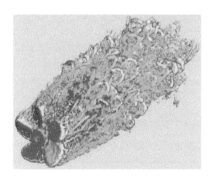

Figure 20.9 *Vorticity patterns past a propeller. The simulation was performed with a 27 discrete-speed entropic LB on a 900^3 grid. Courtesy of S. Chikatamarla (15).*

In 2001, I wrote: *This "much gain no pain" opportunity offered by LBE is probably one of the main assets for its future.*

Fifteen years down the line, this asset stands borne out by many successful applications, on all kinds of parallel architectures including clusters of GPU's (Graphic Processing Units) (see Figs. 20.8 and 20.9). In hindsight, however, the *"much gain no pain"* logo appears simplistic: such top performances don't come from thin air and much care must exercised in order to minimize the overheads due to memory access.

Indeed, since LB is memory intensive, of the order of 20–30 arrays per lattice site, efficient data access and reuse, especially when faced with realistically complex geometries (non-spherical chickens) commands professional attention.

For the case of parallel implementations, recent large-scale simulations with up to 500^3 sites, on heterogeneous clusters with roughly 0.25 Teraflops/s (1 Teraflop = 1000 Gigaflop) peak-performance and a 10 Gb/s infiniband connectivity have shown evidence of better performance of LB versus PS by about a factor 3–5, see (13).

As mentioned, this requires careful optimization of the LB code, and particularly of the strategies aimed at minimizing memory access overheads. To this purpose, data are best organized in terms of energy shells, i.e., populations with the same value of c_i^2, should be stored and accessed contiguously. For instance, to compute the local density of a given site, it is most efficient to proceed by first computing the partial densities of each single shell, and sum them up to form the total density. In equations for the D2Q9 lattice:

$$\rho = \rho_0 + \rho_1 + \rho_2 \tag{20.13}$$

where

$$\rho = f_0, \ \rho_1 = \sum_{i=1}^{4} f_i, \ \rho_2 = \sum_{i=5}^{8} f_i \tag{20.14}$$

are the partial densities of the three energy shells with $c_i =^2= 0$, $c_i^2 = 1$ and $c_i^2 = 2$, respectively.

Another practice which proves useful to minimize data traffic from memory to the processing unit and back is the so-called "fused" strategy, whereby the streaming and collisions steps are performed within a single routine. In practice, this means that the update = stream + collision, from time t to time $t + 1$ at lattice site $\vec{x}$ is performed by directly accessing the neighbor locations $\vec{x} - \vec{c}_i$ at time t. This saves a load/store memory access for each population. On the other hand, in a naive implementation, this requires doubled storage for both arrays at time t (present) and $t + 1$ (future). Such doubled-storage requirement can be avoided by a careful programming of temporary storage preserving the content of the arrays at time t before updating them at time $t + 1$. The price, of course is an increase of the programming complexity.

The reader interested in these highly technical and yet crucial details, is referred to the original papers (16).

20.4 Summary

To sum up, the LB method proves a competitive performer for the direct simulation of incompressible turbulent flows in both two and three dimensions. While its performance on serial computers is comparable with the most consolidated existing methods, typically spectral methods, the main LB asset rests with its high amenability to parallel implementations, especially in the presence of geometries of real-life complexity.

In spite of these inviting credentials, the use of LB for the direct numerical simulation of incompressible turbulence does not seem to have flourished outside a number of academic groups. One possible reason is that the size of current simulations might not have exposed yet a compelling case for LB versus more seasoned numerical methods.

....

REFERENCES

1. A.N. Kolmogorov, "The local structure of turbulence in incompressible viscous fluid for very large Reynolds numbers", *Dokl. Akad. Nauk SSSR*, **30**, 9, 1941.

2. R. Benzi, S. Succi, "Two-dimensional turbulence with the Lattice Boltzmann equation", *J. Phys. A*, 23, L1, 1990.
3. U. Frisch, *Turbulence, the Legacy of A. Kolmogorov*, Cambridge University Press, Cambrdige, 1996.
4. S. Orszag and D. Gottlieb, *Numerical Analysis of Spectral Methods*, SIAM, Philadelphia, 1977.
5. J. Ferziger, "Large eddy simulation", in *Simulation and Modelling of Turbulent Flows*, ICASE/LarC Series in Computational Science and Engineering", p. 109, T. Gatski, M. Hussaini and J. Lumley eds, Oxford University Press, Oxford, 1996.
6. Y. Qian and S. Orszag, "Lattice BGK models for the Navier–Stokes equations: nonlinear deviations in compressible regimes", *Europhys. Lett.*, 21, 255, 1993.
7. D. Martinez, W. Matthaeus, S. Chen and D. Montgomery, "Comparison of spectral methods and lattice Boltzmann simulations of two-dimensional hydrodynamics", *Phys. Fluids*, 6, 1285, 1994.
8. S. Chen, Z. Wang, X. Shan and G. Doolen, "Lattice Boltzmann computational fluid dynamics in three dimensions", *J. Stat. Phys.*, 68, 379, 1992.
9. H. Yu, S.S. Girimaji and L.S. Luo, "Lattice Boltzmann simulations of decaying homogeneous isotropic turbulence", *Phys. Rev E*, 71, 016708, 2005.
10. A. Bartoloni *et al.*, "LBE simulations of Rayleigh–Bénard convection on the APE-100 parallel processor", *Int. J. Mod. Phys. C*, 4, 993, 1993.
11. G. Amati, S. Succi and R. Piva, "Massively parallel lattice-Boltzmann simulation of turbulent channel flow", *Int. J. of Mod. Phys. C*, 8, 869, 1997.
12. R. Benzi, S. Ciliberto, R. Tripiccione, F. Massaioli, C. Baudet and S. Succi, Extended self-similarity in turbulent flows, *Phys. Rev. E*, 48, R29, 1993.
13. S. Ansumali, private communication.
14. M. Namburi, S. Krithivasan and S. Ansumali, "Crystallographic Lattice Boltzmann", *Method Sci. Rep.*, 6, 27172, 2016.
15. S.S. Chikatamarla *et al.*, "Lattice Boltzmann method for direct numerical simulation of turbulent flows", *J. Fluid Mech.*, 656, 298 (2010).
16. A.G. Shet *et al.*, "Data structure and movement for lattice based simulations", *Phys. Rev. E*, 88, 013314 (2013).

EXERCISES

1. Compute the ratio of computational to physical degrees of freedom for a 3D turbulent flow. What do you conclude in the limit of $Re \to \infty$?
2. Prove that the vorticity source/sink term in the vorticity equation vanishes identically in 2D.
3. Derive the Kolmogorov spectrum $E(k) \sim k^{-5/3}$ from the relation (20.2).

Part IV

Lattice Kinetic Theory: Advanced Topics

In Part IV, we present some advanced topics of lattice kinetic theory, namely entropic methods, thermodynamic LB schemes and LB formulations in arbitrary geometries. Each of these developments considerably extends the range and scope of hydrodynamic applications of Lattice Boltzmann theory.

21

Entropic Lattice Boltzmann

The LB with enhanced collisions opened the way to the top-down design of LB schemes, with major gains in flexibility and computational efficiency. However, compliance with the second principle was swept under the carpet in the process, with detrimental effects on the numerical stability of the method. It is quite fortunate that, albeit forgotten, the compliance did not get lost in the top-down procedure. Entropic LB schemes, the object of the present chapter, explain the why's and how's of this nice smile from Lady Luck.

(At equilibrium things work, out of equilibrium, they discover new worlds)

21.1 Rescueing the Second Principle

Like most success stories in computational physics, the LBE theory has gone through a number of lucky strikes. Among others, the fact of exhibiting stable operation in a regime where, in principle, there was no a-priori guarantee of numerical stability.

The point is as follows: by imposing a prescribed form of the local equilibria, the top-down approach described earlier, achieves a dramatic gain in computational efficiency, while still preserving the basic conservation principles. However, the *evolutionary* constraints, as expressed by the second principle and its kinetic underpinning, Boltzmann's *H* theorem went swept under the carpet in the process. Why? Because the local equilibrium was prescribed and attained through relaxation instead of being realized through actual collisions.

This opens a potential liability toward microscopic realizability, hence numerical stability, which we are going to discuss in the sequel.

The Lattice Boltzmann Equation. Sauro Succi, Oxford University Press (2018).
© Sauro Succi. DOI: 10.1093/oso/9780199592357.001.0001

21.1.1   Linear Stability and Time Marching

Let us begin with some elementary considerations on linear stability.

As shown previously, the viscosity of the LB fluid is given by (in lattice units):

$$\nu = c_s^2 \left(\frac{1}{\omega} - \frac{1}{2} \right) \tag{21.1}$$

From this expression, we appreciate that the viscosity is positive in the range

$$0 \le \omega \le 2 \tag{21.2}$$

The lower bound, $\omega = 0$, corresponds to no-collision, i.e., free-streaming, while the upper one is a (singular) zero-viscosity limit. Such positivity range coincides with the linear stability of the explicit Euler time marching for the simple-scalar equation $\dot{f} = -\omega f$, namely

$$f(t + 1) = (1 - \omega)f(t) \tag{21.3}$$

where the timestep is made unit for simplicity.

Numerical stability imposes that, for positive ω, the solution should decay in time, $|f(t + 1)| < |f(t)|$, which implies $|1 - \omega| < 1$, namely the condition (21.2).

Thus, the linear stability of Euler time marching fixes the positivity range of the fluid viscosity. This is an expected and yet reassuring result.

21.1.2   Stability and Realizability

Any kinetic scheme is subject to the so-called realizability constraint, namely

$$0 \le f_i \le 1 \tag{21.4}$$

where we have taken a unit fluid density $\rho = 1$.

In the sequel, we shall focus on the lower bound, which often proves the most demanding one.

For this purpose, let us rewrite the LBGK update in compact form as

$$\hat{f}_i = f_i' \tag{21.5}$$

where

$$\hat{f}_i \equiv f(\vec{x} + \vec{c}_i, t + 1) \tag{21.6}$$

is the spacetime-shifted distribution and

$$f_i' \equiv (1 - \omega)f_i + \omega f_i^e \tag{21.7}$$

is the post-collisional one.

As usual, we have set $\Delta t = 1$ for matter of convenience.

It is of interest to identify the following three distinguished limits:

$$\omega = \begin{cases} 0, & f' = f^e + f^{ne} \equiv f \\ 1, & f' = f^e \\ 2, & f' = f^e - f^{ne} \equiv f^M \end{cases} \tag{21.8}$$

where f^M denotes the mirror of state f under inversion of the non-equilibrium component.

The first case is the plain absence of collision, i.e., free-streaming corresponding to a formally infinite diffusivity. We emphasize that in this limit diffusivity does *not* identify with viscosity, since the latter is a collective property which only makes sense in the hydrodynamic regime. Entropy stays obviously unchanged since there is no interaction.

The last case corresponds to the maximum change of the distribution function compatible with linear stability. As we shall show shortly, this too entails no entropy change, whence the label of "perfect" collision, consistently with the limit of zero viscosity.

Finally, the intermediate case represents the quickest route to local equilibrium and it associates with the maximum entropy change, as detailed below.

The entropy production exhibits a duality between no-collisions and perfect-collisions, while quick-collisions are self-dual, i.e., dual to themselves.

Back to realizability. Realizability imposes that by starting with a non-negative pair $f_i \geq 0$ and $f_i^e > 0$, the shifted $\hat{f}_i$ should also remain non-negative. It is immediately seen that this is guaranteed only in the so-called *under-relaxation* regime:

$$0 \leq \omega \leq 1 \tag{21.9}$$

How about the other, the *over-relaxation* branch of the stability bound? That is,

$$1 < \omega \leq 2 \tag{21.10}$$

By writing (21.7) as

$$f' = f^e - (\omega - 1)f^{ne} \tag{21.11}$$

it is apparent that the more f departs from f^e, the more positivity is endangered by over-relaxation: strong non-equilibrium is the threat.

On the other hand, a back-of-the-envelope calculation shows that owing to resolution constraints, this is precisely the regime LBGK needs to operate for the simulation of high Reynolds flows.

To illuminate the point, let us first write the Reynolds number as

$$Re = \frac{U_{lb}N}{\nu_{lb}} \tag{21.12}$$

where subscript *lb* denotes LB units and N denotes the number of lattice sites per linear dimension.

Let us further recast the relation (21.1) in the form

$$\omega_{lb} = \frac{2}{1 + 6\nu_{lb}} \qquad (21.13)$$

where we have taken $c_s^2 = 1/3$.

Next, let us recall that a well-resolved simulation of a turbulent flow at Reynolds number Re requires of the order of $N = Re^{3/4}$ grid-points per linear dimension (see chapter 20). Given that $U_{lb} \sim 0.1$, the previous expressions show that in order to achieve, say, $Re = 10^4$ with say, $N = 10^3$, (one billion points in three dimensions) one needs $\omega_{lb} = 2/(1 + 0.06) \sim 1.887$, well into the over-relaxation regime.

It is quite remarkable that the old-days LBE with enhanced collisions could indeed function in this potentially highly unprotected regime!

The question is, why?

Because, by a large smile from Lady Luck, the H-theorem, albeit forgotten, did not get lost in the top-down procedure!

This lucky strike was realized only about a decade later, with the development of the entropic version of LB, now known as ELB, for Entropic LB.

In the sequel, we present the basic ingredients of the ELB method, directing the detail-thirsty reader to the original publications.

The *H*-Theorem

The mandate of the ELB is to develop a lattice analog of the H-theorem, namely find a functional $H[f]$ of the discrete populations f_i, such that under the discrete dynamics, the functional can only increase or stay constant, the latter condition marking the attainment of local equilibrium.

Let us begin by considering the H-theorem for the BGK equation in the continuum:

$$\partial_t f + v_a \partial_a f = (1 - \omega)f + \omega g \qquad (21.14)$$

where g is a shorthand for the *attractor* of the collisional operator.

One should *not* hasten to identify g with a local Maxwellian at this stage, but rather regard it as a generic (nonlinear) map of the distribution function onto the attractor itself:

$$g = M[f] \qquad (21.15)$$

The *fixed points* of this mapping, defined by

$$f^* = M[f^*]$$

deliver the local equilibria, i.e., the zero-point of collisional relaxation.

To simplify things further let us consider the case of global, homogeneous equilibria constant in spacetime, i.e., $g = w(v)$.

To this purpose, let us define the departure from the attractor, sometimes also called *residual*,

$$\phi \equiv f - g \qquad (21.16)$$

By definition, the residual ϕ must decay to zero in the course of the evolution taking the system from f to its attractor g.

Since the streaming operator annihilates homogeneous equilibria, the BGK equation takes the simple linear form:

$$\partial_t \phi + v \partial_x \phi = -\omega \phi \qquad (21.17)$$

Upon multiplying both sides by ϕ and integrating in space, and assuming zero flux at the boundaries, one obtains

$$\frac{dH}{dt} = -2\omega H \qquad (21.18)$$

where we have defined the H-functional as

$$H(t) \equiv \frac{1}{2} \int \phi^2(x, t) \, dx \qquad (21.19)$$

This shows that the function $S \equiv -H$ is monotonically increasing in time and can therefore be identified as the entropy of the system.

The inhomogeneous case is clearly more complicated. Nevertheless, as shown in Zwanzig's book, in the general case the BGK equation admits the same H-function as Boltzmann, namely $h[f] = \log f$, with a local-entropy production rate given by

$$\dot{s} = \int (f - g) \log(f/g) \, dv \qquad (21.20)$$

Since the "distance" function is non-negative definite, (for non-negative f and g):

$$d(f, g) \equiv (f - g) \log(f/g) \geq 0 \qquad (21.21)$$

the local entropy production rate is also non-negative definite, $\dot{s} \geq 0$.

The H-theorem for BGK hinges on the orthogonality condition

$$(f - g, \log g) = 0 \qquad (21.22)$$

sometimes also known as logarithmic cylinder condition, because $f - g$ orthogonal to $\log g$ implies that by moving g along its equilibrium subspace, the "vector" f paints a cylinder in Hilbert space (see Figure 21.1).

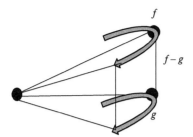

Figure 21.1 *The cylinder condition in Hilbert space. The cylinder is defined as the set of distributions f such that the departure f − g is orthogonal to g.*

The cylinder condition, in turn, derives from the fact that if g is a local Maxwellian, its logarithm depends only on the five microscopic-collision invariants. Consequently, g lives in the submanifold of Hilbert space spanned by the first five Hermite's polynomials, see Fig. 21.1. It can be checked that in the limit $f \rightarrow g$, the logarithmic distance $d(f,g)$ reduces to the standard Euclidean form $(f - g)^2$, which is conducive to an algebraic H-function of the form $h[\phi] = \phi^2$, as in the homogeneous case.

To prepare the H-theorem for the lattice BGK, let us first multiply the BGK equation by $-f$ and integrate in velocity space to obtain

$$\partial_t s + \partial_a s_a = \omega \left[(f,f) - (f,g) \right] \tag{21.23}$$

where

$$s(\vec{x}, t) = -\frac{1}{2}(f,f) \equiv -\frac{1}{2}\int f^2 d\vec{v} \tag{21.24}$$

$$s_a(\vec{x}, t) = -\frac{1}{2}\int v_a f^2 d\vec{v} \tag{21.25}$$

are the local-entropy density and the corresponding-entropy flux, respectively.

In order for the local entropy the production rate $\dot{s} = \omega(f - g, f)$ to be non-negative, the following algebraic-cylinder condition must hold:

$$(f - g, g) = 0 \tag{21.26}$$

so that one can write

$$\dot{s} = \omega(f - g, f - g) \geq 0$$

q.e.d.

A moment's thought reveals that this cannot be true in general, since, unlike its logarithm, the local Maxwellian contains Hermite polynomials at all orders, not just the five microscopic invariants.

However, by identifying g with a quadratic expansion of the local Maxwellian, rather than the full Maxwellian itself, the algebraic-cylinder condition (21.26) can be restored, thus opening the way to a lattice formulation of the H-theorem.

To accomplish this task, a number of technical and conceptual issues have to be addressed, as we are going to discuss in the sequel.

21.2 Lattice *H*-Theorem

We next move on to the fully discrete LBGK world.

Here, things are more complicated because $f_i(\vec{x}, t)$ and $\hat{f}_i \equiv f(\vec{x} + \vec{c}_i, t + 1)$ are two distinct objects, which contribute separately to the local entropy production.

Upon multiplying the LBGK equation once by f_i and once by $\hat{f}_i$ and summing up over all discrete speeds *and* lattice sites, we obtain the following expression for the global entropy production:

$$\Delta S \equiv -\left(H_2(t+1) - H_2(t)\right) = \omega \sum_x \sum_i \left(f_i - g_i\right)\left(f_i + \hat{f}_i\right) \qquad (21.27)$$

As usual, we have assumed that boundary conditions are such to entitle the replacement of $\hat{H}_2 \equiv \sum_x \sum_i \hat{f}_i^2$ with $H_2 \equiv \sum_x \sum_i f_i^2$.

The global entropy production term on the right-hand side of (21.27) has the same sign indeterminacy of its continuum analog, with the additional problem of the presence of the shifted distribution $\hat{f}_i$.

The latter difficulty can be disposed of by rewriting LBGK in the compact form as

$$\hat{f}_i = (1 - \omega) f_i + \omega g_i \qquad (21.28)$$

Substituting this in (21.27) yields

$$\Delta S = -\omega \sum_x \sum_i \left[\omega \phi_i^2 - 2 f_i \phi_i\right] \qquad (21.29)$$

where $\phi_i \equiv f_i - g_i$.

Invoking again the cylinder condition $(g_i, f_i - g_i) = 0$, Expression (21.29) simplifies to

$$\Delta S = \omega (\omega - 2) \sum_x \sum_i \phi_i^2 \qquad (21.30)$$

which is the direct transcription of the continuum counterpart, with the renormalization $\omega \to \omega (\omega - 2)$.

The expression (21.30) shows that discrete target equilibria fulfil a global *H*-theorem, provided ω lies in the range $0 < \omega < 2$, the same stability range of linear stability. This is consistent with the fact the same stability range controls the positivity of the fluid viscosity.

Also to be appreciated is the duality

$$\omega \leftrightarrow 2 - \omega \qquad (21.31)$$

which reflects the entropic equivalence between the infinitely weak (no-collisions) and infinitely strong (instantaneous collisions) regimes discussed earlier. As a anticipated, the self-dual collisions, $\omega = 1$ attain the maximum value of the entropic prefactor $\omega(2 - \omega)$.

Thus, we have now learned that *the H-theorem extends to the over-relaxation branch of the linear stability range*, which is the reason why early LBE's could operate at high-Reynolds number.

This is *not* to say that one can accomplish the dream of zero viscosity, i.e., run the LB simulation all the way up to $\omega = 2$, but explains why one can get pretty close to it, in fact *as close as lattice resolution allows*.

By leveraging the entropic procedure, one can push the viscosity below the limits set by the lattice resolution, i.e., in the subgrid regime, which is in the fact where the ELB shines at its highest.

21.2.1   Entropic LB in the Sub-Grid Regime

In actual practice, nonlinear stability holds up to a thin strip

$$0 < \omega < \omega_{NL}[f] < 2 \tag{21.32}$$

where the threshold $\omega_{NL}[f]$ is an unknown functional of the solution f, via the flow itself and its gradients.

Common practice suggests $\omega_{NL} \sim 2-\epsilon$, where $\epsilon \sim 1/N$, N being the number of lattice sites in one dimension. Note that ϵ identifies with the Lattice-Knudsen number $\Delta x/L$.

Nonetheless, subsequent developments (1; 2; 3) indicated that the actual value of $\omega_{NL}[f]$ can be computed by requiring that a given H-function, or, more generally, Lyapunov functional $\Lambda[f]$, should not decrease under the effect of collisions.

As shown in the chapter 16, the maximum ω compatible with the positivity constraint reads as

$$\omega < \omega_{max} = 1 + |\frac{f^e}{f^{ne}}| = 2 + \left(|\frac{f^e}{f^{ne}}|-1\right) \tag{21.33}$$

For a smooth flow, $|f^{ne}| << f^e$, the constraint (21.33) is always fulfilled since $\omega_{max} > 2$. For very-sharp flows however, ω_{max} is only marginally above 2 and can eventually get smaller than 2, whenever $|f^{ne}| > f^e$, formally yielding to negative viscosity.

It is common ELB practice to express the actual value of the relaxation parameter as a product of two-separate parameters:

$$\omega = \alpha\beta \tag{21.34}$$

with $0 \leq \alpha \leq 2$ and $0 \leq \beta \leq 1$. The first sets the limits where entropy is bound not to decrease, while the latter is then used to regulate the actual value of the viscosity.

The value of α is computed by solving the nonlinear-algebraic problem

$$\Lambda[f] = \Lambda[f^*] \tag{21.35}$$

where star labels the specific post-collisional state, such that entropy is conserved in a collision (isentropic partner).

In other words,

$$f^* = f - \alpha^* f^{ne} \tag{21.36}$$

The value of α is found by solving the isentropic condition (21.35) site by site at each timestep, and sets the fence within which local entropy is bound to not decrease.

Note that the case $\alpha^* = 2$ corresponds to the standard LBGK update, the parameter β being then used to fix the viscosity. In this limit, the isentropic partner simply coincides with the mirror state, $f^* = f^M$.

It should be noted that the ELB is a two-way procedure: not only it increases the viscosity where sharp gradients may go unstable, but it can also decrease it whenever sharp features undergo excessive smearing.

Also to be noted that the ELB viscosity may occasionally turn even negative, $\omega > 2$, thus triggering local instabilities. Provided they don't last "too long," such instabilities can prove beneficial, as they may save small-scales from dissipative death due to overdamping.

This sophisticated, *self-driven*, scenario takes place only wherever and whenever needed, typically in very small critical regions of the fluid.

Everywhere and everyone else, ELB sits quietly under LBGK vests.

21.3   Entropy Minimization Strategies

We next present a few remarks on the interconnection between the existence of an *H*-theorem (nonlinear stability) and the absence of lattice anomalies (consistency).

We have learned previously that Maxwell equilibria necessarily lead to breaking of Galilean invariance in the lattice, because they require isotropy at all-orders in the Mach number.

We also learned that Galilean invariance can be restored to the relevant order by replacing Maxwellian (Fermi–Dirac) distributions with suitable second-order polynomials in the Mach number, and that such polynomial equilibria are in principle consistent with a quadratic *H*-function $H = \sum_i (f_i - g_i)^2$.

Let us now provide some detail on the general mathematical procedure.

The strategy is to minimize the following constrained functional (energy is not included because thermal LBE schemes have yet to be discussed):

$$H_f = \sum_i h(f_i) + A\left[\sum_i f_i - \rho\right] + B_a\left[\sum_i f_i c_{ia} - \mathcal{J}_a\right] \tag{21.37}$$

where $h(\cdot)$ is a generic *convex, positive-definite* functional and A, B_a are Lagrangian parameters.[47]

If such a convex functional can be found, the *local H*-theorem follows straight from eqn (21.28):

$$H_f(t+1) \leq (1-\omega) H_f(t) + \omega H_g(t) \tag{21.38}$$

thus proving the discrete *H*-theorem in local form, see Fig. 21.2.

The global *H*-theorem follows from the local one, with suitable boundary conditions, upon recognizing that the streaming term cannot alter the entropy inside the fluid volume.

Let us now put the entropy-minimization procedure at work to identify the corresponding-discrete equilibria.

The variational requirement min$\{H\}$ implies

$$\frac{\delta H}{\delta f_i} = 0 \quad \rightarrow \quad h'[f_i] = -I_i \equiv -(A + B_a c_{ia}) \tag{21.39}$$

where the prime means derivative with respect to f_i and "$\rightarrow$" means logical implication.

For the specific case of the LBGK operator, it proves convenient to inspect the following family of polynomial *H*-functions:[48]

$$h_p[f] = \frac{f^{p+1} - f}{p}, \quad p > 0 \tag{21.40}$$

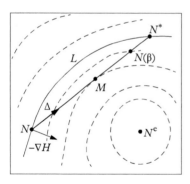

Figure 21.2 *Entropic LBE update. The solid line L is the isentropic curve where $\Lambda(f) = \Lambda(f^*)$ ($N \equiv f$ in this figure). The vector Δ represents the collision operator and the angle between Δ and ∇H is the entropy production inequality ($H \equiv \Lambda$). Point M is the minimum entropy along the segment joining the isentropic partners. The result of the entropic update with (21.34) is represented by the point $N(\beta)$ (the case shown is over-relaxed). A standard LBGK update would be represented by a point along the segment joining N to the local equilibrium N^e. Courtesy of I. Karlin.*

---

[47] A function $f(x)$ is said to be strictly convex if it satisfies the inequality $f(ax + by) \leq af(x) + bf(y)$, with $0 < a, b < 1$, $a + b = 1$. The geometrical meaning is that the segment joining any two points of the graph of f lies below the graph itself.

[48] This is a glorious functional in modern statistical physics, as it relates to many important subjects such as multifractal processes, the replica trick in spin-glass theory and, Tsallis' non-extensive thermodynamics.

Equation (21.39) delivers a whole family of solutions of the form

$$g_i = \frac{1}{p+1}(1 + pI_i)^{1/p} \tag{21.41}$$

It can readily be checked that Boltzmann entropy is recovered as a special case $p \to 0$.

For any integer $p \geq 1$, the equilibria (21.41) represent irrational functions of the argument I_i, which cannot be captured by any finite degree polynomial expansion, apparently suggesting that anomaly-free LBGK polynomial equilibria of any finite degree cannot exhibit an H-theorem (4).

This negative conclusion refers to a "perfect" version of the H-theorem, one that would apply independently of the lattice spacing.

Restricted versions, where compliance with H-theorem is only imposed to the order of accuracy of the LBE method itself (a fairly reasonable stance after all), have indeed been found by Karlin *et al.* (1).

The main point is to enlarge the list of constraints so as to include the hydrodynamic momentum flux tensor $P_{ab} = \rho u_a u_b + p\delta_{ab}$ as well.

Some of the resulting equilibria, like the D2Q9 model are recognizable, some others are not.

It is important to realize that most of these entropies are difficult to analyze because the actual match of the hydrodynamic constraints generally results in a highly nonlinear-multi-dimensional algebraic problem for the Lagrangian parameters, which does not generally deliver closed analytical solutions of these parameters as a function of the hydrodynamic fields.

This stood as a rather serious snag in the way of ELB, hindering further progress in the field for a while.

Early hopes in this direction were provided by Boghosian and collaborators, based on a very elegant-geometrical decomposition of kinetic space into hydrodynamic and kinetic *polytopes* (2).

In the 2001 edition of this book, I wrote: "It is certain that there will be more to learn on this matter in the forthcoming years (6; 7)."

Such substantial developments did indeed take place, as we now proceed to discuss in some detail.

21.4   Modern ELB Developments

Entropic methods have known major developments the last decade, both in terms basic methodology and applications.

Since the material is dense and specialized, in the following we shall be content with a brief survey of the basic ideas, leaving full details to the original papers.

First, a general expression of the lattice H-function, remarkably close to the so-called Kullback-Leibler entropy in information science, has been found, namely

$$H[f] = \sum_{i=1}^{b} f_i \log \frac{f_i}{w_i} + \sum_{k=0}^{N} \lambda_k (M_k - m_k[f]) \qquad (21.42)$$

where the second term on the right-hand side represents the constraints imposed to a list of macroscopic moments M_k, typically, density, momentum and eventually momentum flux. The symbol m_k stands for the microscopic (kinetic) expression of the same moments, i.e., $m_0 = \sum_{i=1}^{b} f_i$ for density, $m_a = \sum_{i=1}^{b} f_i c_{ia}$, $a = 1, 2, 3$ for the current. In general, m_k is a linear map of the discrete populations f_i.

Entropy minimization is achieved by solving the associated variational problem

$$\begin{cases} \dfrac{\delta H}{\delta f_i} = 0 \\[2mm] \dfrac{\delta H}{\delta \lambda_k} = 0 \end{cases} \qquad (21.43)$$

where λ_k denote the variational parameters associated with the constraints.

The solution of variational problem (21.43) delivers *exact* lattice equilibria in the following explicit and *separable* form (8):

$$f_i^{eq} = \prod_{a=1}^{3} \phi(M_a) \chi(M_a)^{\frac{c_{ia}}{c}} \qquad (21.44)$$

where $\phi(M_a)$ and $\chi(M_a)$ are suitable *non-polynomial* shape functions.

For the case of athermal flows, i.e., by imposing only the density and momentum constraints, the analytical solution reads as follows:

$$\phi(M_a) = 2 - \sqrt{1 + M_a^2} \qquad (21.45)$$

and

$$\chi(M_a) = \frac{2u_a/c + \sqrt{1 + M_a^2}}{1 - u_a/c} \qquad (21.46)$$

where $M_a \equiv u_a/c_s$ are the three directional Mach numbers along x, y and z.

The one-dimensional solution is the well-known D1Q3, with the exponent c_{ia}/c taking only the values $(-1, 0, +1)$. Its two- and three-dimensional extensions are the familiar D2Q9 and D3Q27 lattices.

A few comments are in order.

These equilibria are one of the few precious exact results in LB theory. In hindsight, this is due to their inherently one-dimensional nature, separability paving the way to two and three-dimensional extensions. It is interesting to note that separability was the basic argument used by Maxwell to derive his eponymous equilibria, *before* Boltzmann himself rederived them in light of his *H*-theorem!

These equilibria provide a remarkable non-perturbative extension of the standard LBGK equilibria, to which they reduce upon retaining terms up to second order in the local Mach number M_a. They also feature a broader-positivity range, namely $u < c$, $c = \sqrt{3}c_s$ being the lattice light speed.

They show that lattice H-theorems are not restricted to polynomial H-functions and resulting polynomial equilibria.

Finally, the entropy-minimization procedure extends to thermal flows, in which further macroscopic fields, such as the energy, the momentum flux or the energy flux, are added to the list of constraints.

In this case, larger stencils are required, which do not fall within the class of integer multiples of $D1Q3$. Consequently, in order to avoid interpolation, optimal integer approximates of such non-integer lattices have been developed. The reader interested in these specialistic developments are best directed to the original papers.

21.5 Entropic Misunderstandings

Summarizing, the entropic procedure, based on the functional minimization of a *discrete* H-function in Kullback form, leads to an *exact* solution in terms of direct products of one-dimensional, non-polynomial, lattice equilibria.

The discrete speeds are in one-to-one correspondence with the roots of Hermite polynomials, which explains their separability and shows that ELB owes a lot not only to Boltzmann but to Maxwell as well . . . .

By enlarging the list of constraints beyond mass and momentum, the procedure keeps delivering exact, separable, non-polynomial solutions. However, larger stencils are required, which do not fall within the class of integer multiples of $D1Q3$, so that optimal integer approximate lattices have to be used.

One point should be made clear in order to forestall potential misunderstandings on the use of the ELB (9).

The true engine of ELB, the one protecting against numerical instabilities, is not the entropic equilibria, but the entropic estimate leading to a self-adaptive relaxation parameter instead, i.e. eq. (21.35). The use of entropic equilibria alone, in isolation from the entropic estimate of the parameter α discussed previously, cannot and should not be expected to deliver any game-changing benefit as compared to standard LBGK.

Actual practice shows that ELB protects against the development of numerical instabilities in a very subtle and "pre-emptive" way. Whenever the calculation is well resolved, with no danger of local instability, the entropic estimate returns the standard BGK value, $\alpha = 2$. That is, ELB is basically LBGK in disguise, with possibly a minor benefit due to non-perturbative equilibria.

It is only when stability is locally endangered, typically because of strong gradients, that the entropic estimate takes full stage correcting for a self-consistent α, so as to locally steer the local viscosity in such a way to tame instabilities or prevent excessive dissipation, see Fig. 21.3.

And, with a touch of decided elegance, once the danger is over, the ELB kindly bows away returning quietly behind the LBGK curtain.

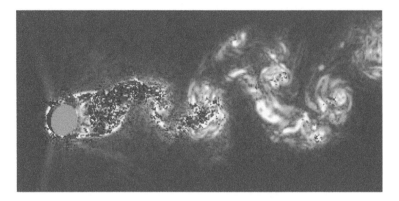

Figure 21.3 *ELB in action for a flow past a cylinder: The red (blue) spots indicate the space location where α < (>)2, respectively As one can appreciate, the α pattern is highly spotty and only covers the nevralgic regions of the flow. Everywhere else ELB is just LBGK in disguise. From I. Karlin et al, Entropy, 17, 8099, (2015).*

Because the viscosity is no longer a fixed parameter, but in fact a "surgical" self-consistent field, some authors have argued that ELB can no longer be classified as direct numerical simulation method. In this author's opinion, such a stance appears rather artificial: all methods based on the direct solution of the Navier–Stokes equations, when stretched to their limits, must resort to some form of "numerical viscosity."

ELB is no exception to the rule, save that its "artificial" viscosity is injected in a most elegant and sensible way, perhaps not coincidentally, due to the fact that ELB is directly informed by the second principle.

It is also inappropriate to dismiss ELB as less efficient than LBGK by comparing the two for flows which are well resolved in the grid.

Indeed, the entropic estimate requires the solution of a nonlinear equation for the parameter α et each lattice site and timestep, which typically requires a few Newton iterations. This is making ELB times more expensive than LBGK by some prefactor between three and five, depending on the flow under consideration.

As explained previously, the comparison is kind of pointless for well-resolved simulations. *The true added value of ELB is for unresolved simulations*, at Reynolds numbers such that the smallest dynamical scales (Kolmogorov scales, see Chapter 20) are not resolved in the grid. Under such conditions, LBGK would typically crash, while ELB continues to provide stable and reliable operation, see Fig. 21.4. Achieving the same stability and accuracy with LBGK would require much finer grids, with a resulting computational overhead largely in excess of the aforementioned factor three to five.

Thus, ELB is basically equivalent in accuracy and more expensive than LBGK for well-resolved simulations. However it permits us to perform unresolved simulations which would require much larger, hence way-more expensive grids, using LBGK. For a detailed discussion of these aspects, see (11).

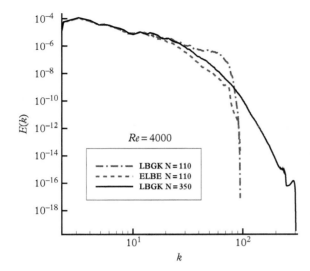

Figure 21.4 *The energy spectrum of a two-dimensional turbulent flow computed with BGK and ELB. It is apparent that ELB with 110 grid points provides a resolved spectrum fairly comparable with the one requiring 350 LBGK grid points. Reprinted with permission from I. Karlin et al, Phys. Rev. E, 84, 068701, (2011). Copyright 2011 by the American Physical Society.*

21.6 Summary

Summarizing, ELB provides two major forward steps: on the theoretical side, it shows that the top-down approach is still compatible with an underlying H-theorem, provided a proper lattice is used. On the practical side, it provides a well-defined procedure to extend the numerical stability beyond the limits imposed by the lattice resolution. This sub-grid regime is indeed the one where ELB outshines LBGK. The operation mode is particularly elegant: the viscosity is corrected only wherever and whenever needed, in a fully self-adaptive operation. Everywhere else, ELB sits silent, virtually indistinguishable from LBGK.

REFERENCES

1. I. Karlin, A. Ferrante and H.C. Oettinger, "Perfect entropies in the lattice Boltzmann method", *Europhys. Lett.*, **62**(3), 1999.
2. B. Boghosian, P. Coveney, J. Yepez and A. Wagner, "Entropic lattice Boltzmann methods", *Proc. R. Soc. London A*, **457**, 717, 2001.
3. H. Chen and C. Teixeira, "H-theorems and origins of instability in thermal lattice Boltzmann models", *Proceedings of the 7th International Conference on Discrete Simulation of Fluids*, Tokyo, July, 1999, North–Holland, Amsterdam; and *Comp. Phys. Comm.* **129**, 21, 2000.
4. A. Wagner, "An H-theorem for the lattice Boltzmann approach to hydrodynamics", *Europhys. Lett.*, **44**(2), 144, 1998.

5. L.S. Luo, "Some recent results on discrete velocity models and ramifications for lattice Boltzmann equation", *Comp. Phys. Comm.*, **129**, 63, 2000.
6. S. Ansumali, I.V. Karlin and H.C. Oettinger, "Single relaxation time model for entropic lattice Boltzmann methods", *Europhys. Lett.*, **63**, 798, 2003.
7. S. Succi, I. Karlin and H. Chen, "The role of the *H* theorem in lattice Boltzmann hydrodynamic simulations", *Rev. Mod. Phys.*, **74**, 1203, 2003.
8. I. Karlin and S.S. Chikatamarla, *Phys. Rev. Lett.*, **97**, 190601, 2006.
9. L.S. Luo *et al.*, "Numerics of the lattice Boltzmann method: Effects of collision models on the lattice Boltzmann simulations", *Phys. Rev. E*, **83**, 056710, 2011.
10. I.V. Karlin, F. Bosch, S. Chikatamarla and S. Succi, "Entropy-assisted computing of low-dissipative systems", *Entropy*, **17**, 8099, 2015.
11. I.V. Karlin, S. Succi and S.S. Chikatamarla, "Comment on Numerics of the lattice Boltzmann method: Effects of collision models on the lattice Boltzmann simulations", *Phys. Rev. E*, **84**, 068701, 2011.

....

EXERCISES

1. Solve the variational problem (21.43) for the case of D1Q3 with only mass and momentum constraints.
2. Compare the D1Q3 entropic equilibria with the one obtained by further imposing the energy constraint.
3. Compare the latter equilibria with the standard D1Q3 expressions.

22
Thermohydrodynamic LBE Schemes

In this chapter we discuss extensions of the LB scheme to thermal flows, where temperature is no longer a control parameter constant throughout the fluid, but a dynamic field evolving in space and time under the drive of energy fluxes within the flow and with the external environment. The extension from athermal to thermal-lattice fluids involves a significant leap of complexity, due to the need of correctly recovering higher-order kinetic moments associated with the energy flux.

In this chapter, we shall describe a number of techniques to accommodate such energy flux within the LB formalism.

(Thermodynamics has many founders but few laws ....)

22.1 Isothermal and Athermal Lattices

Our attention thus far has been confined to the case of *isothermal* fluids, in which the temperature is a constant in space and time throughout the fluid. Many applications, however, involve energy exchanges within the flow and with the external environment. As a result, the fluid temperature becomes a spacetime dependent field, with a dynamics of its own.

To describe thermal flows, the Navier–Stokes equations must be augmented with a dynamic-evolution equation for the internal energy of the fluid. This reads as follows:

$$\partial_t(ne) + \partial_a(neu_a) = \partial_a(n\chi\,\partial_a e) + \dot{q} \tag{22.1}$$

where the internal-energy density is defined as

$$ne = \frac{m}{2}\int (\vec{v}-\vec{u})^2\ f d\vec{v} \tag{22.2}$$

and consequently it relates to the thermal speed as follows:

The Lattice Boltzmann Equation. Sauro Succi, Oxford University Press (2018).
© Sauro Succi. DOI: 10.1093/oso/9780199592357.001.0001

$$e = \frac{mv_T^2}{2} = D\frac{k_B T}{2} \tag{22.3}$$

where D is the number of spatial dimensions and $v_T^2 = D\frac{k_B T}{m}$. In the above, χ is the kinematic energy diffusivity (thermal conductivity) and the last term on the right-hand side is the energy source due to the work done by the momentum flux tensor (stress and pressure) on the fluid. In compact notation:

$$\dot{q} = P_{ab} S_{ab} \tag{22.4}$$

where $S_{ab} = (1/2)(\partial_a u_b + \partial_b u_a)$ is the strain-rate tensor and repeated indices are summed upon.

The eqn (22.1) is best interpreted as the time derivative of the internal energy density, driven by the divergence of the heat flux, which includes both advective and diffusive components, namely

$$Q_a = n e u_a - \partial_a (n\chi \partial_a e) \tag{22.5}$$

This is the low-Knudsen limit of the kinetic heat-flux:

$$q_a = \frac{m}{2} \int v_a (\vec{v} - \vec{u})^2 \, f d\vec{v} \tag{22.6}$$

Being the heat flux a kinetic moment of order *three*, it is immediately concluded that its lattice transcription requires isotropic tensors of order *six*. Such order of isotropy can only be attained by extending the discrete velocity set beyond the first Brillouin cell, i.e., D1Q3, D2Q9 and D3Q27 in lattice parlance.

As a result, thermal LB schemes are typically, but not exclusively, formulated on higher-order lattices.

At the time of this writing, it would be easy to write down the general form of thermal equilibria as a third-order expansion of a local Maxwelian, namely

$$f_i^e = \rho w_i \left(1 + \beta u_i + \frac{\beta^2}{2} q_i + \frac{\beta^3}{6} t_i \right) \tag{22.7}$$

where $\beta = \frac{1}{k_B T}$ is the inverse temperature.

In the above, $u_i = u_a c_{ia}$ and $q_i = u_a u_b (c_{ia} c_{ib} - v_T^2 \delta_{ab})$ are known faces from athermal LB theory, while the cubic (exapole) term

$$t_i = u_a u_b u_c \left[c_{ia} c_{ib} c_{ic} - \frac{1}{\beta} \left(c_{ia} \delta_{bc} + c_{ib} \delta_{ac} + c_{ic} \delta_{ab} \right) \right] \tag{22.8}$$

is the new thermal entry.

Note that in this chapter we shall occasionally make the identification

$$T = c_s^2 \tag{22.9}$$

which is the ideal gas equation of state in natural units $m = k_B = 1$.

Second-order Brillouin lattices, D1Q5, D2Q25 and D3Q125, have sufficient symmetry to fill the six-order isotropy bill in one, two and three spatial dimensions, respectively.

For the pragmatic reader, this could be the operational bottomline.

However, the historical developments of the subject, as well as more recent extensions, have a richer story to tell, which we proceed to illustrate in some detail.

As of today, several families of thermal LBE's can be identified:

- *Extended parametric equilibria*
- *Energy-conserving entropic LB*
- *Double-distribution LB*
- *Hybrid approach*
- *Passive scalar approach*

The first family is the historical one. It counts many variants, all sharing the common feature that the full list of hydrodynamic moments, density, velocity, temperature and heat flux are computed based on a *single* discrete distribution. Hence, they *must* live on higher-order lattices.

The second family is a modern development based on the entropic methods discussed in Chapter 21.

The third family places *two* distinct discrete distributions on a standard (athermal) lattice, one for mass and momentum, and the other for the energy degrees of freedom. In a purely symbolic notation, we may call them particles (*matter*) and phonons (*radiation*). Although somehow artificial, such "matter-radiation" representation proves very convenient because it does not require any higher-order lattice. Over the last decade, it has made proof of significant stability and gained a prominent role in the simulation of thermal LB flows.

The fourth family is called hybrid, as it uses LB only for the fluid density and velocity leaving the solution of the temperature equation to any other numerical method of convenience, typically finite-differences. The rationale of this pragmatic stance is to do away with the extra-complexity associated with higher-order lattices.

Finally, the passive scalar approach is a specific variant which applies to thermal flows within the Boussinesq approximation, i.e., flows in which density changes can be neglected with respect to temperature changes. Among others, an important application is Rayleigh–Bénard convection.

As usual, each of these methods comes with its ups and downs, and, at the time of this writing, they all count a fair share of users within the community. In the sequel, we shall provide the basic ideas behind each of these family of models.

Before doing so, however, a few general considerations on the meaning of temperature in the lattice framework are warranted.

22.2   Temperature in a Discrete World

In classical thermodynamics, temperature is a collective property of macroscopic bodies at equilibrium which governs heat flux between them. From a microscopic viewpoint, temperature is defined through the logarithm of the Boltzmann distribution, $f^e(E) = f^e(0) \, e^{-\beta E}$, namely

$$k_B T = \frac{E}{\log[\frac{f^e(E=0)}{f^e(E)}]} \tag{22.10}$$

This coincides with the kinetic temperature defined through the second-central moment of the distribution function:

$$\int \frac{mc^2}{2} f(\vec{c}) d\vec{c} = \rho D \frac{k_B T}{2} \tag{22.11}$$

where $\vec{c} = \vec{v} - \vec{u}$ is the peculiar velocity.

Note that the kinetic temperature is the same in and out of equilibrium because the above integral *defines* the temperature entering the local equilibrium. Consequently, the kinetic temperature is the same, regardless of whether f or f^e is used to perform the integration.

It is important to bear in mind that the equivalence between the thermodynamic temperature, defined through the Boltzmann logarithm (22.10), and the kinetic one, defined via the second moment (22.11), holds *only* at local equilibrium! More radically, out of equilibrium, thermodynamic temperature looses physical meaning altogether.

Indeed, away from equilibrium, the kinetic temperature stays unchanged by construction, while the thermodynamic one looses physical meaning because the distribution function is no longer exponential and consequently the expression (22.10) delivers a value of T which depends on the actual value of the energy E, hence losing thermodynamic status.

Tricky as this is in the continuum, in a discrete lattice the situation gets even trickier.

Kinetic temperature can still be defined by replacing continuum integrals with discrete sums, but the thermodynamic one is very problematic because the distribution function is a superposition of monochromatic beams, each of which, thermodynamically speaking, carries zero temperature. This is why some authors find it more appropriate to qualify LBE schemes as *athermal* (no-temperature) rather than *isothermal* (constant temperature).

This is more than mere semantics.

Indeed, as pointed out by M. Ernst (1), "*many authors have introduced quantities called 'temperatures' ... which do not coincide with the true temperature in the sense of thermodynamics and statistical mechanics.*" Although Ernst's criticism was directed to cellular automata fluids, it applies to LBE as well, since they are both based on discrete speeds.

The point is that, *in discrete velocity space, the kinetic temperature and the thermodynamic one do not coincide even at local equilibrium.*

This is because, first, the Boltzmann logarithm as applied to the discrete equilibria does not generally return a constant value, so that thermodynamic temperature is ill-defined. Second, even when it does, the reader can check that such is indeed the case for D2Q9, the corresponding temperature does not match the kinetic one.[49]

A subtle description of the ensuing paradoxes and solutions thereof can be found in (2).

22.2.1   Equation of State and Thermodynamic Consistency

In continuum thermodynamics, density, pressure and temperature are related through an equation of state $p = p(\rho, T)$. Such relation is assumed to hold at each point in space and at each time instant, so that only two of these three variables can change independently. In a thermal and compressible flow, pressure is therefore bound to respond to both density and temperature changes, the latter being usually dominant for the case of incompressible fluids.

The simplest equation of state is the one of ideal gases, which we next proceed to derive.

Given the kinetic definition of pressure in an isotropic fluid:

$$p = m \int \frac{c^2}{D} f d\vec{c} \tag{22.12}$$

with $a = x, y, z$, one readily derives the equation of state of an ideal fluid, namely

$$p = \rho v_T^2 = n k_B T \tag{22.13}$$

and the attendant expression for the sound speed:

$$c_s^2 = \frac{\partial p}{\partial \rho}|_T = v_T^2 \tag{22.14}$$

As mentioned previously, thermodynamic consistency requires that changes in temperature, pressure and density be related through the above equations at any point in space and throughout the entire evolution of the flow.

This is not a minor demand for a system living in a lattice, with just a handful of discrete speeds.

A necessary, yet not sufficient, prerequisite for thermodynamic consistency is that these discrete speeds should *not* share the same magnitude (*multi-energy lattices*).

Otherwise, temperature would trivialize to a constant, as we are going to show in the following.

---

[49] To be sure, for D2Q9 the two values are pretty close, 1/3 for the kinetic temperature and $1/(2log(4)) \sim$ 0.36 for the "thermodynamic" one!

Let us consider a discrete distribution function consisting of a set of zero-temperature monochromatic beams:

$$f(\vec{v}) = \sum_{i=1}^{b} f_i \delta(\vec{v} - \vec{c}_i) \tag{22.15}$$

with $|\vec{c}_i| = c$.

Under zero flow, $\vec{u} = 0$, it is readily seen that $\int f(\vec{v}) v^2 d\vec{v} = \sum_i f_i c_i^2 = (\sum_i f_i) c^2 = \rho c^2$. Here $f_i = w_i f(\vec{v}_i)$, where w_i are suitable integration weights.

As a result,

$$k_B T = mc^2$$

In other words, the temperature has no range because all discrete velocity share the same magnitude, with no energy dispersion.

Such a dispersion becomes possible as soon as multiple energy levels become available. For instance, with two distinct levels, $c_1 < c_2$, with densities ρ_1 and ρ_2, respectively, one obtains

$$\frac{k_B T}{m} = (\rho_1 c_1^2 + \rho_1 c_2^2)/(\rho_1 + \rho_2)$$

This clearly exhibits a dynamic range of temperature, between c_1^2 and c_2^2, through the partial densities ρ_1 and ρ_2.

22.3 Thermodynamic Equilibria and Multi-Energy Lattices

The mandate of a thermodynamically consistent LBE is to provide stable operation within the widest possible range of temperatures.

To inspect this issue, it proves convenient to organize thermal LB lattices in terms of *energy shells*, according to a two-index notation:

$$\epsilon_j = \frac{1}{2} c_{ij}^2, \ i = 1, N_j \tag{22.16}$$

where j labels the energy level and i the discrete speed "momentum eigenstate" within energy level j.[50]

Multi-energy lattices have been met before in this book. For instance, the $D2Q9$ model splits into the direct sum of three sub-lattices D2E0, D2E1, D2E2, with energy 0, 1/2, 1, respectively, where the notation Ej means $2\epsilon = j$, see Figs. 22.1 and 22.2.

---

[50] A useful mnemonic is to think of a momentum–energy pair of quantum numbers.

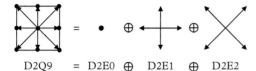

D2Q9 = D2E0 $\oplus$ D2E1 $\oplus$ D2E2

Figure 22.1 *The decomposition of D2Q9 in three energy levels, 0, 1 and 2.*

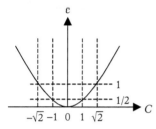

Figure 22.2 *Sketch of the parabolic energy–momentum relation with three discrete energy levels.*

With this definition, the internal-energy density ρe becomes a weighted sum over three energy levels:

$$\rho e = \rho_1 \epsilon_1 + \rho_2 \epsilon_2 + \rho_3 \epsilon_3 \tag{22.17}$$

where ρ_j are the partial densities of the three energy shells.

This does *not* mean that D2Q9 supports thermohydrodynamic flows; in fact, it does not because the heat-flux tensor is not isotropic. It just shows that some athermal LB lattices exhibit a finite-temperature range.

As note above, a given lattice with minimum(maximum) speeds c_{min}, c_{max} supports temperatures in the range

$$\frac{c_{min}^2}{D} < \frac{k_B T}{m} < \frac{c_{max}^2}{D} \tag{22.18}$$

For instance, with reference to D2Q9, $T = 0$ means that all discrete populations have "condensed" into the rest particle state $f_0 = \rho$, $f_i = 0$, $i > 0$, whereas in the "hottest" phase $T = 1$ only the high-energy states $c = \sqrt{2}$ are populated. Pictorially, we could call these the "Bose-Einstein-Condensate" and "Quark-Gluon-Plasma" limits, respectively, for they mimic the distribution function in momentum space of these two extreme opposite states of matter.

It should be clearly borne in mind that these are only realizability constraints, which say a precious little about the actual stability of the corresponding thermal LBE.

Intuitively, we expect safer navigation away from the realizability edges, but there is no way to decide this *a priori*, without first developing a more precise notion of discrete thermodynamic equilibria.

This requires the inspection of the set of thermohydrodynamic constraints to be fulfilled by thermal LBE equilibria, which was precisely the starting point of earliest efforts in the field.

22.4   Extended Parametric Equilibria

At a time where it was not yet realized that lattice equilibria can be derived systematically from a low-Mach expansion of a local Maxwellian, the most natural route to include
thermal (and compressibility) effects was to include higher-order terms in the polynomial expression of the local equilibria. Thermal effects are accounted for by letting the
coefficients of the expansion depend smoothly on the temperature T of the flow.[51]

To widen the range of accessible Mach numbers, as well as to enforce the correct
thermodynamic constraints, Alexander, Chen and Sterling proposed the following $6\mathcal{J}$
parameter (each parameter accounts for $\mathcal{J}$ degrees of freedom, one per energy level)
expansion of the local equilibrium (4):

$$f_{ij}^e = A_j + B_j u_{ij} + C_j u_{ij}^2 + D_j u_j^2 + E_j u_{ij}^3 + F_j u_{ij} u_j^2 \tag{22.19}$$

where $u_{ij} = c_{ija} u_a$ are the covariant components of the flow speed along the discrete
directions of the lattice and $u_j^2 = \sum_i u_{ij}^2$.

The idea is to adjust the $6\mathcal{J}$ Lagrange multipliers $A_j, \ldots, F_j$ in such a way as to recover
the desired set of macroscopic moments. Owing to the wide degree of latitude provided
by this set of parameters, this looks like a rather comfortable situation, as long as one is
able to pick up physically meaningful solutions out of a redundant set of possible ones.

Let us focus on the task of putting this huge freedom at work to impose the set of
usual hydrodynamic constraints:

$$\sum_{ij} f_{ij}^e = \rho \tag{22.20}$$

$$\sum_{ij} f_{ij}^e c_{ija} = \rho u_a \tag{22.21}$$

$$\sum_{ij} f_{ij}^e c_{ija} c_{ijb} = \rho \left(u_a u_b + v_T^2 \delta_{ab} \right) \tag{22.22}$$

$$\sum_{ij} f_{ij}^{ne} c_{ija} c_{ijb} = \rho \nu [S_{ab} - (\partial_c u_c) \delta_{ab}] \tag{22.23}$$

where S_{ab} is the strain rate tensor.

This set must be augmented with the additional constraints ensuring that the internal
energy (temperature) e obeys the correct energy-transport equation.

The first new entry is the *heat flux vector*

$$q_a = \frac{1}{2} \sum_{ij} f_{ij} c_{ij}'^2 c_{ija}' \tag{22.24}$$

---

[51] In the continuum, temperature enters the game in a beautifully self-similar form, via the dimensionless
relative speed $(v - u)/v_T$. This expression encodes a further symmetry of continuum Maxwellian equilibria,
namely scale-invariance under space and time dilatations/contractions: $x \to \lambda x$, $t \to \lambda^h t$, $v \to \lambda^{1-h} v$, for *any*
value of exponent h. Out of local equilibrium this invariance is broken by dissipative effects.

where

$$c'_{ija} = c_{ija} - u_a \tag{22.25}$$

is the speed of the lattice molecules relative to the fluid ("peculiar" speed). Note that we have taken $m = k_B = 1$ unit.

The requirements on the heat flux vector are

$$q_a^e = \frac{1}{2} \sum_{ij} f_{ij}^e c'^2 c'_{ija} = 0 \tag{22.26}$$

$$q_a^{ne} = \frac{1}{2} \sum_{ij} f_{ij}^{ne} c'^2 c'_{ija} = -\chi \, \partial_a T \tag{22.27}$$

where χ is the thermal diffusivity of the fluid.

Manifestly, the set of constraints (22.20)–(22.23), (22.26)–(22.27) is outnumbered by the free parameters at our disposal. Hence, the problem of fixing the discrete thermodynamic equilibria is underdetermined: more solutions than unknowns.

Out of various admissible solutions, Alexander *et al.* propose the following one ($\rho = 1$):

$$A_0 = 1 - \frac{5}{2}e + 2e^2, \; A_1 = \frac{4}{9}(e - e^2) \; A_2 = \frac{4}{9}(e - e^2) \tag{22.28}$$

$$B_1 = \frac{4}{9}(1 - e), \; B_2 = \frac{1}{36}(-1 + 4e) \tag{22.29}$$

$$C_1 = \frac{4}{9}(2 - 3e), \; C_2 = \frac{1}{72}(-1 + 6e) \tag{22.30}$$

$$D_0 = \frac{1}{4}(-5 + 8e), \; D_1 = \frac{2}{9}(-1 + e), \; D_2 = \frac{1}{72}(1 - 4e) \tag{22.31}$$

$$E_1 = -\frac{4}{27}, E_2 = \frac{1}{108} \tag{22.32}$$

This refers to the big-brother of FHP-7, namely a two-dimensional triangular lattice with three energy levels, $\epsilon = 0, 1, 2$, and thirteen speeds:

$$c_0 = 0, \quad c_{jix} = j \cos \theta_i, \quad c_{jiy} = j \sin \theta_i \tag{22.33}$$

$$\theta_i = \frac{2\pi i}{6}, \quad j = 1, 2, \quad i = 1, \dots, 6 \tag{22.34}$$

The transport parameters take the following expressions:

$$\mu = \rho e \left(\tau - \frac{1}{2} \right), \quad \chi = 2\mu \tag{22.35}$$

Note that within this model, the Prandtl number (momentum diffusivity versus heat diffusivity) is frozen to the value $Pr = 1/2$. This is about half the value of ideal gases, and far from other important fluid-heat carriers, for instance liquid metals, which feature $Pr = \nu/\chi \sim 0.01$.

This limitation can be partially removed, see (5), by using a *two-time* relaxation operator of the form

$$C_{\mathrm{BGK},2} = -\left(\frac{f_i - f_i^e}{\tau_1} - \frac{f_{i*} - f_{i*}^e}{\tau_2} \right) \tag{22.36}$$

where the star denotes mirror conjugation, i.e., $(\vec{c}_i + \vec{c}_{i*} = 0)$.

Capabilities and limitations of this two-time relaxation model for heat transfer problems are discussed in (3; 5).

The derivation of the thermohydrodynamic equation of motion from the multi-energy LBGK proceeds along the same general lines described for isothermal flows: multiscale expansion as combined with systematic neglect of terms $O(Kn)$ and $O(Ma^2)$.

This derivation is rather lengthy and will not be reported here. The reader interested in full details is referred to the relevant literature (6; 7). Suffice it to say that, as pointed out by many authors, anomalies are obviously more cumbersome than in the isothermal case, simply because sizeable Mach numbers and temperature dynamics are deeper probes of the discrete fabric of the lattice.

In particular, since the discrete velocities are uniform, i.e., everywhere the same across the lattice, it is not surprising that many anomalies which lay dormant in the constant-temperature case do raise their face as soon as temperature becomes a spacetime dependent field.

Alexander *et al.* did not discuss the numerical properties of their model (positivity, stability and related issues) any further, but moved on to demonstrate its practical viability on a number of test cases, including *(i) isothermal Poiseuille flow to test the fluid viscosity μ as a function of the relaxation parameter τ, (ii) heat transfer across a transversal longitudinal temperature jump* to compute the thermal conductivity χ.

In both cases, the authors reported excellent agreement with theoretical expressions.

In addition they successfully checked that density waves propagate according to the desired expression $c_s^2 = 2e/D$, in compliance with the ideal gas equation of state. They also simulated a Couette shear flow with a temperature gradient between the boundaries and checked against the analytical temperature profile, see Fig. 22.3:

$$e^* \equiv \frac{e - e_0}{e - e_1} = \frac{1}{2}(1 + y) + Br\frac{(1 - y^2)}{8} \tag{22.37}$$

Hot

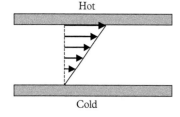

Cold

Figure 22.3   *Thermal Couette flow.*

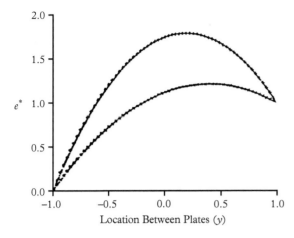

Figure 22.4 *Internal energy (in lattice units) for the Couette flow with heat transfer, for Brinkman numbers Br = 5 and 10. The upper wall is moving with speed u = 0.1 and the lower wall is at rest. The solid lines represent the analytical results (from (4)).*

where y is the normalized distance from the center, e_0 and e_1 are the upper- and lower-wall internal energies, respectively, and Br is the Brinkman number, a measure of dissipative versus conductive heat effects.

The energy profiles for the case $U_1 = 0.1$ when $Br = 5$ and $Br = 10$ are reported in Fig. 22.4 (from (4)).

These tests brought convincing evidence that the Alexander *et al.* thermohydrodynamic LBGK scheme reproduces the basic mechanisms of mass, momentum and energy transfer.

This trailblazing study did not address the thorny issue of *nonlinear* mass, momentum and heat transfer.

Notwithstanding various limitations, Alexander, Chen and Sterling's landmark paper opened the way to the development of thermal LB schemes.

22.4.1 Parametric Equilibria without Nonlinear Deviations

As anticipated, the Alexander *et al.* scheme is exposed to anomalies coming from higher-order terms in the double Knudsen–Mach expansion (5).

To remove these anomalies, Yu Chen (7) proposed higher-order parametric equilibria fulfilling the *full set* of thermohydrodynamic constraints:

$$\rho = \sum_{ij} f^e_{ij} \tag{22.38}$$

$$\mathcal{J}_a = \sum_{ij} f^e_{ij} c_{ija} = \rho u_a \tag{22.39}$$

$$P_{ab} = \sum_{ij} f^e_{ij} c_{ija} c_{ijb} = \rho \left(u_a u_b + v_T^2 \delta_{ab} \right) \tag{22.40}$$

$$Q_{abc} = \sum_{ij} f^e_{ij} c_{ija} c_{ijb} c_{ijc} = \rho \left(u_a u_b u_c + v_T^2 t_{abc} \right) \qquad (22.41)$$

$$K_{ab} = \sum_{ij} f^e_{ij} c_{ija} c_{ijb} c^2 = \rho \left[u_a u_b \left(u^2 + (D+4) v_T^2 \right) + u^2 v_T^2 \delta_{ab} + (D+2) v_T^2 \right] \qquad (22.42)$$

where $t_{abc} \equiv u_a \delta_{bc} + u_b \delta_{ac} + u_c \delta_{ab}$.

We shall refer to Q_{abc} and K_{ab} as to momentum-flux triple tensor (third order) and energy flux tensor (second order).

To ensure isotropy of six-order tensors, this author introduced an $8\mathfrak{F}$ parameter family, fourth-order polynomial expressions of the form

$$f^e_i = A_i + B_i u_i + C_i u^2 + D_i u_i^2 + E_i u_i u^2 + F_i u_i^3 + G_i u_i^2 u^2 + H_i u^4 \qquad (22.43)$$

As for Alexander *et al.* the coefficients $[A_i, \ldots, H_i]$ depend linearly on the fluid density and at most quadratically on the internal energy e, so that the generic coefficient $X_i \equiv [A_i, \ldots, H_i]$ can be cast in the form

$$X_i = \rho \sum_{j=0}^{2} X_{ij} e^j \qquad (22.44)$$

Matching the coefficients X_{ij} to the above set of constraints generates a linear system with $N_k = 24b$ kinetic degrees of freedom (eight parameters per energy level, three energy levels, b discrete speeds each) to match a total of

$$N_c = 1 + D + \frac{1}{2} D (D+1) + \frac{1}{6} D (D+1) (D+2) \qquad (22.45)$$

constraint equations. This means 4, 10, 20 constraints in one, two and three dimensions, respectively.

It is perhaps instructive to compare the number of isothermal and thermal constraints in various dimensions, see Table 22.1.

Table 22.1 *Number of isothermal and thermal constraints in dimension D.*

D	N_c^{iso}	N_c^{th}
1	3	4
2	6	10
3	10	20

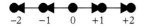

Figure 22.5 *The D1V5 lattice (five speeds, three energies).*

Free parameters and constraints cannot be expected to be in exact balance for a generic lattice, an unsatisfactory aspect of the parametric approach betraying its empirical nature, to which we shall return shortly.[52]

Be as it may, Yu-Chen provides a number of solutions at all dimensions $D = 1, 2, 3$.

In one dimension, he found the so-called five-speed model (D1V5) depicted in Figure 22.5.[53] The parameters (normalized to $\rho = 1$) are given in the Appendix.

In two dimensions the solution is the 16-speed lattice depicted in Figure 22.6 (note the absence of rest particles). The explicit expression of the corresponding parameters is also given in the Appendix.

Finally in $D = 3$ the minimal thermodynamic lattice requires 40 speeds (D3V40), organized as follows:

- Six face centers: $(\pm 1, 0, 0)$ and permutations thereof
- Six far face centers: $(\pm 2, 0, 0)$ and permutations thereof
- Twelve edge centers: $(\pm 1, \pm 1, 0)$ and permutations thereof
- Eight vertices: $(\pm 1, \pm 1, \pm 1)$
- Eight far vertices: $(\pm 2, \pm 2, \pm 2)$

The corresponding set of parameters X_{ji} can be found in Yu Chen's thesis work, where he demonstrates his thermal BGK schemes for mildly supersonic one- and two-dimensional shocks at $Ma \sim 1.1$.

These tests consist of preparing a gas with two distinct sets of thermodynamic conditions (ρ_l, P_l, T_l) and (ρ_r, P_r, T_r) on the left/right side of a separating sect. Upon removing the sect, a shock wave starts propagating rightward out of the high-pressure (left) region, see Fig. 22.7. The density and pressure jumps across the shock front are related by the Rankine–Hugoniot equations (subscripts r, l mean right and left, respectively)

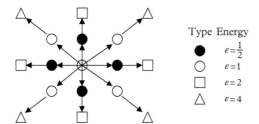

Type	Energy
●	$\varepsilon = \frac{1}{2}$
○	$\varepsilon = 1$
□	$\varepsilon = 2$
△	$\varepsilon = 4$

Figure 22.6 *The D2V16 (16 speeds, 4 energies) thermal lattice.*

---

[52] Beware: a strict N_c versus N_k count would be deceptive because the problem has many symmetries which strongly reduce the effective number of kinetic degrees of freedom.

[53] D1V5 means five discrete velocities in one spatial dimension. It has the very same meaning of D1Q5, but some authors feel like it is more neutral as it does not refer to any author's name.

$$\frac{\rho_l}{\rho_r} = \frac{(\gamma + 1)\, M^2}{(\gamma - 1)\, M^2 + 2} \tag{22.46}$$

$$\frac{T_l}{T_r} = \frac{2\gamma M^2 - (\gamma - 1)\left[(\gamma - 1)\, M^2 + 2\right]}{(\gamma + 1)\, M^2} \tag{22.47}$$

where M is the Mach number, $\gamma = C_p/C_v = (D+2)/D$, as it befits a mono-atomic ideal gas in dimension D.

Excellent agreement with analytical results was reported for various values of the Mach number Ma and other thermodynamic parameters.

These results prove that this thermodynamic LBGK does possess a non-empty working window of thermodynamic parameters also in the nonlinear compressible regime, a remarkable result which furthered the path opened up by Alexander, Chen and Sterling.

However, it still tells very little about the actual size of the stability window in thermodynamic parameter space.

In other words, given a set of parameters for which the method proves stable, one is still left with little feeling for how far one can sail away from this set before incurring numerical troubles, see Figs 22.7 and 22.8.

This mixed state of affairs was pinpointed quite sharply in early work by B. Alder and collaborators, which we now proceed to discuss in some detail.

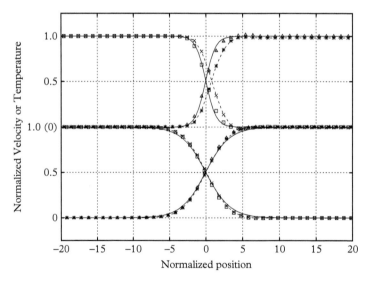

Figure 22.7 *Comparison of the structure of the shock wave fronts. Squares and triangles correspond to the D2V16 model; lines with symbols (dashed line with crosses and stars) are the results of the D2V13 model; solid lines represent the analytical solution. The left/right steps represent the sharp temperature increase, while the right/left step is the velocity increase. The upper half of the graph is for Ma = 1.14 while the lower half corresponds to Ma = 1.02. (After (7), courtesy of Y. Chen.)*

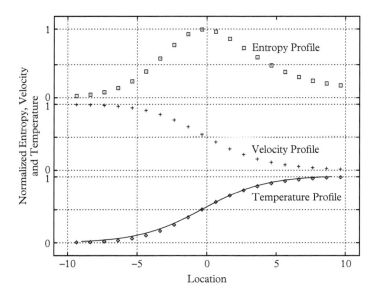

Figure 22.8 *Profiles of the shock wave front. The solid line is obtained by numerical integration of the Navier–Stokes equations, while symbols correspond to the results of the D1V5 model. The Mach number is 1.01. (After (7), courtesy of Y. Chen.)*

22.5   Thermohydrodynamic LB Schemes Get in trouble

McNamara, Garcia and Alder started from Grad's 13-moment approach to fluid dynamics and observed that in order to recover the 13-moment Grad equations, the lowest 26 moments of the discrete distribution should match those of the Maxwell–Boltzmann distribution. As a result, at least 26 discrete speeds are required to fill the bill.

It so happens that a discrete set consisting of 12 speeds $\sqrt{2}$, 8 speeds $\sqrt{3}$ and 4 speed two particles, provides the desired match: 26 degrees of freedom.

Unfortunately, such a minimal scheme proved very unstable for all but the largest transport coefficients, typically of order $O(1)$ in lattice units, far too high for most applications.

Surprise? In hindsight, not really.

We already have clues from athermal fluids that kinetic space, the one spanned by the discrete speeds, ought to be "larger" than the hydrodynamic one, if the correct macroscopic symmetries are to be fulfilled, with some room still left for entropy minimization!

In this light, it is pretty plausible to expect that a minimal kinetic space matching exactly the size of thermohydrodynamic space would leave little left for stability.

A natural move is therefore to extend the set of discrete speeds so as to allow more freedom (as done by Yu Chen). Unfortunately, the authors reported no success in this direction (8).

The best guess is that by adding speeds without a physically informed guideline, brings more freedom to match symmetries, but also higher exposure to numerical instabilities. Some form of GPS is needed to navigate discrete phase space.

A possible way out suggested by McNamara *et al.* is to go just the other way around, namely *reduce*, instead of raising, the number of discrete speeds, in an attempt to *trade thermohydrodynamic consistency on the highest moments for better stability*.

To this purpose, McNamara *et al.* used a 21-speed lattice consisting of:

- 1 particle with energy 0 (rest)
- 6 particles with energy 1/2 (face-centers)
- 8 particles with energy 3/2 (vertices)
- 6 particles with energy 2 (far-face-centers)

With this lattice, the following 21 moments are reproduced:

$$\langle 1 \rangle = \rho \tag{22.48}$$

$$\langle c_{ia} \rangle = \rho u_a \tag{22.49}$$

$$\langle c_{ia}^2 \rangle = \rho \left(u^2 + 2e \right) \tag{22.50}$$

$$\langle Q_{iab} \rangle = 0 = \rho u_a u_b - \frac{1}{3} u^2 \delta_{ab} \tag{22.51}$$

$$\langle c_i^2 c_{ia} \rangle = \rho u_a \left(u^2 + \frac{10}{3} e \right) \tag{22.52}$$

$$\langle c_{ia} c_{ib} c_{ic} - \frac{1}{5} c_i^2 t_{iabc} \rangle = 0 \tag{22.53}$$

$$\langle c_i^2 Q_{iab} \rangle = 3\rho u_a u_b - \rho u^2 \delta_{ab} \tag{22.54}$$

$$\langle c_i^4 \rangle = \frac{20}{3} \rho e^2 \tag{22.55}$$

$$\langle c_i^4 - 5 c_{ix}^2 c_{iy}^2 + c_{ix}^2 c_{iz}^2 + c_{iy}^2 c_{iz}^2 \rangle = 0 \tag{22.56}$$

where each $\langle \cdot \rangle$ denotes thermal equilibrium averages, i.e. $\sum_i f_i^e(\cdot)$.

The first five moments are the usual conserved densities and they are automatically conserved by the collision operator, since they are explicitly included in the local equilibrium.

Entries four and five correspond to the viscous and thermal transport properties. By making Q_{iab} eigenvectors of the scattering matrix (McNamara *et al.* used a matrix LBE instead of LBGK), one generates separate values of the viscosity and thermal conductivity thereby allowing non-unit Prandtl numbers. This completes the list of Grad's 13 moments.

The remaining moments are purely kinetic modes, which should not couple to the Navier–Stokes–Fourier dynamics.

The LBE scattering matrix may be tuned so as to wipe them out as fast as possible, as it was done for ghost modes in early days LBE. The equilibrium component of these kinetic modes does couple, however, to Navier–Stokes–Fourier dynamics and this is the cause of nonlinear deviations, typically scaling like Ma^2.

Since McNamara *et al.* focus on incompressible thermal applications, typically Rayleigh–Bénard convection, these errors are no source of major concern. Still, we are again left with a rather unsatisfactory situation.

Indeed, McNamara *et al.* concluded on a rather disheartening note that they "*have not discovered any advantage to employing LBE over conventional Navier–Stokes solvers for thermal systems.*"

End of the line for LB thermal models?

Fortunately, no.

As we shall see, the ingenious work of many researchers in the field led to several ways out of this impasse.

22.6   Early Attempts to Rescue Thermal LBE

Numerous attempts to improve on the gloomy picture emerging from McNamara *et al.* analysis were put in place in the subsequent years, which we briefly mention in the following.

22.6.1   Tolerance to Realizability Violations

A set of mild pain-easing results were provided by de Cicco *et al.* (9), who showed that positivity of the discrete equilibria, definitely not a sufficient condition for stability, might not be a necessary one either.

We refer to this property as "tolerance to lack of realizability." At each point of the fluid domain let us introduce the realizable hydrodynamic space $\mathcal{H}_\mathcal{R}$ defined as the set of triples $(\rho, \vec{u}, T)$ such that the discrete equilibria are realizable; that is

$$0 < f_i^e < \rho \tag{22.57}$$

In principle, this is a superset of the stable hydrodynamic subspace $\mathcal{H}_\mathcal{S}$ spanned by the triples $(\rho, \vec{u}, T)$ such that the scheme is numerically stable.

The finding by de Cicco *et al.* is that the hydrodynamic behavior can be reasonable even in the presence of *negative* populations (mildly) breaking the realizability constraint (22.57). This property has been confirmed by several other studies; while realizability is highly desirable, it is nevertheless neither necessary, nor sufficient to guarantee numerical stability.

22.6.2   The Kinetic-Closure Approach

Along the spirit of reduced thermohydrodynamics, Renda *et al.* proposed an alternative approach based upon the idea of representing high-energy (secondary) equilibria in terms of low-energy (primary) ones via a *kinetic closure* (10). Primary equilibria are then computed by solving exactly a lower-dimensional system of constraints. The guiding idea is that *secondary equilibria are not allowed to go negative if the primary ones do not.* This cannot secure automatic stability, but certainly makes life harder for numerical errors to drive high-energy populations to negative values. Even though these thermal LBEs never become unstable, they deliver *quantitatively* correct results only for a limited range of values of the thermodynamic parameters.

22.6.3 Non-Space-Filling Lattices

Another interesting proposal is to use non-space-filling lattices, typically octagons offering a higher degree of isotropy (20).

For instance, a 17-speed lattice defined by

$$
\vec{c}_{ji} = j \left(\cos \left(\frac{\pi i}{8} \right), \sin \left(\frac{\pi i}{8} \right) \right), \quad j = 1, 2, \quad i = 0, 1, \ldots, 7
$$

plus one rest particle, can be shown to provide isotropy up to sixth-order tensors (11).

The price is that the corresponding tiling is no longer space-filling, so that particle microdynamics needs to be supplemented with an interpolation step, in order to latch onto the "crystal" grid after streaming.

The authors demonstrate the method for isothermal plane Poiseuille flow and two-dimensional Rayleigh–Bénard convection reporting good quantitative agreement with the previous literature, as well as enhanced numerical stability. However, the method does not appear to have gained much popularity within the community, possibly because interpolation spoils on of the most precious assets of LBE: exact streaming.

22.6.4 Models with Rest Energy

LBE models with pure kinetic energy necessarily lead to the ideal gas relation $\gamma = C_p/C_v = (D + 2)/D$. This hinders gas dynamic simulations in less than three dimensions since the correct value 5/3 is replaced by 3 and 2 in one and two dimensions, respectively.

In addition, specific numerical values such as $\gamma = 1.4$ (air) are not realizable either. It is therefore very desirable to extended thermal models so as to include the capability of a tunable γ parameter.

An intuitive way of achieving this goal is to make allowance for a non-zero rest mass energy ϵ_{0j} for each energetic level j. This corresponds to the following energy–momentum relation:

$$
\epsilon_j = \epsilon_{0j} + \frac{1}{2} c_{ji}^2 \tag{22.58}
$$

The internal gas energy density then becomes

$$\rho e = \sum_j \left(\epsilon_{0j} + \frac{1}{2} c_j^2 \right) \rho_j \tag{22.59}$$

where $\rho_j = \sum_j f_{ji}$ is the partial density of the j-th energy shell.

By allowing distinct rest energies for different energy levels, one can manipulate the equation of state of the LBE gas. It can be shown that by appropriate tuning of the rest energies, the range of density and temperature gradients supported by the model, i.e., the range of nonlinear stability can be extended significantly.

This idea has been successfully demonstrated by Shouxin *et al.* (12). The theoretical arguments are supported by convincing simulations of one- and two-dimensional shock wave propagations at density ratios of order ten, see Fig. 22.9.

In a similar vein, other authors (12) have proposed a new thermal LBE in which the internal energy is modeled by a scalar field using a separate distribution function.

The energy is coupled to the fluid density and momentum by means of an appropriate repartition between resting and moving particles in conventional two speed models. This repartition amounts to making the density of rest particles a suitable function of density

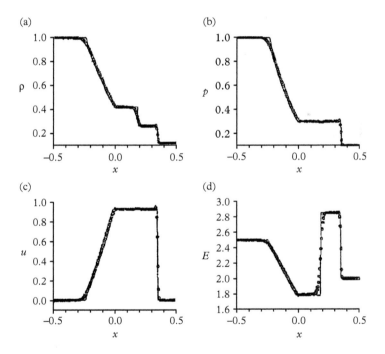

Figure 22.9 *Density, pressure, velocity and energy profiles as obtained by the compressible LBE model by Shouxin* et al. *The solid lines represent the analytical solution. (From (12)).*

and energy. The model does not account for heat dissipation effects but, in principle, it can handle non-ideal thermodynamics.

All of the above represent interesting attempts to escape the problems raised by Alder and Mc Namara, but none of them seems to have turned into a systematic and robust solution to the problem. Other roads have proven more effective, which we now proceed to illustrate in some detail.

22.7   Systematic Third-Order Equilibria

The laborious matching problem associated with the parametric approach can be disposed of by expanding a local Maxwellian to third order. This delivers the expression (22.7), which provides an elegant solution to the sixth-order isotropy issue.

However, many lattices exist which can accommodate sixth–order equilibria, with no unique criterion to decide which one provides the best performance. Nevertheless, by using a variety of high-order lattices, a broad spectrum of hydrothermal flows has been simulated in the last decade.

Typically, in these applications, the temperature T in the local equilibria must kept to a constant value throughout the fluid, on pain of incurring stability problems. Clearly, by the time the temperature in the local equilibria is replaced by a global constant, local energy conservation is ruined.

A pragmatic fix is to compensate with a "restoring" thermal force of the form

$$\vec{F}_{th} = -\nabla[\rho(T - T_0)] \tag{22.60}$$

where T and T_0 are the local and global temperatures, respectively.

This fix saves the day in many applications, but it cannot be expected to be accurate beyond first order. Recently, it has been found that consistency with continuum equations to second order is obtained by shifting the velocity and temperature in the local equilibria, by a force-dependent terms proportional to $F_{th}\Delta t/\rho$. Details can be found in (14)).

While surely mitigating the stability issue to a good extent, the thermal force procedure does not cut to the heart of the thermo-hydrodynamic instability.

To this purpose, an approach with built-in conservation and compliance with the H-theorem is required, as explained in the sequel.

22.8   Entropic Schemes with Energy Conservation

A major development to the LB modeling of thermal with moderate range of temperature variation and mild compressibility effects was proposed by Ansumali and Karlin (15), based on the application of entropic principles discussed in Chapter 21.

The authors start from the observation that the lack of energy conservation within local equilibria of standard athermal LB models, is not only unviable for thermal simulations (of course), but also causes numerical instabilities for the athermal ones. The culprit is a large-bulk viscosity, a direct consequence of lack of energy conservation in the local equilibria. Besides, such non-zero-bulk viscosity also precludes the correct attainment of the hydrodynamic limit for compressible flows. To mend this deficiency, the authors propose to derive lattice equilibria by direct discretization of a corresponding entropy functional, rather than by expanding continuum local equilibria in powers of the Mach number.

More precisely, with reference to the three-dimensional D3Q27 lattice, hence *within the first Brillouin cell*, the H-functional takes the following form:

$$H[f] = \sum_{i=1}^{27} f_i ln\left(\frac{f_i}{W_i}\right) + \alpha\left[\sum_{i=1}^{27} f_i - \rho\right] + \beta_a\left[\sum_{i=1}^{27} c_{ia}f_i - \mathcal{J}_a\right] + \gamma\left[\sum_{i=1}^{27} c_i^2 f_i - K\right] \quad (22.61)$$

where K is twice the kinetic energy and α, β_a and γ are the five Lagrangian parameters associated with the five hydrodynamic fields $\{\rho, \mathcal{J}_a, p\}$, i.e., density, current density and pressure, according to the definition

$$\sum_{i=1}^{27} f_i\{1, c_{ia}, c_i^2\} = \{\rho, \mathcal{J}_a, K \equiv 3p + \mathcal{J}^2/\rho\} \quad (22.62)$$

The extremum condition $\delta H = 0$ yields exponential equilibria of the form

$$f_i^e = W_i e^{-(\alpha + \beta_a c_{ia} + \gamma c_i^2)} \quad (22.63)$$

Inserting this into the constraint relations (22.62) delivers the following elegant expression for the energy-compliant entropic equilibria:

$$f_i^e = \rho W_i(r)\left[1 + \frac{c_{ia}\mathcal{J}_a}{p} + \frac{\mathcal{J}_a\mathcal{J}_{ab}}{2p^2}(c_{ia}c_{ib} - C_i^2(r)\delta_{ab})\right] \quad (22.64)$$

where we have set

$$r \equiv p/\rho \quad (22.65)$$

The weights and sound speed in the above equilibria are given by

$$W_i(r) = (1 - r)^3\left(\frac{r}{2(1 - r)}\right)^{c_i^2} \quad (22.66)$$

where

$$C_i^2(r) = \frac{2r^2 + c_i^2(1/3 - r)}{(1 - r)} \tag{22.67}$$

A few remarks are in order.

First, we note that in the athermal limit $r \to 1/3$, the entropic weights $W_i(r)$ reduce to the standard D3Q27 values and similarly, $C_i^2(1/3) = 1/3$.

Hence energy-compliant entropic equilibria represent a natural extension of their athermal counterpart to the case $p/\rho \neq 1/3$, i.e., the equation of state of an ideal gas with a free-temperature $T \neq 1/3$.

The authors proceed to show that the non-equilibrium momentum-flux tensor also takes the correct expression, namely

$$P_{ab}^{ne} = -\tau p \left[\partial_a \left(\frac{\mathcal{J}_b}{\rho} \right) + \partial_b \left(\frac{\mathcal{J}_a}{\rho} \right) - \frac{2}{3} \partial_c \left(\frac{\mathcal{J}_c}{\rho} \right) \delta_{ab} \right] \tag{22.68}$$

i.e., exhibiting the correct prefactor in front of the divergence to achieve zero-bulk viscosity.

On the same line, the dissipative heat flux is given by

$$q_a^{ne} = -2\tau p \partial_a \left(\frac{p}{\rho} \right) \tag{22.69}$$

which is again the correct continuum expression.

The reader may plausibly wonder: how come this magic, given that the present scheme still lives in the first Brillouin cell, which *cannot* fulfill the sixth-order isotropy requirement?

The answer is in the authors own words:

> It should be noted that the present model does not solve the problem of bulk viscosity and heat conductivity entirely, nor can it be applied to highly compressible flows. However, the domain of validity is wide enough to include such important flows as convection flows and microflows.

Once again, we are faced with a *quasi*-solution, yet one which can handle a non-trivial subset of problem space. The authors proceed to show the validity of the approach by computing a non-equilibrium Couette flow, including finite-Knudsen corrections.

The "consistent Lattice Boltzmann" represents a significant advance: not only it permits us to handle weakly compressible flows with moderate temperature excursions around $p/\rho \sim 1/3$, but it also improves the stability of athermal flows. In a way, it provides an interesting quasi-solution to the stability problem raised by Alder and Mc Namara a decade ahead.

22.9   Of Fluons and Phonons: The Double-Distribution Method

So far, we have been discussing models with a single distribution on either standard or higher-order lattices. Next we present the third family: double distribution models on standard lattices.

Such model with heat dissipation effects was proposed in (16). The main point is that internal energy is described as the density associated with a separate distribution, say $g \equiv g(\vec{x}, \vec{v}; t)$, which we may call "phonons."

By definition:

$$ne = \int g d\vec{v} \tag{22.70}$$

and similarly for the heat flux

$$Q_a = \int g v_a d\vec{v} \tag{22.71}$$

In order for the two relations above to hold true, the phonon distribution must be linked to the particle distribution f via the constraint

$$g = \frac{c^2}{2} f \tag{22.72}$$

where $\vec{c} = \vec{v} - \vec{u}$ is the peculiar velocity, as usual.

The g distribution is then postulated to obey its own LBGK equation:

$$\partial_t g + v_a \partial_a g = -\frac{g - g^e}{\tau_g} - c_a \partial u_a \tag{22.73}$$

where we have set

$$\partial \equiv \partial_t + c_b \partial_b \tag{22.74}$$

The last term on the right-hand side describes the heat dissipation and g^e is related to the particle equilibrium f^e by the condition (22.72), i.e.,

$$g^e = \frac{c^2}{2} f^e \tag{22.75}$$

Equation (22.73) is then solved concurrently with a standard LBGK for f.

The time discretization is obviously more involved than in standard LBEs, due to the time derivative entering the expression of the heat dissipation Q. In fact, a *second-order* time marching is required, the details of which are worth discussing in some detail, since this technique has proven very useful in general, not only for thermal-flow applications.

Before proceeding further, we wish to mention allegedly simpler versions of the DD method have been developed in the subsequent years, see (17).

22.9.1   Second-Order Time Marching

The idea is to march the LB scheme in time using a second-order trapezoidal rule, for the integration of the collision term.

This yields (for simplicity, we illustrate it for f only):

$$\hat{f}_i - f_i = -\frac{\Delta t}{2\tau}(f_i + \hat{f}_i - f_i^e - \hat{f}_i^e) \tag{22.76}$$

where, as usual, we have set

$$\hat{f}_i \equiv f_i(\vec{x} + \vec{c}_i \Delta t, t + \Delta t) \tag{22.77}$$

Note that the integration of the collision operator also proceeds along the trajectory, with both end points contributing the same factor 1/2.

At first sight, eqn (22.76) looks like a major complication, since the right-hand side depends on the actual unknown $\hat{f}_i$ (implicit form).

Fortunately, a local transformation saves the day.

By defining the new distribution:

$$\bar{f}_i \equiv \left(1 + \frac{\theta}{2}\right) f_i - \frac{\theta}{2} f_i^e \equiv f_i + \frac{\theta}{2} f_i^{ne} \tag{22.78}$$

with $\theta = \omega \Delta t$, it is readily checked that the apparently implicit expression eqn (22.76) turns back into an explicit one for the transformed distribution $\bar{f}_i$, i.e.,

$$\hat{\bar{f}}_i - \bar{f}_i = -\frac{\theta}{1 + \theta/2}(\bar{f}_i - \bar{f}_i^e) \tag{22.79}$$

In other words, $\bar{f}_i$ satisfies the same explicit LBGK equation as f_i, only with a modified relaxation rate:

$$\bar{\theta} = \frac{\theta}{1 + \theta/2} \tag{22.80}$$

This is a very nice result, for it permits us to achieve second-order accuracy in time without changing the explicit structure of the LBGK scheme. Note that $\bar{\theta}$ still lies in the range $0 < \bar{\theta} < 2$, as for θ in the standard LBGK.

A similar trick would do for the matrix version too, except that the denominator $1 + \theta/2$ would be replaced by the inverse of the matrix $1 + \Omega \Delta t/2$, which is more laborious to compute but still does not involve any spatial coupling between neighboring lattice sites (locally implicit).

One may argue that the thermo-hydrodynamic variables must be evaluated with the original f and not with $\bar{f}$. However, the very definition eqn (22.78) shows that the two sets of moments obey a linear relation

$$M_k = \bar{M}_k + \frac{\theta}{2}\bar{M}_k^{ne} \tag{22.81}$$

where we have used the property $\bar{f}^e = f^e$, as it stems directly from the definition (22.78).

In other words, conserved quantities are invariant under the transformation $f \rightarrow \bar{f}$, hence they can be computed directly with $\bar{f}$. On the other hand, the non-conserved ones are easily reconstructed according to eqn (22.81).

The scheme was numerically tested for the two-dimensional Couette flows with temperature gradients and Rayleigh–Bénard convection providing good quantitative agreement with the previous literature in both cases. The authors did not mention performance issues, but they did advocate better numerical stability, as compared to previous thermal LBEs.

Several years down the line, these claims appear to be well supported by a large body of applications. As of today, it is fair to say that the DD method has gained a leading role in the LB simulation of thermal flows.

22.10   Hybrid Methods

Simulating thermal flows with an inherently athermal technique, such as LB, might have well resulted into a mission-impossible. It only goes to the credit of the LB community that a variety of ingenious solutions have been worked out, to the point that LB can now be routinely used for thermal calculations, at least as long as compressibility and thermal exchange issues are not overwhelming.

On the other hand, it is also natural to wonder why should one insist in using LB all the way, i.e., for mass, momentum and energy transport, rather than using it for the athermal part of the problem, namely mass and momentum transport alone leaving the solution of the energy equation to any other numerical method of convenience?

Basically, the idea is to solve LB for density and momentum and provide flow velocity to the transport equation for the temperature.

If temperature is just a passive scalar, this is it. In the active case, the temperature is fed back into LB, typically via a thermal correction force $-\nabla\rho(T - T_0)$, T_0 being the reference temperature to be used in the athermal LB equilibria. Eventually, different grids and/or different resolutions can be used for the two set of equations.

In this author's opinion, this hybrid stance makes perfect sense.

In the first place, one should always bear in mind that LB is memory-thirsty, one ends up with tens of discrete populations to ultimately model the evolution of four fields, the fluid temperature and the heat flux.

There must be good motivations to do so.

Potential motivations are not lacking: easy implementation (caveat boundary conditions), parallel amenability, efficiency, stability and likely a larger-timestep: the usual

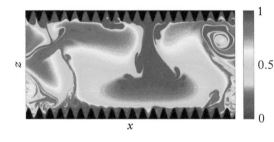

Figure 22.10 *LB simulation of Rayleigh–Benard convection in the presence of wall corrugations. The corrugations can be fine-tuned to optimize the heat transfer across the system. (Courtesy of S. Toppaladoddi and J. Wettlaufer).*

items where LBE may offer a good bargain. Nevertheless, memory considerations and ensuing issues of efficiency associated with memory access should not go forgotten. That is why the hybrid approach should be taken into serious consideration.

22.11 Passive-Scalar Schemes

It is worth stressing that athermal LBE techniques can be successfully employed for flows with *mild* compressibility and dissipative heating effects, where temperature behaves like a passive scalar, with no feedback on the fluid.

A typical example is Rayleigh–Bénard convection within the Boussinesq's approximation, namely $\delta\rho/\rho \ll \delta T/T$:

$$\partial_t T + u_a \partial_a T = \chi \partial_a^2 T + \alpha g \left(T - T_0 \right) \tag{22.82}$$

Here α is the thermal-dilatation coefficient, g the gravitational acceleration, and T_0 a reference temperature. This equation can be efficiently handled by using athermal methods with passive scalars, buoyancy effects being mimicked by external sources.

This approach was pioneered by Massaioli *et al.* (18) and further developed by several other authors (19; 20) for the case of three-dimensional Rayleigh–Bénard turbulence.

Massaioli *et al.* used the historical four-dimensional FCHC lattice, leveraging the fact that the current along the fourth dimension behaves like a passive scalar in the three-dimensional fluid domain. This is a very elegant and compact way to solve passive-scalar dynamics almost for free. However, being tied down to the old-days FCHC lattice, this scheme has gone to a (rather undeserved) oblivion.

Alternatively, one can add a second population to the LB scheme, and treat its density as a passive-scalar fluid temperature. This approach, developed by X. Shan, runs on any LB lattice, and consequently it has met with significant popularity over the last decade, see Fig. 22.10.

22.12 Further Variants for Compressible Flows

Many other variants of thermal LB schemes have been developed to deal with shocks and strong compressibility effects. Typically, these variants import within the LB framework ideas and methods from the vast literature on numerical methods for compressible flows, such as Flux-Corrected Transport and Total-Variation Diminishing methods. Among others, see (21) and the nice review by Aiguo Xu and collaborators (22).

22.13   Boundary Conditions for Thermal Flows

In chapter 17 we have provided a cursory view of the main ideas behind the implementation of various types of boundary conditions for *athermal* flows.

These notions form the basis for elaborating more general schemes dealing of thermal flows, where additional conditions on the temperature and/or heat flux must be imposed at the boundary.

Details are inevitably more involved, due to multi-energy levels. Once again, we face an issue that would warrant way more space than affordable in this book. Here, we can only assume and hope that, with the basic notions learned in this chapter, the interested reader will be able to find their way into the current literature on the subject. Among others, see Li *et al.* ((23), MTR formulation), D'Orazio *et al.*, ((24) double-distribution for open flows), Scagliarini *et al.* ((25), finite-volume formulation).

22.14   Summary

Despite many promising attempts, none of the thermal LBEs presented so far provides a watertight solution to the problem of propagating heat and temperature in a discrete lattice over an *extended* range of values. This is an open frontier of LBE research. At this stage it is very hard to foresee whether we are facing an unforgiving "no go" barrier or if instead as often in the past of the LGA and LBE, some smart intuition will finally save the day.

The above is the verbatim close of this chapter in the 2001 Edition. As the reader can appreciate, with all due restrictions on the intensity of compressibility and viscous heating effects, the situation has undergone significant progress in the last fifteen years.

22.15   Appendix: Parameters of the D1V5 and D2V16 Thermal model

D1V5 model:

$$A_0 = 1 - \frac{5e}{2} + 3e^2, \; A_1 = \frac{4e}{3} - 2e^2, \; A_2 = -\frac{e}{12} + \frac{e^2}{2} \tag{22.83}$$

$$B_1 = \frac{2e}{3} - e^2, \; B_2 = -\frac{e}{24} + \frac{e^2}{4} \tag{22.84}$$

$$C_1 = -\frac{5}{4} + 3e, \; C_2 = -\frac{1}{3} + \frac{e}{2} \tag{22.85}$$

$$D_1 = 1 - \frac{5e}{2}, \; D_2 = -\frac{1}{64} - \frac{5e}{32} \tag{22.86}$$

$$E_1 = -\frac{17}{12}, \; E_2 = \frac{1}{24} \tag{22.87}$$

$$F_1 = \frac{5}{4} \tag{22.88}$$

$$G_1 = -\frac{1}{4}, \; G_2 = \frac{1}{64} \tag{22.89}$$

$$H_0 = \frac{1}{4}, \; H_1 = \frac{1}{12}, \; H_2 = -\frac{1}{48} \tag{22.90}$$

D2V16 model:

$$A_1 = \frac{8}{15} - \frac{2}{3}e + \frac{1}{3}e^2 \tag{22.91}$$

$$A_2 = -\frac{1}{30} + \frac{1}{24}e + \frac{1}{24}e^2 \tag{22.92}$$

$$A_3 = -\frac{4}{15} + \frac{2}{3}e - \frac{5}{12}e^2 \tag{22.93}$$

$$A_4 = \frac{1}{60} - \frac{1}{24}e + \frac{1}{24}e^2 \tag{22.94}$$

$$B_1 = \frac{2}{3}e - e^2 \tag{22.95}$$

$$B_2 = -\frac{1}{24}e + \frac{1}{8}e^2 \tag{22.96}$$

$$B_3 = \frac{1}{4}e \tag{22.97}$$

$$C_1 = -\frac{2}{3} + \frac{5}{6}e \tag{22.98}$$

$$C_2 = \frac{1}{24} + \frac{5}{12}e \tag{22.99}$$

$$C_3 = \frac{1}{6} - \frac{7}{4}e \tag{22.100}$$

$$C_4 = -\frac{1}{96} + \frac{1}{96}e \tag{22.101}$$

$$D_1 = \frac{2}{3} - e \tag{22.102}$$

$$D_2 = -\frac{1}{48} + \frac{1}{16}e \tag{22.103}$$

$$D_3 = \frac{1}{6} - \frac{1}{8}e \tag{22.104}$$

$$D_4 = -\frac{1}{384} + \frac{1}{128}e \tag{22.105}$$

$$E_1 = -\frac{1}{2} \tag{22.106}$$

$$E_2 = 0 \tag{22.107}$$

$$E_3 = -\frac{1}{8} \tag{22.108}$$

$$E_4 = 0 \tag{22.109}$$

$$F_1 = \frac{1}{3} \tag{22.110}$$

$$F_2 = \frac{1}{96} \tag{22.111}$$

$$F_3 = \frac{1}{8} \tag{22.112}$$

$$F_4 = 0 \tag{22.113}$$

$$G_1 = -\frac{1}{6} \tag{22.114}$$

$$G_2 = \frac{1}{96} \tag{22.115}$$

$$G_3 = -\frac{1}{48} \tag{22.116}$$

$$G_4 = \frac{1}{768} \tag{22.117}$$

$$H_1 = \frac{1}{18} \tag{22.118}$$

$$H_2 = -\frac{1}{64} \tag{22.119}$$

$$H_3 = -\frac{1}{32} \tag{22.120}$$

$$H_4 = 0 \tag{22.121}$$

REFERENCES

1. M. Ernst, Temperature and heat conductivity in cellular automata fluids, in *Discrete Models of Fluid Dynamics*, A. Alves ed., p. 186, World Scientific, Singapore, 1991.
2. C. Cercignani, "On the thermodynamics of a discrete velocity gas", *Transp. Theory and Stat. Phys.*, **23**(1–3), 1, 1994.
3. Y. Chen, H. Ohashi and M. Akiyama, "Prandtl number of lattice Bhatnagar–Gross–Krook fluid", *Phys. Fluids*, 7, 2280, 1995.
4. F. Alexander, S. Chen and J. Sterling, "Lattice Boltzmann thermohydrodynamics", *Phys. Rev. E*, **53**, 2298, 1993.
5. Y. Chen, H. Ohashi and M. Akiyama, "Thermal lattice Bhatnagar–Gross–Krook model without nonlinear deviations in macrodynamic equations", *Phys. Rev. E*, **50**, 2776, 1994.
6. C. Teixeira, *Continuum Limit of Lattice Gas Fluid Dynamics*, Ph.D. Thesis, Nuclear Engineering Dept, MIT, 1992.
7. Y. Chen, *Lattice Bhatnagar–Gross–Krook Method for Fluid Dynamics: Compressible, Thermal and Multiphase Models*, Ph.D. Thesis, Dept. of Quantum Engineering and System Science, University of Tokyo, 1994.
8. G. McNamara, A. Garcia and B. Alder, "Stabilization of thermal lattice Boltzmann models", *J. Stat. Phys.*, **81**, 395, 1995.
9. M. de Cicco, S. Succi and G. Bella, "Nonlinear stability of thermal lattice BGK models", *SIAM J. Sci. Comp.*, **21**(1), 366, 1999.
10. A. Renda, S. Succi, I. Karlin and G. Bella, "Thermohydrodynamic lattice BGK equilibria with non-perturbative equilibria", *Europhys. Lett.* **41**(3), 279, 1998.
11. P. Pavlo, G. Vahala and L. Vahala, "Higher-order isotropic velocity grids in lattice methods", *Phys. Rev. Lett.*, **80**(18), 3960, 1998.
12. S. Hu, G. Yan and W. Shi, "A lattice Boltzmann model for compressible perfect gas", *Acta Mech. Sinica.*, **13**, 3, 1997.
13. G. Yan, Y. Chen and S. Hu, "Simple lattice Boltzmann model for simulating flows with shock waves", *Phys. Rev. E*, **59**(1), 454, 1999.
14. M. Sbragaglia, R. Benzi, L. Biferale, H. Chen, X. Shan, and S. Succi, "Lattice Boltzmann method with self-consistent thermo-hydrodynamic equilibria", *J. Fluid Mech.*, **628**, 299 (2009)
15. S. Ansumali and I.V. Karlin, "Consistent lattice Boltzmann method", *Phys. Rev. Lett.*, **95**, 260605, 2005.
16. X. He, S. Chen and G. Doolen, "A novel thermal model for the lattice Boltzmann method in the incompressible limit", *J. Comp. Phys.*, **146**, 282, 1998.
17. Y. Peng, C. Shu, Y.T. Chew, "Simplified thermal lattice Boltzmann model for incompressible thermal flows", *Phys. Rev. E*, **68**, 026701, 2003.
18. F. Massaioli, R. Benzi and S. Succi, "Exponential tails in two-dimensional Rayleigh–Bénard convection", *Europhys. Lett.* **21**(3), 305, 1993.

19. X. Shan, "Simulation of Rayleigh–Bénard convection using a lattice Boltzmann method", *Phys. Rev. E* **55**, 2780, 1997.
20. G. Vahala, P. Pavlo, L. Vahala and N. Martys, "Thermal lattice Boltzmann models for compressible flows", *Int. J. Mod. Phys. C* **9**(8), 1247, 1998.
21. V. Sofonea, A. Lamura, G. Gonnella, "Finite-difference lattice Boltzmann model with flux limiters for liquid-vapor systems", *Phys. Rev. E*, **70**, 04670, 2 2012.
22. Aiguo Xu et al, "Lattice Boltzmann modeling and simulation of compressible flows", *Front. Phys.*, **7**, 582 2012.
23. Like Li, R. Mei, J. F. Klausner, "Boundary conditions for thermal lattice Boltzmann equation method", *J. Comp. Phys.* , **237**, 366, 2013
24. A. D'Orazio, S. Succi, C. Arrighetti, "Lattice Boltzmann simulation of open flows with heat transfer", *Phys. Fluids* **15** 2778. 2003
25. A. Scagliarini, L. Biferale, M. Sbragaglia, K. Sugiyama, and F. Toschi, "Lattice Boltzmann methods for thermal flows: Continuum limit and applications to compressible Rayleigh-Taylor systems", *Phys. Fluids*, **22**, 055101, 2010.

··

EXERCISES

1. Draw the realizability domain of the D1V5 lattice.
2. Compute the temperature for the D2V9 lattice as a function of the partial densities ρ_j, $j = 0, 1, 2$. At a given ρ_0 for which ratio ρ_2/ρ_1 do we get population inversion (negative temperature)?
3. Write the kinematic viscosity of the thermal fluid within the double-distribution model, in terms of $\bar{\theta}$. Can you reach negative viscosity? If yes, how? If not, why?
4. Show that $\bar{f}(t) \sim f(t + \Delta t/2)$.

23

Out of Legoland: Geoflexible Lattice Boltzmann Equations

The LBEs discussed to this point lag behind "best in class" Computational-Fluid Dynamics (CFD) methods for the simulation of fluid flows in realistically complicated geometries, such as those presented by most real-life devices. This traces back to the constraint of working along the light-cones of a uniform spacetime. Various methods have been proposed to remedy this unsatisfactory state of affairs. Among others, a natural strategy is to acquire geometrical flexibility from well-established techniques which can afford it, namely Finite Volumes (FV), Finite Differences (FD) and Finite Elements (FE). Alternatively, one can stick to the cartesian geometry of standard LB, and work at progressive levels of local grid refinement.

In this chapter we shall present the general ideas being both strategies.

If we can't beat them, let's join them!

(B. Alder)

23.1 Coarse-Graining LBE

The merging of LBE with finite volume, difference or element methods is most conveniently formulated within the general framework of *coarse-graining*.

The main idea is to gain geometrical flexibility by coarse-graining the information carried by the *differential* form of LBE (continuum spacetime) and define hydrodynamic observables on a coarse grid of virtually arbitrary shape. The key observation is that the coarse grid *need not be tied down to the symmetries of the underlying fine grid*, where the dynamics of the differential LBE takes place.

The starting point of the coarse-graining process is the differential form of LBE:

$$\partial_t f_i + c_{ia}\partial_a f_i = C_i, \quad i = 1, \ldots, b \tag{23.1}$$

The Lattice Boltzmann Equation. Sauro Succi, Oxford University Press (2018).
© Sauro Succi. DOI: 10.1093/oso/9780199592357.001.0001

where C_i denotes any suitable LBE collision operator. In the following, we shall refer to BGK for simplicity.

Formally, the expression (23.1) represents a set of b hyperbolic Partial Differential Equations (PDEs), which are clearly amenable to standard finite difference/volume/element discretization. The result is a set of bN ordinary differential equations for the generic degree of freedom f_{im} representing the i-th population attached to the spatial node, cell, element labeled by $m = 1, N$.

These equations take the general form:

$$\sum_{m=1}^{N} \left(M_{lm} \frac{df_{im}}{dt} + \vec{c}_i \cdot \vec{S}_{lm} f_{im} \right) = -\omega M_{lm}(f_{im} - f_{im}^e), \quad i = 1, \ldots, b, \quad l = 1, \ldots, N \quad (23.2)$$

where M_{lm}, $\vec{S}_{lm}$ are the grid representations of the identity and gradient operators, respectively.[54]

23.2 Finite Volume LBE

In this section we shall substantiate the abstract eqn (23.2) to the specific case of Finite Volume (FV) discretization.

Finite volumes provide a natural framework to carry on the information-removal procedure associated with coarse-graining, in a way compliant with the basic conservation principles. Moreover, they are very sound from a physical point of view, because the basic observables are defined as *averages over finite regions of space*, which is certainly closer to experimental practice and also smoother from a numerical viewpoint.

Conservation stems from the fact that cell-centered variables change in time under the effect of fluxes through the cell boundaries. By construction, these fluxes obey the identity:

"Flux-out from cell C to neighbor cell C' = - Flux-in from neighbor cell C' to C."

This secures automatic conservation (to machine round-off) of the physically conserved quantities.

For the sake of simplicity, we shall refer to a two-dimensional situation; three-dimensional extensions are conceptually straightforward, although more laborious.

We begin by considering a finite volume in the form of a two-dimensional trapezoid (see Fig. 23.1).

For this purpose, let us introduce a projection operator $\mathcal{P}$, mapping the fine-grain distribution f into the coarse-grained one, F:

$$F = \mathcal{P}f \quad (23.3)$$

---

[54] We recall that the identity operator translates into a unit matrix only if the unknown function is projected upon an orthonormal set of basis functions. This is typically the case for finite differences and volumes but *not* for finite elements, since the latter are generally overlapping in space.

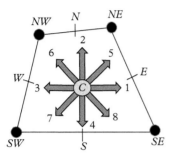

Figure 23.1 *A trapezoidal finite volume in 2D defined by the North-East (NE), North-West (NW), South-West (SW) and South-East (SE) corners. Actual unknowns, the nine populations f_i, are located in the cell centroid C.*

By definition, the projection operator removes information contained in the spatial scales below the size of the finite volumes. In practice, this operator is defined by a spatial integration over the cell Ω_C, and generates the coarse-grained populations $F_i(C)$ at each cell center C:

$$F_i(C) = \frac{1}{V_C} \oint_{\Omega_C} f_i(x,y)\, dx\, dy \tag{23.4}$$

where V_C is the volume of the cell centered in C.

In physical terms, $F_i V_C$ represents the total number of particles in the cell Ω_C traveling along the i-the direction.

Application of $\mathcal{P}$, as combined with explicit Euler forward-time marching in steps, yields (1):

$$F_i(C, t + \Delta t) - F_i(C, t) + \frac{\Delta t}{V_C} \sum_{s=1}^{4} \Phi_{is} = -\omega \Delta t [F_i(C) - \bar{F}_i^e(C)] \tag{23.5}$$

where

$$\Phi_{is} = \vec{c}_i \cdot \oint_{\partial \Omega_s} \vec{n}_s f_i(x,y)\, dA_s \tag{23.6}$$

is the flux across the s-th surface of outward normal $\vec{n}_s$, the index $s = 1,\ldots,4$ running over the four "East, North, West, South" boundaries of the cell Ω_C, of volume V_C, see Fig. 23.1.

The right-hand side is the coarse-grained BGK-collision operator, where

$$F_i^e(C) = \frac{1}{V_C} \oint_{\Omega_C} f_i^e(x,y)\, dx\, dy \tag{23.7}$$

stands for the coarse-grained local equilibrium distribution.

Equation (23.5) is not operational yet because the integrals defining the coarse-grained streaming and collision operators involve the fine-grain distribution f_i.

To this purpose a *reconstruction operator* is required in order to map the coarse populations F back into the fine ones, f:

$$f = \mathcal{R}F \qquad (23.8)$$

Formally, such reconstruction operator is the inverse of the projection operator discussed earlier on, i.e., $\mathcal{R} \equiv \mathcal{P}^{-1}$.

However, this expression is clearly ill defined since, by the definition, the projection operator $\mathcal{P}$ cancels information, hence it cannot have a unique inverse $\mathcal{P}^{-1}$.

As a result, the closed loop "fine-to-coarse-to-fine" does not land on the starting point f, i.e.,

$$f' \equiv \mathcal{R}\mathcal{P}f \neq f$$

The same goes for the reciprocal fine-to-coarse-to-fine loop:

$$F' \equiv \mathcal{P}\mathcal{R}F \neq F$$

In other words, neither $\mathcal{R}\mathcal{P}$ nor $\mathcal{P}\mathcal{R}$ coincides with the identity, so that information is lost in the process of projecting and reconstructing the fine-grain information.

The missing information must therefore be supplied by other means, typically by interpolation. Indeed, the actual form of $\mathcal{R}$ depends on the specific-interpolation scheme.

For the sake of simplicity, in the sequel we shall restrict our analysis to the case of uniform quadrilaterals of sides a and b, the case $a = b = \Delta x = 1$ denoting the standard LB scheme.

23.2.1 Piecewise-Constant Streaming

The simplest reconstruction procedure consists of assigning the same value $F_i(C)$ to all points (x, y) within the cell Ω_C.

This corresponds to a *(piecewise-constant)* (PWL) spatial representation.

The streaming operator for, say, eastward movers reads as follows:

$$F_C(t + \Delta t) = (1 - u)\, F_C(t) + u F_W(t) \qquad (23.9)$$

where

$$u \equiv \frac{c\Delta t}{a}$$

is the advection-Courant number, and W denotes the "West" neighbor cell.

The PWC representation is very simple, but it introduces severe numerical diffusion effects, which make it unviable for all but the lowest-Reynolds flows.

To appreciate the point let us Fourier analyze (23.9), where we set $k \equiv ka$ and $\omega \equiv \omega \Delta t$ for notational simplicity:

$$e^{i[k-\omega]} = 1 + (1-u)\left(e^{ik} - 1\right) \tag{23.10}$$

By writing $\omega = \omega_r + i\gamma$, summing the squares of real and imaginary parts and dividing imaginary over real components, we obtain

$$\begin{cases} e^{2\gamma} = 1 + 2u(1-u)(\cos k - 1) \\ tg\,(\omega_r - k) = \frac{(1-u)\sin k}{(1-u)\cos k + u} \end{cases} \tag{23.11}$$

It is readily seen that in the standard LB limit $u \to 1$, i.e., the exact continuum dispersion relation $\omega_r = k$ and $\gamma = 0$ is recovered.

Non-unit Courant numbers, inevitable in FV-LB formulations, lead to departures from these continuum relations.

In the limit of short wavelengths, $k \ll 1$, $\cos k \sim 1 - k^2/2$, the second of the above relations simplifies to

$$e^{2\gamma} = 1 - k^2 u\,(1-u) \tag{23.12}$$

In the small-dissipation limit $\gamma \ll 1$, this delivers

$$\gamma \sim -\frac{k^2 u\,(1-u)}{2} \tag{23.13}$$

exposing a diffusive correction with a (positive) numerical-diffusion coefficient

$$D_{\text{num}} = \frac{u(1-u)}{2} \tag{}$$

in lattice units.

This numerical diffusion vanishes in the LB limit $u = 1$, and attains its maximum, 1/8 at $u = 1/2$. In the limit $u \ll 1$, $D_{\text{num}} \sim u$, which represents quite a sizeable effect since u cannot be made much smaller than 1, on pain of incurring severe numerical dispersion effects, as it is apparent from the second dispersion relation (23.11).

By and large, not a very encouraging state of affairs.

23.2.2   Piecewise-Linear Streaming

The natural way to improve on the unsatisfactory piecewise-constant representation is to move to a higher-order interpolation.

The next level is *piecewise-linear* (PWL) interpolation, namely (index i dropped for simplicity):

$$f\,(x,y) = F_C + F_x \cdot (x - x_C) + F_y \cdot (y - y_C) + F_z \cdot (z - z_C) \tag{23.14}$$

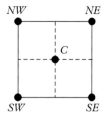

Figure 23.2 *Square cell centered in C and related areas as weighting factors.*

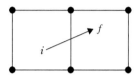

Figure 23.3 *Fine-grid particle crossing the coarse-cell boundary.*

where the coefficients F_x, F_y, F_{xy} must be expressed in terms of the nodal values F_P, $P = [C, E, N, W, S, \dots]$, i.e., the values at the centers of the neighboring cells. This is a statement on the value of the function at the cell center C, together with its partial derivatives $\partial_x F$, $\partial_y F$ and $\partial_{xy} F$.

Upwind ideas are useful for this purpose. For instance, East movers ($i = 1$) on the East boundary read as (see Figs 23.2 and 23.3)

$$f_1(e) = F_1(C) + \left[\frac{F_1(C) - F_1(W)}{x_C - x_W} \right] (x_e - x_C) \tag{23.15}$$

where subscript e refers to the center point e on the "East" surface.

Similar expressions hold for other directions.

Cross derivatives are slightly more complicated.

The complete expression for North-East movers ($i = 5$) is

$$f_5(e) = p \left[F_5(C) + \frac{F_5(C) - F_5(W)}{x_C - x_W} \left(x_e^* - x_C \right) \right] + (1-p) F_5(S) \tag{23.16}$$

where $p = 1 - 1/2a$ and the effective East location $x^*(e)$ can to some extent be treated as a free parameter in order to minimize numerical viscosity.

Some algebra (2) yields

$$x^*(e) = x_C + \frac{a-1}{2p} \tag{23.17}$$

which provides virtually zero numerical viscosity in a uniform mesh of size a (in units where $\Delta x = 1$). Note that the location $x_C + (a-1)/2$ is the position of particles in the fine grid which cross the boundary of the macrocell at half-integer times $t + 1/2$, (see Fig. 23.3).

The correction p has no transparent geometrical interpretation, at least not for this author.

It is also worth observing that while p is irrelevant at large $a \gg 1$, it becomes key at $a = 1$. In this limit, the expression (23.16) reduces to $f_5(e) = (F_5(C) + F_5(S))/2$, and it is left for the reader to prove that by summing up all four contributions from East, North, West and South boundaries, the standard LBE expression for North-East propagation is recovered:

$$F_5(C, t + 1) = F_5(SW, t) \tag{23.18}$$

23.2.3   Piecewise-Linear Collision Operator

Since the fluid field is a linear combination of the discrete populations, the piecewise-linear assumption on f carries over to the velocity field. To the purpose of coarse-graining, the latter is best split into "slow" and "fast" components

$$u_a = U_a + \tilde{u}_a \tag{23.19}$$

where $\tilde{u}_a$ is the piecewise linear fluctuation on top of the background piecewise constant velocity U_a. As a result, the contribution of sub-cell fluctuations takes the form

$$\mathcal{P}\,[\tilde{u}_a \tilde{u}_b] = \left\{ \oint_{\Omega(C)} [x_c - x_c(C)]\,[x_d - x_d(C)]\,dx\,dy \right\} \partial_c U_a \partial_d U_b \tag{23.20}$$

where $a, b, c, d = x, y$.

The cell integration is easily performed analytically, and the partial derivatives of the coarse-grained field are obtained as linear combinations of the nodal values $F(C)$ via interpolated expressions such as (23.14).

A detailed formulation of the piecewise-linear finite volume LBE in two and three dimensions can be found in (3). The calculations are elementary but lengthy and will not be reported here.

We simply remark that the piecewise-linear collision operator becomes weakly non-local through the partial derivatives of the coarse-grain velocity field. This is the price to pay for the significant drop of numerical viscosity achieved by PWL versus PWC (basically recovering the second-order accuracy of the original LBE scheme).

This reduction was first demonstrated in (2) for a simple two-dimensional Poiseuille flow test with a two-grid lattice with 32 grid points.

The first eight points starting from both walls are equispaced whereas the remaining 16 covering the central region have doubled spacing $a = 2$, for a total channel width $W = 8 + 8 + 16 \times 2 = 48$ (fine) lattice units. Along the vertical direction (y) an equispaced lattice $b = a = 1$ is retained.

The pressure gradient was adjusted in such a way as to produce a maximum speed $U_c = 0.02$ at the centerline of the channel, independently of the physical viscosity ν.

This "numerical viscosimeter" permits the computation of the additional viscosity introduced by the numerical scheme, by simply measuring the ratio $0.02/U_c$ where U_c is the value provided by the numerical simulation. Shown in Figs 23.4 and 23.5 is a series of by now Jurassic simulations with varying viscosities using *piecewise-constant* and *piecewise-linear* interpolations, respectively.

From Fig. 23.4, it is appreciated that the analytical result is only attained for very large values of the physical viscosity, $\nu = O(1)$. This is consistent with the theoretical expectations and shows that the PWC finite-volume LBE scheme (FVLBE) is inadequate for all but smallest Reynolds number simulations. Fig. 23.5 shows the dramatic improvement, nearly two orders of magnitude, resulting from piecewise-linear interpolation.

The critical reader may now wonder why, after having stated that the *piecewise-linear* version does not yield *any* numerical diffusion do we still observe departures from analytical results for Poiseuille flow?

The point is that, while it is true that there is no numerical diffusion on each of the two separate uniform sublattices, at the interface there is a definite "kink" in the slope of the velocity profile, which is responsible for numerical errors.

This kink amplitude scales with the ratio $c_s \delta a/\nu$ where δa is the jump between the two resolutions and the sound speed c_s has been inserted for dimensional consistency. It is therefore clear that the effect of the kink is negligible only at viscosities $\nu \gg c_s \delta a$, which clearly demands $\delta a/a \ll 1$.

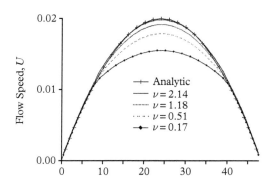

Figure 23.4 *Two-dimensional Poiseuille flow with piecewise-constant finite-volume LBE at various viscosities ν (from (2)).*

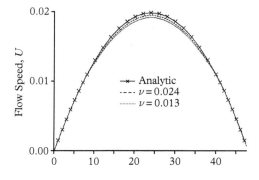

Figure 23.5 *The same case as in the previous figure, with a piecewise-linear finite volume LBE (from (2)).*

Care must therefore be exercised in tailoring the mesh so as to minimize these interface kinks. The earliest finite volume LBE versions were limited to integer jumps of the lattice pitch *a*; subsequently this limitation has been lifted allowing higher flexibility in the grid construction and smoother control of the kink effects.

23.2.4   Piecewise-Parabolic Interpolation

The piecewise-linear finite volume LBE works fine for low–moderate Reynolds flows, but still not adequate for direct numerical simulations of turbulent flows in the range $Re > 10^3$.

The calculation is easy: for a 100^3 grid with $U = 0.1$ (in uniform LBE lattice units) the physical viscosity required to achieve $Re = 10^3$ is around $\nu = 0.01$. With a local stretching $\delta a/a = \epsilon$, the condition for interface kinks to be negligible reads as $\epsilon \ll 1$, which implies a very mild stretching which cannot yield major upsizing of the computational domain with a reasonable number of grid points.

The bottom line is that for state of the art direct finite volume LBE simulations, higher-order interpolates are needed. In fact, second-order interpolation should restore second-order accuracy of the original LBE scheme.

23.3   Unstructured Lattice Boltzmann Equation

Modern finite volume LBE formulations have been introduced in the last decade (4; 5; 6), which achieve higher geometrical flexibility.

The main idea is to use *unstructured* volumes generating grids that do not need to conform to any specific coordinate system, and where the connectivity of the network of nodes can also change from place to place. Protoypical unstructured volumes are triangles and tetrahedra, in two and three dimensions, respectively.

These unstructured Lattice Boltzmann schemes (ULBE for short) integrate the differential form of LBE using a cell-vertex finite-volume technique, in which the unknown populations are placed at the nodes of the mesh and evolve based on the fluxes crossing the edges of the corresponding elements (see Fig. 23.6). It has been shown that ULBE tolerates significant stretching without introducing any appreciable numerical viscosity effect to second order in the mesh size.

This is important for practical applications, since it permits a time accurate description of transitional flows. More precisely, since the mesh is irregular, the negative contribution of propagation viscosity disappears leaving the same expression as in the continuum, i.e.,

$$\nu = c_s^2 \tau \tag{23.21}$$

This implies that, unlike standard LB, small viscosities can only be achieved with correspondingly small timesteps; a key property of LB being sacrificed to the cause of geometrical flexibility.

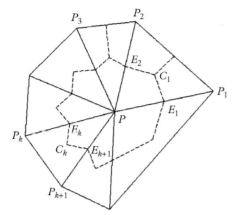

Figure 23.6 *A typical finite volume for the cell-vertex ULBE. The discrete populations at node P are advanced in time based on the fluxes through the surrounding surface (line in 2D) obtained by joining the edge centers E_k with the centers C_k of the surrounding triangles with vertices P, P_k and P_{k+1}. Reprinted with permission from S. Ubertini et al, Progress in Comp. Fluid Dynamics, 5, 84 (2005). Copyright 2005 by Inderscience.*

ULBE methods deliver interesting performance on moderate Reynolds number flows. However, they are subject to stringent timestep limitations, as imposed by the smallest volumes and consequently, their application to high-Reynolds flows simulations appears somehow problematic. Arguably, their best use is in conjunction with standard uniform LBE schemes, whereby ULBE would take care of the boundary layer regions, while standard LBE would be applied in the bulk flow.

At the time of this writing, such coupling just started to appear, Di Ilio et al., *Phys. Rev. E*, 2017 (7).

23.4   Finite Difference LBE

Once it is recognized that the differential form of LBE is basically a set of hyperbolic PDEs, it becomes natural to observe that both time and space derivatives can be discretized in many ways, not just the lock-step light-cone rule $\Delta \vec{x} = \vec{c} \, \Delta t$ characterizing LBE.

With independent time and space discretizations the lattice is set free from the tyranny of symmetry requirements which tie down LBE dynamics and considerable geometrical flexibility can be bought thereof.[55] Of course, such freedom comes at a price, as we shall see in the sequel.

This point of view was first endorsed by S. Chen and co-workers, who developed a number of Finite Difference LBE (FDLBE) methods based on higher-order (Runge–Kutta) time-marching schemes, as combined with various spatial discretization schemes (8; 9; 10).

As a specific example, they used a central finite difference scheme for spatial discretization and a second-order Runge–Kutta (modified Euler) for time marching.

---

[55] One more link with LGA theory being jettisoned ....

The resulting FDLBE schemes take the general form

$$f_i\left(\vec{x}, t + \frac{\Delta t}{2}\right) = f_i(\vec{x}, t) - \frac{1}{2}\Delta t R_i(\vec{x}, t) \tag{23.22}$$

$$f_i(\vec{x}, t + \Delta t) = f_i(\vec{x}, t) - \Delta t\, R_i\left(\vec{x}, t + \frac{\Delta t}{2}\right) \tag{23.23}$$

where

$$R_i(\vec{x}, t) = -c_{ia}\Delta_a f_i - \omega\left(f_i - f_i^e\right) \tag{23.24}$$

and $\Delta_a f_i = [f_i(\vec{x} + \Delta x_a, t) - f_i(\vec{x} - \Delta x_a, t)]/2\Delta x_a$ is the centered partial derivative with respect to x_a.

This FDLBE scheme achieves second-order accuracy in both space and time and has been successfully demonstrated for various flows in bounded geometries, such as two-dimensional Taylor vortex flow, Couette flow with temperature gradient between the walls, and others.

Several finite-difference variants of the LB scheme have been developed in the last decade, not only for the Navier–Stokes equations, but for a broad variety of other linear and nonlinear partial differential equations of mathematical physics. While a general assessment of these FD-LB methods is beyond the scope of this book, here we only wish to reiterate the main conceptual point: finite-difference formulations generally spoil a major property of the basic LB scheme, i.e., exact streaming. Whether this is a price worth paying in view of the resulting gains in geometrical flexibility is an assessment that should made case by case, depending on the problem at hand. Based on the current literature, it appears like FD-LBE represents an interesting option for flows requiring curvilinear coordinates (11).

23.5   Interpolation-Supplemented LBE

Another approach along the line of *grid freedom* is the so-called *ISLBE* (Interpolation-Supplemented Lattice Boltzmann Equation). The idea is still to move the discrete distributions along straight paths $\Delta x_{ia} = c_{ia}\Delta t$, with the important proviso of doing away with the constraint that endpoints of the flights should land on lattice sites.

This is yet another way of setting the control grid free from the symmetry constraints bearing on the microscopic particle motion. The price to pay for this geoflexibility is, as in finite-volume LBE, the need of interpolating between the spatial grid and the particle positions.

It is clear that LBE becomes here a close relative of well-known Particle In Cell (PIC) methods (12), with the LBE highlighting that velocity space is represented by a fistful of discrete speeds. As in PIC methods, the geometry of the spatial control grid can be chosen at will and non-uniform or even non-Cartesian meshes can be adopted. Unlike PIC, however, the particle degrees of freedom carry quasihydrodynamic information

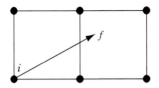

Figure 23.7 *An off-lattice particle flight starting at i and ending at f. The starting point i lies on the lattice, the ending point f does not, and therefore interpolation is needed to bring the information back to the lattice.*

which obviates the need for statistical averaging. Also, the asset of locality is preserved to a great extent, since particles contribute to the grid densities only at the nodes of the cells they belong to.

To minimize numerical viscosity, second-order interpolation is used, see Fig. 23.7.

He *et al.* proved the viability of ISLBE for a two-dimensional flow past a cylinder using cylindrical coordinates (13), at both low- and high-Reynolds number (up to $Re = 10\,000$). Their results compare well with the wide body of existing data and support claims of substantial memory savings (more than an order of magnitude) with respect to standard uniform LBE.

However, this method does not appear to have made into a LB mainstream, although modern semi-lagrangian methods have been developed in the recent literature (14).

23.6   Finite Element LBE

Having discussed the merge of LBE with finite volume and finite difference methods, for the sake of completeness, we shall also mention the possibility of marrying LBE with the third general grid method, namely the Finite Element Method (FEM).

Consistently with the general finite element philosophy, we expand the discrete populations $f_i(\vec{x}, t)$ onto a set of *localized* (finite support) basis functions $\psi_l(x)$ (15):

$$f_i(\vec{x}, t) = \sum_{m=1}^{N_e} f_{im}(t)\, \psi_m(\vec{x}) \tag{23.25}$$

where N_e is the number of finite elements, equal to the number of grid nodes. The basis functions $\psi_m(\vec{x})$ are piecewise polynomials restricted to a local neighborhood of the lattice node $\vec{x}_m$ (whence the nomer *Finite Elements*).

By inserting the finite element expansion (23.25) into the differential LBE (23.1), and taking the scalar product with the generic basis function ψ_l, one obtains a set of ordinary differential equations of the form (23.2), where

$$M_{lm} = (\psi_l, \psi_m) \tag{23.26}$$

$$\vec{S}_{lm} = (\psi_l, \nabla \psi_m) \tag{23.27}$$

In (23.26)–(23.27), the brackets stand for the functional-scalar product $(f, g) \equiv \int f(\vec{x}) g(\vec{x})\, d\vec{x}$.

Each finite-element basis function $\psi_l(\vec{x})$ takes value one at the node $\vec{x}_l$ and zero at every other node, i.e.,

$$\psi_l(\vec{x}_m) = \delta_{lm}$$

which implies

$$f_{il} = f_i(\vec{x}_l)$$

In other words, the unknowns f_{il} represent the value of the function at the node $\vec{x}_l$, see Fig. 23.8.

However, instead of seeking point-wise convergence to such values, the FEM procedure rather focuses on the global convergence ("weak convergence" in math language) in the functional space spanned by the N_e basis functions $\{\psi_l\}$. This is similar to the Hermite expansion in velocity space discussed earlier, with a crucial twist: the finite element basis functions are *compact*, i.e., they are strictly zero outside their spatial support. The numerical benefit is substantial: the basis functions can be made quasi-orthogonal without any sign change, but simply leveraging their compactness. Indeed the mass matrix M_{lm} is non zero only if the supports of the elements ψ_l and ψ_m show a non-zero overlap.

A simple example clarifies the point.

A popular finite element in one spatial dimension is the piecewise linear "hat function" defined as follows:

$$e_l(x) = \begin{cases} \dfrac{x - x_{l-1}}{x_l - x_{l-1}} & x_{l-1} < x < x_l \\[2mm] \dfrac{x_{l+1} - x}{x_{l+1} - x_l} & x_l < x < x_{l+1} \\[2mm] 0 & \text{otherwise} \end{cases} \qquad (23.28)$$

Hence the support of $e_l(x)$ is the interval $I_l \equiv [x_{l-1} < x < x_{l+1}]$. Note that $e_l(x_l) = 1$ and 0 at every other node.

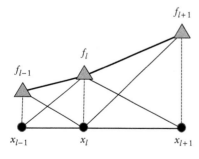

Figure 23.8 *Piecewise linear Finite Element representation of the function $f(x) = f_{l-1}e_{l-1}(x) + f_l e_l(x) + f_{l+1}e_{l+1}(x)$ in the interval $[x_{l-1}, x_{l+1}]$.*

Clearly, the interval I_l overlaps only with its left and right neighbor intervals, I_{l-1} and I_{l+1}, which means that the matrix M_{lm} is tridiagonal, i.e., the only non-zero elements are $M_{l,l-1}$, $M_{l,l}$ and $M_{l,l+1}$.

The same argument applies to the streaming matrix B_{lm}.

The corresponding one-dimensional FE-LBE takes the form of a first-order matrix evolution problem:

$$M_{l,l-1}\dot{f}_{i,l-1} + M_{l,l}\dot{f}_{i,l} + M_{l,l+1}\dot{f}_{i,l-1} = \tag{23.29}$$

$$- v_i[S_{l,l-1}f_{i,l-1} + S_{l,l}f_{i,l} + S_{l,l+1}f_{i,l+1}] - \omega[M_{l,l-1}g_{i,l-1} + M_{l,l}g_{i,l} + M_{l,l+1}g_{i,l-1}] \tag{23.30}$$

where we have set

$$g_{il} \equiv f_{il} - f_{il}^e$$

It can be readily checked that for the case of uniform mesh $x_l = l\Delta x$, the matrix elements reduce to

$$M_{l,l-1} = \frac{\Delta x}{6}, \ M_{l,l} = \frac{4\Delta x}{6}, \ M_{l,l+1} = \frac{\Delta x}{6} \tag{23.31}$$

and

$$S_{l,l-1} = -\frac{1}{2}, \ S_{l,l} = 0, \ S_{l,l+1} = +\frac{1}{2} \tag{23.32}$$

In passing, we note that the mass matrix associated with the hat functions corresponds precisely to the weights of the D1Q3 scheme, and the same holds for higher-dimensional versions.

As a result, the FEMLB equation simplifies to

$$\frac{1}{6}\dot{f}_{i,l-1} + \frac{4}{6}\dot{f}_{i,l} + \frac{1}{6}\dot{f}_{i,l-1} = -\frac{v_i}{2\Delta x}(f_{i,l+1} - f_{i,l-1}) - \frac{\omega}{6}g_{i,l-1} - \frac{4\omega}{6}g_{i,l} - \frac{\omega}{6}g_{i,l-1} \tag{23.33}$$

This is a first order in the time-centered finite-difference matrix problem, for which FEM theory may eventually offer weak convergence to second order in the mesh spacing.

Clearly, this is more complicated than plain LB, and also than the finite-volume or finite-difference variants discussed thus far. Indeed, owing to the non-diagonal mass matrix, FEM-LBE presents us with a *co-evolution* matrix problem.

By co-evolution, we imply the fact that each unknown amplitude f_{il} does not evolve alone, but in conjunction with its left and right neighbors. It is readily recognized that the standard evolutionary form is recovered by introducing the coarse-grained, non-local distribution:

$$\bar{f}_{il} = \frac{1}{6}f_{i,l-1} + \frac{4}{6}f_{i,l} + \frac{1}{6}f_{i,l-1} \tag{23.34}$$

It is also readily checked that such non-local distribution is simply the projection of the full distribution $f_i(x)$ on the basis function $e_l(x)$, i.e.,

$$\bar{f}_{il} = \int_{-\infty}^{+\infty} f_i(x) e_l(x)\, dx$$

This shows that $\bar{f}_{il}$ represents a weighted spatial average over the spatial support of the finite element $e_l(x)$.

It is instructive to recast the expression (23.34) in the following form ($\Delta x = 1$):

$$\bar{f}_{il} = f_{i,l} + \frac{1}{6}[f_{i,l-1} - 2f_{i,l} + f_{i,l+1}] \tag{23.35}$$

This shows that the spatial average corresponds to replacing the original distribution $f(x)$ with its smoother counterpart

$$\bar{f}(x) = f(x) + \frac{\Delta x^2}{6}\frac{d^2 f}{dx^2}$$

In other words, the co-evolution of the original $f(x)$ is replaced by the co-evolution of its smooth associate $\bar{f}(x)$.

The transformation $f \to \bar{f}$ is clearly non-local, and such non-locality is the price to pay to be guaranteed weak convergence, once the proper finite-element basis is chosen.

As compared to finite-volumes, the FEM formulation offers the advantage of a consolidated mathematical background and the existence of powerful convergence theorems. In general, polynomial elements of order p deliver weak convergence to order $p + 1$. It is also more systematic; once the elements are chosen, all the mathematical apparatus follows straight, with little or no room left for discretional choices. In actual practice, however, finite-elements are usually more expensive than finite volumes and finite differences, mainly because they involve a matrix problem even for explicit methods. This is probably the reason why they do not seem to have conquered the mainstream of computational fluid dynamics.

A similar consideration applies to the Lattice-Boltzmann framework; the FE-LB literature appears still comparatively scanty (16).

23.7   Native LBE Schemes on Irregular Grids

We began this chapter with the idea of buying geo-flexibility to LB schemes by marrying them with proved performers in the field, chiefly finite volumes. This is certainly wise, but more daring paths have been explored as well.

The question is: *can we formulate the native LBE microdynamics in an irregular spacetime?*

Irregular (random) lattices are well known in lattice field theory (17), where they are generally credited with a number of virtues, typically regularization of anomalies (the benefits of disorder ...).

The main idea is to define a set of space-dependent discrete speeds $c_{ia}(\vec{x})$ as a local deformation of some suitable standard uniform set c_{ia}^0. The new speeds can be regarded as a "grid flow" with its own speed $c_a = \sum_i c_{ia}$ and deformation tensor $G_{ab} = \partial_a c_b$, both quantities being identically zero in a uniform mesh. The non-Euclidean metric gives rise to inertial forces, a rather undesirable customer for numerical schemes. The main idea is to handle the inertial forces due to the space-dependence of the discrete speeds as it is generally done for any other force, i.e., reabsorb them into properly generalized equilibria. Such generalized equilibria can then be found by entropy minimization principles (18).

This scheme does not seem to have flourished into a competitive method for fluid flows, but it might still have an interest as a discrete kinetic model for flows in disordered and randomly curved media (19).

23.7.1   Implicit LBE Schemes

All schemes discussed so far differentiate on the basis of the way that the space variable is discretized; in all cases the time variable has been treated via explicit discretization. This responds to the criterion of preserving causality in the numerical scheme. On the other hand, once one is willing to forsake the light-cone spacetime discretization, it is rather natural to do away with explicit time integration as well, so as to march in larger steps toward steady-state configurations.

Efforts in this directions have been put forward in the past, but with limited success, mostly on account of the large matrix problem to be solved at each timestep. Moreover, besides mere size issues, the LB matrices do not generally lead to fast convergence of the iterative solvers, due to the hyperbolic nature of the LB equation. At the time of this writing, LB remains largely an explicit scheme also in its finite-volume/difference/elements variants.

23.8   Multigrid Lattice Boltzmann Scheme

Many phenomena of physical and engineering interest exhibit sudden excursions over highly localized regions (boundary layers, shock fronts) which require a correspondingly highly clustered mesh. A popular response to this kind of need in modern computational fluid dynamics is provided by *unstructured* meshes, namely discrete grids in which the number of neighbors of a given node may change from place to place. This allows much stronger distortions of the computational grid, but also requires significantly more complex data structures.

Another popular possibility is provided by *locally embedded* grids, namely grids in which the local connectivity (number of neighbors) stays unchanged, but the lattice spacing is refined or coarsened locally, typically in steps of two for practical purposes.

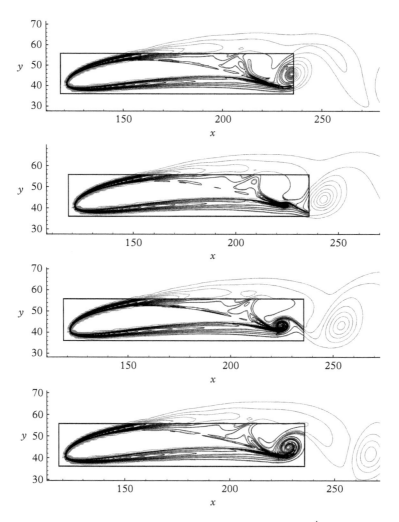

Figure 23.9 *Two-dimensional flow past an airfoil at Re = 10^4 as computed with a multiscale LBE scheme (courtesy of O. Filippova). The grid within the box is refined by a factor of six as compared to the exterior. The four figures represent the time evolution of the vorticity field.*

Local embedding is a specific instance of a more general framework known as *multiscale algorithms*.

Multigrid LBE schemes have been pioneered by Filippova *et al.* (20), who tested and validated them for moderate-Reynolds flows around cylinders and wings (see Fig. 23.9).

Given that local grid-refinement is one of the most popular LB variants for handling complex geometries, a few details of the procedure are worth illustrating, see Fig. 23.10.

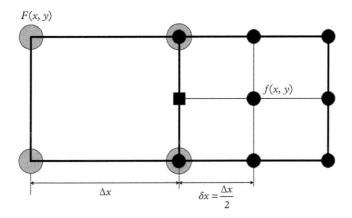

Figure 23.10 *A two-dimensional, two-grid mesh. Large grey and small black circles are the coarse and fine-grained grid nodes, respectively. Double circles are sites common to both coarse and fine grids, squares are fine grid only interface nodes that need to be initialized by interpolation in order to proceed with the update of the interior sites of the fine grid.*

Let us begin by considering two grids, with resolution Δx (coarse) and $\delta x = \Delta x/n$ (fine), with n an even integer, typically $n = 2$ (see Fig. 23.10).

Let the corresponding discrete distributions be F and f, respectively (the discrete velocity index is dropped without loss of generality). Note that the coarse and fine grids have common nodes at the interface. Other options, such as overlapping grids, could be considered as well.

The general multigrid procedure is as follows:

1. *Coarse grid evolution*:
   Evolve $F(t)$ in the coarse grid over a timestep Δt

$$F(x + \Delta x, t + \Delta t) = F' \tag{23.36}$$

$$F' \equiv (1 - \omega_F)F + \omega_F F^{eq} \tag{23.37}$$

where we have set

$$\omega_F = \frac{\Delta t}{\tau_F} \tag{23.38}$$

2. *Coarse to Fine Reconstruction*:
   F is interpolated at boundary nodes at all intermediate times $t' = t, t + m\Delta t/n$, $m = 0, 1, \ldots n - 1$, to provide initial and boundary conditions for the fine-grid evolution.

3. *Fine grid evolution*:
   Using the interpolated values in space and time evolve $f(t')$ on the fine grid, over n timesteps of size $\delta t = \Delta t/n$ each.

$$f(x + \delta x, t + \delta t) = f'(x, t) \tag{23.39}$$

$$f' \equiv (1 - \omega_f)f + \omega_f f^{eq} \tag{23.40}$$

where

$$\omega_f = \frac{\delta t}{\tau_f} \tag{23.41}$$

4. *Fine to coarse projection*:
   The values of F at common nodes are set equal to the corresponding fine-grained values f, so that a new coarse-grain evolution step can be initiated.

5. *Next coarse-grain step*:
   Go to 1.

This general structure is quite transparent, but leaves many important dangling details.

In particular, one has to prescribe the cross-talk protocol between F and f and vice versa, namely provide a mapping of F' as a function of f^e and f' and, conversely, f' as a function of F^e and F'.

This task is accomplished by imposing continuity of mass and momentum, as well as of their fluxes, across the interface between the two grids.

This leads to the following relations between the equilibrium and non-equilibrium components in the coarse and fine grids, respectively:

$$\begin{cases} f^e(I) = F^e(I) \\ f^{ne}(I) = \dfrac{\tau_f}{\tau_F} F^{ne}(I) \end{cases} \tag{23.42}$$

where I denotes the generic interface node.

The former equation is straightforward: since density and flow velocity must be the same at the interface, so do the local equilibria.

The latter stems from the identification of the non-equilibrium distribution with the mean-free path times the gradient of the local equilibria, which is true only to first order in the Knudsen number.

The last step is to establish a the connection between the relaxation times in the two grids.

Equating the viscosity gives

$$\frac{\delta x^2}{\delta t}(\tau_f/\delta t - 1/2) = \frac{\Delta x^2}{\Delta t}(\tau_F/\Delta t - 1/2) \tag{23.43}$$

By imposing the same lattice light speed across the two grids

$$\frac{\Delta x}{\Delta t} = \frac{\delta x}{\delta t} = 1$$

the previous relation simplifies to

$$\omega_f = \frac{\omega_F}{\frac{\omega_F}{2} + n(1 - \frac{\omega_F}{2})} = \frac{\omega_F}{n + (1-n)\frac{\omega_F}{2}} \tag{23.44}$$

Conversely, by swapping $f \leftrightarrow F$ and $n \leftrightarrow 1/n$:

$$\omega_F = \frac{n\omega_f}{(1 - \frac{\omega_f}{2}) + n\frac{\omega_f}{2}} = \frac{n\omega_f}{(1 + (n-1)\frac{\omega_f}{2})} \tag{23.45}$$

The final result is

$$\begin{cases} F' = (1 - \Omega)f^e + \Omega f' \\ f' = (1 - \Omega^{-1})F^e + \Omega^{-1}F' \end{cases} \tag{23.46}$$

where we have set

$$\Omega = n\frac{\omega_f\,(1 - \omega_F)}{\omega_F(1 - \omega_f)} \tag{23.47}$$

A few comments are in order.

First, it is readily checked that in the limit $n \to 1$, i.e., $\Omega \to 1$, one recovers $f' \to F'$, as it should be on matter of consistency.

Second, we observe that the mapping is singular at $\omega_f = 1$, the border between under- and over-relaxation.

Based on the expression (23.44), this occurs at

$$n_{sing} = \frac{\frac{\omega_F}{2}}{1 - \frac{\omega_F}{2}}$$

In other words, starting with $\omega_F > 1$ (over-relaxation) ω_f also remains in the over-relaxation regime only as long as $n < n_{sing}$ (see Fig. 23.11).

For a typical value, say $\omega_F = 1.9$, i.e., $\nu \sim 0.01$ in coarse-grained lattice units, the previous relation gives $n_{sing} = 19$. This allows up to four levels of refinement, $(2^4 = 16)$. With $\omega_F = 1.99$, corresponding to a viscosity $\nu \sim 10^{-3}$ in coarse-grain lattice units, $n_{sing} = 199$, which allows seven levels of refinement $(2^7 = 128)$. These figures indicate

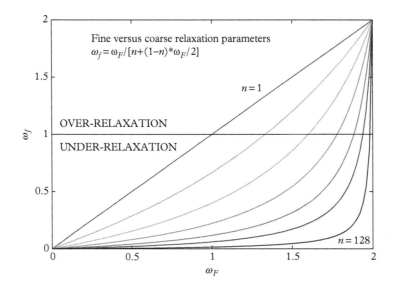

Figure 23.11 *Fine versus coarse relaxation parameter for a sequence of grid refinement levels, $n = 1, 2, 4, 8, 16, 32, 128$. Note that the continuum limit, $n \to \infty$, is singular, as it delivers $\omega_f = 0$ for $0 \leq \omega_F < 2$ and $\omega_f = 2$ for $\omega_F = 2$. In other words, in the continuum limit the only surviving values are the fixed points of the transformation (23.44): $\omega_f = 0$ (no-collisions) and $\omega_f = 2$ (infinitely strong collisions). As we have learned in Chapter 22 on entropic LB, both these extrema leave the entropy unchanged.*

that the singularity constraint is something one can live with, as long as comparatively moderate levels of refinement, say less than ten, are required.

Nevertheless, subsequent work has spawned a number of variants which, besides lifting the singularity can boast improved stability, accuracy and even dynamic adaptivity, i.e., formulations wherein the refined grid moves from place to place in response to the actual structure of the flow.

The matter is steeply technical and the interested reader is kindly directed to the original literature (20; 21; 22; 23).

Inevitably, the complexity of such formulations takes LB a rather long stretch away from the early logo *"Simple Models for Complex Flows,"* *"Complex Models for Complex Flows"* being probably more appropriate. Albeit not terribly exciting, this is after all still better than *"Complex Models for Simple Flows"* . . . .

23.9   Summary

To sum up, while the original LBE is grid-bound and somehow at odds toward realistically complex geometries, many variants are available today that partially mend this weakness.

These variants generally take LBE closer to the realm of computational fluid dynamics and their success depends on the extent to which they manage to handle complex geometries without sacrificing the original LBE assets, simplicity and amenability to parallel computing.

At the time of this writing, local grid refinement seems to offer the most popular deal in this respect, although a general assessment remains difficult to be made, and may change for the future.

....

REFERENCES

1. F. Nannelli and S. Succi, "The lattice Boltzmann equation in irregular lattices", *J. Stat. Phys.*, **68**, 401, 1992.
2. F. Nannelli and S. Succi, "The finite volume formulation of the lattice Boltzmann equation", *Transp. Theory and Stat. Phys.*, **23**, 163, 1994.
3. G. Amati, S. Succi and R. Benzi, "Turbulent channel flow simulation using a coarse-grained extension of the lattice Boltzmann method", *Fluid Dyn. Res.*, **19**, 289, 1997.
4. G. Peng, H. Xi and C. Duncan, "Lattice Boltzmann method on irregular meshes", *Phys. Rev. E*, **58**(4), R4124, 1998.
5. H. Xi, G. Peng and S.-H. Chou, "Finite volume lattice Boltzmann schemes in two and three dimensions", *Phys. Rev. E*, **60**, 3380, 1999.
6. S. Ubertini and S. Succi, "Recent advances of lattice Boltzmann techniques on unstructured grids, Progress in Computational Fluid Dynamics", **5**, 84, 2005.
7. G. Di Ilio, D. Chiappini, S. Ubertini, G. Bella, S. Succi, "Hybrid lattice Boltzmann method on overlapping grids", *Phys. Rev. E*, **95**, 013309, 2017. doi:10.1103/PhysRevE.95.013309.
8. N. Cao, S. Chen, S. Jin and D. Martinez, "Physical symmetry and lattice symmetry in lattice Boltzmann method", *Phys. Rev. E*, **55**, R21, 1997.
9. X. He, L.S. Luo and M. Dembo, "Some progress in lattice Boltzmann method: part I, non-uniform grids", *J. Comp. Phys.*, **129**, 357, 1996.
10. R. van der Sman, *Lattice Boltzmann Schemes for Convection–Diffusion Phenomena; Application to Packages of Agricultural Products*, Ph.D. Thesis, University of Wageningen, Holland, 1999.
11. R. Mei, L.S. Luo and W. Shyy, "An accurate curved boundary treatment in the lattice Boltzmann method", *J. Comp. Phys.*, **155**, 307, 1999.
12. R. Hockney and J. Eastwood, *Computer Simulation Using Particles*, McGraw–Hill, London, 1979.
13. X. He, X. Shan and G. Doolen, "Lattice Boltzmann method on curvilinear coordinates system: vortex shedding behind a circular cylinder", *Phys. Rev. E*, **57**, R13, 1998.
14. B. Dorschner, S. Chikatamarla, and I. Karlin, "Entropic multirelaxation-time lattice Boltzmann method for moving and deforming geometries in three dimensions", *Phys. Rev. E*, **95**, 063306, 2017. doi:10.1103/PhysRevE.95.

15. W. Strang and G. Fix, *An Analysis of the Finite Element Method*, Prentice–Hall, Englewood Cliffs, New Jersey, 1973.
16. Yusong Li, Eugene J. LeBoeuf, and P.K. Basu, "Least-square finite-lement lattice Boltzmann method", *Phys. Rev, E*, **69**, 065701, 2004
17. N.H. Christ, R. Friedberg and T.D. Lee, "Random lattice field theory: general formulation", *Nucl. Phys., B* **202**, 89, 1982.
18. I. Karlin, S. Succi and S. Orszag, "The lattice Boltzmann method on irregular lattices", *Phys. Rev. Lett.*, **82**(26), 5245, 1999.
19. M. Mendoza, S. Succi, H.J. Herrmann, "Flow through randomly curved manifolds", *Sci. Rep.*, **3**, 3106, 2013.
20. O. Filippova and D. Haenel, "Grid refinement for Lattice BGK models", *J. Comp. Phys.*, **147**, 219, 1998.
21. A. Dupuis and B. Chopard, "Theory and applications of an alternative lattice Boltzmann grid refinement algorithm", *Phys. Rev. E*, **67**, 066707 2003.
22. J. Toelke, S. Freudiger and M. Krafczyk, "An adaptive scheme using hierarchical grids for lattice Boltzmann multi-phase flow simulations Computers and Fluids", 35, 820, 2006.
23. D. Yu, R. Mei and W. Shyy, "A multi-block lattice Boltzmann method for viscous fluid flows", *Journal for numerical methods in fluids*, **39**, 99, 2002.

EXERCISES

1. Draw a graph of the coupling coefficients of the piecewise-constant finite volume LBE scheme as a function of the grid size Δx.
2. Estimate the numerical viscosity of piecewise-constant finite-volume LBE.
3. Compute the mass and advection matrix for piecewise-linear finite elements.

24

Lattice Boltzmann for Turbulence Modeling

In this chapter, we present the main ideas behind the application of LBE methods to the problem of turbulence modeling, namely the simulation of flows which contain scales of motion too small to be resolved on present-day and foreseeable future computers.

This chapter is dedicated to the memory of Steven Orszag, a great pioneer of computational fluid dynamics, a wonderful mentor and friend.

24.1 Sub-Grid Scale Modeling

Most real-life flows of practical interest exhibit Reynolds numbers far too high to be directly simulated in full resolution on present-day computers and arguably for many years to come.

This raises the challenge of predicting the behavior of highly turbulent flows without directly simulating all scales of motion which take part to turbulence dynamics, but only those that fit within the computer resolution at hand (see Fig. 24.1).

The name of the game is Turbulence Modeling (TM), the art of modeling the effects of the unresolved scales on the resolved ones.

In principle, this sounds like "Mission Impossible," because, unless one knows the solution beforehand, there is no way to tell *exactly* what effect on the unresolved scales the resolved ones will have. However, as usual, informed guesses (models) can be developed and progressively refined to the point of providing at least a partial answer to the grand-question ask posed.

To this purpose, it proves expedient to split the actual velocity field into large-scale (resolved) and short-scale (unresolved) components, namely

$$u_a = U_a + \tilde{u}_a \tag{24.1}$$

The Lattice Boltzmann Equation. Sauro Succi, Oxford University Press (2018).
© Sauro Succi. DOI: 10.1093/oso/9780199592357.001.0001

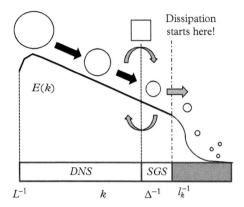

Figure 24.1 *Sketch of the energy spectrum of a turbulent flow. The eddies larger than the grid size Δ are liable to direct numerical simulation (DNS). Since the numerical resolution is not sufficient to reach the Kolmogorov scale l_k, the eddies below the mesh size $l < \Delta$ (subgrid-scales, SGS for short) must be modeled.*

where, by definition, the short-scale fluctuations average to zero

$$< \tilde{u}_a > = 0$$

As a result,

$$U_a = < u_a >$$

brackets denoting some suitable form of averaging.

Since the Navier–Stokes equations are quadratic in the velocity field it is clear that any form of averaging gives rise to new forces due to the non-vanishing quadratic correlations of the short scales.

These "turbulent" forces come in the form of the divergence of the *Reynolds stress tensor*:

$$R_{ab} = \rho \langle \tilde{u}_a \tilde{u}_b \rangle \tag{24.2}$$

The goal of turbulence modeling, sometimes also known as sub-grid scale (SGS) modeling is therefore to express the divergence of this tensor in terms of the resolved field U_a. Symbolically,

$$< NSE[u_a] > = NSE[U_a] + (\partial_b R_{ab})[U] \tag{24.3}$$

where *NSE* stands for Navier–Stokes equations.

24.2 Lattice Boltzmann Turbulence Models

From a fundamental viewpoint, the conceptual path behind hydrodynamic-turbulence modeling can be summarized as follows:

$$f \rightarrow u \rightarrow U \tag{24.4}$$

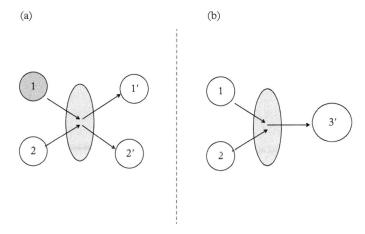

(a)   (b)

Figure 24.2 *Sketch of interactions between small-scale turbulent eddies. On the left side, a scattering interaction $1 + 2 \to 1' + 2'$ leading to no change of the size of the participating eddies. On the right-hand side a merger event $1 + 2 \to 3'$ leading to the emergence of a larger eddy. The latter requires modeling because it leads to the resolved eddy $3'$ out of the two unresolved ones, 1 and 2.*

In other words, by taking the large-scale Chapman–Enskog limit of the Boltzmann equation, one lands on the continuum hydrodynamic description in terms of the flow field u, see Fig. 24.2. For the case of turbulent flows, such field contains strong fluctuations, $\tilde{u}$, and subsequent average over such fluctuations lead to a dynamic equation for the coarse-grained field U. The corresponding lengthscales are

$$l_\mu \to l_k \to \Lambda \to L$$

namely the mean-free path, the Kolmogorov length, the coarse graining length and the global scale L, with the ordering $l_\mu < l_k < \Lambda < L$.

Within the kinetic approach, an alternate path can be conceived, namely

$$f \to F \to U \tag{24.5}$$

In other words, one would coarse grain the Boltzmann equation itself, to produce a large-scale *super-Boltzmann* equation for the distribution F and then apply the Chapman Enskog asymptotic to derive the corresponding large-scale super-hydrodynamics. In general, the super-fields U in (24.4) and (24.5) are not expected to be the same, because the Chapman–Enskog asymptotics does not necessarily commute with spacetime coarse graining taking u into U or f into F. As a result, the kinetic approach offers potential for the formulation of a new class of turbulence models not included in the classical hydrodynamic pattern (24.4).

Based on this, we identify three main families of Lattice Boltzmann TM's, namely:

1. *Hydro-kinetic (HK):* The idea is simply to import existing hydrodynamic TM's within the LB mathematical harness. At first glance, this may sound like a rather pointless "old wine in a new bottle" exercise. However, a subtler inspection reveals that the new bottle can make the old wine better, eventually ....

   The reason is basically that the relaxation mechanism inherent to kinetic theory is more general than the notion of effective viscosity. Indeed, unlike effective viscosity, effective relaxation does not depend on any assumption of scale separation, which is potentially a significant advantage.

2. *Kinetic-hydro (KH):* Models in this class are based on direct coarse graining of the kinetic equation, along the pattern (24.5). As mentioned previously this offers the opportunity of generating new TM's outside the reach of hydrodynamic ones. However, the closure is still imported from hydrodynamics, which is why we label them as kinetic-hydro, instead of fully kinetic.

3. *Fully kinetic (KK):* This is a prospective class of TM's based on coarse-graining of the kinetic equation, with the closure also entirely kinetic in nature. To the best of this author's knowledge, this class is way less developed than the previous two and has not given rise to any practical implementation as yet.

In the sequel, we shall discuss a few examples of the three families presented above.

24.3 Hydro-Kinetic: Eddy-Viscosity Smagorinsky Model

One of the most powerful heuristics behind SGS is the concept of *eddy viscosity*, due to the meteorologist J. Smagorinsky (1924–2005).

This idea, a significant contribution of kinetic theory to fluid turbulence, assumes that the effect of small scales on the large ones can be likened to a diffusive motion caused by random collisions. In other words, small eddies are kinematically transported without distortion by the large ones, while the large ones experience diffusive Brownian-like motion, due to erratic collisions with the small eddies.

Based on this intuition, Smagorinsky's SGS models take the following *algebraic* form (1):

$$R_{ab} = \rho \nu_{eff} \left(|S|\right) S_{ab} \tag{24.6}$$

In (24.6), $S_{ab} = (\partial_a U_b + \partial_b U_a)/2$ is the large-scale strain rate tensor of the incompressible fluid and ν_{eff} is the *effective* eddy-viscosity. The latter is defined as the sum of the molecular plus the turbulent viscosities:

$$\nu_{eff}(S) = \nu_0 + \nu_{tur}(S) \tag{24.7}$$

where S is a suitable scalar contraction of S_{ab} (see below).

A typical expression of the Smagorinsky eddy viscosity is as follows:

$$v_{tur} = C_S \Delta^2 |S|, \quad |S| = \left(\sum_{a,b} 2 S_{ab} S_{ab} \right)^{1/2} \tag{24.8}$$

where C_S is an empirical constant of the order 0.1 and Δ is the mesh size of the numerical grid ($\Delta = 1$ in lattice units).

The rationale of (24.8) is quite transparent: the scale of the unresolved eddies is taken to be the mesh spacing Δ and their representative lifetime is taken to be the inverse magnitude of the strain rate. The ignorance of the details neglected by such a crude yet sensible assumption is channeled into the empirical coefficient C_S (capital, not to be confused with the sound speed c_s).

The stabilizing role of the effective viscosity is equally transparent: whenever large strains develop, R_{ab} becomes large and turbulent fluctuations are damped down, much more effectively than by mere molecular viscosity, a negligible effect at this point.

The Smagorinsky model shines for its simplicity: the structure of the Navier–Stokes equation remains untouched, only with a renormalized effective viscosity, which is readily constructed out of the resolved tensor S_{ab}. Such simplicity comes of course at a price, among others the fact that near walls the damping effects are generally overestimated, so that wall turbulence dies out too fast.

A critical assessment of the Smagorinsky SGS model is beyond the scope of this book; here, we shall simply point out that accomodating Smagorinski-like (*algebraic*) turbulence models of the form $v_{eff} = v_{eff}[S]$ within the LBE formalism is fairly straightforward, as we proceed to discuss in the sequel.

24.3.1 LBE Formulation of the Smagorinsky SGS Model

As noted many times in this book, an important bonus of kinetic theory is that pressure and dissipation emanate from the same quantity: the momentum-flux tensor. As a practical result, both pressure and the strain tensors are available as *local* combinations of the Boltzmann distribution, no need of computing spatial derivatives.

Due to this very nice property, the LB strain tensor reads simply as follows:

$$S_{ab} = \frac{P_{ab}^{ne}}{\rho c_s^2 \tau} \tag{24.9}$$

where

$$P_{ab}^{ne} \equiv \sum_i f_i^{ne} Q_{iab}$$

is the non-equilibrium momflux, as usual.

Owing to the locality of the moment-flux tensor, the implementation of the Smagorinsky model is even simpler than in hydrodynamic codes.

Once the strain tensor S_{ab} is available via (24.9), all we need is to modify the relaxation parameter according to (24.8), so as to obtain the desired expression $v_{eff}(S) = v_0 + C_S\Delta^2 S$, where $S \equiv |S|$ for notational simplicity.

By inverting the standard expression for the LB viscosity,

$$v_{eff} = c_s^2 (\tau_{eff}(S) - \Delta t/2) \tag{24.10}$$

we obtain

$$\tau_{eff}(S) = \frac{\Delta t}{2} + \frac{v_0 + C_S\Delta^2 S}{c_s^2} \tag{24.11}$$

As a result, S is obtained by solving the nonlinear constitutive relation:

$$\tau_{eff}(S)\, S = P \tag{24.12}$$

which results upon contracting the expression (24.9).

In (24.12), we have defined the dimensionless non-equilibrium parameter:

$$P = \frac{\sqrt{P_{ab}^{ne} P_{ab}^{ne}}}{\rho c_s^2}$$

where the indices a, b are summed upon, as usual.

For the specific case of Smagorinsky, this becomes a quadratic equation for S,

$$\tau_0 S + \tau_\Delta^2 S^2 = P$$

where we have defined the molecular and SGS timescales as follows:

$$\tau_0 = \Delta t/2 + v_0/c_s^2$$

and

$$\tau_\Delta \equiv C_S^{1/2}\Delta/c_s$$

The solution of the quadratic equation delivers

$$S = \frac{1}{\tau_\Delta}\left(\frac{-\theta + \sqrt{\theta^2 + 4P}}{2}\right) \tag{24.13}$$

where we have set $\theta \equiv \frac{\tau_0}{\tau_\Delta}$.

Inserting the value of S given by the solution (24.13) into (24.12) completes the LB implementation of the Smagorinsky model.

One can readily check that for small strain rate, $S^2\tau_\Delta^2 \ll S\tau_0$, one recovers the molecular expression $S = P/\tau_0$, whereas in the large strain limit, one obtains $S \sim P^{1/2}/\tau_\Delta$.

For general algebraic models of the form $\nu_{eff} = \nu_{eff}(S)$, the solution of the nonlinear equation (24.12) is usually obtained by standard iteration techniques.

The LB-Smagorinsky model was first used for the simulation of 2D cavity flows at $Re = 10^6$, on a small 256^2 lattice (2). Three-dimensional pipe flows were simulated Eggels and Somers at $Re = 50\,000$ on a 80^3 mesh. Using this model, these authors performed a by then very impressive parallel large-eddy simulation of a turbulent flow in a baffled stirred tank chemical reactor (3). More sophisticated variants of the Smagorinsky approach, eventually including multi-relaxation features, continue to appear in the current literature. For an updated summary, see the very informative books by Guo-Shu and Krueger et al.

24.4   Two-Equation Models

The main virtue of the Smagorinsky SGS model is simplicity: it is an algebraic model which does not imply *any* change in the mathematical structure of the Navier–Stokes equations. However, it is known to cause excessive damping near the walls, where S is highest.

The next level of sophistication is to link the eddy viscosity to the actual turbulent kinetic-energy density (unit-mass density for simplicity):

$$k = \frac{1}{2}\langle \tilde{u}^2 \rangle \tag{24.14}$$

and dissipation

$$\epsilon = \frac{\nu}{2} \sum_{a,b} \langle (\partial_a \tilde{u}_b)^2 \rangle \tag{24.15}$$

These quantities are postulated to obey advection–diffusion–reaction equations of the form (4):

$$\hat{D}_t k - \nabla \cdot (\nu_k \nabla k) = P_k - R_k \tag{24.16}$$

$$\hat{D}_t \epsilon - \nabla \cdot (\nu_\epsilon \nabla \epsilon) = P_\epsilon - R_\epsilon \tag{24.17}$$

where ν_k, ν_ϵ are effective nonlinear viscosities for k and ϵ, respectively, and P_k, R_k and P_ϵ, R_ϵ are local production and removal terms of k and ϵ, respectively (5).

Finally, $\hat{D}_t \equiv \partial_t + \hat{u}_a \partial_a$ with $\hat{u}_a \equiv u_a - \partial_a \nu$, where $\nu \equiv \nu_k$ or $\nu \equiv \nu_\epsilon$ (4). These equations constitute the well-known k–ϵ model of turbulence, still one of the most popular options for engineering applications.

The k–ϵ is easily adapted to the LBE harness too. To this purpose, one defines extra populations carrying k and ϵ as the respective densities (6).[56]

---

[56] By now, the reader should be willing to accept that any physical phenomenon governed by the general triad *convection–diffusion reaction* is amenable to LBE treatment!

A more practical option is to solve only the flow equations with LB, and couple to a standard grid discretization of the $k - \epsilon$ equations. This is basically the hybrid approach mentioned earlier on in connection with thermal flows.

The practical use of the $k - \epsilon$ equations may raise concerns of convergence, due to the nonlinear dependence of the eddy viscosities and stiffness of the source terms as well. In addition, the set-up of proper boundary conditions, especially inlets and outlets in complex geometries, also presents a non-trivial task. These issues are however independent of the LB implementation.

24.5   Reynolds-Averaged Navier–Stokes (RANS)

The $k - \epsilon$ model provides an improvement over Smagorinsky, but it is still liable to criticism. In particular, it does not account for the directional nature of turbulence near solid boundaries. To cope with this problem, a further level of sophistication is introduced, by formulating dynamic equations for the Reynolds stress itself R_{ab}. These equations tend to be rather cumbersome, as they involve higher-rank turbulent tensors and also in connection with the formulation of suitable boundary conditions. Even though they enjoy significant popularity in the CFD engineering community, we shall not delve any further into this option.

24.6   Large Eddy Simulation

Another popular form of subgrid-scale modeling is the so-called large eddy simulation (LES).

The basic idea is to express the resolved velocity field (large eddy) as a spatial convolution (filter) of the actual velocity field, namely

$$U_a^{LES}(\vec{x}; h) = \int_{\Omega} G_\Delta (\vec{x} - \vec{y}) \, u_a(y) \, d\vec{y} \tag{24.18}$$

In (24.18), $G_\Delta(\vec{x} - \vec{y})$ is the kernel of the (homogeneous) filter, typically a shape function localized within a finite region $|\vec{x} - \vec{y}| < \Delta$ of the fluid domain Ω.

Formally, the low-pass filter can be regarded as a projector which eliminates wavelengths shorter than Δ. By acting with the filter upon the Navier–Stokes equations, one generates the equations for the large eddy field, U_a. These look like the Navier–Stokes equations, augmented, as usual, with the divergence of the subgrid stress tensor:

$$T_{ab}^{LES}(\vec{x}) = \int_{\Omega} G_\Delta (\vec{x} - \vec{y}) \, \tilde{u}_a(y) \tilde{u}_b(y) \, d\vec{y} \tag{24.19}$$

where $\tilde{u}_a \equiv u_a - U_a$ is the fluctuating velocity field.

At variance with Smagorinsky, $k - \epsilon$ and RANS, the LES approach is based on a systematic projection formalism. Of course, this does not relinquish the need for a closure, but offers a higher level of mathematical systematicity.

Typically, the subgrid stress tensor splits into three distinct components related to the large–large, large–small and small–small eddies interactions, also known as Leonard, Clark and Reynolds stress tensors, respectively. These can take several forms depending on the closure adopted, the simplest one being again the Smagorinsky expression (24.8) given previously. Modern variants, including dynamic versions whereby the constant C_s is promoted to the status of a self-consistent, spacetime dependent field, also known as Germano models, are also available, see Fig. 24.3.

The main appeal of the LES approach is that it leaves the Navier–Stokes equations formally unchanged, with the sole addition of the subgrid viscosity. As a result, LES can deal with time-dependent turbulence, typically at Reynolds numbers ten times larger than DNS. On the other hand, the use of simple convolution filters, such as flat-top or Gaussian, is adequate for idealized geometries, such as channels, cubes and similar. Extensions to real-life geometries have also become available in the last decade through the resort to complex grid structures.

As noted before, the inclusion of LES subgrid viscosity within the LB framework is pretty straightforward.

LES versions of the LB–MTR formulation have been recently developed by M. Krafczyk and coworkers (7), and more recently by P. Sagaut and collaborators (9; 10).

24.6.1 Kinetic Hydro: Lattice Boltzmann Large-Eddy Simulation

We have noted before that "turbulent forces" arise form coarse-graining the non-linear advection term of the Navier–Stokes equations. A very interesting question, first raised and developed (in Fourier space) by Chen *et al.* (8), is to explore what happens by coarse-graining the Boltzmann equation instead of the Navier–Stokes equations.

Leaving a discussion on the conceptual implications of such a move to the subsequent section, here we illustrate the formal procedure which arises when the coarse graining is performed via spatial filters, as it pertains to LES practices.

The following material inscribes within the family of Kinetic–Hydro TMs.

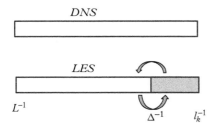

Figure 24.3 *Sketch of the LES procedure. Scales larger than the filter size Δ are simulated, while the shorter ones (in grey) are filtered and modeled. The modeling is particularly sensitive around the filter scale, where the interaction between the filtered and unfiltered scales is strongest (the region encircled by the curvy arrows).*

We begin by defining the coarse-grained (CG) Boltzmann distribution as follows:

$$F = <f>$$

where brackets stand for convolution via LES filter, namely

$$F(\vec{x}, \vec{v}; t) = \int_\omega K_\Delta(\vec{x} - \vec{y}) f(\vec{y}, \vec{v}; t) d\vec{y} \qquad (24.20)$$

where $K_\Delta(\vec{x}, \vec{y}) = K_\Delta(\vec{x} - \vec{y})$ is the kernel of the kinetic filter. In principle, the filter also extends to the time variable, but we confine to pure spatial averaging for simplicity. Note that since both space and time are coarse grained, the velocity variables remains basically untouched.

The CG Boltzmann equation takes the following form:

$$\partial_t F + v_a \partial_a F = C(F) + R \qquad (24.21)$$

where

$$R \equiv <C(f)> - C(F) \qquad (24.22)$$

is the *residual* collision term emerging from filtering the nonlinear interactions in the collision operator.

Formally, the residual can also be viewed as the effect of the commutator between the filtering and collision operators, i.e.,

$$R = (\mathcal{K}\mathcal{C} - \mathcal{C}\mathcal{K})f$$

where $\mathcal{K}$ is the filtering operator, such that, formally $F = \mathcal{K}f$ and $\mathcal{C}$ is the collision operator defined by $\mathcal{C}f = C(f)$. Note that the streaming operator remains unchanged because owing to the homogeneity of the kinetic kernel, filtering commutes with both time and space derivatives.

The problem, as usual, is to express the residual in terms of the coarse-grained distribution F, a task which necessarily calls for some form of closure.

An interesting form of closure, directly patterned after LES practices, is as follows (9). Define a "deconvolution" operator $\mathcal{D}$, such that

$$f_D = \mathcal{D}F$$

tentatively reconstructs the information lost in the filtering procedure.[57] Ideally, the deconvolution operator $\mathcal{D}$ should be the inverse of $\mathcal{K}$, which would correspond to a

---

[57] The reader will notice that the filtering and convolution operators are strict relatives of the projection and reconstruction operators introduced in the finite-volume formulation of LBE.

perfect reconstruction. In practice, one can at best find an approximate pseudo-inverse, such that

$$\mathcal{D}\mathcal{K} = I + O(\Delta^p)$$

where I is the identity, and p is the order of approximation.

In more explicit terms

$$f = f_D + f_E$$

where f_E is an error distribution of order Δ^p.

Specific expressions of the deconvolution operator are part of the LES practice and can be found in the original references.

Once the form of the pseudo-inverse is stipulated, one can construct f_D and use it to perform a perturbative expansion of the residual R.

The procedure goes as follows.

Let us sum and subtract $< C(f_D) >$ in the expression of the residual to obtain

$$R = < C(f_D) - C(F) + < C(f) > - < C(f_D) >$$

This naturally splits into two terms. The first

$$R_1 = < C(f_D > -C(F)$$

needs no modeling because both f_D and F are known.

The second

$$R_2 = < C(f) > - < C(f_D) >$$

does need modeling instead, because f is unknown.

A zeroth-order closure is immediately available, that is

$$R_2 = 0$$

This is not a joke: it tantamounts to ignoring the difference between f and f_D in the collision operator, which is still better than ignoring the difference between f and F.

To proceed to the next order, one rewrites R_2 as follows:

$$R_2 = \mathcal{K}[C(f) - C(f_D)]$$

and expands f around f_D to first order in the error f_E. This yields

$$R_2 = \mathcal{K}[\mathcal{J}(f_D)(f - f_D)] = \mathcal{K}[C'(f_D)(I - \mathcal{D})f]$$

where $\mathcal{J}(f) \equiv \delta C(f)/\delta f$ is the Jacobian of the collision operator. This is still open, for it depends on the unknown f. However, as a first-order approximation, one can replace f with f_D in the previous expression and close the R_2 residual as follows:

$$R_2 \sim \mathcal{K}[C'(f_D)(I - \mathcal{D})f_D] = \mathcal{K}[\mathcal{J}(f_D)(f_D - \mathcal{D}f_D)]$$

By inserting R_1 and this expression for R_2 into the coarse-grained Boltzmann equation (24.21), one finally obtains the desired LES-LB equation in closed form.

The reader can now appreciate that the present model falls within the kinetic-hydro class: filtering is kinetic, but the closure is still based on arguments imported from numerical LES practice.

The formal procedure described has been implemented in a series of interesting papers and found to yield encouraging results for weakly compressible and compressible turbulent flows as well (9; 10). Among other interesting findings, the authors report that a consistent treatment of compressible turbulence requires the LB operator in matrix form.

24.7 The Kinetic Approach to Fluid Turbulence, Again

In Section 24.6, we have illustrated the mathematical procedure of kinetic filtering, as inspired to LES practices. Now, we spend a few comments on the conceptual and practical implications of filtering the kinetic equation first, and (eventually) take the hydrodynamic limit only in a subsequent stage.

The first observation is that *all subgrid models of fluid turbulence based on the coarse-graining of the Navier–Stokes equations necessarily rely on an argument of scale separation between the large and small eddies.* Based on such argument, the effect of the small (un-resolved) eddies on the large (resolved) ones can be described as a diffusive process, whereby the large eddy undergo a sort of Brownian motion under the effect of multiple collisions with the small ones, while the small ones are simply advected away by the large ones. The analogy with kinetic theory is poignant and appealing.

However, unlike molecular fluids, where microscopic and macroscopic degrees of freedom are indeed well separated, the Knudsen number being much smaller than one, turbulent flows exhibit a *continuum* spectrum of scales. This means that, by definition, eddies in the vicinity of the spatial cut off, Δ, simply *cannot* be well separated.

In fact, the largest small eddy (LSE) and the smallest large eddy (SLE) live basically on the same scale, $\Delta(1 \mp \epsilon)$, respectively, see again the curvy arrows in Fig. 24.3.

Hence, the corresponding Knudsen number is

$$Kn \sim \frac{1 + \epsilon}{1 - \epsilon} \sim 1$$

As a result, these eddies are bound to interact strongly with each other and such strong interactions cannot be described in terms of an effective local viscosity.

Starting from this elementary, yet inescapable observation, it has been argued that kinetic theory may offer a unique new angle of attack to the problem of fluid turbulence, namely one in which no assumption of scale-separation is needed. This is because kinetic theory can, in principle deal with arbitrary Knudsen numbers, i.e., flows very far from local turbulent equilibria (11).

As anticipated earlier, the standard coarse-graining procedure goes from the Boltzmann kinetic distribution, f, to the Navier–Stokes description of the fully resolved velocity field u, and then to the large scale subgrid model equations for the coarse-grained velocity U_H, where subscript H stands for hydrodynamic.

Symbolically,

$$f \rightarrow u \rightarrow U_H$$

The main idea is to *revert* this sequence and coarse-grain the Boltzmann equation first, thus generating a coarse-grained "super-Boltzmann" equation for F, and from this super-Boltzmann derive a dynamic equation for the large-scale velocity U_K where subscript K stands for kinetic (see Fig. 24.4).

$$f \rightarrow F \rightarrow U_K$$

One moment's thought reveals why two procedures do not necessarily yield the same result, i.e., $U_H \neq U_K$ in general.

To appreciate the point, let us consider the hierarchy of scales associated with the standard coarse-graining $f \rightarrow u \rightarrow U$, namely:

$$l_\mu < l_k < \Delta \tag{24.23}$$

where l_μ is the molecular mean-free path, l_k is the Kolmogorov length and Δ is the coarse-graining scale, here identified with the grid spacing.

The super-Boltzmann equation contains an effective relaxation time (effective mean-free path), L_μ which incorporates the nonlinear correlations due to kinetic coarse graining.

If such effective mean-free path remains smaller than the grid spacing, $L_\mu < \Delta$, there is no reason to expect any major departure between the hydrodynamic and kinetic coarse graining. If, on the other hand, $L_\mu > \Delta$, then the hydrodynamic limit of the super-Boltzmann equation, $F \rightarrow U_K$, cannot be dealt within the standard Chapman–Enskog procedure.

Under these conditions, which are the rule rather than the exception for very-large eddies sensitive to initial and boundary conditions, the kinetic coarse-graining is expected to bring a new class of turbulence models.

Most real-life turbulent flows fall indeed within this class, since, especially near solid walls, the small eddies are far from being in equilibrium with the large ones.

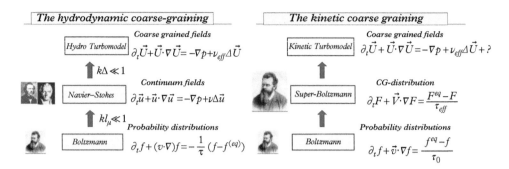

Figure 24.4 *Hydrodynamic(left) and kinetic (right) approach to turbulence modeling. In the hydrodynamic case, one applies coarse graining to the hydrodynamic equations, which in turn emerge from the underlying-Boltzmann equation. In the kinetic approach, coarse graining is applied to the Boltzmann equation first to generate a sort of super-Boltzmann equation, i.e., a kinetic equation which applies to larger-scales than Navier–Stokes hydrodynamics. The super-Boltzmann equation may or may not converge to a hydrodynamic coarse-grained model, depending on the ratio between the coarse-grained mean-free path L_μ and the coarsening scale Δ.*

The analog of the Knudsen number for such flows is the so-called *distortion parameter* defined as

$$\eta = \frac{kS}{\epsilon} \tag{24.24}$$

This parameter is the ratio between the relaxation timescale of turbulent eddies in the bulk, k/ϵ, and near wall, $1/S$, respectively. Whenever such parameter is $\eta \sim 1$, as it is often the case for turbulent flows in real-life geometries, the notion of scale separation goes out of context. It is then only natural to think of an *effective* kinetic equation for the turbulent eddies. The question, of course is how to actually derive one! So far, this has been done by making the effective relaxation time a proper function of the distortion parameter, as we shall detail shortly, see Fig. 24.4.

Before proceeding along this line, let us mention an interesting finding: in order for the kinetic approach to yield a well-defined effective viscosity, the size of the filter must relate to the Knudsen number according to the following condition:

$$\Delta = L\, Kn^{1/2} \tag{24.25}$$

where L is a macroscopic scale of the flow, see Fig. 24.5.

Differently restated, the mean-free path in the coarse-grained kinetic equation must scale like $L_\mu \sim \Delta^2/L$. By choosing the filter "too wide," say $\Delta/L = const$, i.e., Knudsen invariant, the coarse graining is excessive and would contaminate the advective term of the hydrodynamic equations. On the other hand, if the filter is too "narrow," say $\Delta/L = Kn$, then the correction to the molecular viscosity would only appear at higher order, in the form of a second-order Laplacian (hyperviscosity), leaving standard dissipation

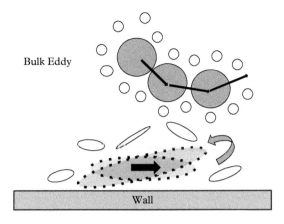

Figure 24.5 *Sketch of equilibrium and non-equilibrium turbulent eddies. The small eddies in the bulk are in dynamic equilibrium with their guiding large eddy, since the latter is not significantly strained. The small eddies near the wall, however, do not settle down to a local equilibrium because the associated large eddies are substantially strained, hence experience fast changes themselves.*

untouched. This analysis is elegant and useful, for it indicates the theoretical fences of the kinetic coarse graining.

In actual practice, however, the effective mean-free path must be *dynamically* linked to the actual turbulent environment, that is the effective relaxation time (or mean-free path) must be promoted to the status of a spacetime dependent dynamic field, responding to the local turbulent environment.

Based upon renormalization group arguments, Yakhot and Orszag proposed the following expression:

$$\tau_{eff} = \tau_0 + C \frac{k}{T} \frac{k/\epsilon}{(1 + \eta^2)^{1/2}} \tag{24.26}$$

where T is the fluid temperature and C is a calibration constant of the order of 0.1.

In (24.26), τ_0 is the bare microscopic relaxation time. Note that the turbulent contribution, the second term at the right-hand side of the expression, reduces to k/ϵ in the limit of weakly strained turbulence, $\eta \to 0$, which holds away from solid walls. The prefactor k/T is due to the fact that the turbulent kinetic energy, k, can be regarded as a sort of "temperature" associated with the turbulent velocity fluctuations. Note that for very large strains $\tau_{eff} \to 1/S$ indicating that even the large eddies are dynamics, thus preventing any form of local equilibrium of the small ones.

24.7.1   Very Large-Eddy Simulation

The lattice super-Boltzmann permits us to implement the so-called very large eddy simulation (VLES) strategy. This is the way Steven Orszag (1943–2011), a giant of computational fluid dynamics and a strong advocate of the kinetic approach, used to refer to it.

VLES consists in focusing the computational power on the largest eddies, those mostly affected by boundary and initial conditions and thus less prone to follow universality arguments à la Kolmogorov, or generalizations thereof. By definition, VLES are dynamic and non-universal, and consequently, they can only be simulated by a numerical method capable of keeping up with their fast time-dependence. Such is indeed the VLES-LBE, which is inherently time-dependent.

Based on this, the LBE-VLES operational procedure is quite transparent. The LBGK equation provides the VLES velocity field to the equations advancing k and ϵ in time. These, in turn, provide the effective relaxation time to the LBGK, according to the expression (24.26), in a self-consistent loop.

Once combined with appropriate boundary conditions, always the main divide between failure and success, this procedure permits us to simulate several turbulent flows in real-life complex geometries, such as cars, trucks and airplanes. In fact, it lies at the heart of the commercial LB software POWERFLOW, see Fig. 24.7.

It is argued that, like LES, VLES-LB can deal with time-dependent turbulence, with the major advantage of having no restrictions on the geometrical complexity. On the other hand, as compared to turbulence models in the $k - \epsilon$ flavor, VLES-LB can deal with larger distortion parameters, $\eta \gg 1$.

24.7.1.1 *Some criticism*

Notwithstanding its successful implementation in POWERFLOW, the kinetic approach to fluid turbulence is still open to a number of criticisms.

First, the use of a (lattice) BGK equation to describe turbulent eddies is a purely heuristic, although well educated, guess: such an equation has not been derived by any rigorous coarse graining of the continuum BGK equation, let alone the Boltzmann equation itself.

Second, the use of the two separate equations for k and ϵ is conceptually unsatisfactory, as these equations should also be derived upon coarse graining of the microscopic kinetic equation, rather than postulated "from the outside." A similar criticism goes for the expression (24.26) of the effective relaxation time, see Fig. 24.6.

Summarizing, the kinetic approach provides a fresh new angle of attack to fluid turbulence, free from the scale-separation restrictions inherent to the very notion of eddy diffusivity, yet, at the time of this writing, a *self-consistent* theory of eddy LBE remains elusive. The subject is important and up for grabs.

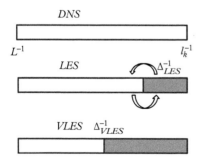

Figure 24.6 *Sketch of the VLES philosophy. A model capable of accounting for non-universal interactions between eddies far from equilibrium, can in principle confine the simulation to a narrower range, toward the large scale end of the spectrum. As a result, it can afford a larger filter size $\Delta_{VLES} > \Delta_{LES}$. This amounts to less computing at a given Reynolds number or, conversely, a larger Reynolds number at a given computational cost.*

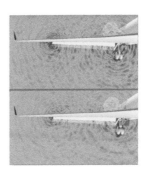

Figure 24.7 *A VLES-LB simulation of a flow past a racing car and details of a VLSE simulation of the landing of a full-scale airliner. Courtesy of EXA Corporation.*

24.8 Fully Kinetic Approach

As mentioned, a fully self-consistent kinetic approach to turbulence models should be based not only on direct coarse graining of the kinetic equations but also on a genuinely kinetic closure, not imported from hydrodynamic TM's. In practice, this means that the residual collisor

$$R = < C(f) > -C(<f>)$$

should be closed based on informed assumptions on the distribution f itself and not on its hydrodynamic moments. For the sake of concreteness, let us refer to the BGK operator, in which case the residual collisor takes the form

$$R = \omega(<f^e(f) > -f^e(F))$$

A fully kinetic closure consists of *deriving* a functional relation between $< f^e(f) >$ and $f^e(F)$. In passing, we note that such kinetic closure identifies a renormalized frequency $\hat{\omega}$, formally defined by imposing the invariance of the collision operator under coarse graining, namely

$$\omega(<f^e(f) > -F) = \hat{\omega}(f^e(F) - F)$$

whence

$$\frac{\hat{\omega}}{\omega} = \frac{<f^e(f) > -F}{f^e(F) - F} \tag{24.27}$$

While it is clear that the right-hand side is a function of the coarse-graining scale Δ, finding a rigorous link between $\hat{\omega}$ and the fluctuating kinetic energy k and energy dissipation rate ϵ appears to be a fairly non-trivial task.

24.8.1   Relation between Kinetic and Hydrodynamic Closures

For the case of the BGK operator it is possible to establish a clearcut formal connection between kinetic and hydrodynamic closures. To this purpose, let us remind the well-known Hermite expansion of the local Maxwellian, namely

$$R = \omega \, e^{-v^2/2} \sum_{n=0}^{\infty} H_n(v)(< u^n > -U^n) \tag{24.28}$$

where tensor indices are omitted for simplicity and we have set $v_T = 1$.

Expression (24.28) informs us that a kinetic closure on the local equilibrium is tantamount to a hydrodynamic closure for the cumulants $< u^n > /U^n$ at all orders.

In the case of the standard lattice BGK with second-order polynomial equilibria, the closure involves the same Reynolds stress tensor as hydrodynamics, namely $< u^2 > -U^2$, hence it does not seem that the kinetic closure may differ substantially from the hydrodynamic one.

Higher-order equilibria, though, may have potential to open genuinely new prospects along these lines.

In the following we portray a speculative route, *with the cautionary disclaimer that what we are going to deal with is unpublished tentative material* (12).

Let us consider again the Hermite expansion of the residual collisor (24.28) and express the molecular velocity in the following "double-fluctuating" form:

$$v = U + \tilde{u} + c \equiv U + \tilde{V} \tag{24.29}$$

where c is the peculiar speed reflecting molecular fluctuations, while $\tilde{u}$ represents the *turbulent* fluctuations. The total fluctuation $\tilde{V} = \tilde{u} + c$ is thus the sum of molecular and turbulent fluctuations.

In this formalim, the closure takes a form of a statement on the probability distribution function (PDF) of the turbulent fluctuations, say $P(\tilde{u})$.

The ensemble average associated to coarse graining may be written as an integral over the PDF, namely

$$F(V) \equiv < f(v) > = \int f(V + \tilde{u}) P(\tilde{u}) d\tilde{u}$$

where $V = U + c$ is the *coarse-grained* molecular velocity. i.e. the molecular velocity which is left once turbulent fluctuations have been filtered out.

As a first guess, one could assume $P(\tilde{u})$ to be a Gaussian with zero mean and variance $< (\tilde{u})^2 > = 2k$, namely

$$P(\tilde{u}) \propto k^{-D/2} e^{-(\tilde{u})^2/2k} \qquad (24.30)$$

in D spatial dimensions.

The heuristic behind this kinetic closure is that *one-point* turbulent fluctuations are Gaussian, with an "temperature" given by turbulent-kinetic energy k.

With such a position, the coarse-grained equilibrium can be computed exactly, as it only involves gaussian integrals. The calculation delivers again a Maxwell–Boltzmann distribution, whose variance is simply the sum of the temperature- and turbulent-kinetic energy:

$$< f^e(v) > \propto < T^{-D/2} e^{-(v-u)^2/2T} > = (T+k)^{-D/2} e^{-(V-U)^2/2(T+k)}$$

in units $k_B = m = 1$.

This looks like a very pretty result and one that explicitly exposes the ratio k/T in the coarse-grained equilibrium. It also makes perfect physical sense, in that turbulent fluctuations give rise to a sort of Doppler broadening of the local equilibrium, which is an expected result.

Unfortunately, it carries no sign of the dissipation rate ϵ, hence it cannot match the expression of the effective eddy-relaxation time given by the expression (24.26). This is where this author left it; hopefully a more proficient reader than this author can take it on from here ....

24.9   Entropic Models: Domesticating the Ghosts

As discussed in Chapter 23, the Entropic LB could be regarded as a sort of Turbulence Model, in that the viscosity is locally adjusted to secure stability based on the second principle. However, this view appears debatable to this author, since it is far from obvious that the local adjustment of viscosity is capable of representing the physics of unresolved eddies on the resolved ones. This said, recent developments, whereby ghosts are "domesticated,"[58] i.e., optimally relaxed based on entropic principles, may change the game (13). Time will tell.

24.10   Wall–Turbulence Interactions

Before closing, we wish to emphasize that the ultimate challenge for the successful implementation of *any* turbulence model is the correct implementation of appropriate boundary conditions describing the interaction of the turbulent eddies with solid walls.

---

[58] This colorful definition is due to Iliya Karlin.

If the bulk is somewhat forgiving, boundaries are usually not.

It is generally agreed that near solid boundaries the grid resolution is not sufficient to capture the fine-scale structure of the flow. Based on this premise, the value of the velocity field is often specified in terms of analytic expressions based on asymptotic solutions of boundary layers. A very popular expression in this respect is the celebrated law-of-the-wall

$$\frac{u(y)}{u^*} = \kappa \, log(y/y^*) + B \tag{24.31}$$

where u^* is a typical amplitude of the turbulent velocity at the wall and $y^* = v/u^*$ is the size of the turbulent layer, typically much smaller than the grid spacing Δ, see Fig. 24.8.

In (24.31), y is the coordinate across the boundary layer, κ the so-called von Karman constant and B is a (more or less) universal parameter. Based on this, one can use the value of $u(y = \Delta)$ obtained by the simulation to compute u^* from the eqn (24.31), which is typically achieved within a few nonlinear iterations.

With u^* at hand, one can estimate the wall stress tensor $\sigma^* = \rho u^{*2}$, hence the turbulent force acting on the populations living on the near-wall cell.

This is just the basic idea, whose actual implementation depends on many crucial details. Indeed, the procedure is conceptually transparent, but subject to a number of practical limitations.

First, the fact that the asymptotic relation (24.31) refers to planar walls with zero curvature. Corrections due to finite curvature must be taken into account, which become

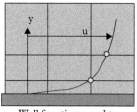
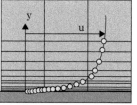

Wall functions used to
resolve boundary layer
———— Boundary layer

Wall functions not used to
resolve boundary layer

Figure 24.8 *Sketch of an undersolved (left) and resolved (right) boundary layer. In the former case, the size near-wall cell is larger than the thickness of the boundary layer, $\Delta > y^*$ and consequently the value of the flow velocity in that cell is taken from analytical asymptotics (wall functions) and matched to the value in the next cell by means of the iterative procedure described in the text. In the latter case, $\Delta < y^*$ and consequently the fluid is simulated all the way to the wall. This requires much higher resolution, and can possibly be realized trough extensive grid refinement near the wall. From www:computationalfluiddynamics.com.au.*

especially demanding in the case of sharp features, such as backward facing steps, where local adverse pressure gradients play a major role on the local structure of the flow. This subject is highly specialized and in constant evolution, hence not suitable for a (non-specialistic) book.

This represents a frontier of LBE research which is likely to keep expanding in the coming years.

24.11 Summary

In summary, the kinetic approach to turbulence modeling presents us with three options of increasing conceptual ambition:

i) Hydro-kinetic models, in which existing TMs are incorporated within the LB mathematical framework;

ii) Kinetic-hydro models, based on the direct coarse graining of the kinetic equation, followed by a hydrodynamic closure;

iii) Fully kinetic models, based on the coarse-graining of the kinetic equation, followed by a kinetic closure.

Harnessing existing hydrodynamic TMs within the LBE formalism, the hydro-kinetic approach appears to be relatively straightforward and efficient.

A few kinetic-hydro TMs based on the direct coarse graining of the kinetic equations, yet closed on hydrodynamic grounds, have been formulated in the last decade. The main appeal of such an approach is its independence of any assumption of scale separation between resolved and unresolved eddies. A fully self-consistent kinetic approach to turbulence modeling, kinetic coarse graining followed by a genuinely kinetic closure remains to be developed.

From the practical standpoint, opinions are mixed.

On the one side, one must concede that the community of turbulence modeling at large does not seem to have subscribed to the LB approach to any massive extent. On the other hand, the fact that the LB approach has made it into the fiercely competitive world of commercial CFD, surely is nothing to be underscored either. In addition, very recent developments from the entropic LB front may lead to significant progress in the coming years.

REFERENCES

1. J. Smagorinsky, "General circulation experiments with the primitive equations I: the basic experiments", *Monthly Weather Rev.*, **91**, 99, (1963).

2. S. Hou, J. Sterling, S. Chen, G.D. Doolen, "Lattice Boltzmann Subgrid Model for High Reynolds Number Flows", *J. of Comp. Phys.*, **118**, 329 (1995).

3. J. Eggels and J. Somers, "Direct and large eddy simulations of turbulent fluid using the lattice Boltzmann scheme", *Int. J. Heat Fluid Flow*, **17**, 307, (1996).

4. B. Launder, "An introduction to single-point closure methodology", in *Simulation and Modeling of Turbulent Flows*, ICASE/LarC Series in Computational Science and Engineering, T. Gatski, M. Hussaini and J. Lumley eds, p. 243, Oxford University Press, Oxford 1996.

5. C. Teixeira, "Incorporating turbulence models into the lattice Boltzmann method", *Int. J. Mod. Phys. C*, **9**(8), 1159, (1999).

6. S. Succi, G. Amati and R. Benzi, "Challenges in lattice Boltzmann computing", *J. Stat. Phys.*, **81**(1/2), 5, (1995).

7. M. Krafczyk, J. Tolke and L.S. Luo, "Large-Eddy simulation with a multiple-relaxation-time model", *Int. J. Mod. Phys C*, **17**, 33, (2003).

8. H. Chen, S. Succi, S. Orszag, "Analysis of subgrid scale turbulence using the Boltzmann Bhatnagar-Gross-Krook kinetic equation", *Phys. Rev. E*, **59** R2527, (1999).

9. Y.H Dong, P. Sagaut, P.S. Marie, "Inertial consistent subgrid model for large-eddy simulation based on the lattice Boltzmann method", *Physics of Fluids* **20**, 035104 (2008).

10. O. Malaspinas and P. Sagaut, "Consistent subgrid scale modelling for Lattice Boltzmann methods", *J. Fluid Mech.*, **700**, 514 (2012).

11. H. Chen *et al.*, "Extended Boltzmann Kinetic Equation for Turbulent Flows", *Science*, **301** 633, (2003).

12. V. Yakhot, many private communications.

13. B. Dorschner, F. Bosch, S.S. Chikatamarla, *et al.*, "Entropic multi-relaxation time lattice Boltzmann model for complex flows", *Journal of Fluid Mechanics*, **801**, 623–651, (Aug 2016).

....

EXERCISES

1. Prove that homogeneous filters commute with spatial derivatives.
2. Analyse the additional terms which emerge if the filter is non homogeneous.

Part V

Beyond Fluid Dynamics: Complex States of Flowing Matter

The ideas and applications discussed thus far have been dealing with flows of "simple" fluids whose flow can be described in terms of the Navier–Stokes equations.

Modern "do smaller–do faster" high-tech applications are placing a pressing demand of quantitative understanding of situations in which even the micro-physical world is not simple any more. Flows with chemical reactions, phase-transitions, multiphase flows, suspensions of colloidal particles, non-Newtonian flows, polymer melts to cite but a few, all call for non-trivial extensions of the Navier–Stokes equations and sometimes to more general equations altogether.

This emerging sector of modern science, often referred to as "complex fluids," or more trendily, "soft matter physics," portrays a multidisciplinary scenario whereby fluid dynamics makes contact with other disciplines, primarily chemistry, material sciences and biology as well.

To us, this may also be denoted as *complex states of flowing matter*.

There is a growing sensation that LBE, and extensions thereof holds a vantage point as a computational framework for the simulation of complex states of flowing matter. Ideally, LBE would fill the gap between fluid dynamics and molecular dynamics, namely the huge and all-important region where fluid dynamics breaks down and molecular dynamics is not yet ready to take over for lack of compute power.

LBE is a good candidate to fill this gap because of its inherent ability to incorporate micro/mesoscopic details into the kinetic theory formalism via suitable external fields and/or equivalent generalizations of local hydrodynamic equilibria: a microscope for fluid mechanics, a telescope for molecular dynamics.

To forestall potential misunderstandings, it is probably useful to state the obvious: *LB is not Molecular Dynamics, and it is not Boltzmann either*! LB cannot solve the Boltzmann equation in any rigorous sense. What it can do, though is to provide useful insights into problems which lie beyond the realm of continuum hydrodynamics and yet do not require a full Boltzmann treatment. It is not necessary to solve the Boltzmann equation to gain insights into moderately non-equilibrium phenomena beyond the Navier–Stokes description. A similar statement holds, mutatis mutandis, for the relation between Lattice Boltzmann and Molecular Dynamics.

What is necessary though is to be aware of the approximate nature of the LB results since, unlike the familiar hydrodynamic scenario, this new landscape offers no rigorous guarantee of convergence.

The next part of this book is therefore devoted to an illustration of LBE methods and applications in the direction of soft matter research. Because this is an open research front, the material is often qualitative in character, the main task being again not to provide full hard-core details, but rather to stimulate new ideas and further activity in this rapidly evolving frontier of LB research.

25

LBE for Generalized Hydrodynamics

In this chapter we shall discuss the main techniques to incorporate the effects of external and/or internal forces within the LB formalism. This is a very important task, for it permits us to access a wide body of generalized hydrodynamic applications whereby fluid motion couples to a variety of additional physical effects, such as gravitational and electric fields, potential energy interactions, chemical reactions and many others.

Don't fight forces: use them

(R. Buckminster Fuller)

25.1 Introduction

Many fluid dynamic applications in modern science and technology deal with *complex* flows, such as flows with chemical reactions, phase transitions, suspended particles and other assorted sources of complexity. For such applications, fluid dynamics couples to a broad range of new physical effects and phenomena which go beyond the framework described by the "plain" Navier–Stokes equations.

This is what we shall refer to as *generalized* hydrodynamics.

It should be emphasized that while hosting a broader and richer phenomenology than "plain" hydrodynamics, generalized hydrodynamics still fits the hydrodynamic picture of *weak departure from suitably generalized local equilibria.*

This class is all but an academic curiosity; for instance, it is central to the fast-growing science of *Soft Matter,* a scientific discipline which has received an impressive boost in the last decades, under the drive of micro and nano-technological developments and major strides in biology and life sciences at large.

To handle this much broader context, generalizations of LBE are required, which incorporate the aforementioned effects. This can be accomplished by either including explicit sources of mass, momentum and energy to the force-free LB, or by shifting the local equilibria by a force-dependent offset in velocity space. We shall refer to the two

The Lattice Boltzmann Equation. Sauro Succi, Oxford University Press (2018).
© Sauro Succi. DOI: 10.1093/oso/9780199592357.001.0001

approaches as to *Driven Hydrodynamics* and *Shifted Hydrodynamics*, respectively. In the sequel we shall proceed to show how to implement both strategies within the LB harness. Subsequent chapters shall deal with specific adaptions of the generalized LBE formalism to flows with chemical reactions, phase transitions and suspended objects. In addition, the extension of the LB methodology to flows at micro- and nanoscales shall also be covered is some detail.

25.2   Including Force Fields

The LBEs discussed so far revolve around two basic ideas:

1. *Construction of suitable discrete local equilibria, consistent with hydrodynamic conservation laws.*

2. *Collisional relaxation to these local equilibria, possibly in compliance with the second principle of thermodynamics.*

This is all we need to generate realistic dynamics of "simple[59]" *quasi-incompressible* fluids. A most natural way of generalizing LBE along the previous directions is to consider the Boltzmann equation with force term, namely

$$\partial_t f + v_a \partial_a f + g_b \nabla_b f = C(f, f) \tag{25.1}$$

where

$$g_b(\vec{x}, t) = F_b(\vec{x}, t)/m \tag{25.2}$$

denotes the acceleration due to a generic force $\vec{F}$. For notational convenience $\partial_a \equiv \partial/\partial x_a$ denotes the spatial derivatives and $\nabla_b \equiv \partial/\partial v_b$ the velocity ones. As usual, Latin indices run over spatial dimensions, $a, b = x, y, z$.

The generalized force $\vec{F} \equiv F_a$, hereafter a function of space and time only is meant to account for *both* external fields such as gravity or electric fields and *self-consistent* interaction forces associated with intermolecular interactions, or any other effective interaction we may wish to include.

In this chapter, we shall be dealing with four basic types of forces, namely:

- *Gradient forces*:

$$F_a(\vec{x}) = -\partial_a V(\vec{x})$$

where $V(\vec{x})$ is a scalar potential.

---

[59] Beware: simple fluids does *not* mean simple flows (turbulence docet!).

- *Contact forces*:

$$F_a(\vec{x}) = -\partial_b T_{ab}(\vec{x})$$

where T_{ab} is the momentum-flux tensor associated with the contact force.

- *Two-body pseudo-potential forces*:

$$F_a(\vec{x}) = \psi(\vec{x}) \int r_a\, G(\vec{x}, \vec{x} + \vec{r})\, \psi(\vec{x} + \vec{r}) d\vec{r}$$

where $\psi(\vec{x}) \equiv \psi[\rho(\vec{x})]$ is a generalized fluid density and $G(\vec{x}, \vec{y})$ is the integral kernel associated with potential energy interactions. Note that the pseudopotential depends on space only via the fluid density, whence the prefix "pseudo," which is consistent with the mesoscopic nature of the force.

- *Electromagnetic (Lorentz) forces*:

$$F_a(\vec{x}, \vec{v}) = \epsilon_{abc} v_b B_c,$$

where ϵ_{abc} is the Levi–Civita tensor implementing the curl operator and B_c is the magnetic field. For simplicity, the electric charge is taken unit and $v \equiv v/c$, c being the speed of light.[60]

The first family typically describes the interaction with gravitational or electrostatic potentials and includes the case of constant forces discussed earlier on in this book to mimic the effects of pressure gradients. It is also of relevance to the case of fluids interacting with self-consistent potentials obeying the Poisson equation, a very important class of problems for electronic transport as well for many biological applications.

The second family is typically associated with the presence of extra-stresses resulting either from fluctuating fields (turbulence models, nanofluids) or potential energy interactions (non-ideal fluids, multiphase flows). Clearly, it contains the first one as a special case of diagonal tensor $T_{ab} = V\delta_{ab}$.

The third family pertains mostly to the treatment of complex fluids within the framework of density functional theory, whereby the (pseudo)force depends on space and time only through the intermediate of the fluid density. Eventually, this includes the second family, in the case where the momentum-flux tensor can be derived from a free-energy functional, as we shall detail.

The common feature between the first three families above is that the force only depends on space and time. While covering a broad variety of complex flows, this leaves aside other important interactions, typically electromagnetic ones. Therefore, the

---

[60] We remind that the Levi–Civita antisymmetric tensor takes value +1 for even permutations of xyz, −1 for the odd ones and 0 otherwise. In particular, it is zero whenever any two indices are the same. As a result, the vector $V_a = \epsilon_{abc} A_b B_c$ is the cross product $\vec{V} = \vec{A} \times \vec{B}$ in coordinate notation.

procedure to implement electromagnetic forces will also be illustrated briefly, although not pursued any further in this book.

In geometrical terms, the force field activates streaming along the velocity dimension of phase space; unsurprisingly, an enormous wealth of new physics opens up by promoting information flow along the velocity extra dimension.

The relevant control parameter is the Froude number defined as

$$Fr = \frac{g\tau}{v_T} \tag{25.3}$$

This measures the shift in velocity space due to the force field in units of the thermal velocity. Likewise, the ratio of Froude to Knudsen number

$$\zeta \equiv \frac{Fr}{Kn} = \frac{g\tau}{v_T} \frac{L}{l_\mu} = \frac{mgL}{k_B T} \tag{25.4}$$

measures the relative strength of potential versus kinetic energy. Clearly, $\zeta \to 0$ designates ideal fluids with no potential energy interactions.

As we shall see, current LB versions for generalized hydrodynamics are restricted to weak interactions with $Fr << 1$, the reason being that they are basically a perturbative extension of the force-free case, $Fr = 0$, see Fig. 25.1.

Indeed, the LB implementation faces with an immediate question: since forces change the molecular velocity, *how can we possibly keep sticking to a formalism which is strictly rooted into a constant-velocity (free-particle) representation?* The same question can be rephrased in more mathematical terms: how can we possibly provide an accurate description of the gradient in velocity space with just a few discrete speeds?

Fortunately, in the limit of sufficiently weak forces, this turns out to be possible *without altering the structure of discrete velocity space*, i.e., without loosing the major perks of the Stream–Collide kinetic paradigm typical of the LB formalism.

In the sequel, we shall provide the main ideas behind this key procedure.

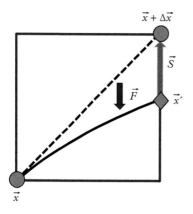

Figure 25.1 *A curved trajectory in the lattice, due to the external force. The straight dashed line represents the force-free, unperturbed trajectory from $\vec{x}$ at time t to $\vec{x} + \Delta\vec{x}$ at time $t + \Delta t$. Due to the force $\vec{F}$ pointing downward, the trajectory bends down ending in $\vec{x}'$ instead of $\vec{x} + \Delta\vec{x}$ at $t + \Delta t$. The source correction term, $\vec{S}$ (vertical arrow) compensates for the curvature effect. This compensation procedure is effective only at small Froude numbers, i.e., when the distance $|\vec{x} + \Delta\vec{x} - \vec{x}'|$ is small compared to the lattice spacing $\Delta\vec{x}$.*

25.2.1 Including Forces with a Force-Free Formalism: Driven Hydrodynamics

We start with the usual observation that all physical quantities of interest are defined in terms of integrals in velocity space.

This implies that, upon integrating by parts, the velocity derivatives can be transferred from the distribution function f to known polynomials (typically, but not necessarily Hermite) which can then be computed analytically. In other words, we need not worry about an accurate discretization of velocity gradients of the distribution function, simply because there is no need of directly computing such gradients (1; 2).

Given any microscopic quantity $\phi(\vec{v})$, the rate of change of the corresponding macroscopic quantity $\Phi(\vec{x}, t)$, due a spacetime dependent force term, reads as follows:

$$\dot{\Phi}(\vec{x}; t) = -\int \phi(\vec{v}) g_a(\vec{x}, t) \nabla_a f(\vec{x}, \vec{v}; t) d\vec{v} = g_a(\vec{x}; t) \int f(\vec{x}, \vec{v}; t) \nabla_a \phi(\vec{v}) d\vec{v} \qquad (25.5)$$

where we have simply taken out the velocity-independent acceleration and applied integration by parts in velocity space. As a result, the distribution function f is left untouched.

The minus sign is due to the fact that the rate of change is formally dictated by the term $-g_a \nabla_a f$ on the right-hand side of the kinetic equation. This sign change responds to a very transparent physical interpretation of the streaming term in velocity space as the result of soft-core interactions, as discussed in Part I of this book.

For the specific case of mass, momentum and momentum-flux, we have

$$\dot{\Phi}_0(\vec{x}; t) \equiv mg_a(\vec{x}; t) \int (\nabla_a 1) \, f d\vec{v} = 0 \qquad (25.6)$$

$$\dot{\Phi}_a(\vec{x}; t) \equiv mg_b(\vec{x}; t) \int (\nabla_b v_a) \, f d\vec{v} = \rho g_a \qquad (25.7)$$

$$\dot{\Phi}_{ab}(\vec{x}; t) \equiv mg_c(\vec{x}; t) \int (\nabla_c v_a v_b) \, f d\vec{v} = \rho(g_a u_b + g_b u_a) \qquad (25.8)$$

where $\rho \equiv mn(\vec{x}; t) = m \int f d\vec{v}$ is the mass density.

In this respect, generalized hydrodynamics is basically hydrodynamics plus external drives, which is why we call it *Driven Hydrodynamics*.

As long as these sources are sufficiently weak, in a sense to be clarified shortly, there is no need to alter the structure of velocity space.

In a pictorial form: propagation along a curved trajectory (curvature being due to the force) can be approximated by propagation along a force-free straight line, plus a correction to the distribution function, which precisely the source term.

Clearly, such approximation only holds as long as the curvature is sufficiently weak.

The same conclusion can be reached in a slightly more sophisticated but perhaps more insightful way, i.e., by absorbing the source term within a suitable generalized equilibrium. An approach that we call *Shifted Hydrodynamics*, since it amounts to shifting the local equilibria by a velocity offset dictated by the force itself. Here is how the story goes.

25.2.2 Shifted Hydrodynamics

Let us begin by moving the force term to the right-hand side, so as to define a generalized BGK collision term:

$$C_F \equiv \omega(f^e - f) - g_a \nabla_a f \tag{25.9}$$

For reasons to become apparent shortly, it proves expedient to evaluate the force term as acting on the local equilibrium, namely

$$C_F \sim C_F^e \equiv \omega(f^e - f) - g_a \nabla_a f^e \tag{25.10}$$

The good news is that the replacement of C_F with C_F^e is no big crime, since it has no impact on the mass, momentum transfer to the fluid by the force field, as we are going to show shortly.

Next, we use a second-order Taylor expansion of the exponential:

$$[e^{g_a \tau \partial_a}] f^e = \left(1 + g_a \tau \partial_a + \frac{\tau^2}{2} g_a g_b \nabla_a \nabla_b \right) f^e + O(Fr^2)$$

where $\tau = 1/\omega$ and we have defined the local *Froude* number as

$$Fr \equiv |g\tau \frac{\nabla_a f^e}{f^e}| \sim \frac{g\tau}{v_T} \tag{25.11}$$

As a result, we can write

$$(1 + g_a \tau \nabla_a) f^e \sim [e^{g_a \tau \partial_a}] f^e - \frac{\tau^2}{2} g_a g_b \nabla_a \nabla_b f^e$$

Given the identity

$$e^{\vec{g}\tau \cdot \nabla_v} f = f(\vec{v} + \vec{g}\tau)$$

and recalling that $\omega\tau = 1$, the shifted collision operator reads

$$C_F^e \equiv -\omega \left[f(v_a) - f^e(v_a + g_a \tau) \right] + \frac{\tau}{2} g_a g_b \nabla_a \nabla_b f^e + O(Fr^2) \tag{25.12}$$

The shifted equilibrium used in actual practice is

$$C_{FS}^e \equiv -\omega \left[f(v_a) - f^e(v_a + g_a \tau) \right] \tag{25.13}$$

which is tantamount to

$$C^e_{FS} \sim C^e_F + \frac{\tau}{2} g_a g_b \nabla_a \nabla_b f^e \tag{25.14}$$

This means that shifted equilibria include the effects of diffusion in velocity space to the second order in the Froude number.

In other words, BGK with shifted equilibria encodes a Fokker–Planck extra diffusion term in velocity space, with diffusivity $D_v = g^2 \tau$.

As it can be readily checked by direct integration, the diffusion term has no impact on mass and momentum conservation. However, it does affect momentum-flux dynamics, usually in the form of a stabilizing effect. Since this term introduces an additional dependence on the relaxation parameter τ, it must be handled with due care.

The condition $Fr \ll 1$ means that this shift is bound to a small fraction of the thermal speed of the Maxwellian distribution: this is the precise meaning of "weak" force, or weak curvature, we have previously alluded to.

25.2.3 Galilean Rescue

From the previous discussion, it looks like the molecular velocity should be shifted by an amount $\vec{g}\tau$, thereby violating the basic LB requirement of working with constant velocities.

Here Galilean invariance comes to a graceful and powerful rescue. Indeed, we know that f^e does not depend on the molecular velocity itself, but rather on the relative velocity $c_a = v_a - u_a$.

This means that the shift can safely be transferred from the molecular speed to the fluid one, that is

$$v_a - g_a \tau - u_a = v_a - (g_a \tau + u_a) \equiv v_a - \tilde{u}_a \tag{25.15}$$

which defines the shifted *fluid* velocity:

$$\tilde{u}_a \equiv u_a + g_a \tau \tag{25.16}$$

This is how Galilean invariance solves the problem of describing a weakly-interacting system via an interaction-free formalism, i.e., free particles moving along straight trajectories and relaxing to a shifted equilibrium.

It is now worth commenting on why it proves so expedient to evaluate the force term on the local equilibrium instead of the distribution function itself.

In the latter case, the shifted operator would look like

$$C_{FS} = -\omega \left[f(v_a - g_a \tau) - f^e(v) \right] + \frac{\tau}{2} g_a g_b \nabla_a \nabla_b f + O(Fr^2) \tag{25.17}$$

This clearly shows that the shift would apply to f, thereby quenching the benefits of Galilean invariance: while local equilibria are explicitly Galilean invariant, the actual distribution is not!

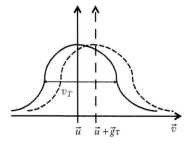

Figure 25.2 *A local Maxwellian centered about $\vec{v} = \vec{u}$ and the shifted local Maxwellian centered about $\vec{v} = \vec{u} + \vec{g}\tau$. The shift must be small as compared to the width of the Gaussian.*

Summarizing, Shifted Hydrodynamics is equivalent to Generalized Hydrodynamics with a drift-diffusion term in velocity space, see Fig. 25.2:

$$S_v = -g_a \nabla_a f + \frac{\tau}{2} g_a g_b \nabla_a \nabla_b f \tag{25.18}$$

Having laid down the basic ideas behind the perturbative approach to generalized hydrodynamics, let us now move on to its implementation within the lattice context, see Fig. 25.2.

25.3   Generalized Lattice Hydrodynamics

The generalized LBGK in the driven-hydrodynamics framework reads as follows:

$$f_i(\vec{x} + \vec{c}_i \Delta t; t + \Delta t) - f_i(\vec{x}; t) = -\omega(f_i - f_i^e) + \Delta f_i \tag{25.19}$$

where $\omega = \Delta t/\tau$ and the source Δf_i is the discrete-velocity representation of the source term due to the force (25.2).

The source term can be regarded as the bias on the i-th population due to a source of mass, momentum, energy or any other macroscopic observable of the form

$$\Phi(\vec{x}, t) \equiv \sum_i \phi(\vec{c}_i) f_i(\vec{x}, t) \tag{25.20}$$

where $\phi(\vec{c}_i)$ is a generic microscopic quantity.

The rate of change in time of the observable Φ due to the source F_i is given by

$$\dot{\Phi}_F(\vec{x}, t) = \sum_i \phi(\vec{c}_i) F_i \tag{25.21}$$

where we have set $F_i \equiv \Delta f_i / \Delta t$.

Depending on its strength and shape in momentum space, this term can result in dramatic changes of the flow dynamics, such as sharp interfaces, phase transitions, chemical reactions and other assorted complexities.

In the simplest instance, F_i is just a set of constants representing the effect of a gravitational field or a constant pressure gradient as discussed earlier in this book. More general situations are readily generated by letting F_i carry a generic spacetime dependence, either explicitly and/or in the form of a self-consistent *mean-field* coupling to macroscopic fields and/or their derivatives:

$$F_i [\vec{x}, t; \Phi (\vec{x}, t), \nabla \Phi, \ldots] \qquad (25.22)$$

Expressions like (25.22) are well suited to encode phenomenological information about the micro–mesoscopic behavior of the system.

On fairly general grounds, one can express F_i as an expansion onto the usual set of kinetic eigenvectors, namely

$$F_i = w_i \left(F_0 + \frac{F_{1,a} c_{ia}}{c_s^2} + \frac{F_{2,ab} Q_{iab}}{2 c_s^4} + \ldots \right) \qquad (25.23)$$

where F_0 is the contribution to the mass density, $F_{1,a}$ to the flow speed, $F_{2,ab}$ to the (traceless) momentum-flux tensor, as given by expressions (25.6–8) for the case of a space dependent force.

This is the lattice version of driven hydrodynamics.

Given that the force $\vec{F}$ is a known function of space and time, the unknown coefficients F_0, $F_{1,a}$, $F_{2,ab}$ can be computed analytically beforehand.

Expression (25.23) is fully operational: by tuning the moments of the source term, one can selectively inject mass, momentum and energy into the fluid in a fairly general and systematic way. Clearly, the forcing scheme must be sufficiently small to preserve the realizability of the kinetic scheme, i.e., do not violate the positivity constraint on the distribution function.

25.3.1   Shifted Lattice Equilibria

The generalized LBGK equation in shifted-hydrodynamic version reads simply as follows:

$$f_i(\vec{x} + \vec{c}_i \Delta t; t + \Delta t) - f_i(\vec{x}; t) = -\omega \left[f_i - f_i^e(\vec{u} + \vec{g}\tau) \right] \qquad (25.24)$$

where

$$f_i^e(\vec{u} + \vec{g}\tau) = w_i \rho \left[1 + \frac{(u_a + g_a\tau)c_{ia}}{c_s^2} + \frac{(u_a + g_a\tau)(u_b + g_b\tau)Q_{iab}}{2c_s^4} \right] \qquad (25.25)$$

Note that this expression includes terms up to τ^2, which must be explicitly coded within the term $F_{2,ab}$ in the Driven Hydrodynamic formulation.

Zero Force   Non Zero Force: Fx

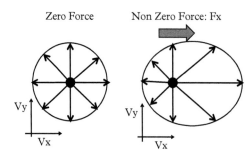

Figure 25.3 *The contour of the distribution function $f(v_x, v_y)$ for the case of D2Q9 lattice, without (left) and with (right) the bias F_i due to the force F_x (thick arrow). Without force, the distribution function is isotropic around the local flow velocity $\vec{u}$, the center point in the figure. The local force promotes the values of the co-flowing populations $(\vec{F} \cdot \vec{c}_i > 0)$ and depletes the counter-flowing ones, $(\vec{F} \cdot \vec{c}_i < 0)$. As a result, the contour turns from a circle to an ellipsoid. If the distortion is sufficiently small, the ellipsoid can be approximated by a circle shifted along the v_x axis by an amount $g_x \tau$.*

It is instructive to perform the calculation explicitly to compare driven versus shifted lattice hydrodynamics.

The shifted equilibria of Driven Hydrodynamics are defined by the identity

$$\omega \tilde{f}_i^e = \omega f_i^e + \Delta f_i \tag{25.26}$$

namely

$$\tilde{f}_i^e = f_i^e + \tau F_i \tag{25.27}$$

Collecting the lattice expansion of both equilibria, and the forcing term, we formally obtain

$$\tilde{f}_i^e = \rho w_i \left[\left(1 + \frac{F_0 \tau}{\rho} \right) + \frac{c_{ia} \left(u_a + \frac{F_{1,a} \tau}{\rho} \right)}{c_s^2} + \frac{Q_{iab} \left(u_a u_b + \frac{F_{2,ab} \tau}{\rho} \right)}{2 c_s^4} + \dots \right] \tag{25.28}$$

A direct calculation shows that

$$F_0 = 0, \; F_{1,a} = \rho g_a, \; F_{2,ab} = \rho (g_a u_b + g_b u_a) \tag{25.29}$$

with $\rho = nm$ and $g_a = F_a/m$.

As a result, we obtain (see Fig. 25.3)

$$\tilde{f}_i^e = \rho w_i \left[1 + \frac{c_{ia}(u_a + u_{F,a})}{c_s^2} + \frac{Q_{iab}(u_a u_b + u_a u_{F,b} + u_b u_{F,a})}{2 c_s^4} + \dots \right] \tag{25.30}$$

where we have set $u_{F,a} \equiv g_a \tau$, the velocity shift over a time-span τ. The expression (25.30) shows that the shift in velocity only applies to first order in the Hermite expansion, the second-order term $O(\tau^2)$ falling short by a factor $u_{F,a} u_{F,b}$.

In passing, we wish to point out that the notion of shift is not limited to the fluid velocity. Indeed, Sbragaglia *et al.* have shown that in order to recover the correct *thermo-hydrodynamic* continuum equations, not only the current but also the temperature needs a force-dependent shift of the form

$$\tilde{T} = T - g^2\tau^2/D \tag{25.31}$$

in D spatial dimensions (3) and $k_B = m = 1$ units.

Many other forcing schemes have been published in the recent years, whose details can be found in the literature.

25.3.2 Multi-Relaxation Time

The force term can also be implemented within the multi-relaxation time (MRT) formalism. Here however shifted equilibria are no longer well defined because they cannot be obtained by inversion of the scattering matrix, since such matrix is singular.

One could of course introduce non-zero eigenvalues also for mass and momentum, but this is not necessary, since, as we have seen in chapter 14, MRT does not make use of discrete equilibria in velocity space anyway.

As a result, the MRT version of generalized hydrodynamics reads simply as follows:

$$C_i = \sum_k [\omega_k(M_k^e - M_k) + S_k]A_i^k \tag{25.32}$$

where A_i^k is the k–th eigenvector in kinetic space and the sum runs on the non-zero ω_k only.

In (25.32), $S_0 \equiv F_0$, $S_a \equiv F_{1,a}$, $S_{ab} \equiv F_{2,ab}$.

A generalized equilibrium can be defined via a corresponding shifted in the various moments, i.e.,

$$\tilde{M}_k^e = M_k^e + S_k/\omega_k$$

The above expression is well defined, since all $\omega_k \neq 0$.

25.4 Lattice Implementation of the Force Term: Details Matter!

The previous section laid down the conceptual procedure to implement external sources within LBE. The details of such an implementation can make a major difference in actual practice.

Several authors have pointed out that lattice corrections due to the finite timestep affect the continuum limit of generalized LB hydrodynamics (1; 2), hence the accuracy of the lattice implementation. For the sake of conciseness, here we refer to the very informative analysis presented by Guo and coworkers (4).

By performing a detailed Chapman–Enskog analysis in the presence of the force term, these authors showed that the linear choice

$$F_0 = 0, \quad F_{1,a} = nF_a, \quad F_{2,ab} = 0 \tag{25.33}$$

leads to deviations of order $\Delta t(\nabla \cdot \vec{F})$ in the continuity equation and $(\tau - \Delta t/2)\nabla \cdot (\vec{u}\vec{F})$ in the momentum equation.

It should be emphasized that the forced Navier–Stokes equations in the continuum are naturally formulated in terms of the τ-shifted flow velocity:

$$\tilde{u}_a = u_a + F_a\tau/m$$

where τ descends from the fluid viscosity. This is quite natural, as there is no Δt in the continuum and the only relevant timescale is τ.

On the other hand, Guo and coworkers show that within the Chapman–Enskog analysis of the forced LB, it is more natural to refer to Navier–Stokes equations in which the flow velocity is shifted by a half timestep, namely

$$\hat{u}_a = u_a + g_a\frac{\Delta t}{2} \sim u_a\left(t + \frac{\Delta t}{2}\right) \tag{25.34}$$

The discreteness errors remain silent only if the force and flow fields are divergence free, which includes the important case of incompressible flows subject to a constant force. However, whenever the force field exhibits generic variations in space and time, these errors cannot be neglected.

The authors also show that inserting quadrupole contributions, i.e.,

$$F_0 = 0, \quad F_{1,a} = nF_a, \quad F_{2,ab} = n(F_au_b + F_bu_a) \tag{25.35}$$

doesn't really help, since the error in the continuity equation is left untouched and the one on the momentum equation is simply turned into $\frac{\Delta t}{2}\nabla \cdot (\vec{u}\vec{F})$.

They go on by discussing a series of other options to conclude for the following optimal one:[61]

$$F_0 = 0, \quad F_{1,a} = nF_a, \quad F_{2,ab} = n\left(1 - \frac{\Delta t}{2\tau}\right)(F_au_b^* + F_bu_a^*) \tag{25.36}$$

where the effective flow field satisfying the Navier–Stokes equations is defined as follows:

$$\rho u_a^* = \sum_i c_{ia}f_i + nF_a\frac{\Delta t}{2} \tag{25.37}$$

---

[61] Unfortunately, the authors use the notation τ for what is in fact $\tau/\Delta t$, which turns out to be a bit confusing.

In other words, the appropriate velocity to be monitored during the evolution of the system is not the plain fluid velocity, $\vec{u}$, but the half-step force-corrected one, $\vec{u}^*$.

Rearranging the terms, the source term can be recast in the following form:

$$F_i = w_i \left(1 - \frac{\Delta t}{2\tau}\right) \left(\frac{\vec{c}_i - \vec{u}}{c_s^2} + \vec{c}_i \frac{\vec{c}_i \cdot \vec{u}}{c_s^4}\right) \cdot \vec{F} \qquad (25.38)$$

To be noted that the prefactor in the above equation is precisely the ratio ν_{LB}/ν between the LB viscosity and the continuum one $\nu = c_s^2 \tau$ indicating that this prefactor traces back to second-order terms in the Taylor expansion of the streaming operator.

The basic difference between Guo's scheme and shifted equilibria is the time shift $\Delta t/2$ instead of τ. The two cannot be the same, other than at zero viscosity, and consequently the lattice shifted equilibria lead to inaccuracies scaling like $(\tau - \Delta t/2)$.

Guo and collaborators state their case by head-on comparison of the various methods for the case of steady-Poiseuille flow and unsteady Taylor–Green vortices showing that the scheme (25.38) provides the closest agreement with analytical results.

For full details, see the original reference.

25.5   Electromagnetic Forces

For the sake of completeness, we provide herewith a brief description of the implementation of electromagnetic forces (Lorentz) within the LB formalism.

To this purpose, let us begin by writing the Lorentz force in coordinate form

$$F_a^L = \epsilon_{abc} v_a B_c \qquad (25.39)$$

in units $q = c = 1$, q being the electric charge and c the speed of light. Note that, at variance with the forces considered this far, the Lorentz force depends explicitly on the molecular velocity $\vec{v}$. Hence, the calculation of mass, momentum and momentum-flux change in time must be performed with some extra care.

The mass change per unit time is given by

$$\dot{\rho}^L = -\epsilon_{abc} B_c \int (\nabla_a v_b)\, f d\vec{v} = 0 \qquad (25.40)$$

where we have integrated by parts and used the relations $\nabla_a v_b = \delta_{ab}$ and $\epsilon_{abc}\delta_{ab} = 0$, due to the antisymmetry of the Levi–Civita tensor.

By the same token, the change of momentum per unit time is given by

$$\dot{\mathcal{J}}_a^L = -\epsilon_{bcd} B_d \int \nabla_b (v_a v_c)\, f d\vec{v} \qquad (25.41)$$

Using the identity $\nabla_b (v_a v_c) = v_a \delta_{bc} + v_c \delta_{ab}$, we obtain

$$\dot{\mathcal{J}}_a^L = -\epsilon_{bcd} B_d \delta_{ab} \int f v_c\, d\vec{v} = -\epsilon_{acd} B_d \int v_c\, f d\vec{v} = \epsilon_{acd} \mathcal{J}_c B_d \qquad (25.42)$$

which is nothing but the macroscopic Lorentz force $\vec{\mathcal{J}} \times \vec{B}$.

Finally, the change per unit time of the momentum flux reads as follows:

$$\dot{P}^L_{ab} = -\epsilon_{cde}B_e \int f \, \nabla_c(v_a v_b v_d) \, d\vec{v} = (\epsilon_{ade}P_{bd} + \epsilon_{bde}P_{ad})B_e \tag{25.43}$$

Recalling the expression of the hydrodynamic tensor, $P_{ab} = p\delta_{ab} + \rho u_a u_b$, and using the antisymmetry of the Levi–Civita tensor, it can be checked that pressure gives no contribution, while the inertial part of the hydrodynamic tensor can be written in vector notation as

$$2\vec{J}(\vec{u} \times \vec{B}) = \vec{u}(\vec{J} \times \vec{B} - \vec{B} \times \vec{J})$$

The attentive reader might have noticed that, even though the Lorentz force depends explicitly on the molecular velocity, the end result for the contribution of mass, momentum and momentum-flux change in time still inscribes within the general relations (25.6–8), which were derived for spacetime dependent forces. How come? The reason is that the Lorentz force is divergence-free in velocity space, i.e., $\partial_a(\epsilon_{abc}v_b B_c) = 0$, hence it can be taken out of the velocity gradient, as if it were a spacetime dependent force only.

25.5.1 Two-Dimensional Magnetohydrodynamics

The expressions simplify considerably when both the flow and the magnetic field live in two spatial dimensions, $\vec{J} = \{J_x(x,y), J_y(x,y)\}$ and $\vec{B} = \{B_x(x,y), B_y(x,y)\}$.

In this case, the Lorentz force is perpendicular to the plane and has no bearing on the momentum equation.

It can also be shown that under the solenoidality condition, $\partial_a B_a = 0$, the Lorentz force can be expressed as the divergence of the *Maxwell tensor*:

$$M_{ab} = -B_a B_b + \frac{B^2}{2}\delta_{ab} \tag{25.44}$$

Based on (25.44), simple algebra shows that in the two-dimensional case, the Lorentz force can be accounted for by simply adding a "magnetic" contribution to the quadrupole component of the local equilibria, namely

$$f^e_i = w_i \rho \left[1 + \frac{c_{ia}u_a}{c_s^2} + \frac{Q_{iab}(u_a u_b - c_{A,a}c_{A,b})}{2c_s^4} \right] \tag{25.45}$$

In the above,

$$c_{A,a} = \frac{B_a}{\sqrt{\rho}}$$

is the Alfven speed, i.e., the typical propagation velocity of magnetic perturbations. The square of the ratio of Alfven to sound speed, or, equivalently, magnetic to thermodynamic pressure:

$$\beta = \frac{c_A^2}{c_s^2}$$

plays a major role in the hydrodynamics of magnetic fluids, and in particular on the confinement of thermonuclear plasmas.

For a detailed treatment of the lattice kinetic theory for magnetohydrodynamic applications, see (5) and references therein.

25.6 Toward Strong Interactions: Hermite Representation

The treatment so far was restricted to the case of Froude numbers well below one. To check this statement, let us just consider the stability condition:

$$|\frac{\Delta f_i}{f_i}| \sim \frac{w_i c_{ia} F_{1,a} \Delta t}{w_i \rho c_s^2} \sim \frac{g \Delta x}{c_s^2} \equiv Fr_\Delta \ll 1$$

where we have used the relations (25.29).

Overcoming such limitation without abandoning the free-particle formalism is not an easy task.

In principle, a possibility is to proceed to higher-order expansions in the lattice Froude number Fr_Δ (closing eyes on convergence).

The formal path can be taken from the very instructive paper by N. Martys *et al.*, where the authors analyze the representation of the forcing term in Hermite polynomials.

The starting point of the analysis is the identity

$$H_n(v) = c_n w^{-1}(v) \, (\nabla_v)^n \, w(v) \tag{25.46}$$

where $c_n = \frac{1}{\sqrt{2\pi}} (-1)^n/n!$ are suitable normalization coefficients and $w(v) = e^{-v^2/2}$.

The above expression is operational: by starting from the ground level $H_0(v) = 1$, all subsequent Hermite polynomials descend upon applying the operator ∇_v to the weight function $w(v)$.

As a reminder, $H_0(v) = 1$, $H_1(v) = v$ and $H_2(v) = v^2 - 1$.[62]

---

[62] These are sometimes called the "probabilist's" Hermite polynomials, as opposed to the "physicist's" one, which feature e^{-x^2} as a weight function. This is a bit weird, since physicist are accustomed to Maxwell–Boltzmann, which includes the factor 1/2 in the exponent.

The usual Hermite expansion, $f(x, v; t) = w(v) \sum_{n=0}^{\infty} M_n(x, t) H_n(v)$, takes then the form

$$f(x, v; t) = \sum_{n=0}^{\infty} c_n M_n(x; t) (\nabla_v)^n w(v) \tag{25.47}$$

The streaming term in velocity space is readily computed as

$$S(x; t) \equiv -g \nabla_v f = -g \sum_{n=0}^{\infty} c_n M_n(x; t) (\nabla_v)^{n+1} w(v) \tag{25.48}$$

By downshifting the index, this also rewrites as

$$S(x; t) = -g \sum_{n=1}^{\infty} M_{n-1}(x; t) c_{n-1} (\nabla_v)^n w(v) \tag{25.49}$$

As a result, the moments of the streaming source are given by

$$S_n(x; t) = -g \frac{c_{n-1}}{c_n} M_{n-1}(x; t) = g \, n M_{n-1}(x; t) \tag{25.50}$$

where we have used the property $c_{n-1} = n c_n$.

This gives $S_0 = 0$, $S_1 = g M_0$ and $S_2 = 2 g M_1$, which is nothing but the one-dimensional version of the relations (25.6–8).

Incidentally, this also shows that the first three terms are the same, regardless of whether they are computed at local equilibrium or with the actual distribution function. This is why it is no big crime to apply the streaming term to the local equilibrium f^e rather than to the actual distribution f.

These relations are not new, but the expression (25.50) illuminates the route toward the description of stronger interactions via higher-order tensors and correspondingly higher-order lattices.

Despite being restricted to weak forces, the LB for generalized hydrodynamics has found a broad spectrum of applications beyond the realm of Navier–Stokes hydrodynamics.

25.7 Summary

Summarizing, the inclusion of force terms within the Lattice Boltzmann formalism proceeds through a systematic procedure, whereby the contribution of the force to the various kinetic moments can be computed and translated into a well-defined source term into the force-free LB (Driven Hydrodynamics). Such source term can be eventually absorbed into a suitably generalized local equilibrium (Shifted Hydrodynamics). Either ways, the basic free-stream and collide structure of the LB scheme is preserved, which is key to its computational efficiency. A few lines of extra-code bring in an entire new world of physics.

The procedure is clearly perturbative in nature, i.e., it applies to sufficiently weak forces, i.e., Froude numbers well below one. In addition, the lattice implementation commands great care in handling spacetime heterogeneities, as they lead to inaccuracies due to the finite size of the lattice timestep. To leading orders in the timestep, such inaccuracies can be reabsorbed in the definition of a half timestep shifted velocity.

With all these caveats in mind, the inclusion of force terms in the LBE has spawned a broad range of applications beyond Navier–Stokes hydrodynamics, as we are going to illustrate in the forthcoming chapters.

. . . .

REFERENCES

1. N. Martys, X. Shan, and H. Chen, *Phys. Rev. E*, **58**, 6855, (1998).
2. J.M. Buick and C.A. Greated, *Phys. Rev. E*, **61**, 5307, (2000).
3. M. Sbragaglia, R. Benzi, L. Biferale *et al.*, *J. Fluid Mech.*, **628**, 299–309, (2009).
4. Z. Guo, C. Zheng, B. Shi, *Phys. Rev. E.*, **65**, 046308, (2002).
5. P. Dellar, *J. Comp. Phys.*, **179**, 95 (2002).

. . . .

EXERCISES

1. Show that the error associated with the transfer of the shift $g\tau$ from f to f^e has zero impact on the momentum imparted to the fluid by the force field.
2. Provide the explicit expression of the above error in terms of non-equilibrium macroscopic moments.
3. Show that the Fokker–Planck diffusion term does not break mass and momentum conservation.

26

Lattice Boltzmann for reactive flows

The dynamics of reactive flow lies at the heart of several important applications, such as combustion, heterogeneous catalysis, pollutant conversion, pattern formation in biology and many others. In general, LB is well suited to describe reaction–diffusion applications with flowing species. In this chapter, we shall provide the basic guidelines to include reactive phenomena within the LBE formalism.

Ignis mutat res

(Fire changes things)

26.1 Chemical Reactions

Reactive flows obey the usual fluid equations, augmented with a reactive source term accounting for species transformations due to chemical reactions (1).

Such term comes typically in the form of a polynomial product of the mass densities of the reacting species.

For a generic reaction $R \to P$ turning N_R reactants into N_P product species according to

$$a_1 R_1 + a_2 R_2 \dots a_{N_R} R_{N_R} \to b_1 P_1 + b_2 P_2 \dots b_{N_P} P_{N_P}$$

the overall reaction rate is given by

$$\dot{\mathcal{R}} = k_{fw}(T) \prod_{r=1}^{N_R} \rho_r^{a_r} - k_{bw}(T) \prod_{p=1}^{N_P} \rho_p^{b_p} \tag{26.1}$$

where the integers a_r and b_p are the stoichiometric coefficients, i.e., the number of molecules involved in the forward reaction $R \to P$ and in the backward one $P \to R$, respectively.

The Lattice Boltzmann Equation. Sauro Succi, Oxford University Press (2018).
© Sauro Succi. DOI: 10.1093/oso/9780199592357.001.0001

The density change per unit time due to this reaction is then $\dot{\rho}_r = -a_r\dot{\mathcal{R}}$ for reactants and $\dot{\rho}_p = b_p\dot{\mathcal{R}}$ for products.

The forward and backward coefficients, k_{fw} and k_{bw}, are usually following an Arrhenius-like relation

$$k_{fw}(T) = k_{\rightarrow}e^{-\frac{\Delta E_{fw}}{k_BT}}, \ k_{bw}(T) = k_{\leftarrow}e^{-\frac{\Delta E_{bw}}{k_BT}} \tag{26.2}$$

where $k_{\rightarrow}$ and $k_{\leftarrow}$ and are attempt rates (inverse time) and $\Delta E_{fw,bw}$ are the corresponding activation energies, i.e., the energy barriers to be overcome in order to transit from R to P and vice versa, see Fig. 26.1. Chemical reactions can exert substantial effects on the dynamics of fluid flows, typically in terms of density changes induced by temperature excursions due to the heat released(absorbed) by exothermic(endothermic) reactions. When such heat release is *fast and strong*, heat exchange and compressibility effects take the main stage in the dynamics of the fluid flow.

If, in the other extreme, reactions are *weak and slow*, then chemistry can be handled as a perturbation on top of inert fluid dynamics.

The progress rate of chemical reactions is typically measured in terms of the diffusive *Damkoehler* number,

$$\mathcal{D}a_D = \frac{L^2}{D\tau_{ch}}$$

where

$$\tau_{ch} \sim 1/\dot{\mathcal{R}}$$

is the typical chemical timescale, L is the global size of the reactor and D is the diffusion coefficient.

The Damkholer number is the ratio between the chemical and diffusion timescales, and, consequently, the condition $\mathcal{D}a_D \gg 1$ configures the so-called *fast-chemistry* regime, sometimes also called *thin-flame regime*, in which chemical reactions are confined within a thin reaction layer of thickness $\delta \sim \sqrt{D\tau_{ch}} \ll L$.

In contrast, slow chemistry tends to occur over the entire domain, $l \sim L$, the so-called *well-stirred* reactor regime characterized by $\mathcal{D}a_D \ll 1$.

A similar reasoning goes for the advective Damkohler number, defined as the ratio between the advective and convective timescale,

$$\mathcal{D}a_A = \frac{L}{U\tau_{ch}}$$

where U is the typical flow velocity.

The energetic impact of a chemical reaction is controlled by the amount of heat released or absorbed in the process of converting reactants into products. This is governed by the energetic gap between the reactant and product states, i.e.,

$$\Delta E = E_R - E_P$$

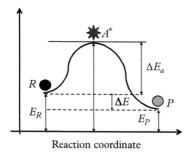

Reaction coordinate

Figure 26.1 *Energetic landscape of an exothermic chemical reaction. The reactant state R transits into an activated complex A^* which subsequently decays into the product complex P. The energy released in the process is $\Delta E = E_R - E_P$. Note that in order to transit from R to P the molecular system must overcome an energy barrier $\Delta E_a = E_{A^*} - E_R$, known as activation energy. Note that the activation barrier for the inverse reaction $P \rightarrow R$, $E_{A^*} - E_P$ is higher than for the direct one.*

The condition $\Delta E > 0$ denotes *exothermic* reactions, in which the excess energy liberated in the reaction is released to the fluid. The opposite condition denotes *endothermic* reactions. Typical chemical reactions feature values of ΔE in the order of fractions of electronvolts (*eV*) corresponding to thousands of Kelvin degrees. We remind that 1 eV ~ 11,000 Kelvin degrees, so that the ambient temperature $T = 300$ Kelvin corresponds to about 1/40 eV. The heat release is typically measured in kilocalories per mole (Kcal/mole): a mole of reactants undergoing a reaction with $\Delta E = 0.1$ eV releases about 2.5 Kcal/mole.[63]

The strength of exothermic reactions is characterized by the "thermal" Damkholer number:

$$\mathcal{D}a_T = \mathcal{D}a \, \frac{\Delta E}{k_B T}$$

where $\mathcal{D}a$ denotes either the diffusive or advective Damkohler number.

Hence, large values of the thermal Damkohler number associate with fast and strong chemical reactions, while small ones characterize slow and weak reactions. Clearly, flows with fast and strong reactions are significantly more difficult to model and simulate (by any numerical method) than weak and slow ones.

The earliest reactive LB schemes were indeed developed for laminar flames with fast chemistry and mild heat release, where compressibility effects can safely be neglected. Over the years, more elaborated compressible LB schemes have also been developed. In this chapter, however, we shall deal only with the former and simpler class of reactive fluids, directing the reader interested in more advanced treatments to the specialized literature.

26.2 Lattice Boltzmann Schemes with Chemical Reactions

Reaction–Diffusion systems have been explored in depth using Lattice Gas Cellular Automata (2). Once reactions take place under flowing conditions, Lattice Boltzmann can offer increased flexibility due to its hydrodynamic capabilities, as we proceed to describe.

---

[63] By definition of Kcal, this is the amount of heat raising the temperature of one liter of water in standard conditions by 2.5 degrees Celsius.

Let us refer to a reactive mixture of $N_S = 4$ miscible species (components), undergoing the following reaction:

$$aA + bB \leftrightarrow cC + dD$$

The reaction rate is given by

$$\dot{\mathcal{R}} = k_{fw}\rho_A^a \rho_B^b - k_{bw}\rho_C^c \rho_D^d \tag{26.3}$$

where k_{fw} and k_{bw} are the forward and backward rates, respectively. The balance between forward and backward reactions defines the conditions of chemical equilibrium:

$$\frac{(\rho_A^e)^a (\rho_B^e)^b}{(\rho_C^e)^c (\rho_D^e)^d} = \frac{k_{bw}}{k_{fw}} \equiv K(T) \tag{26.4}$$

where T is the fluid temperature and superscript "*e*" stands for equilibrium.

To model reaction (26.4), we introduce a *multicomponent distribution function* f_{si}, where the index s runs over the different chemical species $s = 1, \dots, N_S$ participating to the chemical activity of the fluid system.

Chemical reactions are naturally represented as a mass source term $\dot{R}_{si}$, expressing the mass change per unit time of species s propagating along the i-th direction.

Based on the above, the reaction term can be formally written as

$$\dot{R}_s = \frac{\rho_s - \rho_s^e}{\tau_{ch}}$$

which *defines* the effective chemical rate for the given reaction and species.

The reactive LBGK equation for the s-th reactive species takes the form

$$f_{si}(\vec{x} + \vec{c}_i \Delta t, t + \Delta t) = (1 - \omega_s)f_{si} + \omega_s f_{si}^e + \dot{R}_{si}\Delta t \tag{26.5}$$

where we have set $\omega_s = \frac{\Delta t}{\tau_s}$.

The reactive term $\dot{R}_{si}$ must be designed so as to reproduce the correct density change rates $\dot{\rho}_s$ and energy input $\dot{Q}_s$ for each given species s.

In equations:

$$m_s \sum_i \dot{R}_{si} = \dot{\rho}_s \tag{26.6}$$

$$m_s \sum_i \dot{R}_{si} c_{ia} = 0 \tag{26.7}$$

$$m_s \sum_i \dot{R}_{si} \frac{c_i^2}{2} = \dot{Q}_s \tag{26.8}$$

where m_s is the mass of the generic s-th species.

Moreover, since the overall mass and momentum are conserved, the following additional constraints apply:

$$\sum_{s,i} m_s \dot{R}_{si} = 0 \tag{26.9}$$

$$\sum_{s,i} m_s \dot{R}_{si} c_{ia} = 0 \tag{26.10}$$

$$\sum_{s,i} m_s \dot{R}_{si} \frac{c_i^2}{2} = \dot{Q} \tag{26.11}$$

where $\dot{Q}$ is the total heat release/absorption per unit volume in the reactive flow.

It should be noted that the density change rates $\dot{\rho}_s$ and the heat input $\dot{Q}$ are very sensitive functions of the fluid temperature T, typically through an Arrhenius form $e^{-E_a/k_B T}$.

The unknown $\dot{R}_{si}$ can be sought in the parametric form

$$\dot{R}_{si} = A_s + B_s \left(c_i^2 - C_2 \right) \tag{26.12}$$

where $C_2 = \sum_i c_i^2$ serves the purpose of orthogonalizing the basis of lattice kinetic eigenvectors, $E_i^{(0)} = 1$, $E_i^{(a)} = c_{ia}$, $E_i^{(2)} = c_i^2 - C_2$.

Solving the system (26.6–8) yields

$$A_s = \frac{\dot{\rho}_s}{E_1}, \quad B_s = \dot{Q}_s - \frac{\dot{\rho}_s}{E_2} \tag{26.13}$$

where $E_k \equiv \sum_i |E_i^{(k)}|^2$ is the square norm of the k-th eigenvector $E_i^{(k)}$.

The relations (26.13) specify the reactive LBE scheme in full detail.

A few comments are now in order.

First, since $\dot{\rho}_s$ only depends on the local densities $\rho_1, \ldots, \rho_{N_S}$, the reactive LBE scheme preserves its much prized locality in configuration space.

A further comment regards numerical stability, which demands that chemical changes do not spoil the positivity of the fluid density:

$$|\dot{R}_s|\Delta t \ll \rho_s \tag{26.14}$$

This means that the lattice timestep Δt cannot exceed a fraction of the shortest chemical timescale τ_{ch}. Securing this inequality under all thermodynamic conditions is pretty demanding because of the dramatic dependence of the reaction rates on the temperature, which sends $\tau_{ch} \to 0$ in the limit $T \gg E_a/k_B$.

It is therefore clear that in the fast-chemistry regime the advection-diffusion-reaction LB eqn (26.5) must proceed in smaller timesteps than the hydrodynamic solver, while the opposite is true if chemistry is very slow.

It is instructive to inspect some concrete figures.

Consider a reacting fluid in a cubic box 1 m in linear size, and suppose to cover the box with 10^9 lattice nodes (thousand per site) yielding a lattice spacing $\Delta x = 0.1$ cm. Take $c_s \sim 300$ m/s as the sound speed, the resulting $\Delta t \sim 0.1/(3 \times 10^4) \sim 3 \times 10^{-6}$ s.

Such a computation requires approximately $20 \times 4 \times 10^9$ bytes of memory (single precision), namely 80 Gbytes of storage. Assuming 100 floating operations (flops) per site at each timestep, a computer delivering one Gigaflop would take 100 CPU seconds per step, namely 10^5 CPU seconds, roughly a day, for a 1000 step long simulation (a millisecond in real time). These simple estimates indicate that fast chemistry can be handled "head-on" only for very small systems.

A final comment concerns the need of *memory-efficient* multicomponent transport algorithms. Reactive LBE schemes for an N_S species mixture with $2d$ discrete speeds per species in d spatial dimensions require $2dN_S$ arrays. This is a significant memory overhead, readily exceeding the storage requirements of the LB hydrodynamic solver whenever more than two or three species must be accounted for. Hence, the development of memory-lean reactive LB schemes is in significant demand. We shall return to this point shortly.

26.3   Advection-Diffusion-Reaction Equations

In section 26.2, we have illustrated a procedure to include chemical reactions within the LB formalism.

The next step is to model the advection-diffusion-reaction (ADR) equations which describe the evolution of the various species in space and time. We refer here to a model in which the overall fluid momentum evolves according to the Navier–Stokes equations, coupled to N_S evolution equations for the species densities ρ_s.

Such densities obey a corresponding set of continuity equations with chemical sources:

$$\partial_t \rho_s + \nabla \cdot (\rho_s \vec{u}_s) = \dot{R}_s \qquad (26.15)$$

In (26.15) $\vec{u}_s$ is the flow velocity of the s species, which is generally different from the overall barycentric fluid velocity defined as

$$\vec{u} = \frac{\sum_s \rho_s \vec{u}_s}{\sum_s \rho_s} \qquad (26.16)$$

The relative shift $\vec{u}_s - \vec{u}$ is typically expressed in terms of Fick's diffusion:

$$\vec{u}_s - \vec{u} = -D_s \frac{\nabla \rho_s}{\rho_s} \qquad (26.17)$$

giving rise to the following set of advection-diffusion-reaction (ADR) equations:

$$\partial_t \rho_s + \nabla \cdot (\rho_s \vec{u}) = D_s \Delta \rho_s + \dot{R}_s \tag{26.18}$$

The ADR structure emerges from the corresponding LBE much the same way as in the case of the Navier–Stokes equations. However, at variance with the latter, ADR equations do not contain quadratic terms in the fluid velocity. Consequently, the corresponding local equilibria can be truncated to first order in the barycentric velocity, i.e.,

$$f_{i,s}^e = \rho_s w_i \left(1 + \frac{\vec{c}_i \cdot \vec{u}}{c_s^2} \right) \tag{26.19}$$

As a result, the momentum of a given species is not conserved, i.e.,

$$\sum_i f_{i,s}^e \vec{c}_i = \rho_s \vec{u} \neq \sum_i f_{i,s} \vec{c}_i = \rho_s \vec{u}_s \tag{26.20}$$

Projection upon the first two moments for each species gives

$$\partial_t \rho_s + \partial_a \mathcal{J}_{a,s} = 0 \tag{26.21}$$
$$\partial_t \mathcal{J}_{a,s} + \partial_b P_{ab,s} = -\omega_s (\mathcal{J}_{a,s} - \mathcal{J}_{a,s}^e) \tag{26.22}$$

By invoking the usual enslaving argument

$$\mathcal{J}_{a,s} \sim \mathcal{J}_{a,s}^e - \tau_s \partial_b P_{ab,s}^e \tag{26.23}$$

as combined with the ideal equation of state:

$$P_{ab,s}^e \sim \rho_s c_s^2 \delta_{ab} \tag{26.24}$$

one finally obtains

$$\partial_t \rho_s + \partial_a (\rho_s u_a) = c_s^2 \tau_s \Delta \rho_s \tag{26.25}$$

which is the desired AD equation, with diffusivity $D_s = c_s^2 \tau_s$.

Inclusion of second-order terms in the Taylor–Chapman–Enskog expansion delivers the familiar $-1/2$ correction, yielding

$$D_s = c_s^2 (\tau_s - \Delta t/2) \tag{26.26}$$

Finally, we note that the absence of quadratic terms in the local equilibria implies that LB schemes for ADR equations can run on simpler lattices than hydrodynamic ones, usually with just $2d$ nearest-neighbors in d spatial dimensions.

This said, the use of more isotropic grids, possibly combined with second-order equilibria, typically enhances the quality of the numerical results thanks to the higher isotropy, with significant benefits on the quality of the numerical results (3).

26.4 Critical Remarks

As of today, LB schemes for ADR equations come in many variants, such as multi-relaxation time, as well as hybrid finite differences or finite volumes. The reader interested in the details of these developments is kindly directed to the numerous papers available in the original literature, among others (4; 5; 6; 7).

Indeed, LB schemes for ADR equations offer a number of appealing features, such as

i) low numerical dispersion, due to the fact that information moves along the lattice lightcones,

ii) low diffusion coefficients, thanks to the $-\Delta t/2$ shift,

iii) linear scaling of the timestep with the mesh-spacing (whenever weak-compressibility errors can be tolerated),

iv) high amenability to parallel computing in complex geometries.

However, the appealing features should always be critically weighted against the extra-memory required by the $2d$ (or more) arrays, f_{si}, to describe just one physical scalar, namely the density ρ_s.

This issue is made more pressing by modern computers, where the costs of accessing memory are taking an increasing fraction of the CPU time spent on number crunching.

In the sequel, we provide a cursory discussion of a few variants along these lines, namely the hybrid approach, the moment-propagation method (MPM) and the reduced allocation approach.

26.4.1 Hybrid Approach

Nothing new under the sky: as already discussed for the case of thermal flows, the option is simply to integrate the ADR equations with your grid method of choice, leaving the LB for the momentum equation only. Main pro is the use of one array per scalar, the cons are the same we have being discussing all along: timestep size,[64] amenability to parallel

---

[64] This advantage is not obvious, though, as shown by the following simple calculation. The stability condition for an explicit time-marching scheme is $\frac{D\Delta t}{\Delta x^2} < 1/2d$, d being the number of spatial dimensions. In terms of timestep limitations, this means $\Delta t < \frac{1}{2d}\frac{\Delta x^2}{D}$. Next, write the diffusion coefficient as $D = D_{lb}\frac{\Delta x^2}{\Delta t}$, where we have set $D_{lb} \equiv \frac{1}{3}(\frac{\Delta t}{\tau} - \frac{1}{2})$ taking 1/3 for the lattice sound speed squared. Using this expression, the timestep limit rewrites as $\Delta t < \frac{1}{2d}\frac{\Delta t}{D_{lb}}$, which is fulfilled whenever $D_{lb} < \frac{1}{2d}$. Even in the most stringent case, $d = 3$, this gives $D_{lb} < 1/6$, which is indeed fulfilled in many LB applications. Below such value, the LB timestep

computing, complex geometries and so on. The balance between the two is basically a problem-dependent, end-user decision.

26.4.2 Moment-Propagation Method

The idea of the moment-propagation method (8) is to propagate the scalar field $C(\vec{x}; t)$ using the LB populations f_i to define the probability of moving along the i-th link, $i = 0, b$.

In equations:

$$C(\vec{x}, t + \Delta t) = \sum_i p_i(\vec{x} - \vec{c}_i \Delta t, t) C(\vec{x} - \vec{c}_i \Delta t, t) + \frac{\rho_0}{\rho(\vec{x}; t)} C(\vec{x}; t) \qquad (26.27)$$

where

$$p_i(\vec{x}; t) = \frac{f_i(\vec{x}; t) - \rho_0/b}{\rho(\vec{x} - \vec{c}_i \Delta t; t)} \qquad (26.28)$$

In (26.28), the parameter ρ_0 fixes the fraction of $C(\vec{x}; t)$ which remains in place.

A straightforward Taylor expansion of (26.27) delivers the advection-diffusion equation, with a diffusion coefficient (in lattice units):

$$D = \frac{c_s^2}{2} \left(1 - d\frac{\rho_0}{\rho}\right) \qquad (26.29)$$

in d spatial dimensions.

The main merit of the MPM is its deliberate target: memory savings.

On the other hand, this comes with some strings attached. An immediate one is that the propagation probability p_i becomes negative whenever the corresponding population falls below a minimum threshold, $f_i < \rho_0/b$.

This sets a limit on the minimum achievable diffusivity, or, in more technical terms, to the maximum cell-Peclet number defined as

$$Pe_\Delta = \frac{u\Delta x}{D} \qquad (26.30)$$

In fact, high cell-Peclet numbers provide a characteristic figure of merit of the numerical schemes for AD equations. The MPM permits us to reach $Pe_\Delta \sim O(1)$.

---

is not any larger than the one pertaining to an explicit Euler-time marching. The actual LB advantage, however, eventually rests with the opposite limit, i.e. very small diffusivity, $D_{lb} = \epsilon << 1$, while keeping an $O(1)$ timestep $\Delta t = \tau/2 + \epsilon\tau \sim \tau/2$. To the best of this author's knowledge, this matter remains largely unsettled.

LB schemes, on the other hand, can operate at significantly higher cell-Peclet numbers, of the order of $Pe_\Delta \sim 10$ and above. As a result, as mentioned earlier on, LB schemes may indeed offer a valuable option whenever low diffusivity is a high priority.

A detailed analysis of the numerical properties of several ADR-LB schemes can be found in the instructive paper (9).

26.4.3   Presence Index Approach

Caveat: this section is highly speculative in character, its purpose being to stimulate the inventive mind rather than reporting consolidated material.

Consider a fluid mixture of $N_S \gg 2$ species and assume that there is a plausible physical argument to rule out the possibility that more than $N_R \ll N_S$ species occupy the same site at any given time instant. Note that the actual species that occupy a given site at any given time needs not stay the same; in fact it typically changes in time and from site to site.

Iff (double f) this is the case, it is clear that one can allocate storage for the reduced set of N_R species rather than the much larger original set of N_S. For instance, one may posit that in a fluid mixture with, say 30 species, only five can plausibly be at the same site at the same time, thus saving a factor six in memory allocation. Clearly, the list of *which* species occupy a given site at any given time must be updated dynamically in the course of the simulation.

This idea has been first proposed and put into practice by Dupin *et al.*, for the case of *immiscible* mixtures, i.e., multicomponent flows with interfaces (10). Essentially, at each node (x, y) in two-dimensional space, the authors associate an integer "presence index" (PI) p,

$$1 \leq p(x, y, i, r) \leq N_S$$

which codes for the presence of a given species at site (x, y) moving along the i-th lattice link. Note that, at variance with p, the index r runs from 1 to N_R only. The PI permits us to store and perform the LB evolution only on the species which are effectively present at the node (x, y) at time t, thus compressing the storage requirements by a factor $N_R/N_S << 1$. Obviously, this comes with the additional complexity of dynamically updating the PI list at each timestep.

The authors prove nonetheless the viability of this shrewd idea for a number of multicomponent flow applications (10; 11).

Clearly, the application of the PI strategy to the case of *miscible* reactive flows is much more problematic because, by definition, all species can coexist at a given lattice site. However, for some specific flows it might be possible to argue against the simultaneous *substantial* presence of all species at a time at a given lattice site, in which case the PI procedure might then prove beneficial.

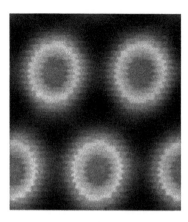

Figure 26.2 *The formation of hexagonal density patterns in an LBE simulation of the Selkov model, from (12).*

26.5 Reactive LBE Applications

Reactive LBEs have been applied to a number of reactive flow simulations, including, among many others:

1. Isothermal reaction–diffusion–advection with no heat release (Selkov model) (12).
2. Two-dimensional methane laminar flame with instantaneous chemistry and significant heat release (13).
3. Low-Mach laminar combustion (14).
4. Various types of combustion systems (15; 16).
5. Fuel-cells and catalytic reactors (17).
6. Many more . . .

The Selkov case is relevant to biology-oriented applications, whereas the other ones are more directed to energy applications, such as combustion engineering.

The Selkov model was addressed by means of a reactive LBGK with an equidistributed mass source of the form (26.12) and temperature treated like a freely tunable parameter. The authors investigated several pattern formation scenarios with various diffusion coefficients, with and without flowing solvent, consistently reporting agreement with literature data, see Figs 26.3.

A very early laminar methane flame application was addressed with a four-dimensional FCHC scheme in which two dimensions are compacted to deliver two additional scalars, the temperature T and the fuel/oxidizer mixture fraction Z, dynamically coupled to the two-dimensional reactive flow, see Figs. 26.4, 26.5 and 26.6. The scheme is thermodynamically consistent only in the limit of infinitely fast chemistry, in which the reaction front becomes infinitely thin and can therefore be described by a single-scalar function contour $Z(x, y, t) = $ constant, providing the geometrical location of the reaction front.

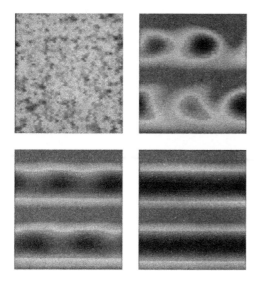

Figure 26.3 *The formation of lamellar density patterns in an LBE simulation of the Selkov model with fluid flow (from (12)).*

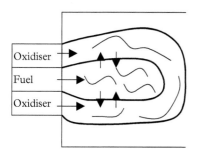

Figure 26.4 *Schematic picture of a diffusion flame. The central region is mostly filled with fresh unburned fuel, while the exterior one is mostly populated by the oxidizer. Upon diffusing at the interface, fuel mixes with the oxidizer and chemical reactions take place.*

The third application refers to a weakly compressible low-Mach flow with temperature dynamics. Finite-time chemistry was accommodated via a suitable generalization of lattice BGK equilibria incorporating the effects of pressure fluctuations. The scheme is enriched with grid refinement techniques which lead to a significantly boost of computational efficiency, see Figs. 26.7 and 26.8.

The fourth type of applications concerns fuel-cell modeling (17) and low-Reynolds flows with heterogeneous surface reactions in porous materials for catalytic reactors (18), see Fig. 26.9. These applications are central to all-important energy savings and environmental sustainability issues.

Besides the applications previously mentioned, many more have appeared in the recent years, whose description goes however beyond the scope of this work.

These applications are well suited to reactive LBE schemes because they need not handle large systems and can easily accommodate surface irregularities, thanks to the geometrical flexibility of the LB method.

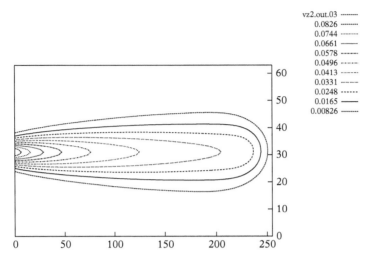

Figure 26.5 *Fuel mixture fraction $Z(x, y)$ contours of a diffusion laminar flame (after (13)).*

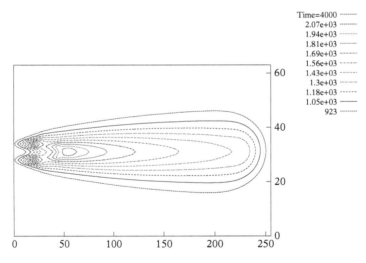

Figure 26.6 *Temperature $T(x, y)$ contours of the laminar flame (after (13)).*

(a)
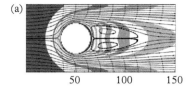

50   100   150

(b)
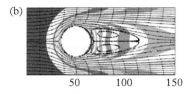

50   100   150

Figure 26.7 *Low Mach reactive flow LBGK calculation. Velocity streamlines and temperature contours past a cylinder. Case (a) finite difference solution of the Navier–Stokes equations with a pressure relaxation method. Case (b) LBGK solution. (After (14), courtesy of O. Filippova.)*

(a)
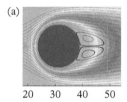

20 30 40 50

(b)
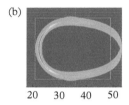

20 30 40 50

Figure 26.8 *Low Mach reactive flow LBGK calculation. Close-up of the temperature and reaction rates. (After (14), courtesy of O. Filippova.)*

Figure 26.9 *Three-dimensional LB simulation of a two-species reactive gas flow across a nanoporous catalytic ingot. From A. Montessori, P. Prestininzi, M. La Rocca and S. Succi, "Effects of Knudsen diffusivity on the effective reactivity of nanoporous catalyst media", J. of Comp. Sci., 17, 377, (2016).*

26.6 Summary

Summarizing, the LB formalism is well adapted to describe advection-diffusion-reaction phenomena. The advantages of the LB representation must nevertheless be critically balanced against the significant memory demand, an important issue in view of the increasing cost of accessing versus processing data in modern computer architectures. To this purpose, the development of memory-lean LB schemes for advection-diffusion-reaction equations appears to be a priority in the field.

With this caveat in mind, the fact remains that reactive slow-flows in complex geometries represent one of the most consolidated applications of LB techniques.

....

REFERENCES

1. F. Williams, *Combustion Theory*, 2nd Edn, Benjamin/Cummings, Menlo Park, (1985).
2. J. P. Boon, D. Dab, R. Kapral and A. Lawniczak, "Lattice gas automata for reactive systems", *Phys. Rep.*, **271**, 2, 58, 1996.
3. S. P. Thampi, S. Ansumali, R. Adhikari, *et al.*, "Isotropic discrete Laplacian operators from lattice hydrodynamics", *J. Comput. Phys.*, **234**, 1–7, 2013.
4. R.G.M Van der Sman and M. Ernst, "Convection-Diffusion Lattice Boltzmann Scheme for Irregular Lattices", *J. Comp. Phys.*, **160**, 766, 2000.
5. I. Rasin, S. Succi and W. Miller, "A multi-relaxation lattice kinetic model for passive scalar diffusion", *J. Comp. Phys.*, **206**, 453, 2005.
6. I. Ginzburg, D. d'Humieres and A. Kuzmin, Optimal stability of advection-diffusion lattice Boltzmann models with Two Relaxation times for positive/negative equilibrium *J. Stat. Phys.*, **139**, 1090, 2010.
7. R. Blaak and P. Sloot, "Lattice dependence of reaction-diffusion in lattice Boltzmann modeling", *Comp. Phys. Comm.*, **129**, 256, 2000.
8. R. Merks, A. Hoekstra and P. Sloot, "The Moment Propagation Method for advection-diffusion in the Lattice Boltzmann method: validation and Peclet Number Limits", *J. Comp. Phys.*, **183**, 563, 2002.
9. B. Servan-Camas and F.T.C. Tsui, "Non-negativity and stability analyses of lattice Boltzmann method for advection-diffusion equation", *J. Comp. Phys.*, **228**, 236, 2009.
10. M. Dupin, T. Spencer, I. Halliday and C. Care, "A many-component Lattice Boltzmann Equation Simulation for Transport of Deformable Particles", *Phil. Trans. Roy. Soc: Math. Phys. and Eng. Sciences*, **362**, 1822, 1885, 2004.
11. T. Spencer, I. Halliday and C. Care, "Lattice Boltzmann equation method for multiple immiscible continuum fluids", *Phys. Rev. E.*, **82**, 066701, 2010.
12. M. Ponce Dawson, S. Chen and G. Doolen, "Lattice Boltzmann computations for reaction–diffusion equations", *J. Chem. Phys.*, **98**, 2, 1514, 1993.
13. S. Succi, G. Bella and F. Papetti, "Lattice kinetic theory for numerical combustion", *J. Sci. Comp.*, **12**, 4, 395, 1997.
14. O. Filippova, "Lattice BGK model for low Mach number combustion", *Int. J. Mod. Phys. C*, **9**, 8, 1439, (1998).

15. K. Yamamoto, X. He and G.D. Doolen, "Simulation of combustion field with the lattice Boltzmann method", *J. Stat. Phys.*, **107**, 367, 2002.

16. S. Meng *et al.* "A simple lattice Boltzmann scheme for combustion simulation", *Comp. and Math. with Applications*, **55**, 1424, 2008.

17. P. Asinari, M. Cali' Quaglia, M. von Spakovsky and B.V. Kasula, "Direct numerical simulation of the kinematic tortuosity of reactive mixture flow in the anode layer of solid oxide fuel cells by the lattice Boltzmann method", *J. of Power Sources.*, **170**, 359, 2007.

18. G. Falcucci, S. Succi, A. Montessori, *et al.*, "Mapping reactive flow patterns in monolithic nanoporous catalysts," *Microfluids and Nanofluids*, **20**, 7 (105), 2016.

EXERCISES

1. Estimate the fastest chemical reaction that can be simulated by a reactive LBE scheme in a cube of side 10 cm discretized with 1000 grid points per side.

2. Estimate the spatial extent of the region where steady-state chemistry takes place (assuming zero net flow).

3. Write a one-dimensional ADR-LB scheme with just three speeds, $c_i = \{-1, 0, 1\}$ and compare its numerical diffusion and dispersion properties against your favored finite-difference scheme.

27

Lattice Boltzmann for Non-Ideal Fluids

In this chapter we discuss the extension of the LB methodology to the case of non-ideal fluids, i.e., fluids in which potential energy sits on a par with kinetic energy. The macroscopic consequences are major, primarily phase-transitions and attendant interface formation, which lie at the heart of the physics of multiphase and multicomponent flows, a branch of the physics of fluids with numerous applications in modern science and engineering.

Have no fear of perfection–you will never reach it

(Salvador Dali')

27.1 Introduction

The dynamics of multiphase flows plays a major role in many areas of applied science and engineering, including oil–water flow in porous media, boiling fluids, liquid metal melting and solidification to name but a few.

The numerical simulation of multiphase flows is a very challenging subject because, in addition to the usual difficulties associated with single-phase motion, it also requires the tracking in time of the interfaces (moving boundaries) between the different fluids (1; 2). The morphology and topology of such interfaces can become fairly complex in the course of time, which makes it pretty hard to provide a quantitative description of the dynamics of multiphase flows. Multiphase and multicomponent flows represent one of the most active fronts of LB research at large, the current literature being characterized by a steadfast outstream of new applications and methodological variants. Covering such a vast and varied landscape would definitely warrant a monograph on its own, an ordeal beyond the scope of this book.

The Lattice Boltzmann Equation. Sauro Succi, Oxford University Press (2018).
© Sauro Succi. DOI: 10.1093/oso/9780199592357.001.0001

As a result, in the sequel we shall present a cursory view of the basic ideas behind the main LBE formulations for multiphase flows leaving the most in-depth details to the current literature and to the very timely and useful monograph by Huang, Sukop and Lu (3).

We shall cover the following families of LB methods for non-ideal fluids:

- *Chromodynamic*
- *Pseudo-potential*
- *Free energy*
- *Finite density*
- *Miscellaneous*

This list alone clearly indicates that no single method has emerged as the "best" LBE solution for multiphase problems. In fact, the very notion of "best" solution is probably ill-posed, since each method may give its best for certain problems instead of others, depending on the parameter regime, scales of motion, boundary conditions and so on. Before plunging into the description of the various multiphase LBE models, for the sake of self-containedness, we provide a brief survey on the physics of non-ideal flows and the associated numerics.

27.2 Non-Ideal Equation of State and Surface Tension

The main distinctive features of the physics of non-ideal fluids and multiphase flows are a *non-ideal equation of state* and *surface tension*, which we are going to discuss in some detail.

27.2.1 Non-Ideal Equations of State

The equation of state (EoS) is the thermodynamic relation between fluid pressure of a given fluid at equilibrium and its density and temperature.

In equations,

$$p = f(\rho, T) \tag{27.1}$$

For an ideal gas with no potential energy, this reduces to the well-known linear relation

$$p = n k_B T = \rho v_T^2$$

which we have encountered many times before in this book.

As we have seen in Part I, in the presence of potential energy, this relation picks up nonlinear density contributions due to many-body interactions.

Of particular interest is the case in which the EoS displays a "loop" in the density-pressure plane, at a given temperature. Such loop corresponds to the existence of three different values of the density (at a given temperature) supporting the same value of the pressure. The lowest density (vapor) and the highest one (liquid) are stable thermodynamic phases, whereas the intermediate one is unstable. This means that the slightest perturbation would take the fluid at such density into either the vapor or the liquid phase. Such spontaneous separation of a homogeneous fluid into a vapor and liquid phase defines a so-called *phase transition*.

The prototypical equation of state supporting phase transitions is the one due to Van der Waals, as we have discussed in the first part of this book. Phase-transitions lie at the heart of the thermodynamics of multiphase fluids, but say little about the hydrodynamics of a multiphase *flow* away from thermodynamic equilibrium (see Fig. 27.1). This latter is governed by the spacetime dynamics of the liquid-vapor interface, which is a very complex non-equilibrium structure connecting the equilibrium bulk phases, typically a complex pattern of liquid droplets or vapor bubbles.

Central to the dynamics of the interface is the transport property known as surface tension, which we next discuss in some detail.

27.2.2   Surface Tension

Surface tension is the macroscopic manifestation of the different attraction experienced by a molecule sitting at a liquid/vapor interface, due to the different density of the two phases (see Figure 27.2).

From a macroscopic point of view, surface tension is defined as the reversible work per unit surface (force per unit length) needed to increase the area A of a given surface S by an amount ΔA:

$$\Delta W = \sigma \, \Delta A \tag{27.2}$$

This is the analog of the bulk relation, $\Delta W = p\Delta V$, which defines the work done on the fluid by the fluid pressure p through a change of volume ΔV. In the sequel, we shall

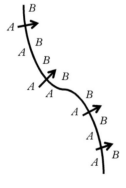

Figure 27.1 *An interface between fluids A and B in a binary A-B mixture. The arrows represent the density gradient normal to the interface, pointing from the light (A) to the heavy (B) fluid. For a single component fluid, A and B stand for the vapor and liquid phase.*

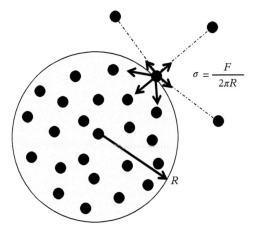

$$\sigma = \frac{F}{2\pi R}$$

Figure 27.2 *Microscopic origin of surface tension. A molecule sitting at the interface between the liquid droplet and the outside vapor experiences stronger attraction toward the interior than the exterior, due to the higher density inside. Surface tension is the energy per unit area (length in the two-dimensional figure) associated with this different attraction.*

refer to a spherical liquid droplet of radius R at a pressure p_l, immersed in its vapor at a pressure $p_v < p_l$ (see Fig. 27.2).

The question is how much work is to be spent on the vapor–liquid system to expand the radius of the liquid droplet from, say, R to $R + \Delta R$ (condensation).

This work is given by $\Delta W = (p_l - p_v)\Delta V$, where $\Delta V = 4\pi R^2 \Delta R$ is the volume change of the droplet. This is the energy supply from the exterior needed to win the action of surface tension, which acts so as to withstand the growth of the liquid surface, see Fig. 27.2 and 27.3.

In equations:

$$(p_l - p_v)\,\Delta V = \sigma\,\Delta A \tag{27.3}$$

Upon spelling out the elementary geometrical factors, we obtain the celebrated Laplace relation:

$$\Delta p = \frac{2\sigma}{R} \tag{27.4}$$

Note that, formally, surface tension acts as a "charge" for curvature-driven changes of pressure, often referred to as capillary pressure. Based on Laplace's relation, capillary pressure dominates the physics of highly curved interfaces, while its effects vanish for flat ones. The Laplace relation lies at the heart of capillary motion, namely the penetration of fluids in small pores. Fluids with large surface tension, such as water ($\sigma = 0.070$ N/m), can penetrate deep in narrow pores, a property which is crucial for life in our planet, trees being an outstanding witness in point.

The propensity of droplets and bubbles to experience significant distortions in a fluid flow is measured by the *Weber* number:

$$We = \frac{\rho U^2 D}{\sigma} \tag{27.5}$$

where ρ is the droplet density, U its macroscopic speed and D its diameter.

By definition, the Weber number measures inertial forces versus capillary ones and can easily take values in excess of 10^3 in engineering applications, such as spray dynamics in diesel engine combustion.

As mentioned earlier on, surface tension stems from the inhomogeneity of the intermolecular interactions at the interface between the two phases.

With reference again to a liquid droplet surrounded by its vapor, let us consider a molecule sitting right at the droplet interface. Since the liquid is denser than the vapor, this boundary molecule interacts with more "liquid" molecules than "vapor" ones and since the intermolecular potential is attractive (on scales above the hard-core repulsion region, typically a few Ångströms), the net result is that the boundary molecule naturally tends to be pulled back into the liquid region. This contrasts with the condition of an internal molecule which, being surrounded by an equal number of molecules in all directions, does not experience any net force on average.

The conclusion is that a surface molecule has an excess of energy (surface energy) with respect to an internal one, the difference representing precisely the work needed to extract the internal molecule and "peel it off" the surface.

The same energy must be supplied to push a vapor molecule inside the liquid droplet, which is yet another way of saying that surface growth involves an energy toll.

Surface tension is a decreasing function of temperature and vanishes at the critical liquid–vapor point, where the two phases become virtually indistinguishable. A derivation of this dependence from first principles implies detailed studies at the molecular level, see Fig. 27.3.

Usually, these will be of no direct concern to us, as long the appropriate physics at the mesoscopic scale can be reproduced via appropriate effective interactions. For our purposes, surface tension can be regarded as a given input parameter in the same league as viscosity, conductivity and other transport properties, except that it depends on potential energy.

The central difference, though, is that while viscosity and conductivity are well-defined concepts also for an ideal gas, surface tension commands the presence of potential energy interactions. Hence, surface tension will be dictated by the stylized model chosen to represent potential energy interactions. This is the major feature that needs to be addressed by all numerical methods for multiphase flows, LB being no exception to the rule.

high σ

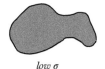
low σ

Figure 27.3 *A droplet of fluid with high surface tension sticks to circular shape much more tightly than a low-surface tension droplet.*

27.3   Numerical Methods for Flows with Interfaces

The numerical simulation of multiphase flows is a challenging subject because, in addition to the usual difficulties associated with single-phase motion, it requires the tracking in time of the interfaces (moving boundaries) between the different fluids, whose shape can become unwieldy complex as time unfolds.

The existing methods to deal with moving interfaces split into basically two main categories:

- *Front-Tracking (FT)*
- *Front-Capturing (FC)*

In FT methods, dynamic degrees of freedom (markers) are attached to the moving interface, and their dynamics is explicitly designed to follow the interface evolution. By doing so, the interface can be tracked fairly accurately, as long as the topology remains sufficiently smooth. If, however, the interface topology becomes very tortuous or undergoes qualitative changes, such as break-up and reconnection, FT run into the typical difficulty of Lagrangian methods, namely ill-conditioning and singularities, due to markers coming too close to each other. These distortions can be healed through grid-rezoning and grid-remapping procedures, which come however with a significant computational overhead.

FC methods, such is LB, solve this problem by defining a data structure throughout the entire computational domain (Eulerian approach). The interface is then located where discontinuities arise.

These methods side-step problems related to large distortions of the interface but, in revenge, they suffer from significant numerical diffusion effects which tend to smear out the interface in the course of the computation.

Many variants of both front-tracking and front-capturing methods have been proposed in the numerical multiphase literature, including hybrid methods aimed at combining the best of the two.

This is a very active and fast moving "front" (forgive the pun) of research, which goes beyond the scope of the present book.

Having clarified that LB belongs to the FC class, it still exhibits a number of peculiar aspects which we now proceed to illustrate.

27.4   LB Schemes for Multiphase Flows: Numerical Challenges

Owing to its kinetic nature, LB offers in principle an interesting opportunity to advance fundamental research in a difficult sector of non-equilibrium thermodynamics, with a broad array of practical fluid dynamic applications. On the other hand, in order to realize such an opportunity, it is necessary to overcome a number of conceptual and technical issues stemming from the discrete nature of the lattice.

Possibly, the most compelling ones are the following two:

- *Low density ratios*
- *Thick interfaces*

Let us dig a bit deeper into these matters.

27.4.1   Low Density Ratios

The transition between the liquid and vapor phases occurs over a very thin interface of at most a few molecular layers. Within such nanometric interface, density may change by several orders of magnitude, say 1 : 1000 for the prototypical case of water-air. This gives rise to very strong interfacial density gradients, whose simulation raises a challenge to any numerical method, no disrespect to lattice Boltzmann. Early LB methods typically attained density ratios around 50 between the heavy and light phase.

The density ratio is often expressed through the so-called *Atwood number*, defined as

$$At = \frac{\rho_l - \rho_v}{\rho_l + \rho_v} \tag{27.6}$$

where subscripts l and v stand for liquid and vapor, respectively.

By definition, the Atwood number ranges from 0 to 1 in the limit $\rho_l/\rho_v \to 1$ and $\rho_l/\rho_v \to \infty$, respectively. As a result, for air-water interfaces $At \sim 0.999$.

The limited range of density ratios achievable by LB methods is mostly due to the onset of spurious currents in the vicinity of the interface, which, in turn result from the lack of symmetry of the underlying lattice. Such spurious currents scale with the interfacial force and consequently they ruin the simulation whenever the density ratio exceeds a given threshold, typically around 50.

In practice, several phenomena of interest show a relatively mild dependence on the density ratio. Consequently, once this ratio is well above 1, valuable information can be gained at smaller values of the Atwood number than the real ones.

Nevertheless, simulating realistic values of Atwood numbers remains a highly-prized goal in multiphase LB research.

27.4.2   Thick Interfaces

As already mentioned, liquid-vapor interfaces are typically molecular thin. The ratio of the interface width, w, to a typical macroscale of the problem, say the droplet radius R(see Fig. 27.5), is known as *Cahn number*:

$$Cn = \frac{w}{R} \tag{27.7}$$

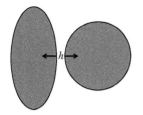

Figure 27.4 *Two approaching droplets can come to a near-contact distance h close to the width of the interface. The near-contact force per unit area, sometimes called disjoining pressure, can be either attractive or repulsive. The former case promotes coalescence and coarsening processes steering the fluid toward the minimum-surface equilibrium state. The repulsive case (shown in the figure) frustrates coalescence and promotes long-lived metastable states with large surface/volume ratios.*

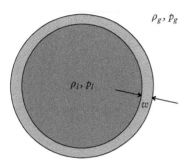

ρ_g, p_g **Figure 27.5** *A finite-width droplet of liquid immersed in a vapor (gas) environment. In most real-life applications, the width w is below a nanometer, leading to Cahn numbers of the order of $Cn \sim 0.001$ and below. In LB simulations, the diffuse interface covers a few lattice sites, which would imply sub-nanometric lattice spacings and prohibitive computational costs. To cope with this problem, LB simulations are typically run at Cahn numbers of the order of $Cn \sim 0.1$, the assumption being that the resulting error be a smooth function of Cn, thus leading to tolerable inaccuracies. Of course, this must be checked very carefully, in order to stave off computer "hallucinations"...*

For the case of flows with high surface to volume ratios, a relevant definition of the Cahn number is

$$Cn = \frac{w}{h} \qquad (27.8)$$

where $h \sim V/S$ is a measure of the gap between near-contact interfaces.

The Cahn number also provides a measure of the strength of interfacial forces which drive the migration of molecules from one phase to another, see Fig. 27.4.

Such forces (per unit volume) can be estimated as

$$F_{int} \propto \frac{\Delta p}{w} \sim \frac{\sigma}{wR} \qquad (27.9)$$

Taking $\Delta p/R$ as a typical strength of the force per unit volume with a uniformly distributed density profile, we see that the actual force at interface is $R/w = 1/Cn$ times larger.

Thick interfaces imply a loss of efficiency as compared to macroscopic front-tracking methods for multiphase flows, which treat the interface as a zero-thickness mathematical surface.

Being a mesoscopic method, LB is generally not aimed at a realistic description of the nanoscale structure of the interface and consequently one would like such interface to take the least number of lattice sites, ideally just one. Unfortunately, this is not compatible with the Eulerian structure of the method, a limitation which is inherent to all *diffuse-interface* methods. Thus, the liquid-vapor interfaces in LB simulations typically span several lattice sites corresponding to a much wider interface than the physical one.

Here one may invoke BA, for "Benevolent Asymptotics": if $Cn \ll 1$ in the application at hand, one may plausibly assume that any Cn sufficiently below 1 may just work nearly as well, as long as the error scales analytically with Cn (the actual definition of BA ...). As we shall see in the sequel, some multiphase fluid problems come indeed with some degree of BA, thereby spawning significant opportunities for LB simulations.[65]

Notwithstanding major progresses in the last decade and some generous help from Nature's benevolence, these items still stand as a continued challenge to this important front of LB research.

27.5 Chromodynamic Models

The first multiphase LB model was introduced by Gunstensen *et al.* (1991) (4), based upon the two-component lattice gas model developed by Rothman and Keller (5).

In these models one begins by introducing two particle distributions, say Red and Blue, f_{iR} and f_{iB}, for the two different fluids. Of course, color is just a mnemonic for different species, say water and oil, or any other attribute allowing to tell the two fluids apart. Each phase is assumed to obey its own LBGK equation:

$$\Delta_i f_{is} = -\omega_s (f_{is} - f_{is}^e) + S_{is}, \quad i = 1, \dots, b, \quad s = \text{Red, Blue} \tag{27.10}$$

The source term S_{is} represents the mesoscopic interaction between the two phases and it is therefore in charge of describing phase separation via surface tension effects. The central quantities in Gunstensen Rothman Zaleski Zanetti treatment are the *color current* K_a and *color gradient* G_a defined as follows:

$$K_a(\vec{x}, t) = \sum_i c_{ia}(f_{iB} - f_{iR}) = K_{aB} - K_{aR} \tag{27.11}$$

and

$$G_a(\vec{x}, t) = \sum_i c_{ia} [\rho_B(\vec{x} + c_{ia}) - \rho_R(\vec{x} + c_{ia})] \tag{27.12}$$

Note that both quantities vanish in color-blind homogeneous regions ($f_{iR} = f_{iB}$) where the two phases are in a perfect balance (perfectly miscible fluids). It can also be shown that single-phase regions ($f_{iB} \cdot f_{iR} = 0$) are color gradient free if both fluids are incompressible. It is natural to define a *color potential* $V \sim G^2$ such that the gradient G_a points in the direction normal to the Blue–Red interface (see Fig. 27.6). Gunstensen *et al.* choose the source term in the heuristic form

$$S_i = A \frac{G_i^2 - kG^2}{G} \tag{27.13}$$

---

[65] The notion of BA has been invoked before in this book, when dealing with flows in porous media, whereby the Knudsen and Mach numbers are artificially inflated to increase the time-step of the simulation.

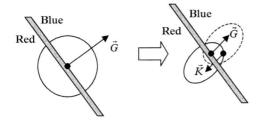

Figure 27.6 *Effect of the chromodynamic multiphase collision rule. The color gradient breaks the Red–Blue internal symmetry (circle) and introduces a preferential direction for Red or Blue particles to move across the interface (ellipses).*

where $G_i = G_a c_{ia}$ is the projection of the color gradient along the i-th direction, $k = bc^2/D$ and A is a free parameter controlling the actual value of surface tension in D spatial dimensions. Since this term does conserve Blue and Red densities separately, it can only set the stage for phase segregation but not promote it in actual terms. For this purpose Gunstensen *et al.* introduce an additional *color redistribution step*, which consists of (counter) aligning the color current K_a with the color gradient G_a in such a way as to minimize the "color energy" (see Fig. 27.6):

$$W = \sum_a K_a G_a \qquad (27.14)$$

This rule lies at the heart of phase-segregating behavior (chemo-taxis).

In passing, we note that such a type of "smell and go" dynamics is commonplace in many other sectors of complex fluid dynamics including polar fluids and biological flows.

The Gunstensen *et al.* multiphase LBE opened the way to a series of applications such as two- and three-dimensional simulations of multiphase flows in porous media.

A more in-depth analysis revealed a few weak sides, however, in particular a certain degree of anisotropy of the surface tension producing a spurious dependence on the interface orientation. In addition, near-interface parasitic microcurrents were reported, which stem from spurious invariants (these parasitic currents affect also other numerical methods!). In addition, the Gunstensen *et al.* scheme is relatively heavy from the computational point of view because of the variational minimization required by the recoloring step. This spurred the development of further models, which are currently receiving increasing attention for the simulation of multicomponent flows with multiple components (6)

27.6   The Shan–Chen Pseudo-Potential Approach

The chromodynamic approach of Gunstensen *et al.* builds on a major abstraction of the physical reality; in fact, the "color force" is nothing but the logical statement that molecules sitting at the interface between, say, dense and light fluid experience a net force driven by the different values of the average intermolecular distance in the two fluids. It is therefore natural to look for more physically-oriented representations, in which these forces are directly encoded as the result of pairwise (pseudo)-molecular interactions.

This is the pseudo-potential approach introduced by X. Shan and H. Chen (7), arguably the most popular LB model for non-ideal fluids to date.

These authors represent the source term in (27.10) as follows:

$$S_{is} = \vec{F}_s \cdot \vec{c}_i \tag{27.15}$$

where the force $\vec{F}_s$ on the s species comes from the following pairwise interaction potential:

$$V_s(\vec{x}) = \sum_{s'} \sum_{i} V_{ss'}(\vec{x}, \vec{x} + \vec{c}_i) \tag{27.16}$$

The physical meaning is clear: at each lattice site $\vec{x}$, the net force experienced by a particle of species s is the sum of the momentum exchanges with particles of all other species in the neighborhood $\vec{x}_i = \vec{x} + \vec{c}_i$ of the site $\vec{x}$ (lattice spacing is made unit).

The pseudo-potential matrix $V_{ss'}(\vec{x}, \vec{y})$ is chosen in a typical *propagator* form

$$V_{ss'}(\vec{x}, \vec{y}) = \Psi_s(\vec{x}) \, G_{ss'}(\vec{x}, \vec{y}) \, \Psi_{s'}(\vec{y}) \tag{27.17}$$

where $G_{ss'}(\vec{x}, \vec{y})$ is the Green's function expressing the pairwise interaction between species s and s' at sites $\vec{x}$ and $\vec{y}$, respectively, and $\Psi_s(\vec{x}) \equiv \Psi[\rho_s(\vec{x})]$ is a a local functional of the fluid density, see Fig. 27.7.

For most practical purposes, this interaction is taken in the simple form

$$G_{ss'}(\vec{x}, \vec{y}) = \begin{cases} G_{ss'} & \text{if } |\vec{y} - \vec{x}| < c_{max} \\ 0 & \text{otherwise} \end{cases} \tag{27.18}$$

where c_{max} denotes the maximum magnitude of the discrete speeds, i.e., $c_{max} = \sqrt{2}$ in the D2Q9 lattice.

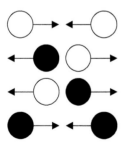

Figure 27.7 *The meaning of the pseudo-potential force. Each lattice population experiences an attractive force from neighbors of the same species and a repulsive one from neighbors of different species. The resulting force is the total sum over the neighborhood.*

The strength of the interaction is controlled by the amplitude parameters $G_{ss'}$ playing the role of effective inverse temperatures, or better said, potential versus kinetic energy. Conventionally, negative $G_{ss'}$ code for attraction and negative $G_{ss'}$ attraction.

A further simplification, due to Shan and Chen permits the simulation of binary fluids with a single species only, i.e., $G_{ss'} = G\delta_{ss'}$, whose density automatically identifies the light and heavy phases (7). In this case, the transition between heavy and light fluid is controlled by the function $\Psi[\rho]$, which serves as a generalized density of the model.

This function is typically taken in the form

$$\Psi(\vec{x}) = \Psi[\rho(\vec{x})] = \rho_0 \left(1 - e^{-\rho/\rho_0}\right) \qquad (27.19)$$

where ρ_0 is a reference density, marking the borderline $\rho_0 = 0(1)$ between light (heavy) fluids, respectively.

The force associated with this pseudo-potential is given by

$$\vec{F}(\vec{x}, t) = G\Psi(\vec{x}, t) \sum_i w_i \Psi(\vec{x}_i, t) \vec{c}_i \qquad (27.20)$$

where we have set $\Psi(\vec{x}, t) \equiv \Psi[\rho(\vec{x}, t)]$ and $\Psi(\vec{x}_i, t) \equiv \Psi[\rho(\vec{x} + \vec{c}_i, t)]$.

Note that $G < 0$ codes for attractive interactions, the only ones considered in the original Shan–Chen model, see Fig. 27.8.

Also to be noted that in the high-density limit $\rho \gg \rho_0$, the function $\psi(\rho)$ tends to a constant, so that the Shan–Chen force vanishes. This is the way the Shan–Chen fluid protects against density pile-up instabilities, notwithstanding the absence of repulsive interactions.

It is readily checked that the Shan–Chen force conserves the total momentum of the system, but not the local one on a site-by-site basis. Thus, this term contributes a source term $n\vec{F}$ to the right-hand side of the Navier–Stokes equations of each species separately.

Of course, the sum over all species yields zero, since the momentum lost on one species must necessarily be gained by another one. This is an important difference with respect to the Gunstensen *et al.* model, in which momentum is conserved locally for each species.

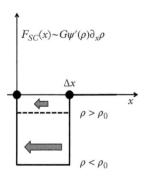

Figure 27.8 *The Shan–Chen force, an attractive interaction confined to the first Brillouin region. The dashed line associates with a higher density, corresponding to a lower force intensity, as per the expression* $\Psi = \rho_0(1 - e^{-\rho/\rho_0})$ *yielding* $\frac{d\psi}{d\rho} = e^{-\rho/\rho_0}$, *an exponentially decreasing function of the density* ρ.

The Shan–Chen force contributes an extra term to the momentum-flux tensor[66]

$$P_{ab} \rightarrow P_{ab}^{\mathrm{NS}} + \frac{1}{2} G \Psi(\vec{x}) \sum_i \Psi(\vec{x}_i) \, c_{ia} c_{ib} \qquad (27.21)$$

where, again, the superscript "NS" denotes the ordinary single-phase Navier–Stokes expression resulting from kinetic energy alone.

By letting Ψ depend on the fluid density ρ according to (33.18), the following non-ideal gas equation of state is obtained:

$$p(\rho) = \rho c_s^2 + \frac{G}{2} c_s^2 \Psi^2(\rho) \qquad (27.22)$$

The reader can readily verify that, by taking $\rho_0 = 1$ for simplicity, such EoS supports a loop structure (see Fig. 27.9) at the critical density

$$\rho_c = \ln 2$$

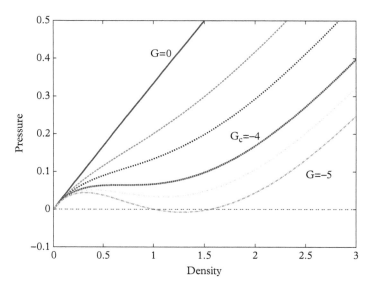

Figure 27.9 *The Shan–Chen equation of state for* $G = 0, -2, -3, -4,$
*$-4.5, -5$ (top to bottom). Below the critical value $G_c = -4$ the curve develops
a loop structure supporting coexistence of a vapor and a liquid phase. Note
that for G sufficiently below G_c, the pressure can attain negative values, a
property which has been used even for cosmological applications (9)!*

---

[66] The exact expression of the lattice momentum-flux tensor is more complicated than (27.21), and can be found in X. Shan, (8). The expression (27.21) provides nonetheless a useful qualitative proxy of the exact one.

provided the coupling strength obeys the inequality

$$G < G_c = -4$$

The appealing feature of this model is that phase separation takes place spontaneously whenever the interaction strength G falls below the critical threshold G_c, thus fitting naturally the physical notion of G as an effective inverse temperature of the system. Indeed, the parameter G is basically the potential energy in units of $k_B T$, hence large ratios (in absolute value) are tantamount to low temperatures.

Such a property was indeed beautifully confirmed by early numerical simulations of binary fluid separation.

Also, the model is easy to use, which explains its growing popularity to this day, despite substantial criticism, to which we shall return in Chapter 28. For a recent informative review of pseudo-potential methods, see (10).

27.7 The Free-Energy Approach

The original Shan–Chen model was formulated directly in terms of a pseudo-potential force, without being derived from a corresponding free energy, hence offering an exposure to thermodynamic inconsistency. Moreover, being a one-parameter model, the non-ideal equation of state and surface tension cannot be changed independently.

A significant step forward in the direction of improving on both liabilities discussed was taken by Swift, Osborne and Yeomans (11). These authors incorporated the equilibrium pressure tensor, also known as Korteweg tensor, for a non-ideal fluid directly into a suitably extended collision operator.

By doing so, the fluid is instructed to reach the right thermodynamic equilibrium directly under the guiding action of the free-energy functional.

The Swift–Osborne–Yeomans method builds on the Van der Waals formulation of a two-component isothermal fluid. The basic object of the theory is the Cahn–Hilliard free-energy density functional (free energy per unit volume) defined as

$$\mathcal{F}[\rho] = \int \left[\frac{k}{2} (\nabla \rho)^2 + f(\rho) \right] d\vec{x} \qquad (27.23)$$

In (27.23), the first term is the energy penalty paid to build and sustain a density gradients, whereas the second term is the bulk contribution controlling the equation of state.

The non-local pressure relates to $\mathcal{F}$ through the Legendre transformation:[67]

$$P = \rho \frac{d\mathcal{F}}{d\rho} - \mathcal{F} = P_0 - k\rho \nabla^2 \rho^2 - \frac{1}{2} k |\nabla \rho|^2 \qquad (27.24)$$

---

[67] We recall that the Helmholtz free energy F of a fluid of volume V is defined as $F = U - TS$ where U is the internal energy and S the entropy. The Gibbs free energy of the same fluid is given by $G = F + PV$. For an isothermal process ($dT = 0$), we have $dF = -P\,dV$ and $dG = V\,dP$.

where

$$p_0 \equiv \rho \mathcal{F}' - \mathcal{F} \tag{27.25}$$

is the equation of state of the fluid (prime stands for derivative with respect to density), see Figs 27.8 and 27.9. The full pressure tensor in a non-uniform fluid includes an off-diagonal component

$$P_{ab} = p\delta_{ab} + k\partial_a\rho\partial_b\rho \tag{27.26}$$

where the second term is related to interfacial surface tension effects.

How does one encode this pressure tensor in the equilibrium distribution?

The recipe is to add non-local terms to the discrete equilibria. In particular, for the seven velocity FHP lattice, Swift–Osborne–Yeomans proposed the following parametric expression:

$$f_i^e = A + Bc_{ia}u_a + Cu^2 + Dc_{ia}c_{ib}u_au_b + F_ac_{ia} + G_{ab}c_{ia}c_{ib} \tag{27.27}$$

The Lagrangian parameters A, B, C, D, F_a, G_{ab} are prescribed by the usual conservation of mass, momentum, momentum flux tensor constraints, with P_{ab} given by (27.26). These relations prove sufficient to compute the parameters as a function ρ *and its spatial derivatives*, thus solving the problem of identifying a proper free energy for the multiphase LB system.

Swift–Osborne–Yeomans demonstrated their model via 2D simulations based on the Van der Waals fluid free-energy density $\Psi_{\mathrm{vdw}} = \rho T \ln \rho/(1 - \rho^b) - a\rho^2$.

Like their predecessors, they tested their scheme against Laplace's law:

$$P_{\mathrm{in}} - P_{\mathrm{out}} = \frac{\sigma}{R} \tag{27.28}$$

where "in/out" refer to inner/outer pressure of a bubble of radius R (Fig. 27.10). Additional tests referred to the dispersion relation of capillary waves (Fig. 27.11).

They also computed the coexistence curve between the two phases for several values of the fluid temperature reporting excellent agreement with thermodynamic theory (Fig. 27.11 and 27.12).

The authors advocated significant reduction of the velocity fluctuations near the interface and satisfactory isotropy of the surface tension.

As a further bonus, they also proposed to study wetting phenomena by including a suitable external chemical potential $\mu(\vec{x})$.[68] Gradients of the chemical potential act as an effective thermodynamic force $\vec{F}_t = -\rho\nabla\mu/3$ and can be readily included as an extra source term on the right-hand side of LBE. By letting a nonzero μ only at the walls with a different intensity for the two species, the affinity of the wall to the two fluids can be varied in a simple and physically appealing way. In particular, by tuning the range of

---

[68] The chemical potential measures the energy change with the number of molecules in the system, namely $\mu = \partial U/\partial N$.

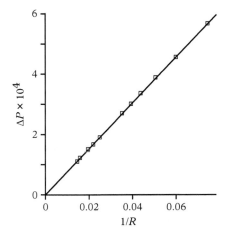

Figure 27.10 *Pressure difference between the inside and the outside of a spherical domain of radius R, as a function of 1/R (test of Laplace's law). The solid line is the exact result for a flat interface (R → ∞) (from (11)).*

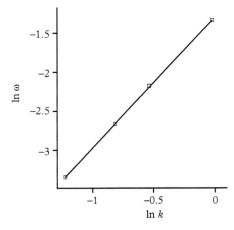

Figure 27.11 *The dispersion curve for capillary waves on an interface. The best fit line has a slope 1.6 ± 0.05, slightly in excess of the theoretical value 3/2. This departure is attributed to curvature effects (from (11)).*

the chemical potential, one can construct diffuse boundaries along the lines of molecular dynamics. According to the authors, these diffuse boundaries significantly soften the pathological orientation issues associated with sharp crystallographic boundaries, see Fig. 27.10.

Potential liabilities to be signaled are: (i) density gradients in the local equilibria are exposed to higher-order lattice artifacts close to sharp interfaces; (ii) analysis of the energy equation reveals that the condition of a constant temperature cannot be satisfied, so that thermodynamic consistency is not achieved. The actual impact of these liabilities is problem dependent and hard to judge *a priori*, but the model certainly has significant value to it.

In the last decade, the free-energy method has found a wide body of interesting applications, especially for the numerical simulation of complex microfluidic flows on heterogeneous and patterned surfaces.

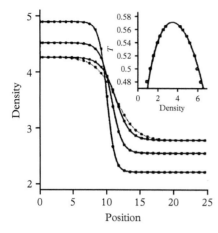

Figure 27.12 *Equilibrium density profiles normal to a flat interface for a Van der Waals fluid for the three highest values of temperature shown on the coexistence curve (inset). The solid lines are numerical solutions of the continuum thermodynamic equations, while the points are from the lattice Boltzmann simulations (from (11)).*

27.8   Free-Energy Finite-Difference Methods

As shown in the previous section, the *free-energy* approach derives from a density functional theory, namely a Cahn–Hilliard mixing energy density formulation for an isothermal system.

In terms of the liquid-phase concentration, with subscript g standing for the gas phase:

$$C = \frac{\rho_l}{\rho_l + \rho_g} \tag{27.29}$$

the Cahn–Hilliard free-energy density functional reads as

$$E\left[C, \nabla C\right] = E_0\left(C\right) + \frac{\kappa}{2}\left|\nabla C\right|^2$$

where E_0 is the bulk contribution and κ is the gradient parameter controlling the interface contribution.

The bulk energy can be rewritten as

$$E_0\left(C\right) \approx \beta C^2\left(C - 1\right)^2$$

where β is a constant fixing the free-energy barrier between the equilibrium states, $C = 0$ (gas=vapor) and $C = 1$ (liquid).

The same parameter accounts for the non-ideal bulk pressure through the Legendre transform:

$$p_0(C) = C\frac{\partial E_0}{\partial C} - E_0(C) \tag{27.30}$$

The two free parameters β and κ provide separate control of the surface tension and interface thickness, respectively:

$$\sigma = \frac{\sqrt{2\kappa\beta}}{6}, \qquad \delta = \sqrt{\frac{8\kappa}{\beta}} \tag{27.31}$$

The internal force representing the non-ideal gas effects reads as follows:

$$\vec{F} = -\nabla(p_0 - \rho c_s^2) + \rho\kappa\nabla\nabla^2\rho \tag{27.32}$$

where the first term on the right-hand side is the gradient of the non-ideal (excess) pressure, while the second one encodes surface tension effects.

Note that the latter involves third-order spatial derivatives.

A variant of the FE approach, due to Lee and collaborators, (12) builds upon direct finite-differencing of the non-ideal force (Free–Energy–Finite–Difference, FEFD) using second neighbors for the third-order spatial derivatives.

From a technical viewpoint, FEFD differs significantly from the previous FE approach, where finite differencing was used only in an ancillary form, i.e., to compute density gradients in the generalized local equilibria. Moreover, as we shall see, FEFD explicitly involves second-neighbor interactions in order to discretize third-order space derivatives.

We refer to Lee's implementation (13), which evolves *pressure* instead of density.

Consequently, the discrete distribution function is defined as follows:

$$g_i = f_i c_s^2 + \left(p_1 - \rho c_s^2\right)\Gamma_i(0) \tag{27.33}$$

where f_i is the usual discrete particle distribution, as defined in standard LB theory. In (27.33)

$$\Gamma_i(\mathbf{u}) = f_i^{eq}(\vec{u})/\rho$$

and p_1 a scalar pressure to be defined later.

This distribution is characterized by the following local equilibrium:

$$g_i^{eq} = w_i\left[p_1 + \frac{\vec{c}_i\cdot\vec{u}}{c_s^2} + \frac{(\vec{c}_i\cdot\vec{u})^2}{2c_s^4} - \frac{(\vec{u}\cdot\vec{u})}{2c_s^2}\right] \tag{27.34}$$

Use of Crank–Nicolson time marching suggests the following change of variables:

$$\bar{g}_i = g_i + \frac{\Delta t}{2\tau}\left(g_i - g_i^{eq}\right) - \frac{\Delta t}{2}\left(\vec{c}_i - \vec{u}\right)\cdot\left[\nabla(\rho c_s^2\left(\Gamma_i - \Gamma_i(0)\right)) - C\nabla\mu\,\Gamma_i\right] \tag{27.35}$$

and

$$\bar{g}_i^{eq} = g_i^{eq} - \frac{\Delta t}{2} (\vec{c}_i - \vec{u}) \cdot \left[\nabla(\rho c_s^2 (\Gamma_i - \Gamma_i(0)) - C\nabla\mu \Gamma_i \right] \tag{27.36}$$

In (27.36) $\mu = \partial E_0/\partial C$ denotes the chemical potential.

A second-order integration in time (Crank–Nicolson) delivers the following LBE for pressure field:

$$\bar{g}_i (\vec{x} + \vec{c}_i\Delta t, t + \Delta t) - \bar{g}_i (\vec{x}, t) = -\frac{1}{\tau + \Delta t/2} (\bar{g}_i - \bar{g}_i^{eq}) (\vec{x}, t) \tag{27.37}$$

$$+\Delta t (\vec{c}_i - \vec{u}) \cdot \left[\nabla(\rho c_s^2 (\Gamma_i - \Gamma_i(0)) - C\nabla\mu\Gamma_i \right] (\vec{x}, t)$$

The same procedure can be applied to the concentration C, by introducing the auxiliary distribution:

$$h_i = \frac{C}{\rho} f_i, \quad h_i^{eq} = \frac{C}{\rho} f_i^{eq}$$

This distribution is shown to obey the following LBE:

$$\bar{h}_i (\vec{x} + \vec{c}_i\Delta t, t + \Delta t) - \bar{h}_i (\vec{x}, t) = -\frac{\Delta t}{\tau + \Delta t/2} (\bar{h}_i - \bar{h}_i^{eq}) (\vec{x}, t) \tag{27.38}$$

$$+\Delta t (\vec{c}_i - \vec{u}) \cdot \left[\nabla C - \frac{C}{\rho c_s^2} (\nabla p_1 + C\nabla\mu) \right] \Gamma_i(\vec{x}, t)$$

$$+\Delta t \nabla \cdot (M\nabla\mu) \, \Gamma_i(\vec{x}, t) \tag{27.39}$$

where the modified equilibrium distribution $\bar{h}_i$ and its equilibrium are calculated as in (27.35) and (27.36), namely

$$\bar{h}_i^{eq} = h_i^{eq} - \frac{\Delta t}{2} (\vec{c}_i - \vec{u}) \cdot \left[\nabla C - \frac{C}{\rho c_s^2} (\nabla p_1 + C\nabla\mu) \right) \Gamma_i (\vec{x}, t) - \frac{\Delta t}{2} \nabla \cdot (M\nabla\mu) \, \Gamma_i (\vec{x}, t) \tag{27.40}$$

In (27.40), M is the mobility, a transport parameter controlling the rate of convergence to the equilibrium. The composition, the hydrodynamic pressure and the fluid momentum are calculated by taking the zeroth and the first moments of the modified particle distribution function:

$$C = \sum_i \bar{h}_i + \frac{\Delta t}{2} \nabla \cdot (M\nabla\mu) \tag{27.41}$$

$$\rho c_s^2 \vec{u} = \sum_i \vec{c}_i \bar{g}_i - \frac{\Delta t}{2} C \nabla \mu \qquad (27.42)$$

$$p_1 = \sum_i \bar{g}_i + \frac{\Delta t}{2} \vec{u} \cdot \nabla \rho c_s^2 \qquad (27.43)$$

Equation (27.41–43) is clearly nonlinear. However, due to the slow variation of the chemical potential on the timescale of a single timestep, C at time t can be updated with the value of μ at the previous timestep. The density and the relaxation time are calculated as local functions of the composition:

$$\rho(C) = C\rho_1 + (1 - C)\rho_2, \quad \tau(C) = C\tau_1 + (1 - C)\tau_2 \qquad (27.44)$$

where subscripts 1 and 2 refer to $t - \Delta t$ and t, respectively.

As is apparent from equations (27.41)–(27.44), Lee's model is significantly more laborious and computationally demanding than its forerunners. In addition, the original version was affected by lack of mass conservation due to the different numerical treatment of the $\vec{c}_i \cdot \nabla C$ and $\vec{u} \cdot \nabla C$ terms.

However, it leads to a significant reduction of the spurious currents, thus permitting to access a broader range of density ratios. Whether such bonuses overweigh the computational overhead must be decided case by case. Be as it may, the FEFD method has proven capable of dealing with a broader range of parameters, i.e., Reynolds, Weber and Atwood numbers, than its predecessors, which has earned it a visible role, especially for macroscale engineering applications, as opposed to micro- and mesoscale applications of both pseudo-potential and free-energy methods.

27.9 Attempts toward Thermodynamic Consistency

The multiphase models presented so far include temperature only as a static parameter, no self-consistent dynamics is allowed, on pain of incurring thermodynamic inconsistencies. As usual, this difficulty is not merely technical.

The point is that the self-consistent inclusion of energy conservation necessarily involves potential energy, hence the two-body radial distribution function, $\rho_{12} = \int f_{12} \, d\vec{v}_1 \, d\vec{v}_2$:

$$E = E_{\text{kin}} + E_{\text{pot}} = \int f(\vec{x}, \vec{v}) \frac{m}{2} v^2 \, d\vec{v} + \int \rho_{12}(\vec{x}, \vec{x} + \vec{r}) V(\vec{r}) \, d\vec{r} \qquad (27.45)$$

This is a daunting task since the two-body distribution function carries a lot of information, thirteen independent variables in three dimensions! Failing a full solution of the two-body Liouville equation, approximate recipes must be devised, as we are going to briefly illustrate in the sequel.

27.9.1 Lattice Enskog Equation

In an early paper, L.S. Luo suggested that thermodynamic consistency could be achieved by going back to the Boltzmann equation for dense (non-dilute) gases, the time-honored Enskog equation, which we have discussed in the Part I of this book.

The author argues that there is no consistent way of accounting for non-local interactions, without including finite size (hence finite density) effects as well. The proposal is then to solve the Boltzmann–Enskog equation, exactly the same way one does with the Boltzmann equation for dilute gases, namely by projecting onto a Hermite basis and subsequently evaluate the kinetic moments by numerical quadrature (14). Although interesting, the idea does not seem to have been pursued any further.

27.9.2 Two-Body Liouville Equation

A more general analysis has been presented by Xe and Doolen starting from an elegant treatment of the two-body Liouville equation in the continuum (15).

In this paper, the contributions of short-range repulsion and long-range attraction are analyzed separately. This analysis makes a nice connection with the two-body Liouville theory, but again, no numerical validation is provided in the original paper. A straighforward LB implementation of the resulting equations is often reported to suffer numerical instabilities. The situation improves by resorting to more sophisticated implementations making use of two separate distribution functions and local implicit timestepping. However, a stable numerical implementation of the He–Doolen approach remains an open issue.

27.9.3 Merging LB with Dynamic Density Functional Theory

Discrete kinetic equations for dense fluids have been developed by Melchionna and Marini–Bettolo, via an elegant merge between density functional theory and LB numerical implementation techniques (16).

These authors do not explicitly refer to the Enskog equation, but formulate directly a non-local BBGKY equation for the two-body distribution. The effective one-body force in their equation reads as follows:

$$\vec{F}(\vec{r}, \vec{v}) = - \int \nabla_r V(\vec{r} - \vec{r}') f_{12}(\vec{r}, \vec{r}'; \vec{v}, \vec{v}') d\vec{r}' d\vec{v}' \tag{27.46}$$

where $V(\vec{r} - \vec{r}')$ is the pair potential and f_{12} the two-body distribution.

Following a consolidated practice in the molecular theory of fluids, the pair potential is then split into a short-range repulsive and long-range attractive components. The short-range component is shown to lead to a revised Enskog-like term.[69] Long-range

---

[69] We remind that the Revised Enskog Theory is an extension of the standard Enskog theory which ensures compliance with irreversible thermodynamics for the case of fluid mixtures. In actual practice, it is based on a non-homogeneous representation of the two-component radial distribution function $g_{ss'}(\vec{r}, \vec{r}')$.

interactions are handled within the Random Phase Approximation and lead to soft-core forces similar to Shan–Chen, plus additional drag and dissipative terms similar to the ones appearing in the Langevin formulation of dense fluids.

Thus, instead of *replacing* non-ideal soft-core forces, like those appearing in the free-energy and pseudo-potential methods, with finite-size Enskog-interactions, the authors include both elements within a unified mathematical framework.

Having laid down the theoretical background within the framework of dynamic density functional theory, the authors proceed to discretize the equations according to standard LB practices. In particular, velocity integrals and short-range collisions are performed using typical LB stencils, while long-range interactions are handled with higher-order stencils extending over several neighbors.

Various applications of this model to nanofluidics and electrorheology of single-component and multi-component fluids are available in the literature.

27.10   Assorted Multiphase LB Approaches

In the last decade, several alternative approaches have appeared, which combine ideas from the basic multiphase approaches previously discussed with assorted techniques from other numerical methods for multiphase flows.

27.10.1   Projection Methods

Inamuro and collaborators developed a method based on two distribution functions, one for the order parameter $\phi = \rho_l/\rho$ and one for the velocity field (17).

The two LBGK's feature extended equilibria which resemble the ones of the free-energy approach, in the sense that they contain matching parameters and spatial derivatives of the macroscopic fields. However, spatial derivatives are computed with high-order stencils involving several layers of lattice sites. In addition, the exact incompressibility condition, *div* $\vec{u} = 0$, is enforced by solving the corresponding Poisson equation for the pressure, whence the name of projection methods, since the fluid velocity is projected onto the space of solenoidal fields. The Poisson equation is then solved by an iterative procedure based again on a LB scheme.

The Inamuro method is computationally demanding but leads to impressive results, such as rising bubbles in three-dimensional flows with density ratios very close to $1 : 1000$ (see Fig. 27.13).

27.10.2   Alternative Equations of State

It should be noted that the Shan–Chen equation of state supports a larger sound speed in the vapor than in the liquid phase, which is clearly unphysical and leads to numerical instabilities in the presence of large density gradients at the interface. It is therefore of prime interest to explore different equations of state, free from such liability.

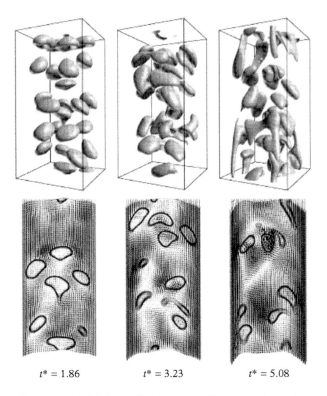

$t^* = 1.86$ $t^* = 3.23$ $t^* = 5.08$

Figure 27.13 *Rising bubbles in a three-dimensional two-phase fluid with the projection method. Reprinted from T. Inamuro et al, J. of Comp. Phys., 128, 628 (2004) with permission from Elsevier.*

Yuan *et al.* inspected a broad array of EoSs, mostly in the form of cubic and non-cubic variants of the Van der Waals EoS (18).

Given the excess pressure corresponding to the desired EoS, they define the corresponding generalized density as follows:

$$\psi[\rho] = \sqrt{\frac{2p^*(\rho)}{Gc_s^2}} \tag{27.47}$$

From there on, the standard Shan–Chen algorithm can formally be applied.

Typical EoS considered in this work are deformations of the cubic Van der Waals EoS in the form

$$p = \frac{\rho c_s^2}{1 - b\rho} - a\frac{\alpha(T, w)\rho^2}{1 + c\rho - d\rho^2} \tag{27.48}$$

where $\alpha(T, w) = (1 + q(w))(1 - (T/T_c)^{1/2})$, q being a quadratic polynomial in the adjustable parameter w, known as acentric factor, and $T_c = \frac{8a}{27b}$ being the VdW critical temperature. The critical density and pressure are $\rho_c = 1/3b$ and $p_c = 8a/27b^2$, respectively. Clearly, the VdW case is recovered by setting $\alpha = 1$ and $c = d = 0$. Note that this class of EoS only alters the attractive term of the VdW equation. The authors consider further variants in which the repulsive branch is also modified.

A particularly interesting variant is the Carnahan–Starling (CS) EoS:

$$p_{CS} = \rho c_s^2 \left[\frac{1 + r + r^2 - r^3}{(1 - r)^3} \right] - a\rho^2 \tag{27.49}$$

where $r = b\rho/4$. Unlike previous ones, the CS EoS modifies the hard-core repulsion term, leaving the attraction term unchanged.

The authors examine the coexistence curves and associated spurious currents and provide a wealth of insightful information concluding that the CS EoS with $a = 0.4963\,T_c^2/p_c$ and $b = 0.187\,T_c/p_c$ offers distinct advantages in terms of reduced spurious currents, hence broader range of density ratios.

In particular, at $T/T_c = 0.53$, the CS EoS reaches up to a density ratio $\rho_l/\rho_g \sim 1360$, with maximum spurious currents of about 0.09, to be contrasted with a Shan–Chen density ratio $\rho_l/\rho_g \sim 58$ at $T/T_c \sim 0.59$ and comparable levels of spurious currents.

Even though the authors do not mention it explicitly, the likely reason for such favorable behavior is a smaller density gradient in the gas phase at a given value of the density ratio, due to a smaller T_c, hence a smaller sound speed at a given T/T_c.

Because of this, the CS equation of state has gained significant popularity in the last decade.

27.10.3 Adaptive Equations of State

Based on the insight that large density gradients in the light phase are the main source of instability, Colosqui and coworkers proposed the use of an *adaptive equation of state*, which would dynamically adjust only the unstable branch of the EoS so as to place the enhanced density gradient mostly on the dense phase (19).

This requires a dynamic self-adaptive spinodal branch[70] which is computed on-the-fly during the evolution.

Since the spinodal branch is unstable, hence experimentally inaccessible, this strategy does not affect the physically measurable component of the EoS.

The simplest customized EoS is a piecewise linear relation of the form:

$$p(\rho) = \begin{cases} C_V \rho, & 0 < \rho < \rho_1 \\ C_1 \rho, & \rho_1 < \rho < \rho_M \\ C_2 \rho, & \rho_M < \rho < \rho_2 \\ C_L \rho, & \rho_2 < \rho \end{cases} \tag{27.50}$$

---

[70] The spinodal points are defined by the condition $\frac{\partial P}{\partial \rho} = 0$.

with $C_L > C_V$, in order to enforce a higher sound speed in the liquid phase than in the vapor one.

In (27.50), V and L are the vapor and liquid coexistence points, 1 and 2 are the spinodals, and M denotes an intermediate and adjustable density-pressure point within the spinodal branch.

Such point controls the perturbation to the piecewise linear EoS connecting 1 and 2 through a straight line.

As a result, the density and pressure of the mid-point $M = (\rho_M, p_M)$, or, equivalently, the spinodal slopes

$$C_1 = \frac{p_M - p_1}{\rho_M - \rho_1} > 0$$

and

$$C_2 = \frac{p_2 - p_M}{\rho_2 - \rho_M} < 0$$

provide the adjustable parameters of the procedure.

The optimization procedure consists of finding optimal values of (C_1, C_2) obeying the equilibrium condition on the chemical potential, namely

$$\int_{x_L}^{x_V} F_x(x)\,dx = k_B T \, log\frac{\rho_L}{\rho_V} \tag{27.51}$$

where x_L and x_V denote the end-points across the interface toward the liquid and vapor bulk values, respectively, and T is the fluid temperature.

It can be shown that perturbations with $C_1 > 0$ and $C_2 < 0$ place a larger share of the pressure gradient on the liquid side, and vice versa. The former option helps stability, since numerical instabilities are typically triggered by oscillations leading to negatives values of the vapor phase.

Numerical simulations of bubble formation in a periodic box showed stable operation under density ratios close to $1 : 1000$.

27.10.4   Merging LB with Flux-Limiting Techniques

The instabilities at the low-density end of the interface are essentially part of a more general and notorious family of numerical instabilities, known as "Gibbs phenomena,"

These are spurious oscillations which result from the lack of high-frequency, short wavelength modes not resolved by the actual grid.

This is a long-standing issue in computational fluid dynamics, especially in connection with sharp front propagation, for which a whole class of methods, known as Flux-Corrected-Transport (FCT) has been developed and constantly refined over the last four decades.

Basically, the FCT idea is to set up a feedback mechanism on the mass flux, so as to prevent it from exceeding the threshold leading to the onset of Gibbs phenomena. This is a feedback-control procedure based on the usual assumption that missing wavelengths would act on the resolved ones via a sort of diffusion process leading to a smoothing of the interface.

In the LB scenario, the use of CFT ideas is still relatively scanty.

Sofonea, Gonnella, Lamura and collaborators have embedded flux-limiting ideas in the form of additional force terms, specifically designed in such a way as to reduce spurious currents. These authors report successful implementations for the case of single interfaces (20). Similar techniques have been extended by A. Xu and coworkers to the case of compressible flows (21).

Similar to all the methods described in this last section, success is encouraging but still in need of consolidation in terms of assessing the robustness of these methods across a wide range of conditions. Whether such miscellaneous methods will make into a mainstream, only time will tell. Given the level of sophistication of CFT-like methods, it is this author's opinion that the merge of LB with such techniques may well result in a significantly improved new class of numerical schemes for multiphase flows.

27.10.5   Entropic Method for Multiphase Flows

Very recently, the entropic method has been extended to multiphase flows (22). Although this finding is too recent to be sensibly commented here, it may offer a potential complement/alternative to the hybrid methods mentioned previously. In fact, based on the citation rate of (20), it seems like the entropic method holds promise to become a major player in the field in the coming years.

27.11   Miscellaneous Multiphase LBE Applications

As anticipated, the application of LB methods to non-ideal fluids has met with burgeoning growth in the last decade. In light of this growth, the next examples look decidedly under-representative and a bit naive as well.

Nonetheless, we leave them nearly untouched, because, i) their pedagogical value stands intact, ii) the body of recent applications is way too vast to be sensibly represented by anything short of a dedicated monograph.

For a recent but necessarily incomplete review, see (23) and, above all, the recent monograph by Huang, Sukop and Lu (3).

Among many others, successful multiphase LBE applications include:

- Spinodal decomposition of binary and ternary mixtures (24; 26)
- Liquid–gas two domain growth (27)
- Rupture and coalescence of fluid interfaces

Figure 27.14 *Phase separation in a binary LBE fluid. (Courtesy of J. Yeomans.)*

Typical phase separation patterns and the growth of their size in time are shown in Figs 27.14 and 27.15.

The reported advantages of the LB approach for this type of problem are:

- *Access to mesoscales virtually off-limits for molecular dynamics*
- *Handy fine-tuning of the free energy to pick up the desired equilibrium as an attractor of the relaxation dynamics*

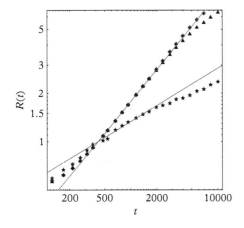

Figure 27.15 *Mean domain size of the two separated phases as a function of time. (Courtesy of J. Yeomans.)*

As mentioned before, Lee's approach is credited to achieve a significant reduction of spurious currents even at density ratios as high as $\sim 1 : 1000$ (29).

As an example, Figure 27.16 shows the spike and bubble penetrations in a Rayleigh–Taylor instability, as computed by Chiappini *et al.* using the FEFD method (30). The initial exponential growth, the rising bubble of the light fluid and the spikes of denser fluid moving in the opposite direction, as well as the superficial wave breaking at a later stage of the simulation are all well visible. The flow field is qualitatively consistent with the typical Rayleigh-Taylor instability dynamics, experimentally and numerically observed by various authors.

Many more applications are nowadays available in the LB literature, an ocean that cannot be sailed here. With proper thanks to Brin and Page, Google comes to a good rescue, and much better so does the recent monograph by H. Huang et al. (3).

27.12 Summary

As of today, many options are available for the LB simulation of non-ideal fluids and multiphase flows. Among them, the Shan–Chen pseudo-potential method has unquestionably gained the most popular position, mostly on account on its conceptual transparence and computational simplicity. First-generation free-energy models are also in extensive use, mostly for microfluidics and soft-matter research. Both methods are rather limited in parameter space, but serve well the purpose of gaining semi-quantitative insight into highly complex flows at a comparatively minor effort, both in simulation and programming time. Subsequent developments have managed to extend the parameter regimes, although sometimes at the expense of simplicity. Notwithstanding a number of limitations, LB schemes for multiphase flows have gained major popularity in the last decade. In some precious cases, they have been used to predict new phenomena (31) and even new types of soft materials (32). In general, they represent one of the richest offsprings of the original LB theory, see also (3).

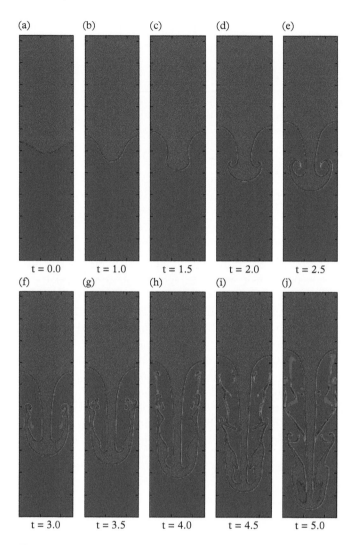

Figure 27.16 *Richtmeyer-Meskov instability as computed with the FEFD method. The main parameters are Re = 2048, We = 12288, At = 0.5 on a 256 × 1024 grid. Reprinted from D. Chiappini et al, Comm. in Comp. Phys., 7, 423, (2010) with permission from Cambridge University Press.*

....

REFERENCES

1. A. Prosperetti and G. Tryggvason, *Computational Methods for Multiphase Flow*, Cambridge University Press, Cambridge, 2005.
2. R. Scardovelli and S. Zaleski, "Direct numerical simulation of free surface and interface flow", *Ann. Rev. of Fluid Mech.*, **31**, 567, 1999.
3. H. Huang , M. Sukop and X. Lu, *Multiphase Lattice Boltzmann Methods: Theory and Application, 1st Edition,* Wiley and Blackwell, NY, 2015.
4. A. Gunstensen and D. Rothman, "Lattice Boltzmann studies of immiscible two phase flows through porous media", *Phys. Rev. A*, **43**, 4320, 1991.
5. D. Rothman and J. Keller, "Immiscible cellular automaton fluids", *J. Stat. Phys.*, **52**, 1119, 1988.
6. S. Leclaire, *et al.*, "Generalized three-dimensional lattice Boltzmann color-gradient method for immiscible two-phase pore-scale imbibition and drainage in porous media." *Physical Review E*, **95.3**, 033306, 2017.
7. X. Shan and H. Chen, "Lattice Boltzmann model for simulating flows with multiple phases and components", *Phys. Rev. E* 47, 1815, 1993.
8. X. Shan, "Pressure tensor calculation in a class of nonideal gas lattice Boltzmann models Lattice Boltzmann Simulation of non-ideal Fluids", *Phys. Rev. E*, 77, 066702, 2005.
9. D. Bini, A. Geralico, D. Gregoris and S. Succi, "Dark energy from cosmological fluids obeying a Shan-Chen nonideal equation of state", *Phys. Rev. D*, **88**, 063007, 2013.
10. L. Chen, Q. Kang, Y. Mu, Y.L. He and W.Q. Tao, "A critical review of the pseudo-potential multiphase lattice Boltzmann model: methods and applications", *Int. J. of Heat and Mass Transfer*, **76**, 210, 2014.
11. M. Swift *et al.*, "Lattice Boltzmann Simulation of Non-ideal Fluids", *Phys. Rev. Lett.*, 75, 830, 1995.
12. T. Lee and C.L. Lin, "A stable discretization of the lattice Boltzmann equation for simulation of incompressible two-phase flows at high density ratio", *J. Comp. Phys.*, **206**, 16, 2005.
13. T. Lee, "Effects of incompressibility on the elimination of parasitic currents in the lattice Boltzmann equation for parasitic fluids", *Computers and Mathematics with Applications*, **58**, 987, 2009.
14. L. S. Luo, "Unified theory of lattice Boltzmann models for non-ideal gases", *Phys. Rev. Lett.* **81**, 1618, 1998.
15. X. He and G. Doolen, "Thermodynamic Foundations of Kinetic Theory and Lattice Boltzmann Models for Multiphase Flows", *J. Stat. Phys.*, **107**, 309, 2002.
16. S. Melchionna and U. Marini Bettolo, "Kinetic theory of correlated fluids: From dynamic density functional to Lattice Boltzmann methods", *J. Chem. Phys.*, **131**, 014105, 2009.

17. T. Inamuro *et al.*, "A lattice Boltzmann method for incompressible two-phase flows with large density differences", *J. Comp. Phys.*, **128**, 698, 2004.

18. P. Yuan and L. Schafer, "Equations of state in a Lattice Boltzmann model", *Phys. Fluids*, **18**, 042101, 2006.

19. C. Colosqui *et al.*, "Mesoscopic simulation of non-ideal fluids with self-tuning of the equation of state, *Soft Matter*, **8**, 3798, 2012.

20. V. Sofonea, A. Lamura, G. Gonnella and A. Cristea, "Finite-difference lattice Boltzmann model with flux limiters for liquid-vapor systems", *Phys. Rev. E*, **70**, 046702, 2004.

21. G. Yanbiao, A. Xu *et al.*, "Discrete Boltzmann modeling of multiphase flows: hydrodynamic and thermodynamic non-equilibrium effects", *Soft Matter*, **11**, 5336, 2015.

22. I. Karlin *et al.*, A. Mazloomi, S. Chikatamarla and I. Karlin, "Entropic Lattice Boltzmann Method for Multiphase Flows", *Phys. Rev. Lett.*, **114**, 174502, 2015.

23. G. Falcucci *et al.*, "Lattice Boltzmann methods for multiphase flow simulations across scales", *Comm. in Comp. Phys.*, **9**, 269, 2011.

24. W. Osborne, E. Orlandini, J. Yeomans and J. Banavar, "Lattice Boltzmann study of hydrodynamic spinodal decomposition", *Phys. Rev. Lett.*, **75**, 4031, 1995.

25. A. Lamura, G. Gonnella and J. Yeomans, "A lattice Boltzmann model of ternary fluid mixtures", *Europhys. Lett.*, **45**, 314, 1999.

26. F. Alexander, S. Chen and D. Grunau, "Hydrodynamic spinodal decomposition: growth kinetics and scaling", *Phys. Rev. B*, **48**, 634, 1993.

27. D. Grunau, S. Chen and K. Eggert, "A lattice Boltzmann model for multiphase fluid flows", *Phys. Fluids A* **5**, 2557, 1993.

28. D. Grunau, T. Lookman, S. Chen and A. Lapedes, "Domain growth, wetting and scaling in porous media", *Phys. Rev. Lett.*, **71**, 25, 4198, 1993.

29. X. He, S. Chen and R. Zhang, "A lattice Boltzmann scheme for incompressible multiphase flow and its application in simulation of Rayleigh-Taylor instability", *J. Comput. Phys.*, **152**, 642, 1999.

30. D. Chiappini, G. Bella, S. Succi *et al.*, "Improved Lattice Boltzmann Without Parasitic Currents for Rayleigh-Taylor Instability", *Comm in Comp. Phys.*, 7, 423 2010.

31. Liu, Yahua; Andrew, Matthew; Li, Jing; *et al.*, "Symmetry breaking in drop bouncing on curved surfaces", *Nature Comm.*, 6, 10034 2015.

32. K. Stratford, R. Adhikari, I. Pagonabarraga, JC Desplat, ME Cates, "Colloidal jamming at interfaces: a route to fluid-bicontinuous gels", *Science*, 309, 2198–2201 2005.

....

EXERCISES

1. Compute the critical pressure of the Shan–Chen equation of state and compare with the spinodal values.
2. Plot the Shan–Chen equation of state for several values of G. At which value G does pressure go negative?
3. Compute the Weber number of a raindrop falling at 1 m/s.

28

Extensions of the Pseudo-Potential Method

In this chapter we provide an account of subsequent extensions of the Shan–Chen pseudo-potential method, including more elaborated potentials which extend beyond the first Brillouin cell. These extensions permit to lift a number of limitations of the original model and considerably expand its scope and range of applications.

Things are fluid in this world, and if you don't remain fluid, you get lost in the sauce

(Talib Kweli)

28.1   Introduction

In Chapter 27, we have discussed a variety of LB techniques for non-ideal fluids. As usual, each method comes with its ups and downs, but actual evidence shows that the Shan–Chen (SC) model has enjoyed increasing popularity over the years. Interestingly, such popularity stands in the face of a fair amount of substantial criticism. In this chapter, we first revisit the Shan–Chen model in some more detail and discuss ways out of the criticism. Subsequently, we shall discuss the extension of the SC technique to the case of multi-range potentials extending beyond the first Brillouin cell. As we shall see, this extension proves effective in softening many of the weaknesses of the original formulation, thereby considerably expanding its scope and range of applications.

28.2   Shan–Chen Revisited

Let us begin by considering again the expression of the Shan–Chen pseudo-potential force:

$$F_a(\vec{x}) = G\Psi(\vec{x}) \sum_i w_i r_{ia} \Psi(\vec{x}_i), \quad a = x, y, z \tag{28.1}$$

The Lattice Boltzmann Equation. Sauro Succi, Oxford University Press (2018).
© Sauro Succi. DOI: 10.1093/oso/9780199592357.001.0001

where the notation is as follows:

$$\Psi(\vec{x}) \equiv \Psi[\rho(\vec{x})], \ \Psi(\vec{x}_i) \equiv \Psi[\rho(\vec{x} + \vec{r}_i)], \ \vec{r}_i = \vec{c}_i \Delta t$$

Taylor expansion of the expression of eq. (28.1) delivers

$$F_a(\vec{x}) = \Psi \sum_i G_i r_{ia} \left\{ 1 + r_{ib}\nabla_b + \frac{r_{ib}r_{ic}}{2}\nabla_b\nabla_c + \frac{r_{ib}r_{ic}r_{id}}{3!}\nabla_b\nabla_c\nabla_d \right.$$
$$\left. + \frac{r_{ib}r_{ic}r_{id}r_{ie}}{4!}\nabla_b\nabla_c\nabla_d\nabla_e + \ldots \right) \right\} \Psi \tag{28.2}$$

where we have set $\Psi \equiv \Psi(\vec{x})$ and $G_i \equiv G w_i$ for notational simplicity.

As usual, odd-order terms give zero contribution, so that, to fourth order, we are left with

$$F_a(\vec{x}) = G \left(\frac{< r_a r_b >}{2}\nabla_b\Psi^2 + \Psi \frac{< r_a r_b r_c r_d >}{4!}\nabla_b\nabla_c\nabla_d\Psi \right) \tag{28.3}$$

where $< r_a r_b > \equiv \sum_i w_i r_{ia} r_{ib}$ and similar for the fourth-order term. The expansion exhibits an elegant diagrammatic representation, as reported in Figure 28.1.

On the assumption that both second- and fourth-order bracket tensors are isotropic, which is indeed the case for hydrodynamic LB lattices, the first term is readily recognized to provide the gradient of the non-ideal bulk pressure, sometimes known as excess pressure, namely

$$p^\star(\rho) = \frac{Gc_s^2}{2}\Psi^2(\rho) \tag{28.4}$$

in lattice units $\Delta t = 1$.

After some algebraic manipulations, the fourth-order term gives rise to divergence of the following momentum-flux tensor:

$$p_{int}\delta_{ab} + \hat{P}_{ab}(1 - \delta_{ab}) \tag{28.5}$$

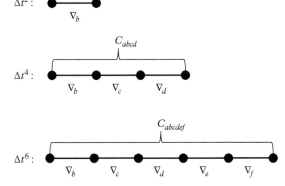

Figure 28.1 *Diagrammatic representation of the Taylor expansion of theShan-Chen force, up to sixth-order terms. The connectivity tensors are defined as follows:* $C_{ab} = \sum_i w_i c_{ia} c_{ib}$, $C_{abcd} = \sum_i w_i c_{ia} c_{ib} c_{ic} c_{id}$, $C_{abcdef} = \sum_i w_i c_{ia} c_{ib} c_{ic} c_{id} c_{ie} c_{if}$.

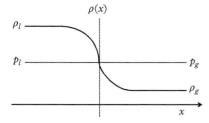

Figure 28.2 *The structure of the interface. The density undergoes a steep drop from the liquid to the vapor(gas) phase across the interface. The P_{xx} component remains constant across the interface and takes the same value in the liquid and gas phase. The interface width scales inversely with the surface tension.*

where

$$p_{int} = \frac{Gc_s^4}{4}\left[\Psi\Delta\Psi - \frac{1}{2}(\nabla\Psi)^2\right] \tag{28.6}$$

and

$$\hat{P}_{ab} = (1 - \delta_{ab})\frac{Gc_s^4}{4}\nabla_a\Psi\nabla_b\Psi \tag{28.7}$$

Both terms conspire to determine the spatial structure of the interface and vanish in the bulk (see Fig. 28.2).

In particular, the off-diagonal tensor $\hat{P}_{ab}$ fixes the surface tension through the relation

$$\sigma = \int_{-\infty}^{+\infty}(p_t - p_n)dn \tag{28.8}$$

where subscripts t and n denote the pressure component tangential and normal to the surface, respectively.

As a result, we see that the second-order term in the Taylor expansion controls the non-ideal equation of state, while surface tension and the structure of the interface depend on fourth-order terms.

All in all, the overall momentum-flux tensor takes the final form

$$P_{ab} = (p_{bulk} + p_{int})\delta_{ab} + \hat{P}_{ab} \tag{28.9}$$

where

$$p_{bulk} = \rho c_s^2 + \frac{Gc_s^2}{2}\psi^2(\rho) \tag{28.10}$$

and p_{int} and $\hat{P}_{ab}$ are given by the expressions (28.6) and (28.7).

How about higher-order terms?

These do not map into any well-recognizable macroscopic property, such as surface tension, but affect capillary forces which arise when two interfaces come to near-contact (sometimes also known as dispersion forces). As we shall see, these terms are exposed to the infamous *spurious currents* discussed in Chapter 27; hence they deserve special attention.

Before venturing into the discussion of spurious currents, let us first dig a bit deeper into the Shan–Chen equation of state and surface tension.

28.2.1 Equation of State

As mentioned in Chapter 27, the excess pressure gives rise to a critical density and pressure, defined by the usual conditions:

$$\frac{\partial p}{\partial \rho}\big|_c = 0, \quad \frac{\partial^2 p}{\partial \rho}^2 \big|_c = 0 \tag{28.11}$$

With the choice $\Psi(\rho) = 1 - e^{-\rho}$, a simple calculation delivers the critical values of density and pressure, namely

$$\Psi_c = 1/2, \quad \rho_c = log2, \quad G_c = -4, \quad p_c = (log2 - 1/2)/3 \sim 0.063 \tag{28.12}$$

where we have set $\rho_0 = 1$ for simplicity.

In the low-density region, $\rho \to 0$, $\Psi \to \rho$ the SC EoS reduces to

$$p = \rho c_s^2 + \frac{Gc_s^2}{2}\rho^2 \tag{28.13}$$

This is the low-density, soft-core dominated region of the VdW equation of state, $p - a/V^2 = nkT$.

For attractive potentials, with $G < 0$, such quadratic EoS is readily seen to go unstable at $\rho > \rho_G \equiv 1/|G|$, as the sound speed becomes imaginary beyond such density ($\partial_\rho p < 0$). This instability, a sort of "gravitational collapse" reflects the untamed density pile up due to purely attractive interactions.

In the VdW equation this instability is tamed by short-range repulsion, which becomes dominant above the hard-core density $\rho > 1/b$. As discussed in Chapter 27, the SC model does not cater for any repulsive interactions. The reason is practical: hard-core interactions are deliberately written off the Shan–Chen picture, because they entail very strong forces which would impose impractically small timesteps.

Instead, the instability is tamed by the specific functional form of the generalized density $\Psi(\rho)$ and notably by the fact that such generalized density saturates to a constant value in the high-density limit $\rho \gg \rho_0$. A constant Ψ implies zero force, which also means that once the density goes significantly above ρ_0, the SC force fades away, thereby preventing the system from going unstable.

Borrowing from quantum chromodynamics, this can pictorially be associated with a form of *asymptotic freedom*, meaning by this that Shan–Chen "molecules" stop interacting as they come sufficiently close, roughly speaking below a distance $r_0 \sim \rho_0^{-1/3}$.

In this high-density limit, the SC EoS reduces to

$$p = \rho c_s^2 + \frac{Gc_s^2}{2}\rho_0^2 \tag{28.14}$$

This shows that at high density, the SC EoS regains a linear trend, with the *same* slope (square sound speed) as the ideal gas branch, plus a constant offset.[71]

This is a far cry from the high-density behavior of the VdW EoS, which exhibits a diverging pressure as the hard-core density is approached. This discrepancy is of no concern, as long as the SC model is operated well below such hard-core density, as it is indeed the case for most hydrodynamic purposes.

Indeed, hard-core repulsion is responsible for the small-scale structure of liquids, while thermodynamic properties are mostly dictated by the soft-core attractive tails.

Nevertheless, a serious liability must be pointed out: according to the SC EoS, sound propagates faster in the vapor than in the liquid phase, which is clearly unphysical. A direct calculation yields (with $\rho_0 = 1$):

$$c_s^2(\rho) \equiv \frac{\partial p}{\partial \rho} = c_s^2 (1 + G\psi (1 - \psi))$$

Since $G < 0$ and $\psi (1 - \psi) > 0$, $c_s^2(\rho) < c_s^2$ for any finite $\rho > 0$, q.e.d.

As discussed in Chapter 27, this is a serious threaten to numerical instability, especially at high-density ratios.

28.2.2   Interface Pressure and Surface Tension

The interface pressure, p_{int}, is the inhomogeneous contribution which arises in the vicinity of the interface in order keep the normal component of the fluid pressure tensor constant across a planar interface, thereby securing its mechanical stability.

For a two-dimensional flat interface developing along the x direction and uniform along y, the condition for mechanical stability reads simply as follows:

$$\partial_x P_{xx} = 0 \tag{28.15}$$

with boundary conditions at infinity (bulk phases):

$$P_{xx}(x \to -\infty) = P_{xx}(x \to -\infty) = p_{bulk} \tag{28.16}$$

where

$$p_{bulk} = \rho_l c_s^2 + G\Psi_l^2/2 = \rho_g c_s^2 + G\Psi_g^2/2 \tag{28.17}$$

is the common value of the gas (vapor) and liquid bulk pressure.

The presence of the interface pressure configures (28.15) as a nonlinear partial differential equation, whose solution delivers the actual profile of $\Psi(x)$ across the interface, typically a hyperbolic tangent connecting the two bulk phases through a thin interface of width inversely proportional to the surface tension.

---

[71] This is reminiscent of the so-called "bag model" equation of state of hadronic matter, quarks and gluons, where the constant term associates with the vacuum fluctuations.

28.2.3 Surface Tension

As discussed in the previous section, the surface tension is measured by the mismatch between the normal and tangential components of the fluid pressor across the interface. Some algebra delivers the following expression:

$$\sigma = -G\frac{c_s^4}{4} \int_{-\infty}^{+\infty} (\partial_x \Psi)^2 \, dx \tag{28.18}$$

An order of magnitude estimate gives

$$\sigma \sim \frac{Gc_s^4}{4} \frac{(\delta\Psi)^2}{w}$$

where $\delta\Psi$ is the jump at the interface and w the interface width.

Typical values for Shan–Chen simulations are:

$G \sim -5$, $c_s^4 = 1/9$, $\delta\Psi \sim \Psi_l \sim 1$, $w \sim 10$, yielding $\sigma \sim 10^{-2}$ in lattice units.[72] Since the only free-parameter is the coupling strength G, whose range of variation is limited, say $-6 < G < -4$, on account of numerical stability, the range of variation of the surface tension is correspondingly rather narrow.

This the consequence of a very basic limitation of the SC model: with a single parameter, the equation of state and the surface tension cannot be changed independently, which is again patently unphysical.

28.3 Shan–Chen Limitations and How to Soften Them

Having revisited the SC model in some more detail, let us summarize the main criticisms and then move on to show how they have been mitigated over the past decade, leading to a new generation of pseudo-potential models which is proving very useful for the simulation of flows of great complexity, such as foams and emulsions (1).

Here come the main "baddies":

1. *The model is thermodynamically inconsistent, as it does not derive from a thermodynamic free-energy*

2. *The equation of state and surface tension cannot be changed independently*

---

[72] It is of interest to estimate what this value means in physical units. Let $\sigma = \sigma_{LB}\frac{k_B T}{\Delta x^2}$ be the surface tension in physical units, with σ_{LB} the one in LB units. Given that $k_B T \sim 4 \times 10^{-21}$ Joule, taking $\Delta x = 10^{-9}$ m, the value $\sigma_{LB} = 1$ corresponds to $\sigma \sim 4 \ 10^{-3}$ N/m, which is an order of magnitude below the typical values of many fluids, particularly water ($\sigma = 0.07$) N/m. It thus appears like with standard Shan–Chen, sub-nanometric spacings are needed to recover realistic values of the surface tension in physical units. As is often the case with LB, however, the target are not the physical properties in physical units, but rather the dimensionless groups expressing the relative strength between concurrent mechanisms. We shall deal with this matter in more detail in chapter 29.

3. *The sound speed is higher in the vapor than in the liquid phase*
4. *The model is unstable beyond density ratios of the order of 30–50*
5. *The interface is too thick, as it covers of the order 5–10 lattice sites*

The first limitation is that mechanical stability and thermodynamic stability (Maxwell's area rule) are not compatible, unless the generalized density is taken in the form $\Psi[\rho] = \rho_0(1 - e^{-\rho/\rho_0})$. On the other hand, continuum free-energy models depend on the physical density ρ, hence compatibility with such models would command $\Psi \equiv \rho$, no trick. Strictly speaking, this defeats the very purpose of the method, which builds on the idea of replacing the physical density ρ with the generalized density $\Psi(\rho)$ in order to tame density pile-up instabilities. So, it looks serious.

The next two limitations stand from the fact that SC is a one-parameter model.

The fourth is mostly associated with spurious currents, i.e., currents which owe their existence to the lack of lattice symmetry beyond fourth order. In this respect they are not unique to LB but pertain basically to any discretization method.

The fifth limitation is very practical; like in any other diffuse-interface method, the interphase boundaries cover a sizeable number of grid points, between, say five and ten. For macroscopic flows, in which the interface is treated as a mathematical surface of zero thickness, this is a patent waste of resolution. On the other hand, one may argue that resolving the interface may bring genuine microscopic physics to the table. In principle, this would very intriguing, save for the fact are nanometer thick, hence would command overly expensive sub-nanometric LB simulations.

All of these appears quite an imposing array of criticism for a method which has gained so much popularity over the years! How come?

As with most success stories, this comes from a precious combination of ingenuity and generous smiles of Lady Luck.

Let us next go over to the list of remedies.

For convenience, we shall leave the first limitation at the end.

28.4   Surface Tension Independent of the Equation of State: Multirange Pseudo-Potentials

A most natural move to lift the second SC limitation is to endow the model with more than just one free parameter. Typically, this amounts to introduce some form of repulsion and set its range and strength independent of the attractive one. In the sequel, we discuss one such model which has proven pretty effective to this purpose, the *multi-range* pseudo-potential method (2).

The basic idea of the multirange method is to introduce a second force, acting beyond the first Brillouin region (belt or shell for simplicity). Let us consider the case in which

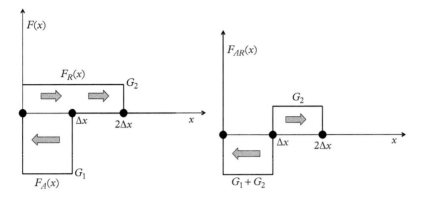

Figure 28.3 *Scheme of a short-range attractive, mid-range repulsive force $F_{AR}(x)$. The attractive force F_A of strength $G_1 < 0$ acts on the first shell only, $0 < x < \Delta x$, while the repulsive one F_R is active on both shells $0 < x < 2\Delta x$, with strength $G_2 > 0$. In this figure, $G_2 + G_1 < 0$, and the combination of the two results in the force $F_{AR}(x)$ shown in the right panel.*

such second force acts on both the first and second shells (see Fig. 28.3), with a separate coupling strength G_2.

As a result, the first shell experiences a cumulative coupling $G_1 + G_2$, while the second shell hosts the G_2 coupling alone. By suitably tuning the magnitude of the two parameters G_1 and G_2, the extended model can include both attractive and repulsive interactions. In explicit terms:

1. $G_1 + G_2 > 0$, $G_2 > 0$: RR, short-range repulsion, mid-range repulsion
2. $G_1 + G_2 > 0$, $G_2 < 0$: RA, short-range repulsion, mid-range attraction
3. $G_1 + G_2 < 0$, $G_2 > 0$: AR, short-range attraction, mid-range repulsion
4. $G_1 + G_2 < 0$, $G_2 < 0$: AA, short-range attraction, mid-range attraction

By the usual Taylor expansion, we obtain

$$p^\star \propto (G_1 + G_2)C_1^2 + G_2 C_2^2 \tag{28.19}$$

$$\sigma \propto -(G_1 + G_2)C_1^4 - G_2 C_2^4 \tag{28.20}$$

where we have used the shorthand $C_k^p \equiv \sum_i w_{ik} c_{ik}^p$, $k = 1, 2$ denoting the two shells and $p = 2, 4$.

These relations show that, in principle, the excess pressure and surface tension can now be changed independently by proper tuning of the parameters G_1 and G_2.

Here, we wish to call the reader's attention on the fact that by working in the AR region, i.e., attraction in the first shell and repulsion in the second one, surface tension can be taken down to very small values, $\sigma \ll 1$ in lattice units. This region, inaccessible to standard Shan–Chen, is key to model the rheology of very complex flows with high-surface/volume ratios, such as foams and emulsions.

Mathematically, we note that $G_2 > 0$, i.e., second-belt repulsion, contributes indeed a negative surface tension. As a result, the repulsive barrier lowers the surface tension of the fluid and the cost of maintaining interfaces within the flow.

In particular, the double-belt SC potential gives rise to long-lived mestastable states in the form of multi-droplet configurations, which prove capable of surviving coalescence much longer than in the plain SC scenario.

The reason is intuitively clear: when two droplets come in near-contact, in the SC scenario there is little story: the attractive potential promotes a rapid merger into a single and larger droplet. The merger is particularly effective whenever the droplets are well separated in size.

This is the way the system attains its minimum surface time-asymptotic condition: a single spherical drop.

When second-belt repulsion is switched on, the picture changes dramatically: the droplets must now overcome an energy barrier before they can merge, thereby advancing the coarsening process toward the minimum surface/volume equilibrium configuration.

If there is sufficient kinetic energy to overcome the repulsive barrier, the merge will indeed occur. Failing sufficient energy, however, the repulsive barrier cannot be overcome and small droplets can happily shun the fatal attraction of the larger ones, see Fig. 28.4, 28.5 and 28.6.

The result is the occurrence of metastable states characterized by long-lived multi-droplet configurations which delay the coarsening process leading to the single-droplet minimum S/V configuration.

This qualitative scenario sets the stage for the study of soft-glassy materials, i.e., soft matter systems exhibiting long-time relaxation and nonlinear rheology. A systematic study of the two-belt SC models in the (G_1, G_2) parameter plane can be found in Falcucci *et al.* (3).

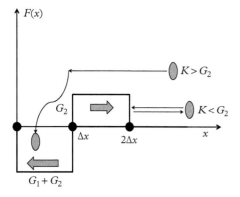

Figure 28.4 *Sketch of the multi-range survival scenario. Droplets with sufficient kinetic energy to overcome the repulsive barrier, $K > G_2$ are "swallowed" by the attractive potential and merge with the attracting droplet (coarsening). The less energetic ones, with $K < G_2$, are bounced back and escape the "fatal attraction" of the large eddies.*

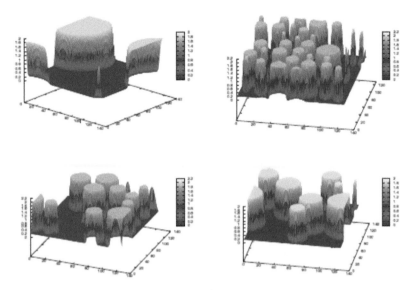

Figure 28.5 *A metastable multi-droplet density configuration computed with the AR two-belt Shan–Chen potential. From G. Falcucci et al., EPL 82, 24005 (2008).*

As usual, some limitations still remain; for instance, it has been noted (4) that the equation of state and surface tension cannot be tuned independently across the entire parameter space, because the values of G_1 and G_2 affect the density of the coexisting phases. More specifically, any non-zero G_2 affects the density ratio of the coexisting phases on top of the relations (28.19), leading to residual correlation between the density ratio and the surface tension. This impacts on the minimum surface tension attainable by the model, with ensuing limitations on a number of engineering applications, such as spray injection in diesel engines. Remedies against these limitations are proposed in (4), see Fig. 28.4.

Notwithstanding such limitations, the multirange option has proven fairly beneficial in several respects and contributed a significant boost to the range of applications of the pseudo-potential method.

28.5 Vapor Versus Liquid Sound Speed: Alternative Equations of State

The next limitation, faster sound in vapor than in liquid, can be disposed of by turning to two-parameters equations of state, such as those discussed in Chapter 27. As commented earlier on, a lower sound speed in the vapor translates into smaller interface forces on the vapor side, the one mostly exposed to numerical instabilities. It is reasonable to assume that the enhanced stability observed with Van-der-Waals like EoS,

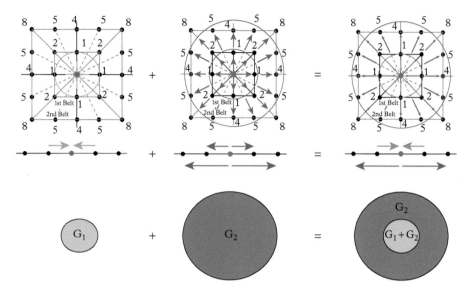

Figure 28.6 *Typical example of a double-belt lattice. The first of strength G_1 acts on the first belt, consisting of the usual D2Q9 neighbors. The second force, with strength G_2, acts on both belts, the D2Q9 one and the second belt consisting of 16 neighbors, for a total of 25. The integer labels stand for the energy magnitude c_i^2. The rightmost panel represents the net sum of the two forces, namely a standard Shan–Chen with strenghth $G_1 + G_2$, plus a second belt force of strength G_2. By playing with the sign and amplitude of G_1 and G_2 different regimes can be attained, with a separate control of the equation of state and surface tension. From (3).*

such as Carnahan–Starling, is due precisely to this feature. It is worth noting that the actual implementation of any equation of state can proceed along the same lines as for the Shan–Chen model. To this purpose, one defines the generalized density $\psi(\rho)$ associated with the chosen EoS by simply letting

$$\psi(\rho) = \sqrt{\frac{2p^*(\rho)}{c_s^2}}$$

where $p^*(\rho)$ is the excess pressure associated with the given EoS at hand. Note that we have set $G = 1$, since this parameter is now immaterial, the parametric dependence being incorporated within the expression of the excess pressure.

28.6   Large Density Ratios and Spurious Currents

Typical values of the Shan–Chen liquid-vapor coexistence densities are of the order of $\rho_l \sim 2$ and $\rho_g \sim 0.05$, corresponding to density ratios around 30–50.

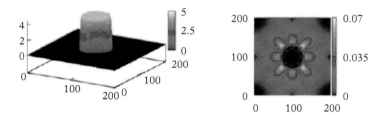

Figure 28.7 *A liquid droplet and the associated spurious currents. The eightfold symmetry of the underlying D2Q9 lattice is clearly visible. From (3)*

Attempts to extend such ratio necessarily command lower vapor densities, since in the SC model, the liquid density cannot grow much above $\rho_l \sim 2$ without entering the unreliable high-density branch of the SC equation of state.

This facilitates the appearance of spurious currents at the interface triggering density fluctuations in the vapor phase, which eventually lead to spurious currents, unphysical negative densities and numerical disruptions, see Figs 28.7.

Clearly, the details of the numerical implementation may affect significantly the instability scenario and many antidotes have been proposed in the literature.

These antidotes are fairly disparate in nature; some appeal to different numerical discretizations, some others propose more general potentials and/or equations of state. Since they are all fairly technical in character, here we simply mention the informing principle leaving the details to the original publications.

28.6.1 Alternative Discretizations

Some authors maintain that slightly different implementation of the Shan–Chen force may lead to significant stability enhancement.

For instance, Kupershtock proposes the following variant (5):

$$F_a = A\Psi \sum_i G_i \Psi_i r_{ia} + (1 - A) \sum_i G_i \frac{\Psi_i^2}{2} r_{ia} \qquad (28.21)$$

where we have set $G_i = G w_i$.

In (28.21), the first term on the left-hand side is the standard Shan–Chen scheme, whereas the second one is a non-local approximation of the gradient.

The authors discuss the benefits of choosing an optimal value of the interpolation coefficient, not necessarily in the natural range $0 < A < 1$, and report very large density ratios, of the order of 10^6 or more, for negative values of A. However, no information on the magnitude of spurious currents is provided. For a recent and informative comparison between the Shan–Chen and Kupershtock approaches, see (6).

28.6.2 Local Grid Refinement

Spurious currents are known to recede at increasing spatial resolution. Consequently, local grid refinement is a failsafe recipe to increase the stability of multiphase LB simulations. However, grid refinement adds significantly to the complexity of the implementation.

Moreover, for the case of non-ideal fluids, it faces with genuinely new problems. In particular, matching the equilibrium and non-equilibrium components of the distribution function at the coarse/fine interface with just a single coupling parameter is clearly impossible because such matching requires two distinct values of the surface tension in the coarse and fine grids.

The multi-range Shan–Chen model can in principle circumvent the problem.

However, as mentioned, the implementation is laborious and, to the best of the author's knowledge, to date it only works if the coarse/fine numerical interface does not overlap with the liquid/vapor physical interface. This means that the refined grid must dynamically adapt to the interface, a highly demanding task in the presence of complex interfaces.

28.6.3 Multi-Range Potentials, Again

The second shell of the multi-range SC model has been reported to bring along benefits in terms of spurious currents, mainly because of the higher (8th)-order isotropy of the lattice.

Details of higher-order lattices up to 16th-order isotropy can be found in (7). Since spurious currents are due to the lack of symmetry of the lattice, it is natural to expect that two-belt potentials should lead to some benefits in terms of density ratios as well.

Falcucci *et al.* presented a thorough study of the two-belt Shan–Chen force in the parameter plane (G_1, G_2) and showed that a significant broadening of the coexistence diagram, with tolerable levels of spurious currents, can be achieved by a proper combination of double-attractive potentials.

This spawns the conceptual possibility of using *adaptive potentials*, i.e., different values of G_1 and G_2 in the vicinity of the interface and away from it. In particular, one may think of adapting the EoS with a given surface tension, or vice versa. However, to the best of this author's knowledge no investigation along these lines has been performed to date.

In passing, we wish to mention that multirange SC models for binary mixtures have permitted to gain new insights into the non linear rheology of complex fluids and soft amorphous materials, such as foams and emulsions. In particular, it has been shown that a proper tuning of the attraction/repulsion parameters for the two species permits us to realize not only small surface tension, but also positive *disjoining pressure*, thereby allowing the simulation of flows with extremely long-lived and complex interfaces, see Fig. 28.8.

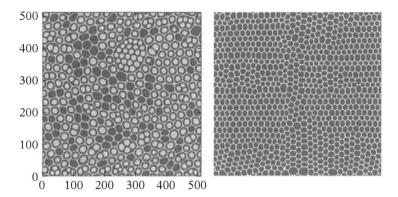

Figure 28.8 *A pictorial example of an emulsion-like (left) and a foamy-like (right) fluid configuration. The left panel represents the density of fluid A in a binary mixture A+B, yellow(blue) coding for high(low) density. The complex configuration of droplest A in fluid B and vice versa is a long-lived metastable configuration, which owes its existence to the competition of short-range attraction and long-range repulsion. The right-hand side represents a foam-like density configuration, with cells of gas (deep blue) surrounded by thin liquid channels (white). None of the two can be represented by a standard Shan–Chen model. From M. Sbragaglia et al., EPL, 91, 14003 (2010). Courtesy of M. Sbragaglia.*

28.7   Thick Interfaces

As noted many times in this book, one of the most appealing features of the LB approach to multiphase fluids is the spontaneous emergence of the interface from the underlying mesoscopic physics. On the other hand, since the dynamics is grid bound, the interface must necessarily extend over a few lattice sides by mere arguments of numerical diffusion.

Let us analyze the point in little more detail.

By letting $\Delta\rho$ the density change across a single-lattice spacing Δx, a condition for the interface to be well resolved is

$$\frac{\Delta\rho}{\bar{\rho}} \ll 1 \tag{28.22}$$

where $\bar{\rho}$ can be taken as the average between the liquid and gas phase, hence about $\rho_l/2$, since $\rho_g \ll \rho_l$. On the other hand, the density jump across the interface is, by definition, $\rho_l - \rho_g \sim \rho_l$.

By writing $\Delta\rho \sim \Delta x\, \nabla\rho$ and $\rho_l - \rho_g \sim w\, \nabla\rho$, w being the width of the interface, one readily obtains

$$\tilde{w} \equiv \frac{w}{\Delta x} \gg 1 \tag{28.23}$$

Actual practice shows that $\tilde{w} \sim 10$.

Here, one must come clean on the scale and purpose of the LB simulation, a subject which touches at some basic issues which we are to discuss a bit further in the sequel.

28.7.1 Weakly Broken Universality

For a macroscopic calculation, in which there is no point nor wish to resolve the spatial structure of the interface, $w/\Delta x \sim 10$ is just a bad waste of resolution. The reader should note that the name of ten in three dimensions is thousand, so the waste is substantial.

Thus, for macroscopic flows, one would rather turn to methods designed in such a way as to achieve the correct jump within just a single lattice spacing, $w = \Delta x$. There exists a whole class of numerical methods for sharp interfaces specifically designed to achieve this goal. Such methods typically enforce a numerical limiter on the mass flux across the interface and some LB versions equipped with flux-limiters seem indeed to result in sharper interfaces.

If, on the other hand, the simulation is intended to resolve the structure of the interface, which is of decided interest in micro- and nanofluidic applications, then one must confront the fact that physical interfaces are basically sub-nanometer wide. Hence, in order to resolve them, mesh spacings in the order of Armstrongs are needed taking LB to basically the same spatial resolution as Molecular Dynamics (MD).

Here, we are faced with a natural question: under such conditions, what's the point of using multiphase LB instead of MD?

Before elaborating on this delicate item, let us forestall potential misunderstandings by stating a Lapalissian "theorem".[73]

Lattice Boltzmann is not Molecular Dynamics!

Having clarified the obvious, let us now proceed to spell out the idea of Weakly Broken Universality (WBU) with the aid of some numbers. Let L be the typical scale of interest, say the width of a micro-channel hosting the multiphase flow. With $L = 1$ mm, the ratio w/L (Cahn number) is of the order of $Cn = 10^{-6}$. Let us further assume an LB mesh spacing $\Delta x = 100$ nm, whence $w = 1$ mm, the LB Cahn number is $Cn_{LB} = w/L = 10^{-3}$, three orders of magnitude larger than the physical one, but still a small number, much smaller than one.

The basic question then shifts to the sensitivity of the physical phenomenon in point to the actual value of the Cahn number. It is plausible to expect that at least *some* aspects of the multiphase flow, say the overall speed of propagation of the interface, would be relatively indifferent to molecular details, hence not show any appreciable change in going from 10^{-6} to 10^{-3}.

This is what we call *Weakly-Broken Universality*, to contrast it with the unbroken universality of continuum Navier–Stokes hydrodynamics.

---

[73] A byname for obvious, from "La chanson de La Palisse," B. de la Monnoye (1641–1728).

Universality is broken whenever the scale separation between molecular and hydrodynamic degrees of freedom is not complete, so that additional descriptors and associated transport parameters, are required to capture the relevant phenomena.

Matching these parameters one-to-one may simply prove computationally unviable for LB. However, whenever the dependence of such phenomena on the previously mentioned parameters is sufficiently smooth, which we take here as a definition of WBU, then very small numbers can be replaced by just small ones without spoiling the essential physics.

Under such WBU conditions, LB is a wonderful tool to compute orders of magnitude faster and larger than molecular dynamics.

Not because LB would contain the same physics as MD, which most certainly does not (see Lapalisse again), but simply because, *for that given phenomenon,* MD is a computational overkill.

We have seen WBU in action before; the Mach number is often artificially blown up to march in much larger timesteps. Promoting very small numbers to just small ones (where small means distinctly below one) can save major amounts of computational power: this is where LB shines high!

WBU may be regarded as a generalization of what we have previously called Benevolent Asymptotics (BA). WBU implies that the number of additional descriptors of non-universal behavior is small and *more importantly,* that i) the associated dimensionless groups are very small (much smaller than 1) and ii) act smoothly on the system. Under such conditions, very small numbers can be replaced by small ones (just below 1), thereby providing major computational gains without compromising physical insight.[74]

Therefore, it is precisely the degree of WBU available for a given problem which dictates how close LB can get to MD for that problem.

WBU is by no means a given, much less so a physical principle, it is instead a generous gift that mother nature may or may not want grace us with. Arguably, in soft matter this gift comes by more frequently than one would expect, and this is the main reason why LB can prove useful for problems which would normally be regarded as "MD-only."

A few concrete examples may serve illustrating the point.

28.7.1.1 *Capillary filling*

Under a suitable set of approximations, the motion of an advancing liquid front in a capillary channel is given by the so-called Washburn–Lucas law (8; 9; 10):

$$z^2(t) = V_{cap}Lt \tag{28.24}$$

where $z(t)$ is the tip of the front, $V_{cap} = \sigma/\mu$ is the capillary speed, σ is the surface tension and μ is the dynamic viscosity. Being derived under the assumption of zero interface thickness, the Washburn law holds in the limit of zero Cahn number.

---

[74] WBU is sometimes like skating on thin ice, it might be dangerous, but it can take you far.

For a channel width of, say, $L = 1$ mm, the actual Cahn number is about 10^{-6}, which is arguably a close enough proxy of zero. However, LB simulations of liquid penetration into a vapor-filled capillary showed that the Washburn law is satisfactorily reproduced at values of $0.01 < w/L < 0.1$, hence within LB reach, see Fig. 28.9.

Of course, there is more to capillary filling than the position of the front, but these results indicate nonetheless that the Washburn law is fairly insensitive to the Cahn number. As a result, LB can at the very least provide a useful starting point for more complex investigations, before the relay baton is handed over to MD.

28.7.1.2 *Super-hydrophobicity*

Another example of WBU is super-hydrophobicity, as first discussed in Sbragaglia *et al.* (7). In a nutshell, super-hydrophobicity is the apparently paradoxical phenomenon by which a thin vapor film develops within nanoscale cavities, so that the upper-lying liquid can flow on the corrugated surface with less friction than on the smooth one (11). Super-hydrophobicity is central to many microfluidic applications and has been studied in detail by MD simulations.

It was found that some of these details can be reproduced *quantitatively* by a suitably tuned Shan–Chen model. In particular, the pressure drop between the vapor and liquid phase as a function of the nanocavity depth compares with MD results within a few percent accuracy.

With a $h = 10$ nm deep nanocavity, the inner Cahn number w/h is about $1/30$ and the outer one, $w/H \sim 1/1000$, H being the height of the corrugated microchannel, see Fig. 28.10. Both numbers are within reach of (large-scale) LB simulations with $\Delta x \sim 0.3$

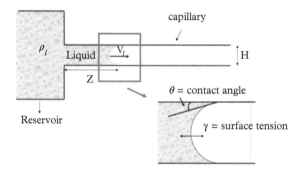

Figure 28.9 *Sketch of a capillary filling experiment. A liquid at the density ρ_l flows left to right from the reservoir to the capillary channel pulled by the capilary forces at the fluid-solid interface and dragged by ordinary friction with the wall. Under suitable approximations, the front centerline, $z(t)$, evolves according to the Washburn–Lucas's law (28.24). Readapted from F. Diotallevi et al., Lattice Boltzmann simulations of capillary filling: Finite vapor density effects, Eur. Phys. J. Special Topics, 171, 237–243 (2009), EDP Sciences, Springer-Verlag 2009. DOI: 10.1140/epjst/e2009-01034-6*

nm, featuring a computational Cahn number around 1/3. The good agreement between MD and LB points again to a weak dependence of the pressure drop on the Cahn number.

Thus, at least *some* aspects of capillary flows and superhydrophic phenomena do enjoy the perks of WBU.

On the other hand, there are many other phenomena where atomistic details do persist up to supramolecular scales, in which case WBU simply does not hold. Under such circumstances, molecular dynamics is no overkill and LB must gently bow away.

One such problem might occur in the study of capillary filling of micro-channels with random nanoscale wall corrugations (12; 13).

The near-wall profile of the flow may depend on geo-chemical details of the texture of the solid wall in a way which cannot be captured by a mean-field picture based upon transport parameters, such as surface tension, contact angles and apparent slip length. This is still an open and hotly debated topic in current microfluidic and soft matter research.

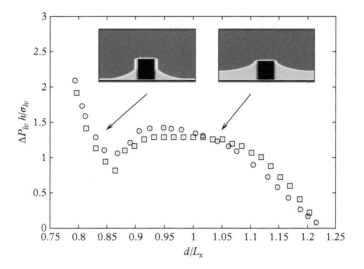

Figure 28.10 *Superhydrophic effect: a liquid (dark) flowing on top of his vapor (light), as triggered by the presence of a grooved surface. The normalized pressure drop, $\Delta P_{lv} h/\sigma_{lv}$, between the two bulk phases is shown as a function of the normalized distance d/L_x, where d is the distance of the top of the groove from the upper wall, while L_x is the size of the periodic cell. LB (squares) results are plotted against MD ones (circles) for the same contact angle $\theta \sim 160^{\circ}$. The two insets represent the density configuration at the onset of the wetting/ dewetting transition (right) and for a wetted configuration (left). A remarkable agreement between LB and MD is apparent. The inner Cahn number is $w/H \sim 1/30$, h being the height of the corrugation and w the width of the interface. From (7)*

28.8   Thermodynamic Consistency: Shan–Chen from Pseudo-Free Energy

One of the earliest criticism to the SC model pointed to its lack of a rigorous theoretical foundation, i.e., a free-energy principle.

Notwithstanding such criticism, the SC method has merrily continued to provide useful insights for more than two decades. As it is often the case for methods which work even though they in principle shouldn't, it was later realized that SC had more theoretical basis to it than originally anticipated.

To begin with, it has been realized that the SC model does stem from a *quasi*-free energy principle, where the *quasi* prefix underscores the fact that the actual distance from a genuine free-energy principle is often negligible for most practical purposes.

The pseudo-free energy (PFE) associated with the SC model reads as follows (14)

$$\mathcal{L}[\rho, \Psi] = f(\rho) + \frac{\kappa}{2}(\nabla\Psi(\rho))^2 \qquad (28.25)$$

Note that this PFE comes in hybrid form, the bulk contribution depends on the physical density ρ, while the interface term depends on the generalized density $\Psi(\rho)$.

It is of course possible to recast it in terms of the physical density alone:

$$\mathcal{L}[\rho, \nabla\rho] = f(\rho) + \frac{\kappa}{2}(\Psi')^2(\nabla\rho)^2 \qquad (28.26)$$

where $\Psi' \equiv d\Psi/d\rho$.

In this "canonical" representation, the mixed PFE is equivalent to a standard FE, with a density-dependent surface parameter:

$$\kappa(\rho) = \kappa\Psi'^2(\rho) \qquad (28.27)$$

Following the usual free-energy formalism, one can compute the pressure tensor as the second-order functional derivative of the PFE with respect to the density gradient, and define the associated force as the divergence of such pressure tensor (for details see the original publication).

The detailed calculations show that starting from the PFE eqn (28.25), or, equivalently, (28.26), one obtains precisely the Shan–Chen force, plus an extra contribution in gradient form, namely

$$\vec{F} = \vec{F}_{SC} - \nabla V_{extra} \qquad (28.28)$$

In (28.28), the extra potential is given by

$$V_{extra} = \frac{G}{2}\lambda(\rho)\Delta\rho \qquad (28.29)$$

with $\lambda(\rho) = \Psi - \rho\Psi'$, the Legendre transform of the generalized density, which clearly vanishes in the thermodynamic limit $\Psi \to \rho$.

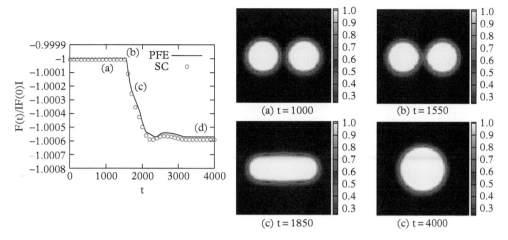

Figure 28.11 *Time evolution of the Pseudo-Free energy (28.25) associated with the bare SC forcing and the corrected one, eqn (28.28), for the case of a droplet merger event. As one can see from the leftmost panel, the two quantities show very little departure from each other. From M. Sbragaglia et al., EPL 86, 24005 (2009).*

The key observation is that, to most practical purposes, the extra-force contributes very little to the total integral expressing Maxwell area's rule. This means that the SC force provides a fairly reasonable quasi-solution to the Maxwell's area rule.

Moreover ensuing numerical simulations provided clear evidence that the SC pseudo-free energy shows genuine free-energy behavior, i.e., monotonic time-decay toward its asymptotic configuration.

Indeed, simulations of phase-separation with and without the extra-force term show no appreciable departures from each other, see Fig. 28.11.

Summarizing, while it is formally true that the SC pseudo-potential does not stem from a free-energy, it is also true that the SC force derives from a pseudo-free energy. Most importantly, this latter is found to be numerically close to the correct free energy. In hindsight, this is probably one of the main theoretical reasons behind the success of the Shan–Chen method.

A word of caution is in order, though.

The above PFE was derived in the continuum and even though the numerical simulations do not seem to indicate any major overturn of the continuum picture due to lattice discreteness, a fully discrete expression of the lattice PFE remains to be found.

28.9 Summary

The development of modern LB methods for multiphase fluids is a rich subject and one which is still in flux. Here, I can only offer a personal, hence necessarily biased, view of the state of affairs (15).

Many brilliant ideas have come to light in the last decade from many different angles; for a nice informative account see the recent review (16) and especially the recent monograph by Huang, Sukop and Lu (17).

As typical of LB, a basic distinction can be made between numerically-oriented versus physics-oriented approaches. Higher-order discretizations and local-grid refinement belong to the former category, whereas modified EoS and multi-range potentials inscribe to the latter. In this author's opinion, some form of symbiotic growth is possibly the best ace to spawn further strides toward the simulation of complex interfaces and strong density contrasts under a broader range of parameters than currently attainable. This will require a lot of ingenuity, hard numerical work and, sure enough, further smiles from Lady Luck.

....

REFERENCES

1. R. Benzi, S. Chibbaro and S. Succi, "Mesoscopic Lattice Boltzmann Modeling of Flowing Soft Systems", *Phys. Rev. Lett.*, **102**, 026002, 2009.
2. G. Falcucci *et al.*, "Lattice Boltzmann models with mid-range interactions", *Comm. Comput. Phys.*, **2**, 1071–84, 2007.
3. G. Falcucci *et al.*, "Lattice Boltzmann simulations of phase-separating flows at large density ratios: the case of doubly-attractive pseudo-potentials", *Soft Matter*, **6**, 4357-4365, 2010.
4. Q. Li and K.H. Luo, "Achieving tunable surface tension in the pseudo-potential lattice Boltzmann modeling of multiphase flows", *Phys. Rev. E.*, **88**, 053307, 2013.
5. L. Kupershtok, D.A. Medvedev, D.I. Karpov, "On equations of state in a lattice Boltzmann method", Comput. *Math. Appl.*, **58**, 965–74, 2009.
6. Q. Li, K. Luo, X. Li, "Forcing scheme in pseudopotential Lattice Boltzmann model for multiphase flows", *Phys. Rev. E.*, **86**, 016709, 2012.
7. M. Sbragaglia *et al.*, "Surface roughness-hydrophobicity coupling in microchannel and nanochannel flows", *Phys. Rev. Lett.*, **97**, 204503, 2006.
8. P.G. De Gennes, F. Brochard-Wyart and D. Quere, "Capillarity and wetting phenomena", Springer Verlag, 2003.
9. E. Washburn, "The dynamics of capillary flow", *Phys. Rev.*, **17**, 273, (1921)
10. R. Lucas, "Ueber das Zeitgesetz des Kapillaren Aufstiegs von Flussigkeiten", *Kolloid-Z*, **23**, 15, 1918.
11. S. Wang, L. Jiang, L. "Definition of superhydrophobic states", *Advanced Materials*, **19**, 3423, 2007.
12. H. Kusumaatmaja, C. Pooley *et al.*, "Capillary filling in patterned channels", *Phys. Rev. E.*, **77**, 067301, 2008.
13. J. Hyvaluoma and J. Harting, "Slip flow over structured surfaces with entrapped microbubbles", *Phys. Rev. Lett.*, **100**, 246001, 2008.

14. M. Sbragaglia *et al.*, "Continuum free-energy formulation for a class of lattice Boltzmann multiphase models", *EPL*, **86**, 24005, 2009.

15. G. Falcucci *et al.*, "Lattice Boltzmann methods for multiphase flow simulations across scales", *Comm. Comput. Phys.*, **9**, 269–96, 2011.

16. Q. Li, K. Luo, Q. Kang, *et al.*, "Lattice Boltzmann methods for multiphase flow and phase-change heat transfer", *Progress in Energy and Combustion Science*, **52**, 62, 2016.

17. H. Huang , M. Sukop and X. Lu, Multiphase Lattice BoltzmannMethods: Theory and Application, 1st Edition, Wiley and Blackwell, NY, 2015.

....

EXERCISES

1. Show that the sound speed associated with the Shan–Chen EoS is always higher in the gas than in the liquid phase.

2. Discuss the difference between the Shan–Chen and the Kuperstock force.

3. Recast the expression (28.21) in terms of symmetric and non-symmetric non-local terms $\Psi_i^{\mp} \equiv \frac{\Psi(\vec{x}+\vec{c}_i) \mp \Psi(\vec{x}-\vec{c}_i)}{2}$.

29

Lattice Boltzmann Models for Microflows

The Lattice Boltzmann method was originally devised as a computational alternative for the simulation of macroscopic flows, as described by the Navier–Stokes equations of continuum mechanics. In many respects, this still is the main place where it belongs today.

Yet, in the last decade, LB has made proof of a largely unanticipated versatility across a broad spectrum of scales, from fully developed turbulence, to microfluidics, all the way down to nanoscale flows. Even though no systematic analog of the Chapman–Enskog asymptotics is available in this beyond-hydro region (no guarantee), the fact remains that, with due extensions of the basic scheme, the LB proves capable of providing valuable insights into the physics of flows at micro and nanoscales.

This does not mean that LBE can solve the actual Boltzmann equation or replace Molecular Dynamics, but simply that it can offer useful insights into some flow problems which cannot be described within the realm of the Navier–Stokes equations of continuum mechanics. In this chapter, we provide a cursory view of this fast-growing front of modern LB research.

If you want a guarantee, buy a toaster

(C. Eastwood)

29.1 Basics of Microfluidics

Micro and nanometric flows play a crucial role on many emerging applications in modern engineering (1) and consequently a deeper understanding of the physics of fluids at micro and nanoscales is paramount to these applications. From both practical and conceptual standpoints, the distinctive feature of micro/nanoflows is the substantial

The Lattice Boltzmann Equation. Sauro Succi, Oxford University Press (2018).
© Sauro Succi. DOI: 10.1093/oso/9780199592357.001.0001

increase of surface/volume (S/V) effects as a result of their reduced size. The immediate consequence is a major enhancement of dissipative versus inertial effects (low Reynolds flows), which configures microfluidics as an arena hosting the competition between dissipative effects, pressure drive and capillary forces triggered by substantial S/V ratios. This also places a major emphasis on the role of fluid-wall interactions, which is where microphysics takes stage.

Making abstraction of capillarity for the moment, let us briefly review the zero-Reynolds scenario governed by the competition between pressure and dissipation, as formally described by the incompressible Stokes equations:

$$\begin{cases} \mu \Delta \vec{u} = -\nabla p \\ \nabla \cdot \vec{u} = 0 \end{cases} \tag{29.1}$$

with appropriate initial and boundary conditions.

In (29.1), $\vec{u}$ is the flow velocity, p the flow pressure and μ the dynamic viscosity. Due to the high S/V ratios, it is clear that boundary conditions are paramount to the overall dynamics of the microflow. In particular, they bear heavily on the propensity of the fluid to flow across microchannels on an affordable energy budget.

A simple calculation makes the point.

By adopting the standard no-slip boundary conditions,

$$\vec{u}_{y_{wall}} = 0 \tag{29.2}$$

where y_{wall} is the cross-flow wall coordinate, the Stokes equations across a planar channel deliver the well-know Poiseuille parabolic profile:

$$u_x(y) = 4U_c \frac{y}{H} \left(1 - \frac{y}{H} \right), \ u_y = u_z = 0 \tag{29.3}$$

where

$$U_c = \frac{1}{8} \frac{\Delta p}{L} \frac{H^2}{\mu} \tag{29.4}$$

is the centerline speed and H and L are the channel height and length, respectively, see Fig. 29.1. The solution holds in the limit of infinite channel width, but this is rather immaterial to our present purposes, since the exact solution for a finite-width channel would only add mathematical complications without affecting the qualitative message.

The flow rate across the channel is then given by

$$Q = \bar{U}HW \tag{29.5}$$

where $\bar{U} = 2U_c/3$ is the average speed and W the channel width, which, by the argument, we shall take much larger than L and H.

Collecting all together, we obtain

$$Q = \frac{1}{12} \frac{\Delta p}{L} \frac{H^3 W}{\mu}$$

(29.6)

In general,

$$Q = kA \frac{\Delta p}{L} \frac{H^4}{\mu}$$

(29.7)

where $A = W/H$ is the transversal aspect ratio (width/height) and k is a geometry-dependent dimensionless numerical factor.

This shows that, at a given pressure gradient $\Delta p/L$, and at a given aspect ratio A, the flow rate scales with the *fourth* power of the channel height, H. This power-four law is quite general, for it results from the combination of inverse curvature, the inverse Laplacian, scaling like H^2 and the channel cross section $S = HW$.

For a three-dimensional channel of width W, the S/V is ratio is given by

$$S/V = \frac{2HL + 2WL}{LHW} = \frac{2}{W} + \frac{2}{H}$$

Since most microfluidic devices are characterized by large aspect ratios,

$$W/H \gg 1$$

this relation reduces to $S/V \sim 2/H$, showing that channel height can be taken as a representative measure of the inverse surface to volume ratio, see Fig. 29.1.

Thus, upon halving the channel height, a 16-fold increase of the pressure gradient is required to keep the flow rate Q constant across the channel.

This presents a very steep dissipative barrier to microscale fluid motion, which explains why the pursuit of ways out of this barrier constitutes one of the leit motifs of modern microfluidic research.

As we shall see, most of these ways out focus on strategies aimed at facilitating relative motion of the fluid with the respect to the wall, i.e., at promoting slip flow.

These strategies are usually based on optimized geo-chemical treatment of the wall surface, via coating, etching and various other techniques. Although details may change

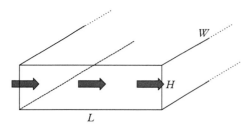

Figure 29.1 *Sketch of a microchannel flow with $H \ll L \ll W$.*

from case to case, most of these technological stratagems make strong reliance on non-equilibrium effects, whence the importance of describing fluid flow beyond the Navier–Stokes framework.

29.2   Non-Hydrodynamic Behavior Beyond Navier–Stokes

One of the central issues raised by microfluidic gas flows is the lack of separation between the molecular scale, the mean-free path, and the smallest geometrical size of the device, say H transverse to fluid motion.

In other words, non-negligible Knudsen numbers

$$Kn = \frac{l_\mu}{H} \sim 1$$

Since the mean-free path scales inversely with the fluid density, sizeable Knudsen numbers usually arise in connection with rarefied gases, as they occur, say in the outer atmosphere. In microdevices, however, rarefaction is not a requisite: the mean-free path of air in standard conditions is about 100 nm; hence micrometric devices are generally exposed to non-equilibrium finite-Knudsen effects.

Indeed, microscale flows exhibit a number of non-hydrodynamic effects, which escape a quantitative description in terms of continuum macroscopic equations. These include non-zero flow speed at the wall (slip flow), anomalous Knudsen layers in the vicinity of the wall, where non-equilibrium effects dominate the scene. As we shall see, similar non-equilibrium effects also occur in connection with the motion of thin liquid/vapor interfaces in capillaries.

29.2.1   Slip-Flow and Knudsen Layers

A central issue of microfluidics is the potential breakdown of the continuum representation whenever the size of the domain becomes comparable with the molecular mean-free path. This issue is mostly relevant for gaseous microflows, where l_μ is typically of the order of 0.1 μm, so that Knudsen numbers around $Kn \sim 1$ are easily reached. Less so in liquids, where molecules move just a fraction of nanometer away from their equilibrium configuration. Yet, finite-Knudsen issues may still become relevant whenever the liquid develops nanometric interfaces and/or thin films near solid walls.

The breakdown of the continuum approximation implies that the flow is no longer close to local-thermodynamic equilibrium, so that, in principle, the full Boltzmann equation, or a fully atomistic approach, should be used.

Since both options are extremely intensive from the computational point of view, the most common strategy is to enrich the Navier–Stokes equations with generalized boundary conditions, allowing for slip flow at the wall. It should be noted that the possibility of slip flow was contemplated by Navier himself as early as in 1823 (2), who postulated a direct proportionality between the slip velocity and the gradient of the velocity at the wall. By definition, the proportionality constant is precisely what we call slip length l_s. For a one-dimensional flow across parallel plates at a distance h, the Navier slip boundary adds a slip-term l_s/h, on top of the parabolic Poiseuille bulk profile $1 - (y/h)^2$. One may therefore argue that the notion of slip flow is perfectly compatible with continuum hydrodynamics. This is certainly true for the case of macroscopic flows, in which the slip term $l_s/h \ll 1$. However, for micro- and nanoflows where $l_s/h \sim 1$, this is no longer the case, since the slip term becomes dominant over the bulk contribution. Moreover, under such conditions, it is not obvious that the bulk component still obeys a parabolic flow profile, because $l_s/h \sim 1$ means that non-equilibrium is as strong as equilibrium, thus contradicting the low-Knudsen assumption underneath the continuum picture. Also to be noted, that a similar case may occur even if the nominal Knudsen number $Kn = l_\mu/h \ll 1$ provided the slip length far exceeds the molecular mean-free path l_μ, so that $l_s/h \sim 1$ even though $l_\mu/h \ll 1$. This may indeed occur in nanoflows with h of the order of a few nm, $l_\mu \sim 0.1 - 1$ nm and l_s of the order of tens to several hundreds of *nm*. Boundary conditions are paramount to the physics of fluids at any scale, and even more so for the case of microflows. From a macroscopic point of view, boundary conditions reflect the molecular interactions taking place at the fluid-solid interface. To this regard, the non-slip boundary conditions simply encode the idea that fluid molecules in the immediate vicinity of a solid wall do not move relative to the wall, because the interactions with the wall molecules are strong enough to suppress any relative motion. Intuitively, the point is that at the molecular scale, the geometrical irregularities of the wall surface would trap the liquid molecules and grind them to a halt. Alternatively, strong attraction between fluid and solid molecules may eventually lead to a similar scenario.

Yet, it should be appreciated that the *non-slip boundary conditions do not follow from any basic principle of Newtonian dynamics*. In fact, they should rather be viewed as a *postulate* of continuum mechanics based on the intuitive picture given here.

On the other hand, it has long been recognized that several fluids, such as methane or polymer melts, do exhibit a net flow relative to the wall itself, a phenomenon known as slip flow.

In equations

$$u_s \equiv u_{y_{wall}} - u_{wall} \neq 0 \tag{29.8}$$

where $u_{y,wall}$ denotes the fluid speed at the wall location and u_{wall} is the wall speed itself. u denotes the flow speed tangent to the wall.

Slip motion has a major impact on the overall mass-flow rate, since it softens the fourth-order dependence on H as given in eqn (29.7), thus leading to potentially major energy savings.

A quantitative measure of slip motion is the *slip length*, l_s, defined as the extrapolated distance from the wall at which the fluid speed matches exactly the wall speed (zero, for the case of a static wall). In equations

$$l_s = |\frac{u}{\partial_y u}|_{y_{wall}} \tag{29.9}$$

Substantial slip flow is defined by the condition that the slip length be significantly larger than the mean-free path, i.e.,

$$l_s \gg l_\mu$$

For instance, in water microflows, l_s can reach up to ten or even hundreds of nanometers on some specific types of solid walls, such as pristine graphene (fast water transport in carbon nanotubes).

The slip length is generally proportional to the molecular mean-free path. However, depending on the nature of the fluid and the geo-chemical properties of the surface, the law of proportionality may eventually become pretty involved leading to sizeable deviations from hydrodynamics in the vicinity of the wall, see Fig. 29.2.

As the Knudsen number increases, deviations from the Poiseuille profile become apparent, the so-called Knudsen layer, which may extend within the bulk flow.

Under such conditions, a full kinetic treatment is required.

Perhaps, the most immediate macroscopic manifestation of slip flow is the so-called *Knudsen paradox*.

This refers to the non-monotonic behavior of the mass flow across a microchannel as a function of the Knudsen number.

According to the Poiseuille solution, the total flow across the channel scales like the inverse viscosity, hence the inverse mean-free path, that is like $1/Kn$.

On the other hand, a wide body of evidence indicates that above a given threshold, roughly around $Kn \sim 0.1$, the flow starts to increase at increasing Knudsen number, *as if* the effective viscosity of the fluid would *decrease* upon increasing the mean-free path, whence the (apparent) paradox.

The analytical expression of the total flow across the channel in the vicinity of $Kn < 1$, but not above, reads as follows (see Fig. 29.3):

$$Q = \frac{a}{Kn} + bKn + cKn^2 \tag{29.10}$$

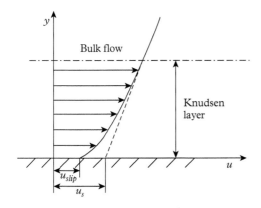

Figure 29.2 *Knudsen layer and apparent slip length for the Kramers problem, i.e., a gas filling half-space $y > 0$ and driven by a constant shear at $y \rightarrow \infty$. The velocity profile extrapolated from the bulk flow gives a non-zero velocity u_s at the wall. The actual velocity profile in the Knudsen layer departs from the bulk-extrapolated values and leads to a different slip-velocity $u_{slip} < u_s$, sometimes referred to as "microscopic" slip velocity. Reprinted with permission from J. Meng, Y. Zhang, Phys. Rev. E 83, 036704, (2011). Copyright 2011 by the American Physical Society.*

where b and c are numerical coefficients carrying the first and second-order deviations from Poiseuille flow. At very large Knudsen, $Kn \gg 1$, the direct dependence on Kn turns into a much weaker $log(Kn)$ form, which is not captured by the expression (29.10). The linear and quadratic terms at the right-hand side of eq. (29.10) reflect the dependence of the slip length, namely the off-wall distance at which the flow profile falls to zero, on the mean-free path.

The inverse dependence on Kn reflects the *collective* behavior of the fluid molecules (strongly interacting regime), whereas the direct dependence signals the onset of *single-particle* behavior (weakly interacting regime).

To appreciate this point, let us write the Poiseuille velocity as follows:

$$u = -\frac{\nabla p}{\rho} \tau_d$$

where $\tau_d \propto H^2/\tau_c$ is the diffusive hydrodynamic timescale. The latter scales *inversely* with the collisional timescale, τ_c, whence the inverse dependence of the hydrodynamic speed on the Knudsen number.

Next, consider instead a single-particle Langevin-like equation:

$$\dot{v} = -\eta v - \nabla p/\rho$$

where η is the friction coefficient, i.e., the (inverse) dissipative timescale, $\eta = 1/\tau_c$.

The steady-state solution

$$v = -\frac{\nabla p}{\rho} \tau_c$$

shows that the molecular speed is *directly* proportional to τ_c, hence to the Knudsen number.

The argument is far from being rigorous (the use the collective force $-\nabla p/\rho$ in the single-particle Langevin equation is pretty questionable), but conveys nonetheless the point: *the Knudsen paradox is a macroscopic manifestation of the crossover from collective to individual molecular dynamics.*

Hence, in order to capture such crossover, one needs to take a step beyond the Navier–Stokes equations.

29.3 Multiphase Flows

The examples given refer to deviations from continuum theory due to non-equilibrium affects associated with finite Knudsen numbers. Such kind of deviations are typical of the rarefied gas regime, but similar effects can also be observed in multiphase flows, especially near liquid-vapor interfaces, where microscopic physics takes full stage, see

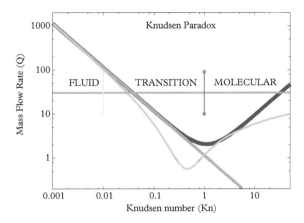

Figure 29.3 *Schematic view of the Knudsen paradox, namely the dependence of the mass flow rate Q on the Knudsen number Kn. Three main regimes can be identified, the Fluid regime for Kn < 0.01, the Transition regime 0.01 < Kn < 1 and the Molecular regime Kn > 1. The fluid regime is characterized by an inverse dependence on the Knudsen number Q = a/Kn, where a is a geometry-dependent coefficient (a = 1/8 for planar Poiseuille flow). In the Transition regime the corrections due to single-particle motion start to become sizeable and cause a deviation from the hydrodynamic law (straight line). Beyond a critical value of the order of $Kn_c \sim 1$, single particle contributions become dominant, leading to an increase of the mass flow rate with the Knudsen number. At very large Knudsen, in the Molecular regime, single particle contributions turn to a much slower (logarithmic) growth, as reported in the lower-lying curve (only the molecular branch of this curve carries physical meaning).*

Fig. 29.3. Indeed, as repeatedly mentioned previously, interfaces are typically subnanometric thick; hence the local Knudsen number in their vicinity is easily of order unity

$$Kn \sim l_\mu |\frac{\partial_y \rho}{\rho}| \sim 1$$

where the coordinate y runs across the interface.

29.3.1   The Moving Contact Line

A typical example is the motion of a liquid-vapor front in a capillary channel.

This is known as *capillary imbibition*, a process whereby a liquid front penetrates a capillary channel, typically filled with vapor, under the drive of capillary forces arising at the contact point of the liquid-solid interface with the solid wall.

The condition of mechanical equilibrium of the contact line between the three phases, liquid, gas and solid, reads as follows:

$$\sigma_{SG} = \sigma_{SL} + \sigma_{LG}\cos\theta \tag{29.11}$$

where $\sigma_{\alpha,\beta}$ is the surface tension between phase α and β, and θ is the (static) contact angle, see Fig. 29.4.

The contact angle dictates the capillary pressure acting on a liquid/vapor interface, as expressed by the Young–Laplace equation:

$$p_{cap} = \frac{2\sigma}{R} = \frac{4\sigma\cos\theta}{H} \tag{29.12}$$

where σ is the surface tension and $R = \frac{H}{2}\cos\theta$ is the radius of curvature of the meniscus, i.e., the surface of an ideal sphere of radius R intersecting the plane walls of a channel of height H.

The contact angle is a material-dependent property which measures the affinity of the fluid to the solid, i.e., its propensity to wet the solid surface.

The range $0 < \theta < 90^0$ defines *hydrophilic* surfaces, namely good wettability, liquid-solid attraction or weaker liquid-solid than liquid-liquid repulsion (The lofty rule "I like A, because i dislike it less than B").

The complementary range $90^o < \theta < 180^o$ defines *hydrophobic* surfaces corresponding to poor wettability, liquid-solid repulsion or stronger liquid-solid than liquid-liquid attraction.

Based on the Young–Laplace equation, on hydrophobic surfaces the capillary forces drive the front motion against dissipation because the liquid molecules do not like the company of the solid ones. On hydrophilic surfaces, the opposite is true. Thus, hydrophobicity is generally perceived as the friend of low dissipation and high droplet mobility.

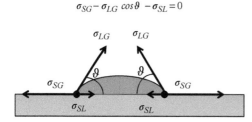

Figure 29.4 *Definition of the constant angle of a liquid droplet (L) on a solid substrate (S), surrounded by its vapor or gas (G). The figure reports the case of a hydrophilic droplet with $\theta < 90^o$. For a static droplet, the left and right contact angles are the same, but for moving droplet this is no longer the case because the motion of the droplet alters its shape resulting in different contact angles at the receding and advancing fronts.*

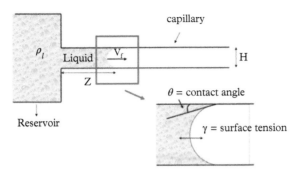

Figure 29.5 *A typical set-up of capillary imbibition. The liquid in the reservoir is pulled into the micro-channel under the drive of capillary forces acting from left to right along the tangent to the flow profile defined by the contact angle. As the liquid progresses into the micro-channel, z = z(t), where small z denotes the centerline coordinate of the front, inertia grows because of the growing mass of the liquid, while the capillary forces remain the same. In the long term, this leads to a vanishing imbibition speed (see main text).*

Understandably, the study of droplet motion on material surfaces is paramount to a wide host of applications in micro and nanotechnology, see Fig. 29.5.

The motion of liquid-vapor fronts on smooth surfaces is governed by the so-called Washburn–Lucas equation, which we have seen before in Chapter 28. This is essentially Newton's law, as applied to the fluid moving in the channel under the combined effect of capillarity and viscous dissipation.

The equation of motion of the liquid front advancing in the vapor-filled channel reads as follows:

$$\frac{d}{dt}(\rho HWz\dot{z}) = 2\sigma\cos\theta - C\eta W\frac{z\dot{z}}{H} \tag{29.13}$$

where $\dot{z}$ is the speed of the front located at $z(t)$, H and W are the channel height and width, and η the dynamic viscosity of the fluid.

Finally, C is a geometry dependent constant, typically of order ten, which is computed by identifying $\dot{z}$ with the central speed of the Poiseuille profile.

Note that the vapor is not included in the momentum balance, based on the assumption $\rho_v/\rho_l \ll 1$ (small Atwood number).

At steady state, (29.13) delivers

$$z\dot{z} = V_{cap}\frac{2H}{C}\cos\theta \tag{29.14}$$

where we have defined

$$V_{cap} = \frac{\sigma}{\eta} \tag{29.15}$$

as the capillary speed, i.e., the typical speed of propagation of capillary waves at the interface.

A simple quadrature, with the initial condition $z(t = 0) = 0$, yields

$$z(t) = \left(V_{cap}\frac{H}{C}\cos\theta\right)^{1/2} t^{1/2} \tag{29.16}$$

The above expression, also known as Washburn's law, shows that the front advances according to a *diffusive, $t^{1/2}$* law; hence its long-time dynamics is much slower than ballistic motion.

It should be pointed out that the square root time dependence is by no means the result of diffusion, but rather of a mass which grows linearly with the front coordinate. The practical outcome is that the velocity of the front vanishes asymptotically as $t^{-1/2}$, i.e., in the long term, the liquid virtually stops moving.

This motivates a significant body of applied research aimed at finding ways to speed up the front propagation via chemico-geometrical treatment of the solid walls.

It should be borne in mind that the motion of liquid-vapor fronts, or droplets on non-smooth, nanocorrugated surfaces is significantly more complicated than the idealized Washburn–Lucas picture.

For one thing, the contact angle of the moving front may depart significantly the static value, depending on the so-called *Capillary number*:

$$Ca = \frac{V}{V_{cap}} = \frac{V\eta}{\sigma} \tag{29.17}$$

where V is the speed of the capillary front.

Moreover, the interaction of the interface with the nanocorrugation is quite complex, as the contact angle undergoes sudden changes due to the abrupt change of orientation of the surface, e.g., 90° from a flat horizontal to a flat vertical wall). Such asperities may even prevent the front from moving altogether, a phenomenon called *front pinning*.

All of these complications speak clearly for the potential deviations from the Weakly–Broken–Universality scenarios already discussed.

Let us now return to singularity issues. Clearly, the very fact that the liquid-vapor interface moves along the microchannel implies that the standard non-slip boundary condition cannot hold true. Formally, this is signaled by the divergence of the stress tensor at the contact point with the solid wall, sometimes called the *Heracles paradox*, a colorful metaphor to indicate that it would take a formally infinite amount of energy to move the front ahead against the non-slip boundary condition.

Clearly, the singularity is not real, but just a signal of the breakdown of the continuum picture. Indeed, it is readily disposed of by relaxing the non-slip condition and replacing it with a more general one allowing a net relative motion between the fluid and wall.

Formally, the simplest generalization reads as follows:

$$(u + l_s \partial_y u)|_{y=0} = 0 \tag{29.18}$$

where u is the fluid velocity tangent to the wall located at $y = 0$, and the subscript y indicates the normal to the wall. In (29.18), l_s is the slip length discussed previously, i.e., the distance normal to the wall at which the velocity takes zero value. Clearly, the limiting case $l_s \to 0$ restores the non-slip boundary condition.

The notion of slip length is very handy, for it permits us to stick to the continuum formalism, if only with a different boundary condition, von Neumann versus Dirichlet.

The problem with this kind of generalized boundary condition is that, at a macroscopic level, l_s is a purely empirical input parameter, which must necessarily be imported from experiment or from an underlying microscopic theory.

By mere dimensional arguments, one posits:

$$l_s \sim l_\mu \tag{29.19}$$

i.e., the slip length should be of the same order of the molecular mean-free path. A more refined closure takes the form:

$$l_s = l_\mu f(l_\mu; Ca) \tag{29.20}$$

where $f(Ca) \to 1$ in the limit $Ca \to 0$. This expression is typical Weakly-Broken-Universality in action: physics beyond Navier–Stokes which can be encoded via general relations between dimensionless parameters.[75]

However, the situation is usually more complicated than this. In particular, in the presence of net motion, the contact angle acquires a dynamic component, whose departure from the static one depends not only on the Capillary number, but also on a series of other parameters, such as the intensity of the wall shear stress and/or the details of the wall geo-chemical composition. Thus a more realistic closure would look like

$$\theta_{dyn} - \theta = f(Ca, l_s/l_\mu \dots) \tag{29.21}$$

where ... stands for further parameters, possibly beyond the WBU framework.

As already mentioned, more microscopic treatments, such as molecular dynamics or the solution of the full-Boltzmann equation for gaseous flows, are well positioned to deliver the right answer. Unfortunately, both are extremely demanding from the computational point of view; hence they often fall short of reaching scales of experimental interest.

This is where a stylized mesoscopic approach, such as *lattice Boltzmann*, may make a genuine contribution, provided one can prove the viability of LB beyond the strictly hydrodynamic realm.

This is a challenging issue, which we next proceed to discuss in some more detail.

---

[75] This author is not aware of any dimensionless number for l_s/l_μ, but it could well be called slip-Knudsen number ...

29.4   LB without Chapman–Enskog: An Eluded No-Go?

The use of LB for microflows is relatively recent and it cannot be said to have enjoyed a smooth. To be sure, it was rather categorically ruled out (6; 7), mostly on account of the lack of a rigorous asymptotic back-up, à -la Chapman–Enskog.

While formally unquestionable, this no-go has been disclaimed has been soon disclaimed by a number of simulations, which have flourished into a substantial stream of publications in the subsequent years.

Besides a number of ad-hoc solutions, the main ingredients which take LB down to microscale flows are essentially the following ones:

1. *Higher-order lattices*

2. *Kinetic boundary conditions*

3. *Filtering of kinetic modes (Regularization)*

Higher-order lattices are needed to equip LB with beyond-hydrodynamic information, i.e., reproduce higher-order moments describing non-equilibrium effects. Kinetic boundary conditions are essential to properly account for fluid-wall interactions beyond the hydrodynamic no-slip description. Filtering and/or fine-tuning of the kinetic modes, especially in combination with the two ingredients mentioned previously, may also prove pretty effective to minimize the effect of higher-order kinetic modes (ghosts) on transport phenomena.

This can be achieved by using either MTR or the regularized LBGK methods discussed previously in this book. The latter (and simpler) option, seems to provide satisfactory results across the full range of Knudsen numbers, from the hydro regime at $Kn \to 0$ all the way up to $Kn = 10$ and beyond, see (11).

Finally, the discovery of some *exact solutions* also added significant evidence that higher-order LB can go beyond hydrodynamics, at least under specific set-ups, such as plane channel and Couette flows.

Before we move on to a more detailed discussion, let us first attempt a more general approach.

Kinetic theory lies between the continuum and atomistic worlds. At its inception, the LB method was explicitly targeted to the former, and has indeed been developed basically in the form of an alternative to Navier–Stokes solvers for macroscopic flows. In this territory, it enjoys the perks of a rigorous mathematical background, mainly Chapman–Enskog asymptotics. For some authors, this still is the only "legal" use of LB, the point being that numerical results alone do not speak for the validity of the method. While formally legitimate, we do not feel like this concern should not necessarily be turned into a categoric no-go. For many authors, this writer included, the hunting ground is much wider than the Chapman–Enskog fence and may eventually extend to micro and even nanoscale fluid flows, provided due care is exercised. Unquestionably, this task faces with a number of conceptual and practical issues.

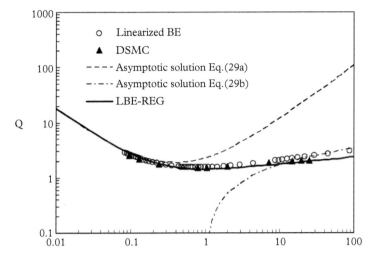

Figure 29.6 *Mass flow Q across a channel at different Knudsen numbers in abscissas. Linearized BE: Numerical solution of linearized Boltzmann equation; DSMC: Direct Simulation Monte Carlo; eq. (29a) in the original reference (11): Low-Knudsen asymptotic solution; eq. (29b) in the original reference (11): High-Knudsen asymptotic solution; LBE-REG: Regularized LBE, with higher-order 21-speed lattice and kinetic boundary conditions. From (11).*

First, as previously mentioned, the non-hydrodynamic phenomena occurring in microfluids cast a shadow on the possibility of using LB for microfluidics altogether. The technical rationale is that standard LB models do not provide enough symmetry in velocity space to quantitatively describe the evolution of higher-order moments of the distribution function, i.e., the ones carrying the non-hydrodynamic information. Differently restated, there is no firm theoretical back-up behind, and consequently the use of LB in this "unprotected" region is basically an uncontrolled approximation. Fortunately, not all uncontrolled approximations are doomed to turn into "computer hallucinations", and indeed this view has proven decidedly over-restrictive, see Fig. 29.6, as we proceed to detail in the following.

29.5   Early Minimal Models: Density-Dependent Relaxation

The earliest LB simulation of microflows was performed by Nie *et al.* (12). These authors pursued a minimalistic approach to microfluidics, by leaving LB completely unchanged, except for fixing the dynamic viscosity, rather than the kinematic one, to a constant $\mu = \rho v = const.$

This is motivated by the fact that microfluidic flows usually take place in long-thin channels, where density changes along the streamwise coordinate cannot be ignored (experimental set-ups feature typical 100 : 1 aspect ratios and density drops $\rho_{out}/\rho_{in} \simeq 2$).

Apart from this, LB was kept completely unchanged and the Knudsen number was raised by simply taking large values of the relaxation time, according to the usual expression of the kinematic viscosity,

$$
\nu = c_s^2 \left(\tau - \frac{\Delta t}{2} \right)
$$

By imposing $\rho \nu = \mu_0 = const.$, the effective relaxation parameter acquires the following dependence on the fluid density:

$$
\tau(\rho) = \frac{\Delta t}{2} + \frac{\mu_0}{\rho c_s^2}
$$

At solid walls, the standard bounce-back boundary condition was used.

Notwithstanding such drastically simplified strategy, the authors (12) report excellent agreement with experimental data, both in terms of the slip-velocity, density and pressure profiles.

However, such successful match with the experimental data requires an adjustment factor, a, in the definition of the Knudsen number:

$$
Kn = a c_s (\tau - \Delta t/2)/H
$$

where H is the channel height.

The fact that such a minimal micro-LB does yield satisfactory agreement with (some) experimental data is pleasing on one side, but puzzling nonetheless.

First of all, the use of bounce-back with large values of τ, say above one, is questionable to say the least, because bounce-back is supposed to enforce non-slip boundary conditions, while large values of τ take LB in the non-hydrodynamic regime where slip flow is a fact.

Indeed, it is known from analytical solutions of LBGK in straight channel geometries (13) that bounce-back boundary conditions lead to an artificial dependence of the slip length on the Knudsen number,

$$
\frac{l_s}{H} = a_0(\epsilon) + a_1(\epsilon, \tau) \, Kn + a_2(\epsilon, \tau) \, Kn^2
$$

where $\epsilon \equiv 1/N$ is the inverse number of lattice sites across the flow (a numerical Knudsen number, interfering significantly with the physical one).

On the other hand, continuum kinetic theory gives a linear dependence (Maxwell relation) plus quadratic corrections:

$$
\frac{l_s}{H} = 1.15 Kn + 0.92 Kn^2
$$

Thus, it is literally true that LBGK with bounce-back boundary conditions is *not* adequate for microflow simulation.

This said, for moderate Knudsen numbers, say $Kn \sim 0.1$, and sufficiently large lattices, say $N > 100$, the numerical data do not differ significantly. This explains why a simple rescaling of the Knudsen number may sometimes recover satisfactory numerical results.

More elaborated forms of effective relaxation parameters carrying an explicit dependence on the Knudsen number, as well as on the actual distance from the wall have been employed by a number of authors.

Typically, the effective, local relaxation time is taken in the following form:

$$\tau_{eff}(y) = \frac{\tau}{1 - \psi(y; Kn)} \tag{29.22}$$

where τ is the bare relaxation time and ψ a universal function of the cross-flow coordinate y and the Knudsen number.

By definition, such function must obey the consistency condition $\psi \to 0$ in the limit $y \to \infty$, so as to recover the bare value τ in the bulk. Positive values of ψ in the vicinity of the wall eventually lead to slip flow due to an increased relaxation time. Space-dependent relaxation parameters are reminiscent of the so-called "wall functions" used to describe the near-wall behavior of turbulent flows. They represent an informed guess on the shape of the profile, typically drawing from asymptotic analytical solutions supplemented with numerical accomodation coefficients.

As anticipated earlier on, these must be regarded as ad-hoc solutions, which permit us to capture the Knudsen layer, eventually, but surely do not cut to the heart of the non-equilibrium physics behind microfluidics. To this purpose, a new generation of LB's must be developed, as we shall detail in the sequel.

29.6 High-Order Lattices

A formal theory of LB methods beyond Navier–Stokes, based upon higher-order lattices, was developed in a remarkable paper by X. Shan *et al.*, (14). Here, the authors develop the discrete analog of Grad's expansion in Hermite polynomials, with full details on its specific implementation on a series of higher-order lattices associated with different numerical quadrature rules.

A natural question arises: given that higher-order LB (HOLB) fall into Grad's footsteps, why would HOLB succeed where Grad's approach would generally not? This is the typical kind of question capable of igniting highly opinionated discussions within the LB community.

For those with a strong penchant for no-gos, HOLB are still insufficient to describe fluids beyond Navier–Stokes. To a point, they have it right: indeed, HOLB alone is not enough. For others, however it does, provided the other ingredients are put in place. Here, we try to offer a hopefully balanced view, yet overtly more in empathy with the optimistic stancers than with no goers.

Three arguments can be brought up to this purpose (15), namely

i)   Counting of Degrees of Freedom,
ii) Structural Organization (Linear Multiscalars versus Nonlinear Tensors)
iii) Realizability of Kinetic Boundary Conditions

29.6.1   Counting Degrees of Freedom: Multi-Scalars versus Tensor Fields

First, a mere counting argument: HOLB's provide a larger set of effective degrees of freedom to work with. The discrete speeds of HOLB lattices are typically of order 40 or more, hence many more than the 13 Grad moments (3) (16) (see Fig. 29.7). This permits us to represent third-order anisotropic moments, such as Q_{xxy} and similar, which are not included in Grad's hierarchy. In other words, instead of contracting the energy flux triple tensor $Q_{abc} \equiv < v_a v_b v_c >$ to a single vector $Q_c = \sum_a v_a^2 v_c$, one retains the double tensor $Q_{ac} \equiv < v_a^2 v_c >$, without summing upon the a subscript to form the kinetic energy.

This is tantamount to retaining three separate energy-flux vectors, $\vec{Q}_a = < v_a^2 \vec{v} >$, $a = x, y, z$, corresponding to the flux of the three "directional" kinetic energies v_x^2, v_y^2, v_z^2, respectively. Since this introduces another six extra fields, one might appropriately name it the Grad-19 formulation.

From equilibrium, the three components of this multi-scalar energy define three distinct pseudo-temperatures T_x, T_y, T_z, which eventually reunite into a single temperature $T_x = T_y = T_z = T$ under near-equilibrium conditions.

These anisotropic terms play an important role near the boundaries, and their neglect represents a source of inconsistency in Grad's treatment of boundary conditions.

Higher-order lattices offer sufficient room for such extra-components, which is a good start. However, this is not sufficient per-se, unless suitable boundary conditions are

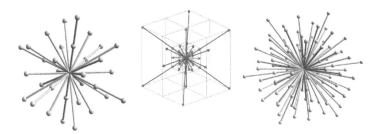

Figure 29.7 *Example of 3d HOLB lattices with 33,41 and 93 discrete speeds, respectively. Many others can be obtained by pruning the mother D3Q125 lattice, consisting of $5^3 = 125$ discrete velocities. Reprinted with permission from A Montessori,* Phys Rev E *92, 043308 (2015). Copyright 2015 by the American Physical Society.*

also put in place. This is precisely where the lattice formulation makes a key contribution: while it appears very unwieldy to devise well-posed boundary conditions for tensor fields of the Grad formulation, as well as its regularized versions, the *multi-scalar* lattice formulation lends itself to a conceptually transparent and mathematically well-posed formulation of the boundary conditions. The point is always the same: the information moves along straight lines.

For a LB treatment of Grad's boundary conditions on open flows, see (17).

29.6.2 Structural Organization: Multiscalars versus Tensors

Second, besides counting, which is in itself a necessary but not sufficient condition for beyond Navier–Stokes behavior, an equally decisive point regards the *structural organization* of the discrete degrees of freedom, i.e., the multi-scalar set of populations versus the tensor hierarchy of the kinetic moments.

To this regard, it is worth noting that LB enjoys a failsafe representation at *both* ends of the Knudsen spectrum; $Kn \to 0$, the Chapman–Enskog protected region, and $Kn \to \infty$, the free-streaming regime.

In the latter, the LB provides an *exact* representation of the free-streaming solution of the initial value problem $f(\vec{x}, \vec{v}_i; t = 0) = f_0(\vec{x}, \vec{v}_i)$, namely:

$$f(\vec{x}, \vec{v}_i; t) = f_0(\vec{x} - \vec{v}_i t, \vec{v}_i) \tag{29.23}$$

Of course, this only applies along the "highway" directions defined by the discrete velocities $\vec{v}_i$.

It is important to note that this property simply *cannot* hold for whatever method based on the hierarchical moment representation of the distribution function, whereby the time change of a given kinetic moment is driven by the divergence of the upper and lower neighbors in the kinetic ladder.

In equations (one dimension for simplicity):

$$\partial_t \rho_n + \partial_x \mathfrak{J}_n = C_n \tag{29.24}$$

where the *n*-th order current is given by

$$\mathfrak{J}_n = \int v f(v) H_n(v) dv = \rho_{n+1} + (n-1)\rho_{n-1} \tag{29.25}$$

the generalized density ρ_n being the *n*-th order kinetic moment.

The above relation a direct consequence of the Hermite recursion relation:

$$H_{n+1}(v) = v H_n(v) - (n-1) H_{n-1}(v) \tag{29.26}$$

The expression (29.24) provides a highly structured hierarchy of PDE's, which can be solved quite effectively by various techniques. However, none of these techniques enjoys the perk of an exact computational representation of the transport terms (advection), because the moments move along *material*, hence flow-dependent, streamlines, instead of straight molecular streamlines (the LB lightcones).

The same point can be made by observing that the moments at time t:

$$\rho_n(x; t) = \int H_n(v) f_0(x - vt, v) dv = \sum_{m=0}^{\infty} (-t)^m \partial_x^m \int H_n(v) v^m f_0(x, v) dv \qquad (29.27)$$

involve an infinite hierarchy of space derivatives of the moments at time $t = 0$.

For all its apparent simplicity free-streaming excites a full (infinite) hierarchy of Hermite modes in Hilbert space! *All* of these modes are strictly required to keep a *complete* record of the initial distribution $f_0(x, v)$: perfect memory, non-equilibrium in full action!

This structural difference between material transport (advection-diffusion) and kinetic free-streaming becomes more and more crucial as the Knudsen number is increased. As discussed many times in this book, this is possibly the most essential property of the LB method.

29.6.3 Realizable Kinetic Boundary Conditions

The third point relates to boundary conditions. Indeed, these properties hold in the bulk and would not prove decisive for practical purposes, unless suitable boundary conditions are formulated, taking proper care of fluid-wall interactions. It is precisely the realizability of these kinetic boundary conditions which permits us to extend higher-order LB to the beyond Navier–Stokes territory. Let us expand this important point into some detail.

29.7 Diffuse Boundary Conditions

As already pointed out many times in this chapter, non-equilibrium flows must be left free to exchange momentum with the solid walls. This is the mandate of kinetic boundary conditions.

An efficient set of kinetic boundary conditions for low-Knudsen flows has been formulated by Ansumali and Karlin (18). These authors developed a lattice transcription of the time-honored Maxwell's full-accomodation model in rarefied gas dynamics, known also as diffuse boundary conditions (19).

The idea is that molecules impinging on the wall lose track of their incoming speed. Consequently, they are reinjected into the fluid along a random direction, with a velocity drawn from a Maxwellian at the local-wall speed and temperature.

In equations (time is omitted because the fluid-wall interaction is assumed to be instantaneous):

$$f(\vec{v}) v_n = \int_{v_n' < 0} f(\vec{v'}) v_n' B(\vec{v'} \to \vec{v}) d\vec{v'}, \quad (v_n > 0) \qquad (29.28)$$

where

$$\vec{v} = \vec{c} - \vec{u}_w \tag{29.29}$$

is the molecular speed relative to the wall moving at velocity $\vec{u}_w$ and $v_n = \vec{v} \cdot \hat{n}$, $\hat{n}$ being the solid to fluid unit normal to the wall (oriented from fluid to wall).

Finally, $B(\vec{v}' \to \vec{v})$ is the fluid-wall scattering kernel, obeying the mass conservation and detailed balance conditions:

$$\int_{v_n > 0} B(\vec{v}' \to \vec{v}) d\vec{v}' = 1 \tag{29.30}$$

$$|v_n'| f^{eq,0}(\vec{v}') B(\vec{v}' \to \vec{v}) = |v_n| f^{eq,0}(-\vec{v}') B(-\vec{v}' \to -\vec{v}) \tag{29.31}$$

where the subscript 0 denotes local equilibrium at zero-wall speed, $\vec{u}_w = 0$. For the case of random reinjection, $B(\vec{v}' \to \vec{v}) = \frac{1}{4\pi v^2}$, which clearly satisfies both conditions above.

Using the these properties, after some algebra, one can express the reflected distribution function in terms of the impinging one, through the following equation:

$$f(\vec{c}) = f_w^{eq}(\vec{c}) \frac{\int_{v_n' < 0} |v_n'| f(\vec{c}') d\vec{c}'}{\int_{v_n' < 0} v_n' f_w^{eq}(\vec{c}') d\vec{c}'} \tag{29.32}$$

which holds for $v_n > 0$, the reflected distribution from solid back to fluid.

In (29.32), the subscript w indicates that equilibria are evaluated at the wall parameters $\{\rho_w, \vec{u}_w, T_w\}$.

With these prescriptions, all quantities in the right-hand side of eqn (29.32) are known, and this equation provides operational knowledge of the reflected distribution function, see Fig. 29.8.

The authors go on and carry these expressions from continuum to discrete velocity space. To this purpose, they make reliance on the Hermite–Gauss quadrature described earlier, and come up with the following expression for the reflected populations:

$$g_i = \frac{\sum_j v_{jn}' g_j^{eq}(\vec{c}_i; u_{wall})}{\sum_j v_{jn}' g_j^{eq}(\vec{u}_w)} \tag{29.33}$$

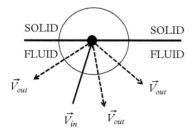

SOLID   SOLID

FLUID   FLUID

$\vec{V}_{out}$ $\vec{V}_{out}$

$\vec{V}_{in}$ $\vec{V}_{out}$

Figure 29.8 *Maxwell's diffuse boundary condition. A molecule impinging on the wall (solid arrow) is re-injected back along a random direction, independent of the impinging one (dashed arrows). For elastic walls, the magnitude of the velocity is conserved, so that energy is conserved while momentum is not.*

where $\vec{v}'_i = \vec{c}_i - \vec{u}_w$, and we have defined the following rescaled distribution:

$$g_i \equiv w_i \rho \, \frac{f_i}{f^{eq,0}(\vec{c}_i)} \qquad (29.34)$$

where f^{eq} denotes the local equilibrium in continuum velocity space, so that g_i absorbs this ratio.

The Ansumali–Karlin boundary conditions exhibit a number of appealing properties. First, they preserve the positivity of the distribution at boundary nodes, i.e., if the incident distribution is positive, the reflected one is guaranteed to be positive too. Second, they readily extend to more general scattering kernels, such as those allowing a blend of slip and reflection.

Interestingly, the standard bounce-back condition is recovered only in the case of the three-velocity one-dimensional model. This is the only case in which the local equilibrium depends only on the actual distribution and its mirror partner. Be as it may, the Ansumali-Karlin diffuse boundary conditions have gained a prominent role in the framework of LB microfluidics.

29.7.1   Partial-Accomodation Kinetic Boundary Conditions

An empirical generalization of the bounce-back rule, allowing for partial slip, besides reflection at the wall was introduced in (20), in the footsteps of previous work in the context of LGCA's (21)).

To discuss the slip-reflection rule, let us remind that the non-slip boundary condition, $\vec{u}_w = 0$, is typically imposed by reflecting the outgoing populations back into the fluid domain, via the bounce-back rule, formally

$$B(\vec{v}, \vec{v}') = \delta(\vec{v} + \vec{v}')$$

A useful generalization consists of making allowance for a mix of bounce back and specular reflections, see Fig. 29.9:

$$f_{\searize}(x, y, H + 1) = r f_{\searrow}(x + 1, y, H) + s f_{\nearrow}(x - 1, y, H) \qquad (29.35)$$

where $0 \le r \le 1$ is the reflection coefficient and $s = 1 - r$ is the slip coefficient. In (29.35), the arrows denote the direction of propagation of the corresponding distributions. This three-site interaction is more complicated than pure bounce-back, but it permits us to model free-slip motion at the solid interface.

Within this approach, the reflection(slip) coefficient must be regarded as a free parameter to be calibrated through comparison with experimental data. In particular, the amount of slip flow is a very sensitive function of s in the near vicinity of $s = 1$, where it diverges like $(1 - s)^{-1}$.

The realizability of the reflect-slip boundary conditions has been proven analytically by Sbragaglia and Succi (22), who showed that it corresponds to a lattice transcription of the popular Cercignani–Lampis (CL) scattering kernel (4). We remind that the CL kernel is a generalization of Maxwell's diffuse boundary conditions, in which two distinct coefficients are introduced to allow a separate mechanism for tangential and normal relaxation at the wall boundary.

Sbragaglia and Succi also show how to incorporate partial accomodation, i.e., the possibility for the reflected particle to carry less energy than the incident one, as a result of inelastic collisions with the solid wall.

The specific form of the discrete boundary kernel for the D2Q9 model is as follows:

$$B_{ij} = \begin{pmatrix} r + aw_2 & aw_2 & s + aw_2 \\ aw_1 & r + s + aw_1 & aw_1 \\ s + aw_2 & aw_2 & r + aw_2 \end{pmatrix}$$

where a is the accomodation coefficient fulfilling the normalization condition:

$$r + s + a = 1 \tag{29.36}$$

In Fig. 29.9, $w_1 = 1/9$ and $w_2 = 1/36$ are the standard weights of the D2Q9 scheme. Clearly, in the limit $a \to 0$, this reduces to the slip-reflect condition (29.35). This model can reproduce experimental data up to second order in the Knudsen number and has been successfully extended to heterogeneous conditions, i.e., with (r, s, a) depending on the local position on the wall. Such an extension permits us to compute slip flows over micro-patterned surfaces in excellent agreement with analytical solutions.

Summarizing, Ansumali and Karlin identified the correct framework to formulate lattice boundary conditions for non-zero Knudsen numbers. Their model was confined to the case of full accomodation, although the authors clearly recognized its applicability to more general conditions. Indeed, generalizations to partial-accomodation were formulated first on an empirical basis in (20), and subsequently backed up by analytical work in (22).

29.7.2 Volumetric Boundary Conditions

The kinetic boundary conditions described pertain to regular boundaries aligned with the lattice.

General boundary conditions capable of dealing with arbitrary complex geometries have been proposed by H. Chen (23) and successfully implemented in the commercial LB software POWERFLOW. These are based on a volumetric formulation of the LB

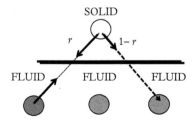

Figure 29.9 *Boundary conditions with partial reflection, as per eqn (29.35). Reprinted with permission from A Montessori, Phys Rev E 92, 043308 (2015). Copyright 2015 by the American Physical Society.*

scheme, which permits us to handle the fluid-wall interactions in terms of coupled dynamics of volume and surface elements (*voxels and surfels*, respectively). These boundary conditions can handle the complex boundary layers associated with turbulent flows and have later been adapted to microflows reproducing the slip flow, the Knudsen paradox and Knudsen layers (10).

Since these boundary conditions are explicitly designed to handle boundary layers, it is not surprising that they can handle slip-flow situations too. The formulation is significantly more elaborate, and we refer the detail-avid reader to the original paper (23).

29.7.3 Non-Equilibrium Boundary Conditions

Besides the kinetic boundary conditions illustrated before, another class of non-equilibrium boundary conditions for open flows is worth mentioning.

They rely upon extrapolation not only of the flow field, but also of its gradient, namely the stress tensor. These boundary conditions do not necessarily involve higher-order lattices, but illustrate the realizability of Grad's program on a discrete lattice.

The incoming populations at the outlet boundary nodes ("missing data" in the authors terminology) are set as follows (31):

$$f_i(B) = f_i^{eq}[\rho(F), u(F)] + f_i^{ne}[\sigma(F)] \qquad (29.37)$$

where B and F stand for the boundary-fluid pairs of lattice sites connected by the interpolation scheme. In (29.37), ρ, u and σ stand for the fluid density, velocity and stress tensor, vector and tensor indices being relaxed for notational simplicity.

In the actual simulation, a three-dimensional flow past backward facing step, both equilibrium and non-equilibrium components contain terms only up to second order, i.e., Grad's expression for an athermal flow.

Significant enhancement of the accuracy in the prediction of the reattachment length past the step is reported, as compared to the case where non-equilibrium data (σ) are not included. The indication is that inclusion of non-equilibrium terms according to expressions like (29.37) and generalizations thereof to account for higher-order fields provides a *lattice realizable* implementation of Grad's program. Of course, such a statement ought to be proved case by case, but the realizability of kinetic boundary conditions appears to be more than an occasional lucky strike.

29.8 Regularization

One further point regards the Regularization procedure. As discussed earlier, this consists of filtering out non-hydrodynamic moments, the ones fueling non-equilibrium, right after the streaming step.

In equations:

1. *Stream:* $f_i(x, t) = f_i'(x - c_i \Delta t, t - \Delta t)$

2. *Regularize*: $h_i(x, t) = \mathcal{P}f_i(x; t)$
3. *Collide*: $f'_i(x, t) = (1 - \omega)h_i(x; t) + \omega h_i^{eq}(x; t)$

In the calculation, $\mathcal{P}$ denotes a projector in kinetic space, filtering out the non-hydrodynamic (ghost) modes, i.e.,

$$\mathcal{P}f_i = h_i \tag{29.38}$$

$$(1 - \mathcal{P})f_i = g_i \equiv f_i - h_i \tag{29.39}$$

By hydrodynamic modes, we imply here density ρ, momentum $\mathcal{J}_a$ and momentum-flux, P_{ab} for standard LB, with the addition of the third-order energy-flux tensor Q_{abc} for the case of HOLB.

All remaining moments classify under the ghost rubric, i.e., the kinetic moments lacking a direct macroscopic interpretation.

This hydro-ghost representation follows in the steps of Grad's 13-moment generalized hydrodynamics, with two crucial twists:

i) It contains a larger set of kinetic moments
ii) These moments are organized by the symmetry of the discrete lattice rather than by the Hermite ladder in Hilbert space

The benefits associated with both items have been commented before, essentially i) offers more degrees of freedom than Grad-13, in particular the non-isotropic components Q_{xxy} and similar, which are relevant to the near-wall physics, and ii) These degrees of freedom propagate error-free, no less of memory, along the discrete lightcones. To dig a bit deeper into these items, let us remind that the hydrodynamic component receives contribution both from conserved ($\rho, \mathcal{J}_a$) and transport (P_{ab}) modes, so that it features a non-zero equilibrium and non-equilibrium components:

$$h_i = h_i^e + h_i^{ne}$$

The ghosts, on the other hand, are purely non-equilibrium objects:

$$g_i^e = 0, \ g_i = g_i^{ne}$$

hence they only affect transport modes and higher order kinetic modes. The whole point of the Regularization procedure is to annihilate g_i after streaming, so as to minimize their impact on transport modes.

In formal terms:

$$\hat{h}_i + \hat{g}_i = (1 - \omega)(h_i + rg_i) + \omega h_i^e$$

where hat denotes shifted distributions at $(\vec{x} + \vec{c}_i \Delta t, t + \Delta t)$. In this, $r = 1$ denotes the standard BGK, while $r = 0$ stands for the Regularized version. In this formalism, g_i are the ghosts generated by the streaming at $t - \Delta t$, which are then filtered out before collision, while $\hat{g}_i$ are the ghosts generated by streaming at time t.

While still under exploration, the Regularized LB has shown several encouraging evidences that it can help reproducing at least the global features of finite-Knudsen flows, such as the transition from collective hydrodynamics to single-particle molecular regimes in finite Knudsen flows between parallel plates (see Fig. 29.12).

29.9   Exact Solutions

We started this chapter by acknowledging that LB for microfluidics does not enjoy the rigorous mathematical back-up of Chapman–Enskog asymptotics. We also commented on the fact that this lack of rigorous asymptotics has exposed a major point of criticism toward the applicability of LB to microfluids. While such statements had a formal point for standard LB lattices, they appear less tenable toward the HOLB schemes described, see Fig. 29.10. One can take a further step further along this direction: if only in special cases, HOLB enjoy more than rigorous asymptotics, namely *exact* solutions at finite-Knudsen numbers.

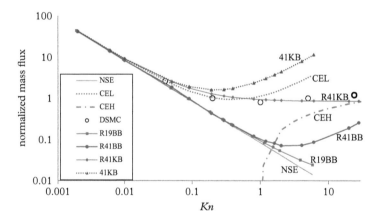

Figure 29.10 *The mass flow rate as a function of the Knudsen number for a two-dimensional channel flow. The open circles refer to the DSMC solution, which is taken as a benchmark reference. CEL and CEH refer to analytical asymptotic solutions obtained by Cercignani in the low (CEL) and high (CEH) regimes, respectively. NSE reports the Navier–Stokes solution, while the remaining curves refer to assorted variants of the LB method: KB refers to kinetic boundary conditions versus bounce back (BB). The former allow for slip flow at the wall, while the latter does not. The prefix R denotes the Regularized version and finally 19 and 41 refer to the number of discrete velocities. As one can see, the R41KB model comes very close to DSMC. From (15).*

Ansumali, Karlin and coworkers have been able to exhibit an exact solution to the hierarchy of nonlinear lattice Boltzmann (LB) kinetic equations for stationary planar Couette flow at non-vanishing Knudsen numbers (24). Using the diffuse boundary conditions described, these authors derived closed-form solutions for all higher-order moments. They further performed convergence tests against exact results and DSMC, (24), see Fig. 29.11.

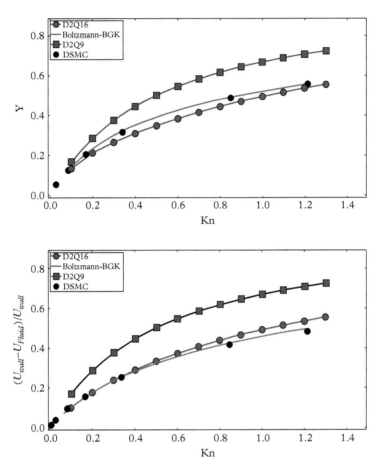

Figure 29.11 *Relative slip flow at the wall as a function of the Knudsen number. The figure highlights the significant improvement obtained with the higher-order D2Q16 lattice as compared with standard D2Q9. The solution of the fully kinetic BGK as well Direct Simulation Monte Carlo results are shown as a reference. In the left panel $Y = \frac{1}{\Delta U}\frac{du_x}{dy}|_0$, ΔU is the relative velocity of the plates and $y = 0$ denotes the wall location. Reprinted with permission from S. Ansumali et al, Phys Rev Lett. 98, 124502 (2007). Copyright 2007 by the American Physical Society.*

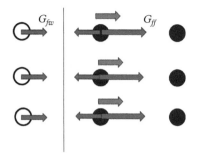

Figure 29.12 *Sketch of hydrophobic walls, as obtained by setting $G_{ff} < G_{fw} < 0$, i.e., stronger fluid-fluid than fluid-wall attraction. The result on the layers of molecules adjacent to the wall (thin vertical line) is a net repulsion from the wall.*

The results indicate that the LB hierarchy with larger velocity sets does approximate kinetic theory beyond the Navier–Stokes level. Most interestingly, these authors pointed out that the use of diffuse boundary conditions based upon discrete Boltzmann populations rather than kinetic moments, is key to the realizability of the LB approach, as opposed to the standard Grad's 13-moment method. In a subsequent study, Yudistiawan, Ansumali and Karlin (5) went one step further and showed quantitative accuracy of the standard D2Q9 model equipped with kinetic boundary conditions, with the slip-flow solution of Couette and Poiseuille flow by Cercignani (4). Although such exact solutions only hold for specific set-ups, they indicate that HOLB equipped with kinetic boundary conditions does carry quantitative non-hydrodynamic information.

29.10   Partial Summary

Despite major progress in the last decade, the theory of LB beyond Navier–Stokes is still in need of consolidation. While it is clear that HOLB plus kinetic boundary conditions are essential ingredients to recover quantitative agreement with the Boltzmann equation at finite Knudsen, up to, say, $Kn \sim 1$, systematic convergence criteria remain to be formulated. For instance, Kim *et al.* reported lack of improvement at increasing order of accuracy of Hermite quadratures (25), which points to a dependence on the specific lattice. On the other hand, other authors (15) report fairly good agreement also in terms of flow profiles by using all three ingredients, namely, HOLB, kinetic boundary conditions *and* Regularization. It is quite possible that Regularization, which is lattice specific (ghost fields are not universal), may explain the lack of systematic convergence reported by Kim *et al.* As mentioned previously, at the time of this writing, a fully fledged theory of HOLB with kinetic boundary conditions and Regularization is still awaiting for a systematic formulation. In particular, it would be highly desirable to elucidate the formal connections with modern versions of Regularized Grad-13 theory.

29.11   Multiphase Microflows

The matter discussed so far concerns the use of LB for single-phase fluids out of equilibrium, such as rarefied gases.

Modern technology has moved large strides in the direction of micro and nanofluidics applications involving multiphase flows, say liquid-vapor interfaces in capillary channels. This has proved to be one of the fastest growing sectors of LB research in the last decade, mostly based on the use of the various multiphase models described in Chapters 27 and 28.

These models are typically complemented with specific variants that best adapt the problem at hand. Here, "best adapt" means the usual LB strategy: inject the required amount of specificity of fluid-wall interactions, without incorporating full molecular details. The territory previously referred to as to *Weakly Broken Universality*.

In the sequel, we shall provide a cursory view of some of the LB variants designed to chart the WBU territory.

29.11.1 Fluid-Wall Potentials

A microscopic approach to account for fluid-wall interactions consists of equipping LB with non-local fluid-wall potentials.

A typical wall-fluid force reads as follows:

$$V_{fw}(y) = G_{fw} \chi (y/r_w)$$

where y is the normal-to-wall coordinate.

The r_w is the range of the fluid-wall potential, typically a few lattice sites, G_{fw} is the coupling strength and $\chi(y)$ is a fast-decreasing function of the distance from the wall, i.e., $\chi \to 0$ for $y \gg r_w$.

The inclusion of explicit fluid-wall potentials leads to the following slip-flow scenario: the repulsive wall-fluid forces (hydrophobic interaction) leads to a depletion the fluid layer near the wall. Due to this density drop, the dynamic viscosity also drops down, leading to an acceleration of the near-wall fluid, as compared to the bulk flow.

For sufficiently large values of the fluid-wall potential, i.e., $G_{fw}/k_B T \gg 1$, this scenario was shown to yield slip-lengths l_s, well in excess of the intermolecular range r_w, i.e.,

$$l_s \gg r_w \sim l_\mu$$

This is in qualitative agreement with experimental observations of apparent slip lengths of order of $10\,nm$ or higher.

However, it should be noted that the use of fluid-wall potentials with an explicit dependence on the spatial coordinates raises an issue of consistency with the mesoscopic nature of LB. As discussed previously, LB combines naturally with mean-field *pseudo-potentials* stemming from density-dependent functionals. The other face of the same medal is an issue of spatial resolution, since in real-life applications the range of the potential is typically below a nanometer, thus requiring sub-nm LB spacings, hence the inclusion of statistical fluctuations.

29.11.2   Fluid-Wall Pseudo-Potentials

A rich variety of complex microscale flows is currently being simulated with LB schemes for non-ideal fluids, especially the pseudo-potential and free-energy versions. Here, the versatility of LB and its aptness at capturing WBU regimes shines high. A full account of the wide body of work in this direction is way beyond the scope of this book. Hence, in the following we only sketch a few ideas.

For instance, by introducing separate fluid-fluid and fluid-wall pseudo potentials with different coupling strengths, G_{ff} and G_{fw}, respectively, it is possible to code for both hydrophobic and hydrophilic interactions, (see Fig. 29.12).

As an example, by using the Shan–Chen attractive pseudo potentials, the condition $|G_{ff}| > |G_{fw}|$, with both G's negative to code attraction results in a net repulsion of the near-wall fluid layer, thereby implementing a hydrophobic wall.

By the same token, the reverse inequality codes for hydrophilic walls.

The same qualitative result can be obtained by using the same coupling strength everywhere in the fluid, but tuning the strength of the fluid-wall interactions by setting the value of the density ρ_W entering the Shan–Chen force between the wall and the adjacent fluid layer.

Interestingly, these models are *realizable*, i.e., by tuning the interaction strength, it is possible to recover basically the full range of contact angles 0–180°. This is by no means a given.

Therefore, once the LB parameters are calibrated so as to reproduce the desired contact angle, they can be used for larger scale simulations, closer to scales of experimental relevance.

29.11.3   Heterogeneous Free-Energy Models

In the free-energy versions, a similar effect can be attained by catering for space-dependent heterogeneous coefficients in the local equilibria.

For instance, by letting the chemical potential vary along the wall, it is possible to describe complex droplet interactions with chemically heterogeneous surfaces, micro-patterned surfaces, and many other phenomena of great interest in modern microfluidics and soft-matter research (26; 27). This permits us to handle many problems for which molecular dynamics would be a computational overkill, in the sense that the essential physics can still be encoded within transport parameters, such as the surface tension and contact angles, with no need of specifying full atomistic detail. In general, this approach appears to have met with significant success for the Lb simulation of a variety of complex multiphase microflows.

29.12   Further Miscellaneous Developments

Based on the arguments given in this chapter, one may conclude that LB, equipped with higher-order lattices, kinetic boundary conditions and Regularization *can* indeed handle non-equilibrium phenomena beyond Navier–Stokes hydrodynamics. Likewise, by suitable formulations of non-ideal fluid-fluid and fluid-wall interactions, it can also describe complex microfluidics effects beyond the continuum Navier–Stokes picture.

The developments mentioned are the ones which permit modern LB to handle the physics of microfluids in the WBU regime.

Several items for future development still remain, some of which we briefly comment upon in the following.

29.12.1   Spurious Knudsen Layers

As shown long ago by Cornavin *et al.*, (28), solid walls not aligned with the lattice grid are known to excite spurious Knudsen layers, which may contaminate the physical solution far inside within the flow. Due to the lack of symmetry in velocity space, these spurious layers are highly anisotropic. The question remains as to whether a sufficiently accurate discretization, jointly with a volumetric formulation of the LB scheme, and possibly higher-order lattices, may curb the problem. Some evidence in the affirmative is indeed available, but much more validation work is required to put these observations on a systematic basis.

29.12.2   High-Order Boundary Conditions

Another practical question concerns what kind of high-order lattice should be used for highly non-equilibrium flows. At the moment, a large variety of options is available, but no consolidated winner has emerged as yet. The question becomes even more urgent with regard to the implementation of boundary conditions, especially solid walls and outlets, a subject which is still in flux at the time of this writing.

29.12.3   Non-Uniform Grids

Most microfluidic devices are characterized by very large aspect ratios, $A = L/H$, L being the length (centimeters) and H being the width, typically of the order of tens-hundreds of microns, leading to $A = 10^2 \div 10^3$. This implies that the efficient simulation of a full microfluidic device is best performed with highly anisotropic and stretched lattices with $\Delta x >> \Delta y$, x and y being the streamwise and crossflow coordinates. While several LB variants dealing with non-cartesian or even more general lattices are available, the majority of microfluidic LB applications still runs on uniform cartesian lattices. It is quite likely that future LB microfluidics will shift increasingly toward more advanced lattices.

29.12.4   Thermal Microflows

Since heat exchange is a crucial phenomenon in many microfluidic applications, it is clear that the formulation of thermal effects in microfluidic LB schemes is an important item on the agenda.

Once again, several developments have taken place among multiple directions, far too many for a complete coverage.

Interesting extensions of the Maxwell model for gas-surface interactions, known as Langmuir boundary conditions, have been developed, which are apparently able to handle Knudsen layers, as well as temperature jumps.

The idea is to express the slip velocity, as well as the thermal jump, as a linear combination of the velocity (temperature) of the wall and the one in the first vapor layer next to it, i.e.,

$$u_s = (1 - \alpha)u_w + \alpha u_v \tag{29.40}$$

$$T_s = (1 - \alpha) T_w + \alpha T_v \tag{29.41}$$

where the coefficient α is a function of the Knudsen number and the fluid pressure

$$\alpha = \frac{\beta p / k_b T_w}{1 + \beta p / k_b T_w} \tag{29.42}$$

Here, p is the fluid pressure, T_w the wall temperature and $\beta \propto 1/Kn$.

Clearly, in the limit $Kn \to 0$, one has $\alpha \to 1$, so that the non-slip boundary condition is recovered.

For simplicity, we have taken the same accomodation coefficient, α, for velocity and temperature, but in general, two separate coefficients are required.

With the values of the slip velocity and temperature, one can construct the incoming populations through linear combinations of equilibrium and non-equilibrium distributions at the wall and in the near-wall gas layer. For more details, see (32) and references therein.

These boundary conditions have been used in conjunction with the double-distribution thermal LB scheme described earlier in this book. They have been shown to reproduce analytical solutions for near-wall velocity and temperature profiles up to $Kn \sim 0.1$ using the standard D2Q9 lattice.

29.12.4.1 *Thermal Entropic Models*

Another route to thermal microfluids is provided by thermal entropic LB schemes (29).

By thermal entropic, it is implied that, besides mass and momentum, the total energy density, $e = \rho(v_T^2 + u^2)/2$ also takes part to the list of constraints in the entropy minimization procedure.

This takes LB off-lattice, because the discrete velocities are no longer integer multiples of each other.

By combining ELB with the Ansumali–Karlin boundary conditions, the authors demonstrate satisfactory compliance with the Knudsen paradox. While thermal microflows have indeed been simulated at moderate Knudsen numbers and simple geometries, much remains to be done for the case of strong non-equilibrium and thermal effects in complex geometries. For a very recent thorough effort along these lines, see (30) and references therein.

29.13 Summary

The "LB beyond Chapman–Enskog" research line has witnessed major progress over the last decade.

Initially based on purely heuristic attempts, the subject has flourished into a substantial stream of new research based on the synergistic contribution of multiple concurrent items.

First, a systematic theory of higher-order lattices following in the footsteps of discrete-velocity implementation of Grad's program, has been developed, including the discovery of some exact solutions. These advances have been matched by the development of new kinetic boundary conditions providing a lattice counterpart of those used in the continuum kinetic theory of rarefied gases. Such kinetic boundary conditions have proven instrumental for the realization of Grad's program on higher-order lattices. Both items have been further strengthened by the development of Regularization procedures aimed at minimizing the effects of non-hydrodynamic ghosts on the transport modes. The concurrency of these three streamlines has resulted into a major extension of the standard LB theory, as we knew it, say ten years ago.

At the same time, workers in the field have come up with a number of ingenuous variants of the original LB schemes for non-ideal fluids, which have found extensive use in the WBU sector of microfluidics.

For those who feel like kinetic theory is not complete until it is shown to converge to a given set of partial differential equations, the hard-core Chapman–Enskog line, all of this still may fall short of providing a convincing case for LB beyond Navier–Stokes.

While this stance deserves serious consideration, the amount of numerical and also theoretical evidence in favor of LB beyond Navier–Stokes, commands major attention too.

As a purely personal opinion, possibly the most far-reaching bottomline of the HOLB approach is simply this: *the lattice organization of degrees of freedom is more transparent and effective than the moment hierarchy in Hilbert space.* What I mean by this is that the nonlinear tensor field theory which HOLB *eventually* converges to, is likely to look much more involved and unwieldy than HOLB itself, with substantial consequences on its realizability, especially in connection with boundary conditions for strongly confined microflows.[76] Future success of the HOLB line will hinge on two major items: i) A firm theoretical control on whether or not WBU applies to the microfluidic problem at hand, ii) The implementation of HOLB methods within efficient computational tools for highly confined microflows.

None of the two is going to hang low from the tree, but the pay-off for (the WBU sector of) microfluidics might be a truly major one.

---

[76] This is not to say that the pursuit of a such a tensor field theory is pointless, quite the opposite. It simply means (to this author) that there is no reason to put the HOLB agenda on hold till the day comes, when such theory is worked out in full detail. The two avenues should be pursued concurrently with mutual benefit for each other.

REFERENCES

1. A. Beskok and G.E. Karniadakis, *"Microflows, Fundamentals and Simulation"*, Springer Verlag, Berlin, 2001.
2. C.L.M.H. Navier, *Memoirs de lAcademie Royale des Sciences de lInstitut de France*, **1**, 414–16 1823.
3. J. Meng and Y. Zhang, "Gauss-Hermite quadratures and accuracy of lattice Boltzmann models for nonequilibrium gas flows", *Phys. Rev. E*, **83**, 036704 2011.
4. C. Cercignani and M. Lampis, "Kinetic models for surface interactions", *Transport Theory and Stat. Phys.*, **1**, 101 1971.
5. W. P. Yudistiawan, S. Ansumali and I. V. Karlin, "Hydrodynamics beyond Navier-Stokes: the slip flow model", *Phys. Rev. E*, **78**, 016705 (2008)
6. L.S. Luo, Comment on "Discrete Boltzmann Equation for Microfluidics", *Phys. Rev. Lett.*, **92**, 139401-1 2004.
7. M. Junk, A. Klar and L.S. Luo, "Asymptotic analysis of the lattice Boltzmann equation", *J. Comp. Phys.*, **210**, 676 2005.
8. F. Toschi and S. Succi, "Lattice Boltzmann method at finite Knudsen numbers", *Europhys. Lett.*, **69**, 549 2005.
9. Y. Zhang, R. Qin, and D.R. Emerson, "Lattice Boltzmann simulation of rarefied gas flows in microchannels", *Phys. Rev. E*, **71**, 047702 2005.
10. Y. Zhou, R. Zhang, I. Staroselsky, H. Chen, W Kim and M. Jhon, "Simulation of micro and nano-scale flows via the lattice Boltzmann method", *Physica A*, **362**, 68 2006.
11. X.D. Niu, S.A. Hyodo, T. Munekata and K. Suga, "Kinetic Lattice-Boltzmann method for microscale gas flows: Issues on boundary condition, relaxation time and regularization", *Phys. Rev. E*, **76**, 036711 2007.
12. X.B. Nie, G.D. Doolen and S.Y. Chen, "Lattice-Boltzmann simulations of fluid flows in MEMS *J. Stat. Phys.*, **107**, 279 2000.
13. X. He *et al.*, "Analytic solutions of simple flows and analysis of nonslip boundary conditions for the Lattice Boltzmann BGK model", *J. Stat. Phys.*, **87**, 116 1997.
14. X. Shan, X. Yuan and H. Chen, "Kinetic theory representation of hydrodynamics: a way beyond the NavierStokes equation", *J. Fluid Mech.*, **550**, 413 (2006)
15. A. Montessori *et al.*, Higher order Lattice Boltzmann method for non-equilibrium flows, *Phys. Rev. E*, **92**, 043308, 2015.
16. R. Zhang, X. Shan, H. Chen, "Efficient kinetic methods for fluid simulation beyond the Navier-Stokes equation", *Phys. Rev. E*, **74**, 046703 2006.
17. S. Chikatamarla, S. Ansumali and I.V. Karlin, "Grad's approximation for missing data in lattice Boltzmann simulations", *Europhys. Lett.*, **74**, 215 2006.
18. S. Ansumali and I.V. Karlin, "Kinetic boundary conditions in the lattice Boltzmann method", *Phys. Rev. E*, **66**, 026311 2002.
19. J.C. Maxwell, "On stresses in rarefied gases arising from inequalities of temperature", Philos. *Trans. R. Soc.* London **170**, 231 1879.

20. S. Succi, "Mesoscopic modeling of slip motion at fluid-solid interfaces with heterogeneous catalysis", *Phys. Rev. Lett.*, **89**, 064502 2002.

21. P. Lavallé, J.P. Boon and A. Noullez, "Boundaries in lattice gas flows", *Physica D* **47**, 1, 1 1991.

22. M. Sbragaglia and S. Succi, "Analytical calculation of slip flow in lattice Boltzmann models with kinetic boundary conditions", *Phys. of Fluids*, **17**, 093602 2005.

23. H. Chen, "Volumetric formulation of the lattice Boltzmann method for fluid dynamics: basic concept", *Phys. Rev. E*, **3**, 58 1998.

24. S. Ansumali *et al.*, "Hydrodynamics beyond Navier-Stokes: Exact Solution to the Lattice Boltzmann Hierarchy", *Phys. Rev. Lett.*, **98**, 124502 2007.

25. S.H. Kim, H. Pitsch and I. D. Boyd, "Accuracy of high-order lattice Boltzmann methods for microscale flows with finite Knudsen numbers", *J. Comp. Phys.*, **227**, 8655, 2008.

26. H. Kusumaatmaja *et al.*, "Drop dynamics on chemically patterned surfaces", *Europhys. Lett.* **73**, 740, 2006.

27. J. Hyvaluoma and J. Harting, "Slip flow over structured surfaces with entrapped microbubbles", *Phys. Rev. Lett.*, **100**, 246001, 2008.

28. R. Cornubert, D. d'Humieres and D. Levermore, A Knudsen layer theory for lattice gases, *Physica D*, **47**, 241, 1991.

29. M. Mazloomi, S. Chikatamarla, and I. Karlin, "Entropic Lattice Boltzmann Method for Multiphase Flows", *Phys. Rev. Lett.*, **114** 174502, 2015.

30. N. Frappolli, "Entropic lattice Boltzmann models for thermal and compressible flows", ETHZ Thesis, 2017.

31. S. S. Chikatamarla, S. Ansumali, I. V. Karlin, "Grad's approximation for missing data in lattice Boltzmann simulations", *EPL (Europhysics Letters)*, **74** (2), 215–254, 2006.

32. J. Wang, L. Chen, Q. Kang, S. S. Rahman, "The lattice Boltzmann method for isothermal micro-gaseous flow and its application in shale gas flow: a review". https://arxiv.org/pdf/1508.01562.

....

EXERCISES

1. Compute the gain in flow rate due to a slip velocity equal to 1/10 of the central velocity U_c. What is the corresponding gain upon halving the channel height?

2. Compute the capillary rise of water ($\sigma \sim 0.07\,N/m$) on a cylindrical capillary of radius 1 *mm*.

3. Compute the geometrical coefficient C in the Washburn law, eq. (29.16) for the case of a planar Poiseuille flow.

30

The Fluctuating Lattice Boltzmann

Fluid flow at nanoscopic scales is characterized by the dominance of thermal fluctuations (Brownian motion) versus directed motion. Thus, at variance with lattice Boltzmann models for macroscopic flows, where statistical fluctuations had to be eliminated as a major cause of inefficiency, at the nanoscale they have to be summoned back. In this Chapter, we discuss this "nemesis of the fluctuations" and describe the way they have been inserted back within the LB formalism. The result is one of the most active sectors of current Lattice Boltzmann research.

I frequently hear music in the very heart of noise

(G. Gershwin)

30.1 Nanoscale flows

The study of fluid flow at nanometric scales has witnessed major progress in the last decade, mostly under the impressive drive of nanotechnological advances, spawning numerous applications where fluid dynamics meets with allied disciplines, such as material science, chemical engineering, biology and medicine.

By definition, the dynamics of nanoscale flows is highly exposed to molecular fluctuations, due to the small volumes involved (1 cubic nanometer of water contains about 30 molecules). In principle, this rules out a continuum description, and even though such description may sometimes persist down to nanometric scales, the numerical simulation of nanoflows is usually demanded to atomistic methods, most notably molecular dynamics (MD). Yet, whenever molecular specificity is not the dominant factor, one may plausibly argue that a continuum, or a mesoscopic mean-field picture, enriched with fluctuations, may still capture the essentials of nanofluid motion at a much lower computational cost than MD.

This is precisely the framework which the fluctuating lattice Boltzmann equation (FLBE) described in this Chapter inscribes within.

The Lattice Boltzmann Equation. Sauro Succi, Oxford University Press (2018).
© Sauro Succi. DOI: 10.1093/oso/9780199592357.001.0001

The reader may notice some irony here, or, better said, the "nemesis" of fluctuations: LB was invented precisely to get rid of statistical fluctuations and now we need them back [77] ! Indeed, the study of nanoflows with FLBE represents one of the most active areas of current LB research.

The study of nanoscale flows becomes particularly rich in the case where such flows contain suspended bodies, such as colloids, cells or biological molecules. In this case, the fluid acts as a *solvent* for the *solute* molecules, typically much larger and heavier than the solvent ones. A typical example of great relevance to many engineering applications, soft-matter and biology are colloidal suspensions (1).

The numerical simulation of suspensions is traditionally handled by mesoscopic stochastic particle methods, such as the Brownian–Stokesian dynamics, in which the effects of nanoscopic scales on the mesoscopic ones are represented via stochastic terms in the equations of motion (Langevin equations, see Part I).

Apart from direct particle-particle collisions, which become dominant at high concentrations, the suspended particles interact through the intermediate of the surrounding solvent via long-range correlations, also known as hydrodynamic interactions (HI).

The HI configure a computationally demanding classical many-body problem.

At their physical root, these long-range forces emerge from the coherent superposition of short-range interactions between the suspended macro-particles and the fluid molecules. However, a fully molecular treatment is faced with a daunting scale separation between the solvent and solute molecules, typically nanometers and below versus microns. Since these are diffusion dominated, near-zero Reynolds number flows, the gap in time is even more daunting than the one in space, easily six decades and more.

This fuels a major stream of modern computational fluid-dynamics aimed at developing multiscale methods to bridge these gaps.

Clearly, it would be unrealistic to expect that LB can take this all, in fact, no single method can. However, one may realistically try to put the dual particle-field nature of LB at work in the hope of making a fresh contribution to the gap problem discussed above.

To handle scale separation, two substantial extensions of the LBE methodology are called for:

1. *Incorporation of fluctuating forces,*
2. *Handling of fluid–solid moving boundaries.*

In this chapter, we shall focus on the former leaving the latter to Chapter 31.

---

[77] Personal unglorious note: back then, Tony Ladd, the main character of the FLBE plot, was kind enough to email me for an opinion on the potential use of reintroducing fluctuations back into the then-newborn LB scheme. Much to my regret, my reply was a dumb "what for?". Being the smart person he is, Tony wisely ignored my dumb reaction and went on to develop one of the most productive offsprings of LB to this day.

30.2   The Fluctuating LBE

As discussed in Part I of this book, flows at nanometric scales are fully exposed to the discrete nature of the underlying molecules. As a result, particle fluctuations can no longer be ignored and take a dominant role in the dynamics of the nanoscale fluid.

As a result of their interaction with the surrounding solvent molecules, suspended particles experience a systematic drag, as well as a random "kicks" due to thermal fluctuations.

The effects of these fluctuations grow with decreasing size and mass of the suspended particles, according to

$$\tilde{a} = \frac{k_B T}{m_p R} \tag{30.1}$$

where $\tilde{a}$ is the random acceleration experienced by a particle of mass m_p and radius R, suspended in a solvent fluid at temperature T, see Fig. 30.1.

For a typical colloid of radius $R = 1\,\mu$m and a density comparable to water, the mass m_p is of the order of 10^{-15} kg, corresponding to a random acceleration of about 1 m/s^2.

This is typically several orders of magnitude larger than the inertial acceleration,

$$a_I = \frac{u^2}{R}$$

where u is the directed motion of the colloid. Typically, such directed motion from the application of an external potential $V(x)$, so that, in the high-damping limit relevant to nanofluidics, one has

$$u(x) \sim -\frac{1}{m_p \gamma} \frac{\partial V(x)}{\partial x}$$

γ being the friction coefficient due to particle collisions with the solvent molecules, see Fig. 30.2.

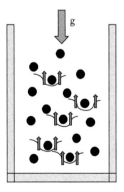

Figure 30.1 *A colloidal suspension under the effect of gravity. The colloidal particles fall to the bottom under the effect of their own weight. In the process, they displace the surrounding fluid, which exerts a drag force (upward arrows) opposing the downfalling motion. The displacement of the surrounding fluid is also responsible for long-range hydrodynamic correlations indicated by the wavy lines in the figure.*

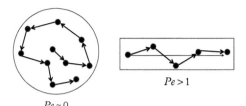

Figure 30.2 *Brownian motion at near-zero Pèclet number (left) and directed motion at Pe > 1 (right).*

The relative importance of directed versus Brownian motion for mass transport is conveniently measured by the Péclet, the accent on e is acute number,

$$Pe = \frac{uR}{D}$$

where

$$D = v_T^2/\gamma$$

is the diffusion coefficient of the colloidal particle in the solvent.

Most colloidal dispersions at equilibrium feature $Pe \ll 1$.

Given that mass scales like R^3, the random acceleration grows like the fourth power of the inverse radius, which explains why thermal fluctuations play a major role on the motion of nanoscale bodies.

The inclusion of such fluctuations and their coupling to the motion of suspended particles is a central subject of the theory of hydrodynamic motion at micro and nanoscales, a subject known as *fluctuating hydrodynamics*.[78]

The basic idea of fluctuating hydrodynamics (2) is that thermally induced fluctuations can be represented as random fluctuations in the fluxes of the conserved quantities.

In particular, the momentum exchange due to random interactions is cast in the form of a fluctuating stress tensor, which adds to the standard stress tensor of the Navier–Stokes equations.

The fluctuating LBE (FLBE) is designed according to the top-down principles of lattice kinetic theory, namely a minimal kinetic equation recovering the equations of fluctuating hydrodynamics upon coarse graining the microscopic degrees of freedom in a discrete lattice.

This is a canonical way of taking kinetic theory closer to molecular dynamics.

By the Lapalisse theorem mentioned earlier in chapter 29, we know that the two cannot be made equal, and the best one can hope for is to develop an overlap region. The goal of FLBE is thus to maximize this overlap (see Fig. 30.3).

We shall focus our attention on the momentum flux tensor P_{ab}, whose divergence yields the force experienced by the fluid element.

---

[78] Fluctuating hydrodynamics can be regarded as a coarse-grained version of molecular hydrodynamics, i.e. the study of the collective motion groups of molecules at the nano and microscale, see J.P. Boon and S. Yip, *Molecular Hydrodynamics*, Mc Graw Hill, NY, 1980.

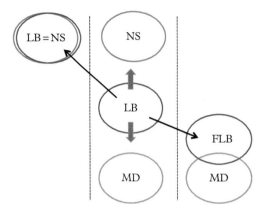

Figure 30.3 *Lapalisse theorem: LB lies in-between Navier–Stokes (NS) and Molecular Dynamics (MD) as "infrared" and "ultraviolet" limits, respectively (middle). The former (left) can be reached asymptotically via Chapman–Enskog, the latter(right) cannot. FLB aims at maximizing the MD–LB overlap.*

Since this tensor is explicitly tracked by LBE, it appears natural to implement fluctuations directly at the level of this tensor, rather than as a volume force.

Mathematically, this codes into a supplementary source term $\tilde{S}_i$ (the tilde stands for fluctuating) on the right-hand side of LBE, namely

$$f_i(\vec{x} + \vec{c}_i, t + 1) - f_i(\vec{x}, t) = -A_{ij}\left(f_j - f_j^e\right) + \tilde{S}_i \tag{30.2}$$

Since $\tilde{S}_i$ contributes only to the stress tensor, the fluctuating source takes the form (Latin indices summed upon):

$$\tilde{S}_i = A\, Q_{iab}\, S_{ab} \tag{30.3}$$

where the amplitude A is pinned down by the requirement of Gaussian statistics of the two-time correlator:

$$\langle \tilde{S}_{ab}\,(\vec{x}_1, t_1)\, \tilde{S}_{cd}\,(\vec{x}_2, t_2)\rangle = A\delta\,(\vec{x}_1 - \vec{x}_2)\,\delta\,(t_1 - t_2)\left(\delta_{ac}\delta_{bd} + \delta_{ad}\delta_{bc} - \frac{2}{3}\delta_{ab}\delta_{cd}\right) \tag{30.4}$$

The variance is adjusted in such a way as to define the temperature of the system, via the *fluctuation–dissipation* relation:

$$A = \frac{1}{3}\rho k_B T\left[1 - (1 + \lambda)^2\right] = \frac{2\eta k_B T}{\lambda^2} \tag{30.5}$$

where λ is the leading nonzero eigenvalue of the scattering matrix and η is the dynamic viscosity of the fluid.

The latter expression is not straightforward and follows from the solution of a discrete Langevin equation, details of which can be found in Ladd's pioneering paper (3).

Ladd brought solid evidence that both over- and under-relaxation, $-1 < \lambda + 1 < 0$ and $0 < \lambda + 1 < 1$, respectively, yield excellent agreement between theory and numerical experiment.

The expression (30.5) fills the first task on the way toward a LB for nanofluids, namely the formulation of an explicit expression of the fluctuating hydrodynamic stress tensor.

30.3   Some FLBE Liabilities

The early theory of fluctuating LB discussed is exposed to a number of liabilities and limitations, primarily:

i) *Lack of equipartition at small scales,*

ii) *Weak fluctuations (thermal speed much smaller than the sound speed).*

In the sequel, we shall dig a bit deeper into the issues and see how they can possibly be solved or at least mitigated.

30.4   Thermal Equipartition at All Scales

A direct way of testing equipartition is to check that the translational and rotational kinetic energy of the suspended bodies be an exact match with the thermal energy of the fluctuations, namely

$$\frac{1}{N}\sum_{p=1}^{N}\frac{m_p V_p^2}{2} = \frac{1}{N}\sum_{p=1}^{N}\frac{m_p \Omega_p^2 R_p^2}{2} = k_B T \qquad (30.6)$$

where Ω_p denotes the angular velocity of the p-th particle.

The early fluctuating LBE simulations showed that the translational and rotational degrees of freedom of the solid suspension come only into an approximate equilibrium with the underlying fluid, thus leading to small-scale violations of the Fluctuation-Dissipation Theorem (FDT).

The culprit is lack of energy conservation, which prevents complete equipartition between the modes of the system.

More precisely, it was found that the mean translational and rotational energy are almost equal, but typically some twenty percent lower than the solvent fluctuations. This lack of equipartition is theoretically disturbing and also practically, as it hinders the measurement of temperature dependent transport parameters such as the viscosity of the suspension.

Adhikari *et al.* noticed that, even for an isothermal system, the FDT relations as given in Ladd's original scheme are strictly valid only in the continuum limit $k \to 0$. Since such limit cannot be reached in any actual simulation, small-scale features of the flow are particularly exposed to FDT violations.

To restore FDT at all scales, Adhikari *et al.* proposed to extend the stochastic source to *all* non-conserved modes including the higher-order kinetic ones, the ghosts (4), see Fig. 30.4.

The rationale for such a move is that ghost modes act as an internal reservoir, eventually absorbing the fluctuating thermal energy supplied to the stress tensor.

Formally, one writes

$$\tilde{S}_i \equiv \tilde{S}_i^T + \tilde{S}_i^G = \frac{1}{2c_s^4} Q_{iab}\tilde{S}_{ab} + \sum_{k=n_h+1}^{b} w_i E_{ki}\tilde{S}_k \tag{30.7}$$

where $n_h = 1 + D + D(D+1)/2$ is the number of hydrodynamic (conserved+transport) modes in D spatial dimensions.

In (30.7), E_{ki} is the i-th component of the k-th eigenvector of the collision matrix (or any suitable basis vector in kinetic space) and $\tilde{S}_k$ is the source contribution to the corresponding k-th kinetic moment.

Note that the index k runs only within the ghost sector, $n_h + 1 \leq k \leq b$, since the $(1 + D)$ conserved modes are excluded and the $D(D + 1)/2$ components of the stress tensor are already taken into account by the first term on the right-hand side.

The fluctuating sources $\tilde{S}_k$ must obey their own FDT relations. Upon expressing the discrete distribution as the sum of a mean component $\bar{f}_i$ and a fluctuating one, $f_i = \bar{f}_i + \delta f_i$,

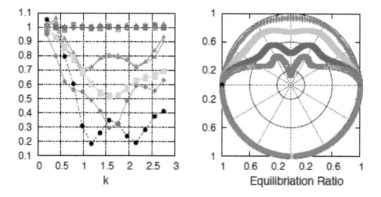

Figure 30.4 *Left panel: equilibration of the density (circle), momentum (triangle) diagonal (square) and off-diagonal (diamonds) components of the stress tensor as a function of the wave number for the D3Q15 lattice. Open versus solid symbols refer to implementations with and without ghost forcing. In the latter case, equipartition only holds below $k \sim 0.5$, while in the former it extends above $k \sim 2$. Note that $k = 1$ means a wavelength 2π longer than the lattice spacing $\Delta x = 1$. The right panel reports the magnitude of the Fourier transform of the horizontal current, $|\mathcal{J}_x(\vec{k})|$ at various $k = 0.62, 1.25, 1.88, 2.51$, top-to-bottom, as a function of the angle θ for the D2Q9 model with $\tau = 1$. Here θ defines the direction of the wavevector, i.e., $\vec{k} = k(\cos\theta, \sin\theta)$. The ghost-forced version reported in the lower plane shows very nice equilibration at all angles, as opposed to the non-forced one (upper plane). From R. Adhikari et al, EPL 71, 473, (2005).*

the solution of the linearized LBE leads to the following FDT relation for the ghost component of the fluctuating source:

$$\tilde{S}_i^G = \theta \sqrt{\rho\omega(2-\omega)} \sum_{k=h+1}^{b} w_i N_{k,i} r_k \tag{30.8}$$

where $N_{k,i}$ is the k-the orthonormal kinetic eigenvector, and

$$\theta = \frac{kT}{mc_s^2} \tag{30.9}$$

is the ratio of thermal to sound speed squared, $\omega = \Delta t/\tau$ and r_k is a uniformly distributed random number with zero mean and unit variance. The notation here has been slightly modified with respect to the original papers.

First, we refer to a single-relaxation time operator, second the kinetic eigenvectors are bi-orthonormal with respect to the weights w_i, namely

$$\sum_i w_i N_{k,i} N_{l,i} = \delta_{k,l} \tag{30.10}$$

$$\sum_k w_i N_{k,i} N_{k,j} = \delta_{i,j} \tag{30.11}$$

These scalar products are convenient, as they ensure that the ghosts do not project upon the local equilibrium.

The use of a collision matrix would simply turn the factor $\omega(2-\omega)$ into $\omega_k(2-\omega_k)$ within the sum over the various modes. The MRT version proves indeed beneficial in actual simulations.

The use of orthonormal eigenvectors leads to some change in the numerical factors. For the D2Q9 lattice (3–3–3 Conserved-Transport-Ghost modes), Duenweg and Ladd provide the full list of un-normalized eigenvectors, as reported in Tables 30.1–3, where $e_k = \sum_i w_i E_{k,i}^2$ is the norm of the un-normalized kinetic vector.

These are readily ortho-normalized via a simple rescaling:

$$N_{k,i} = \sqrt{\frac{w_i}{e_k}} E_{k,i} \tag{30.12}$$

Table 30.1 *D2Q9: Conserved Sector*

k	$E_{k,i}$	e_k
0	1	1
1	c_{ix}	1/3
2	c_{iy}	1/3

Table 30.2 *D2Q9: Transport Sector*

k	$E_{k,i}$	e_k
3	$3c_i^2 - 2$	4
4	$2c_{ix}^2 - 2c_i^2$	4/9
5	$c_{ix}c_{iy}$	1/9

Table 30.3 *D2Q9: Ghost Sector*

k	$E_{k,i}$	e_k
6	$(3c_i^2 - 4)c_{ix}$	2/3
7	$(3c_i^2 - 4)c_{iy}$	2/3
8	$9c_i^4 - 15c_i^2 + 2$	16

The same authors also provide the following list for the D3Q19 lattice ($4 - 6 - 9$ Conserved-Transport-Ghost modes), see Tables 30.4–6.

The conserved eigenvectors are the usual ones, just 1_i and c_{ia}, while transport (viscous) eigenvectors are linear combinations of the well-known projectors Q_{iab}. The ghost eigenvectors are less transparent. In principle, they can be obtained by a standard Gram–Schmidt orthogonalization, paying attention to discarding linear dependencies due the lattice identities, such as $c_{ia}^3 = c_{ia}$ and $c_{ia}^4 = c_{ia}^2$, ($a = x, y, z$) discussed earlier in chapter 14.

However, the physical meaning remains pretty opaque.

For the D2Q9 lattice, the ghost modes take the form of a second-order pseudo-current:

$$\tilde{\mathcal{J}}_a = \sum_i \tilde{\rho}_{2i} c_{ia}$$

where

$$\tilde{\rho}_{2i} \equiv 3c_i^2 - 4$$

Table 30.4 *D3Q19: Conserved sector*

k	$E_{k,i}$	e_k
0	1	1
1	c_{ix}	1/3
2	c_{iy}	1/3
3	c_{iz}	1/3

Table 30.5 *D3Q19: Transport sector*

k	$E_{k,i}$	e_k
4	$c_i^2 - 1$	2/3
5	$3c_{ix}^2 - c_i^2$	4/3
6	$c_{iy}^2 - c_{iz}^2$	4/9
7	$c_{ix}c_{iy}$	1/9
8	$c_{iy}c_{iz}$	1/9
9	$c_{ix}c_{iz}$	1/9

Table 30.6 *D3Q19: Ghost sector*

k	$E_{k,i}$	e_k
10	$(3c_i^2 - 5)c_{ix}$	2/3
11	$(3c_i^2 - 5)c_{iy}$	2/3
12	$(3c_i^2 - 5)c_{iz}$	2/3
13	$(c_{iy}^2 - c_{iz}^2)c_{ix}$	2/9
14	$(c_{iz}^2 - c_{ix}^2)c_{iy}$	2/9
15	$(c_{ix}^2 - c_{iy}^2)c_{iz}$	2/9
16	$(3c_i^4 - 6c_i^2 + 1)$	2
17	$(2c_i^2 - 3)(3c_{ix}^2 - c_i^2)$	4/3
18	$(2c_i^2 - 3)(c_{iy}^2 - c_{iz}^2)$	4/3

plus a fourth-order pseudo density:

$$\tilde{\rho}_4 \equiv \sum_i (9c_i^4 - 15c_i^2 + 2)$$

The prefix "pseudo" indicates that the ghost densities are not positive definite. For instance, since in both D2Q9 and D3Q19, $c_i^2 = \{0, 1, 2\}$, we have that $\tilde{\rho}_{2,i} = \{-4, -1, 2\}$ for rest particles, nearest neighbor and next-nearest neighbor particles, respectively.

For the D3Q19, the Duenweg–Ladd representation consists of two set of currents associated with two distinct second-order pseudo densities (Hermite level 3), plus three fourth-order pseudo densities (Hermite level 4).

As a mathematical curiosity, for the D2Q9 it is possible to find a "dual" representation with just one fourth-order ghost pseudo density and its associated current.

This is given by

$$\tilde{\rho}_{4i} = \frac{9}{2}c_i^4 - \frac{15}{2}c_i^2 + 1$$

Interestingly, this ghost density is obtained from the fluid density by just swapping the D2Q9 weights $w_i = \{0, 4/9, 4/36\}$ with their ghost partners, $\tilde{w}_i = \{1, -2, 4\}$, such that $\sum_i \tilde{w}_i = 0$.

That is, $\tilde{\rho} = \sum_i f_i$, $\tilde{\rho}_{4i} = \sum_i \tilde{w}_i p_i$, where we have defined the ghost weights as: $\tilde{w}_i = \rho_{4i} w_i$ and $p_i \equiv f_i/w_i$ are the reduced populations.

Dual representations are available for D3Q16 and D3Q25 lattices, but not for D3Q19. For details see Adhikari *et al.* (5).

While duality surely adds to the elegance and transparency of the ghost representation, so far it has made no compelling evidence of being superior to other representations.

As an aside, it may be of interest to note that the dual density and currents relate to gradients of the distribution in velocity space.

Indeed, by performing a quadrature of the identity $\int_{-\infty}^{\infty} \frac{df}{dv} dv = 0$, we obtain $\sum_i \tilde{w}_i p_i = 0$ for any arbitrary sequence p_i, which implies $\sum_i \tilde{w}_i = 0$.

To the best of this author's knowledge, the connection between ghost modes and velocity gradients of the distribution function remains largely unexplored.

30.5   Theoretical Foundations of the Fluctuating LB

The theoretical foundations of Ladd's formulation and its subsequent extension by Adhikari et al have been laid down in a series of elegant and useful papers by a number of authors (4; 6; 7). In passing, these developments also illuminate the relation between temperature and sound speed in FLBE, thereby clarifying the reasons why the former is much smaller than the latter, which results in a major limitation for its practical use.

In the sequel, we provide a brief survey of the main ideas behind these important developments.

30.6   FLB from Equilibrium Statistical Mechanics

As mentioned, FLBE is a literal nemesis of fluctuations; swept under the carpet at the very inception of LB to deal efficiently with macroscopic flows, fluctuations had to be summoned back for reinsertion in order to handle nanoscopic ones.

The reinsertion, as developed in original Ladd's version was typical top-down design: compliance with fluctuating hydrodynamics.

In the recent years, the more canonical bottom-up approach has also been brought to existence, mainly thanks to Duenweg and Ladd's derivation of the FLBE from a generalized lattice-gas model allowing for arbitrary occupation numbers along each discrete direction.

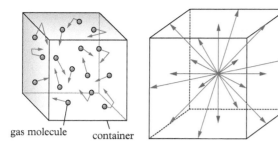

Figure 30.5 *The basic LB cell as a thermodynamic subsystem. In a real fluid (left) molecules move along any direction and any magnitude (below the speed of light), while in the LB cell (right) velocities are "frozen" to the discrete velocities.*

More precisely, each lattice cell is treated as a thermodynamic subsystem composed of $\Delta N = n\Delta x^3$ particles of mass m at temperature T (see Fig. 30.5).

By treating ΔN as a random variable, subject to Poisson statistics, the authors derive a FLBE which is consistent with both fluctuating hydrodynamics (top-down) and equilibrium statistical mechanics (bottom-up). Consistently with the foundations of statistical mechanics, the noisy source is formulated in terms of stochastic collisions based on a Monte Carlo process obeying detailed balance.

Within this procedure, the thermalization of all kinetic modes follows as a natural consequence of the detailed balance condition, thus placing Ladd's original formulation as well as subsequent extensions, on a solid theoretical basis.

Incidentally, this formulation also sheds light on practical corrections to Ladd's original scheme. For instance, the authors observe that also the amplitude of the ghost-noise term must receive a correction of the order of $(1-\gamma)^2$, γ being the relaxation parameter associated with shear and bulk stresses.

Physically, this is because the FDT relations describe a system at a more coarse-grained *time*scale than fluctuating hydrodynamics itself. Consequently, the $\delta(t-t')$ term in the FDT is replaced by a finite-memory term over a timestep Δt, namely $1/\Delta t$. Since the time integral of the time correlator must be the same, a renormalization prefactor is required in order to obtain the same macroscopic viscosity.

This is only a cursory account of an elegant but not-so-easy subject. The interested reader is kindly directed to the original publications.

30.7 The Boltzmann Number

The FLBE is shown to represent the most probable state of this generalized lattice-gas, the thermal fluctuations about the equilibrium state being dictated by the number of particles in the discrete states, as it is the case for actual molecules.

The relative intensity of the fluctuations, proportional to $1/\sqrt{\Delta N}$, is controlled by what the authors name the *Boltzmann number* defined as

$$Bo = \left(\frac{\theta}{n\Delta x^3}\right)^{1/2} \tag{30.13}$$

For most fluids $\theta \equiv v_T^2/c_s^2 \sim O(1)$, $\theta = 1$ corresponding to the ideal gas limit. The denominator is the number of molecules represented by a single LB particle, once we stipulate a number density $n_{lb} = 1$ in lattice units, which is always possible in an incompressible fluid.

The (squared) Boltzmann number can also be recast in the alternative form:

$$Bo^2 = \theta \left(\frac{d}{r_0}\right)^3 \left(\frac{r_0}{\Delta x}\right)^3 \tag{30.14}$$

where d is the interparticle distance and r_0 is the range of the intermolecular potential.

The second term on the right-hand side is the granularity parameter (see Part I) controlling the intensity of non-diluteness, while the third one is a direct measure of the spatial coarse graining, hence LB specific.

By definition, the standard non-fluctuating LB corresponds to the macroscopic limit

$$Bo \to 0 \tag{30.15}$$

To be noted that in a liquid, where $d/r_0 \sim O(1)$, and even more so in a dilute gas, $d/r_0 \gg 1$, the smallness of the Boltzmann number is entirely in charge of spatial coarse graining, i.e., $r_0/\Delta x \ll 1$.

In a nanofluid however, the latter condition is no longer guaranteed.

For macroscopic flows, the Boltzmann number is pretty small indeed, typically of the order of the inverse square root of the Avogadro number.

Take a millimeter cube of water, about 30 molecules per cubic nanometer: with $\Delta x = 10^{-3}$, $d/r_0 \sim 1$, and $\theta \sim 1$, we obtain $Bo \sim 10^{-10}$.

In microfluidics, the Boltzmann number gets larger: with $\Delta x = 10^{-6}$ m, we obtain $Bo \sim 10^{-5}$, still much smaller than 1, no point for FLBE.

Nanofluidics, though, tells another story; with $\Delta x = 10^{-9}$ m, $Bo \sim 0.2$.

By pushing LB even further down, to atomistic scales, $\Delta x = 0.1$ nanometers, see (8), the Boltzmann number may even get larger than one: the LB particles become "quarky", i.e., instead of representing a large collection of molecules, they represent a *fraction* of molecule!

Clearly, this is a very wildly fluctuating regime, for which the notion of FLBE as a weak perturbation on top of LBE is hardly tenable. In fact, at such sub-nanometric scales, the very notion of distribution function becomes shaky because of many-body density correlations (See Part I of this book).

The question is: *how can we possibly use FLBE altogether at such small scales?*

Here, mass and temperature come to some partial rescue, let us see how.

30.7.1   Thermal versus Inertial Mass

As anticipated, FLBE simulations work at $v_T^2 \sim 10^{-4} \div 10^{-3}$, much smaller than $c_s^2 \sim O(1)$, both in lattice units, for otherwise numerical instabilities arise due to the stochastic source being too strong for the purpose of numerical stability.

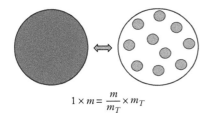

$$1 \times m = \frac{m}{m_T} \times m_T$$

Figure 30.6 *Variance reduction through the notion of thermal mass. One particle of mass m (dark circle) is statistically equivalent to $\frac{m}{m_T} \gg 1$ particles of mass $m_T \ll m$ (gray circles). Note that the radius is the same and mass is coded by the darkness of the circle.*

This constraint is conveniently analyzed in terms of the "thermal mass" introduced by Duenweg and Ladd (9; 10), namely

$$m_T \equiv \frac{k_B T}{c_s^2} = m\theta \tag{30.16}$$

where m is the mass of the solvent molecules, i.e., $m/m_p \ll 1$.

Given that the inertial mass is typically set to $m = 1$ in LB units (see Appendix on LB units), if the thermal mass were to coincide with the inertial one, i.e., $m_T = m = 1$, one would indeed obtain $v_T^2/c_s^2 \sim O(1)$ in lattice units.

Since in actual practice this ratio is of the order 10^{-3} at most, the thermal mass is much smaller than one, thus ensuring the condition $Bo \ll 1$ even when fluctuations in particle number are of order $O(1)$.

In practical terms, it is like replacing one particle of mass m with $m/m_T \gg 1$ particles of mass m_T, a procedure with many similarities to variance reduction techniques popular in Monte Carlo simulations (see Fig. 30.6).

30.8 FLBE from Fluctuating Kinetic Theory

Like for standard LB, the fluctuating LBE can also be derived by direct discretization of a corresponding fluctuating Boltzmann equation.

In another remarkable paper, following upon the lines of fluctuating kinetic theory (11), Gross *et al.*, cast the fluctuating discrete Boltzmann equation in terms of a suitable Onsager–Machlup theory of linear fluctuations.

The statistical properties of the noise term are fixed by invoking a fluctuation-dissipation theorem based on fluctuating *kinetic theory*, rather than Landau–Lifshitz fluctuating hydrodynamics.

The central object of fluctuating kinetic theory is the Klimontovich one-particle distribution defined as

$$\tilde{f}(\vec{x}, \vec{v}; t) = \sum_{p=1}^{N} \delta\left[\vec{x} - \vec{x}_p(t)\right] \delta\left[\vec{v} - \vec{v}_p(t)\right] \tag{30.17}$$

where the index p runs over the N particles in the system.

The corresponding Boltzmann distribution function is obtained upon averaging over an ensemble of particle trajectories, i.e.,

$$f(\vec{x}, \vec{v}; t) = \langle \tilde{f}(\vec{x}, \vec{v}; t) \rangle \tag{30.18}$$

By assuming ergodicity, this is equivalent to integrating upon a small but finite volume of phase space $\Delta\vec{x}\Delta\vec{v}$ and over a finite-time interval Δt.

By these simple definitions, it is apparent that the fluctuating LB is a coarse-grained *lattice Klimontovich equation*, in which the many-body correlations are represented through a Gaussian noise.

Schematically,

$$\tilde{f} \to \tilde{f}_g \to f \tag{30.19}$$

where the first arrow implies replacement of fluctuations with Gaussian noise, while the second is the $Bo \to 0$ limit of zero fluctuations.

This shows that the FLBE is inherently a theory of weak fluctuations.

It also indicates a potential route toward a Lattice Klimontovich equation, namely a fluctuating LB with *strong fluctuations and non-Gaussian noise*, a largely unexplored topic to which we shall return shortly.

As a matter of fact, since the Boltzmann equation is nonlinear, the resulting fluctuating distribution function $\tilde{f}$ is no longer a Gaussian stochastic process. However, the fluctuations $\delta f = \tilde{f} - f$ obey a linearized Boltzmann equation which, as shown by Fox and Uhlenbeck back in the 70s, can be formally recast in terms of a Gaussian Markov process.

These authors derived a fluctuating Lattice Boltzmann equation, which, in their own words, "*provides an efficient way to solve the equations of fluctuating hydrodynamics for ideal and non-ideal fluids alike.*"

The advocated advantage of treating fluctuations at the kinetic rather than the hydrodynamic level is that the kinetic formalism provides a unified framework to deal with ideal and non-ideal gases at the same time.

To the best of this author knowledge, the issue of how to deal with strong fluctuations, potentially hampering realizability of the distribution function, remains however open.[79]

Readers interested in an in-depth understanding of this fascinating subject are best directed to the original papers.

30.9   Numerical Stability and Computational Considerations

As previously discussed, stability considerations prevent the thermal energy density nk_BT from being comparable to the fluid pressure ρc_s^2, since this would entail "kicks"

---

[79] For promising recent developments on this topic, see M. R. Parsa and A. Wagner, "Lattice gas with molecular dynamics collision operator," *Phys Rev E*, **96** (1), 013314, Jul 21, 2017. DOI: 10.1103/PhysRevE.96.013314.

of order $O(1)$, which would ruin the realizability of the FLBE scheme, whence the constraint:

$$v_T^2 \ll c_s^2 \qquad (30.20)$$

Complying with such a strong constraint, say $v_T^2/c_s^2 \sim 10^{-3}$, would clearly collapse the timestepping procedure. For an elegant treatment of the mesh and timestep limitations of FLBE, see (12).

Applications based on the fluctuating LBE with suspended particles are obviously more demanding than standard LB ones. Apart from the cost of integrating the particle dynamics, a significant overhead stems from the computation of the fluid-solid interactions, which requires a significant amount of computational work at each particle-fluid interface, including grid-to-particle and particle-to-grid interpolations. Moreover, the implementation of the stochastic force requires a random number at each lattice site and timestep, which costs several floating point operations.

A few simple estimates help put things in perspective.

A cube of side L lattice sites along each dimension, with N suspended particles of diameter D, leaves an interparticle gap:

$$h = d - D \qquad (30.21)$$

where

$$d = L/N^{1/3} \qquad (30.22)$$

is the mean interparticle distance (center to center, see Fig. 30.7).

Hence, a moderate-size simulation with, say $L = 200$, can host of the order of $N \sim 10^3$ particles of diameter $D = 10$, corresponding to a gap $h = 20 - 10 = 10$.

A large-scale simulation with $L = 10^3$, on the other hand, can easily host millions of well-resolved particles; with $N = 10^6$ and $D = 50$, the relation (30.21) yields $h = 100 - 50 = 50$.

In general, the condition that the gap be well resolved, i.e., $h \gg 1$ (in lattice units), implies the following limitation on the number of particles:

$$N \ll N_{max} = \frac{L^3}{(1 + D)^3} \sim (L/D)^3 \qquad (30.23)$$

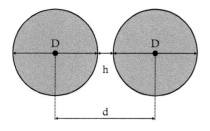

Figure 30.7 *Two colloidal particles of diameter D at a near-contact separation h = d − D smaller than D. In the figure, d is the mean interparticle distance.*

Or, conversely,

$$L \gg L_{min} = (1 + D)N^{1/3} \sim DN^{1/3} \tag{30.24}$$

The overall computational timescales like

$$t_{CPU} \sim t_F L^3 + t_P N_P + t_X N_P D^2$$

corresponding to the fluid-fluid, particle-particle and fluid-particle components, respectively. The coefficient t_F is the CPU time to update a single-lattice fluid cell without any particle, t_P the time cost of updating a particle without any surrounding fluid, while the third term collects the cost of fluid-particle, particle-fluid coupling, is actually the most demanding component as the particle concentration is increased. In the above expression, we have taken D^2 on the assumption that the particle-fluid cross-talking region scales like the area of the fluid-particle interface. However, the actual form of this term depends on the specific implementation (see chapter 31).

A notable aspect is the linear scaling of this term with the number of fluid particles, which stems from the local nature of the fluid-particle interaction, as opposed to long-range interactions characterizing stochastic particle methods, such as Brownian dynamics.

This linear dependence is key in allowing FLB simulations with a very large number of particles, up to hundreds of millions, which are out of reach for long-range particle methods.

Indeed, back in the mid 90s, Ladd reported simulations with 64 000 particles on mid-range workstations, significantly higher than typical values for Brownian dynamics simulations of that time. Nowadays, upper edge, massively parallel LB simulations with billions of cells and hundreds of million particles have been become available.

A word of caution is in order, though. Due to long-range hydrodynamic correlations, the box size L cannot be chosen based on purely static geometric criteria, such as the relation (30.24).

Owing to long-range hydrodynamic correlations, finite-size effects play a major role on the physics of suspensions, the result being that the box size L must scale with higher powers than $N^{1/3}$, thus leading to superlinear computational cost in the number of particles (13).

A final remark concerns the cost of *accessing* data versus *processing* them. As noted previously, the LB overall performance is increasingly more dependent on the former than the latter and consequently any semi-quantitative performance assessment should take proper account of memory organization issues.

30.10   Loose Ends and Future Developments

The theory of FLBE has witnessed significant advances in the last decade, both on practical and foundational fronts. However, several open items still remain. The subject is not an easy one, as it presents a rather inextricable tangle of in-depth technicalities and fundamental physical issues, which we now briefly comment upon.

30.10.1 Validity of FDT Under Confinement

One of the most fundamental physics issues is the validity of the FDT, as we know it, for highly confined fluids, most notably in the vicinity of confining walls. We have already noted that FDT relations are based on the assumption of Gaussian fluctuations, hence, strictly speaking, only hold for near-equilibrium fluids.

On the other hand, it is known that equilibrium conditions are generally violated in the vicinity of solid walls, where the fluid flow necessarily develops strong spatial inhomogeneities.

Deviations from equilibrium can be pretty drastic; the motion of solid objects of nanometric size in the vicinity of solid walls is known to develop hydrodynamic singularities. For instance, the dynamic viscosity of a nanosphere impinging on a solid wall shows a divergence of the viscosity perpendicular to the wall, as the distance goes to zero. In other words, Stokes's law:

$$\vec{F}_d = 6\pi \eta R \vec{V} \tag{30.25}$$

expressing the force experienced by a sphere of radius R moving at speed $\vec{V}$ in a solvent of dynamic viscosity η holds only in free space. Once the sphere gets in the vicinity of a planar wall, the drag force shows a significant, anisotropic increase, due to the hydrodynamic pressure resulting from the interaction with the planar wall (see Fig. 30.8).

This enhancement develops into a formal singularity in the limit of zero distance from the wall. The particle then experiences a local viscosity that depends on the distance to the wall and on the direction of motion, i.e., parallel versus perpendicular. The analytical description of the viscosity increase in the parallel direction was first developed by Faxen in 1922 and has been the object of intense molecular dynamics simulations in the recent past.

As usual, the singularity is lifted by replacing the continuum equations with an atomistic representation, but the specifics of this regularization still make the object of active investigation.

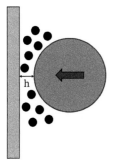

Figure 30.8 *A colloidal particle impinging on a flat wall. When the distance h becomes comparable with the size of the solvent molecules, continuum hydrodynamics goes under question. In particular, the dynamic viscosity perpendicular to the wall exhibits a R/h singularity in the limit h/R → 0.*

Under such conditions, the dynamic viscosity acquires a non-isotropic and singular dependence on the distance from the wall. Under more drastic conditions, particularly in strongly sheared nano-flows, the FDT might even fail altogether. Since FLBE is as good as the FDT statement it incorporates, it can only be trusted in conditions where FDT is failsafe (14).

30.10.2   FLBE Under Strong Confinement

Second, there remain a number of technical issues regarding the validity of FLBE itself, even in confined flows where the FDT is not under question.

For instance, Wagner *et al.* have recently pointed out that the standard FLB formulation violates Galilean invariance, in that the correlators show a dependence on the net-flow velocity (15). They propose an ingenious antidote against such violations, namely velocity-dependent multi-relaxation. In other words, the relaxation matrix would pick up a local and instantaneous dependence on the flow field. The authors recognize the "prohibitive" cost of such solution, but hasten to reassure the reader that the dependence needs not be machine-accurate, after all. Hence, it might eventually be implemented through pre-wired look-up tables with consequent major computational savings.

30.10.3   Handling Strong Fluctuations

As mentioned many times by now, the current FLB cannot handle strongly fluctuating regimes with $\frac{|\delta f|}{f} \sim 1$, unless the timestep is made prohibitively small. The main problem is broken realizability by strong fluctuating sources.

A radical way to cope with this problem might be to take the nemesis of fluctuations in earnest and go back to "primeval" LB models, such as the early Mc Namara-Zanetti version, if not a full many-body LGCA.

As discussed previously, such schemes feature built-in realizability and should therefore be able to handle much larger fluctuations than LB's with prescribed equilibria, either in LBGK or MRT form. Of course, efficiency issues must be carefully inspected, but with an open mind, since reasons that caused the decline of these models some thirty years ago need not hold true today. Another, less radical, option is to explore fluctuating entropic models and, perhaps, higher-order lattices too.

However, at the time of this writing, these are mere speculations; substantial work is needed to assess their viability.

30.11   Summary

Summarizing, the fluctuating Lattice Boltzmann equation permits us to reinstate the main effects of molecular fluctuations within the single-particle Lattice Boltzmann formalism. Initially developed in a pure top-down approach from fluctuating hydrodynamics, its theoretical foundations have lately been significantly elucidated and consolidated.

Significant work is still required to enhance its stability against strong fluctuations and also to place it on a solid theoretical basis for strongly confined nanoflows.

Such developments have potential to turn FLBE into a precious "tool of discovery" for non-equilibrium confined nanoflows at spatial and timescales not accessible to molecular dynamics simulations. All of the above touches at very basic topics at the crossroad between applied nanofluidics and fundamentals of non-equilibrium statistical mechanics of small systems far from the thermodynamic limit (16).

REFERENCES

1. G. Bossis and J. Brady, "Stokesian Dynamics", *Annu. Rev. Fluid Mech.*, **20**, 111, 1998.
2. L. Landau, E. Lifshitz, *Fluid Mechanics*. London: Butterworth-Heinemann, 1995.
3. A.J.C. Ladd, "Numerical simulations of particulate suspensions via a discretized Boltzmann equation". Part 1. Theoretical Foundations, *J. Fluid Mech.*, **271**, 285 1994.
4. R. Adhikari, K. Stratford, M. Cates and A. Wagner, "Fluctuating Lattice Boltzmann", *Europhys. Lett.*, **71**, 473, 2005.
5. R. Adhikari, S. Succi, "Duality in matrix lattice Boltzmann models", *Phys. Rev. E*, **78**, 066701, 2008.
6. B. Duenweg, A.J.C. Ladd, "Lattice Boltzmann simulation of soft matter systems", *Adv. Polym. Sci.*, **221**, 89, 2009.
7. M. Gross, R. Adhikari and M.E. Cates, "Thermal fluctuations in the lattice Boltzmann method for nonideal fluidsbibitem", *Phys. Rev. E* **82**, 056714, 2010.
8. J. Horbach and S. Succi, "Lattice Boltzmann versus molecular dynamics simulation of nanoscale hydrodynamic flow", *Phys. Rev. Lett.*, **26**, 924503, 2006.
9. B. Duenweg, A.J.C. Ladd, "Statistical mechanics of the fluctuating lattice Boltzmann equation", *Phys. Rev. E* **76**, 036704 2007.
10. B. Duenweg, U. Schiller and A.J.C. Ladd, "Progress in the understanding of the fluctuating lattice Boltzmann equation", *Comp. Phys. Comm.*, **180**, 605, 2009.
11. R. Forrest and G. Uhlenbeck, "Contributions to Nonequilibrium Thermodynamics. II. Fluctuation Theory for the Boltzmann Equation", *Phys. Fluids*, **13**, 2881 1970.
12. B. Duenweg, U. Schiller and A. Ladd, "Statistical mechanics of the fluctuating lattice Boltzmann equation", *Phys. Rev. E*, **76**, 036704, 2007.
13. B. Duenweg and A. Ladd, "Lattice Boltzmann Simulations of Soft Matter Systems", *Adv. Polym. Sci.*, **221**, 89, 2009.
14. H. Basagouglu, S. Melchionna, S. Succi and V. Yakhot, "Fluctuation-Dissipation relation from a FLB-LBGK model", *Europhys. Lett.*, **99**, 64001, 2012.

15. G. Kaheler and A. Wagner, "Fluctuating ideal-gas lattice Boltzmann method with fluctuation dissipation theorem for nonvanishing velocities", *Phys. Rev. E* 87, 063310 2013.
16. U. Seifert, "Stochastic thermodynamics, fluctuation theorems and molecular machines", *Reports on Progress in Physics*, 75 (12), 126001, 2012.

....

EXERCISES

1. Show that the Peclet number can be expressed as the product of the diffusive time across the colloid and the local-shear rate u/R.
2. Compute the time and space partial derivatives of the Klimontovich distribution. How do they connect to each other?
3. Write a 1d FLBE for a colloidal trapped into an optical potential, $V(x) = \frac{Kx^2}{2}, -L/2 < x < L/2$ and find out the highest-ratio $k_B T/KL^2$ consistent with numerical stability.

31

LB for Flows with Suspended Objects: Fluid–Solid Interactions

In the recent years the theory of the fluctuating LB, as it was proposed and developed by A.J.C. Ladd in the early 90s, has undergone major developments, both at the level of theoretical foundations and practical implementation. In this chapter we provide a cursory view of such developments, with special focus on the general formulation of fluid-solid interactions within the Lattice Boltzmann formalism.

Fluids tell solids how to bend, solids tell fluids how to move.

(readapted from John Wheeler)

31.1 Introduction

In Chapter 30, we have discussed the fundamentals of the fluctuating lattice Boltzmann formulation, namely the inclusion of statistical fluctuations within the LB formalism. We have also mentioned that this development is mostly aimed at micro- and nanoscale flows with suspended bodies, say colloidal particles, cells, polymers and macromolecules of various sorts.

Clearly, the rheological behavior of these suspensions is highly affected by the way the suspended particles interact with the fluid and among themselves. From the mathematical and computational standpoint, this configures a technically thick issue, namely the treatment of fluid-solid moving boundaries, in a more macroscopic-oriented context also known as *fluid-structure interactions* (FSI). In the sequel, we shall provide a description of a number of methods which have been developed to include FSI within the LB formalism. In particular, we shall cover the case of both *rigid* and *deformable* bodies, which are both vital to many applications in science and engineering.

As to the methodology, we shall refer to two main classes of methods:

1. *Collision-based Coupling* (CbC)
2. *Force-based Coupling* (FbC)

The Lattice Boltzmann Equation. Sauro Succi, Oxford University Press (2018).
© Sauro Succi. DOI: 10.1093/oso/9780199592357.001.0001

In the former the fluid structure (solvent-solute in soft-matter language) interactions are described through explicit collisions with the suspended body, whereas in the latter, such interactions take place through the exchange of action and reaction forces between the fluid and the suspended bodies.

It should be noted that in the collision methods only the fluid is loaded with thermal fluctuations, since momentum-conserving collisions are automatically in charge of transferring the fluctuating forces to the particles. In force-based methods, though, fluctuations are explicitly added to both fluid and particles, in order to ensure momentum conservation in compliance with fluctuation-dissipation theorem (FDT).

31.2   Ladd's Collision-Based Coupling Method

The first model including FSI within the (fluctuating) LB was presented in the trailblazing work by Tony Ladd (1).

In the original Ladd's method, the solid particle (colloid) is represented by a closed surface $\mathcal{S}$, for simplicity the surface of a rigid sphere in the following. When painted in the lattice, the surface $\mathcal{S}$ is replaced by a staircase approximation Σ, defined by the set of lattice links cut by $\mathcal{S}$. As a result, each particle p is associated with a staircased sphere of surface Σ_p (see Fig. 31.1).

The accuracy of this representation is only $O(1/R)$, R being the radius of the sphere in lattice units, so that, in actual practice, the simulations always deal with rough spheres, which is generally a source of numerical inaccuracy, but sometimes it is not so unrealistic condition, provided the numerical roughness $1/R$ can be made comparable to the physical one.

The next important question is how to deal with the lattice sites within the solid sphere.

In Ladd's formulation, both exterior and interior regions are filled with fluid. This simplifies the technical procedure while leaving the physics basically unaffected, as long as the *solid fraction*

$$\phi = \frac{4\pi \sum_{p=1}^{N} R_p^3}{3L^3} \tag{31.1}$$

remains negligible, $\phi \ll 1$, as it is the case for dilute suspensions.

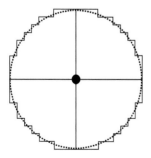

Figure 31.1 *Staircase approximation (solid) of a spherical particle (dotted), in two space dimensions. The lattice rugosity $1/R$ fades away in the "inflationary" limit $R \to \infty$.*

In (31.1), L denotes the side of a cubic box hosting N suspended particles of radius R_p.

The solid–fluid interaction proceeds as follows.

Having placed solid nodes in the middle point of a lattice link, we identify mirror pairs of directions say $(i, \bar{i})$ colliding head-on along the cut link. These mirror pairs are then treated via a straightforward generalization of the bounce-back rule taking into account the relative motion of the spherical particle to the surrounding fluid.

The velocity at a boundary site belonging to a particle p, and carrying the solid–fluid interaction across the $(i, \bar{i})$-th link, is given by

$$\vec{u}_b = \vec{v}_p + (\vec{r}_b - \vec{r}_p) \times \vec{\omega}_p \tag{31.2}$$

In (31.2), $\vec{r}_b = \vec{r}_s + \frac{1}{2}\vec{c}_i$ is the location of the boundary node along the i-th link emanating from the solid site $\vec{r}_s$ and connecting to the fluid site $\vec{r}_i = \vec{r}_s + \vec{c}_i$. All coordinates are relative to the center of the p-th particle, located at position $\vec{r}_p$ and moving with translation and angular velocities $\vec{v}_p$ and $\vec{\omega}_p$, respectively, see Fig. 31.2.

The timestep is made unit for simplicity.

This velocity sets the bias between colliding pairs:

$$f_i\left(\vec{r}_s + \vec{c}_i, t + 1\right) = f_i\left(\vec{r}_s + \vec{c}_i, t'\right) + 2\rho w_i u_{bi} \tag{31.3}$$

$$f_{\bar{i}}\left(\vec{r}_s, t + 1\right) = f_i\left(\vec{r}_s, t'\right) - 2\rho w_i u_{bi} \tag{31.4}$$

where t' denotes post-collisional states and we have set

$$u_{bi} = \frac{\vec{u}_b \cdot \vec{c}_i}{c_s^2}$$

Note that these rules reduce to the usual bounce-back conditions for a solid at rest, $\vec{v}_p = \vec{\omega}_p = 0$.

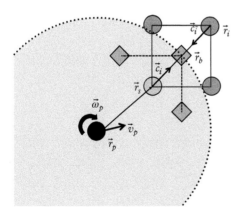

Figure 31.2 *Head-on collisions along a link cutting through the staircased surface (diamonds). The ideal surface is represented by the dotted line. In the figure $\vec{r}_i = \vec{r} + \vec{r} + \vec{c}_i$ and the discrete boundary nodes (diamonds) lie inbetween the lattice sites (circles).*

These collision rules produce a net momentum transfer between the fluid and the solid site:

$$\vec{F}_i\left(\vec{r}_b, t + \frac{1}{2}\right) = 2\vec{c}_i\left[f_i(\vec{r}_s, t') - f_i(\vec{r}_i, t') - 2\rho w_i u_{bi}\right] \tag{31.5}$$

The net force acting upon particle p is obtained by summing over all boundary sites $\vec{r}_b$ and associated interacting links, namely

$$\vec{F}_p = \sum_{\vec{r}_b, i \in \Sigma_{p,b}} \vec{F}_i(\vec{r}_b) \tag{31.6}$$

where $\Sigma_{p,b}$ denotes the set of links emanating from the boundary site $\vec{r}_b$ which connect with a fluid site by cutting across the lattice surface $\Sigma_{p,b}$,

Similarly, the total torque $\vec{T}_p$ is computed as

$$\vec{T}_p = \sum_{\vec{r}_b, i \in \Sigma_{p,b}} \vec{F}_i(\vec{r}_b) \times \vec{r}_b \tag{31.7}$$

We are now in a position to advance the particle position, speed $\vec{v}_p$ and angular momentum $\vec{\omega}_p$, according to Newton's equations of motion:

$$\begin{cases} \dfrac{d\vec{r}_p}{dt} = \vec{v}_p, \\[2mm] m_p\dfrac{d\vec{v}_p}{dt} = \vec{F}_p \\[2mm] I_p\dfrac{d\vec{\omega}_p}{dt} = \vec{T}_p \end{cases} \tag{31.8}$$

where m_p and I_p are the particle mass and moment of inertia, respectively.

For the purpose of stability, Ladd recommends to advance in time according to a leap-frog scheme:

$$\begin{cases} \vec{r}_p(t) = \vec{r}_p(t-2) + 2\vec{v}_p(t), \\[2mm] \vec{v}_p(t+1) = \vec{v}_p(t-1) + \dfrac{2}{m_p}\vec{F}_p(t) \\[2mm] \vec{\omega}_p(t+1) = \vec{\omega}_p(t-1) + \dfrac{2}{I_p}\vec{T}_p(t) \end{cases} \tag{31.9}$$

where

$$\vec{F}_p(t) = \frac{\vec{F}_p(t-1/2) + \vec{F}_p(t+1/2)}{2}$$

and

$$\vec{T}_p(t) = \frac{\vec{T}_p(t - 1/2) + \vec{T}_p(t + 1/2)}{2}$$

consistently with the fact that solid–fluid collisions take place at half-integer times.

This smoothing average is key to damping out short-scale torque oscillations triggered by discreteness effects at the particle–fluid interface.

By a similar token, the fluid speed is smoothed out by a three-point average:

$$\vec{u}(r, t) = \frac{\vec{u}(r, t - 1) + 2\vec{u}(r, t) + \vec{u}(r, t + 1)}{4} \tag{31.10}$$

This implies a memory of the overload ideas because three-time levels need to be kept track of, but according to Ladd, the gain is worth the pain.

This set of equations takes into account the full many-body hydrodynamic interactions, since the forces and torques are computed with the actual flow configuration, as dictated by the presence of *all N particles simultaneously*.

The FLBE coupled to the particle dynamic equations (31.8) provide the promised set of equations describing the self-consistent motion of the suspended particles within the fluid solvent.

Note that the fluctuations are injected into the fluid only, and dynamically transmitted to the particle through explicit collisions with the solute particle, see Fig. 31.3. This is a major departure from stochastic particle methods, such as Langevin and Brownian dynamics, in which the noise acts directly on the particle dynamics as a proxy of actual solute-solvent collisions. A major computational consequence is that Ladd's scheme only involves local operations, thereby achieving, in principle, *linear* computational complexity in the number of particles. As commented previously, the price is small timesteps. In addition, the method is actually laborious, as it requires to keep track of the dynamic connectivity between boundary nodes and the associated interacting fluid sites.

31.2.1 Numerical Tests

In a series of thorough papers, Ladd performed an array of validation tests, including (see Figs. 31.5–8):

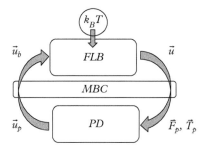

Figure 31.3 *Schematic CbC coupling between the fluctuating LB (FLB) for the fluid solvent with the particle dynamics (PD) for the colloids. The coupling is fully in charge of the moving boundary conditions (MBC), which promote the exchange of momentum and torque from the fluid to the particle ($\vec{F}_p$, $\vec{T}_p$) and vice versa. Formally, the particle is driven by the fluid via force and torque and reacts back to the fluid via the boundary velocity $\vec{u}_b$. Note that fluctuations ($k_B T$) are injected into the solvent alone and then transmitted to the particles through the boundary collisions.*

Brownian Dynamics   LB Dynamics

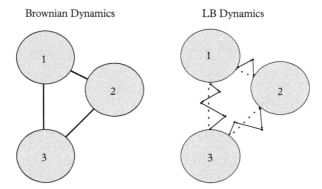

Figure 31.4 *Sketch of the basic difference between N-body Brownian dynamics and local LB dynamics. In Brownian dynamics, the three colloids 1, 2, 3 interact simultaneously with a strength inversely proportional to their center to center distance (thickness of the connecting solid lines). In the LB dynamics, the three colloids interact through direct collisions with the solvent particles as they move and collide in the solvent. For the figure in point, one, three and two (the number of kinks in the trajectories connecting the particles) solvent-solvent collisions for 1-2 , 1-3 and 2-3 colloid interactions, respectively.*

- Flow past a row of cylinders
- Flow past a cubic lattice of spheres
- Time-dependent hydrodynamic interactions (isolated moving sphere)
- Short-time velocity autocorrelation
- Short-time dynamics, prior to onset of Brownian motion

A detailed description of these tests can be found in the original papers. Here we just report some pictorial evidence.

These careful studies show excellent agreement with theoretical and previous numerical results. However, surface roughness may introduce slight deviations from continuum results (smooth surfaces) which fade away approximately like $O(1/R)$, in lattice units.

A further comment on the stochastic fluctuating LBE dynamics is in order.

A key task of fluctuating LBE is to regain non-Boltzmann correlations which fell alongside when moving from LGCA to LBE.

A probing test for the validity of this approach consists of measuring the decay of the linear and angular velocity autocorrelation functions:

$$A_v(t) = \langle v_p(t)\, v_p(0) \rangle \tag{31.11}$$

$$A_\omega(t) = \langle \omega_p(t)\, \omega_p(0) \rangle \tag{31.12}$$

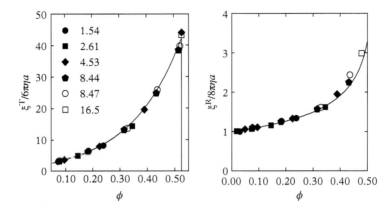

Figure 31.5 *Longitudinal and transversal drag of a cubic lattice of spheres as a function of the volume fraction. The solid lines are accurate solutions of the Stokes equations. (After (2), courtesy of A. Ladd.)*

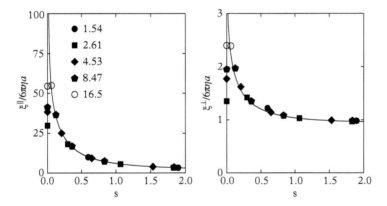

Figure 31.6 *Longitudinal and transversal drag between a pair of moving spheres as a function of the separation s. The solid lines are solutions of the Stokes equation in the same geometry. (After (2), courtesy of A. Ladd.)*

As is well known from the famous Alder–Wainwright molecular dynamics experiment (3), the hydrodynamic interaction between the moving particle and the surrounding flow induced by the particle motion itself promotes the persistence of the particle trajectory along the initial direction, hence a memory effect which invalidates Boltzmann's assumption of molecular chaos, see Fig. 31.5–6.

Mathematically, these non-Boltzmann memory effects manifest themselves through the algebraic decay of fluctuations, as opposed to purely exponential Boltzmann decay of uncorrelated motion, see Fig. 31.6. In particular, in three dimensions, $A_u(t)$ is expected to decay like $t^{-3/2}$.

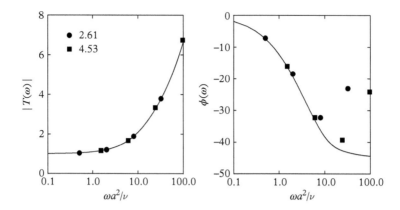

Figure 31.7 *Torque and phase lag of a rotating sphere as a function of the reduced frequency $\omega R^2/\nu$. Solid lines are taken from expressions given in Landau–Lifshitz. (After (2), courtesy of A. Ladd.)*

This algebraic decay and the ensuing long-lasting memory effects have indeed been reproduced in numerical experiments by Ladd (2) (the effect is measured at sufficiently short times, so as to rule out finite size effects).

The simulations also show that the normalized autocorrelation falls off in time exactly like the steady decay of the translational and rotational velocities of the sphere, thus proving that the fluctuating LBE obeys the fluctuation–dissipation theorem and reinstates (at least part of) the non-Boltzmann physics inaccessible to standard LBEs, see Fig. 31.7.

31.3 Improving on Ladd's Model

The application of LB techniques to flow with suspended bodies has experienced a burgeoning growth in the last decade. It comes therefore as no surprise that the original Ladd's method has been put under intense scrutiny and gone through to a series of major upgrades, see Fig. 31.8. As anticipated earlier, these methods can be grouped into two main classes:

i) *Collision-based Coupling* (CbC) and

ii) *Force-based Coupling* (FbC).

The CbC class follows the basic strategy of Ladd's model, namely describe the particle-fluid interactions via explicit collisions between the LB populations and the sites on the moving boundary. The FbC class, on the other hand, describes the fluid-particle interactions via suitable force fields.

Collision-based methods are *topological*, i.e., they require knowledge of the connectivity of the interacting sites (active links).

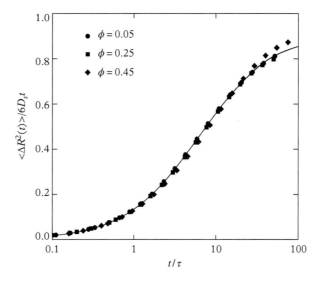

Figure 31.8 *Scaled mean-square displacement $\Delta R^2/6D_s t$ at short pre-Brownian times versus reduced time $t^* \equiv t/\tau(\phi)$, where $\tau(\phi)$ is a universal function of the solid fraction. Simulation for 128 spheres (solid symbols) are shown at various packing fractions Φ. (After (2), courtesy of A. Ladd.)*

The force-based methods, on the other hand, are *metric*, i.e., they identify the list of interacting nodes based on their relative distance. Programming-wise, the latter are simpler, as they require less bookkeeping.

The details, which may differ considerably even within the same class of methods, tend to be highly technical in nature. Therefore, in the following we shall provide only a cursory account, directing the reader keen of specifics to the original publications.

A very valuable guide, in this respect, is provided by the nice review of Aidun *et al.*, explicitly targeted to recent developments of LB techniques for complex flows (4). A similar statement goes for the very informative review of Duenweg-Ladd (8) on the LB methods for soft matter applications[80], and especially the recent book by Krueger et al.

31.3.1 Dealing with Impulsive Forces: the Cover-Uncover Transition

For matter of simplicity, Ladd's method treats the solid particle as a liquid with the same density of the solvent and a much higher viscosity. The idea is that the dynamics of the internal fluid relaxes to a rigid body motion within a short timescale, of the order of

---

[80] I am very grateful to these authors, for their papers and books provide me with a handy "escape-route" whenever my knowledge of the subject falls below the required standard ...,

$\tau_p \sim R_p^2/\nu_p$, ν_p being the kinematic viscosity of the colloidal particle. So, if all is well, the error is ephemeral and dissolves after a short transient.

Clearly this works well only as long as, i) the particle and the fluid have similar densities, which is indeed the case in many neutrally buoyant suspensions, ii) the suspension is very dilute, so that the particle contribution to the mass-momentum balance is negligible. Let us dig a bit deeper into these matters.

More precisely, the mass contained in a cubic cell of side d, the interparticle distance, is given by

$$M_c = \rho_p V_p + \rho_f V_f$$

where $V_p = 4\pi R_p^3/3$ is the volume of the particle, $V_c = d^3$ is the volume of the cubic cell and $V_f = V_c - V_p$ is the volume of the fluid in the cell.

Ladd's treatment is appropriate as long as the cell mass remains close to mass of the same fluid cell *without* any particle, namely i.e., $M_f^0 = \rho_f d^3$.

In equations,

$$gIP \equiv \frac{M_c - M_f^0}{M_f^0} = \frac{|\rho_p - \rho_f|}{\rho_f} \frac{4\pi R_p^3}{d^3} << 1 \qquad (31.13)$$

With a minor twist from Schiller's original definition (7) which applies to a lattice cell of side Δx instead of d, we call this ratio the global *Immersion Parameter*.

From (31.13), it is appreciated that $gIP \to 0$ for $\rho_p \to \rho_f$ as well as in the limit $R_p/d \to 0$, proving the assertion.

For dilute suspensions the issue is minor, but at high concentrations it becomes substantial even for particles with solvent-like density.

Technically, this bears directly on the so-called *cover/uncover* procedure, i.e., the change of "state" of fluid nodes which become "solid" upon being covered by the particle as it moves across the solvent, and conversely, those that transit from solid to fluid state upon being uncovered by the particle motion.

Let us expand a bit further on this technical but crucial topic.

Consider an interior site at time t which gets uncovered at time $t + \Delta t$ (see Fig. 31.9); the question is how to define the discrete distribution function at the newly uncovered site transiting from solid to fluid status.

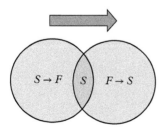

Figure 31.9 *Sketch of the cover/uncover procedure for a right-moving particle. Lattice sites in the region past the particle motion (left) must turn from solid to fluid, while those ahead of the particle undergo the reverse, fluid to solid, transition.*

Prior to the $S \rightarrow F$ transition, the site might be regarded as virtually filled with an equilibrium distribution with the local-particle density ρ_p and the local-fluid speed $\vec{u}_s = \vec{v}_p + \vec{\omega}_p \times \vec{r}_s$, $\vec{r}_s$ being the solid site position with respect to the particle centermass $\vec{r}_p$.

After uncovering, no unique recipe is available to fill in the newborn fluid site, but a reasonable option is to use the equilibrium distribution with the density and velocity taken as an average from a local neighborhood of fluid notes.

If the interior fluid is coupled to the particle motion the cover/uncover problem is basically no issue because the newly covered/uncovered nodes get automatically the right density and a near-exact velocity. If this is not the case, the average fluid velocity, say $\vec{u}_{av}$, is unlikely to match the solid-site velocity $\vec{u}_s$ and consequently the uncovered sites experience an impulsive force of the form

$$\vec{F}_{S \rightarrow F} = \rho_p(\vec{u}_{av} - \vec{u}_s)\frac{\Delta x^3}{\Delta t} \tag{31.14}$$

where the lattice-spacing $\Delta x = 1$ and the timestep $\Delta t = 1$ have been exposed for the sake of dimensional clarity.

A reciprocal expression holds for the covered sites, undergoing the fluid-to-solid transition. By taking $\rho_p = 1$ and $\vec{u}_{av} - \vec{u}_s \sim \vec{v}_p$, the expression (31.14) is readily seen to take values of the order $\vec{F}_{S \rightarrow F} \sim v_p \sim 10^{-3} \div 10^{-2}$, hence pretty substantial in lattice units.

Therefore, sudden covering/uncovering forces may easily generate substantial spurious pressure pulses which need to be kept at bay, especially for non-dilute suspensions where the pulse decay may easily exceed the interparticle collision time.

Several recipes have been put in place to this purpose.

Essentially, they all amount to some form of buffering in space and time. Higher-order time-marching schemes, such as fourth-order Runge–Kupta, or eventually implicit time stepping, have shown evidence of beneficial effects in terms of damping down the impulsive force eqn (31.14).

Also useful have proven space-buffering strategies wherein the particle surface is turned into a shell of finite-thickness $w > 1$ lattice units. The corresponding nodes are set to an intermediate state between the solid and fluid ones. The benefit is that the solid-fluid and fluid-solid transitions are smoothed out by a factor $1/w$, i.e., no node can transit from one state to another in a single timestep, but it takes w steps on average. Additional benefits may result from including non-equilibrium terms in the newly covered/uncovered nodes. On a technical side, such buffering techniques lean heavily on non-local spacetime interpolations, which add significantly to the complexity of a scheme which is already rather complex on its own. Clearly, we are far cry from the early-days motto *"Simple Models of Complex Fluids"*

31.3.2 Concentrated Suspensions: Near-Contact Problems

Another major issue, which is common to all numerical methods for suspended flows is the treatment of near-contact particle encounters, as they occur especially in non-dilute suspensions.

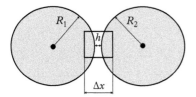

Figure 31.10 *Sub-grid near contact. The two colloidal particles come to a distance smaller than the grid size (square box). At such distances, lubrication forces, scaling like the inverse distance, can no longer be neglected.*

The problem is particularly acute whenever the boundary-to-boundary distance, h, becomes smaller than the lattice spacing (sub-grid near contact):

$$h < \Delta x \tag{31.15}$$

In the first place, the very notion of hydrodynamic motion within a subgrid distance h goes under question. In addition, it is clear that under the condition $h < \Delta x$, the same fluid node is shared by two particles; hence it must satisfy simultaneously *two* moving boundary constraints, leading to a potentially ill-posed formulation, see Fig. 31.10.

Worst yet, sub-grid near the contact can take place with no intermediate lattice nodes whatsoever, in which case using the standard bounce-back condition with two separate particle velocities leads to violations of global mass conservation.

Global mass conservation can be restored by bouncing nodes with the average velocity of the two nearby particles, but local violations still remain.

Local conservation can be restored by treating interior nodes as fluid ones and performing the stream-collide LB dynamics at both sides of the interface. For details, see Aidun–Clausen's review (4) and Kruger's et al. book.

As the reader may well appreciate, this entails significant bookkeeping to track of the time-changing connectivity of the fluid-solid interaction.

Other recipes have been put in place to ease out the procedure.

Some authors augmented the fluid-particle dynamics with lattice analogs of lubrication forces in the form of two-body forces. For instance,

$$\vec{F}_{12} = 6\pi\eta \left(\frac{R_1 R_2}{R_1 + R_2} \right)^2 \left(\frac{1}{h} - \frac{1}{h_c} \right) \hat{r}_{12}\hat{r}_{12} \cdot \vec{v}_{12} \tag{31.16}$$

where R_1 and R_2 are the particles radii, h is the size of the gap between the two spheres, h_c is a corresponding cutoff, $\hat{r}_{12}$ is the unit vector joining the centers of the particles and $\vec{v}_{12}$ is the relative centermass speed. Note that the lubrication forces scale like R/h; hence they diverge in the (non-hydrodynamic) limit $h/R \rightarrow 0$.

The form, (31.16), first proposed by Ladd, has been subsequently extended and modified to account for gap curvature and other higher-order effects.

By and large, these recipes prove pretty effective, and permit us to handle deep sub grid near contacts, down to $h \sim 0.01R$.

Nevertheless, an accurate and efficient treatment of sub-grid near encounters remains an outstanding issue in modern LB research, and one in constant need of improvement, especially in connection with multiscale applications of increasing complexity.

31.3.3   Higher-Order Boundary Conditions

As noted earlier, boundary conditions at solid boundaries split into two general families: *link-based* and *node-based*. The former are credited for second-order precision, as they place the interaction at mid-lattice locations. However, on-site bounce-back is often not too bad, since errors averaged over the entire boundary may often benefit from cancellations. Be as it may, both classes need to be extended so as to incorporate the relative motion of the particle to the fluid.

31.3.3.1   *Bouzidi-Lallemand bounce-back rule*

For link-based methods, the standard bounce-back must be generalized so as to accommodate boundaries that do not fall at the midpoint of the link. Typically, this implies that the only a fraction of the fluid population is bounced back, and that fraction is a function of the geometrical parameters of the boundary.

Such parameters might either be the solid fraction in the cell surrounding a lattice node, or other interpolation methods.

These methods interpolate from fluid nodes adjacent to the FBN (Fluid-Boundary Node), as reported in Fig. 31.11 for the case of linear interpolation (9).

As discussed in the chapter 17, q is the boundary distance defined as a fraction of the total link distance, $e_i = c_i \Delta t$. The Bouzidi-Lallemand boundary condition is flexible and efficient: it applies to generic moving boundaries wherein the coefficient q becomes a spacetime dependent field $q(\vec{x}; t)$. Indeed, this boundary condition has gained major popularity for LB simulations with curved and/or moving boundaries.[81]

It should also be reminded that its use in conjunction with the BGK single-time-relaxation operator is subject to the usual pathologies, namely the wall location depends on the actual value of the relaxation parameter τ. Such corrections, largely negligible for macroscopic flows, cannot be passed out in the case of nearby particle encounters.

For such situations, the common practice is to switch to the multi-relaxation time-collision operator.

Finally, problems of mass leakage are sometimes signaled, which this author is candidly unable to comment in any reasonable depth, since the detail dependence of these matters allows no substitute for first-hand experience. Just a word of caution for the reader.

31.4   Force-Based Coupling Methods for Fluid-Structure Interactions

Fluid-structure interactions are a consolidated mainstream in structural engineering, typically a happy marriage between finite-volumes for the fluid and finite-elements for the solid. Lately, they have been gaining major focus also in soft matter research, mostly in connection with their numerous LB applications to biological flows.

In the sequel, we shall focus mostly on this latter direction.

---

[81] For the record, it should be mentioned that the Bouzidi–Lallemand condition is rooted into the work of Ginzburg and D'Humières, described previously.

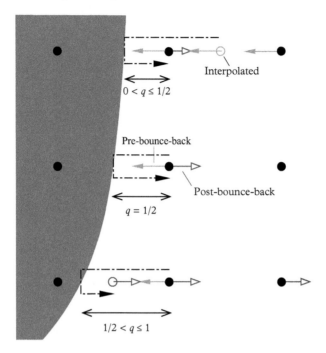

Figure 31.11 *The Bouzidi-Lallemand scheme. Top: The boundary is located at less than mid-link and thus bounce-back proceeds from an interpolated distribution (open circle), calculated from neighboring distributions (second and third black circles from the left). Middle: The boundary is located exactly at mid-link and therefore no interpolation is required. Bottom: The boundary is located beyond mid-link, thus the post-bounce-back takes place at an intermediate location (open circle) and then interpolated to the nearest fluid boundary node (second black circle from the left).*

31.4.1 Flexible Bodies: Polymers in LB Flows

Ladd's method trailblazed a series of schemes tracing extended *rigid* bodies within LB flows. Besides solid spheres, another example of great practical importance is given by *flexible* strings and their immediate application to polymer-solvent solutions.

Ahlrichs and Dunweg were the first to combine LB with force-based Langevin-like stochastic particle dynamics (SPD) models for polymer chains (10).

The two schemes are coupled via a frictional force proportional to the speed of the monomer (*bead* in computational parlance) relative to the local fluid speed, plus a stochastic component for the fluctuations. The friction coefficient is treated like a free tunable parameter subject to the fluctuation-dissipation theorem. The polymer is assumed to be formed by a chain of beads interacting via a pair potential $V(r)$ implementing excluded-volume effects, the effective size of the point-like monomers.

This comes in a typical Lennard–Jones form

$$V_{LJ}(r) = 4V_0 \left[\left(\frac{R}{r} \right)^{12} - \left(\frac{R}{r} \right)^6 \right]$$

(31.17)

where r is the separation between any two monomers of radius R and V_0 is the depth of the potential well.

The binding interactions are modeled via a spring-like nearest-neighbor potential of the form

$$V_{bind}(r_p, r_{p\mp 1}) = \frac{\kappa}{2} \left(r_{p,p\mp 1} - r_0 \right)^2$$

(31.18)

where $r_{p,p\pm 1}$ is the separation along the polymer chain of the two neighbors of the p-th bead, r_0 is the equilibrium distance and κ is the spring stiffness.

Each bead obeys the stochastic Newton equations for *point-like* particles:

$$\begin{cases} \dfrac{d\vec{r}_p}{dt} = \vec{v}_p, \\ m_p \dfrac{d\vec{v}_p}{dt} = \vec{F}^{int} + \vec{F}^h_p \end{cases}$$

(31.19)

where

$$\vec{F}^{int}_p = -\nabla_p (V_{LJ} + V_{bind}) + \vec{F}^h_p$$

is the interparticle force and $\vec{F}^h_p$ is the hydrodynamic force exerted by the fluid on the p-th particle.

These equations can be integrated in time with any of the efficient methods available from the stochastic particle dynamics(SPD) literature, typically with a shorter timestep than LB, since the polymer dynamics is usually significantly faster than the solvent one. A typical value is

$$\frac{\Delta t_{SPD}}{\Delta t_{LB}} \sim 1/10$$

The key issue is how to evaluate the force $\vec{F}^h_p$ exerted by the fluid on the p-th particle and the corresponding back reaction from the particle to the fluid.

The simplest recipe is to assume linear proportionality with the bead–fluid relative speed, namely

$$\vec{F}^h_p = -\eta R_p \left(\vec{v}_p - \vec{u}_p \right)$$

(31.20)

where η is the fluid dynamic viscosity.

The local-flow speed $\vec{u}_p$ at the bead location can be evaluated through a simple interpolation from the lattice grid points (G2P: Grid-to-Particles):

$$\vec{u}_p = \sum_g W_{gp} \vec{u}_g \qquad (31.21)$$

where the sum runs over all fluid grid points. In practice, this sum is typically confined to the lattice cell which contains the bead in point.

The G2P interpolation coefficients W_{gp} depend on the chosen interpolation scheme. A simple equipartition, $W_{gp} = 1/4$ (for the case of two spatial dimensions) proves often adequate.

Next we need is to compute the back-reaction force on the fluid.

A straightforward procedure is to interpolate this force on each of the grid nodes surrounding the bead and bias the corresponding populations according to the any of various schemes discussed in this book to impart momentum to the LBE populations.

In equations,

$$\vec{F}_g = \sum_p W_{pg} \vec{F}_p \qquad (31.22)$$

where the sum runs over all particles in the system. In practice such sums runs only within the dual cell centered around the grid position $\vec{r}_g$ (see Fig. 31.12–13).

The particle-to-grid (P2G) interpolation matrix W_{pg} is the adjoint of W_{gp}.[82]

Once the force on the generic grid node g is defined, the directional share on the i-th population is computed according to the standard expression:

$$\Delta f_i = \rho w_i \frac{\vec{F}_g \cdot \vec{c}_i}{c_s^2} \Delta t \qquad (31.23)$$

where we have reinstated the explicit dependence on the timestep for reasons to be detailed shortly. The expression (31.23) corresponds to a low-order implementation of the forcing term, just for purpose of illustration.

At each lattice site, the populations are then updated according to the usual procedure, namely

$$f_i = f_i + \Delta f_i \qquad (31.24)$$

where f_i at the right-hand-side is the unforced discrete distribution after the stream-collide steps.

The combination of LBE plus molecular dynamics supplemented with the grid-to-particle and particle-to-grid interpolation (31.21)–(31.22) yields a self-consistent scheme to evolve the polymer within the LB flow.

Ahlrichs and Dunweg demonstrated their method for a series of deterministic and stochastic experiments. The latter revealed a basic failure of the LBE–SPD scheme to recover the fluctuation dissipation theorem.

In order to repair this inconsistency, the authors introduced a stochastic force $\tilde{F}_a$, whose covariance is adjusted to the temperature according to the usual fluctuation–dissipation relation

---

[82] The adjoint satisfies the double orthonormality conditions $\sum_g W_{pg} W_{gq} = \delta_{pq}$ and $\sum_p W_{pg} W_{hp} = \delta_{gh}$, which secure momentum conservation.

$$\langle \tilde{F}_a\,(t)\,\tilde{F}_b\,(t') \rangle = 2k_B T \eta R \; \delta_{ab}\, \delta\,(t - t') \tag{31.25}$$

By adding this stochastic force to the friction force (31.20) and to the LB as well, the authors were able show that the corresponding Langevin dynamics achieves full compliance with the fluctuation dissipation theorem. For full details see the original reference.

The coupled LB-SPD dynamics proceeds through the following four basic steps:

1. *G2P: interpolate the fluid velocity from Grid to Particles (Gather) at time t.*

2. *PM: Particle Mover, compute the force on the particle and move it from t to t + 1*

3. *P2G: Interpolate the force from Particles to Grid (Scatter) at time t + 1*

4. *FM: Fluid Mover, advance the LB distributions from time t to t + 1 with the force $\vec{F}_g$*

These are schematically depicted in Fig. 31.12–15.

A schematic view of the force-based LB-SPD coupling scheme is reported in Fig. 31.16.

These multiscale techniques have been used extensively for biofluidic simulations, such as the translocation translocation through nanopores and biological membranes (11) and, more recently, the effects of hydrodynamic correlations on protein aggregation (12).

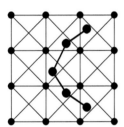

Figure 31.12 *A polymer chain swimming in an LBE flow.*

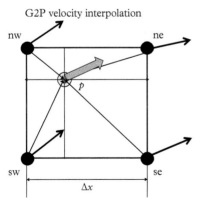

Figure 31.13 *Interpolating the particle velocity from the grid to the monomer location p (gray dot) within the LB cell. The summation runs over the four north-east (ne), north-west (nw), south-west (sw) and south-east (se) grid nodes. In the simplest instance one can take $W_{gp} = 1/4$ for all $g = 1, 4$. Another possibility is the "nearest takes it all" option, $W_{gp} = 1$ for the nearest g, namely nw in this figure, and $W_{gp} = 0$ for all other grid points. Yet another option is bilinear interpolation, in which $W_{gp} = 1 - A_{gp}/A$, A_{gp} being the area of the square defined by g and p and A is the area of the cell. Once the velocity $\bar{u}_p$ is known, the friction force can be evaluated via the eqn (31.20).*

P2G force interpolation

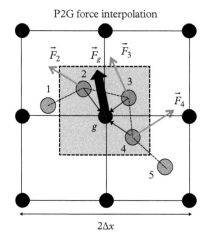

$2\Delta x$

Figure 31.14 *P2G force interpolation. The forces acting on the beads belonging to the dual control-cell centered at the grid location g are collected into the force $\vec{F}_g$. In this picture only beads labeled $2, 3, 4$ contribute to the force $\vec{F}_g$.*

G2P2G velocity-force gather-scatter interpolation

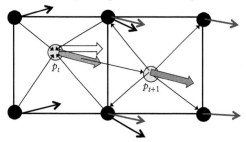

Figure 31.15 *Geometrical sketch of the full LB-SPD coupling. The particle p at time t receives the interpolated velocity from the fluid, $\vec{u}_p$, (G2P) which forms the viscous force $\vec{F}_p^h$. The particle is then advanced from t to $t+1$ by its equation of motion and finally transmits the force to the neighboring fluid grid-points (P2G), which can in turn be advanced to the next timestep. In computer parlance, the G2P and P2G interpolations are sometimes called Gather and Scatter, respectively.*

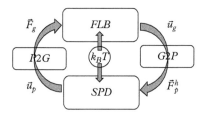

Figure 31.16 *Logical sketch of the LB-SPD coupling. To be noted that both fluid and particles are subject to stochastic forcing.*

31.4.2   Force-Based Coupling: the "Raspberry" Model

The Ahlrichs–Duenweg procedure previously described readily extends to higher-dimensional objects, typically two-dimensional membranes.

A simple way of dealing with fluid-membrane coupling is to treat the membrane as a set of point-like off-lattice beads connected trough flexible springs. The beads move along Lagrangian trajectories under the effect of the viscous drag from the fluid and the shape-restoring forces of the neighboring nodes in the membrane.

It should be noted that within this model, the membrane is permeable, in that a non-zero mass flow usually takes place across the boundary. Hence it cannot serve as a realistic model for the solid particles, such as colloids. However, at virtually zero Reynolds number (Stokes flow), this is arguably not a serious concern. For a very thorough analysis and modern refinements of the raspberry models see (13; 14).

31.4.3 A Few Words of Comment

Before moving further along, let us recollect a few basic points of the Duenweg–Ahlrichs force-based LB-SPD versus collision-based Ladd's method for colloidal particles. In the latter, the fluid sticks to the solid particle, in that the scheme is designed to prevent any relative motion between the two at the solid surface. The coupling is conservative because momentum is conserved in each and every collision and consequently the particles need no stochastic input since momentum fluctuations are transmitted self-consistently via the fluid-solid collisions. The coupling is topological and demands knowledge of dynamic connectivity of the fluid-solid interaction, which implies a significant algorithmic bookkeeping. Time marching generally proceeds with the same timestep for both fluid and particle components.

In the force-based LB-SPD model, the solid is permeable and the fluid generally slips on it. *Since the fluid-particle coupling is dissipative, both fluid and particles must receive a stochastic input.* The coupling is force based; hence it requires only the metric of the interactions, as combined with interpolations to manage the grid-to-particle and particle-to-grid information transfer. Time marching needs not proceed on the same timescale, the SPD component being usually advanced in smaller steps than the LB fluid. It is important to be aware that both interpolation procedures (short-scales) and finite-size effects (large-scales) concur to change the effective size of the particle as compared to its geometrical value. For instance, the effective radius of a spherical particle of bare radius R is given by:

$$\frac{1}{R_{eff}} = \frac{1}{R} + \frac{c}{\Delta x} - \frac{C}{L}$$

where c and C are positive constants and L is the size of the simulation box.

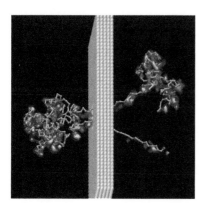

Figure 31.17 *Translocation of a double-stranded biopolymer through a membrane. The reddish color codes for hydrodynamic interactions. Reprinted with permission from M. Bernaschi et al, Nanoletters 8, 1115 (2008). Copyright 2008 American Chemical Society.*

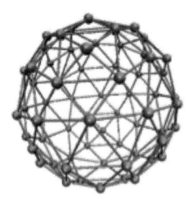

Figure 31.18 *Raspberry representation of the colloidal membrane. The nodes on the surface of the sphere interact with the surrounding fluid according to the Langevin equation (31.20). Courtesy of J. Horbach.*

To be noted that short-scale corrections make the effective radius smaller than the bare one, because they associate with backflow effects kicking the particle from the backside, thus decreasing dissipation, hence its effective radius.

Finite-size corrections, on the other hand, have just the opposite effect: they cut-off large-scale modes which would otherwise contribute to the particle mobility, hence the absence of these modes increases dissipation, i.e. the effective radius.

For a detailed description of these "renormalization" effects, see the review by Duenweg and Ladd.[83]

I am highly indebted to Burkhardt Duenweg for clarifying this point to me.

31.5   The Immersed Boundary Method

The Immersed Boundary Method (IBM for short, nothing to do with International Business Machines!) is a very elegant and powerful technique to deal with motion of deformable bodies moving within the flow. It is originally due to C. Peskin, in connection with the modelling of cardiovascular hemodynamics (25).

The basic idea of IBM is to treat the fluid (solvent) in Eulerian form, hence well suited to LB, and the immersed body as an off-grid Lagrangian structure (membrane) moving along with the fluid.

The main appeal of IBM is *dynamic adaptivity*, the membrane degrees of freedom move along Lagrangian trajectories, thus setting the computation free from expensive grid refinement around the immersed body.

The membrane is typically given a parametric representation of the form

$$\vec{X} = \vec{X}(\xi, \eta) \tag{31.26}$$

---

[83] This author is very indebted to Burkhard Duenweg for clarifying these matters to him.

where $0 \leq \xi \leq 1$ and $0 \leq \eta \leq 1$ are the Lagrangian coordinates describing the floating membrane $\mathcal{M}$ and $\vec{X} = (X, Y, Z)$ is the three-dimensional coordinate vector constrained to the membrane of equation $Z = Z(X, Y)$.

The fluid-membrane interaction takes place as follows: the membrane acts as a boundary, hence singular, source of force on the fluid, whose motion, in turn *fixes* the velocity on the membrane itself. Here, "fixes" means that the membrane moves exactly at the same local velocity of the fluid, i.e., no slip.

The force exerted by the membrane at the fluid location $\vec{x}$ at time t is expressed by a double integral over the membrane of the force $\vec{F}_m$, emanating from each single point $\vec{s} = (\xi, \eta)$ of the membrane, namely

$$\vec{F}_f(\vec{x}; t) = \int_{\mathcal{M}} \vec{F}_m(\vec{X}(\vec{s}); t)) \delta[\vec{x} - \vec{X}(\vec{s}; t)] d\vec{s} = \vec{F}_m(\vec{X}; t) \tag{31.27}$$

where $\mathcal{M}$ denotes the membrane manifold.

It is worth noting that, in the above, $\vec{F}_m$ is the force per unit area on the membrane, so that the total integral over the membrane provides the force acting on the fluid. Note that $\vec{F}_m$ in (31.27) includes the Jacobian of the transformation $\vec{s} = (\xi, \eta) \rightarrow \vec{X} = (X, Y, Z(X, Y))$.

The membrane force $\vec{F}_m$ is computed as the divergence of the elastic stress tensor of the membrane, which is taken as an input from elasticity theory. The details of this interaction may change considerably depending on the material, but in general, they imply a non-local and sometimes also non-liner coupling between different regions of the membrane, which are transmitted to the fluid via the relation (31.27)

On the assumption that each point of the membrane moves at the same speed of the fluid, one writes

$$\vec{u}_m(\vec{s}; t) = \int_{\mathcal{V}} \vec{u}_f(\vec{x}; t) \delta[\vec{x} - \vec{X}(\vec{s}; t)] d\vec{x} \tag{31.28}$$

where $\mathcal{V}$ is the entire volume of the fluid.

Under the same assumption, the equation of motion of the membrane is given by

$$\frac{d\vec{X}(\vec{s}; t)}{dt} = \vec{u}_m(\vec{s}; t) \tag{31.29}$$

The equation of motion of the membrane is dynamically coupled to the fluid equation, using a Lagrangian-particle method for the former and an Eulerian one (LB in our case) for the latter.

This set of simultaneous equations is generally solved by iteration, until the condition (31.29) is met.

In fact, such a condition is most appropriately regarded as a Lagrangian constraint on the fluid equations, which take the following general form:

$$
\begin{cases}
\dfrac{D\vec{u}_f}{Dt} = -\nabla p + \nu \Delta \vec{u}_f + \int \vec{F}_f(\vec{x}; t) \\[2ex]
\dfrac{d\vec{X}}{dt} = \int \vec{u}_m(\vec{X})
\end{cases}
\tag{31.30}
$$

In (31.30), D/Dt is the material derivative, and we have set $\vec{u}_m(\vec{X}; t) \equiv \vec{u}_m(\vec{s}; t)$.

The generic IBM algorithm proceeds through the following four basic steps:

1. *Compute the membrane force on the fluid, $\vec{F}_f$, eqn (31.27), at time t*
2. *Solve the fluid equations for the fluid velocity $\vec{u}_f$ from t to t + Δt*
3. *Compute the membrane velocity, $\vec{u}_m$, eqn (31.28), at time t + Δt*
4. *Advance the membrane position from t to t + Δt via eqn (31.29)*

This is the simplest instance, whereby the system (31.30) is solved explicitly, i.e. the fluid is advanced from time t to $t + \Delta t$ using the fluid force at time t and the membrane is advanced with either the velocity at time t or the newly updated velocity at time $t + \Delta t$.

For strongly coupled problems, this must replaced by an implicit procedure, whereby the coupled system (31.30) gives rise to a full nonlinear matrix problem, so that steps one to four must be iterated to convergence, each iteration requiring the solution of a linearized matrix problem. The procedure is computationally demanding but very robust and systematic.

31.5.1   Discrete Dirac's Delta: Smoothed Particles

Central to the accuracy and efficiency of the IBM procedure is a suitable smooth representation of the Dirac's deltas and the ensuing particle-to-grid and grid-to-particle interpolations.

The former step is achieved by replacing the Dirac's delta by finite-support smooth shape functions (also known as *smoothed particles*). The shape functions come in many sorts, but they all must share a few general properties, such as:

$$
\int_{\mathcal{V}} W_h(\vec{r} - \vec{r}')\, d\vec{r}' = 1, \quad \int_{\mathcal{V}} W_h(\vec{r} - \vec{r}')\vec{r}'\, d\vec{r}' = \vec{r}
$$

for mass and momentum conservation, respectively see,

In addition, the shape function must recover the Dirac's delta in the limit of zero support, namely

$$
lim_{h \to 0} W_h(\vec{r} - \vec{r}') = \delta(\vec{r} - \vec{r}')
$$

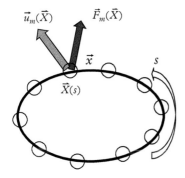

Figure 31.19 *The basic idea of the IBM scheme. The membrane (a closed string in the two-dimensional figure) exerts a force $\vec{F}_f(\vec{x})$ on the fluid and the fluid motion imposes the membrane velocity $\vec{u}_m(\vec{X})$. The membrane location $\vec{X}$ and the fluid location $\vec{x}$ sit on top of each other due to the Dirac's delta constraint. In the numerical implementation, the cross-talking region between the fluid and the membrane is smoothed out and extends over a few lattice sites depending on the chosen shape function (see text and Fig. 31.20). The fluid location $\vec{x}$ and the membrane location $\vec{X}$ are kept separate in the picture, but the Dirac's delta brings them on top of each other, a property which does not hold in the lattice representation (see Fig. 31.20).*

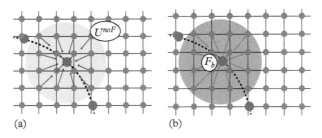

(a)   (b)

Figure 31.20 *The basic IBM interpolations: the smoothed particle imports the velocity from the fluid sites (left) and exports the force back (right). Reprinted with permission from A. Amiri et al, Phys Rev E 89 053312 (2014). Copyright by the American Physical Society*

The simplest choice is the piece-wise constant function (in one dimension for simplicity):

$$W_h(x) = \begin{cases} 1/h, & |x| < h/2 \\ 0, & |x| \ge h/2 \end{cases} \tag{31.31}$$

This is ill-suited to practical computations, since it suffers a discontinuity at the particle edge.

A better option is provided by the piecewise linear function:

$$W_h(z) = \begin{cases} 1 - |z|, & |z| < 1 \\ 0, & |z| \ge 1 \end{cases} \tag{31.32}$$

where we have set $z \equiv x/h$.

This is a popular choice for finite-element simulations, but not very suitable to IBM because the derivatives are discontinuous at the edge.

Higher-order shape functions are required. Among others, a popular choice are irrational splines of the form

$$
W_h(z) = \begin{cases} \dfrac{1}{8h}(3 - 2z + \sqrt{1 + 4z - 4z^2},\ 0 \le |z| < 1 \\[2ex] \dfrac{1}{8h}(5 - 2z - \sqrt{-7 + 12z - 4z^2},\ 1 \le |z| < 2 \\[2ex] 0,\quad |z| \ge 2 \end{cases} \tag{31.33}
$$

It is a simple exercise to check that such shape function goes to zero with its first and second derivatives at the particle edge $|z| = 2$. It should also be noticed that each smooth particle of the form (31.33) covers four h units, to be contrasted with the two of piecewise linear functions and just one for piecewise constants. Of course, $h \ge 1$ in lattice units.

With such finite-size and smooth particle representation of the Dirac's delta, the convolution integrals (31.27–28) are replaced by the corresponding discrete sums:

$$
\vec{F}_f(t) = \sum_{m=1}^{N_m} \vec{F}_m(t)\, W_{mf}(t)
$$
$$
\vec{u}_m(;t) = \sum_{f=1}^{N_f} \vec{u}_f(t)\, W_{fm}(t) \tag{31.34}
$$

where N_f is the number of fluid grid points, N_m the number of points discretizing the membrane while $W_{mf}(t)$ and $W_{fm}(t)$ are the discrete representations of the Dirac's delta connecting the membrane to the fluid and vice versa. Note that they depend on time through the Lagrangian motion of the membrane.

The IBM procedure is elegant and general, as it is not committed to any specific fluid solver, as long as an Eulerian method is used. Moreover, the membrane force, which in the original formulation derived from continuum elasticity theory, can be replaced by atomistic-oriented versions, such as beads and springs, in the framework of soft-matter simulations.

All of these points make IBM well suited to coupling with LB. Full details of the numerical LB-IBM procedure can be found in the book by T. Krueger et al.

31.5.2   Merging LB with IBM: Some Applications

One of the earliest LB–IBM mergers is due to Feng *et al.*, who simulated the sedimentation of weakly deformable colloidal particles (16). In the author's words, "This novel method preserves the advantages of LBM in tracking a group of particles and, at the same time, provides an alternative and better approach to treating the fluid-solid boundary conditions."

Further IBM-LB work deals with validation benchmarks of flows past static and moving cylinders (17; 18). The application of IBM–LB for macroscopic fluid-structure problems appears to be a fast moving front of modern LB research.

IBM–LB studies can be profitably undertaken at smaller scales as well, particularly in the context of biofluidics and blood flow micro-circulation problems. J. Zhang and collaborators used IBM–LB to simulate the motion deformable liquid capsules and its application to blood flows (19).

These authors performed two-dimensional simulations of single and multiple red blood cells in shear and channel flows and reported successful reproduction of several hemo-rheological features, such as tank-treading motion, cell migration from the vessel wall, slipper-shaped cell deformation, cell-free layers and others.

Another remarkable biofluidic implementation is due to Melchionna (20), who, similarly to Zhang *et al.*, represented colloidal particles by means of truncated-cosine smooth particles of the form

$$
W_h(x) = \begin{cases} \dfrac{1}{2h}\left(1 + \cos\left(\dfrac{\pi x}{h}\right)\right), & 0 \le |x| \le h \\ 0, & |x| \ge h \end{cases}
\tag{31.35}
$$

Melchionna *et al.* implemented red-cells as rigid ellipsoids based on three-dimensional anisotropic extensions of (31.35), and reported multiscale hemodynamic LB simulations in realistic human arteries with up to one billion lattices sites and several hundred millions solid bodies, thus reaching physiological hematocrit concentrations, near the physical scale of red blood cells, i.e., about ten microns (21) see, Fig. 31.21.

Such significant advantages do not come for free, since the volume integral involves of the order of h^3 lattice sites per colloid, typically about $4^3 = 64$ in practical applications.

According to the author, this computational overload is more than compensated by the benefits of smoothness.

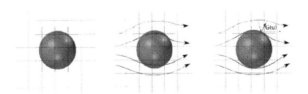

Figure 31.21 *Smooth particle representation of the colloidal particle. The particle-particle and particle-fluid forces result from a convolution over the volume of the particle. These forces (G in the figure) are also in charge of excluding the fluid from the particle volume; hence they are mostly aptly regarded as a dynamic constraint on the fluid motion. Courtesy of S. Melchionna.*

Further interesting IBM-LBM work can be found in Dupuis *et al.* (22), where non-slip boundary conditions on a non-grid conforming boundary are applied to the case of flows past cylinders. For a recent assessment of the LB-IBM method see the recent paper I. Mountrakis, E. Lorenz, A.G. Hoekstra, Revisiting the use of the immersed-boundary lattice-Boltzmann method for the simulation of suspended particles, Phys Rev E 96, 013302 (2017).

For a recent account of LB simulations of complex fluid-fluid interface dynamics, the reader is kindly directed to the informative review by Krueger and collaborators (23), as well as the recent monograph G. Falcucci and A. Montessori, Lattice Boltzmann Modeling of Complex Flows for Engineering Applications (2018), http://iopscience.iop.org/book/978-1-6817-4672-.

31.6 External Boundary Force: A Unified Framework for Force-Based Methods

As noted before, by stipulating that the membrane must move with the same local velocity of the fluid, the IBM method implements a built-in non-slip condition In discrete terms, this means that the membrane is represented as a collection of point-like *markers*, i.e., beads of zero mass which present no inertial response to the fluid motion. As a result, the IBM method is mostly suited to problems in which the membrane and the fluid move moreless on the same timescale.

Some authors have noted that the non-slip condition can also be attained within the context of a force-based SPD method by treating the fluid at the membrane location as a discrete collection of beads with mass

$$m_f = \rho_f \Delta x^3$$

ρ_f being the fluid density in the local cell of volume Δx^3.

At each discrete location $\vec{r}_p$ on the membrane, one defines the slip velocity as follows:

$$\vec{w}_p = \vec{v}_p - \vec{u}(\vec{r}_p) \tag{31.36}$$

The time-marching scheme is then designed in such a way that upon starting with a zero-slip at time $t = 0$, the same condition remains true at any subsequent discrete time

$$\vec{w}_p(t = 0) = \vec{w}_p(n\Delta t) = 0, \ n = 1, 2 \ldots.$$

The associated *External Boundary Force* (EBF) method is then shown to provide a second-order accurate time-discretization of the following set of stochastic particle equations:

$$\begin{cases} m_p \frac{d\vec{v}_p}{dt} = -\gamma(\vec{v}_p - \vec{u}_p) - \vec{\xi}_p - (1 - \mathcal{I})\vec{F}_p^{int} \\ m_f \frac{d\vec{u}_p}{dt} = \gamma(\vec{v}_p - \vec{u}_p) - \vec{\xi}_p + \mathcal{I}\vec{F}_p^{int} \end{cases} \tag{31.37}$$

where we have set $\vec{u}_p \equiv \vec{u}(\vec{r}_p)$ and $\vec{F}_p^{int}$ denotes the force experienced by the p-th bead as a result of the interaction with the other beads. Note that it is precisely these interactions that would prevent the no-slip condition on timescales longer than m_p/γ. In (31.37),

$$\mathcal{I} = \frac{m_p}{m_p + m_f}$$

is the *local immersion parameter* (7). As one can see, such parameter controls the distribution of the internal force between the fluid and the particles in a given cell. Interestingly, this parameter smoothly interpolates between the IBM and the force-based raspberry methods, which are attained in the limit $\mathcal{I} \to 0$ and $\mathcal{I} \to 1$, respectively. Thus, in a way, the EBF method provides an elegant unification between IBM and raspberry methods.

31.6.1 Time Marching

All along this chapter we have devoted comparatively little attention to the details of the time-marching schemes to advance the coupled fluid-solid system of equations. This is motivated by the high density of crucial details characterizing the subject, which are best appreciated by reading through the original publications. Nevertheless, we wish to call the reader attention on the fact that such details affect the physics of the problem and, as shown by Schiller, they also permit us to devise a unified picture of the various force-based methods through the aforementioned local immersion parameter.

31.7 Eulerian–Eulerian Multicomponent Models

Before closing, we wish to mention an interesting alternative to IBM–LB for the simulation of deformable bodies within fluid flows.

The idea is to represent the deformable body as a coherent "droplet" of a dense liquid navigating within a light one. Technically, this is achieved by formulating multicomponent LB models with forcing terms, explicitly designed in such a way as to reproduce the correct dynamics of the deformable membranes, say vesicles, capsules and similar biological objects (24).

These methods classify as grid-grid ("Eulerian–Eulerian"), as opposed to the grid-particle ("Eulerian–Lagrangian") nature of the IBM-LB procedure. The appeal of the Eulerian–Eulerian approach is the fact of being a fully-integrated LB scheme, with no need of expensive grid-to-particle and particle-to-grid interpolations.

However, the design of the proper forces implementing the desired interfacial dynamics is by no means is a simple task, because such forces must comply with a number of local and global dynamic constraints, including i) constant length of the (two-dimensional) closed membrane, ii) preferential local curvature, iii) bending rigidity, and iv) adjustable interface compressibility.

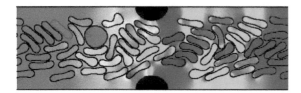

Figure 31.22 *Snapshot of pressure-driven motion of a collection of 58 vesicles in a constrained microchannel using the Eulerian–Eulerian multicomponent approach. Courtesy of I. Halliday.*

For the case of a two-dimensional droplet of dense fluid A within a light fluid B, the authors proposed the following expression (24):

$$\vec{F} = \frac{1}{2}\nabla\phi \, \kappa \left\{ [\sigma - \alpha_0(1 - L/L_0)] - \frac{b}{2}\kappa(\kappa^2 - \kappa_0^2) - b\frac{\partial^2\kappa}{\partial s^2} \right\} \qquad (31.38)$$

where $\phi = (\rho_A - \rho_B)/(\rho_A + \rho_B)$ is the phase field locating the interface at $\phi = 0$, κ is the local preferential curvature, σ the surface tension, α_0 the interfacial compressibility, b the bending rigidity, L the membrane length and s the curvilinear coordinate along the membrane. This force can be derived from a suitable free-energy functional.

As one can readily appreciate, the interfacial compressibility tends to keep the interface length at its reference value L_0 and the bending rigidity tends to maintain the local curvature around κ_0. In the limit $\alpha_0 \to 0$ and $b \to 0$, the force reduces to a pure gradient of the phase field, $\frac{1}{2}\kappa\sigma\nabla\phi$, with no control on the length and curvature of the interface.

A representative picture of red-blood cells motion based on the formulation is given in Fig. 31.22.

To be noted that this force is purely normal; hence it has no effect on the tangential degrees of freedom of the membrane. The latter, however, play a crucial role on the motion of the membrane on other substrates, the so-called *tank-treading* motion, typical for instance, of red blood cells motion on vascular walls.

Indeed, the authors report variations of the tangential speed v_t along the interface as large as 50%, in substantial breach of the tank-treading constraint $v_t = const$.

Thus, the tank-treading constraint requires additional forces, both normal and tangential to the interface.

Since the subject is still highly in flux, we leave it here directing the reader keen of the latest developments on this subject to the original authors.

31.8 Summary

As shown in this chapter, the LB modeling of fluid-structure interactions, both at macroscopic and microscopic scales is a very vibrant subject of ongoing research. Many options are available, with many technical variants which adapt to a number of different applications. In general, finite-size smooth representations seem to offer more simplicity original collision-based procedures, since the book-keeping of the fluid-particle interactions appears to be simpler and possibly more efficient as well. On the other hand, by dispensing with explicit fluid–particle coupling, the Eulerian–Eulerian approach holds a major potential in terms of computational efficiency The range of applications is wide-open and leaves enough room for many further developments and applications in soft matter and biofluidics as well.

REFERENCES

1. A.J.C. Ladd, "Numerical simulations of particulate suspensions via a discretized Boltzmann equation. Part 1: Theoretical Foundations", *J. Fluid Mech.*, **271**, 285 1994.
2. A.J.C. Ladd, "Numerical simulations of particulate suspensions via a discretized Boltzmann equation. Part 2: Numerical results" *J. Fluid Mech.*, **271**, 311 1994.
3. B. Alder and T. Wainwright, "Phase Transition for a Hard Sphere System", *J. Chem. Phys.* **27**, 1208 1957.
4. C. Aidun and J. Clausen, "Lattice Boltzmann method for complex flows", *Annu. Rev. Fluid Mech.*, **42**, 439, 2010.
5. B. Duenweg, A.J.C. Ladd, "Lattice Boltzmann simulation of soft matter systems", *Adv. Polym. Sci.*, **221**, 89, 2009.
6. B. Duenweg, A.J.C. Ladd, "Statistical mechanics of the fluctuating lattice Boltzmann equation", *Phys. Rev., E* **76**, 036704 2007.
7. B. Duenweg, U. Schiller and A. Ladd, "Statistical mechanics of the fluctuating lattice Boltzmann equation", *Phys. Rev. E,* **76**, 036704, 2007.
8. B. Duenweg and A. Ladd, "Lattice Boltzmann Simulations of Soft Matter Systems", *Adv. Polym. Sci.*, **221**, 89, 2009.
9. A. Bouzidi and P. Lallemand, "Momentum transfer of a Boltzmann-lattice fluid with boundaries", *Phys. Fluids*, **13**, 3452 2001.
10. P. Ahlrichs and B. Duenweg, "Simulation of a single polymer chain in solution by combining lattice Boltzmann and molecular dynamics", *J. Chem. Phys.*, **111**, 8225 1999.
11. M. Fyta, S. Melchionna, E. Kaxiras and S. Succi, "Multiscale coupling of molecular dynamics and hydrodynamics: application to DNA translocation through a nanopore", Multiscale Modeling and Simulation **5**, 1156–73 2006.
12. M. Chiricotto, S. Melchionna, P. Derreumaux, F. Sterpone, "Hydrodynamics effects on beta-amyloid peptide aggregation", *J. Chem. Phys.*, **145**, 035102 2016.
13. L. P. Fischer, T. Peter, C. Holm, and J. de Graaf, "The raspberry model for hydrodynamic interactions revisited. I. Periodic arrays of spheres, and dumbbells", *J. Chem. Phys.*, **143**, 084107, 2015.
14. J. de Graaf, T. Peter, L. Fischer and C. Holm, "The raspberry model for hydrodynamic interactions revisited. II. Periodic arrays of spheres and dumbbells". *J. Chem. Phys.*, **143**, 084108, 2015.

15. T.T. Pham, U. Schiller, J.R. Prakash and B. Duenweg, "Implicit and explicit solvent models for the simulation of a single polymer chain in solution: Lattice Boltzmann versus Brownian dynamics", *J. Chem. Phys.*, **131**, 164114, 2009.
16. Zhi-Gang Feng, E. Michaelides, "The immersed boundary-lattice Boltzmann method for solving fluid-particles interaction problems", *J. of Comp. Phys.*, **195**, 602, 2004.
17. X. Niu, C. Shu, Y.T. Chew and Y. Peng, "A momentum exchange-based immersed boundary-lattice Boltzmann method for simulating incompressible viscous flows", *Phys. Lett. A*, **354**, 173, 2006.
18. A. De Rosis, G. Falcucci, S. Ubertini, F. Ubertini and S. Succi, "Lattice Boltzmann analysis of fluid-structure interaction with moving boundaries", *Comm. in Comp. Phys.*, **13**, 3, 823, 2013.
19. J. Zhang, P.C. Johnson and A.S. Popel, "An immersed boundary lattice Boltzmann approach to simulate deformable liquid capsules and its application to microscopic blood flows", *Phys. Biol.*, **4**, 285, 2007.
20. S. Melchionna, "Incorporation of smooth spherical bodies in the Lattice Boltzmann method", *J. of Comp. Phys.*, **230**, 3966, 2011.
21. M. Bernaschi *et al.*, "MUPHY: A parallel MUltiPHYsics/Scale code for high-performance biofluidic simulations", *Comp. in Phys.* **180**, 1495, 2009.
22. A. Dupuis, P. Chatelain and P. Koumoutsakos, "An immersed boundary lattice-Boltzmann method for the simulation of the flow past an impulsively started cylinder", *J. Comp. Phys.*, **227**, 4486, 2008.
23. T. Krueger, S. Frijters, F, Guenther, B. Kaoui and J. Harting, "Numerical simulations of complex fluid-fluid interface dynamics", *Eur. Phys. J. Special Topics*, **222**, 177, 2013.
24. I. Halliday, S. Lishchuk, T. Spencer *et al.*, "Local membrane length conservation in two-dimensional vesicle simulation using a multicomponent lattice Boltzmann equation method", *Phys. Rev. E*, **87**, 023307, 2016.
25. CS, C. Peskin, "Numerical-Analysis of Blood Flow in Heart," *Journal of Computational Physics*, **25** (3), 220–252, 1977.

....

EXERCISES

1. Check the dimensional consistency of the eqn (31.25).
2. Compute the mass matrix $M(x, y) = \int W_h(x-z) W_h(z-y) dz$ for the case of piecewise-constant and piecewise-linear shape functions.
3. The same as Number 2 for Gaussian shape functions: what is the main qualitative difference between the piecewise and Gaussian cases?

Part VI

Beyond Newtonian Mechanics: Quantum and Relativistic Fluids

The lattice Boltzmann schemes discussed so far, all relate to classical fluids obeying classical Newtonian mechanics, without any relativistic or quantum effects.

Neither of these two restrictions is inherent the LB formalism, which can be extended to the case of relativistic as well as quantum fluids in a comparatively simple manner.

This final part of the book deals precisely with these two major subjects.

In passing, we wish to note that, besides relativity and quantum physics, the LB harness and kinetic theory in general, can serve as a very effective functional generator of a broad variety of linear and non-linear partial differential equations of mathematical physics, such as Maxwell's electromagnetism, Korteweg De Vries, Kuramoto-Sivashinsky, various types of non-linear Schrödinger equations and many others.

Covering this ground will take this book beyond reasonable size limits, and consequently the reader interested to these developments is kindly directed to the current literature.

We are now ready to proceed with quantum and relativistic LB.

32

Quantum Lattice Boltzmann (QLB)

> The lattice Boltzmann concepts and applications described so far refer to classical, i.e., non-quantum, physics. However, the LB formalism is not restricted to classical Newtonian mechanics and indeed an LB formulation of quantum mechanics going by the name of quantum LB (QLB) has been in existence for more than two decades. Even though it would be far-fetched to say that QLB represents a mainstream, in the recent years it has captured some revived interest, mostly on account of recent developments in quantum-computing research. In this chapter we provide an account of the QLB formulation: stay tuned, LBE goes quantum!

> *We're ugly, but we have the quantum physics* . . . .
>
> (readapted from Janis Joplin)

32.1 Fluid Formulation of Quantum Mechanics

Intriguing analogies between quantum mechanics and fluid mechanics have been pointed out since the earliest days of quantum theory (1). The common tenet still is that these analogies are purely formal in character and do not bear upon the basic physics of quantum phenomena. A less common view maintains instead that quantum mechanics and notably its philosophical hard core, Heisenberg's uncertainty principle, are nothing but a mirror of our ignorance of the underlying (hidden) microscopic physical level. This leads to the intriguing theory of "hidden variables" which can be traced back to the famous Einstein–Podolski–Rosen paradox and subsequent Bohm's non-local quantum mechanics (2; 3).

This debate has gained major momentum in recent years, mainly under the impact of mind-boggling experiments in the field of quantum optics which provide increasing support to the idea of quantum entanglement as an inescapable non-local feature of quantum mechanics. For the sake of completeness, we just remind that two quantum systems A and B (the popular Alice and Bob characters in virtually all quantum tales)

The Lattice Boltzmann Equation. Sauro Succi, Oxford University Press (2018).
© Sauro Succi. DOI: 10.1093/oso/9780199592357.001.0001

are entangled whenever the product state $C = A \otimes B$ contains states which cannot be obtained as a direct product of single eigenstates of A and B, separately.

In a more formal language, let

$$\phi^A(x) = \sum_i a_i \phi_i^A(x)$$

and

$$\phi^B(y) = \sum_j b_j \phi_j^B(y)$$

be the wavefunctions of A and B, and

$$\phi^C(x, y) = \sum_{i,j} c_{ij} \phi_i^A(x) \phi_j^B(y)$$

the wavefunction of the product state C.

Entanglement is described by the condition

$$c_{ij} \neq a_i b_j$$

Apart from the mathematical definition, the consequences are profound and often highly counterintuitive.

In particular, entangled states can be spectacularly non-local; take two entangled photons, next to each other, one spin-up and the other spin-down. Let them move far apart and then flip the spin of any of the two: as a result, the spin of the other is seen to flip too, with a delay much shorter than the light flight-time between them. This faster-than-light communication, until recently mostly the ground for wild (yet serious) speculations and movie scifi, is receiving increasing experimental support: entanglement and non-locality seem to be deeply rooted into the foundations of quantum mechanics (4), beyond the famous pragmatic stance "shut-up and calculate"!

With some regret, this fascinating topic goes beyond the scope of this work and we must therefore leave it here, turning instead to a pragmatic point: what can the fluid-quantum analogy do for us in terms of numerical modeling of quantum mechanical phenomena?

The question is perfectly legitimate because, regardless of its philosophical implications, the fluid analogy provides an intuitive and physical sound basis to develop numerical methods for time-dependent quantum mechanics. In particular, it appears reasonable to wonder whether and to what extent can the advantages brought about by LBE in fluid dynamics be exported to the context of quantum mechanics.

Before we put forward our discrete kinetic theory version of the analogy, it is useful to provide a cursory survey of the main ideas behind the analogy itself. To this end, a short recap of basic notions of quantum mechanics is in order.

32.2 The Fluid Formulation of the Schrödinger Equation

Let us begin with the Schrödinger equation for a non-relativistic quantum particle of mass m in an external potential $V(\vec{x})$:

$$i\hbar\,\partial_t\Psi = \left[-\frac{\hbar^2}{2m}\Delta + V(\vec{x})\right]\Psi \tag{32.1}$$

where $\Psi(\vec{x}, t)$ is the wavefunction of a material particle of mass m.

Upon multiplying (32.1) by the complex conjugate Ψ^*, and the complex conjugate of (32.1) by Ψ and then subtracting, we obtain

$$i\hbar\,\partial_t|\psi|^2 = -\frac{\hbar^2}{2m}\left(\Psi^*\Delta\Psi - \Psi\Delta\Psi^*\right) \tag{32.2}$$

A simple integration by parts of the right-hand side delivers the following conservative form, in coordinate notation:

$$\partial_t|\Psi|^2 + i\frac{\hbar}{m}\partial_a\left(\Psi\partial_a\Psi^* - \Psi^*\partial_a\Psi\right) = 0 \tag{32.3}$$

which is readily recognized as a familiar continuity equation for a quantum "fluid" with density

$$\rho = |\Psi|^2 \tag{32.4}$$

and current density

$$\mathcal{J}_a = i\frac{\hbar}{m}\left(\Psi\partial_a\Psi^* - \Psi^*\partial_a\Psi\right) \tag{32.5}$$

For the sake of the analogy with fluid motion, the wavefunction is normalized to the mass of the particle, namely

$$\int \rho(\vec{x}, t)\,d\vec{x} = m$$

Similarly, by multiplying by $\partial_a\Psi^*$ eqn (32.1) and its complex conjugate by $\partial_a\Psi$, then summing up, we obtain the evolution equation for the current density in fluid-conservative form:

$$\partial_t\mathcal{J}_a + \partial_b P_{ab} = 0 \tag{32.6}$$

where the momentum flux tensor of the quantum fluid is given by

$$P_{ab} \equiv P\delta_{ab} = \left[\frac{\hbar^2}{2m^2} (\Psi^* \Delta \Psi + \Psi \Delta \Psi^*) + \frac{V}{m}|\Psi|^2 \right] \delta_{ab} \tag{32.7}$$

The first two terms on the right-hand side will soon be shown to correspond to the kinetic energy plus a genuinely quantum potential, while the third term is a classical term resulting from the external potential. The quantum potential may also be regarded as a sort of non-local pressure, driven by quantum rather than thermal fluctuations. The above formulation configures the quantum wavefunction as an *inviscid, irrotational* and *compressible* fluid.

The inviscid character of the quantum fluid reflects the reversible nature of the Schrödinger equation, a diffusion equation in *imaginary* time.

The quantum analogy reveals itself in its full splendor in the so-called eikonal representation (5):

$$\Psi = Re^{i\theta} \tag{32.8}$$

where both the amplitude R and the phase θ carry a spacetime dependence.

Upon inserting this expression into eqn (32.5), we obtain

$$\mathcal{J}_a = \frac{\hbar}{m} \partial_a \theta \equiv \rho u_a \tag{32.9}$$

which defines the "macroscopic flow speed" u_a associated with the wavefunction. Here "macroscopic" means free from quantum fluctuations.

Likewise, by inserting (32.8) in (32.7), we obtain:

$$P = \rho \left(\frac{u^2}{2} + \frac{V}{m} + \frac{Q}{m} \right) \tag{32.10}$$

where:

$$Q \equiv -\frac{\hbar^2}{2m} \frac{\Delta R}{R} \tag{32.11}$$

is the famous quantum potential advocated by Bohm and co-workers to support the picture of quantum mechanics as an inherently non-local description of the microscopic world. Note, that the quantum potential represents the fluctuating component of kinetic energy due to the quantum fluctuations, as opposed to the systematic component provided by the first term at the r.h.s. of eq. (32.10). It is of some interest to note that the quantum potential bears a formal analogy with surface tension. To appreciate the point, just write $R = \rho^{1/2}$ and check by direct calculation that

$$Q = -\frac{\hbar^2}{2m}\left[\frac{1}{2}\frac{\Delta\rho}{\rho} - \frac{1}{4}\left(\frac{\nabla\rho}{\rho}\right)^2\right]$$

Apart from a prefactor $1/\rho^2$, this is formally the inhomogeneous interface term in the diagonal component of a classical non-ideal fluid pressor, with the important twist of a *negative* sign. Indeed, the quantum mechanical potential drives *loss of coherence*, which is just the opposite effect of classical (positive) surface tension.

The eikonal formulation naturally invites the picture of a cloud (fluid) of particles moving about their center of mass with a spatial distribution $\rho(\vec{x}, t)$, and a systematic speed given by the spatial gradient of the phase $\theta(\vec{x}, t)$ according to the expression (32.9) and a fluctuating speed associated with the square root of the quantum potential. In this respect, a formal analogy with fluid turbulence is also pretty inviting, although we shall not pursue it any further in this book.

Hard to resist the appeal of the celebrated de Broglie's "pilot wave" picture: the classical trajectory is just the average over the cloud of "particles" (hidden variables) in much the same way as fluid density and velocity represent the average over an ensemble of molecular trajectories.

So much for the analogy in the continuum world. What about the discrete lattice world? Interestingly enough, once transposed into the language of the lattice world, this analogy becomes even *more* compelling

In fact, the lattice formulation naturally calls for an "upgrade" from the non-relativistic Schrödinger equation to its relativistic associate, the Dirac equation.

Symbolically, the analogy goes as follows: *DE: Dirac Equation, SE: Schrödinger Equation, LBE: Lattice Boltzmann Equation, NSE: Navier–Stokes Equation,*

$$\boxed{\begin{array}{l} DE \rightarrow SE \\ LBE \rightarrow NSE \end{array}} \qquad \begin{array}{r}(32.12)\\(32.13)\end{array}$$

Incidentally, the physical $DE \rightarrow SE$ side of analogy is known since long as Foldy–Wouthuysen transformation (6), an elegant technique to derive quasi-relativistic approximations of the Dirac equation upon expanding in v/c as a smallness parameter.

Before venturing into the task of substantiating this analogy, for the sake of self-containedness, let us proceed with a short recap of the Dirac equation.

32.2.1 Relativistic Quantum Mechanics: the Dirac Equation

The Dirac equation is the "Wunderkind" of the marriage between the two high points of modern physics: relativity and quantum theory. It builds upon a formal requirement of Lorentz invariance on quantum mechanics, which implies a symmetric balance between space and time derivatives. Such a balance is manifestly broken by the Schrödinger equation, in which time derivatives appear at first order with the spatial ones at second order. The foresighted reader might already have a clue to the analogy: like the Schrödinger equation, the Navier–Stokes equation shows the same imbalance. And similarly for the Dirac equation, in LBE the space and time derivatives are well balanced, both first order.

But let us go on with the Dirac equation.

As is well known, Erwin Schrödinger (1887–1961) derived his equation by formulating an operator transcription of the non-relativistic Hamiltonian:

$$E = H(p, q) = \frac{p^2}{2m} + V(q)$$

with the quantum-mechanical correspondence rules:

$$\hat{q}_a \leftrightarrow q_a, \ \hat{p}_a \leftrightarrow -i\hbar\partial_a, \ \hat{E} \leftrightarrow i\hbar\partial_t \tag{32.14}$$

where hat denotes the operators.

Paul Dirac (1902–84) endeavored to do exactly the same starting however from the relativistic expression of the energy, namely (5):

$$E = \pm\sqrt{m^2 c^4 + p^2 c^2} \tag{32.15}$$

In doing so, he realized that the only way to attach operator meaning to the square root on the right-hand side was to allow the particle momentum to become a 4 × 4 matrix, namely

$$(p_a)_{jk} = -i\hbar\alpha_{jk}^a\partial_a, \ a = x, y, z$$

where the discrete indices j, k run from 1 to 4 and the Latin index is left unsummed upon.

As a consequence, the scalar wavefunction ψ of the non-relativistic Schrödinger equation is promoted to a four-component complex wavefunction $\psi_j, j = 1, 4$, also known as *four-spinor* (see the short Appendix at the end of this chapter).

The discrete indices describe internal degrees of freedom, conventionally associated with the internal angular momentum of particle, best know as spin $\vec{s}$.

According to quantum mechanics, a particle with spin s admits $2s + 1$ measurable components along a spatial axis. Thus, for electrons, which feature $s = 1/2$ there are only two components. By taking the spatial direction aligned with the particle momentum $\vec{p}$, there are therefore two states, spin-up (aligned) and spin-down (counter-aligned). How about the other two components? Since the energy comes from a square-root, in principle it can take both positive and negative signs. While the negative branch sounded like total nonsense at the time of Dirac's finding, it did not take long to recognize by experimental evidence that it is no less physical than the positive branch: the name of the game being *antimatter*. In other words, for each electron with energy-momentum $(E, \vec{p})$, there is an anti-electron with energy-momentum $(-E, -\vec{p})$, both with spin-up and spin-down components, for a total of four, see Fig. 32.1.

The standard representation of the Dirac equation in atomic units ($\hbar = c = 1$) reads as follows (5):

$$\partial_t\psi_j + \left(\alpha_{jk}^x\partial_x + \alpha_{jk}^y\partial_y + \alpha_{jk}^z\partial_z + im\beta_{jk}\right)\psi_k = 0 \tag{32.16}$$

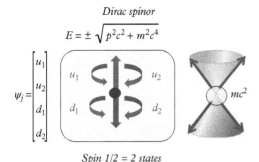

Dirac spinor

$$E = \pm \sqrt{p^2c^2 + m^2c^4}$$

$$\psi_j = \begin{bmatrix} u_1 \\ u_2 \\ d_1 \\ d_2 \end{bmatrix}$$

Spin 1/2 = 2 states

Figure 32.1 *The geometrical representation of a relativistic wavefunction with spin s = 1/2. The energy-momentum relation is represented by the upper and lower cones standing for particles (red) and antiparticles (blue), respectively. Each of the two comes with a spin-up (u_1, d_1) and spin-down option (u_2, d_2), for a total of four components. In the absence of magnetic coupling, the spin-up and spin-down components are indistinguishable and the wavefunctions reduces to a bispinor.*

Here ψ_j is a complex 4-*spinor* describing a particle–antiparticle pair with spin $\pm 1/2$. As mentioned earlier on, relativistic invariance implies that each particle with energy-momentum (E, p_a) and wavefunction ψ_j comes with a corresponding anti-particle of opposite energy-momentum $(-E, -p_a)$ (parity symmetry) and conjugate wavefunction ψ_j^*.

The streaming of this spinning pair is described by the three 4×4 complex matrices α_{jk}^a while the "collisions" are in charge of the mass matrix β.[84]

The former are explicitly given in terms of the 2×2 Pauli matrices σ^a, $a = x, y, z$:

$$\alpha^a = [\sigma^a, 0; 0, -\sigma^a] \tag{32.17}$$

where

$$\sigma^x = [0, 1; 1, 0], \ \sigma^y = [0, -i; i, 0], \ \sigma^z = [1, 0; 0, -1] \tag{32.18}$$

The "collision" matrix is given by

$$\beta = [0, I; I, 0] \tag{32.19}$$

where I is the 2×2 identity matrix and spinorial indices have been dropped for simplicity, see Figs 32.1 and 32.2.

Similarly to the Schrödinger equation, the Dirac equation can also be written in terms of fluid motion. However, since we are dealing with a four-component wavefunction, the relativistic picture points to a *four-component fluid mixture*.

---

[84] Massless fermions are known as Weyl fermions, after the German mathematician Hermann Weyl (1885–1955), who first discussed them in 1929. Weyl fermions were once thought to describe neutrinos, but no more since neutrinos have recently been found to be massive. Weyl fermions might prove overly interesting for new electronic applications because, being massless, they do not suffer any backscatter. Weyl quasiparticles were discovered very recently in exotic crystals with special topological properties.

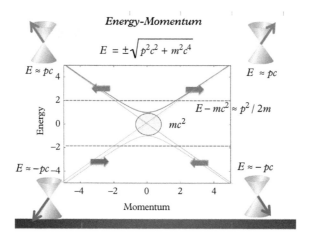

Figure 32.2 *The energy-momentum relation for a relativistic particle of mass m. For each particle with energy-momentum (E, p) (red arrow on the top-right), there is an antiparticle with energy-momentum $(-E, -p)$ (arrow bottom-left). According to the relativistic expression of the velocity $v = \frac{\partial E}{\partial p} = pc^2/E$, the particle and its antiparticle move at the same velocity. This is consistent with the fact that the antiparticle associated to wavepacket $\psi_{(E,p)}(x, t) = e^{+i(px-Et)/\hbar}$ is represented by the complex-conjugate wavepacket $\psi^*_{(E,p)}(x, t) = e^{-i(px-Et)/\hbar} = \psi_{(-E,-p)}(x, t)$. This tight link is particularly apparent for the case of massless particles, m = 0, for which $E = \pm pc$. Right-moving particles and antiparticles belong to the same straight line $E = pc$, while left-moving particles and antiparticles lie on the other bisectrix $E = -pc$.*

32.2.2   The Majorana Representation

To unfold this picture, it proves expedient to recast the Dirac equation into a form where all streaming matrices, sometimes called Weil matrices, become real (see (5)). This is the so-called *Majorana form*, after the italian physicist Ettore Majorana (1906–38).[85]

---

[85] Majorana's asked himself whether Dirac's relativistic quantum wave mechanics could be formulated without resorting to complex numbers. His transformation answers in the affirmative. Given that the wavefunction is real, i.e., $\psi^* = \psi$, this means that Majorana particles are their own anti-particles! For many years, this was considered a sort of mathematical curiosity, but recent developments in high-energy physics and quantum condensed matter, have pointed to the possibility that Majorana particles may indeed exist and be realized in the lab. Majorana was a very singular and mysterious personality. In March 1939 he embarked in a mysterious trip from Naples to Palermo, arrived, then boarded a ship back to Naples and disappeared without a trace. The actual fate of Majorana stands unresolved and still makes the object of disparate hypotheses. For a beautiful account of these fascinating matters see the very enjoyable paper "Majorana returns" by Frank Wilczek (7). From Wilczek quoting Fermi:

> "There are many categories of scientists: people of second and third rank, who do their best, but do not go very far; there are also people of first-class rank, who make great discoveries, fundamental to the development of science. But then there are the geniuses, like Galileo and Newton. Well, Ettore Majorana was one of them."

This was from Fermi, not particularly known for flightiness or overstatement.

The Majorana representation is obtained via the following unitary transformation:

$$\psi_j \rightarrow \psi_j = \frac{1}{\sqrt{2}}(\alpha^y_{jk} + \beta_{jk})\psi_k$$

In a compact four-dimensional notation ($\mu = 0$ for time and $\mu = 1, 2, 3$ for the three spatial dimensions), this yields

$$W^\mu_{jk}\partial_\mu\psi_k = M_{jk}\psi_k, \quad \mu = 0, 3 \tag{32.20}$$

with

$$W^0_{jk} = \delta_{jk}, \quad W^1_{jk} = -\alpha^x_{jk}, \quad W^2_{jk} = \beta_{jk}, \quad W^3_{jk} = -\alpha^z_{jk}$$
$$M_{jk} = -im\alpha^y_{jk} + (qV + A^a \mathcal{J}_a)\delta_{jk}$$

Here $qV + \mathcal{J}_a A^a$ is the interaction of the elementary charge q with an external electromagnetic field described by the 4-vector potential (V, A^a). Since all matrices here are real, except for α^y which is purely imaginary, the Majorana representation of the Dirac equation involves only real matrices.

A scalar product of eqn (32.20) with ψ_j^* yields the desired set of continuity equations:

$$\partial_t\rho_j + \partial_a\mathcal{J}_j^a = S_j, \quad j = 1, 4 \tag{32.21}$$

where

$$\rho_j = \psi_j^*\psi_j$$

is the partial density of the j-th fluid,

$$\mathcal{J}_j^a = \psi_j^*\alpha^a_{jk}\psi_k$$

is the corresponding current density and

$$S_j = i\psi_j^* M_{jk}\psi_k$$

is a mixing term coupling the different components of the relativistic mixture. Note that in the previous expressions only the index k is summed upon.

Unitarity, read norm conservation, implies

$$\sum_j S_j = i\sum_{jk}\psi_j^* M_{jk}\psi_k = 0$$

This is automatically secured by the anti-Hermitian character of the mass matrix: $M_{kj} + M_{jk}^* = 0$.

As promised, the fluid analogy comes by *more* naturally than in the non-relativistic case, because the Dirac equation only involves first-order derivatives.

Another comforting feature is that the external interaction is easily accommodated within a formal redefinition of the mass matrix, without compromising the local nature of the theory. The fluid interpretation of the Dirac equation is equally transparent: four types of *spinning particles* stream in space and once they reach the same spacetime location, they interact via the "scattering matrix" M_{jk}. Again, a decided flavor of LBE.

A qualitative difference with classical particle motion is apparent, though. Classical particles have no internal structure, and consequently a type-1 particle at location x at time t with speed v propagates to $x + v\Delta t$ at time $t + \Delta t$ and it is still entirely of type 1, no mixing.

A relativistic particle, however, undergoes mixing during free propagation because the rotation around the direction of motion mixes up the four spinorial components. This is why the streaming matrix is generally non-diagonal, echoing the fact that spin is not an ordinary vector. This suggests that the discrete spacetime of a relativistic particle should be represented by a four-link network each link carrying a distinct spinorial state (8).

This "quantum lattice theory" is less of a joke than it might seem. It has been realized recently that lattice formulations of field theory based upon spinning particle motion may offer potential advantages over more popular techniques such as path integration (9). In a nutshell, this is because in quantum lattice models "*instead of seeking discretized versions of the Hamiltonian or the Lagrangian, a discretized version of the evolution operator is introduced*" (9). In fact, what this author finds is that "*the rotation group, the Lorentz group and spin emerge automatically in the continuum limit from unitary dynamics on a cubic lattice.*" The reader fond of more details is directed to the original reference.

32.2.3   Dirac to Schrödinger: the Enslaving Approximation

Having briefly reviewed the Dirac equation, we now proceed to substantiate the analogy DE : SE = LBE : NSE. As a first step in this direction, there is a useful lesson to learn by inspecting the way Schrödinger equation is obtained as a long wavelength (low energy) limit of the Dirac equation. We shall see in a moment that this involves a sort of enslaving approximation which is formally very similar to the low-Knudsen expansion taking the Boltzmann equation into the Navier–Stokes equations.

The formal parallel emerging from this analogy is

$$Kn = \frac{l_\mu}{l_M} \leftrightarrow \beta = \frac{v}{c} = \frac{\lambda_c}{\lambda_b} \tag{32.22}$$

where l_μ is the classical mean-free path, l_M is a typical coherence length of the macroscopic fluid, β is the relativistic particle to light speed ratio, $\lambda_c = \frac{\hbar}{mc}$ and $\lambda_b = \frac{\hbar}{mv}$ are the Compton and De Broglie wavelengths, respectively. In other words, the ratio of the material speed to the speed of light plays the role of a relativistic Knudsen number.

To make the argument quantitative, let us consider the 1D version of the DE in Maiorana form

$$\begin{cases} \partial_t u_{1,2} + c\partial_z u_{1,2} = +\omega_c d_{2,1} \\ \partial_t d_{1,2} - c\partial_z d_{1,2} = -\omega_c u_{2,1} \end{cases} \tag{32.23}$$

where

$$\omega_c = \frac{mc^2}{\hbar}$$

is the Compton frequency, often identified with the particle mass in atomic units.

In (32.23), (u_j, d_j) represent a pair of up/down moving bi-spinors with spin up/down (1, 2). They propagate independently (the streaming matrix α^z can always be diagonalized independently of the other two) and mix through the "scattering" effect due to the mass matrix.

Next, let us introduce slow(fast) modes by mixing u, d as follows:

$$\phi^{\pm} = \frac{1}{\sqrt{2}} (u \pm id) e^{i\omega_c t} \tag{32.24}$$

Spinorial indices have been relaxed, which is harmless in the absence of magnetic fields.

The slow and fast modes evolve according to

$$\begin{cases} \partial_t \phi^+ + c\partial_z \phi^- = 0, \\ \partial_t \phi^- - c\partial_z \phi^+ = 2i\omega_c \phi^- \end{cases} \tag{32.25}$$

Now comes the "enslaving" assumption, whose meaning in the context of quantum mechanics will be clarified shortly:

$$|\partial_t \phi^-| \ll 2m |\phi^-| \tag{32.26}$$

This delivers $\phi^- = \frac{\lambda_c}{2i}\partial_z \phi^+$.

By inserting this into eqn (32.25) we obtain

$$\partial_t \phi^+ = i\frac{\hbar}{2m}\partial_z^2 \phi^+ \tag{32.27}$$

which is precisely the sought-after one-dimensional Schrödinger equation for a free particle of mass m.

What is the lesson here?

The relativistic motion implies that any particle of energy-momentum (E, p_a) is associated with an antiparticle with opposed energy-momentum $(-E, -p_a)$. The symmetric combination of the two gives rise to a smooth, emergent field, whereas the antisymmetric combination defines a low-amplitude, high-frequency mode which decouples from the system dynamics in the limit $\beta \to 0$.

More precisely, inspection of eqn (32.27) reveals that the ratio of the amplitudes $|\phi^-/\phi^+|$ scales like β^2, whereas the frequency ratio goes like $1/\beta$.

This means that, as β goes to zero, the antisymmetric mode becomes smaller in amplitude and faster in frequency, so that it finally becomes unobservable on timescales longer than $1/\omega^-$.

This is precisely the enslaving picture of kinetic modes in (discrete) kinetic theory.

The scenario is exactly the same, except for a key difference: kinetic theory describes dissipative phenomena in which adiabatic elimination irons out the initial conditions: transient modes die out, never to return. Quantum mechanics is reversible and fast modes never die out: they just oscillate so fast that any observation on timescales longer than their period of oscillation, would be unable to detect them.

But they are still there and more resolved (higher-energy) measurements could always bring them back to light again.

Another interesting remark concerns the symmetry breaking induced by a nonzero mass m. If m is made zero, the up and down walkers do not see each other and go across with no interaction, the result being the wave equation for photons. Manifestly this is a singular limit which cannot be described by the Schrödinger equation (diffusion coefficient goes to infinity). Any nonzero mass causes "collisions" which slow down the wavepackets and confer on them a subluminal speed $v < c$ characterizing material versus massless particles.

32.2.4   The Interacting Case

Electromagnetic interactions are readily included within the Dirac equation by the well-known generalization of the four-momentum:

$$p^\mu \to p^\mu + qA^\mu$$

where $p^\mu = -i\hbar\partial_\mu$, $A^\mu = (V, \vec{A})$ is the four-dimensional vector potential and q is the electric charge of the particle. This rule immediately informs us that electromagnetic interactions can be formally brought to the right-hand side and incorporated within a generalized mass matrix. For a simple one-dimensional case with purely electrostatic interactions, $(\vec{A} = 0)$, the Dirac equation reads as follows:

$$\begin{cases} \partial_t u_{1,2} + c\partial_z u_{1,2} = +\omega_c d_{2,1} + ig u_{2,1} \\ \partial_t d_{1,2} - c\partial_z d_{1,2} = -\omega_c u_{2,1} + ig d_{1,2} \end{cases} \tag{32.28}$$

where $g = qV/\hbar$ is the coupling frequency of the potential. Self-consistent potentials, such as those arising in connection with the nonlinear Schrödinger equation, are easily accommodated by making g a function of the local density $\rho = u^2 + d^2$, with many interesting applications in quantum condensed matter (Bose-Einstein condensation) and modern cosmology.

32.3   The Quantum LBE

After all this preparation, we are finally in the position to reformulate the basic analogy in quantitative terms.

This is based on the following position: *The discrete velocities $\vec{v}_i$ are the analog of the discrete particle spin states s_j* (8; 10; 11).

From this assumption it follows that:

1. *The discrete population f_i is the analog of the 4-spinor ψ_j.*
2. *The scattering matrix A_{ij} is the analog of the mass matrix M_{jk}.*

Position one follows from the observation that all LBE schemes work with a set of discrete speeds coming in pairs like spins. In quantum mechanics, the spin is intrinsically discrete since it can only take a few quantized values, all intermediate ones being unstable. That is real-world quantum physics.

In LBE the particle speed is made artificially discrete; even though it could take on any value in the continuum, we know that the correct macroscopic physics does not ask for more than a few properly chosen discrete values.

In quantum mechanics spin quantization is a genuine physical effect due to the impossibility of measuring momentum and position at a time. Likewise, the spin, which has dimensions of angular momentum $\vec{r} \times \vec{p}$, can only be measured along a single direction, while its projection over the remaining two is left undetermined.

In LBE, speed quantization is a convenient numerical device, with no physical motivation.

The analogy is tempting, but a minute's thought reveals two major stumbling blocks, related to a *dimensional* and a *structural* mismatch, respectively, between LBE and DE.

The dimensional mismatch refers to the fact that while the four-spinor (we consider $s = 1/2$ throughout) has always four components in *any* spatial dimensions, the discrete distribution f_i is a *multi-scalar* set of b real functions, where b strongly depends of the dimensionality. In one spatial dimension the analogy is real because the spin can always be aligned with the momentum, the so-called helicity representation. In this representation, a quantum D1Q2 would correspond to spin $s = 1/2$, while D1Q3 would associate with $s = 1$ and so on. In a way, LB schemes with(without) rest particles would associate with bosons(fermions), respectively.

In higher dimension though, this is no longer possible because the spin cannot be measured along more than one spatial direction: the *helicity $h = \vec{s} \cdot \vec{p}$* is no longer a good quantum number.

The structural mismatch concerns the streaming operator: while in LBE this is always diagonal in momentum space, there is *no way* the three Weil matrices can be diagonalized simultaneously. Again, this is intimately related to the quantum nature of the spin variable: spinors are neither vectors nor multi-scalars.

This looks like a very serious stumbling block in more than one spatial dimensions.

Remarkably, the block can be circumvented, as we are going to discuss later. Before doing so let us first comment on the important issue of discrete time marching.

32.4 Time Marching

The procedure to march quantum LBE in time deserves some attention, as it presents some peculiarity due to the imaginary character of time in the Schrödinger equation. In the sequel, we shall stick to the one-dimensional case, with no loss of generality.

Direct Fourier analysis shows that an explicit light-cone marching, such as the one used for fluid LBEs, is unconditionally unstable due to the imaginary nature of the diffusion coefficient.

A simple and effective way out of this problem, is to march eqns (32.28) with an implicit Crank–Nicolson method, whereby the right-hand side of both eqns are evaluated at mid-times as follows:

$$\begin{cases} u^> - u = +\mu \left(\dfrac{d^< + d}{2} \right) + i\gamma \left(\dfrac{u^> + u}{2} \right) \\[2mm] d^< - d = -\mu \left(\dfrac{u^> + u}{2} \right) + i\gamma \left(\dfrac{d^< + d}{2} \right) \end{cases} \tag{32.29}$$

where we have set: $u^> \equiv u(z + \Delta z, t + \Delta t)$, $d^< \equiv d(z - \Delta z, t + \Delta t)$
    In (32.29) and,

$$\mu = \omega_c \Delta t \equiv m/m_p$$

is the "relaxation parameter" of the QLB scheme, which can also be interpreted as the ratio of the physical mass m to the "Planck" mass

$$m_p = \frac{\hbar/c}{c\Delta t}$$

associated with the discrete spacetime.
    Likewise, $\gamma = g\Delta t$, where $g = V_0 e/\hbar$ is the electrostatic coupling frequency.
    The scheme (32.29) is linearly implicit, but it does not pose a problem since we can easily invert it analytically site by site to obtain the following *explicit* form:

$$\begin{cases} u^> = pu + qd, \\[2mm] d^< = -qu + pd \end{cases} \tag{32.30}$$

where

$$p = \frac{1 - \Omega^2/4}{1 + \Omega^2/4 - i\gamma}, \quad q = \frac{\mu}{1 + \Omega^2/4 - i\gamma} \tag{32.31}$$

and

$$\Omega^2 = \mu^2 - \gamma^2$$

The reader can check by direct calculation that $p^2 + q^2 = 1$ for any μ and γ.
    This is the final form of the quantum Lattice Boltzmann (QLB) equation in one spatial dimension, see Fig. 32.3.
    Valuable insights on the main properties of the QLB equation can be gained by analyzing the dispersion relation associated with the QLB scheme, eq. (32.30).

The QLB scheme

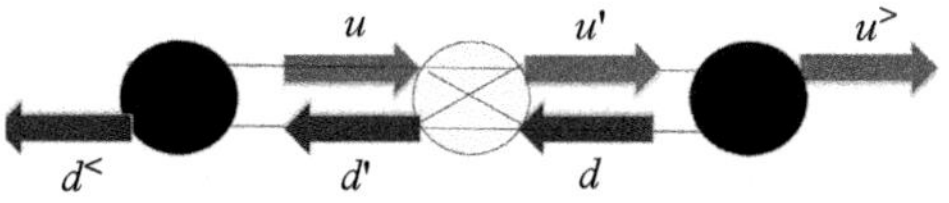

$$u^> \equiv u(z+1,t+1) = + pu(z,t) + qd(z,t) \equiv u'$$
$$d^< \equiv d(z-1,t+1) = - qu(z,t) + pd(z,t) \equiv d'$$

Figure 32.3 *The stream-collide representation of the one-dimensional QLB scheme. The tight analogy with classical LB is apparent, although with an anti-symmetric quantum collision matrix $Q = [p, q; -q, p]$, to respect unitarity. We explicitly display two separate connections for the up and down spinors, to stress the fact that, unlike classical particles, they mix even in the absence of explicit interactions, but just by the very fact of carrying a non-zero mass. Note that in $d > 1$, the up and down spinors do mix even without carrying mass!*

32.4.1   The QLB Dispersion Relation and Lorentz Invariance

Owing to its linearity, the QLB scheme can be analyzed in full by Fourier-decomposition. By expanding the spinors u and d in plane waves, (tilde denotes the Fourier-transform):

$$u(z;t) = \sum_{k,\omega} \tilde{u}(k,\omega)e^{-i(kz-\omega t)}, \ d(z;t) = \sum_{k,\omega} \tilde{d}(k,\omega)e^{-i(kz-\omega t)}$$

we obtain the linear system:

$$\begin{cases} (e^{-i(k\Delta z-\omega\Delta t)} - p)\tilde{u} - q\tilde{d} = 0 \\ q\tilde{u} + (e^{-i(-k\Delta z-\omega\Delta t)}) - p)\tilde{d} = 0 \end{cases} \quad (32.32)$$

Zeroing the determinant and recalling that $p^2 + q^2 = 1$, simple algebra delivers the following dispersion relation:

$$cos(\omega\Delta t) = pcos(k\Delta z) \quad (32.33)$$

This has to be compared with the exact dispersion relation, expressing Lorentz invariance in continuum spacetime. For a massive free particle:

$$\omega^2 - c^2 k^2 = \omega_c^2 \quad (32.34)$$

Apparently, the two are very different, but in fact they are much closer than it seems at a first glance.

To appreciate the point, let us expand (32.33) to second order in the time-step Δt, to obtain:

$$1 - \frac{\omega^2 \Delta t^2}{2} - p\left(1 - \frac{k^2 \Delta z^2}{2}\right) = 0 + O(\Delta t^2) \tag{32.35}$$

namely

$$\omega^2 - pk^2 c^2 = 2(1-p)\omega_p^2 \tag{32.36}$$

where, for convenience, we have defined the "Planck" frequency $\omega_p \equiv 1/\Delta t$ and used the lightcone condition $c = \frac{\Delta z}{\Delta t}$.

This is a Lorentz-deformed dispersion relation, with a modified light speed $c\sqrt{p}$ and a modified mass $2(1-p)m_p$, $m_p \equiv m\omega_p/\omega_c$ being the lattice "Planck" mass.

For non-interacting particles ($\Omega = \mu$), the effective speed and mass of the wavepacket take the following form:

$$c(\mu) = c\left(\frac{1 - \mu^2/4}{1 + \mu^2/4}\right)^{1/2}, \quad m(\mu) = \frac{m}{\sqrt{1 + \mu^2/4}} \tag{32.37}$$

This shows that in discrete time, the wavepacket slows down and stops propagating altogether at $\mu = 2$, beyond which the velocity becomes imaginary, signaling the impossibility of further propagation.

The loss of mass to the lattice, on the other hand, proceeds smoothly up to virtually infinite values of the parameter μ.

For the case of massless (Weyl) particles, $\mu = 0$ and $p = 1$, both effects remain silent and what we get is *exactly* the continuum dispersion relation.

This is not a perturbative result, since for $p = 1$ the discrete dispersion relation delivers $\omega \Delta t = \pm k\Delta z + 2n\pi$. Leaving aside the "alias" modes with $n > 0$, this gives precisely $\omega = \pm kc$, i.e., $\omega^2 = c^2 k^2$.

For massive particles, the picture is a bit less rosy, but still fairly acceptable.

To second order in $\mu \equiv \omega_c \Delta t$, we have $p = 1 - \mu^2/2$, so that, given that $\mu = \omega_c \Delta t$, (32.38) delivers

$$\omega^2 - k^2 c^2 = \omega_c^2 \tag{32.38}$$

which is still the exact continuum dispersion relation, although this time-only second-order accurate in $\omega_c \Delta t$.

At higher orders, both the propagation speed and the mass receive further corrections from the lattice.

32.4.1.1   *Non-relativistic regime*

The previous analysis includes the non-relativistic limit as the special case characterized by the condition

$$|\omega - \omega_c| \ll \omega_c$$

Under such conditions, the continuum dispersion relation (32.28) reduces to

$$2(\omega - \omega_c)\omega_c = k^2 c^2$$

which delivers

$$\omega - \omega_c = \frac{k^2 c^2}{2\omega_c} = \frac{\hbar}{2m} k^2$$

This is precisely the dispersion relation of the Schrödinger equation for the shifted frequency $\omega' \equiv \omega - \omega_c$. This is automatically obtained as the low-frequency branch of the QLB discrete dispersion relation, see Fig. 32.4.

Before concluding, we observe that the QLB scheme is *unconditionally* unitary and stable, because $p^2 + q^2 = 1$ for any value of the timestep Δt.

This is a remarkable property for an *explicit* solver of real-time quantum wave dynamics, although not unique to LB.[86]

Let us now proceed to provide some numerical evidence of the QLB viability for a few simple benchmark problems.

32.5 Numerical Tests in One Space Dimension

The earliest quantum LB scheme were validated in a series of one-dimensional textbook calculations, including (i) *free particle propagation*, (ii) *harmonic oscillator* and (iii) *scattering from a rectangular barrier* (10). Subsequently, the scheme has also been demonstrated for simple cases of nonlinear Schrödinger equations of direct relevance to Bose–Einstein condensation (11), best known as Gross-Pitaevski equation.

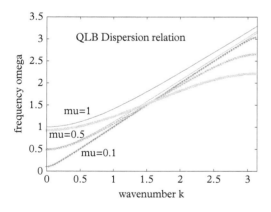

Figure 32.4 *The QLB dispersion relation for $\mu = 0.1, 0.5, 1.0$. For $\mu = 0.1$, no appreciable departure from the continuum case (solid line) is visible at any wavenumber k. The case $\mu = 0.5$ shows no appreciable loss of mass, and slow-down effect, i.e. downward bending of the curve, starting at about $k \sim 2$. This is of no serious concern, since $k = \pi$ corresponds to a wavelength $2\Delta x$, the shortest that can be represented in the lattice. Finally, the case $\mu = 1$ shows a tiny but visible mass reduction and a substantial bending already from about $k \sim 0.5$, implying that wavelengths of the order of 3 lattice spacings experience a sizeable artificial slow-down.*

[86] Stable explicit schemes for the Schrödinger equation can be obtained by integrating the real and imaginary parts of the wavefunction as two separate unknowns, following the formal analogy with classical Hamiltonian systems.

32.5.1 Free Particle Motion

The free motion of a quantum particle in the absence of any external potential is characterized by the well-known phenomenon of loss of coherence; that is, the spatial extent of the wavefunction grows indefinitely in time until any form of coherence is lost in dull uniformity. This phenomenon is best analyzed for the case of the so-called *minimum uncertainty wavepacket*, a Gaussian packet centered at $x = x_0$ propagating at speed $u_0 = p_0/m$ (5):

$$\psi(z,0) = \left(2\pi\delta_0^2\right)^{-1/4} e^{ip_0 z/\hbar} e^{-(z-z_0)^2/4\delta_0^2} \tag{32.39}$$

Denoting by $\delta^2(t)$ the variance of the wavefunction at time t, defined as $\delta^2(t) = \int \Psi^2(z;t)z^2 dz$, the analytical expression for this loss of coherence reads as:

$$\delta^2(t) = \delta_0^2 + \frac{D_q^2 t^2}{\delta_0^2} = \delta_0^2 \left(1 + \frac{t^2}{t_{dec}^2}\right) \tag{32.40}$$

where $D_q = \hbar/2m$ is the quantum diffusion coefficient and $t_{dec} = \delta_0^2/D_q$ is the decoherence time.

Note that, at variance with classical diffusion, $\delta \sim t^{1/2}$, the expression (32.40) implies a ballistic loss of coherence, $\delta \sim t$, in the asymptotic regime $t \gg t_{dec}$.[87] The analytical expression (32.40) has been checked for a series of simulations with different values of δ_0, $u_0 = p_0/m$ and number of grid points N.

A typical set of results is shown in Figs 32.5 and 32.6.

These tests prove that quantum LBE reproduces accurately the analytical results as long as the spatial uncertainty δ is well resolved, typically above 16 lattice units.[88]

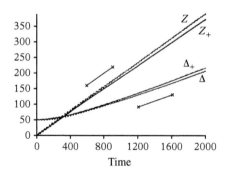

Figure 32.5 *Mean displacement (Z) and variance of the wavefunction (Δ) as a function of time for $\beta = 0.2$ and $\mu = 0.1$. The suffix + indicates the average over the density of the "hydrodynamic" mode. From (10).*

[87] For an electron, $D_q = \hbar/m_e \sim 10^{-4}\ m^2/s$. Hence, by taking $\delta_0 \sim \lambda_b \sim 10^{-10}$ m corresponding to an electron speed of about $c/100$, one estimates $t_{dec} \sim 10^{-16}$ s, one tenth of the typical chemical bond-forming/breaking time. This clearly shows that electrons decohere well within the time it takes to "jump" from one molecule to another, hence they are quintessential quantum objects. By contrast, the decoherence time of a human being can be estimated of the order of 10^{36} seconds, more than the age of the Universe, *squared*, hence they are quintessential quantum particles.

[88] This figure may be contrasted to the 1–2 lattice units required by fluid applications. This seems to indicate that non-relativistic quantum wave behavior on the lattice requires an order of magnitude more lattice sites than fluid behavior. The author does not have any cogent explanation for this.

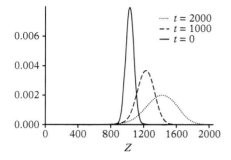

Figure 32.6 *Probability distribution of the "hydro-dynamic" mode at three successive instants in time. Parameters are the same as in Fig. 32.5 from (10).*

Narrower packets perforce contain high wavelengths which fail to obey the adiabaticity assumption underlying the non-relativistic limit and possibly experience the discreteness effects discussed in section 32.4, as well.

32.5.2 Harmonic Oscillator

The next test is the familiar harmonic oscillator, a particle bound in a quadratic harmonic potential:

$$V_H(z) = V_0 \frac{(z - z_0)^2}{4L^2} \tag{32.41}$$

As is well known, a wavepacket with variance $\delta_0 = \sqrt{D_q \Omega}$, $\Omega^2 = 8V_0/mL^2$, behaves like a classical particle bouncing back and forth in the potential with a speed $u_H = \delta_0 \Omega/2\pi$ with no loss of coherence.

The coherent motion of such a wavepacket has been reproduced by quantum LBE for a series of values of the parameters V_0, L and N. A typical result is shown in Fig. 32.7.

Again, quantitative match with analytical result is obtained only for sufficiently broad packets. Packets narrower than 16 lattice units are slightly distorted by relativistic, non-adiabatic, motion of the shortest wavelengths.

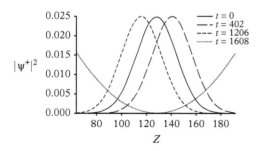

Figure 32.7 *Wavefunction in a harmonic potential at $t = 0$, $T/4$, $3T/4$ and T, where $T = 1608$ is the oscillation period in lattice units. The vertical axis displays the probability density $|\psi^+|^2$ of the "hydrodynamic" mode. From (10).*

32.5.3   Scattering Over a Rectangular Barrier

In the same paper, the quantum LBE has been tested on a simple scattering problem over a rectangular potential of the form

$$V_s(z) = \begin{cases} V_0, & -w/2 < z < w/2 \\ 0, & \text{otherwise.} \end{cases} \qquad (32.42)$$

A typical manifestation of quantal effects is a finite probability for the quantum particle to go through the potential barrier even if is energy E is lower than the potential height V_0 (*quantum tunneling*).

Under tunneling conditions, an incident plane wave $\psi_{\text{in}} = Ae^{ik_{\text{in}}x}$ at $x = -w/2$ develops transmitted and reflected components of the form

$$\psi(z) = Te^{ik_t z} + Re^{-ik_r z} \qquad (32.43)$$

where the reflected and transmitted wavenumbers are given by

$$k_t = \frac{\sqrt{2m(E - V_0)}}{\hbar}, \quad k_r = \frac{\sqrt{2mE}}{\hbar} \qquad (32.44)$$

respectively.

Inside the classically forbidden region $-w/2 < z < w/2$, the wavenumber of the transmitted wave becomes purely imaginary, corresponding to the well-known exponential decay due to the barrier itself: $T/A \sim e^{-2|k_t|w}$.

The existence of a reflected and transmitted waves has been clearly detected in a series of simulations with different values of the parameters V_0, w and N, see Fig. 32.8.

To sum up, these very early tests provided evidence of the viability of quantum LBE, at least in one dimension. The scheme performed efficiently and provided *stability* and *unitarity* at the same time, a very valuable property for an explicit scheme. As commented before, this is related to the peculiar light-cone spacetime marching technique inherent to quantum LB.

32.6   Multi-Dimensional QLB

As discussed earlier, the extension of the QLB to more than one spatial dimension faces with a fundamental limitation: the spin can no longer be kept aligned with the particle momentum,[89] and consequently it can no longer be identified with the discrete velocities of the QLB scheme.

This limitation was circumvented by augmenting the stream-collide dynamics with a "rotation" step, aimed at keeping the spin aligned with the particle momentum along the

---

[89] Multi-dimensional quantum lattice schemes, not based on the spin-speed analogy, though, have been developed by a number of authors (12; 13).

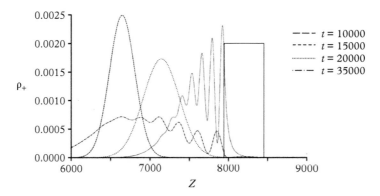

Figure 32.8 *Probability density of the "hydrodynamic" mode at different time instants for the case of the scattering against a rectangular barrier (solid line). The exponential decay within the barrier is clearly visible, and so are reflected waves in front of it. From (10).*

new direction of propagation. The evolution of the wavefunction in D spatial dimensions then splits in D one-dimensional QLB steps. The mathematical justification for such "uncorrelated" strategy being provided by a specific instance of *operator-splitting*, known as Trotter–Suzuki formula:

$$e^{(A+B)\Delta t} = e^{A\Delta t}\, e^{B\Delta t} + C\Delta t^2 + O(\Delta t^2) \tag{32.45}$$

where

$$C = [A, B] \equiv AB - BA \tag{32.46}$$

is the commutator of the two matrices A and B.

The multi-dimensional case is best discussed by starting from the most compact formulation of the QLB scheme, namely:

$$\hat{\psi} = \psi'$$

where $\hat{\psi} \equiv S\psi$ and $\psi' \equiv Q\psi$ are the spacetime shifted and post-collisional spinors, respectively.

The three-dimensional QLB scheme reads as follows:

1. *Motion along x:* $\psi(x + \Delta x, y, z; t + \Delta t) = P_x \psi(x, y, z; t)$
   (a) Rotate ψ with R_x^{-1}
   (b) Collide with $Q_x \equiv R_x^{-1} Q R_x$
   (c) Stream with diagonal S_x
   (d) Rotate back with R_x

2. *Motion along y:* $\psi(x + \Delta x, y + \Delta y, z; t + \Delta t) = P_y\psi(x + \Delta x, y, z; t + \Delta t)$
   (a) Rotate ψ with R_y^{-1}
   (b) Collide with $Q_y \equiv R_y^{-1}QR_y$
   (c) Stream with diagonal S_y
   (d) Rotate back with R_y
3. *Motion along z:* $\psi(x + \Delta x, y + \Delta y, z + \Delta z; t + \Delta t) = P_z\psi(x + \Delta x, y + \Delta y, z; t + \Delta t)$
   (a) Rotate ψ with R_z^{-1}
   (b) Collide with $Q_z \equiv R_z^{-1}QR_z$
   (c) Stream with diagonal S_z
   (d) Rotate back with R_z

Note that since one of the three streaming matrices can always be taken diagonal, the scheme requires in fact only two rotations steps. The specific form of the rotation matrices X, Y, Z can be found in the original papers.

Summarizing, the full 3D QLB propagator can be written as a direct product of three 1D propagators:

$$P = P_zP_yP_x$$

where each 1D propagator comes in the compact form:

$$P_a = R_aS_aQ_aR_a^{-1}, \quad a = x, y, z$$

This procedure was formally devised since the very inception of the QLB scheme (8), but its practical implementation had to await the mid 2000s, with the thesis work by Silvia Palpacelli (14). Subsequent investigations by P. Dellar et al (15; 16) proved crucial for the correct implementation of the three-dimensional scheme. The latter authors computed a few benchmark problems, such as free-particle motion and ground-state wavefunctions in harmonic potentials, both in two and three spatial dimensions providing numerical evidence of the viability of QLB in multiple spatial dimensions.

32.6.1   Quantum LBE: Move, Turn and Collide

The quantum LBE is a bit more cumbersome than its classical counterpart, the only extra charge being the "rotation" and "antirotation" steps needed to keep the spin aligned with the speed. Apart from the anti-symmetry of the collision matrix, this is the only trace left of the quantum nature of the moving particle. It should noted that quantum LBE bears many similarities with other quantum lattice schemes discussed in the recent (12; 13) and not so recent (17; 18) literature.

What sets it apart from all these schemes is the fact of insisting on the diagonal representation of the Weil matrices, so as *to retain as much as possible the notion of classical trajectories*. In fact, the "rotation" operator can formally be interpreted as an "internal scattering" between particle–antiparticle states, thus leaving the concept of quantum trajectory still well defined.

In a pictorial sense, we might say that while classical particles just "stream and collide," quantum particles, somehow like swimmers need to take a somersault each time they have to move along a new spatial dimension. They "stream, turn and collide"! Again, the "turn" step is a necessity induced by the internal structure of the relativistic particle.

32.7 QLB is a Genuine Dirac Solver

The study of the convergence of the QLB results versus exact solutions showed that the error is generally small even at low resolution, but does not decay to zero as fast as it should (second order) at increasing resolution.

In particular, the wavefunction always shows some high-frequency, small-amplitude oscillations, which do not die out at increasing resolution.

Dellar and collaborators realized that these oscillations do not die because ... they *should not* die, since they represent the fast relativistic component of the Dirac spinor, the famous "Zitterbewegung," a prototypical signature of relativistic quantum mechanics.

Zitterbewegung in QLB was demonstrated by Dellar and collaborators, who performed detailed comparisons between the solution of the Dirac equation using QLB versus spectral methods.

These authors also showed that QLB does converge to Dirac better than to Schrödinger, for all parameter regimes within and beyond the enslaving approximation inherent to the non-relativistic Schrödinger equation, see Fig. 32.9.

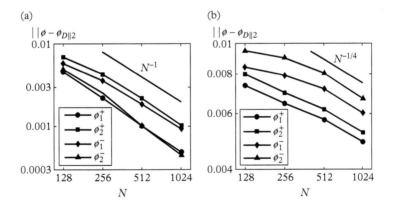

Figure 32.9 *QLB vs pseudo-spectral solution of the Dirac equation (reference solution). The error versus the reference solution as a function of the grid resolution for all four components of the relativistic bi-spinor. Panel (a): early time, Panel (b) late time. From P. Dellar et al, Phil. Trans Roy Soc 046700, (2010), by permission of the Royal Society.*

Although perhaps unsurprising in hindsight, this represents a major conceptual step forward, since it implies that *QLB provides a unified treatment of both non-relativistic and relativistic quantum wave problems* (19).

This opens up a new class of quantum-relativistic applications for QLB.

For instance, it is known that, due to the special symmetries of the honeycomb lattice, electrons in graphene behave like relativistic massless particles. As a result, graphene offers, for the first time ever, the possibility of testing experimentally the so-called *Klein paradox*, namely the perfect transmission (literally zero reflection) of relativistic quantum wavepackets across potential barriers.

Roughly speaking, the condition for a relativistic particle of mass m and charge q to undergo zero-reflection across a barrier of height V reads as follows:

$$qV > mc^2 \qquad (32.47)$$

This was known since the early days of relativistic quantum mechanics, but never tested experimentally since, even for particles as light as electrons, it requires energies of the order of 0.5 MeV. In graphene, however, the effective mass of the electron is much smaller than the bare electron mass, typically two orders of magnitude, depending on density and temperature conditions, so that the condition (32.47) can in principle be met, with major implications on the electric conductivity and other transport properties of utmost relevance for practical applications. As a result, QLB can be used to study electron transport in graphene. Preliminary work in this direction has been performed in the recent years (20), see Fig. 32.10, and much more is expected for the future.

32.8 The Lattice Wigner Equation

We began this chapter by discussing the fluid dynamic formulation of quantum mechanics and went on by positing an analogy between the relativistic Dirac equation and discrete kinetic theory. There is however another formulation of quantum mechanics which is even closer to kinetic theory, known as Wigner's phase-space formulation (21),

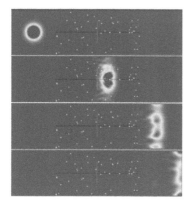

Figure 32.10 *Relativistic electron wavefunction in graphene sample with impurities (dots) in graphene. The wavepacket on the left travels within the sample and gets distorted upon impinging on the impurities. Reprinted with permission from S. Palpacelli et al, Int. J. Mod. Phys C, 23, 1250080 (2012).*

after the hungarian polymath Eugene Wigner (1902–95). This is based on the Wigner function defined as follows (in $d = 1$ for simplicity):

$$W(x, p; t) = \frac{1}{2\pi \lambda_b} \int \psi^* \left(x - \frac{r}{2}; t \right) e^{-i\frac{pr}{\hbar}} \psi \left(x + \frac{r}{2}; t \right) \, dr \qquad (32.48)$$

It is immediatly apparent that the Wigner function bears a striking similarity with Boltzmann's probability distribution $f(x, v; t)$. On the other hand, since in quantum mechanics position and momentum cannot be measured simultaneously to an arbitrary precision, it is clear that the Wigner distribution must present some basic difference as compared to a classical probability distribution. Indeed, it can be readily checked that the Wigner distribution is generally not positive definite: due to quantum interference effects it can develop negatives which betray its non classical nature, see Fig. 32.11.

The Wigner formulation appears particularly well suited to study the emergence of classical behavior in the limit $\hbar \rightarrow 0$, or, better said, in the limit where the De Broglie length becomes vanishingly small as compared to any classical lengthscale, say the mean-free path.

In equations:

$$Kq = \frac{\lambda_b}{l_\mu} \rightarrow 0$$

where we have defined Kq as a *quantum Knudsen number*, see Fig. 32.11.

It is indeed apparent that in the limit $\lambda_b \rightarrow 0$, the Wigner transform (32.48) returns $W(x, p; t) \rightarrow \psi^*(x)\psi(x) = \rho(x)$, which is clearly non-negative definite. By expanding

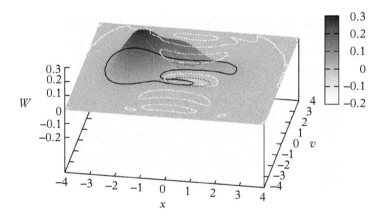

Figure 32.11 *The Wigner function associated with an electron tunneling through a double-well potential. Non-classical regions of negativity are clearly visible. Reprinted from O. Furtmaier et al, J. Comp. Phys. 305, 1015 (2016) with permission from Elsevier.*

the exponential kernel in (32.48) in powers of r/λ_b, an infinite series of terms of alternating sign is obtained, whose positivity is no longer guaranteed due to the interference between the backward wavefunction $\psi^*(x-r/2)$ and forward wavefunction $\psi(x+r/2)$.

The Wigner transform exhibits a number of very useful classical-like properties, which hold at any value of the quantum Knudsen number.

For instance, by integrating in momentum space, we obtain

$$\int W(x,p;t)\,dp = \rho(x)$$

precisely like a classical distribution. The analog integral in configuration space delivers the density in momentum space:

$$\int W(x,p;t)\,dx = \rho(p)$$

By the same token, it possible to define kinetic moments of higher order, like in classical kinetic theory.

It is therefore natural to assume that it should also obey a similar kinetic equation.

This is indeed true, with a crucial twist, though, as we are going to comment below.

The (collisionless) Wigner distribution can be shown to obey the following collisionless kinetic equation:

$$\frac{\partial W}{\partial t} + \frac{p_a}{m}\frac{\partial W}{\partial x_a} + \sum_{l=1,3,5\ldots} \hbar^{l-1}\, F_a^{(l)}(x)\, \frac{\partial^l W}{\partial p_a^l} = 0 \tag{32.49}$$

where

$$F_a^{(l)}(x) = \frac{\partial^l V(x)}{\partial x_a^l}$$

is the l-th order "force" driven by quantumness at order l.

Note that the term $l = 1$ is completely classical, so that quantum effects take stage at second order and higher in the quantum Knudsen number. These effects couple to spatial derivatives of the classical potential from order three and above, thus showing that the harmonic oscillator is immune from quantum interference. Indeed, it can be shown that the Wigner transform of a Gaussian wavepacket remains positive definite.

As previously mentioned, the Wigner formulation of quantum mechanics is particularly well suited to study the semiclassical limit in which quantum effects, notably interference, are relatively mild. A typical use in the study of electron transport in semiconductors.

Lately, the Wigner formalism has been strongly revived for fundamental investigations of the most puzzling aspects of quantum mechanics, such as the Einstein–Podolski–Rosen paradox, quantum entanglement and the other most counterintuitive manifestations of quantum non-locality.

While such subjects lie beyond the scope of this work, they clearly state a case of major interest for the development of powerful techniques to solve the Wigner equation in the strong interference limit.

From the mathematical viewpoint, this is a pretty formidable task, mostly an account of the quantum "forces" described above, which involve high-order derivatives in velocity space.

It is therefore of interest to find out what lattice-kinetic theory might possibly offer on this important table.

Lattice versions of the Wigner equations have been recently investigated using high-order lattices to accommodate the high-order derivatives in momentum space.

At the time of this writing, however, they have not shown any visible advantage over methods based on spectral Hermite expansions (22), but the subject is still in embryonic state indeed very recent studies seem to indicate the possibility of formulating a lattice Wigner equation along the footsteps of LB indeed very recent studies indicate the possibility of formulating a lattice Wigner equation along the footsteps of the LB method, see S. Solorzano et al, The Lattice Wigner Equation, Phys. Rev. E, 97, 013308 (2018).

Much further studies are required to consolidate these fascinating matters, further studies are definitely required to illuminate these fascinating matters.

32.9   Summary

Summarizing, we have presented a Lattice Boltzmann formulation of Dirac's quantum relativistic wave equation which naturally reduces to the Schrödinger equation in the non-relativistic limit. This formulation is based on a formal analogy between the quantum spin and classical discrete momentum which is exact in one spatial dimensions. It extends to higher dimensions through a generalization of operator splitting securing the alignment between the spin and the discrete momentum. This is introduces an error which is within the same order, second, of the lattice discretization. Unitarity, on the other hand, holds exactly regardless of the size of the timestep. The resulting quantum LB scheme inherits the efficiency and amenability to parallel computing of its classical counterpart. Finally, we have also hinted at lattice realizations of the semi-classical Wigner equation.

32.10   Appendix: Spinors

Spinors are geometrical entities characterized by the way they transform under spatial rotations. For instance, we know that upon a rotation of an angle α of the cartesian plane, the coordinate of a vector $P \equiv (x, y)$ turn into $P' \equiv (x' = x\cos\alpha - y\sin\alpha, y' = x\sin\alpha + y\cos\alpha)$.

This shows that full rotations with $\alpha = 2\pi$ and integer multiples thereof leave the vector unchanged. Like vectors and tensors, spinors transform linearly upon rotations of Euclidean space. Unlike vector and tensors, however, they do not come full circle back upon full rotations of 2π and integer multiples thereof.

In particular, under a 2π rotation, a bi-component spinor $\xi = (\xi_1, \xi_2)$ turns into its mirror image (ξ_2, ξ_1) More generally, upon a rotation of an angle α, a spinor of order n rotates only by an angle α/n. On purely pictorial grounds, *we can think of a spinor as of an object with n heads, which turn each into the next one, upon a rotation 2π*. Thus, the previous example of a bicomponent spinor with $n = 2$, can be thought as a head-tail card, where the head turns into the tail and vice versa, upon a 2π rotation of the plane. Likewise, a spinor of order $n = 4$ can be pictured as a four-arm structure, East, North, West, South, (ENWS), turning into South, East, North, West, (SENW) upon a 2π rotation of the plane.

In other words, it takes a rotation of 8π in order for the original ENWS configuration to come full circle back.

These strange objects, first introduced in mathematics back in 1913 by the french mathematician Elie Cartan (1869–1951), have known a major boost of popularity in physics after the discovery of particles carrying internal angular degrees of freedom, best known as *spin*. In particular, a particle of spin s is described by a (complex) spinor of order $n = 2s + 1$.

The most popular example is the electron, which features $s = 1/2$, and is therefore described by a two-component spinor. In fact, as per Dirac's paramount achievement, by a four-component spinor, associated with the electron and its anti-electron!

Clearly, we are here at a major pinnacle of breathtaking resonance between abstract mathematical beauty and physical depth.

Although the algebra of spinors has been worked out in great detail, their deep meaning remains somewhat obscure, as cogently expressed by the Sir Michael Atiyah comment:

> No one fully understands spinors. Their algebra is formally understood but their general significance is mysterious. In some sense they describe the square-root of geometry and, just as understanding the square root of −1 took centuries, the same might be true of spinors.

....

REFERENCES

1. E. Madelung, Quanten theorie in hydrodynamischer form, *Z. Phys.*, 40, 322, 1927.
2. D. Bohm, *Quantum Theory*, Englewood Cliffs, New Jersey, 1951.
3. A. Messiah, *Quantum Mechanics*, North–Holland, Amsterdam, 1962.
4. A. Einstein, B. Podolski, N. Rosen, Quantum-mechanical description of physical reality, *Phys Rev.*, 47, 777, 1935.
5. L. Landau and E. Lifshitz, *Non-Relativistic Quantum Theory*, Pergamon Press, Oxford, 1960.
6. L.L.Foldy and S.A. Wouthuysen, On the Dirac theory of spin 1/2 particles and its non-relativistic limit, *Phys. Rev.*, 78, 29 1950.
7. F. Wilczek, Majorana returns, *Nature Phys.*, 5, 614, 2009.

8. S. Succi and R. Benzi, "Lattice Boltzmann equation for quantum mechanics", *Physica D*, **69**, 327, 1993.

9. I. Bialynicki–Birula, Weil, Dirac and Maxwell equations on a lattice as a unitary cellular automata, *Phys. Rev. D*, **49**, 12, 6920, 1994.

10. S. Succi, "Numerical solution of the Schrödinger equation with discrete kinetic theory", *Phys. Rev. E*, **53**(2), 1969, 1996.

11. S. Succi, "Lattice quantum mechanics: an application to Bose–Einstein condensation, *Int. J. Mod. Phys. C*, **9**(8), 1577, 1998.

12. B. Boghosian and W. Taylor, "Quantum lattice gas models for the many body Schrödinger equation", *Int. J. Mod. Phys. C*, **8**(4), 705, 1997.

13. D. Meyer, "Quantum lattice gases and their invariants", *Int. J. Mod. Phys. C*, **8**(4), 717, 1997.

14. S. Palpacelli and S. Succi, "The quantum lattice Boltzmann equation: Recent developments", *Commun. Comput. Phys.*, 4, 9801007, (2008).

15. P. Dellar, D. Lapitski, "Convergence of a three-dimensional quantum lattice Boltzmann scheme towards solutions of the Dirac equation", *Phil. Trans. Roy. Soc.*, 369 (1944), 046706, 2010.

16. P. Dellar, D. Lapitski, S. Palpacelli and S. Succi, "Isotropy of three-dimensional quantum lattice Boltzmann schemes", *Proc. Rev. E*, **83**, 046706, 2011.

17. R. Feynman and A. Hibbs, *Quantum Mechanics and Path Integrals*, McGraw–Hill, New York, 1965.

18. K. Kulander ed., *Comp. Phys. Comm.* **63**, 1991, special issue on time-dependent methods for quantum dynamics.

19. F. Fillion, M. Mendoza, H. Herrmann, S. Palpacelli and S. Succi, "Formal analogy between the Dirac equation in its Majorana form and the discrete-velocity version of the Boltzmann kinetic equation", *Phys. Rev. Lett.*, **111**, 160602, 2013.

20. M. Mendoza, H. Herrmann and S. Succi, "Preturbulent regimes in graphene flows", *Phys. Rev. Lett.*, **106**, 156601, 2011.

21. E. Wigner, "On the quantum correction for thermodynamic equilibrium", *Phys. Rev.*, **40**, 749, 1932.

22. O. Furtmaier, S. Succi, M. Mendoza, "Semi-spectral method for the Wigner equation", *J. Comp. Phys.*, **305**, 1015, 2016.

··

EXERCISES

1. The QLB dispersion relation yields an imaginary propagation speed for $\mu > 2$. Comment on this result versus the stability condition $0 < \omega < 2$ of the classical LB.

2. Write down the quantum potential for the case of a free-particle.

3. Show that the Wigner transform of a Gaussian wavepacket is positive definite.

33

QLB for Quantum Many-Body and Quantum Field Theory

In Chapter 32 we have expounded the basic theory of quantum LB for the case of relativistic and non-relativistic wavefunctions, namely single-particle quantum mechanics. In this chapter, we shall discuss extensions of the quantum LB formalism to the overly challenging arena of quantum many-body problems and quantum field theory, along with an appraisal of prospective quantum computing implementations.

Hey, scientist, leave the cat alone . . . .

(readapted from Pink Floyd).

33.1 The Quantum *N*-Body Problem

Solving the single particle Schrödinger, or Dirac, equation in three dimensions is a computationally demanding task. This task, however, pales in front of the ordeal of solving the Schrödinger equation for the quantum many-body problem, namely a collection of many quantum particles, typically nuclei and electrons in a given atom or molecule.

This is the central goal of computational quantum chemistry, a cutting edge of modern computational science.

The target is the *N*-body Schrödinger equation:

$$i\hbar\, \partial_t \Phi = \hat{H}_N \Phi \equiv \left[-\sum_{n=1}^{N} \frac{\hbar^2}{2m} \Delta_n + V\left(\vec{R}_N\right) \right] \Phi \tag{33.1}$$

where $\Phi \equiv \Phi(\vec{R}_N, t)$ is the *N*-body wavefunction, $\vec{R}_N \equiv \{\vec{r}_1 \dots \vec{r}_N\}$ is the position vector in 3*N*-dimensional configuration space $\mathcal{R}_{3N}$ and $\vec{r}_n = (x_n, y_n, z_n)$ are the single-particle spatial coordinates in ordinary three-dimensional space.

The Lattice Boltzmann Equation. Sauro Succi, Oxford University Press (2018).
© Sauro Succi. DOI: 10.1093/oso/9780199592357.001.0001

Finally, $V(\vec{R}_N)$ is the interparticle potential, typically in a two-body form:

$$V\left(\vec{R}_N\right) = \sum_{n=1}^{N} \sum_{m>n} V\left(|\vec{r}_n - \vec{r}_m|\right)$$

Note that for large bio-molecules, such as proteins, N can easily get in the range of several thousands or even hundred thousands ....

In sheer mathematical terms, this means solving a (linear) partial differential equation in hundreds of thousands of dimensions, a task which is clearly unfeasible with standard numerical techniques. More precisely, with a grid resolution of N_g grid points per spatial dimension, the number of discrete degrees of freedom of the Hilbert space for a system of N quantum particles in D spatial dimensions is given by

$$\mathcal{N}_H = (N_g^D)^N \tag{33.2}$$

This shows the exponential computational barrier presented by the quantum many-problem: a lowly $N = 10$ particle system, in $D = 3$ spatial dimensions, with just $N_g = 10$ grid points per dimension delivers $\mathcal{N}_H = 10^{30}$ degrees of freedom! This exponential complexity barrier stands unforgivingly tall in the way of any grid-based method, to the point that in his Nobel lecture, Walter Kohn (1923–2016)referred to the many-body wavefunction Φ as to as an uncomputable, hence unphysical construct.

Nevertheless, given the paramount importance of the quantum many-body problem in chemistry, material science and biology, the search of approximate numerical methods to "bypass" this exponential barrier, makes one of the most challenging mainstreams of computational chemistry (1).

33.2   Numerical Quantum Many-Body Methods

In the following we provide a very cursory view of the main tools of the trade for the numerical quantum many-body problem (2). Even though they differ significantly in spirit and detail, all of these method share a common feature: spend the computational resources only on the "active" regions of Hilbert space where the wavefunction takes on substantial values, leaving the unpopulated ones (the overwhelming majority) suitably disattended.

33.2.1   Quantum Monte Carlo Methods

One of the most popular techniques for the quantum many body problem is the quantum Monte Carlo method. This method comes in many variants, whose description lies clearly beyond the scope of this book. Here, we shall simply mention the Diffusion Monte Carlo method (DMC) for many-body *ground-state* calculations. This consists of treating the imaginary-time N-body Schrödinger equation as a diffusion-reaction equation in $3N$-dimensional space $\mathcal{R}_{3N}$. Since grids are unviable due to exponential memory requirements, the DMC moves $N_w \ll \mathcal{N}_H$ grid-free "random-walkers" in $3N$-dimensional space, based on stochastic rules designed in such a way as to recover the diffusion equation in the limit $N_w \to \infty$. The obvious advantage is that the walkers tend to sample only the regions of Hilbert space where the wavefunction is substantial leaving the empty ones unattended.

In imaginary time, the potential $V(\vec{R}_N)$ leads to the disappearance of walkers, which must constantly be re-injected in order to secure norm conservation,

$$\int_{\mathcal{R}_{3N}} |\Phi|^2 (\vec{R}_N) d\vec{R}_N = N$$

Being Monte Carlo, the DMC method is very slowly convergent, but has nonetheless been able to provide groundbreaking insights into the physics of quantum many body systems. In particular, it was used to perform the first pioneering simulation of the ground state of a system of electrons (the electron gas) (3), and continues to be improved and refined to this day.

33.2.2   Variational Monte Carlo

An alternative class of Monte Carlo methods is the Variational Monte Carlo (VMC). As suggested by its name, the idea is to minimize the energy functional:

$$E[\Phi] = < \Phi | \hat{H}_N \Phi > \tag{33.3}$$

where the bra-ket notation stands for scalar product in Hilbert space.

The procedure is based on the variational theorem, according to which the exact ground state of the Schrödinger equation minimizes the (33.3) energy functional. As a result, any other (trial) wavefunction is bound to deliver a higher value of the energy, $E_T \geq E_0$.

The trial function carries a given dependence on the spatial variables and also depends on a small number $P \ll N$ of variational parameters $\Lambda \equiv \{\lambda_1 \ldots \lambda_P\}$, which are varied in such a way as to minimize the expectation value of the energy $E_T(\lambda) = < \Phi_T | \hat{H} \Phi_T >$. This minimization proceeds via the solution of the following set of P constraint equations:

$$\frac{\delta E_T(\lambda)}{\delta \lambda_p} = 0, \ p = 1, P$$

The advantage is that an educated guess on the functional form of $\Phi_T(\vec{R}_N; \lambda)$ permits us to reduce dramatically the parameter space.

The flip-side is that fast convergence is very sensitive to the quality of the trial function. In other words, in order to obtain accurate results, a great deal of foreknowledge must be injected into the choice of the trial function. Whenever such foreknowledge is available, the VMC can prove a very effective quantum many-body solver.

33.2.3 Matrix Eigenvalue Methods: Exact Diagonalization

Another mainstream technique to solve the quantum many-body problem consists in addressing the time-independent version of the Schrödinger equation (33.1), namely

$$\hat{H}_N \Phi = E\Phi \tag{33.4}$$

The task is then "simply" to solve the corresponding numerical eigenvalue problem.

This is more easily said than done, as it involves the diagonalization of monster matrices of rank N_g^{3N}, N_g being the number of grid points per spatial dimension. As noted before, even a lowly $N_g = 10$ grid, with just $N = 10$ particles delivers an unfeasible numerical eigenvalue problem of size 10^{30}.

In one spatial dimension, very clever ad-hoc techniques can save the day and occasionally provide even preciously rare *exact* solutions (5). On the numerical side specialized techniques have also been developed that provide highly accurate solutions for large number of particles in one spatial dimension. However, these techniques heavily lean on the sequential nature of one-dimensional space, hence their extension to $d > 1$ appears highly non trivial, although not unfeasible either in view of the latest developments, such as Tensor Networks theory (see R. Orus, A practical introduction to tensor networks: Matrix product states and projected entangled pair states, ANNALS OF PHYSICS 349, 117–158 (2014)).

33.2.4 Dynamic Minimization

An alternative class of methods is based on the idea of minimizing the energy functional (33.3) by means of iterative methods, such as steepest descent or conjugate-gradient techniques. Let $\{\phi_n\}$ a collection of discrete degrees of freedom describing the N-body wavefunction Φ_N, the minima would be attained by "rolling" the ball in Hilbert space according to some gradient dynamics, say a simple steepest-descent of the form:

$$\frac{d\phi_n}{dt} = -\frac{\partial E[\phi_n]}{\partial \phi_n} \tag{33.5}$$

A particularly elegant and successful form of *dynamic minimization* is the Car–Parrinello ab-initio molecular dynamics (6).

In a nutshell, the Car–Parrinello introduces a *unified* Lagrangian for both nuclear and electronic degrees of freedom,

$$\mathcal{L}_{CP} = \mathcal{L}(\psi_e, R_I) \tag{33.6}$$

where ψ_e stand for the electronic degrees of freedom and R_I are the nuclear coordinates. The electronic degrees of freedom obey effective one-particle Schrödinger-like equations known as Kohn-Sham equations, to be detailed shortly. The nuclear degrees of freedom, being much slower, are handled through classical molecular dynamics. which are assumed to obey classical-Newton mechanics, since ions are much heavier than electrons.

The CP dynamics features a key property: in the process of minimizing the CP Lagrangian functional (33.6), the electrons land exactly on the Born–Oppenheimer hypersurface associated with the instantaneous nuclear coordinates (Born–Oppenheimer approximation, another form of the enslaving principle discussed many times before in connection with the hydrodynamic limit of the LB theory). As a result, the "magic" Car–Parrinello algorithm permits us to track ions and electrons at the same time, whence its nomer *ab-initio* molecular dynamics. On the other hand, in order to stick to the Born–Oppenheimer surface, the electrons must be assigned a smaller mass than their physical one, thereby imposing very small timesteps (this is vaguely reminiscent of the low Mach-number constraint in Lattice Boltzmann theory).

It should be appreciated that the CP method belongs to the conceptual framework of Density Functional Theory, whereby the ground state of the quantum many-body system is described solely in terms of the electron density distribution in ordinary three-dimensional space, $n(\vec{r})$. This is where the dramatic reduction of complexity, from ultra-dimensional Hilbert space to ordinary three-dimensional space comes from. The strength of electronic density functional theory is that such dramatic reduction sits on a very solid theoretical cornerstone, namely the Kohn–Hohenberg theorem to be discussed shortly, as it provides a constructive framework for LB implementations.

33.2.5   Quantum LB as the "Perfect" Lattice Monte Carlo?

After this very cursory and partial view of the tools of the trade, a natural question arises: *what could (quantum) lattice Boltzmann techniques possibly contribute to this challenging forefront of computational science?*

Since the N-body Schroedinger is formally a diffusion-reaction equation in imaginary time, one might speculate that a classical LB scheme for reaction-diffusion could possibly bring something on this table. A similar statement would go for quantum LB in real time. In a way, LB would work as a Monte Carlo method with perfect efficiency since no move is rejected in the stream-collide LB dynamics: the Perfect Monte Carlo!

Albeit true in principle, this rosy idea of the Perfect Monte Carlo is computationally doomed by the exponential complexity barrier mentioned before: totally unviable for more than a fistful of particles. The stopper, as usual is exponential memory demand.

One could imagine a sophisticated series of hierarchical grids focusing resolution only on the relevant regions of the Hilbert space. Even so, the dimensional curse remains unforgiving: just a minimal 3 grid points per dimension for 10 particles in three spatial dimensions requires $3^{30} \sim 10^{14}$ grid points: no way.

I wish I had better news to report, but until the day dawns when an off-grid LB is invented, I'm afraid that this is where we stand[90]. So much for the full hard-core N-body problem: fortunately, other routes may carry some better news, as we are going to briefly discuss in the sequel.

---

[90] Some ten years ago, I received a phone-call from a Californian investor, who wished to pursue the idea of "perfect" Monte Carlo. Much to my regret, despite some short but intense mumbling, I could do no better than convincing him that the dimensional curse could not be beated. Helas, my claim to fame & fortune went shattered on mountain Dimensional Curse . . . .

33.3   Electron-Density Functional Theory

Most many-body problems be they classical or quantum are most effectively solved by casting them, whenever possible, into a suitable *effective* one-body form. This entails drastic approximations, which often turn out to capture much more physics than they were initially supposed to.

The N-body Schroedinger equation offers a spectacular case in point: there exist indeed an *exact* mapping between the N-body Schroedinger equation and a set of N *effective* one-body Schroedinger like equations, best known as Kohn-Sham equations, after Walter Kohn and Lu Jeu Sham (7).

The Kohn–Sham formulation builds on a fundamental theorem proven in the mid 60s by Pierre Hohenberg and Kohn (8).

The theory states is that the ground-state energy of a many-electron wavefunction (nuclei are regarded as classical particles on account of their large mass) is uniquely determined by the electronic density in *ordinary* three-dimensional space, namely, $E = E[n(\vec{r})]$, where

$$n(\vec{r}) = \sum_{j=1}^{N_{orb}} |\phi_j|^2 (\vec{r})$$

is the electron density and ϕ_j are one-particle electron orbitals.

The ground-state energy can then be obtained by summing up the single particle orbital energies, which in turn are obtained by solving the Kohn–Sham equations:

$$\hat{H}_j^{KS} \phi_j = E_j \phi_j, \; j = 1, N_{orb} \tag{33.7}$$

The Kohn–Sham Hamiltonian consists of four contributions:

$$\hat{H}_j^{KS} = -\frac{\hbar^2}{2m} \Delta_j + V_{ext}(\vec{r}) + e^2 \int \frac{n(\vec{r'})}{|\vec{r'} - \vec{r}|} \, d\vec{r'} + V_{ex}[n] \tag{33.8}$$

The first two contributions are the usual kinetic energy and external potential operators, including the electron-ion interactions, while the third one relates to the self-consistent Hartree–Fock potential associated to the electronic cloud.

Finally, the fourth one is an effective "exchange" energy functional which collects all the unknown effects of N-body interactions.

The key point is that, albeit unknown, such an effective functional of the electron density is *proven* to exist! As a result, the ground-state energy of a fictitious system of *near-independent* electrons moving in such a potential is *exactly the same* ground-state energy of the interacting system!

By near-independent, we mean that each of the KS equations are coupled only through the total electron density and therefore they can be solved independently using the total electron density obtained, say, by a previous iteration.

Describing how such a magic comes about is beyond the scope of this book. Here, we shall simply remark that density functional theory leans heavily on the intuitive picture of a quantum many-body system as a backbone of ions glued together by a very mobile *electronic fluid.*

In this respect, it certainly puts a premium on efficient real space solvers for one-particle Schrödinger-like equation, both in the time-independent (ground-state) and time-dependent (excited states) form. Whence potential scope for LB techniques.

33.3.1 Kinetic Approach to Electronic Density Functional Theory

Very recently, the previous program has been implemented in actual simulations of some simple molecules showing competitive performance with the state of the art methods for electronic structure simulations (9) (see Fig. 33.1).

Essentially, a classical LB for diffusion-reaction coupled to a Poisson solver for the interaction potential, is applied to each of the one-body KS equations.

While a massive amount of careful work is needed to assess the effective import of this LB–KS approach, one may hope that the use of LB techniques may open new angles of attack to electronic structure simulations. Particularly interesting from the theoretical viewpoint appears the prospect of a kinetic formulation of the Hohenberg–Kohn theorem. At the moment, however, this prospect remains to be realized.

33.4 Lattice Boltzmann for Quantum Field Theory

The quantum many-body Schroedinger equation describes non-relativistic systems consisting of large but finite number of particles, electrons and nuclei. Electrons and nuclei are stable and therefore their number does not change in time. The relativistic framework presents a very different picture: since particles come in pairs with their antiparticles, they can annihilate and turn into light (photons) and vice versa. Hence the number of particles is not conserved and a statistical theory of quantum fields, (quantum field theory, QFT for short) is necessary to describe the dynamics of the quantum relativistic system.

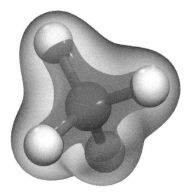

Figure 33.1 *The methane molecule as computed by the LB-KS method. Reprinted with permission fom M. Mendoza et al, Phys Rev Lett 113, 096402 (2014). Copyright 2014 by the American Physical Society.*

33.4.1 Second Quantization and Heisenberg Interaction Representation

QFT is most conveniently formulated in the language of *second quantization*, a formalism centered upon the so-called *generation and annihilation* operators.

The generator $\hat{a}_n^+$ materializes a particle in the n-th quantum state and its conjugate $\hat{a}_n$ annihilates it. Note that "particle," here means an eigenstate described by an associated eigenfunction $\phi_n(x)$. However, in contrast to the Schroedinger formulation, no action is taken on the wavefunction itself, but only on the occupation numbers of the corresponding eigenstate, whence the nomer "second" quantization.

It is indeed clear that suitable combinations of these operators permit us to describe the dynamics of the quantum field in terms of transitions between the various eigenstates.

For instance, the composite operator $\hat{a}_n^+\hat{a}_m$ moves a particle from quantum state m to quantum state n and its conjugate $\hat{a}_m^+\hat{a}_n$ does just the converse.

The action of the previous operators on a quantum wavefunction with n boson particles, denoted by $|n>$ in Dirac's bra-ket notation, is defined by the following relations:

$$\begin{cases} \hat{a}|n> = \sqrt{n} \,|n-1> \\ \hat{a}^+|n> = \sqrt{n+1} \,|n+1> \end{cases} \tag{33.9}$$

From (33.9), one readily derives

$$\begin{cases} \hat{a}^+\hat{a} = n \\ \hat{a}\hat{a}^+ = n+1 \end{cases} \tag{33.10}$$

The former is also denoted as "number operator" $\hat{n}$, for its effect on the state $|n>$ is simply to multiply it by the number of particles, i.e., $\hat{n}|n> = n\,|n>$. From the above relations (33.10) one derives the fundamental commuting rule for bosons:

$$[\hat{a}, \hat{a}^+] = \hat{a}\hat{a}^+ - \hat{a}^+\hat{a} = 1$$

As an example, for a two-state boson wavefunction with n_0 particles in state $|0>$ (ground state) and n_1 particles in the first excited state $|1>$, with $n_0 + n_1 = n$, we have

$$\hat{a}_0|n_0, n_1> = \sqrt{n_0}|n_0 - 1, n_1>$$

and

$$\hat{a}_1^+ |n_0 - 1, n_1> = \sqrt{n_0(n_1 + 1)} \,|n_0 - 1, n_1 + 1>$$

so that

$$\hat{a}_1^+\hat{a}_0 \,|n_0, n_1> = \sqrt{n_0(n_1 + 1)} \,|n_0 - 1, n_1 + 1>$$

In other words, the combined effect of the two operators is to raise one particle (excitation) from the ground state to the first excited state, see Fig. 32.2.

Note that we have said nothing about the spatial structure of the wavefunction. If $\phi_0(x)$ and $\phi_1(x)$ denote the eigenfunctions of the zeroth- and first-order energy levels, respectively, a typical two-body bosonic wavefunction reads as follows:

$$\Phi(1,2) = c_{00}\phi_0(1)\phi_0(2) + c_{01}\phi_0(1)\phi_1(2) + c_{10}\phi_0(2)\phi_1(1) + c_{11}\phi_1(1)\phi_1(2)$$

where the arguments 1 and 2 denote the spatial coordinates of the bosons while subscripts 0, 1 refer to the quantum states energetically available to the system. The amplitudes $|c_{nm}|^2$, $n, m = 0, 1$, represent the probability to occupy the corresponding eigenstates. The previous expression automatically embeds the boson particle-exchange symmetry $\Phi(1,2) = \Phi(2,1)$. The generalization to the case of N bosons over M quantum states is the so-called *Pfaffian* of the matrix $\Phi_{nm} \equiv \phi_m(x_n)$, $n = 1, N$, $m = 1, M$, the symmetric analog of the determinant in which all permutations, odd and even, carry a plus sign.

The relations (33.9) deliver the commutation relation:

$$[\hat{a}_n, \hat{a}_n^+] = 1 \tag{33.11}$$

It can also be checked that commutators between different particle states are identically zero

$$[\hat{a}_n, \hat{a}_m] = [\hat{a}_n, \hat{a}_m^+] = 0 \tag{33.12}$$

This holds for bosons, fermions obeying the same rules, with commutators replaced by anti-commutators $\{A, B\} = AB + BA$. This apparently innocent sign change bears a world of difference between the statistical and dynamical behavior boson and fermions. In particular, the *anti-commuting* relation $\{A, A\} = 2A^2 = 0$ indicates that fermionic operators obey a pretty peculiar algebra, known as Grassmann algebra, after the prussian polymath Herrmann Grassmann (1809-1877), in which the square of a number can be zero even if the number is not. In fact, a Grassmann number can be literally defined as the non-zero square root of zero! This is a close relative of the cross product between two vectors, although a much more general one. Grassmann variables form a well-defined algebra, equipped with all the set of rules that permit to perform actual computations. However, such computations are typically much harder than usual ones based on "ordinary" numbers. In the first place, electronic computers, as we know them, are not instructed to deal directly with anti-commuting Grassmann numbers. Weird as it seems, Grassmann algebra sets a very fertile mathematical background for the quantum theory of fermionic fields.

Having defined the basic algebra of the second-quantization operators, we need to prescribe their dynamic evolution.

According to the spirit of second quantization, the operators themselves, rather than the wavefunction, evolve in space and time.

Since an operator is represented by an infinite-dimensional matrix in Hilbert space, second quantization is basically a formalism for the dynamics of infinite-dimensional matrices.

Within this matrix formulation, physical observables are computed in the *Heisenberg's interaction representation* according to which the expectation value of a given operator $\hat{A}$ is given by

$$A(t) = < \psi(0)|\hat{A}(t)|\psi(0) > \equiv \int \psi^*(x;0)\hat{A}(t)\psi(x;0)dx \qquad (33.13)$$

where x denotes the position in d spatial dimensions, see Fig. 33.2.

This operator evolves in time according to Heisenberg's equation:

$$i\hbar\partial_t\hat{A} = [\hat{H},\hat{A}] \qquad (33.14)$$

where the second-quantized Hamiltonian is given by

$$\hat{H} = \sum_i \hat{a}_i^+ \epsilon_i \hat{a}_i \qquad (33.15)$$

ϵ_i being the energy of the i-th quantum eigenstate.

This is Heisenberg's matrix formulation of quantum mechanics, as opposed to Schroedinger's wavefunction formulation.

Heisenberg's operator equation (33.14) is an evolution equation for the infinite-dimensional matrix $\hat{A}$:

$$\hat{A} = \sum_{ij} \hat{a}_i^+ A_{ij} \hat{a}_j \qquad (33.16)$$

whose elements are defined as follows:

$$A_{ij} = \int \phi_i^*(x)\hat{A}\phi_j(x) \; dx \qquad (33.17)$$

where $\phi_i(x)$ denotes the i-th basis function, out of a complete set in Hilbert.

Heisenberg's formulation is more suitable to relativistic quantum field theory because it is naturally suited to describe transitions between quantum states with a *variable* number of excitations.

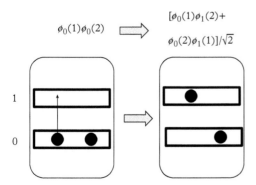

Figure 33.2 *Graphical representation of the effect of generation and annihilation operators upon a system of two bosons, 1 and 2, with two energy levels, denoted by 0 and 1. The figure represents the transition from $|2,0>$ to $|1,1>$ under the effect of the operator $\hat{a}_1^+ \hat{a}_0$. Note that the $|1,1>$ state at the right represents the superposition of two states, $\frac{1}{\sqrt{2}}[\phi_0(1)\phi_1(2) + \phi_0(2)\phi_1(1)]$.*

33.4.2   Quantum Fields in Spatially Extended Systems

For the case of the spatially extended quantum many-body systems, one needs a collection second-quantized operators at each spatial location.

It proves expedient to cast these operators in the following form:

$$\begin{cases} \hat{\psi}(x) = \sum_n \psi_n(x)\hat{a}_n \\ \hat{\psi}^+(x) = \sum_i \psi_n^*(x)\hat{a}_n^+ \end{cases} \tag{33.18}$$

where the spatial coordinate x plays the role of a *continuum* index. In other words, at each spatial location x one may imagine to place a "tower" of quantum states spanned by the index n in (33.18).

The operator $\hat{\psi}^+(x)$ generates a particle (excitation) at position x, through the sum over all possible quantized states, each contributing a weight $\psi^*(x)$, while its complex conjugate, $\hat{\psi}(x; t)$ does just the opposite.

As a result, the non-local operator

$$\hat{K}(x, y) \equiv \hat{\psi}^+(y)\hat{\psi}(x)$$

annihilates a particle at position x and generates a new one at position y, the net result being the hopping of the particle from x to y.

The reader will hardly miss a very close analogy with the streaming operator of lattice kinetic theory, except that here "particles" are in fact elementary excitations of the quantum field, see Fig. 33.4.

The corresponding Heisenberg's equation reads as follows:

$$i\hbar\partial_t\hat{\psi} = [\hat{H}, \hat{\psi}] \tag{33.19}$$

where the second-quantized Hamiltonian of the extended system depends on the specific physics in point.

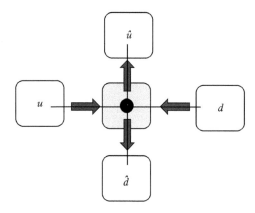

Figure 33.3 *Symbolic representation of the stream-collide for the (infinite)-dimensional matrices u and d. The squares stand for a finite-dimensional representation of the matrices. The stream-collide dynamics must be such to preserve the equal-time commutation relations.*

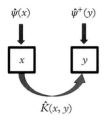

$\hat{\psi}(x)$ $\hat{\psi}^+(y)$

x y

$\hat{K}(x, y)$

Figure 33.4 *Graphical representation of the hopping operator $\hat{K}(x, y)$ associated with kinetic energy. A particle is taken away from site x and deposited at site y, resulting in net motion from x to y.*

A particularly important second-quantized model for quantum many-body lattice systems is the Hubbard–Hamiltonian:

$$\hat{H} = k \sum_{x,y} \hat{\psi}^+(y)\hat{\psi}(x) + g \sum_x V[\hat{\psi}^+(x)\hat{\psi}(x)] \tag{33.20}$$

In (33.20), the first term on the right-hand side is the kinetic hopping from site to site, while the second one is associated with on-site potential energy interaction. The parameter k/g measures the ratio of kinetic to potential energy. Note that in the quantum world, particles can tunnel through potential barriers, which is why our k is often called t, for tunneling.

The Hubbard–Hamiltonian plays a major role in the study of highly correlated quantum many-problems, such as electrons in superconductors and other exotic states of matter where quantum correlations play a defining role.

At skim value, it thus appears like lattice quantum field theory could indeed be formulated as a *lattice kinetic theory of operators*, i.e., infinite-dimensional matrices in Hilbert space. In broad strokes, this lattice kinetic theory is supposed to propagate infinite-dimensional matrices along the lattice according to the rules of quantum field theory.

Propagate means, as usual, stream and collide: thus, the question is, how can such matrices possibly "propagate and collide," in compliance with the rules of quantum field theory?

As a matter of fact, since commutators are the cornerstone of operator dynamics, such compliance is tantamount to fulfilling the appropriate commuting rules at any time along the simulation. Thus, besides the usual requirement of unitarity, the quantum field operators have to obey the equal-time commutation relations (ETCR). For the one-dimensional case:

$$C(x, y; t) \equiv \hat{\psi}(x; t)\hat{\psi}(y; t) \mp \hat{\psi}(y; t)\hat{\psi}(x; t) = 0 \tag{33.21}$$

$$C^{**}(x, y; t) \equiv \hat{\psi}^*(x; t)\hat{\psi}^*(y; t) \mp \hat{\psi}^*(y; t)\hat{\psi}^*(x; t) = 0 \tag{33.22}$$

$$C^*(x, y; t) \equiv \hat{\psi}^*(x; t)\hat{\psi}(y; t) \mp \hat{\psi}^*(y; t)\hat{\psi}(x; t) = i\hbar\delta(x - y) \tag{33.23}$$

where $\mp$ refers to bosons and fermions, respectively.

As we shall see, matching such constraints turns out to be possible, at least in one spatial dimensions.

33.4.3 LB for (1 + 1) Quantum Field Theory

The operator approach to numerical quantum mechanics and quantum field theory is not new: in fact, it was proposed no less than three decades ago by Bender and collaborators in the context of finite-element discretization of the operator equations (10). In the sequel, we show the way that these ideas can be transferred to the LB framework.

Formally, the QFT–LB scheme reads exactly like the standard QLB, except that the discrete wavefunctions ψ_j and ψ_j^* are now replaced by second-quantized operators.

For an interaction-free QFT in one spatial dimension (Majorana representation):

$$\begin{cases} \partial_t \hat{u} + c\partial_x \hat{u} = +\omega_c \hat{d} \\ \partial_t \hat{d} - c\partial_x \hat{d} = -\omega_c \hat{u} \end{cases} \tag{33.24}$$

where we have set $\hat{\psi}_j = (\hat{u}, \hat{d})$ for the up and down moving operators, respectively (in the sequel, up=right, down=left).

These can be discretized in the lattice according to the same rules as QLB yielding a standard transfer-matrix relation:

$$\hat{\psi}_j(x; t + \Delta t) = \sum_{k,y} T_{jk}(x - y; \Delta t)\hat{\psi}_k(y; t) \tag{33.25}$$

The transfer matrix $T_{jk}(x - y)$ is localized to the neighborhood defined by the given lattice kinetic scheme, in our case just the two nearest-neighbors.

In explicit terms:

$$\begin{cases} \hat{u}(x, t + \Delta t) = (1 + p)\hat{u}(x - \Delta x, t) + q\hat{d}(x - \Delta x, t) \\ \hat{d}(x, t + \Delta t) = -q\hat{u}(x + \Delta x, t) + (1 + p)\hat{d}(x + \Delta x, t) \end{cases} \tag{33.26}$$

In symbolic 2×2 matrix notation:

$$T^{QLB}(x, y) = S^> \delta(y - x + \Delta x) + S^< \delta(y - x - \Delta x) + Q\delta(x, y) \tag{33.27}$$

where

$$S^> = \begin{pmatrix} 1 & 0 \\ 0 & 0 \end{pmatrix}, \quad S^< = \begin{pmatrix} 0 & 0 \\ 0 & 1 \end{pmatrix}, \quad Q = \begin{pmatrix} p & q \\ -q & p \end{pmatrix}$$

are the stream-right, stream-left operators and quantum LB collision matrices, respectively.

A graphical representation of the QLB transfer matrix is given in Fig. 33.5.

The (1+1) QFT-LB scheme (33.26) has been shown to be compatible with the commutator constraints (33.21), thus proving the consistency of the QFT-LB scheme (11). Once the effect of the operator $\hat{\psi}(x; t)$ is known at time t, the effect of the time-advanced operator $\hat{\psi}(x; t + \Delta t)$ is also known, via the expression (33.26). As a result, by advancing the operators in time, one can compute any average observable based on the expression (33.13), i.e., with no need of advancing the wavefunction itself.

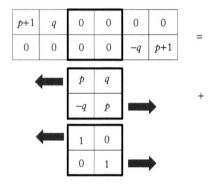

Figure 33.5 *The transfer matrix restriction to the three sites* $[x - 1, x, x + 1]$. *Note that the operators are stored site-wise, i.e.,* $[\hat{u}(x - 1), \hat{d}(x - 1); \hat{u}(x), \hat{d}(x); \hat{u}(x+1), \hat{d}(x+1)]$ *and so on (we take* $\Delta t = \Delta x = 1$ *for simplicity). The composition rule of the transfer matrix is quite simple: take the* 2×2 *quantum collision matrix* $Q(x, x)$ *at site* x *and shift the first row* $[p, q]$ *leftward of two units, from* x *to* $x - 1$. *Then shift the second row* $[-q, p]$ *to the right, from* x *to* $x + 1$. *Same with the unit matrix to account for the streaming and sum the two: the result is the transfer matrix* $T(x, y)$ *for* $y = x - 1, x, x + 1$.

Having brought up this encouraging result we hasten to add that, at the time of this writing, the viability of the QFT-LB scheme remains confined to linear QFT's in one spatial dimension. Relaxing such limitation might open an entirely new field of applications to the QLB formalism, such as high-Tc superconductivity, exotic quantum materials and many other exciting phenomena in the physics of highly correlated quantum many-body systems (12).

An interesting effort in this direction has been recently put forward by Mendl, who developed a LB formulation of a matrix-valued quantum Boltzmann equation originally derived from the Hubbard model (13). Although it is too early to formulate any judgement, the method might prove useful to simulate quantum states of matter, such as complex spin systems and other spintronics phenomena.

33.5 Quantum LB for Quantum Computing

It has been pointed out for a long time (14; 15; 17) that quantum lattice algorithms constitute appealing candidates for prospective implementation on quantum computers. To illustrate the point let us first recall a few basic ideas about quantum computing.

33.5.1 Bits and Qubits

Electronic computers are based on bits, i.e., binary digits which encode just two possible states and outcomes: *yes (1) or no (0)*, tertium non datur.

Quantum bits, qubits for short, encode a graded response encompassing all possible states between yes and no, actually any real number between zero and one. More precisely, any couple of real numbers (p, q), $0 < p, q < 1$, $p + q = 1$, expressing the probability of "yes" (p) and "no" (q) (18; 19).

Any two-level quantum system is a potential candidate for qubit implementation. To appreciate the point let us consider a quantum superposition of the spin-down ($|0 >$) and spin-up ($|1 >$) states of a spin 1/2 fermion:

$$\psi(x) = c_0 |0 > + c_1 |1 >$$

where c_0 and c_1 are complex amplitudes and $|0 > \equiv [1, 0]^T$ and $|1 > \equiv [0, 1]^T$ are the basis states (superscript T stands for transpose).

The probabilities of finding the particle with spin-down(up) are $p_0 = |c_0|^2$ and $p_0 = |c_1|^2 = 1 - p_0$, respectively. Since p_0 and $p_1 = 1 - p_0$ can take any value in the range $[0, 1]$, this two-level system can store an infinite amount of information, i.e., the qubit can be prepared in any of the virtually infinite states labelled by p_0. It should be borne in mind that while the qubit can be prepared in a virtually infinite number of superposition states, it cannot be used to *transmit* more than one classical bit of information. This is because any detection process will necessarily collapse the system to one of the two classical cases $|0 >$ or $|1 >$, with probability p_0 and p_1, respectively.

This is the famous Schroedinger's cat, which is neither dead or alive as long as one does not open the box to find out.

Once the box is opened for observation read the measurement in real life, the poor feline precipitates in one of the two classically observable states, dead or alive.[91]

Bottomline, to paraphrase Pink Floyd: "hey, scientist, leave the cat alone"! But scientists did not quite leave the poor cat alone and made instead the study of quantum entanglement one of the most brazing frontiers of modern physics.

Next consider a *register* consisting of two qubits sitting on neighboring lattice locations, $\psi(x)$ and $\psi(x + 1)$. The system is now described by four basis states, namely:

$|00 >$: down-down

$|10 >$: up-down

$|01 >$: down-up

$|11 >$: up-up

The generic two-site wavefunction describing this two-qubit systems takes the form:

$$\psi(x, x + 1) = c_{00} |00 > + c_{01} |01 > + c_{10} |10 > + c_1 1 |11 >$$

---

[91] Some argues that a third state exists upon opening the box: the poor cat is simply furious!

which requires $2^2 = 4$ complex coefficients, whose magnitude square gives the probability of finding the 2-qubit system in one of the four pure states.

Likewise, a Q-qubit system requires 2^Q complex amplitudes, each delivering the probability of finding the quantum system in one of the 2^Q pures states $|0 \ldots 0 > , \ldots |1 \ldots 1 >$, where each ket counts Q bits.

Thus, Q qubits encode 2^Q classical states, thus offering a potential strategy to defeat the dreadful exponential complexity plaguing the quantum many-body problem and QFT as well. The name of this magic is *quantum entanglement,* namely the subtle web of non-local correlations which sustain the Q-qubit configuration.

However, in order to fulfill the promise, the quantum correlations encoding the 2^Q classical systems must survive long enough to allow the system to perform useful calculations. Unfortunately, this turns out to be exceedingly challenging in practice, because quantum correlations are very fragile and *decohere* real fast on a macroscopic scale (20).[92]

This is ultimately the major stumbling block toward the practical implementation of quantum computers, see Fig. 33.6.

So much for the state representation; how about the time evolution?

Quantum computing proceeds through the application of *unitary* operations. This is no surprise, given that quantum systems evolve in time through the application of the unitary propagator:

$$\hat{U}(t, t') = e^{\frac{i\hat{H}}{\hbar}(t'-t)} \tag{33.28}$$

Note that this is a reversible operation, as it satisfies the group property:

$$\hat{U}(t, t') = \hat{U}^{-1}(t', t)$$

The QLB, as we have discussed it in this book, is a second-order (Crank-Nicolson) approximation of $\hat{U}(t, t + \Delta t)$:

$$\hat{U}_{QLB} = (1 - i\hat{H}\Delta t/2)^{-1}(1 + i\hat{H}\Delta t/2) \equiv (1 + i\hat{H}_{QLB}\Delta t/2) \tag{33.29}$$

The identity on the right-hand side highlights that the QLB formulation can be kept explicit form, as per the expression (33.27), hence suitable for quantum computing.

---

[92] The decoherence time between two quantum states a distance d apart, is roughly given by $\tau_D \sim \tau_R \lambda_b/d$, where τ_R is a typical classical relaxation time and $\lambda_b = \hbar/mv_T$ is the thermal De Broglie length. For macroscopic objects, say two particles of mass 1 gram, placed 1 cm apart at $T = 300$ Kelvin degrees, the ratio τ_D/τ_R is about 10^{-40}! Thus, even taking the age of the Universe for τ_R, would still leave us with $\tau_D \sim 10^{-23}$ seconds. This speaks clearly for the reasons why quantum computing is so hard to achieve in a macroscopic world...Nevertheless, technological advances are regularly reported in the current literature. In addition, the possibility that quantum coherence may play a role in some biological phenomena is currently a subject of active investigation (16).

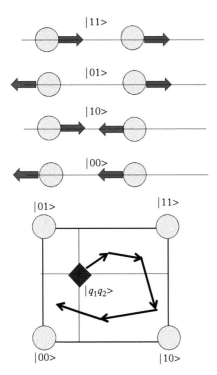

Figure 33.6 *Upper panel: the lattice configuration associated with a two-qubit system. In the figure up=right and down=left. Lower panel: the two-qubit system (diamond inside) can take any position in the plane defined by the four pure states and can fill the plane in time. A classical system, on the other hand, can only jump among the four corners of the plane, with probability $p_{00}, p_{01}, p_{10}, p_{11}$.*

33.5.2   Hamiltonians and Circuits

The basic idea behind quantum computers is to realize the circuitry capable of imitating, or emulating, the unitary evolution driven by the quantum propagator $\hat{U}(t, t')$. Here again, the language of second quantization comes very handy in.

Consider for instance the boolean operator "NOT," which reverts the bit states, $|0>$ to $|1>$ and vice versa (spin-flip). It is readily checked that the corresponding second-quantized matrix is given by $\begin{pmatrix} 0 & 1 \\ 1 & 0 \end{pmatrix}$.

Indeed,

$$\begin{pmatrix} 0 & 1 \\ 1 & 0 \end{pmatrix} \begin{bmatrix} x \\ y \end{bmatrix} = \begin{bmatrix} y \\ x \end{bmatrix} \tag{33.30}$$

Likewise the generation operator is represented by $\hat{a}^{+} = \begin{pmatrix} 0 & 0 \\ 1 & 0 \end{pmatrix}$ and one can readily check that this is the matrix analog of $\hat{a}^{+}|0 >= |1 >$, namely

$$\begin{pmatrix} 0 & 0 \\ 1 & 0 \end{pmatrix} \begin{bmatrix} x \\ y \end{bmatrix} = \begin{bmatrix} 0 \\ x \end{bmatrix} \tag{33.31}$$

Thus, the state $(x = 1, y = 0) \equiv |0 >$ turns into $(x = 0, y = 1) \equiv |1 >$, whereas $|1 >$ turns into the unobservable nihil, $\emptyset \equiv (0, 0)$.

By the same token, the annihilation operator writes

$$\hat{a} = \begin{pmatrix} 0 & 1 \\ 0 & 0 \end{pmatrix} \tag{33.32}$$

This turns (x, y) into $(y, 0)$, i.e., $|1 >$ into $|0 >$ and $|0 >$ into nihil.

It is therefore clear that the second-quantized unitary operator

$$U(t, t + \Delta t) = e^{\frac{i\Delta t}{\hbar} \sum_n \epsilon_n \, \hat{a}_n^+ \hat{a}_n}$$

can be realized by suitable gates implementing products of annihilation-generation operators.

Remarkably, the QLB propagator is also unitary, hence it can be implemented in terms of shift and quantum-collision gates associated with the matrices $S^>$, $S^<$ and Q defined in eqn (33.27).

For a detailed discussion of lattice quantum computers, see (17).

33.6   Quantum Random Walks

The discussion thus far has exposed two routes to quantum physics, Schroedinger's wavefunction formulation and Heisenberg's operator representation.

Either ways, the intimate probabilistic character of quantum physics is inherent to the formalism, with no explicit inclusion of stochastic elements, see Fig. 33.7.

Yet another route to quantum mechanics is to invoke stochasticity at the outset. This is of course a long story, but here we shall focus just on a single realization of the stochastic route to quantum mechanics, namely *quantum random walks*, QRW for short.

QRWs are the quantum counterpart of classical random walks for particles which cannot be precisely localized in both space and momentum due to quantum uncertainity (21). Classical random walks are defined in terms of probabilities to take a jump of a given length left or right (in one spatial dimension), whereas QRWs are defined in terms of probability amplitudes, based on complex coefficients. The physics

$$\begin{bmatrix} u^> \\ d^< \end{bmatrix} = \hat{U}_{QLB} \begin{bmatrix} u \\ d \end{bmatrix}$$

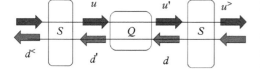

Figure 33.7 *The basic QLB "circuitry" for quantum computing implementation via the streaming and collision gates. Here $\hat{U}_{QLB} = \hat{S}\hat{Q}$. The upper and lower link-lines implement the right and left-streaming operators, $S^>$ and $S^<$, respectively.*

of QRWs is extremely rich, due to a host of new phenomena disclosed by quantum interference effects.

In a recent time, QRWs have known a major surge of interest, due to their conceptual and practical import as discrete realizations of stochastic quantum processes, whose continuum limit recovers a broad variety of relativistic quantum wave equations in flat and curved spaces (22). Moreover, QRWs are amenable to a number of experimental realizations, say ion traps, photonic lattices and other types of optical devices. As a result, they hold promise of playing an important role in many areas of modern physics and quantum technology, such as quantum computing, fundational quantum mechanics and biological physics.

In the following, we show that QRW can be placed in one-to-one correspondence with the QLB scheme.

33.6.1   QLB is a Quantum Random Walk

A $(1 + 1)$-dimensional quantum walk for a pair of complex wavefunctions (bi-spinor) obeys the following discrete spacetime evolution equations (discrete map):

$$\begin{cases} \psi_1(z + \Delta z; t + \Delta t) = B_{11}\psi_1(z; t) + B_{12}\psi_2(z; t) \\ \psi_2(z - \Delta z; t + \Delta t) = B_{21}\psi_1(z; t) + B_{22}\psi_2(z; t) \end{cases} \tag{33.33}$$

where the matrix elements define a $SU(2)$ operator parametrized by the three Euler angles θ, α, β:

$$B_{11} = e^{i\alpha}\cos\theta, \quad B_{12} = e^{i\beta}\sin\theta \tag{33.34}$$

$$B_{21} = -e^{-i\beta}\sin\theta, \quad B_{22} = e^{-i\alpha}\cos\theta \tag{33.35}$$

In the previous calculations, the amplitudes $\psi_\pm$ code for the probability of the quantum walker of moving up(down) along the lattice $z_j = j\Delta z$.

Equations (33.33) present a very rich structure, which has been shown to recover a variety of important quantum wave equations, as soon as the gauge angles are allowed to acquire a spacetime dependence (22).

It is readily recognized that the QLB scheme for a free massive particle falls within the scheme (33.33), under the condition

$$\alpha = \beta = 0, \quad \cos\theta = p(\mu)$$

where $p(\mu) = \frac{1-\mu^2/4}{1+\mu^2/4}$, $\mu = \omega_c\Delta t = m/m_p$ being the dimensionless mass in lattice units discussed in Chapter 32.

It should be noticed that, to second order in μ, one can write $p(\mu) \sim \cos(\mu)$, indicating that μ can be treated indeed as an angle, such that $\theta = \mu$.[93]

---

[93] The use of angles to define the relative strength of the interaction has a time-honored tradition in theoretical physics, especially high-energy physics. Perhaps the most famous example in point is the Cabibbo's angle,

This equivalence holds only in the limit $\mu \ll 1$ but it can be shown that, upon replacing Crank–Nicolson time marching with exact-exponentiation, the identification is exact for any μ (23).

So much for free massive particles.

For the case of purely electrostatic interactions, the coefficients of the quantum collision matrix read as follows:

$$
\begin{cases}
p(\mu, \gamma) = \dfrac{1 - \Omega^2/4}{1 + \Omega^2/4 - i\gamma} \\[2mm]
q(\mu, \gamma) = \dfrac{\mu}{1 + \Omega^2/4 - i\gamma}
\end{cases}
\tag{33.36}
$$

where $\gamma \equiv V\Delta t/\hbar$ is the electrostatic coupling strength and we have set $\Omega^2 = \mu^2 - \gamma^2$. Here, the identification goes as follows:

$$
\begin{cases}
tan(\theta) = \dfrac{\mu}{1 - \Omega^2/4 + \gamma^2} \\[2mm]
tan(\alpha) = tan(\beta) = \dfrac{\gamma}{1 + \Omega^2/4}
\end{cases}
\tag{33.37}
$$

Similar considerations extend to the electromagnetic case and also curved space, see (22; 24).

The bottomline is that QLB, as originally proposed back in 1993 (25), was a quantum walk … without knowing it! In fact, a pretty early one, given that the paper (25) was posted in November 1992, just one month after Ahronov *et al*, (21), the paper officially marking the birth of QRW's.

Be as it may, this shows a second and independent connection between QLB and quantum computing.

33.7 Quantum Simulators for Classical Kinetic Theory

Recent progress in ion-trap and superconducting circuit experiments have shown that these platforms may offer a very appealing environment for quantum information and simulation tasks. By and large, quantum simulators have been used almost exclusively

after the italian physicist Nicola Cabibbo (1935–2010), whose angle θ_C measures the relative probability that *down* and *strange* quarks decay into *up* quarks. The Cabibbo-rotated quarks take the form $d_C = V_{ud}d + V_{us}s$ and $s_C = V_{cd}d + V_{cs}s$, where the Cabibbo's angle is defined as $tan(\theta_C) = \left|\frac{V_{us}}{V_{ud}}\right|$. Its numerical value is $\theta_C \sim 13.02^o$. The 2×2 Cabibbo unitary matrix takes exactly the form (33.33). Cabibbo's angle, dated 1963 was later generalized by Makoto Kobayashi and Toshihide Maskawa into what became to be known as CKM matrix, accounting for CP (charge-parity) violations observed in weak interactions. The 3×3 CKM matrix has proved highly successful to interpret high-energy experiments on weak and strong interactions, to the point of becoming a cornerstone of the standard model of elementary interactions. This major accomplishment garnered the 2008 Nobel Prize in physics, but guess what, not to CKM, but NKM instead, where N stands for Yoichiro Nambu. Why C had to be taken off the widely acknowledged CKM and replaced by N (who largely deserved Nobel on his own) remains a mistery, bordering the outcry.

to tackle quantum problems, in line with Feynman's celebrated paradigm that it takes quantum computers to simulate quantum physics (26).

This include the quantum simulation of highly correlated fermionic systems, fermionic-bosonic models and lattice gauge theories. On the other hand, circuit quantum electrodynamics (QED) setups can nowadays host top-end quantum information protocols, such as quantum teleportation and topological phase transitions (27). These quantum devices are approaching the complexity required to simulate *both classical and quantum* nontrivial problems.

Here, we briefly hint at the possibility of quantum simulation of classical lattice Boltzmann dynamics, as recently proposed by Mezzacapo *et al.* (28).

To this purpose, let us begin by casting classical kinetic theory in second-quantized form.

33.7.1   Second-Quantized Kinetic Theory

In Chapter 32 we have briefly discussed the second-quantization formulation of relativistic quantum mechanics. Although especially well suited to relativistic problems, such formalism may prove useful also in the context of non-relativistic quantum mechanics and eventually *classical*-kinetic theory as well.

To elucidate the point, let us begin by considering the explicit expression of the generation and annihilation operators for the paradigmatic case of the quantum harmonic oscillator, governed by the following Hamiltonian

$$\hat{H} = -\frac{\hbar^2}{2m}\partial_x^2 + \frac{m}{2}\omega^2 x^2$$

in one space dimension without loss of generality.

Next apply the well-known correspondence rules

$$\hat{x} = x, \quad \hat{p} = -i\hbar\partial_x \tag{33.38}$$

so that the Hamiltonian takes its classical form, only with position-momentum replaced by the corresponding operators

$$\hat{H} = \frac{\hat{p}^2}{2m} + \frac{1}{2}m\omega^2\hat{x}^2$$

Following textbook material, the generation-annihilation operators associated with this Hamiltoniand read as follows:

$$\begin{cases} \hat{a} = \hbar\omega\left(\frac{x}{L} + L\partial_x\right) \\ \hat{a}^+ = \hbar\omega\left(\frac{x}{L} - L\partial_x\right) \end{cases} \tag{33.39}$$

where $L = \sqrt{\frac{2\hbar}{m\omega}}$ is the typical lengthscale of the quantum-oscillator wavefunction.

Inverting the relations (33.39) delivers

$$
\begin{cases}
\hat{x} = \dfrac{L}{2\hbar\omega}(\hat{a} + \hat{a}^{+}) \\[2mm]
\partial_x = \dfrac{1}{2\hbar\omega}(\hat{a} - \hat{a}^{+})
\end{cases}
\tag{33.40}
$$

Let us next consider a *linear* BGK kinetic equation of the form

$$
\partial_t f + v\partial_x f = -\Omega(f - f^{eq})
\tag{33.41}
$$

where it is understood that the local equilibrium is a linear map of the distribution, say

$$
f^{eq} = \hat{M}f
$$

where $\hat{M}$ is a local constant operator in space and time.

Let us further define the associated Liouvillean operator

$$
\hat{L} \equiv -v\partial_x - \Omega(I - \hat{M})
$$

where I is the identity. By a formal parallel between the Schroedinger equation, the Liouville operator defines an associated kinetic Hamiltonian:

$$
\hat{H}_K = i\hbar\hat{L}
$$

Taking all together, the kinetic Hamiltonian takes the following second-quantized form:

$$
\hat{H}_K = i\hbar\left[\frac{v\hat{L}}{\hbar\omega}(\hat{a} - \hat{a}^{+}) + \Omega(I - \hat{M})\right]
\tag{33.42}
$$

The mandate is now is to quantum simulate the quantum-kinetic Hamiltonian just the way one does with "genuine" quantum Hamiltonians. This implies the identification of a specific mapping between the Hamiltonian (33.42) and a concrete implementation protocol on actual (real-life!) quantum devices (see Fig. 33.8).

This is a non trivial task even for those in the know, much more so for this author. But the good news is that, at least for linear transport problems, the task can indeed be accomplished.

33.7.2 Non-Unitary Hamiltonians

Leaving the details to the original publication (28), here we conclude with a few comments on a general issue attached to the previous program: true quantum Hamiltonians describe reversible motion, the kinetic Hamiltonian generates irreversible dynamics. More specifically, irreversibility of the kinetic Hamiltonian is built-in within the non-unitary collision matrix Ω: how to deal with this basic difference as compared to quantum systems?

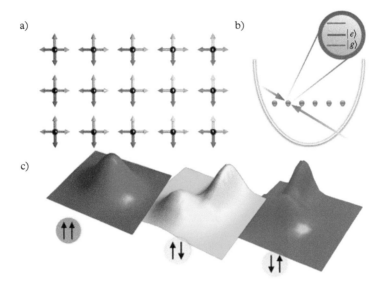

Figure 33.8 *The conceptual set-up of the quantum simulation of a classical LB scheme. The fluid density on a two-dimensional lattice can be simulated via normal motional modes and internal levels of a set of trapped ions (b), (c). Superposition of two motional modes entangled with pseudo-spin states can encode velocity distributions along different lattice directions. From A. Mezzacapo et al, Sci. Rep. 5, 13153 (2015).*

The trick appeals to a simple property: *the sum of two unitaries is no longer unitary in general.*

Based on this observation, one can express a non-unitary matrix as a weighted sum of two unitary ones.

For the case in point, the collision propagator

$$\hat{C} = e^{-\Omega(I-\hat{M})\Delta t}$$

is represented as a weighted sum of two unitary operators:

$$\hat{C} = \hat{U}_1 + \alpha \hat{U}_2$$

By construction, its norm obeys the inequality $1 - \alpha \leq |C| \leq 1 + \alpha$, as it befits an irreversible process, the lower and upper side of the inequality corresponding to stable and unstable dynamics, respectively. The two unitaries $\hat{U}_1$ and $\hat{U}_2$ must then be chosen in such a way as to keep the system on the stable side of the inequality. This is tantamount to securing negative interference between the two unitaries, which turns out to be possible through a specific choice of their eigenvalues. Full details in (28).

33.7.2.1 Heterogeneous collision operators

A major challenge is also posed by heterogeneous operator in which either the collision frequency Ω or the equilibrium map $\hat{M}$ acquire a dependence on space and/or time, either through explicit heterogeneity or by non-linearity, the latter being the case of the Navier Stokes equations. Under such conditions, the kinetic Hamiltonian acquires an explicit and most often non-linear dependence on the operator $\hat{x}$, which also needs to be encoded in the proper quantum circuitry. To the best of this author's knowledge, at the time of this writing, this problem stands unsolved, thereby setting a severe constraint to the applicability of the second-quantized kinetic approach to realistic classical transport and fluid problems.

Before concluding, we wish to emphasize that, at the moment, the quantum simulation of classical LB is not meant to compete with classical implementations, but simply to establish a proof of principle for the future. Such proof of principle fills the missing QC box in the four-by-four matrix:

- CC (Classical computers for Classical physics)
- CQ (Classical computers for Quantum physics)
- QC (Quantum computers for Classical physics)
- QQ (Quantum computers for Quantum physics)

To date, CC and CQ are the consolidated mainstreams, QQ is a burgeoning niche and QC is possibly the least explored of the four. It is reassuring to observe that the LB technique, classical and quantum is able to cover all four paradigms.

33.8 Summary

Summarizing, we have shown that the quantum LB formalism can be extended to quantum many-body problem as well as to (some simple) quantum field theories.

In addition, it is also well suited to the paradigm of quantum computing and matches the requirement of quantum walks. These are very exciting forefronts of modern science, which might prove capable of meeting with otherwise unmanageable demands of computational resources. Whether or not a lattice kinetic approach to the previous problems is going to make a major inroad into these exciting fields remains entirely to be seen. The potential seems to be there, but only systematic and thorough work in the coming years will tell.

···

REFERENCES

1. W. Kohn, "Nobel lecture: electronic structure of matter–wave functions and density functionals", *Rev. Mod. Phys.*, **71**, 1253, 1999.

2. J.M. Thijss, *"Computational Physics"*, Cambridge University Press, Cambridge, 1999.

3. D. Ceperley and B. Alder, "Ground State of the Electron Gas by a Stochastic Method", *Phys. Rev. Lett.*, **45**, 566, 1980.

4. D.M. Ceperley and B.J., *Computational Physics*, Cambridge University Press, Cambridge, 1999.

5. E. Lieb and W. Liniger, "Exact analysis of an interacting boson gas 1. General solution and ground state", *Physical Review*, **130**, 1605, 1963.

6. R. Car and M. Parrinello, "Unified approach for molecular dynamics and density functional theory", *Phys. Rev. Lett.* **55**, 22, 2471, 1985.

7. W. Kohn and L. Sham, "Self-consistent equations including exchange and correlation effects", *Phys. Rev.*, **140**, 1133, 1965.

8. P. Hohenberg and W. Kohn, "Inhomogeneous electron gas", *Phys. Rev.*, **136**, B864, 1964.

9. M. Mendoza, S. Succi and H.J. Herrmann, "Lattice kinetic approach to electron density functional theory", *Phys. Rev. Lett.* **113**, 096402, 2014.

10. C. Bender and D. Sharp, "Solution of Operator Field Equations by the Method of Finite Elements", *Phys. Rev. Lett.*, **50**, 1535, 1983.

11. S. Succi, "Lattice Boltzmann for quantum field theory", *J. Phys. A: Math. and The.*, **40**, F559, 2007.

12. S. Sachdev, "Quantum Phase Transitions", Cambridge University Press, Cambridge, 2nd edition 2011.

13. C. Mendl, "Matrix-valued quantum lattice Boltzmann method", *Int. J. of Mod. Phys. C*, **26**, 1550113, 2015.

14. B. Boghosian and W. Taylor, "Quantum lattice gas models for the many body Schrödinger equation", *Int. J. Mod. Phys. C*, **8**(4), 705, 1997.

15. D. Meyer, "Quantum lattice gases and their invariants", *Int. J. Mod. Phys. C*, **8**, 4, 717, 1997.

16. N. Lambert, Y-N Chen, Y-C Cheng *et al.*, "Quantum biology", *Nat. Phys.*, **9**, 10, 2013.

17. J. Yepez, and B. Boghosian, "An efficient and accurate quantum lattice-gas model for the many-body Schrodinger wave equation", *Comp. Phys. Comm.*, **146**, 280 2002.

18. D. P. di Vincenzo, "Quantum computing", *Science*, **270**, 255, 1995.

19. M. Nielsen and I. Chuang, "Quantum computation and quantum information", Cambridge University Press, Cambridge, 2011.

20. W. Zurek, "Decoherence and the transition from quantum to classical, revisited", *Los Alamos Science*, **27**, 1–25, 2002.

21. Y. Aharonov, L. Davidovich and N. Zagury, "Quantum random walks", *Phys. Rev. A* **48**, 1687, 1993.

22. S. Succi, F. Fillion and S. Palpacelli, "Quantum lattice Boltzmann is a quantum random walk", *EPJ Quantum Technology* **2**, 12, 2015.

23. F. Fillion, M. Mendoza, H. Herrmann, S. Palpacelli and S. Succi, "Formal analogy between the Dirac equation in its Majorana form and the discrete-velocity version of the Boltzmann kinetic equation", *Phys. Rev. Lett.*, **111**, 160602, 2013.

24. P. Arrighi, S. Arrighi and M. Forets, "Quantum walks in curved spacetime", *ArXiv: 1505.07023v1 [quant-ph]* (May 2015); in G. Di Molfetta, M. Brachet, and F. Debbasch, *Physica A: Statistical Mechanics and its Applications* **397**, 157 2014.
25. S. Succi and R. Benzi, "Lattice Boltzmann equation for quantum mechanics", *Physica D*, **69**, 327, 1993.
26. R. P. Feynman, "Simulating Physics with Computers", *Int. J. of Theor. Phys.*, **48**, 467, 1982.
27. R. Blatt, I. Bloch, I. Cirac and P. Zoeller, "Quantum simulation: an exciting adventure, Special Issue on Quantum Simulation", *Annalen der Physik* **525**, A153, 2013.
28. A. Mezzacapo, M. Sanz, L. Lamata, I. Egusquiza, S. Succi and E. Solano, "Quantum simulator for transport phenomena in fluid flows", *Scientific Reports* **5**, 1353, 2015.

....

EXERCISES

1. Discuss the connection between anti-commuting Grassmann variables and Pauli exclusion principle.
2. Show that the NOT operator writes simply as $\hat{a} + \hat{a}^{+}$.
3. Find the expression of the annihilation-generation operators for a generic potential $V(x) = v_2 x^2 - v_4 x^4$.

34

Relativistic Lattice Boltzmann (RLB)

Relativistic hydrodynamics and kinetic theory play an increasing role in many areas of modern physics. Besides their traditional arenas, astrophysics and cosmology, relativistic fluids have recently attracted much attention also within the realm of high-energy and condensed matter physics, mostly in connection with quark-gluon plasmas experiments in heavy-ion colliders and electronic transport in graphene. In this chapter we describe the extension of the lattice Boltzmann formalism to the case of relativistic fluids.

Logic can take you from A to Z: imagination can take you anywhere.

(A. Einstein)

34.1 Relativistic Mechanics

Einstein's theory of relativity stands out as one of the pinnacles of intellectual achievement in the whole history of mankind. By stating its dependence on the state of motion of the observer, it destitutes time from the Newtonian throne of absoluteness.

More precisely, since the speed of light is finite and the same in any frame of reference, the time interval between two events, as measured by a moving observer is *not* the same as the one measured by an observer at rest.

Mathematically, this is expressed by the Lorentz's spacetime transformations (1):

$$
\begin{cases}
x' = \dfrac{x - \beta ct}{\sqrt{1 - \beta^2}} \\[2mm]
y' = y, \quad z' = z \\[2mm]
ct' = \dfrac{ct - \beta x}{\sqrt{1 - \beta^2}}
\end{cases}
\tag{34.1}
$$

where primed quantities refer to space and time as measured by an observer in motion along direction x with constant velocity V, with respect to an observer at rest (the lab).

The Lattice Boltzmann Equation. Sauro Succi, Oxford University Press (2018).
© Sauro Succi. DOI. 10.1093/oso/9780199592357.001.0001

In (34.11),

$$\beta = \frac{V}{c} \tag{34.2}$$

is the ratio of the material speed to the speed of light, c.

It is readily checked that the Lorentz transformations reduce to standard Galilean translations, $\vec{x}' = \vec{x} - \vec{V}t$ and Newton's absolute time $t' = t$, in the non-relativistic limit $\beta \to 0$. Also to be noted is the symmetry between space x and ct, i.e., time appears to all effects and purposes like a fourth spatial dimension.

Third, the speed of light is the same in both frames, reflecting the experimental finding which provided the stepping stone for Einstein's theory of special relativity.

Based on the Lorentz transformations, the velocities can be shown to add up according to the following *nonlinear* rule:

$$\vec{v} = \frac{\vec{V} + \vec{v}'}{\sqrt{1 + \frac{\vec{V} \cdot \vec{v}'}{c^2}}} \tag{34.3}$$

Thus, velocities are not simply additive and the classical addition rule $\vec{v} = \vec{v}' + \vec{V}$ is recovered only in the non-relativistic limit $V/c \to 0$. This also informs that while momentum is a relativistic invariant, velocity is not.

Also to be noted that, according to (34.3), the material speed cannot exceed c, i.e., the light speed fixes an upper bound to the velocity of any material particle, an upper bound which is reached only in asymptotic limit $V \to c$.

Lorentz kinematics permits us to derive the laws of relativistic mechanics, which are based on the energy-momentum relation:

$$E = \mp\sqrt{p^2 c^2 + m^2 c^4} \tag{34.4}$$

where p is the magnitude of the momentum and m is the *rest mass* of the particle, see Fig. 34.1.

A few comments are in order.

First, a particle at rest ($p = 0$) still possesses an energy $E = mc^2$, the celebrated Einstein's equation, which reflects the energy carried by the particle when it is not moving (the internal energy carried by the nuclei).

As the particle moves, kinetic energy adds its contribution. For slow, non-relativistic, motion, $p \ll mc$, a second-order expansion delivers the familiar non-relativistic expression:

$$E - mc^2 \sim \frac{p^2}{2m} + O(p^2/m^2 c^2) \tag{34.5}$$

However, higher terms at all orders appear in the reminder, which can no longer be neglected as the particle speed approaches the speed of light.

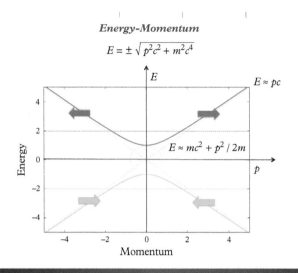

Figure 34.1 *Relativistic energy-momentum relation. Both positive (particles) and negative (antiparticles) branches are represented. The light-cones of massless particles are the straight light-dotted lines bisecting the four quadrants of the energy-momentum plane. Note that the arrows denote the sign of velocity, not of momentum. Based on the relativistic expression of the velocity $v = pc^2/E$, particles in the bottom-left branch move rightward and those in in the bottom-right move leftward, as indicated by the green arrows.*

In the opposite *ultra-relativistic* limit, $p \gg mc$, the expression (34.4) delivers

$$E \sim pc \tag{34.6}$$

This no longer shows any trace of the particle mass, and it is exactly obeyed by massless particles, such as the photon.

Finally, we observe that in relativistic mechanics both signs of the square root make sense, the negative branch corresponding to *anti-particles*, as first predicted by P.A.M Dirac (1902–1984) in the context of relativistic-quantum mechanics.

It should be appreciated that the rest energy is usually pretty large; even for the tiny electron it amounts to about 0.5 MeV (half a million electronVolts), which corresponds to a temperature of about 5 billion Kelvin degrees.

This is why relativistic mechanics remains hidden to matter in standard conditions.

The equations of motion of relativistic mechanics are readily derived from the energy relation and read as follows:

$$\begin{cases} \dfrac{d\vec{x}}{dt} \equiv \dfrac{\partial E}{\partial \vec{p}} = \vec{p}/m \\[3mm] \dfrac{d\vec{p}}{dt} \equiv -\dfrac{\partial E}{\partial \vec{x}} = \vec{F}(\vec{x}) \end{cases} \tag{34.7}$$

Note that the momentum can still be written as $\vec{p} = m(v)\vec{v}$, provided the "moving mass" acquires a dependence on the velocity:

$$m(v) = m/\sqrt{1-\beta^2} \tag{34.8}$$

This shows that the "moving mass" becomes infinite in the limit $v \to c$, providing a dynamic rationale for the speed of light as an upper bound to the material particle speed. It would take an infinite force to attain the asymptotic speed $v = c$. We hasten to add that the notion of "moving mass" is only a metaphor the only physical mass being the rest one.

With these basic notions of relativistic mechanics, we next move on to a brief description of relativistic kinetic theory.

34.2 Relativistic Kinetic Theory

Relativistic kinetic theory plays a crucial role in many areas of modern physics, such as astrophysics and cosmology, high-energy physics, e.g., jets emerging from the core of galactic nuclei or gamma-ray bursts, quark-gluon plasmas produced in heavy-ion collisions. With the recent discovery of graphene, it also bears on many problems related to the electronic transport in graphene samples and other exotic forms of quantum materials.

Despite these many applications, the extension of the LB formalism to relativistic flows had to await some twenty years, the first relativistic LBE model by Mendoza and collaborators having made its appearance only in 2010 (2).

Before discussing the relativistic LB, a brief reminder of the basic notions of relativistic kinetic theory are in order (for more details see (3; 4)).

Relativistic kinetic theory lives in an eight-dimensional *phase-spacetime* labeled by the four-dimensional coordinates

$$\begin{cases} x^\mu = (\vec{x}; ct) \\ p^\nu = (\vec{p}c; E) \end{cases} \tag{34.9}$$

where the four-dimensional indices $\mu, \nu = 0$ label time (energy) and $\mu, \nu = 1, 3$ run over the usual six-dimensional phase-space $(\vec{x}, \vec{p})$.

Note that momentum is used instead of velocity, because the latter is not invariant under the Lorentz transformations.

These are so-called *controvariant* coordinates.

To form scalars, we also need their *covariant* counterparts defined as (for a Minkowski metric):

$$\begin{cases} x_\mu = (ct, -\vec{x}) \\ p_\nu = (E, -\vec{p}c) \end{cases} \tag{34.10}$$

The transformation from one to the other takes place as follows: $V_\mu = g_{\mu\nu} V^\nu$ and vice versa, $V^\mu = g^{\mu\nu} V_\nu$, where the metric tensor, $g_{\mu\nu}$, is defined as follows:

$$ds^2 = g_{\mu\nu} dx^\mu dx^\nu$$

In this calculation, ds is the arclength of the four-dimensional *worldline* $x^\mu = x^\mu(s)$.

For the case of Minkowski, spacetime $g_{\mu\nu}$ is a diagonal matrix with entries $\{1, -1, -1, -1\}$. In explicit terms:

$$ds^2 = c^2 dt^2 - dx^2 - dy^2 - dz^2 \tag{34.11}$$

The condition $ds^2 \geq 0$ denotes the so-called "time-like" intervals, while the opposite case $ds^2 < 0$ is referred to as "space-like." The border case $ds^2 = 0$ defines the so-called "light-cones," which play a central role in LB theory.

It is often useful to express the energy, momentum and velocity in dimensionless units as follows:

$$E = mc^2\gamma, \ \vec{p} = mc\gamma\vec{\beta}, \ \vec{v} = c\vec{\beta} \tag{34.12}$$

where

$$\gamma = \frac{1}{\sqrt{1-\beta^2}} \tag{34.13}$$

is the relativistic Lorentz factor.

The dimensionless triplet $\{\beta, \gamma, \gamma\beta\}$ stands for velocity, energy and momentum in natural units of c, mc^2, mc.

They are related as follows:

$$\beta = \frac{\mu}{\sqrt{1+\mu^2}}, \ \gamma = \sqrt{1+\mu^2}, \ \mu = \beta\gamma \tag{34.14}$$

Non-relativistic mechanics is recovered in the limit $\beta \to 0$, or, equivalently, $\gamma \to 1$. It is worth noting that the most telling metric for relativistic effects is γ rather than β. In fact, inspection of the relation (34.13) shows that sizeable values of $\gamma > 1$ are attained only with β pretty close to 1.

Relativistic particles are sometimes classified as follows:

Non-Relativistic (NR): $\gamma \to 1$
Weakly Relativistic (WR): $1 \leq \gamma \leq 2$
Strongly Relativistic (SR): $\gamma > 2$
Ultra-Relativistic (UR): $\gamma \gg 2$

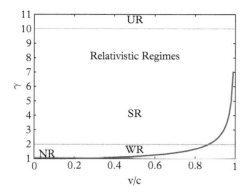

Figure 34.2 *The relation $\gamma = \gamma(\beta)$ and the corresponding NR, WR, SR and UR regimes. Note that it takes pretty substantial values of β before the UR regime is entered, typically $\beta > 0.9$.*

The WR condition denotes the prevalence of rest energy mc^2 over the kinetic one $E_K = (\gamma - 1)mc^2$, while the opposite is true in the SR regime. Note that the borderline, $\gamma = 2$, corresponds to $\beta = \sqrt{3}/2 \sim 0.85$, a very sizeable fraction of the speed of light! The UR regime connotates the overwhelming prevalence of kinetic over rest energy, as measured by the ratio (inverse temperature in rest energy units):

$$z \equiv \frac{mc^2}{k_B T} \ll 1 \tag{34.15}$$

By definition, in the UR regime $z \to 0$, the rest mass can be totally ignored. In this limit,

$$\mu \gg 1, \; \beta \to 1, \; \gamma \to \mu \tag{34.16}$$

as opposed to the non-relativistic limit

$$\mu \ll 1, \; \beta \to 0, \; \gamma \to 1 \tag{34.17}$$

We hasten to add that the notion of temperature here is purely kinetic and cannot be identified with the equilibrium temperature of standard thermodynamics because, unlike the non-relativistic case, relativistic energy does not split into the sum of a macroscopic and thermal components. This is again a result of the non-separability of the energy-momentum relation for massive particles. The UR regime is particularly relevant to astrophysics, where Lorentz factors as high as $\gamma \geq 10^2 \div 10^4$ can be encountered, a very hot subject in point being the case of gamma rays bursts from some types of supernova explosions, see Fig. 34.2.

34.3 Relativistic Boltzmann Equation

The relativistic Boltzmann equation (RBE) reads as follows:

$$\frac{\partial}{\partial x^\mu}(p^\mu f) + \frac{\partial}{\partial p^\nu}(F^\nu f) = Q(f,f) \equiv \frac{mc^2}{\lambda} C(f,f) \tag{34.18}$$

where $f \equiv f(x^\mu, p^\nu)$ is the relativistic distribution function in full eight-dimensional phase-spacetime and $F^\nu = \frac{dp^\nu}{dt}$ is the four-dimensional force[94]. In (34.18), we have exposed the dimensional group mc^2/λ, (energy/length), where λ is the mean-free path, for reasons of dimensional transparency.

By explicitly separating space and time as well as energy and momentum, the eq. (34.18) takes the form:

$$\frac{1}{c}\frac{\partial(Ef)}{\partial t} + \frac{\partial(cp^a f)}{\partial x^a} + \frac{\partial(\vec{F}\cdot\vec{v}f)}{\partial E} + \frac{\partial(F^a f)}{\partial p^a} = Q(f,f)$$

Latin indices running over the three spatial dimensions.

Let us consider the force-free case for simplicity, $F^\nu = 0$, so that the third and fourth terms at the l.h.s. drop apart.

Dividing by E/c and using the relation $v^a = \frac{p^a c^2}{E}$, the left-hand side reduces to the standard non-relativistic streaming operator $\partial_t f + \partial_a(v^a f)$, with

$$v^a = \frac{p^a c^2}{E} = \frac{p^a/m}{\sqrt{1 + (p/mc)^2}} \tag{34.19}$$

The relativistic BE takes then the following non-relativistic look:

$$\partial_t f + \partial_a(v^a f) = \frac{C(f,f)}{\tau} \tag{34.20}$$

where

$$\tau = \frac{\lambda}{c}\frac{E}{mc^2}$$

is the collisional mean-free time.

Note that the distribution function f still depends on the particle energy-momentum and represents a fully legitimate relativistic equation in non-relativistic vests. Moreover, v^a is a nonlinear function of the particle momentum, according to the relation (34.19).

In order to move one step closer toward a non-relativistic lookalike, it is common to integrate over the energy shell, i.e., define a non-relativistic-looking distribution as

$$f_3(\vec{p}, \vec{x}; t) = \int f(E, \vec{p}, \vec{x}; t) dE \tag{34.21}$$

Since for stable particles and quasi-particles energy and momentum are one-to-one related, the energy dependence of f factors out a Dirac's delta distribution $\delta[E - E(p)]$. This trivializes the integration to the following on-shell expression:

$$f_3(\vec{p}, \vec{x}; t) = f[E(p), \vec{p}, \vec{x}; t] \tag{34.22}$$

---

[94] In explicit terms, $F^0 = \frac{dE}{dt} = \vec{F}\cdot\vec{v}$ and $F^a = \frac{dp^a}{dt}$.

In the sequel we shall refer to f_3 as simply to $f(\vec{x}, \vec{p}; t)$.

So much for the force-free relativistic streaming. Let us next consider the relativistic collision operator.

Like for the non-relativistic case, this is generally a very complicated non-linear integral operator in momentum space. Happily enough, like for the non-relativistic case, microscopic details are not needed for hydrodynamic purposes and consequently this term is typically replaced by simplified models pretty much in the same spirit as the non-relativistic BGK.

However, relativistic BGK comes in *two flavors*, the so-called weakly relativistic Marle model

$$C_M = \frac{1}{\tau_M}(f^{eq} - f) \tag{34.23}$$

where $\tau_M = \lambda/c$ is a relaxation time-constant, and the Anderson–Witting model:

$$C_{AW} = \frac{1}{\tau_{AW}}(f^{eq} - f) \tag{34.24}$$

where the Anderson–Witting relaxation time is given by $\tau_{AW} = \frac{mc^2\lambda}{U^\mu p_\mu}$, U^μ being the four-dimensional fluid velocity, so that $\tau_{AW} \sim \lambda/U$.

The Marle model looks exactly the same as BGK, and indeed it can be used only for weakly relativistic fluids. Whenever relativistic effects become significant, the Anderson–Witting model must be employed.

It is important to recognize that the local equilibrium f^{eq} in both collision operators must necessarily reflect the symmetries of relativistic mechanics, hence it exhibits significant departures from the non-relativistic Maxwell–Boltzmann local equilibria.

These equilibria are known as Maxwell–Juettner, or simply Juettner, distribution, after Ferencz Juettner (1878–1958), who first studied and proposed them in 1911, within the context of relativistic kinetic theory. Since the Juettner distribution plays a key role in the formulation of the relativistic LBE, in the sequel we analyze it in some more detail.

34.4 Relativistic Equilibria: the Juettner Distribution

Relativistic equilibria have made the object of much debate since the early days of relativistic kinetic theory, until recent times. In the end, a rather broad consensus seems to have emerged on the fact that they are described by the following Juettner distribution:

$$f^{eq} = Ae^{-\frac{p^\mu u_\mu}{k_B T}} \tag{34.25}$$

where A is a normalization constant and u_μ the (dimensionless) Lagrangian multiplier associated with the macroscopic flow of momentum.

This is best recast in dimensionless form as follows:

$$f^{eq} = f_{\bar{y}}(p) \equiv A e^{-z\gamma\,\gamma_u \beta^\mu u_\mu} \tag{34.26}$$

where we have set

$$\beta^\mu = (1, \vec{v}/c) \tag{34.27}$$

$$u_\mu = (1, -\vec{u}/c) \tag{34.28}$$

and

$$\gamma_u = \frac{1}{\sqrt{1 - u^2/c^2}} \tag{34.29}$$

Note that for ordinary matter, $z = mc^2/k_B T$ is a huge number, in the order of 10^7.[95] Also to be noted that $k_B T$ is here taken as a measure of the average kinetic energy, which exceeds the rest energy in the ultra-relativistic regime discussed before. As a result, the notion of temperature and thermal speed, $v_T = \sqrt{k_B T/m}$ is purely notational.

The five Lagrangian parameters A and u_μ are fixed by the definition of the macroscopic variables, i.e., the number flux and the energy-momentum flux.

In covariant notation,

$$N_\mu = c \int f p_\mu \frac{d\vec{p}}{p_0} \tag{34.30}$$

$$T_{\mu\nu} = c \int f p_\mu p_\nu \frac{d\vec{p}}{p_0} \tag{34.31}$$

Note that the integrals run over $d\vec{p}/p_0$, which is a relativistic invariant, as opposed to $d\vec{p}$ (see (3; 4)).

By performing the algebra, one obtains the following explicit expressions for the macroscopic moments.

Number density:

$$N_0 \equiv nc = c \int f d\vec{p} \tag{34.32}$$

Number density current:

$$N_a \equiv n\vec{u} = c \int f \vec{\beta} d\vec{p} \tag{34.33}$$

Energy density:

$$T_{00} \equiv ne = c \int f\gamma\, d\vec{p} \tag{34.34}$$

---

[95] At room temperature $k_B T \sim 1/40$ eV, hence for electrons at room temperature we have $z \sim 0.5 \; 10^6/(1/40) = 2 \; 10^7$.

Energy flux:

$$T_{0a} = T_{a0} = c \int f \gamma \beta_a d\vec{p} \tag{34.35}$$

Energy-momentum flux:

$$T_{ab} = T_{ba} = c \int f \gamma \beta_a \beta_b d\vec{p} \tag{34.36}$$

The Juettner distribution exhibits a number of properties worth some specific comment. Like its non-relativistic analog, it depends on two macroscopic parameters, the inverse temperature z the flow speed β_u, both in natural units, mc^2 and c, respectively.

In the double non-relativistic limit

$$\frac{k_B T}{mc^2} \equiv 1/z \to 0, \quad u/c \to 0 \tag{34.37}$$

the Juettner distribution recovers the local Maxwell–Boltzmann equilibria. However, away from such limit, it develops a fairly different shape in momentum and, even more so, in velocity space.

In the rest frame $u_a = 0$, it simplifies to a global equilibrium

$$f^{eq,0} = A e^{-z\gamma} \tag{34.38}$$

where the number density constraint fixes the prefactor:

$$A = \frac{n}{4\pi} \frac{z}{K_2(z)} \tag{34.39}$$

$K_2(z)$ being the modified Bessel function (see (3; 4)).

The following asymptotic expansion

$$K_2(z) = \sqrt{\frac{\pi}{2}} z^{-1/2} e^z \left(1 + \frac{1}{2z} + \dots \right) \tag{34.40}$$

is useful to analyze the "low-temperature" non-relativistic limit $z \to \infty$.

Indeed, in such limit, taking into account (34.40), the expression (34.38) returns

$$f^{eq,0} \propto z^{3/2} e^{-(\gamma-1)z} \tag{34.41}$$

which recovers the standard Maxwell–Boltzmann distribution upon setting $\gamma - 1 \sim \beta^2/2$.

In the opposite UR limit, $z \ll 1$, we leave to the reader as an exercise to show that the Juettner distribution in velocity space, i.e., $f_J(v) = f_J(p)|dp/dv| = \gamma^{-3} f_J(p)$, develops a double-humped structure, a critical temperature of the order of $T_c = mc^2/k_B$ and tends

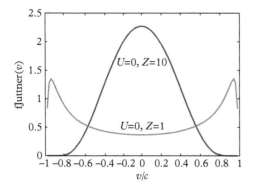

Figure 34.3 *Examples of velocity space Juettner equilibria at rest, $u/c = 0$, for $z = 1$ and $z = 10$. The bimodal structure at $z = 1$ is apparent.*

to a pair of Dirac deltas, centered upon $v = \mp c$ in the limit $z \to 0$ see Fig. 34.3. This is the only way the system can manage to accommodate the double constraint of an increasing energy and a finite velocity.[96]

So much for the rest frame.

In the laboratory frame, where $u_a \neq 0$, the Juettner distribution shifts toward its most probable velocity in the vicinity of $v = u$, much like a non-relativistic shifted Maxwellian. However, besides such a shift, it develops a non-zero skewness in the direction of the macroscopic motion. As the macroscopic flow is increased, such skewness becomes dominant over the non-relativistic shift.[97]

This is the consequence of the fact that the shifted Maxwellian reflects Galilean invariance, a symmetry which simply does not belong to the relativistic context.

These considerations illustrate the fact that, away from the non-relativistic region, $\beta \to 0$ and $z \to \infty$, Juettner equilibria differ drastically from their Maxwell–Boltzmann counterparts. More precisely, the Juettner distribution transits from a classical Gaussian $e^{-p^2/2}$ to an exponential form $e^{-|p|}$, where p is now measured in units of the thermal momentum $p_T = k_B T/c$.

Not unexpectedly, the probability of finding particles at many standard thermal deviations is much higher in relativistic than in Newtonian mechanics.

As we shall see, this has significant implications for the formulation of the relativistic LBE.

---

[96] The following socio-economic analogy cannot be resisted. By injecting unlimited energy into a collection of individuals than cannot run faster than a given upper bound velocity, implies that the middle-class runners ($v = 0$ in a system globally at rest) are totally suppressed and only the extreme runners ($v = \pm c$) can survive, since they are the only ones who can carry a virtually infinite amount of energy. A pretty effective recipe for extinction indeed...

[97] The skewness denotes the third-order moment $\mathcal{S} = <v^3> - <v>^3$. A positive skewness signals depletion of counterflow movers ($v < 0$) movers with a corresponding enhancement of the co-flow movers ($v > 0$). This occurs through a deformation of the distribution function and not just a rigid shift from $<v> = 0$ to $<v> = u$, where we have assumed $u > 0$.

34.5   Relativistic Hydrodynamics

As for the non-relativistic case, the equations of relativistic hydrodynamics can be either postulated "top-down" on macroscopic grounds based on symmetry arguments, or derived bottom-up from the microscopic Boltzmann equation.

With reference with the simplest instance of the Marle BGK equation, integration upon momentum delivers the two conservation equations for the number density and current,

$$\partial_\mu N^\mu = 0 \tag{34.42}$$

and for the energy-momentum density and current density

$$\partial_\mu T^{\mu\nu} = 0 \tag{34.43}$$

These equations are purely formal until a specific form $T^{\mu\nu}$ is prescribed. By mere symmetry arguments, one writes

$$T^{\mu\nu} = (p + \epsilon)u^\mu u^\nu - pg^{\mu\nu} \tag{34.44}$$

where p is the fluid pressure, $\epsilon = ne$ is the energy density and $g^{\mu\nu}$ is the Minkowski metric tensor. In (34.44) $u^\mu = dx^\mu/ds$ is the unit quadrivector tangent to the fluid worldline.

In vector notation:

$$u^\mu = \gamma_u(1, \vec{u}/c) \tag{34.45}$$

and

$$u_\mu = \gamma_u(1, -\vec{u}/c) \tag{34.46}$$

so that $u_\mu u^\mu = \gamma_u^2(1 - u^2/c^2) = 1$. In other words, the four-velocity is a unit vector in (3+1)-dimensional spacetime.

The expression (34.44) refers to the gas of a perfect relativistic fluid with no dissipative effects, the analog of the Euler equations.

The relativistic fluid equation $\partial_\mu T^{\mu\nu}$, can be written in many different forms. Of particular interest, given its similarity with the non-relativistic analog is the following one:

$$\partial_t(\gamma_u h - p) + \partial_a(\gamma_u h u_a) = 0 \tag{34.47}$$

$$\frac{\gamma_u h}{c^2}(\partial_t u_a + u_b \partial_b u_a) = -\partial_a p + u_a \partial_t(p/c^2) \tag{34.48}$$

where

$$h = ne + p \tag{34.49}$$

is the enthalpy per unit volume of the fluid and e is the energy.

It can be checked that in the non-relativistic limit

$$h/c^2 \to \rho, \ \gamma_u \to 1 \tag{34.50}$$

the previous equations reduce to the standard non-relativistic Euler equations.

The extension to the dissipative case is by no means a trivial matter and still somehow under debate. The main point is that dissipative processes cannot be represented by Laplacian operators, as this would introduce an infinite speed of propagation, clashing with the relativistic nature of the fluid. To obviate this problem, dissipative processes are represented through second-order derivatives in both space and time, leading to wave-like equations, also known as telegraphers' equation.

In the sequel, we shall not delve any further into this important topic, but simply observe that kinetic theory is free from such a problem because it comes naturally with space and time on the same footing. This does not mean that dissipation is automatically correct in kinetic formulations of hydrodynamics, but simply that *causality* is not an issue.

34.6   Relativistic LBE

At a first glance, there are several reasons to argue that the lattice kinetic implementation of relativistic hydrodynamics should come by more naturally than its non-relativistic counterpart.

The first consideration is that relativity imposes a *finite* speed on physical grounds, hence it appears naturally suited to the inherently finite-speed LB formalism. Second, kinetic theory comes by construction in covariant language, with space and time on the same footing from scratch, as in relativity.

Third, one notes that the standard LB exhibits exactly the same equation state of ultrarelativistic matter, namely

$$p = \frac{1}{3}\rho c^2 \tag{34.51}$$

Given the relation

$$\beta = \frac{u}{c} = \frac{u}{c_s}\frac{c_s}{c} = \frac{Ma}{\sqrt{3}} \tag{34.52}$$

it is readily appreciated that (34.51) implies that weakly compressible flows, at, say $Ma \sim 0.1$, are inherently weakly relativistic, $\beta \sim 0.1/\sqrt{3} \sim 0.06$.

So much for the qualitative hints.

A closer look at the mathematical details reveals however a number of major hurdles.

First, unlike Maxwell–Boltzmann, the Juettner distribution *is not separable* in the three spatial dimensions, due to the irrational dependence $\gamma = (1 - \beta^2)^{-1/2}$, a direct consequence of Lorentz invariance. This implies that the analogue of expansions in Hermite polynomials are not immediately available.

Second, the macroscopic equations for the number density and the energy-momentum density seem to belong to two distinct and disconnected kinetic ladders. The former, $\partial_\mu N^\mu = 0$ involves kinetic moments of $\{1, \vec{\beta}\}$, while the latter, $\partial_\nu T^{\mu\nu} = 0$ involves $\{\gamma, \gamma\vec{\beta}, \gamma\vec{\beta}\vec{\beta}\}$.

Again, given that for massive particles γ is an irrational function of β, there is no way of merging the two ladders into a single one, described by a sequence of ascending powers of β.

In hindsight, both difficulties can be (partially) circumvented, even though they nonetheless affected the chronological development of RLB (the blame is directly on this author ...).

34.7 Top-Down Approach: RLB by Moment Matching

Due to the lack of a direct analogue of the Hermite expansion of Juettner equilibria, the first RLB was developed in the same way as the standard LB, namely by top-down compliance with relativistic hydrodynamics, via moment matching.

Moreover, in response to the two-ladders problem previously mentioned, the first RLB was based on *two* distribution functions, one, f_i, for the particle number, and the other, h_i, for the particle energy. This is in very close analogy with the double-distribution thermal LB schemes described previously in this book.

The macroscopic target equation look as follows.

Conservation of particle number:

$$\partial_t(n\gamma) + \partial_a(n\gamma u_a) = 0 \tag{34.53}$$

Conservation of particle energy momentum:

$$\begin{cases} \partial_t(h\gamma^2 - p) + \partial_a(h\gamma^2 u_a) = 0 \\ \partial_t(h\gamma^2 u_a) + \partial_b(h\gamma^2 u_a u_b + p\delta_{ab}) = \\ \partial_b\{\eta[\partial_a(n\gamma u_b) + \partial_b(n\gamma u_a) + \partial_c(n\gamma u_c)\delta_{ab}]\} \end{cases} \tag{34.54}$$

In (34.54), $\gamma \equiv \gamma_u = (1 - u^2/c^2)^{1/2}$, $h = p + \rho c^2$ is the enthalpy density and η is the kinematic viscosity.

In full analogy with the non-relativistic scheme, one lays down the following relativistic LBGK in Marle form:

$$\begin{cases} f_i(\vec{x} + \vec{c}_i\Delta t; t + \Delta t) - f_i(\vec{x}; t) = \dfrac{\Delta t}{\tau}(f_i^{eq} - f_i) \\[2mm] h_i(\vec{x} + \vec{c}_i\Delta t; t + \Delta t) - h_i(\vec{x}; t) = \dfrac{\Delta t}{\tau}(h_i^{eq} - h_i) \end{cases} \tag{34.55}$$

The RLB scheme seeks a discrete relativistic equilibrium matching the following hydroynamic constraints:

Particle number:

$$\sum_i f_i^{eq} = \sum_i f_i = n\gamma \tag{34.56}$$

Particle energy-momentum:

$$\sum_i h_i^{eq} = \sum_i h_i = h\gamma^2 - p \tag{34.57}$$

$$\sum_i h_i^{eq} c_{ia} = \sum_i h_i c_{ia} = (h\gamma^2 - p)u_a \tag{34.58}$$

Solving the corresponding algebraic matching problem using the standard D3Q19 lattice delivers the following relativistic local equilibria:

$$f_i^{eq} = n\gamma w_i(1 + 3\beta_i) \tag{34.59}$$

and

$$h_i^{eq} = 3w_i\left[\frac{p}{c^2} + h\gamma^2\left(\beta_i + \frac{3}{2}\beta_i^2 - \frac{1}{2}\beta^2\right)\right], \quad i > 0 \tag{34.60}$$

$$h_0^{eq} = 3w_0\left[h\gamma^2\left(1 - \frac{(c^2 + 2)p}{h\gamma^2 c^2} - \frac{1}{2}\beta^2\right)\right], \quad i = 0 \tag{34.61}$$

In (34.60–61), $\beta_i = \frac{\vec{u}\cdot\vec{c}_i}{c^2}$ and $\beta^2 = u^2/c^2$.

The weights w_i are the same as in the standard D3Q19 lattice.

To be noted that $c = \Delta x/\Delta t$ is now intended to be the *actual* speed of light, so that the edge-center links of the D3Q19 lattice carry faster-than-light particles (*tachyons*) with $c_i = \sqrt{2}c$. Albeit unphysical, this does not cause any numerical problem, as long as u/c remains well below one, i.e., in the weakly relativistic regime.

The previous relativistic equilibria carry a strong similarity with their non-relativistic counterparts, in that they are second-order polynomials of the fluid speed. The numerical details, however, are not the same, as they must reflect the relativistic nature of the macroscopic equations. For more details, the reader is kindly directed to the original publications.

34.7.1   Dissipative Hydrodynamics

By carrying out the Chapman–Enskog expansion, one obtains the familiar expression for the shear viscosity

$$\eta = \frac{\gamma h}{3}c^2\left(\tau - \frac{\Delta t}{2}\right) \tag{34.62}$$

In the relativistic context, such an expression should be taken for no more than a tunable form of numerical viscosity. The reason is that the exact form of dissipative terms in relativistic hydrodynamics is still open to a large extent, even though very recent developments may shed some new light on this subject (5). We shall briefly return on this point in the forthcoming chapter.

34.8   Early RLB Applications

The earliest RLB scheme was first validated against existing relativistic simulations of shock-wave propagation in quark-gluon plasmas. Later, it was also applied to the simulation of electron flows in graphene samples, as briefly discussed below.

34.8.1   Shock Wave in Quark-Gluon Plasmas

As a validation study, the RLB described in the previous section has been applied to the study of shock-wave propagation in quark-gluon plasmas (QGP). This is a very hot (literally!) problem in modern subnuclear and high-energy physics, with potentially far reaching cosmological implications for the origin of our Universe.

This fascinating problem warrants a few extra-lines of comment.

Quark-gluon plasma is the state of matter which is supposed to result from the liberation of quarks from hadronic matter (protons and neutrons,) once the temperature is brought above a critical value, of the order of $T_c \sim 400$ Mev$/k_B$, corresponding to about $4 \cdot 10^{12}$, i.e. four trillions Kelvin degrees. This is the hottest temperature ever achieved on Earth (at the Relativistic Heavy-Ion Collider in 2010), which is about $250,000$ times hotter than the center of the Sun! It is also the temperature of the Universe approximately one microsecond after the Big-Bang and actually the highest in the known Universe...

As a note of curiosity, the other extreme, i.e. the lowest temperature ever achieved, a Bose-Einstein condensate realized at MIT in 2003 is in the order of one nanokelvin.

These extreme temperatures, which are typical of the early Universe, have indeed been achieved in terrestrial lab experiments, such as the Relativistic Heavy Ion Collider (RHIC) at Brookhaven, USA and the Alice (A Large Ion Collider Experiment) experiment at CERN.

The protons emerge from such collisions with an energy comparable to their rest mass, and consequently they undergo a deconfining transition, which eventually liberates their subnuclear components, i.e., quark and gluons.

According to a cornerstone of modern physics, i.e., *asymptotic freedom*, quarks at sufficiently high energies stop interacting, hence they should behave like free particles, i.e., in thermodynamic terms, like a high-Knudsen gas of free particles.

To the contrary, the experiments reveal that the quark-gluon plasma behaves like a strongly interacting fluid, with a mean-free path in the order of femtometers (10^{-15}) meters, i.e., the range of strong interactions.

In fact, the quark-gluon plasma features one of the smallest values of the ratio η/s, s being the entropy density, ever observed in nature, a so-called *quasi-perfect* fluid. It comes close to saturating the celebrated AdS-CFT (Anti DeSitter Conformal Field Theory) duality lower bound:

$$\frac{\eta}{s} \geq \frac{1}{4\pi}\frac{\hbar}{k_b} \tag{34.63}$$

There is therefore a strong interest in understanding the equilibrium (equation of state) and non-equilibrium (transport coefficients) of this fascinating state of matter.

The study of shock waves can provide valuable information on both fronts. First, by measuring the sound speed, c_s, one gains direct access to the ratio c_s/c as a function of energy, which is tantamount to knowing the equation of state of the quark-gluon plasma.

Second, since the occurrence of shock waves is only possible below a given value of the ratio η/s, the observation of such shocks can be taken as an evidence of very low viscosity.

The tool of the trade to investigate these matters are either low-viscous relativistic hydrodynamic codes or, more fundamentally, fully kinetic Boltzmann equation for parton dynamics (BAMPS: Boltzmann Approach for Multi-Parton Scattering). Both methods, particularly BAMPS are computationally demanding.

Interestingly, the shock speed in quark-gluon plasmas is a moderate fraction of the speed of light, typically $u/c \sim 0.3$ corresponding to a Lorentz factor $\gamma_u \sim 1.1$. As a result, there appears to be scope for RLB as a minimalistic, and yet quantitative, approach to the problem.

Indeed RLB has proven capable of reproducing with quantitative accuracy the velocity and pressure profiles of shock waves at typical quark-gluon-plasmas conditions see Fig. 34.4.

RLB was also found to compute an order of magnitude faster than hydrodynamic codes and several orders faster than BAMPS.

This is not surprising, given its minimalistic nature. However, it surely offers a fresh new option to investigate quark-gluon plasma hydrodynamics whenever microscopic details of quark-gluon interactions are not essential for the hydrodynamic evolution of the system.

34.8.2   Astrophysics: Supernova Explosions

The RLB formalism is not restricted to one spatial dimensions.

As a pictorial example of a three-dimensional application to a very different context, shock waves from supernova explosions, is shown in Figure 34.5.

Although no in-depth analysis has been pursued on the quality of such simulations as compared with state-of-the-art computational astrophysics, these figures indicate that RLB might be useful to study relativistic flows in the presence of solid bodies.

Of course, a realistic effort along this line should include the development of suitable boundary conditions, a completely uncharted territory at the time of this writing (the calculation in Figure 34.5 was performed using bounce-back boundary conditions).

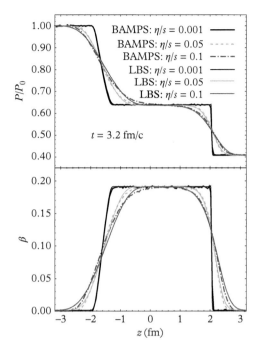

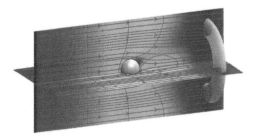

Figure 34.4 *Shock profiles and corresponding beta factor, as computed by RLB (denoted by LBS, for Lattice Boltzmann Solver, in the figure) and the full Boltzmann solver (BAMPS) for different values of the ratio η/s. Reprinted with permission from M. Mendoza et al, Phys Rev Lett 105 014502 (2010). Copyright 2010 by the American Physical Society.*

Figure 34.5 *Density contours and velocity streamlines of a relativistic shock flow impinging against a solid interstellar obstacle, after a supernova explosion. Reprinted with permission from M. Mendoza et al, Phys Rev Lett 105 014502 (2010). Copyright 2010 by the American Physical Society.*

34.8.3 Electron Flows in Graphene

The RLB has also been applied to study of relativistic electron flows in graphene, another very hot topic in modern physics. It is indeed known that, thanks to the special symmetry of the honeycomb lattice, electrons in graphene behave like near-massless excitations obeying Dirac dynamics. Under suitable approximations, it has been speculated that such excitations can be treated like a relativistic fluid, whose quantum nature can be conveyed entirely within the expression of the effective viscosity. The RLB can then be used as a LB solver for this relativistic fluid and explore the consequences of this assumption. Among other things, it has been pointed out that fixed impurities of micrometric size might behave as solid obstacles giving rise to pre-turbulent phenomena at $Re \sim 10 \div 100$.

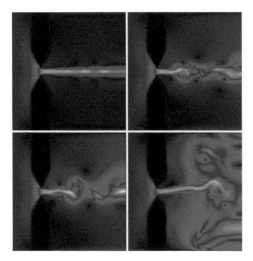

Figure 34.6 *Instability of an electron microjet in graphene. The top and bottom panels refer to subsequent snapshots for the relativistic and non-relativistic cases, respectively. It is noted that relativistic effects delay the jet instability. Reprinted with permission from M. Mendoza et al, Phys Rev Lett 106 156601 (2011). Copyright 2010 by the American Physical Society.*

It was also speculated that such pre-turbulent wakes should leave measurable signatures in the form of hundreds-MHz oscillations in the current signals out of graphene samples. Evidence of hydrodynamic signatures in graphene flows have been confirmed by very recent experiments (8), see Fig. 34.6.

34.9 Strongly Interacting Fluids, AdS-CFT Duality and Super-Universality

Strongly interacting fluids (SIF), such as quark-gluon plasmas, electron flows in graphene and Bose–Einstein lattice condensates have gained tremendous attention in the recent years, mainly under the impulse of major experimental breakthroughs in condensed matter and high-energy physics. They also unveiled profound connections between fluid mechanics, condensed matter, high-energy physics and even string theory (9; 10).

These fluids appear to attain the closest known approximation to a *perfect fluid*, in that they minimize the famous AdS–CFT (Anti de Sitter, Conformal Field Theory) viscosity bound (11):

$$B \equiv 4\pi \frac{\eta}{s} \frac{k_B}{\hbar} \geq 1 \tag{34.64}$$

where η is the dynamic viscosity and s the entropy per unit volume (entropy density is more relevant than number density in a relativistic context). This inequality was derived from the celebrated duality between $(d + 1)$-dimensional gravity and conformal field theory in dimension d.

Since they live on the boundary ("brane" in stringy parlance) of the gravitational domain, these fluids are called "holographic" (12). Although not rigorous, this inequality appears to be fulfilled by all known fluids, with values of B ranging from 400 for water, to about ten for Helium-three, down to about three for graphene electrons and one for quark-gluon plasmas.

It is of interest to recast the ADS-CFT bound in a form more familiar to the fluid-dynamicist. By writing the dynamic viscosity as $\eta = nmv = nmv_T l_\mu$ and recalling that $\lambda_B = \hbar/mv_T$, we obtain

$$B = 4\pi \frac{l_\mu}{\lambda_B}$$

In equivalent terms:

$$l_\mu > \frac{\lambda_B}{4\pi}$$

indicating that quantum physics sets a lower bound to the mean-free path, hence to the Knudsen number.

At a first glance, holographic fluids stand like the graveyard of kinetic theory. Indeed, due to the very strong interactions, it is argued that not only Boltzmann kinetic theory, but the very notion of quasi-particle as a weakly interacting collective degree of freedom, would lose meaning for holographic fluids. Differently restated, since the mean-free path becomes comparable or smaller than the De Broglie length, one is faced with an irreducible quantum many-body problem. Amazingly, hydrodynamics continues to hold for holographic fluids as disparate as super-hot and dense quark-gluon plasmas and super-cold dilute atomic gases, which lie some thirty orders of magnitude apart in density and over twenty. *This is a truly spectacular manifestation of Universality!*, which we may call "Super-Universality"!

Does this "holo-hydrodynamics without kinetic-theory" picture spoil the possibility of using lattice kinetic methods? In our opinion, the opposite is true.

Holo-hydrodynamics only adds to the top-down approach described previously, namely *design* LB schemes based on the symmetries of the macroscopic (field theory) target, rather than *deriving* them *ab-initio* from microscopic dynamics. This point of view has been recently expressed in adamant terms by Paul Romatschke, in the delightful paper: (Do nuclear collisions create a locally equilibrated quark-gluon plasma?, Eur. Phys. J. C, 77, 21, 2917). We quote verbatim *What is hydrodynamics? The equations of hydrodynamics can be derived using a multitude of approaches. Some assume the system to be closed to thermal equilibrium, others assume a weakly coupled microscopic particle description (kinetic theory). In my opinion, the most general derivation of hydrodynamics follows the approach of Effective Field Theory (EFT). According to this viewpoint, hydrodynamics is the EFT of long-lived, long-wavelength excitations consistent with the basic symmetries of the underlying system. .... The above EFT derivation does at no point invoke the presence of an underlying-based, kinetic description of the matter.* This is -literally- the spirit of the top-down LB approach! Impressive QLB simulations of quantum turbulence in Bose–Einstein condensates already exist (13) and one may also hope that LB formulations in curved manifolds should be able to simulate efficiently holographic flows.

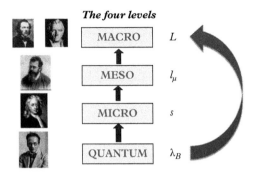

Figure 34.7 *The four levels of quantum statistical physics: Quantum-Micro-Meso-Macro (QM^3) and the associated lengthscales: De Broglie wavelength λ_B, classical size of the molecule (interaction range) s, mean-free-path l_μ and macroscale L. The standard ordering is $\lambda_B < s < l_\mu < L$. In this case, the macroscopic limit is obtained gradually, i.e., from quantum many-body to classical microscale Newtonian mechanics, to Boltzmann's kinetic theory and finally macroscopic hydrodynamics. In exotic states of matter, and typically SIF fluids, this gradual process is broken and the system can jump to the macroscopic level without ever traversing the intermediate stages. This is what we call "Super-Universality". In macroscopic manifestations of quantum effects, the ordering is violated and the De Broglie length "jumps" above s and l_μ, thereby providing a direct connection between the quantum world and the macroscopic level governed by continuum hydrodynamics. In strongly interacting AdS-CFT fluids, it is argued that the mean-free path gets sufficiently close to the De Broglie wavelengths to hamper a kinetic-theory description based on long-lived quasiparticles. Once again, the systems jumps directly from quantum many body to a macroscopic hydrodynamic level. Here hydrodynamics must be taken in its broadest meaning, i.e., as a large-scale limit of an underlying quantum mechanics and quantum field theory.*

34.9.1 Caveat

Classifying graphene and quark-gluon plasmas under the same "quasi-perfect" SIF umbrella as quark-gluon plasmas is formally correct, since they both feature a mean-free path comparable with the De Broglie length. On the other hand, from the strict fluid-dynamic point of view, this may turn out to be very confusing. The point is that while quark-gluon plasmas feature indeed ultrashort mean-free path, of the order of femtometers, depending on density and temperature conditions, graphene may exhibit long mean-free paths, corresponding to a kinematic viscosity about four orders of magnitude larger than water! So, graphene near-saturates the AdS-CFT bound because of the extremely small mass of the electrons, which results in a correspondingly small *dynamic* viscosity. The mean-free path, however is not small at all, it is actually comparable with the size of the current graphene samples!

34.10 Summary

The Lattice Boltzmann formulation of relativistic hydrodynamics, a silent subject for over twenty years, has known a vigorous boost of activity in the last decade. The results are encouraging, as they show that RLB may offer an appealing option, not only over fully kinetic formulations such as BAMPS, but also versus hydrodynamic codes.

A lot remains to be done, in terms of validation and development of suitable boundary conditions. For a wonderful account of the progress of relativistic hydrodynamics, see the recent review (6).

....

REFERENCES

1. L. Landau, E. Lifshitz, The classical theory of fields, 3rd revised english version, Pergamon Press, Oxford, (1971).
2. M. Mendoza, H.J. Herrmann and S. Succi, "Fast Lattice Boltzmann Solver for Relativistic Hydrodynamics", *Phys. Rev. Lett.*, **105**, 014502, 2010.
3. R. Liboff, *Kinetic Theory: Classical, Quantum and Relativistic Descriptions*, Wiley and sons, New York, 2002.
4. C. Cercignani and G.M. Kremer, *The relativistic Boltzmann Equation: Theory and Applications*, Birkhause, Basel, 2002.
5. A. Gabbana, M. Miller, S. Succi and R. Tripiccione, Towards a unified lattice kinetic scheme for relativistic hydrodynmics, *Phys. Rev.*, 95 (5), 2017.
6. P. Romatschke and U. Romatschke, Relativistic Fluid Dynamics Out of Equilibrium Ten Years of Progress in Theory and Numerical Simulations of Nuclear Collisions, arXiv:1712.05815v1 [nucl-th] 15 Dec 2017.
7. M. Miller, H.J. Herrmann and S. Succi, "Pre-turbulent regimes in graphene flows", *Phys. Rev. Lett.*, **106**, 156601, 2011.
8. D.A. Bandurin *et al.*, "Negative local resistence caused by viscous electron backflow in graphene", *Science*, 351, 1055, 2016.
9. S. Sachdev and M. Mueller, "Quantum criticality and black holes", *J. Phys.: Condens. Matter*, 21, 164216, 2009.
10. J. Maldacena, "The large-N limit of superconformal field theories and supergravity", *Int. J. of Theor. Phys.*, 38, 1113, 1999.
11. G. Policastro, D.T. Son and A. Starinets, "Shear viscosity of strongly coupled N= 4 supersymmetric Yang-Mills plasma", *Phys. Rev. Lett.*, 87, 081601, 2001.
12. A. Adams, P. Chesler, H. Liu, "Holographic vortex liquids and superfluid turbulence", *Science*, 341 (6144) 368–72 2013.
13. J. Yepez *et al.*, "Superfluid turbulence from quantum Kelvin wave to classical Kolmogorov cascades", *Phys. Rev. Lett.*, **103**, 084501 2009.
14. D. Hupp M. Miller, I. Bouras, S. Succi and H.J. Herrmann, Relativistic Lattice Boltzmann method for quark-gluon plasma simulations, *Phys. Rev. D*, 84, 125015, 2011.

....

EXERCISES

1. Prove the relations (34.12).
2. Show that the Lorentz transformations leave the line element ds^2 invariant.
3. Prove that $f_{\tilde{\jmath}}(v) = \gamma^{-3}(v) f_{\tilde{\jmath}}(p(v))$.

35

Advanced RLB models

The relativistic LB scheme described in Chapter 34 is based on the top-down approach and limited to weakly relativistic fluids. In this chapter, we present a systematic derivation of relativistic LB based on the continuum kinetic theory. The resulting scheme can handle relativistic flows with Lorentz factors up to order ten, thereby considerably extending the scope of the method. In addition, we also discuss the extension of the LB scheme to generalized coordinates for the simulation of flows on curved manifolds along with preliminary attempts to numerical relativity.

When you are courting a nice girl, an hour seems like a second. When you sit on a red-hot cinder, a second seems like an hour. That's relativity,

(A. Einstein).

35.1 Relativistic LBE from Relativistic Kinetic Theory

The moment-matching formulation described in Chapter 34 suffers from a few basic drawbacks: first, the use of two distributions, f and h, implies a non-physical diffusivity of the number density (due to the finite value of τ). Moreover, the energy density, which should diffuse, is conserved instead, through the coupling with the momentum equation.

At low speed, these two inconsistencies remain silent, but as speed and temperature are raised, they become a source of concern.

As a result, it is highly desirable to develop a RLB, "bottom-up" i.e., from the proper phase-space discretization of the corresponding kinetic equation in the continuum. We shall dub this KRLB, for kinetic RLB.

As anticipated previously, this task proves significantly more involved than its non-relativistic analog, a difficulty which can be traced to the lack of a natural polynomial basis for Juettner equilibria.

The Lattice Boltzmann Equation. Sauro Succi, Oxford University Press (2018).
© Sauro Succi. DOI: 10.1093/oso/9780199592357.001.0001

However, by using the absolute (zero speed) Juettner equilibrium, and by neglecting the rest energy (ultra-relativistic approximation), a basis of orthogonal lattice polynomials can be worked out and shown to reproduce the dynamics of relativistic flows, with Lorentz factors up to $\gamma \sim 10$ (see (1; 2)).

To this purpose, however, higher-order lattices are required, particularly if dissipative phenomena have to be accounted for. Since this subject is highly technical and still in flux, here we provide only the basic ideas directing the detail-thirsty reader to the current literature.

35.1.1   Relativistic Basis Functions

The basic idea is to expand the local equilibria using a relativistic extension of the Hermite basis, with the Juettner global equilibrium as a weight function, namely

$$f^{eq}(\vec{x}, \vec{p}) = W_{\mathcal{J}}(p) \sum_{n=0}^{\infty} M_n^{eq}(\vec{x}) \mathcal{J}_n(\vec{p}) \tag{35.1}$$

where

$$W_{\mathcal{J}}(p) = A(z) e^{-z\gamma} \tag{35.2}$$

is the Juettner equilibrium in the rest frame and $\mathcal{J}_n(\vec{p})$ denotes the (yet unknown) n-th order relativistic polynomial. Note that in our notation, $p \equiv |\vec{p}|$ is the magnitude of the momentum in ordinary three-dimensional space.

As noted before $\gamma = \sqrt{1 + \mu^2}$ is an irrational function of the dimensionless momentum $\mu = p/mc$, which represents a major inconvenience for the lattice discretization. However, for ultrarelativistic particles, $z \to 0$, the Juettner weight simplifies to

$$W_{\mathcal{J}}(p) = e^{-p/p_T}$$

where $p_T = k_B T/c$ is the thermal momentum.

This expression is much more amenable to analytical calculations and ensuing lattice manipulations.

Indeed, this is the weight function of the classical Laguerre polynomials, the analogue of Hermite for positive-definite unbounded variables such as the momentum $0 < p < \infty$.

A systematic procedure to construct relativistic polynomials to the desired order goes as follows.

First, define the basic set of five generating polynomials up to order $r = 1$:

$$\mathcal{G}^{(1)} = \{1, p^0, p^a c\} \equiv \{1, E, \vec{p}c\} \tag{35.3}$$

Higher-order polynomials are generated by Gram–Schmidt orthogonalization, using $W_{\mathcal{J}}(p)$ as an orthogonalizing weight, namely

$$\mathcal{J}_{nm} \equiv \int \mathcal{J}_n(\vec{p}) w_{\mathcal{J}}(p) \mathcal{J}_m(\vec{p}) \frac{d\vec{p}}{p^0} = \delta_{nm} \tag{35.4}$$

Note that since we use ultra-relativistic weights, in (35.4) $p = E/c$.

To the desired third order, the procedure delivers 30 linearly independent polynomials listed in Fig. 35.1. Given that we are dealing with UR particles for which $mc^2/k_B T \to 0$, all expressions in Fig. 35.1 are normalized in thermal units, i.e., $p^0 = p^0/(p_T c)$ and $p^a \equiv p^a/p_T$, $a = x, y, z$, where $p_T = k_B T/c$ is again the thermal momentum.

This completes the task of providing an orthonormal basis of relativistic polynomials in continuum energy momentum, i.e., the stepping stone for the development of a corresponding lattice kinetic formulation.

35.1.2   Relativistic Lattice Polynomials

The next step is to translate the procedure on the lattice.

To this purpose, the integrals are turned into discrete sums, and the discrete orthogonality relations take the form

$$\int \mathcal{J}_n(p^\mu) w_{\mathcal{J}}(p^0) \mathcal{J}_m(p^\mu) \frac{d\vec{p}}{p^0} = \sum_i w_i \mathcal{J}_n(p_i^\mu) \mathcal{J}_m(p_i^\mu) = \delta_{nm} \tag{35.5}$$

As usual, the goal is to find a suitable set of discrete nodes $p_i^\mu = (p_i^0, \vec{p}_i)$ and associated weights w_i, such that in (35.5) constraints are fulfilled *exactly* up to the desired order. On top of this, we shall further require that the nodes lie on a regular lattice, so as to preserve the essential property of LB, that is *exact streaming*.

In equations:

$$\vec{v}_i = (l_i, m_i, n_i) c_{lb} \tag{35.6}$$

where (l_i, m_i, n_i) is a triplet of integers measuring the discrete velocities in units of the lattice speed

$$c_{lb} = \frac{\Delta x}{\Delta t}$$

Finally, since we are dealing with UR particles, we shall also require that the magnitude of all discrete velocities be in exact match with the speed of light, i.e.,

$$|\vec{v}_i| = c \tag{35.7}$$

for each discrete velocity $i = 1, b$.

Order	Polynomial $\mathcal{J}_k$	k
0th	1	0
1st	$\dfrac{p^0-2}{\sqrt{2}},\dfrac{p^x}{\sqrt{2}},\dfrac{p^y}{\sqrt{2}},\dfrac{p^z}{\sqrt{2}}$	1, 2, 3, 4
2nd	$\dfrac{(p^0-6)p^0+6}{2\sqrt{3}},\dfrac{(p^0-4)p^x}{2\sqrt{2}},\dfrac{(p^0-4)p^y}{2\sqrt{2}}$	5, 6, 7
	$\dfrac{(p^0-4)p^z}{2\sqrt{2}},\dfrac{-(p^0)^2+(p^x)^2+2(p^y)^2}{4\sqrt{2}},\dfrac{3(p^x)^2-(p^0)^2}{4\sqrt{6}}$	8, 9, 10
	$\dfrac{p^xp^z}{2\sqrt{2}},\dfrac{p^yp^z}{2\sqrt{2}},\dfrac{p^xp^y}{2\sqrt{2}}$	11, 12, 13
3rd	$\dfrac{1}{12}(p^0-6)^2p^0-2,\dfrac{((p^0-10)p^0+20)p^x}{4\sqrt{5}}$	14, 15
	$-\dfrac{(p^0-6)((p^0)^2-3(p^x)^2)}{24},\dfrac{5(p^x)^3-3(p^0)^2p^x}{24\sqrt{5}}$	16, 17
	$\dfrac{((p^0-10)p^0+20)p^y}{4\sqrt{5}},\dfrac{(p^0-6)p^xp^y}{4\sqrt{3}},\dfrac{p^xp^yp^z}{4\sqrt{3}}$	18, 19, 20
	$\dfrac{(p^0-6)((p^x)^2+2(p^y)^2-(p^0)^2)}{8\sqrt{3}}$	21
	$\dfrac{p^x((p^x)^2+2(p^y)^2-(p^0)^2)}{8\sqrt{3}}$	22
	$\dfrac{p^y(-3(p^0)^2+3(p^x)^2+4(p^y)^2)}{24\sqrt{2}},\dfrac{((p^0-10)p^0+20)p^z}{4\sqrt{5}}$	23, 24
	$\dfrac{(p^0-6)p^xp^z}{4\sqrt{3}},-\dfrac{p^z((p^0)^2-5(p^x)^2)}{8\sqrt{30}},\dfrac{(p^0-6)p^yp^z}{4\sqrt{3}}$	25, 26, 27
	$\dfrac{(p^0-6)p^yp^z}{4\sqrt{3}},\dfrac{p^z(-(p^0)^2+(p^x)^2+4(p^y)^2)}{24\sqrt{2}}$	28, 29

Figure 35.1 *The discrete polynomials up to third order. Reprinted with permission from M. Mendoza et al, Phys Rev D, 87, 065027 (2013). Copyright 2013 by the American Physical Society.*

Taken all together, we obtain

$$|\vec{v}_i|^2 = (l_i^2 + m_i^2 + n_i^2)c_{lb}^2 = c^2 \tag{35.8}$$

namely

$$l_i^2 + m_i^2 + n_i^2 = \frac{c^2}{c_{lb}^2} \equiv R^2 \tag{35.9}$$

where R is an integer measuring the diameter of the sphere where the discrete velocities are located.

The above is a Diophantine equation for the integer triplet (l_i, m_i, n_i).

It clearly shows that by identifying the lattice speed with the actual physical speed of light, $c = c_{lb}$, the only admissible integer triplet is $(\pm1, 0, 0)$ and permutations thereof, the six nearest neighbors of a cubic lattice. Manifestly, this falls short of fulfilling the moment constraints (35.5) at order two and above.

It turns out that the minimal solution at order two is $R^2 = 9$, which delivers the following set of 30 discrete velocities (see Fig. 35.2):

- $(\pm3, 0, 0)$ and permutations thereof (6 velocities)
- $(\pm1, \pm2, \pm2)$ and permutations thereof (24 velocities)

An important detail remains to be pointed out. We have seen before that in order to give RLB a non-relativistic look, the energy in front of the time-derivative has to be factored out, leading to the term c/E in the collision operator. Such term cannot be arbitrary, because the discrete four momentum must obey the condition $p^\mu p_\mu = E^2 - p^2 c^2 = m^2 c^4$. Hence, for UR particles, $E/c = p$, the energy is the magnitude of the three-dimensional momentum over the speed of light.

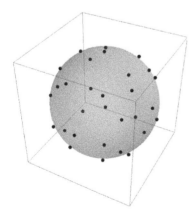

Figure 35.2 *The 30 discrete velocities recovering second-order KRLB. Reprinted with permission from M. Mendoza et al, Phys Rev D, 87, 065027 (2013). Copyright 2013 by the American Physical Society*

The integration over the energy variable is performed in spherical coordinates using the zeroes of the following third-order polynomial:

$$P_3(p) = p(p-6)^2 - 24 = 0$$

Such roots provide the admissible values for the discrete energy, the numerical result being $p_1 \sim 0.936$, $p_2 = 3.305$ and $p_3 = 7.759$. Note that this quadrature is confined to third order since we are targeting a second-order kinetic formulation.

The whole shebang consists of three energy shells, each containing the 30 discrete velocities described previously, for a total of 90.

The analogy with energy-momentum quantization is artificial but nonetheless intriguing, or perhaps intriguing but nonetheless artificial ....

35.1.3   Third-Order KRLB with Improved Dissipation

The KRLB scheme can be furthered to match third-order moments, as it is required to provide a more accurate description at high speed flows and dissipative effects. In this case, larger stencils involving hundreds of discrete speeds, must be used. A typical set of discrete velocities living on a sphere of radius $R = \sqrt{41}$ is shown in Fig. 35.3.

The corresponding discrete velocities are as follows:

- $(\pm 1, \pm 2, \pm 6)$ and permutations thereof (48)
- $(\pm 0, \pm 4, \pm 5)$ and permutations thereof (24)
- $(\pm 3, \pm 4, \pm 4)$ and permutations thereof (24)

for a total of 96 discrete velocities. The integration on the energy shells now requires the zeroes of a fourth-order polynomial, whose roots are found to be $p_1 = 0.743$, $p_2 = 2.572$, $p_3 = 5.731$ and $p_4 = 10.95$, for a total of $96 \times 4 = 384$. However, since many weights can be set to zero without compromising the solution, one ends up with "only" 128 discrete velocities, a number comparable with higher-order non-relativistic lattices.

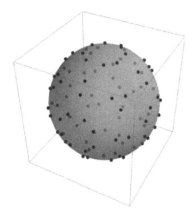

Figure 35.3 *The 96 discrete velocity quadrature points for the third-order KRLB. Reprinted with permission from M. Mendoza et al, Phys Rev D, 87, 065027 (2013). Copyright 2013 by the American Physical Society.*

35.1.4   KRLB Equilibria

With lattice basis functions at hand, the local equilibria can be constructed following a standard expansion up to the third order moments:

$$f_i^{eq}(\vec{x}, \vec{p}_i; t) = \sum_{k=0}^{3} w_i \mathcal{J}_{ik} M_k^{eq}(\vec{x}, t) \tag{35.10}$$

where the kinetic moments M_k^{eq} are available from continuum kinetic theory, see (3).

Although conceptually straightforward, the algebra is rather laborious, as the reader may readily appreciate by glancing at the expressions shown Fig. 35.4.

This completes the derivation of relativistic LB schemes from continuum relativistic kinetic theory.

35.1.5   Kinetic RLB schemes for non-zero mass (finite-temperature) particles

As pointed out several times, the KRLB schemes discussed so far, are built upon lattice polynomials based on the ultra-relativistic assumption $z \to 0$, for which the energy-momentum relation reduces to $E = pc$, leading to a drastic simplification of the algebraic procedure.

On the other hand, many relativistic fluid applications of major interest, such as the time evolution of the early Universe or the dynamics of the quark-gluon plasma following upon a heavy-ion collision, occur at non-zero values of the inverse temperature parameter $z = mc^2/k_B T$, due to rapid decay of the temperature in response to the expansion of the system after the big-bang (or little-bang in the case of heavy-ion collisions). For instance, the temperature of the early Universe is estimated to fall from $T_P = 10^{32}$ Kelvin degrees at the Planck time, $t_P = 10^{-43}$ seconds, to $T_B = 10^{13}$ Kelvin at the baryogenesis time $t_B = 10^{-6}$ seconds, when the quark-gluons plasma started to condense into baryonic matter, i.e. protons, neutrons and their antiparticles. For the case of a proton, with $mc^2 \sim 1$ GeV, corresponding to 10^{13} Kelvin degrees, z goes from about $z_P = 10^{-19}$ to about $z_B = 1$ (see Fig. 35.5). These figures speak clearly for the need of a relativistic LB scheme based on non-zero mass particles at finite temperature. This task has been undertaken very recently by Gabbana *et al.* (4). Using rather sophisticated algebraic procedures and lattices, these authors have succeeded in generating finite mass (temperature) KRLB schemes which work in the range $0.1 \leq z \leq 100$, thus covering an appreciable fraction of the thermal history of the Universe, from the QGP regime onwards in time.

The scheme has been validated by comparing the expression of the dynamic viscosity as a function of z against analytical results from the Chapman-Enskog and Grad's methods. To be noted that, unlike the non-relativistic case, these two methods do *not* lead to the same dependence $\eta = \eta(z)$, not even in the UR limit. More precisely, by writing the dynamic viscosity in the form (lattice units):

$$\eta(z) = kPf(z)(\tau - 1/2)$$

$$f_i^{\text{eq}} = \frac{nw_i}{4T} \left[p_i^{02} \left(T^2 \left(2U^{02} - U^{x2} - U^{y2} - 1 \right) - 2TU^0 + 1 \right) \right.$$

$$+ 2p_i^0 \left(T \left(T \left(U^0 \left(p_i^x U^x + p_i^y U^y + p_i^z U^z - 4U^0 \right) + 1 \right) \right. \right.$$

$$\left. - p_i^x U^x - p_i^y U^y - p_i^z U^z + 7U^0 \right) - 4 \right) + T^2 \left(p_i^{x2} \left(-U^{02} + 2U^{x2} + U^{y2} + 1 \right) \right.$$

$$+ 2p_i^x U^x \left(p_i^y U^y + p_i^z U^z - 4U^0 \right) + p_i^{y2} \left(-U^{02} + U^{x2} + 2U^{y2} + 1 \right) + 2p_i^y U^y \left(p_i^z U^z - 4U^0 \right)$$

$$\left. + 8U^0 \left(U^0 - p_i^z U^z \right) - 2 \right) + 2T \left(5 \left(p_i^x U^x + p_i^y U^y + p_i^z U^z \right) - 8U^0 \right) + 12 \right]. \tag{B1}$$

For the case of the third-order moment expansion, we repeat the same procedure, using all the polynomials ($k = 0, \ldots, 29$). This leads to the following expressions:

$$f_i^{\text{eq}} = \frac{nw_i}{12T} \left[p_i^{03} \left(TU^0 - 1 \right) \left(4U^{02} - 3 \left(U^{x2} + U^{y2} + 1 \right) \right) - 2TU^0 + 1 \right)$$

$$- p_i^{02} \left(T^3 \left(-2U^{02} \left(3p_i^x U^x + 3p_i^y U^y + 2p_i^z U^z \right) \right. \right.$$

$$+ \left(U^{x2} + U^{y2} + 1 \right) \left(3p_i^x U^x + 3p_i^y U^y + p_i^z U^z \right) + 36U^{03} - 6U^0 \left(3U^{x2} + 3U^{y2} + 4 \right) \right)$$

$$+ 3T^2 \left(2U^0 \left(p_i^x U^x + p_i^y U^y + p_i^z U^z \right) - 22U^{02} + 7 \left(U^{x2} + U^{y2} \right) + 9 \right)$$

$$- 3T \left(p_i^x U^x + p_i^y U^y + p_i^z U^z - 14U^0 \right) - 15 \right) - 3p_i^0 \left(T^3 \left(U^{03} \left(p_i^{x2} + p_i^{y2} - 24 \right) \right. \right.$$

$$- U^0 \left(p_i^{x2} \left(2U^{x2} + U^{y2} + 1 \right) + 2p_i^x U^x \left(p_i^y U^y + p_i^z U^z \right) \right.$$

$$+ p_i^y \left(p_i^y U^{x2} + 2p_i^y U^{y2} + p_i^y + 2p_i^z U^y U^z \right) - 12 \right) + 12U^{02} \left(p_i^x U^x + p_i^y U^y + p_i^z U^z \right)$$

$$- 2 \left(p_i^x U^x + p_i^y U^y + p_i^z U^z \right) \right) + T^2 \left(p_i^{x2} \left(-U^{02} + 2U^{x2} + U^{y2} + 1 \right) \right.$$

$$+ 2p_i^x U^x \left(p_i^y U^y + p_i^z U^z - 11U^0 \right) + p_i^{y2} \left(-U^{02} + U^{x2} + 2U^2 + 1 \right) + 2p_i^y U^y \left(p_i^z U^z - 11U^0 \right)$$

$$- 22p_i^z U^0 U^z + 56U^{02} - 14 \right) + 2T \left(6 \left(p_i^x U^x + p_i^y U^y + p_i^z U^z \right) - 25U^0 \right) + 20 \right)$$

$$+ T \left(p_i^{x3} T^2 U^x \left(-3U^{02} + 4U^{x2} + 3U^{y2} + 3 \right) \right.$$

$$+ p_i^{x2} T \left(3 \left(U^{02} - 2U^{x2} - U^{y2} - 1 \right) \left(-p_i^y TU^y + 6TU^0 - 7 \right) \right.$$

$$+ p_i^z TU^z \left(-U^{02} + 4U^{x2} + U^{y2} + 1 \right) \right) + 3p_i^x U^x \left(T \left(T \left(p_i^{y2} \left(-U^{02} + U^{x2} + 2U^{y2} + 1 \right) \right. \right. \right.$$

$$+ 2p_i^y U^y \left(p_i^z U^z - 6U^0 \right) - 12p_i^z U^0 U^z + 24U^{02} - 4 \right) + 14p_i^y U^y + 14p_i^z U^z - 48U^0 \right) + 30 \right)$$

$$+ p_i^{y3} T^2 U^y \left(-3U^{02} + 3U^{x2} + 4U^{y2} + 3 \right) + p_i^{y2} T \left(p_i^z TU^z \left(-U^{02} + U^{x2} + 4U^{y2} + 1 \right) \right.$$

$$+ 3 \left(6TU^0 - 7 \right) \left(U^{02} - U^{x2} - 2U^{y2} - 1 \right) \right) + 6p_i^y U^y \left(T \left(2T \left(-3p_i^z U^0 U^z + 6U^{02} - 1 \right) \right. \right.$$

$$+ 7p_i^z U^z - 24U^0 \right) + 15 \right) + 6p_i^z U^z 2T^2 \left(6U^{02} - 1 \right) - 24TU^0 + 15 \right)$$

$$\left. - 24U^0 \left(T^2 \left(2U^{02} - 1 \right) - 5TU^0 + 5 \right) \right) - 30 \left(T^2 - 2 \right) \right]. \tag{B2}$$

Figure 35.4 *The second- and third-order KRLB equilibria. Checking the correctness of this expression is truly for the brave! Reprinted with permission from M. Mendoza et al, Phys Rev D, 87, 065027 (2013). Copyright 2013 by the American Physical Society.*

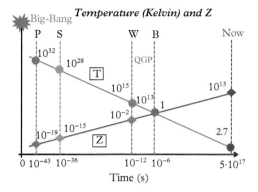

Figure 35.5 *The thermal evolution of the Universe, from the big-bang (t = 0 seconds) to the present day (t ∼ 5·10^{17} seconds). The figure serves a purely illustrative purpose, hence it is out of scale. The first vertical line from the left (P), stands at the Planck time (t = 10^{-43} seconds), when the estimated temperature is about 10^{32} Kelvin degrees, corresponding to a Planck energy of about 10^{19} GeV. The second bar at t = 10^{-36} seconds (S), marks the separation of the nuclear strong force from the electroweak one, at a temperature around 10^{28} K^o, corresponding to about 10^{15} GeV. The third bar, at t = 10^{-12} seconds (W), marks the separation of the weak force from the electromagnetic one, at an energy scale of about 100 GeV, the characteristic energy scale of the Standard Model. The fourth bar at t = 10^{-6} seconds (B), marks the baryogenesis, namely the formation of protons and neutrons from the "condensation" of the plasma of quarks and gluons. Finally, the fifth and last bar (NOW), marks the present time, about 14.5 billion years after the big-bang, with a cosmic background temperature of 2.7 Kelvin degrees. This is the thermal history of the Universe, as we know it. The temperature (upper line) drops of 32 orders of magnitude from the bing-bang to the present time, and the inverse dimensionless temperature z (lower line), grows by the same factor, from 10^{-19} at Planck time to 10^{13} at the present time. The quark-gluon plasma state lives in a narrow but crucial window, between the third and the fourth bar, in the time-slice between 1 picosecond and 1 microsecond, corresponding to the range 0.01 < z < 1.*

with k a numerical factor, P the pressure and $f(z)$ a shape function with the property $f(0) = 1$, Chapman-Enskog predicts $k = 4/5$, while Grad gives $k = 2/3$. Which one of the two methods should be trusted, remains somewhat open in the current literature. The numerical results obtained by running the finite-mass KRLB for selected test-case dissipative flows, such as the relativistic Taylor-Green vortex and one-dimensional shock-wave propagation, provide neat evidence in favor of the Chapman-Enskog method. The finite-mass KRLB scheme has been subsequently tested against BAMPS for the one-dimensional Riemann problem at $\eta/s = 0.1$, and found to provide satisfactory agreement not only for density and velocity profiles, but also for the pressure tensor and the heat flux vector (see Fig. 4 of reference 4).

35.2 Applications of Kinetic RLB

As a check of consistency, the kinetic-based RLB has been tested on the mildly-relativistic quark-gluon shock waves discussed in Chapter 34. As one can appreciate from Fig. 35.6, KRLB achieves much better agreement with BAMPS, as compared to RLB.

For higher $\beta \sim 0.7$, the mildly relativistic LB is known to fail, specifically, it develops near-discontinuities with no counterpart in full Boltzmann simulations. Hence, this presents an excellent testground for KRLB.

A typical shock-wave KRLB simulation at $\beta \sim 0.7$ is shown in Figure 35.7

The third-order kinetic RLB has been tested against BAMPS and shown to provide a satisfactory agreement not only on pressure, density and flow speed, but also on more demanding quantities, such as the viscous-pressure tensor and heat flux. Further tests and numerical comparisons can be found in the original paper (1).

35.3 RLB in Spherical Coordinates

For the sake of completeness, we wish to mention that other versions of KRLB exist, which do not insist on formulating the method on an integer "ijk" cartesian lattice. In

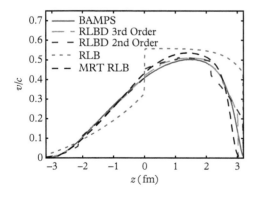

Figure 35.6 *The velocity profile of a shock wave in quark-gluon plasma, computed with RLB both single- and multi-relaxation time, RLB second and third order, as compared with BAMPS. In the figure RLBD stands for RLB with dissipative terms, actually equivalent to KRLB. The benefits of second- and third-order KRLB are clearly visible. Reprinted with permission from M. Mendoza et al, Phys Rev D, 87, 065027 (2013). Copyright 2013 by the American Physical Society.*

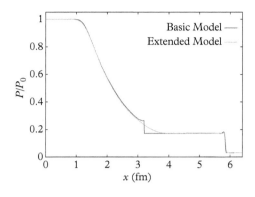

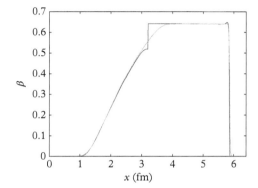

Figure 35.7 *Pressure (top) and velocity (bottom) profiles of a shock wave at β ∼ 0.7, using RLB versus second-order KRLB. As one can see, the artificial kick is basically removed by the KRLB formulation.* Reprinted with permission from M. Mendoza et al, Phys Rev D, *87, 065027 (2013)*. Copyright 2013 by the American Physical Society.

particular, Romatschke *et al.*, (5) developed a KRLB scheme in spherical coordinates, where the discrete velocities are expressed via their magnitude and the two azimuthal and polar angles, i.e., (see Fig. 35.8)

$$f(\vec{p}_i) \equiv f(p_i, \theta_i, \phi_i)$$

The advantage of such formulation is that spherical quadratures come more naturally off the shelf and require less discrete velocities to match the corresponding kinetic moments in the continuum.

The downside is that the discrete velocities are not uniform, i.e., they depend on the location in energy-momentum space, hence they no longer support exact free streaming. This leads to complications in the actual implementation of the scheme, as well as to a loss of accuracy, due to the need of interpolating. The spherical KRLB has nevertheless proven capable of simulating non-trivial relativistic benchmarks, such as exact solutions in relativistic Milne spacetimes. Full details in (5).

35.4   Lattice Boltzmann in Curved Manifolds

Many systems in nature present spatial curvature, called curved space due to the presence of massive bodies in astrophysics and cosmology, soap films or geometric confinement constraining the motion of particles moving on curved manifolds flow between two rotating cylinders and spheres, hemodynamics through deformable vessels, fusion plasmas and flow through porous media. These geometric constraints force the fluid to move along curved trajectories, leading to the upsurge of non-inertial forces which can exert major effects on the dynamics of the fluid flow. The study of these effects phenomena represents a very active area of fluid-dynamic research and it interfaces with allied disciplines, such as material science, soft-matter and biology.

Whence, major scope to developed a lattice kinetic theory in curved manifolds.

35.4.1   Hydrodynamics in Curved Manifolds

The study of physical phenomena in curved manifolds has a long and distinguished tradition in both classical and quantum physics. Moreover, tensor calculus on manifolds makes a major subject of mathematical analysis (6), see Fig. 35.8. In the following, we sketch out the essential notions that are needed to formulate of a "curvy" LB.

Curved manifolds are characterized by a spacetime dependent metric tensor $g_{ab}(x^c)$ and the associated Christoffels' symbols Γ^c_{ab}. Latin indices run over arbitrary coordinates in three spatial dimensions, since we do not consider General Relativity at this point.

The metric tensor measures the distance between two points $P \equiv (x^a)$ and $P' \equiv P + dP \equiv (x^a + dx^a)$ on the manifold $\mathcal{M}$ described by the controvariant coordinates x^a, $a = 1, 3$.

Such distance is measured by the line element:

$$ds^2 = g_{ab} dx^a dx^b$$

where sum over repeated indices is implied, as usual.

The Christoffel symbols measure the change of direction of a vector V^a moving tangentially to the manifold from P to $P + dP$ (see Appendix at the end of this chapter). In equations:

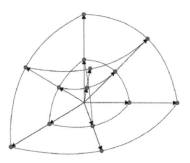

Figure 35.8 *Kinetic RLB in spherical coordinates. It is found that* 27 *discrete speeds are sufficient to recover second-order kinetic moments. Reprinted with permission from P. Romatsche et al, Phys Rev C, 84, 034903, (2011). Copyright 2011 by the American Physical Society.*

$$\Gamma^{c}_{ab} = \frac{1}{2}g^{cd}\left(\frac{\partial g_{da}}{\partial x^{b}} + \frac{\partial g_{db}}{\partial x^{a}} - \frac{\partial g_{ab}}{\partial x^{d}}\right)$$

The Christoffels's symbols reflect the connectivity and curvature of the manifold and consequently they express the inertial forces experienced by a particle whose motion is constrained to the manifold. Of major relevance to describe this motion is the covariant derivative of a vector, defined as

$$\nabla_{a}V^{b} = \partial_{a}V^{b} + \Gamma^{b}_{ca}V^{c}$$

where Γ^{b}_{ca} is the Christoffel tensor, carrying information on the connectivity and curvature of the manifold.

Once the basic notions of differential calculus in manifolds are absorbed, the hydrodynamic equations of motion in curved spaces can obtained by simply replacing partial derivatives with the covariant ones, i.e.,

$$\partial_{a} \to \nabla_{a}$$

Such replacement is sometimes called *minimal coupling* between matter and spacetime, since ∇_{a} smoothly recovers ∂_{a} in the limit where the metric tensor g_{ab} reduces to the flat cartesian metric.

After some algebraic manipulations, the continuity reads as follows:

$$\partial_{t}\rho + \nabla_{a}(\rho u^{a}) = 0 \tag{35.11}$$

Likewise, the momentum equation is given by

$$\partial_{t}(\rho u^{a}) + \nabla_{b}P^{ab} = 0 \tag{35.12}$$

where the momentum-flux tensor takes the following form:

$$P^{ab} = Pg_{ab} + \rho u^{a}u^{b} - \mu(g^{cb}\nabla_{c}u^{a} + g^{ac}\nabla_{c}u^{b} + g^{ab}\nabla_{c}u^{c}) \tag{35.13}$$

μ being the dynamic viscosity. The previous expressions are "minimally coupled," i.e., they recover the standard Navier–Stokes equations in the limit of cartesian manifolds, $g^{ab} = \delta_{ab}$.

Even so, the complication is very substantial because *the covariant derivatives couple the metric nonlinearly and non-locally to the fluid and to the metric itself.*

In other words, the tangle between nonlinearity and non-locality associated with self-advection in the Navier–Stokes equations in flat spaces, here extends to matter as well, through the covariant derivatives. On the other hand, as we shall see in a moment, kinetic theory in curved space remains free from this tangle, potentially offering even more advantages than the corresponding cartesian version.

35.4.2   Kinetic Theory in Curved Manifolds

Similarly to hydrodynamics, kinetic theory can also be formulated directly in curved spacetime.

The Boltzmann equation on a curved manifold reads as follows (7):

$$\partial_t f + v^a \frac{\partial}{\partial x^a} f + \Gamma^a_{bc} p^b p^c \partial_{p^a} f = C(f,f) \tag{35.14}$$

where Latin indices run over the three spatial coordinates. The left-hand side reflects the geodesic equations of motion on a curved manifold, namely

$$\begin{cases} \dfrac{dx^a}{dt} = p^a/m, \\[2mm] \dfrac{dp^a}{dt} = \Gamma^a_{bc} p^b p^c \end{cases} \tag{35.15}$$

where $p^a = mv^a$ are the controvariant momenta.

In other words, the term $F^a = \Gamma^{bc}_a p^b p^c$ represents the inertial forces acting on the particle due to the curvature of the manifold.

Note that even though the inertial forces depend on both space and momentum coordinates, they can still be kept outside the momentum derivative because they are divergence-free in momentum space.

While the streaming probes the structure and topology of the curved manifold, as reflected by the inertial forces, the collisions are local, hence less affected by metric complications.

In particular, the collision term can be taken in the usual Marle form, with the following generalized equilibria:

$$f^{eq} = \frac{\rho \sqrt{g}}{(2\pi v_T^2)^{3/2}} e^{-\frac{(v^a - u^a) g_{ab} (v^b - v^A)}{2 v_T^2}} \tag{35.16}$$

where g is the determinant of the metric tensor. In other words, the manifold makes itself felt only through the metric tensor in the definition of the kinetic energy and the metric prefactor $\sqrt{g}$, both terms being local in spacetime.

35.4.3   Lattice Boltzmann in a Curved Manifold

Equation (35.16) can be given a suitable lattice representation and integrated according to a standard *cartesian*-LB scheme, based on the discretization of the *controvariant* coordinates v^a. It is important to appreciate that, by definition, the controvariant coordinates are free from metric factors, hence well suited to cartesian-looking formulations, such as Lattice Boltzmann. Another bonus of the controvariant formulation is to provide a body-fitted representation of non-cartesian boundaries which lie within the *logically* cartesian grid, with no need of staircasing, as in the standard LB scheme.

For instance, in cylindrical coordinates, with $x^1 = r$, $x^2 = \theta$, $x^3 = z$, the contravariant velocity components are simply $v^1 \equiv v^r = \dot{r}$, $v^2 \equiv v^\theta = \dot{\theta}$ and $v^3 \equiv v^z \equiv \dot{z}$. Hence they can be discretized in a uniform lattice in the (v^r, v^θ, v^z) *logical* space, even though the *physical* velocity space is non cartesian.

The LB in controvariant coordinates reads as follows:

$$f_i(x^a + v^b \Delta t, v^b, t + \Delta t) - f_i(x^a, v^b; t) = \frac{\Delta t}{\tau}(f_i^{eq}(x^a; t) - f_i(x^a; t)) + F_i \Delta t \qquad (35.17)$$

where the local equilibria are given by

$$
f_i^{eq} = w_i \rho \sqrt{g} \Bigg[\frac{5}{2} + 2u_i + \frac{c_i^a g_{ab} c_i^b}{2} - \frac{c_i^a c_i^a}{2} + \frac{u_i^2}{2} - \frac{g^{aa}}{2} - \frac{u^a u^a}{2} + \frac{u_i^3}{6} \\
+ \frac{1}{2}\left(u_i - c_i^a g_{ab} c_i^b - c_i^a c_i^a\right) - \frac{c_i^a u^a u^b u^b}{2} - \frac{u_i}{2}(g_{aa} - 3) - u^a g_{ab} u^b \Bigg]
$$

In the above, we have set $u_i \equiv c_i^a u^a$ and $u^a \equiv u^a/c_s$.

The inertial forces read as follows:

$$
F_i = w_i \rho \sqrt{g} \Bigg[(g^{ab} - \delta^{ab} + u^a u^b)\left(\frac{1}{2} c_i^a c_i^b c_i^c F_c - \frac{1}{2} c_i^c F^c \delta^{ab} - \frac{1}{2} c_i^a F^b - \frac{1}{2} c_i^b F^a \right) + c_i^a F^a \\
+ c_i^a u^a c_i^b F^b - u^b F^b \Bigg]
$$

where $F^a \equiv F^a/c_s$, again for notational simplicity.

Full details can be found in (8).

While the structure of the curvy LB remains formally cartesian, the metric and curvature clearly show up in the details of the equilibria and local inertial forces.

The "curvy" LB scheme has been validated for some representative classical fluids in non-cartesian geometries, such as the Couette flow between two rotating cylinders and also extended to the case of flow between rotating spheres, see Fig. 35.9.

Lately, it is has also been applied flow through *campylotic* media, i.e., media with randomly distributed curvature (9).

These developments are still embryonic, but they may open new perspectives for the simulation of relativistic and also non-relativistic flows in non-trivial geometries, as well as for fluid-structure simulations.

To be stressed again, the substantial advantage over the fluid dynamic representation is that nonlinearly coupled covariant derivatives disappear to be replaced by inertial forces and the metric tensor in local equilibria. This advantage, however, is just beginning to be explored and only time can tell whether and to what extent it can actually be realized in practical simulations, see Fig. 35.10.

This will depend essentially on the ability of mastering the inertial forces and the numerical instabilities which may arise once these forces become substantial at lattice scale.

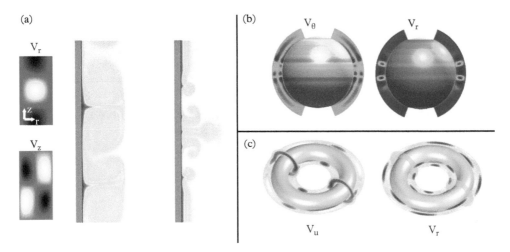

Figure 35.9 *Couette flow between rotating cylinders and spheres (right) and relativistic Taylor instability (left), as computed with the curvy LB. From (8).*

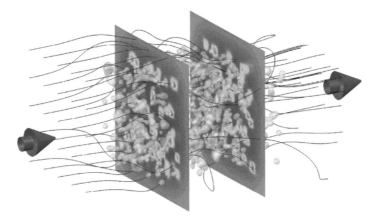

Figure 35.10 *Flow past a campylotic media, i.e., a media with localized random spots of curvature. From (9).*

35.5   Future Developments: Fluid-Structure Interactions

The curvy LBE presented so far deals with *static* manifolds which do not change in time and experience no reaction from the fluid moving along them. Besides general gravitation, a host of important fluid-structure problems do not conform to this approximation, as they require a dynamic geometry responding self-consistently to the fluid motion. In principle, this can be accommodated by extending the forcing methods discussed previously in this book to the curvy version of LB, i.e., by describing fluid motion in the manifold dictated by the dynamics of the structure. To the best of the author's knowledge, this important front is entirely unexplored.

35.6   General Relativity

Another major extension of the curvy LB are general relativistic flows, in which the motion of fluid matter is self consistently coupled to the Einstein equations of general gravitation (EEs). We remind that these read as follows:

$$R_{\mu\nu} - \frac{1}{2}Rg_{\mu\nu} + \Lambda g_{\mu\nu} = \frac{8\pi G}{c^4}T_{\mu\nu} \qquad (35.18)$$

where $R_{\mu\nu}$ is the Ricci curvature tensor, R the Ricci scalar curvature, Λ the cosmological constant, G the universal-gravitational constant and $T_{\mu\nu}$ is the matter tensor.[98]

The Einstein's equations are driven by the material fluid via the stress tensor, which acts as a source for the metric curvature.

The fluid, on the other hand, obeys the usual conservations equations:

$$\nabla_\nu T^{\mu\nu} = 0 \qquad (35.19)$$

whereby the metric field reacts back on the fluid motion through the covariant derivatives (minimal-coupling scenario).

In spite of its formal transparency, equations (35.18) and (35.19) form a system of highly complicated nonlinear tensor equations, which sets one of highest bars in computational physics altogether (10). In particular, given that the EEs are known to generate singular solutions, such as the prototypical black holes, even for relatively simple spacetimes, it is not hard to imagine the kind of complexity which may arise due to the coupling with the fluid motion.

To this regard, it is worth observing that the Einstein equations are highly nonlinear wave equations, which under suitable coordinate representations and appropriate kinematic constraints (10), can be cast in the form of nonlinear, inhomogeneous, hyperbolic equations. Thus, in principle at least, they are liable to an LB treatment. The prospective advantage, as usual is the disentanglement of nonlinearity and nonlocality associated with the uplift from four-dimensional spacetime to eight-dimensional momentum-energy-spacetime.

Just to provide an explicit example of the kind of difficulty met by the LB formulation of EEs, let us consider one the few precious metric spaces which admit exact solutions, namely Gowdy spacetimes (11).

---

[98] The cosmological constant term is often brought to the right-hand side and interpreted as the tensor contribution of vacuum fluctuations, i.e., $T_{\mu\nu}^{(V)} \equiv -\frac{\Lambda c^4}{8\pi G}g_{\mu\nu}$. This variant of the EEs takes the even more compact form, $G_{\mu\nu} = T_{\mu\nu} + T_{\mu\nu}^{(V)}$, where $G_{\mu\nu} \equiv R_{\mu\nu} - \frac{1}{2}Rg_{\mu\nu}$ is the so-called gravitational tensor.

Gowdy spacetimes are relevant to the propagation of gravitational waves in closed universes.

The Gowdy line element for a T^3 topology (three-dimensional torus) reads as follows:

$$ds^2 = e^{-\frac{\lambda}{2}} e^{-\frac{\tau}{2}} (e^{-2\tau} d\tau^2 + d\theta^2) + d\Sigma^2$$

where τ is the proper time, $0 < \theta < 2\pi$ is a cyclic coordinate and $d\Sigma$ is the surface element described by two additional cyclic coordinates and two parametric functions $P(\tau, \theta)$ and $Q(\tau, \theta)$.

Such functions obey the following set of one-dimensional nonlinear wave equations:

$$\begin{cases} \ddot{P} - e^{-2\tau} P'' = -e^{-2P}(\dot{P}^2 - e^{-2\tau} Q'^2) \\ \ddot{Q} - e^{-2\tau} Q'' = +2e^{P}(\dot{P}\dot{Q} - e^{-2\tau} P' Q') \end{cases} \tag{35.20}$$

where dot and prime mean derivative with respect to τ and θ, respectively. These equations must be supplemented with two independent first-order equations for the third parametric function $\lambda(\tau, \theta)$.

The left-hand side of (35.20) is a pair of linear wave equations with a time-asymptotically vanishing speed $e^{-\tau}$. The right-hand side, on the other hand, presents an exponential local nonlinearity in P and a quadratic non linearity in both space and time gradients. While an LB formulation of the nonlinear terms can arguably be found, the main problem for LB is the time-dependent speed, which commands the use of adaptive time stepping.

35.6.1   LB for Numerical Relativity

Notwithstanding the difficulties discussed in 35.6, an LB for numerical relativity has been very recently implemented (12). The implementation is based on the so-called Z4 formulation (13), which casts the EEs in the form of first-order hyperbolic, nonlinear flux equations, plus constraints, thereby leveraging the aforementioned analogy with fluid dynamics.

Leaving all details to the original publication, here we just mention that the Z4 formulation evolves the following set of fields:

$$\{\alpha, \beta, \gamma_{ab}, \theta_a, Z_a, K_{ab}\}$$

for a total of $1 + 1 + 6 + 3 + 3 + 6 = 20$. The first eight fields stem from the line element

$$ds^2 = \alpha dt^2 - \gamma_{ab}(dx^a + \beta^a dt)(dx^b + \beta^b dt)$$

while K_{ab} is the extrinsic curvature tensor.

The extra variable $Z^\mu \equiv (\theta, Z^a)$ is an auxiliary current, whose symmetrized gradient, $\nabla_\nu Z^\mu + \nabla_\mu Z^\nu$ acts as a dynamic constraint to the EEs, which are exactly satisfied whenever $Z_\mu = 0$.

The authors report successful completion of "apple-to-apple tests," such as analytical solutions of an expanding flat universe and other nonlinear wave propagation tests. However, they also point out mid-term numerical instabilities, which develop as a result of some discrete distributions going negative, possibly due to insurgence of strong metric forces.

Incidentally, the authors also report excellent parallel performance indicating that one of the major assets of the LB technique does indeed carry over to the context of numerical relativity.

The scheme is in its infancy and clearly necessitates major validation and development work before a serious assessment of its computational competitivity can be made.

35.6.2   Hybrid Scenarios for Numerical Relativity

An interesting remark by a leading figure in numerical relativity is that the toughest cookie of the Einstein-fluid system is often the fluid part (14).

Based on this remark, and leaving aside the inherent intellectual challenge of finding an LB formulation of EEs, from a practical viewpoint, it might prove more cost-effective to focus the LB technology on the fluid equations alone, leaving the Einstein equations to the highly advanced methods developed by the numerical relativity community.

Such a tantalizing "marriage" remains to be celebrated, though ....

35.6.3   Kinetic Theory of spacetime

Finally, before closing, we wish to point out that the most far-reaching import of a prospective LB for EEs might rest with its conceptual rather than computational value. Indeed, such LB might set a pointer toward the existence of an underlying kinetic theory beneath the Einstein equations, a literal kinetic theory of spacetime. This falls in the line of modern developments in classical and quantum gravity. Indeed, it has been shown that under pretty general conditions, the Einstein equations can be formulated in terms of equilibrium thermodynamics relations, such as the proportionality of entropy and heat $dS = T\delta Q$ (15). In this perspective, the Einstein equations can be regarded as an equation of state of spacetime. If the Einstein equations are the mathematical expression of the continuum thermodynamics of spacetime, then it is natural to speculate that an underlying kinetic theory of spacetime should also exist.

Along the same line, it is also natural to speculate that non-equilibrium kinetic fluctuations might bear a parallel with quantum fluctuations of spacetime at the Planck scale, on top of a smooth metric background obeying Einstein's equations. What form such hypothetical kinetic equations of spacetime should take is totally unknown to this author, but the magic of such prospect is hard to escape.

35.7   Summary

Summarizing, we have presented a formulation of the relativistic LB based on continuum relativistic kinetic theory. Such an extension permits us to handle relativistic flows with substantial Lorentz factors, of the order of ten, and also to provide a more accurate account of dissipative phenomena relevant to quantum condensed matter and cosmology. We have also shown how to extend the relativistic LB to the case of curved manifolds. Extensions of the present methodology to dynamic and self-consistently coupled manifolds may prove useful for the simulation of fluid-structure problems. In addition, they might be coupled to Einstein's equation solvers to produce self-consistent solutions of general gravitation problems in astrophysics and cosmology.

35.8 Appendix: Covariant and Controvariant Coordinates

Scalars, vectors and tensors have a well-defined geometrical meaning regardless of the systems of coordinates used to represent them. In fact, this coordinate invariance is the very reason why they are so useful!

Coordinates, on the other hand, may chosen in many different ways, some being decidedly more convenient than others, depending on the problem at hand. For instance, a problem with spherical symmetry is best handled in polar than cartesian coordinates. Since coordinates are not unique, it is important to assess the way that they transform in going from one representation to another. It turns out that this defines basically two families of coordinates: *Controvariant* and *Covariant*.

Given a basis set $\vec{e}_a$ a $=1,3$, any vector $\vec{V}$ in three-dimensional space can be expressed as follows:

$$\vec{V} = \sum_{a=1}^{3} V^a \vec{e}_a$$

The coefficients of this decomposition V^a are the controvariant coordinates.

Note that the basis set needs be neither orthogonal nor normalized to unity.

Given the vector $\vec{V}$ and the basis set, the controvariant coordinates are obtained as follows:

$$V^a = \frac{\vec{V} \times \vec{e}_b \cdot \vec{e}_c}{V_{abc}}$$

where $V_{abc} = \vec{e}_a \times \vec{e}_b \cdot \vec{e}_c$ is the volume of the parallelepiped define by the three basis vectors, clearly non-zero for a regular basis. Under a change of coordinates $x^a \rightarrow x'^a = f^a(x^b)$, the controvariant coordinates transform as follows:

$$dx'^a = \frac{\partial f^a}{\partial x^b} dx^b$$

The covariant coordinates, on the other hand, are defined by the projection of the generic vector $\vec{V}$ on the basis vectors:

$$V_a = \vec{V} \cdot \vec{e}_a$$

Based on their definition, the two sets of coordinates are related as follows:

$$V^a = \vec{e}_a \cdot \vec{e}_b V_b = g_{ab} V_b$$

where

$$g_{ab} = \vec{e}_a \cdot \vec{e}_b$$

is the covariant metric tensor. Its controvariant version g^{ab} is just its inverse, i.e., $g_{ac} g^{cb} = \delta_a^b$.

These notions carry on to the context of differential calculus on manifolds.

Let us consider a point $\vec{P}$ on a three-dimensional manifold $\mathcal{M}$ described by the generalized coordinates x^a, $a = 1, 3$. Next, let us consider a change in position from $\vec{P}$ to $\vec{P}' = \vec{P} + d\vec{P}$, we have

$$d\vec{P} = \frac{\partial \vec{P}}{\partial x^a} dx^a$$

The partial derivatives define a basis set in the manifold $\mathcal{M}$

$$\vec{e}_a = \frac{\partial \vec{P}}{\partial x^a}$$

so that the displacement writes

$$d\vec{P} = \vec{e}_a dx^a$$

In other words, dx^a are the controvariant coordinates of the displacement $d\vec{P}$. The gradient, on the other hand, is a covariant tensor.

By expressing $\vec{P}$ in cartesian notation as

$$\vec{P} = X\vec{e}_x + Y\vec{e}_y + Z\vec{e}_z$$

we obtain

$$\frac{\partial P}{\partial x^a} = \frac{\partial X}{\partial x^a}\vec{e}_x + \frac{\partial Y}{\partial x^a}\vec{e}_y + \frac{\partial Z}{\partial x^a}\vec{e}_z$$

The distance between $\vec{P}$ and $\vec{P}'$ is given by

$$ds^2 = d\vec{P} \cdot d\vec{P} = \vec{e}_a \cdot \vec{e}_b \, dx^a dx^b = g_{ab} dx^a dx^b$$

This scalar quantity is an invariant, same value in any system of reference, as it results from the contraction of the covariant tensor g_{ab} with the controvariant displacement tensor $dx^a dx^b$. This is a general rule: invariants are zeroth-order tensors obtained by contracting controvariant and covariant tensors of equal order. The simplest case is the first-order differential: $df = \frac{\partial f}{\partial x^a} dx^a$.

A few elementary examples may help the cause of clarity.

In cartesian coordinates, $\vec{e}_x = (1,0,0)$, $\vec{e}_y = (0,1,0)$, $\vec{e}_z = (0,0,1)$, the metric tensor is the unit matrix $g_{ab} = \delta_{ab}$.

For the case of polar coordinates (r, ϕ) in the plane:

$$\begin{cases} x = rcos\phi \\ y = rsin\phi \end{cases} \tag{35.21}$$

we have $\vec{e}_r = (\partial x/\partial r)\hat{i} + (\partial y/\partial r)\hat{j} = cos(\phi)\hat{i} + sin(\phi)\hat{j}$, and $\vec{e}_\phi = (\partial x/\partial \phi)\hat{i} + (\partial y/\partial \phi)\hat{j} = -rsin(\phi)\hat{i} + rcos(\phi)\hat{j}$. Hence $\vec{e}_r$ is the unit vector pointing radially outward, while $\vec{e}_\phi$ is orthogonal to it, with magnitude r.

The metric tensor is therefore $g_{rr} = 1, g_{r\phi} = g_{\phi r} = 0, g_{\phi\phi} = r^2$.

This means that the effect of a change $d\phi$ on the distance ds^2 depends on the spatial location of the basis: the farther from the origin the larger the effect on the distance. This space dependence of the basis function offers great geometrical flexibility, but it also represents a source of mathematical complexity.

....

REFERENCES

1. M. Mendoza, I. Karlin, S. Succi and H.J. Herrmann, "Relativistic Lattice Boltzmann with Improved Dissipation", *Phys. Rev. D*, 87, 065027, 2013.
2. F. Mohseni, M. Miller, S. Succi, and H.J. Herrmann, "Relativistic lattice Boltzmann model for ultra-relativistic flows", *Phys. Rev. D*, 87, 083003 2013.
3. C. Cercignani and G.M. Kremer, *The relativistic Boltzmann Equation: Theory and Applications*, Birkhause, Basel, 2002.
4. A. Gabbana, M. Mendoza, S. Succi and R. *Tripiccione*, Kinetic approach to relativistic dissipation, *Phys. Rev. E*, 96, 023305 (2017).
5. P. Romatsche, M. Mendoza and S. Succi, "A fully relativistic lattice Boltzmann algorithm", *Phys. Rev. C*, 84, 4903, 2011.
6. R.L. Bishop, S.I. Goldberg, *Tensor analysis on manifolds*, Dover unabridged, Mac Millan 1968.
7. R. Liboff, *Kinetic Theory: Classical, Quantum and Relativistic descriptions*, Wiley and sons, New York, 2002.
8. M. Mendoza, J.D. Debus, S. Succi, and H.J. Herrmann, "Lattice kinetic scheme for generalized coordinates and curved spaces", *Int. J. Mod. Phys. C*, 25, 1441001, 2014.
9. M. Mendoza, S. Succi, and H.J. Herrmann, "Flow through randomly curved manifolds", *Scientific Reports*, 3, 3106, 2013.

10. T.W. Baumgarte and S.L. *Shapiro Numerical Relativity: solving Einstein's equations on the computer*, Cambridge University Press, Cambridge, 2010.

11. R. Gowdy, Gravitational waves in closed universes, *Phys. Rev. Lett.*, **27**, 826 1971.

12. T. Ilseven and M. Mendoza, "Lattice Boltzmann model for numerical relativity", *Phys. Rev. E*, **93**, 023303, 2016.

13. C. Bona, J. Masso, E. Seidel and J. Stela, New formalism for numerical relativity, *Phys. Rev. Lett.*, **75**, 600 1995.

14. F. Pretorius, private communication.

15. T. Jacobson, Thermodynamics of spacetime: the Einstein equation of state *Phys. Rev. Lett.*, **75**, 1260 1995.

EXERCISES

1. Compute the metric tensor in spherical coordinates.
2. Compute the Christoffels symbol in polar coordinates.
3. Show that the inertial forces have zero divergence in momentum space.

36

Coda

I am conscious of being only an individual struggling weakly against the stream of time. But it still remains in my power to contribute in such a way that, when the theory of gases is again revived, not too much will have to be rediscovered.

L. Boltzmann

36.1 The Question

In the previous book, "The Lattice Boltzmann Equation (for fluid dynamics and beyond)", I closed the book by raising the (kind of jesty) question: "who needs LBE"?.

Fifteen years later and close to thirty down the full line, the same question stands again, possibly in less jesty vests, namely: _What did we learn through lattice Boltzmann? Did LB make a real difference in our understanding of the physics of fluids and flowing matter in general?_ As usual, views can be widely disparate and, more often than not, quite opinionated as well . . . . Here, I wish to offer my own (biased) view, without any presumption of being right just because of a long duty on the camp.

36.1.1 Biblio-Numbers

To the risk of staring shallowness, I shall begin by playing a plain bibliographic argument. Indeed, if biblio-numbers are anything to go by, scientific databases send off a neat message: several thousands papers, some hundred and-fifty thousand cumulative citations, with an h index above 150, as of March 2017.

Of course, it would be foolish to claim that biblio-numbers carry the full story, for progress in knowledge and understanding surely does not scale with the number of papers, sometimes not even _monotonically_ with it! Yet, when papers score in the many thousands, it is not unreasonable to assume that something beyond "run of the mill" might have happened.

Having acknowledged the quantitative impact, the next question is: what fueled this growth in the first place?

Here, I believe that much of the answer is captured by a witty sentence that goes to Tony Ladd's credit, namely besides being conceptually simple, : _LB cannot be too wrong._

The Lattice Boltzmann Equation. Sauro Succi, Oxford University Press (2018).
© Sauro Succi. DOI: 10.1093/oso/9780199592357.001.0001

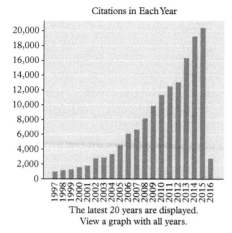

Citations in Each Year

The latest 20 years are displayed.
View a graph with all years.

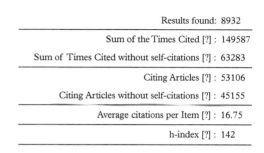

Results found: 8932

Sum of the Times Cited [?] : 149587

Sum of Times Cited without self-citations [?] : 63283

Citing Articles [?] : 53106

Citing Articles without self-citations [?] : 45155

Average citations per Item [?] : 16.75

h-index [?] : 142

Figure 36.1 *The ISI citation record of LB, as of March 2016. This was obtained by typing "Lattice Boltzmann," which delivers eventually some spurious extra-contributions. However, since it also neglects others, I have assumed that the two effects balance somehow. I was unable to update the data because, as of February 2017, LB counts more than 10,000 papers, a threshold above which ISI no longer delivers the Citation Report feature providing the picture above . . . .*

36.1.2   LB Cannot Be Too Wrong

This statement fully resonates with my own perception of the subject. LB can compute across many scales of motion, from fully developed turbulence around an airplane, all the way down to biopolymer translocation through biological membranes, and lately further down to quark-gluon plasmas. At any of these scales, one can possibly find specific (specialized) methods which may eventually compute more accurately than LB.

Yet, the information provided by LB usually comes by *handier and faster*; maybe not perfect, granted, but rarely "too wrong," in Ladd's sense.

For those, like this author, who feel like computer simulation is more for insight than for hard numbers, this is certainly a major plus. Of course, this is not to downplay the role of quantitative benchmarking, which is a most indispensable activity to stave off "computer hallucinations." But the major contribution of LB, in my opinion rests mostly with its broadband flexibility across virtually all walks of the physics of fluids and its interfaces with its (many) allied disciplines.

This speaks for breadth, but how about depth?.

36.1.3   Breakthroughs

Besides routinely use for a broad spectrum of complex flow problems, there are, in this author's opinion, a few precious instances in which LB has made a palpable difference. Maybe not in terms of major discoveries (no Higgs bosons or gravitational waves here ...), but in terms of disclosing insights that would have been much less accessible on count of computer resources, human time, and often both.

At the *macroscale*, fully scalable, multiscale hemodynamics on millions of cores in realistically complex geometries is definitely a case in point. On a different context, the emergence of LB-based commercial software for aerodynamic design should also be regarded as major achievement, as any reader with some experience of the fierce competition in the field of industrial computational fluid dynamics, may well attest.

As discussed mostly in the "Beyond Navier-Stokes" part, LB has spawned a number of precious insights on the behavior of complex multiphase and multicomponent *microscale* flows, sometimes unveiling new basic phenomena or even novel materials.

Finally, LB simulations are providing a cost-effective tool to gain new insights into the physics of *nanoscale* fluids with suspended bodies. In particular, these simulations are helping to unravel the subtle interplay between geometrical confinement and hydrodynamic correlations and their effects on basic biological phenomena, such as biopolymer translocation or protein folding and aggregation.

None of the above comes for free: if it did, it would not be credible.

Instead, hard labor has to be invested to deal with complex boundaries, watch lattice artifacts at near-grid scales, protect numerical stability, secure consistency of particle-LB couplings, and so on and so forth.

But the claim is that, once this toil is all said and done, the result is outstanding and sometimes even possibly unique.

36.1.4   Weakly Broken Universality and Chimaera Simulation

No less importantly, LB is playing a precious role as a gap-bridger between the micro and macroscopic description of complex states of flowing matter.

Not that it would replace Molecular Dynamics or Direct Simulation Monte Carlo, which it surely cannot, but in the sense of providing a *computational chimaera*, capable of capturing Weakly Broken Universality (WBU), namely complex states of flowing matter which challenge the continuum description and yet do not demand full Boltzmann or atomistic detail (1). Spotting WBU instances at the frontiers between fluids and allied disciplines is, in this author's opinion, a significant contribution to the advancement of the, *physics of flowing matter across scales*, from both conceptual and computational viewpoints see Fig. 36.2.

36.2   Boltzmann and Computers

In a most delightful paper on the occasion of Boltzmann's 150th, Christoph Dellago and Harald Posch ask an intriguing question:

What would have Boltzmann done with a computer?

The authors argue, quite convincingly in my opinion, that Boltzmann would have most certainly appreciated the value of computers as "tools of discovery," especially for statistical mechanics.

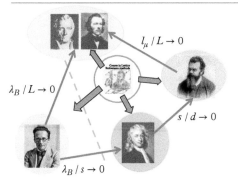

LB: Chimaera simulation

$l_\mu / L \to 0$

$\lambda_B / L \to 0$

$s / d \to 0$

$\lambda_B / s \to 0$

Figure 36.2 *A pictorial view of the LB-Chimaera strategy. By suitable "mutations," the LB can approach Boltzmann, Newton and even the quantum level (Schrödinger). The upward route to Navier–Stokes is the standard and can be taken to the very end, i.e., LB can be made "equal" to Navier–Stokes. However, the path to Boltzmann and Newton cannot be completed in full, i.e., LB cannot be made equal to Boltzmann or Newton. Yet, it can serve as a useful "proxy," whenever the Boltzmann or Newton descriptions are not needed in full, which is precisely the WBU ground.*

(c) Bernhard Reischl

Figure 36.3 *Boltzmann at computational work.   Courtesy of Bernhard Reischl.*

Consequently, the authors go on, he would have invented molecular dynamic to test the basic assumptions behind his equation.

In the same speculative vein, and with an unashamed pinch of wishful thinking, I tend to imagine that he would have invented Lattice Boltzmann as well, see Fig. 36.3 Moreover, given his well-documented appreciation for the practical outcomes of scientific theories, I also dare to imagine that he would have taken a lot of fun and scientific profit at it!

The coda has changed, but not its essence. No differently than in 2001, I feel like closing this second book with the same tribute of gratitude to Ludwig Boltzmann. Now, as ever, it remains true that not a single bit of this subject would have ever come into existence without his work.

Thanks Ludwig, wherever you are.

This closes this long book; for good or for worst, the time has come to lower the curtains. To say it with Mark Knopfler

And it's what it is, it's what it is now.

· ·

REFERENCES

1. S. Succi, Chimera simulation for complex states of flowing matter, Special Issue on "Multiscale Modeling at the Physics-Chemistry-Biology interface," *Phil. Trans. Roy. Soc.*, 374, 20160226, (2016).
2. C. Dellago and H. Posch, "Realizing Boltzmann's dream: computer simulations in modern statistical mechanics, Boltzmann's legacy," G. Gallavotti, W. Reiter and J. Yngvason eds, *ESI lectures in Mathematics and Physics* (2010).

37

Notation

Quantitative computations in kinetic theory develop quickly into a thick jungle of tensors. Compact notation is a key survival tool through this jungle, whence its frequent use throughout this book. To orient the reader, hereafter we provide a few notational examples which should hopefully guide the navigation through the math-denser sections of this book.

37.1 Vectors and Tensors

Vectors shall be denoted either through standard notation, $\vec{V}$, or via the component notation V_a, with Latin indices labeling the spatial dimensions, $a = x, y, z$. Either ways, in three-dimensional cartesian coordinates, both notations mean the following:

$$\vec{V} \equiv V_a = (V_x, V_y, V_z) \tag{37.1}$$

Similarly, second-order tensors shall be denoted either by $\overset{\leftrightarrow}{T}$ or T_{ab}.
In explicit matrix notation,

$$\overset{\leftrightarrow}{T} \equiv T_{ab} = \begin{bmatrix} T_{xx} & T_{xy} & T_{xz} \\ T_{yx} & T_{yy} & T_{yz} \\ T_{zx} & T_{zy} & T_{zz} \end{bmatrix} \tag{37.2}$$

Moving on to triple-order tensors already reveals the advantage of the component notation, which carries seamlessy to T_{abc}, while the matrix notation would necessitate a rather cumbersome triple matrix.

Higher dimensions dramatically add to the point.

In standard lattice Boltzmann theory, for quasi-incompressible, a-thermal flows, tensors up to order four are required, which sets a compelling case for the coordinate notation. Thermal flows, and recent developments in LB theory beyond hydrodynamics

The Lattice Boltzmann Equation. Sauro Succi, Oxford University Press (2018).
© Sauro Succi. DOI: 10.1093/oso/9780199592357.001.0001

require tensors up to order six or more, in which case the survival strategy mentioned above is an absolute must.

37.2 Scalar Products and Tensor Contractions

As is well known, vectors and tensors can be contracted to form lower-order tensors, down to order zero, i.e., *scalars*. A typical contraction is the scalar product of two vectors

$$s = \vec{V} \cdot \vec{W} = \sum_a V_a W_a = V_x W_x + V_y W_y + V_z W_z \tag{37.3}$$

The very same token applies to the scalar product of tensors, for the case of rank two:

$$s = \vec{\vec{T}} : \vec{\vec{U}} \equiv \sum_{ab} T_{ab} U_{ab} \tag{37.4}$$

Explicitly

$$s = T_{xx}U_{xx} + T_{xy}U_{xy} + T_{xz}U_{xz} + T_{yx}U_y + T_{yy}U_{yy} + T_{yz}U_{yz} + T_{zx}U_{zy} + T_{zy}U_{zy} + T_{zz}U_{zz} \tag{37.5}$$

Unless explicitly indicated otherwise, the symbol $\sum$ is omitted, according to Einstein's notation (general relativity sets another pressing demand for survival across tensorial jungles ...).

According to this notation, one writes simply $s = T_{ab}U_{ab}$.

37.3 Gradients, Divergence, Curl and All That

A similar reasoning applies to differential operators.

Gradient

$$\vec{\nabla} \equiv \partial_a = (\partial_x, \partial_y, \partial_z) \tag{37.6}$$

where we have used the common shortand

$$\partial_a \equiv \frac{\partial}{\partial x_a} \tag{37.7}$$

Divergence of a vector (a scalar):

$$s = \vec{\nabla} \cdot \vec{V} \equiv \partial_x V_x + \partial_y V_y + \partial_z V_z \tag{37.8}$$

In Einstein's notation, simply $s = \partial_a V_a$.

Divergence of a second-order tensor (a vector):

$$\vec{V} = \vec{\nabla} \cdot \vec{\vec{T}} \tag{37.9}$$

In explicit component notation,

$$V_x = \partial_x T_{xx} + \partial_y T_{xy} + \partial_z T_{xz}$$

$$V_y = \partial_x T_{yx} + \partial_y T_{yy} + \partial_z T_{yz}$$

$$V_z = \partial_x T_{zx} + \partial_y T_{zy} + \partial_z T_{zz}$$

In Einstein's notation, $V_a = \partial_b T_{ab}$.

Sometimes we shall need the double divergence of a second-order tensor, namely the scalar:

$$s = \vec{\nabla}\vec{\nabla} : \vec{\vec{T}} \equiv \sum_a \sum_b \partial_a \partial_b T_{ab} \tag{37.10}$$

In explicit notation,

$$s = \partial_{xx} T_{xx} + \partial_{xy} T_{xy} + \partial_{xz} T_{xz} +$$

$$\partial_{yx} T_{yx} + \partial_{yy} T_{yy} + \partial_{yz} T_{yz} +$$

$$\partial_{zx} T_{zx} + \partial_{zy} T_{zy} + \partial_{zz} T_{zz}$$

In Einstein's notation, $s = \partial_a \partial_b T_{ab}$.

Finally, for completeness, we mention the curl of a vector

$$\vec{\omega} = \vec{\nabla} \times \vec{V} \tag{37.11}$$

In explicit component notation,

$$\omega_x = \partial_y V_z - \partial_z V_y$$

$$\omega_y = \partial_x V_z - \partial_z V_x$$

$$\omega_z = \partial_x V_y - \partial_y V_x$$

In compact component notation,

$$\omega_a = \epsilon_{abc} \partial_b V_c \tag{37.12}$$

where ϵ_{abc} is the Levi–Civita tensor taking value one for even permutations of $a \neq b \neq c$, −1 for odd ones and zero otherwise.

....

REFERENCE

1. R. Aris, *Vectors, Tensors and the Basic Equations of Fluid Mechanics*, Dover unab., New York, 1989.

Appendices

They promised me Mars, instead I have got Facebook.

(B. Aldrin)

A.1 The Pseudospectral Method

The pseudospectral method is based on the idea of solving the Navier–Stokes equations by means of Fourier transforms in reciprocal (wavevector) space. For the sake of simplicity we shall refer to the two-dimensional case, where fluid equations are most conveniently recast in the *vorticity–streamfunction* representation. Vorticity has already been introduced as the curl of the velocity field, and in two dimensions it reduces to the z component ω, a pseudo-scalar. The streamfunction ψ is defined by the condition that surfaces $\psi =$ constant define lines tangents to the velocity field u, v:

$$\partial_y \psi = u \tag{A.1}$$
$$\partial_x \psi = -v \tag{A.2}$$

From the definition, the streamfunction follows from vorticity via a static Poisson equation $\Delta \psi = \omega$.

The pseudospectral method is based on a discrete Fourier representation of both vorticity and streamfunction:

$$\omega(x, y, t) = \sum_{m,n} \omega_{mn}(t)\, e^{-i\left(k_{x,m}x + k_{y,n}y\right)} \tag{A.3}$$

$$\psi(x, y, t) = \sum_{m,n} \psi_{mn}(t)\, e^{-i\left(k_{x,m}x + k_{y,n}y\right)} \tag{A.4}$$

$$k_{x,m} = \frac{2\pi m}{L}, \quad k_{y,n} = \frac{2\pi n}{L}, \quad m, n = 0, 1, \ldots, N-1 \tag{A.5}$$

The vorticity equation reads

$$\partial_t \omega + \mathcal{J}(\omega, \psi) = \nu \Delta \omega \tag{A.6}$$

where $\mathcal{J}(f, g)$ is the Jacobian operator $\mathcal{J}(f, g) \equiv \partial_x f \partial_y g - \partial_x g \partial_y f$.

Upon Fourier transforming (A.6), one obtains

$$\partial_t \omega_{mn} + \text{FT}\left[\mathcal{J}\left(\omega, \psi\right)\right] = -\nu k^2 \omega_{mn} \tag{A.7}$$

where the symbol FT stands for Fourier Transform. Equation (A.7) is very appealing because all space derivatives are turned into algebraic operators according to $\partial_a = -ik_a$, $a = x, y$. However, the FT of the Jacobian involves a lengthy convolution between the vorticity and streamfunction transforms. Such a convolution would require $O(N^2)$ operations, thus considerably taxing the computational efficiency.

The pseudospectral method of Orszag–Patterson turns around this problem by first computing the Jacobian products in real space and only once these are available, taking the FT of these products. This cuts down complexity to $O(N \log_2 N)$.

In the process spurious modes, known as aliases, are introduced. They correspond to wavevectors $\vec{k}'$, $\vec{k}''$ outside the resolved range, which couple back to the resolved spectrum via nonlinear beatings $\vec{k}' \pm \vec{k}''$. These aliases are usually removed by cutting away all modes with magnitude $k > k_N/\sqrt{2}$, although more sophisticated and accurate procedures are available too.

The pseudospectral method is very accurate and very fast for periodic flows, the latter property being related to the availability of the celebrated Fast Fourier Transform (FFT) algorithm, which computes the Fourier transform, in principle an N^2 operation, in $N \log_2 N$ steps.

A.2   Computer Performance

Computer performance for scientific applications is commonly expressed in *Flops/s*, i.e., Floating point operations per second.

This is basically the time it takes to multiply two (32 or 64 digits) floating-point numbers. For nearly five decades such processing speed has increased at the amazing pace of roughly a factor two every year and a half, a scaling regime famously known as "Moore's law," after Intel's co-founder Gordon Moore (1929), a pioneer of the modern computer era, who predicted such astounding rise back in the mid 70s.

Such a continued exponential growth has little precedent in the technological/industrial context: a factor two every 1.5 years means a factor $2^{10} \sim 10^3$ every 15 and a factor $2^{30} \sim 10^9$ every 45. That's basically where we stand today, from the Megaflops (a million Flops/s) of the mid 70s, to the tens of Petaflops (one Petaflops/s = one billion millions Flops/s) of most powerful current-day supercomputers. Such a breathtaking ride (at a similar pace, today's cars would move faster than light!) has been fueled by the relentless miniaturization of electronic chips and increasing speed of the associated circuitry.

However cranking the clock has come to its limits, not because of fundamental restrictions in further reducing the switching time (the time it takes to flip a bit on and off) below the nanosecond or so, but because of more practical, yet fundamental, issues of *heat consumption*.

Electrons in modern silicon travel close to the speed of light, hence they move some 30 cm in a nanosecond (corresponding to 1 GHz clock time). This is still a long distance in microlectronics units and therefore ballistic delays are not a concern at such clock rates. But heat consumption is.

Faster commuting times demand higher voltage and heat-consumption scales quadratically with the applied voltage. As a result, cooling down the circuitry in the increasingly narrower spaces dictated by faster clock time presents the electronic industry with an extremely daunting thermal barrier. That is why computer manufacturers have turned to a complementary/alternative strategy: concurrent processing, also known as *parallel computing*, which we shall briefly discuss in the following.

A.2.1   Flops, MLUPS and Bytes per Flop

Before turning to parallel computing, let us note that in the recent years the LB community seems to have embraced its own performance metric, namely *MLUPS*, standing for *Million Lattice Updates Per Second*.

What does this mean? If your LB code updates a million sites' lattice in one CPU second, then it performs at one MLUPS, this is what it means.

The typical performance of a plain LB code is around ten MLUPS, give or take. This figure is less mysterious than it seems.

In Flops units, a standard LB code takes about 200 Flops/site and consequently, on a computer delivering one Gflop/s (your laptop is not far from this), this LB code would update a million lattice sites about five times in a second, i.e., five MLUPS. So, all is well between Flops and MLUPS.

We hasten to add that in modern times, the limiting factor is less and less the processing speed itself but rather the cost of accessing data. Take the streaming step of the LB scheme: its performance in MLUPS is virtually infinite, since streaming involves no floating-point operation at all, but only memory shifts.

The latter, though, definitely take time too, and quite appropriately, MLUPS acknowledge this simple yet important fact. Indeed LB streaming has been measured to feature about 30 "equivalent MLUPS," which is better than the ten MLUP mentioned above, but definitely not infinity! The cost of accessing memory must be watched carefully in LB applications, also because it is going to be more and more demanding as the size of the problems to be solved is increased.

A critical figure to be watched from close is BPF, the Bytes to be accessed per Flop to be executed. A plain LB code takes about 200 Flops/site and requires about 20 arrays, i.e., 80 Bytes/site in single precision and 160 in double precision. Hence, on a single-site basis, LB features a BPF of the order of 0.5 Byte/Flop. This is smaller than the BPF of explicit finite-difference schemes, which typically feature about a one Flop/variable, hence a BPF of the order of four to eight Bytes/Flops.

However, things are a little bit more articulated than that.

The cost of accessing memory is very sensitive to the specific type of memory where this data must be accessed from.

Fast memories, also known as *caches* are designed to keep up with the CPU, i.e., sustain a BPF of order unity. This makes caches expensive, hence limited in size, typically tens of Megabytes.

Large applications demand way more memory, and keeping all data in-cache is literally impossible, whence the data access issue.

In this respect, it should be kept in mind that *computer memory is one dimensional,* which means that the element, say, $f(i, j, k + 1)$ of a 3d array of size N^3 lies $4 \times N^2$ bytes away from the physically contiguous element $f(i, j, k)$. This is the Fortran convention, the innermost index runs first, but any other convention would face the same issue, only on a different index.

Such 1D memory distance, known as *stride* in computer jargon, is thus seen to score in the tens of Mbytes for large-size problems, say $N \sim 1000$.

LB is particularly exposed to the stride issue because, with about 20 arrays stored at each lattice site, the stride between the element (i, j, k) of the first array and the element $(i, j, k + 1)$ of the last one is twenty times larger, i.e., $20 \times 4N^2$ bytes.

With $N = 10^3$ this gives 80 Mbytes, well above the cache size.

These simple considerations explain why optimal memory access is one of the most critical issues in optimizing the performance of large-scale LB codes.

A.3　A Primer on Parallel Computing

As discussed above, parallel computing is a powerful antidote against the thermal barrier quenching further decrease of the clock rates of single processors.

This explains why parallel computing plays a central role in the landscape of high-performance computing. The idea is over two thousand years old, as it endorses the time-honored Roman principle *divide et impera* (divide and conquer). To solve a huge problem, break it down in smaller ones, solve them concurrently and then glue the partial solutions back together, to recompose the global solution.

Ideally, upon using P computers each solving a $1/P$-th of the original problem, time-to-solution can be cut down by a factor P, and sometimes even more (superlinear speed-up).

To make this blue-sky scenario work for real, a series of conceptual and practical hurdles must be overcome.

The primary issues are the following ones:

- *Fraction of Parallel Content*
- *Communication to Computation Ratio*
- *Load Balance*

A.3.1　Fraction of parallel content

It is a fact of life that not all sections of a given algorithm are necessarily amenable to parallelization. Take the simple problem of summing N sumbers using $P = N/K$ processors where K is an integer number.

Each processor can compute its share of K numbers, happily unaware of all others. However, in the end, all partial sums computed by each processor in isolation, must be summed together to produce the final result. Summing the partial sums is a task for just one processor, two being already one too many.

This is a so-called "serial bottleneck."

Such bottlenecks may seriously deteriorate the parallel performance.

In fact, it is simple matter to realize that by letting f the fraction of work that can be performed in parallel, the time spent using P processors is

$$T(P) = T(1)\left(\frac{f}{P} + (1-f)\right) \tag{A.8}$$

Consequently, the benefit of using P computers instead of just one (parallel speed-up) is given by

$$S(P) \equiv \frac{T(1)}{T(P)} = \frac{1}{f/P + (1-f)} \tag{A.9}$$

This simple expression shows that, even in the limit of an infinite number of processors $P \to \infty$, the asymptotic speed-up flat-tops at

$$S(\infty) = 1/(1-f)$$

Unfortunately, this is rather flat function of f except in the close vicinity of $f = 1$.

For instance, with only a lowly ten percent of the serial code, this expression shows that no matter how many computers we can afford, the maximum speed-up cannot exceed a factor ten.

Therefore ten processors is the threshold above which parallel computing becomes a mere waste.

These simple and yet far-reaching considerations are known as Amdahl's law, after Gene Amdahl (1922–2015) a major pioneer of computer architecture design.

LBE is well-placed *vis-à-vis* Amdahl's law because most of the LB work comes from the collision step, which is fully local. This means that there is often scope for hundreds of processors and more. In fact, as discussed in the main text, highly specialized parallel implementations of LB hemodynamics have reached parallel speed-up of the order of millions! However, achieving such performance requires highly professional strategies to curb the cost of memory access and communication, as we are going to detail in the sequel.

A.3.2   Communication to computation

The next constraint is communicativity.

As the simple example of summing N numbers already shows, at some point the independent action of the P processors needs be coordinated to glue together a sensible answer.

In the real world, this takes time—the time to communicate results, either among other processors or to a supervisor, a glorious word for a slower machine doing nothing but collecting other processors' work (aptly named "slaves") and perform the comparatively lightweight duty of summing the partial sums.

In more democratic environments the supervisor may well be sharing the slaves' task as well.

Whatever the arrangement, the fact remains that communication should be kept to a minimum because it constitutes an unproductive overhead (the cell-phone syndrome, well captured by the slogan "while talkers do the talking, builders do the building" ...).

Such an overhead is accrued by technological issues: interprocessor communication is usually significantly slower than intra-processor calculations, easily a factor ten.

The figure of merit here is the so-called "Communication to Computation Ratio," (CCR) for short. CCR can be measured in Bytes transmitted off-processor over in-processor executed Floating Point Operations (Flops), i.e., Bytes/Flops. Numerical schemes with low CCRs, well below one, are dear to parallel computers.

How does LB do versus CCR? As discussed earlier on, on a single-site basis LB scores rather well, below one BPF, one Byte per Flop. To go from BPF to CCR, one needs to know the amount of data exchanged by a given processor (*core* in modern computers) versus the amount of computational work executed within the processor (core). This is a Surface/Volume factor, whose specific value is highly dependent on the geometry and the specific implementation of the communication network.

For an ideal case of a cubic domain with n^3 lattice sites assigned to a single processor, the surface (communication) to volume (computation) ratio is $6/n$, yielding a CCR of the order of $0.5 \times 6/n = 3/n$ Bytes/Flop.

With $n \sim 30$, this gives a fairly reasonable CCR, around 0.1.

Of course, extreme parallel implementations involving several hundred thousands, if not millions of processors, may well require much smaller CCRs.

Without entering (crucial) details, we just wish to mention that under these extreme conditions, the only way to offset the communication costs is to "hide" them, by overlapping communication tasks with computational ones. In other words, while waiting for the data required from other processors (cores), each processor (core) strives to perform work with no causal dependencies, i.e. which can be done without waiting for the above data.

Even without delving into any detail, the reader may well appreciate that such overlapping between communication and computation requires highly sophisticated data-dependency analysis and parallel programming techniques, whose description goes way beyond the scope of this Appendix.

Before quitting this section, it is perhaps of some use to pause on how far parallel computing appears to stand from the "social-network" attitude presenting "unlimited communication" as the ultimate good for mankind. To the point of maintaining, with Mark Zuckerberg, that *connectivity is a human right*!

I can't help quoting Bill Gates sober comment: *When I hear that connectivity is the most important issue to build a better world, I believe they are making jokes of me and I wonder: have you ever been in a poor country?*

To say it more succinctly with the astronaut Aldrin: "They promised me Mars, instead I have got Facebook."

A.3.3   Load Balancing

Finally, we come to the issue of load balancing.

Assuming a homogeneous parallel computer, namely all processors work at the same speed, it is clear that to avoid wasteful delays, they should all get the same amount of work, on pain of being pace-made by the slowest one. And for a heterogeneous one, each processor should receive work in proportion to its processing power. The pain is that parallel performance is dictated by the slowest one.

Load balancing is a rather trivial matter with regular geometries, where a simple geometric domain decomposition assures balanced performance. It can become fairly tricky when the geometry is complex and possibly even changing in time (think of piston-valve motion in engine combustion, or deformable bodies within fluid flows). Again, LBE with its simple data structure is well positioned to steer clear of load balancing issues for many practical applications. However, if geometry is complicated, no free lunch, professional load balancing strategies musty be deployed.

A.3.4   Again on Thermal Barriers

The most powerful supercomputer, as of November 2016, was the Chinese Taihulight. In order to deliver its peak performance of 93 Petaflops, Taihulight demands about 15 MW corresponding to a fairly remarkable six Gflop/Watt and, incidentally, at a cost of roughly 300 Mflops/US dollar.

Albeit technologically remarkable, the prospects for further growth opened by these figures are thermally daunting: upscaling to the grand-challenge of one Exaflop computer (a billion billions Flops/s) speaks of tens of Gigawatts, more than a nuclear plant!

With a touch of bitter irony, many have noted the humbling contrast with the human brain, which just demands a modest 25 Watts (after all, the brain is not made of silicon)....

It really sounds like computer industry is at a turning point.

A.4   From Lattice Units to Physical Units

In this book, we have repeatedly emphasized the fact that LBE is most often directed to "synthetic" matter studies, meaning by this that the focus is more on dimensionless numbers than on the actual values of physical properties. This is practical for theoretical studies, but it may easily lead to embarrassing mumbles in response to the request of communicating results in physical units, the good, old centimeters, seconds and grams!

In the following, we remind the basic elementary steps which permit conversion of the result of LBE simulations in actual physical units. For the sake of convenience, we shall refer to the Centimeter-Gram-Second (CGS) system.

A.4.1 Space

Most LBE simulations assume the lattice spacing Δx as the space unit simply: Therefore, upon using N grid points to represent, say, a linear length of L centimeters, the resulting length unit is simply:

$$S_l \equiv \Delta x = \frac{L}{N} \quad \text{(cm)} \tag{A.10}$$

In equivalent terms,

$$\Delta x = \Delta x_{lb} S_l$$

where $\Delta x_{lb} = 1$ by definition, since this is the lattice spacing in LB units.

This shows that, at variance with most numerical methods, the units are not fixed by a physical quantity, but by the chosen resolution. This may cause some headache before one gets used to it, but it is a perfectly well-defined convention.

A.4.2 Time

In a similar way, the time unit of LBE simulations is fixed to the lattice time step.

Its physical value can be defined via the sound speed as follows $C_s = c_s \Delta x / \Delta t$, where the capital denotes the physical value.

The result is

$$S_t \equiv \Delta t = \frac{c_s}{C_s} \Delta x \quad \text{(seconds)} \tag{A.11}$$

Like for space, we have

$$\Delta t = \Delta t_{lb} S_t$$

where $\Delta t_{lb} = 1$ by definition, since this is the lattice timestep in LB units.

For Navier–Stokes fluids, it is often more expedient to fix the lattice spacing based on the fluid viscosity, through the expression $\nu = \nu_{lb} \frac{\Delta x^2}{\Delta t}$, where ν is the kinematic viscosity in physical units and ν_{lb} is the lattice one.

This delivers

$$S_t \equiv \Delta t = \frac{\nu_{lb}}{\nu} \Delta x^2 \tag{A.12}$$

Note that upon keeping the ratio of the physical to LB viscosity at a fixed value, the time-unit scales quadratically with the length unit.

A.4.3 Mass

Since lattice particles are typically given a unit mass, the mass unit is

$$S_m = m \quad \text{(grams)}, \tag{A.13}$$

where m is the physical mass of the fluid molecules.

All other physical quantities can then be derived by the knowledge of the three scale factors S_l, S_t, S_m.

It is of some interest to discuss the scale factor for the number of particles, so as to clarify the often asked question: how many physical molecules does an LBE particle represent?

The count goes as follows.

Let n the number density, i.e., the number of physical molecules per cubic centimeter. Recalling that a lattice cell occupies a volume of S_l^3 cubic centimeters, we obtain

$$S_N = \frac{n S_l^3}{n_{lb}} \tag{A.14}$$

where n_{lb} is the density in LB units. With the typical choice $n_{lb} = 1$, S_N is simply the number of physical molecules in a single LB cell. For macroscopic flows this is typically a large number; for instance, an LB simulation of air flow at a spatial resolution of 1 millimeter delivers $S_N \sim 3 \times 10^{16}$.[99]

As a result, the relevant mass to be used in the conversion to physical units is not the mass m of the single molecule, but rather the mass contained in the LB cell, namely

$$S_M = m S_N = \rho \Delta x^3$$

where ρ is the mass density in physical units, and we have taken $n_{lb} = 1$.

The physical units of any other quantity follow from the three above.

---

[99] We remind that the relevant quantity for the number density in the kinetic theory of gases is the Loschmidt number, namely the number of ideal gas molecules in a cubic meter, which at 0 centigrades and 1 atmosphere is about 2.7×10^{25}, also known as *amagat*, after the French physicist Emil Amagat (1841–1915).

Part VII

Hands-On

In this section, we provide a sketchy description of a minimal LB warm-up computer program. This is a kind of Jurassic piece, yet still a working one. Many versions in your modern language of choice can be found on the web, such as the open source Palabos software (www.palabos.org). Fortran and C(C++) are still recommended whenever performance is not on the byside.

> I have never seen a smart error.
>
> Daan Frenkel (on the joys and pains of programming...)

H.5 Routine List

The fluid lives in a two dimensional lattice labeled by the discrete indices $\{1 \leq i \leq N_x, 1 \leq j \leq N_y\}$. The discrete distribution, however, is defined on an augmented space, consisting of the fluid domain plus a framing set of points (buffers), which are used to implement boundary conditions. As a result we have: $f_i(x, y) = f(ip, i, j)$, with $ip = 0, NP$, the number of populations, and $i = 0, N_x + 1$ and $j = 0, N_y + 1$.

H.5.1 Input

This routine reads off the main input parameters, namely, the initial density, velocity and the relaxation frequency, ω, fixing the lattice fluid viscosity. The code refers to a D2Q9 scheme with discrete velocities numbered as follows: 0 for rest particles, $1 \div 4$ for velocity $c = 1$, and $5 \div 9$ for velocity $c = \sqrt{2}$.

H.5.2 InitHydro

This routine initialises the macroscopic hydrodynamic fields, namely the fluid density $\rho(i, j) = \rho_0$ and the two fluid velocity components $u(i, j) = u_0$ and $v(i, j) = v_0$, along the x and y axes. The current version sets constant values for simplicity, but thre is no restriction for general space-dependent profiles.

H.5.3 Equil

This routine computes the equilibria based on the values of the macroscopic fields initialized in InitHydro. The calculation is kind of "historical", as it reflects the way it was first coded by Francisco Higuera back in 1989 for the very first LB simulation ever! Admittedly, it is a bit involved, but it provides a useful exercise for the reader to check that it corresponds to the standard expression given in the book.

H.5.4 InitPop

The discrete populations are initialized with the equilibroum values. $f(ip, i, j) = f^{eq}(ip, i, j)$. This is a common choice, but other options are certainly available.

H.5.5 Pbc

Fully periodic boundary conditions.

H.5.6 Mbc

Having initialized the lattice fluid, we are ready to cycle in time. The first action is to fill up the buffer nodes with values from the fluid domain, $\{1 \le i \le N_x, 1 \le j \le N_y\}$, Here, 'mbc' stands for mixed, namely periodic inlet and outlet, and bounce-back top and bottom (solid walls, with no-slip boundary condition).

H.5.7 Move

The streaming step. Note the trick of running the loop index *against the direction of motion*. Failing to do so, would consistently overwrite the current memory location, with the result of moving the -same- population across the full lattice! The trick permits to use the same array $f(ip, i, j)$ for both time levels t and $t + 1$.

H.5.8 Hydrovar

Computes the hydrodynamic fields based on the updated populations. These are needed to compute the equilibria at the next time cycle.

H.5.9 Colli

The collision step. The zenith of BGK simplicity: the BGK relaxation in one line! (in fortran 77 we have five for loops, but in fortran 90 it would be -literally- one line).

H.5.10 Force

Adds the source term due to external forces. For the case in point, this is a constant force along the x direction, surrogating a constant pressure gradient. The force is calibrated so as to produce a Poiseuille profile of central velocity u_f.

H.5.11 Obst

A simple example of thin-plate obstacle within the flow. Above a critical Reynold's number, it generates recirculating patterns.

H.5.12 Diag0D

Zero-dimensional diagnostics, i.e. total mass and momentum.

H.5.13 Movie

Velocity profiles for (gnuplot) animation

H.5.14 Profil

One-dimensional output: the velocity profiles across the flow.

H.5.15 Config

Dumps the actual configuration for check-and-restart runs.

H.6 Warm-up program

```
css      Lattice BGK simple start-up code in D=2
css      along with the book:
css      The Lattice Boltzmann equation
css      for fluid dynamics and beyond: Oxford Univ. Press, 2001
css      Slightly readapted for the 2018 version of the book
css      Author: Sauro Succi

css      Disclaimer:
css      The code is a simple warm-up, bug-freedom not guaranteed
css      The author declines any responsibility
css      for any incorrect result obtained via this code.
c
c        D2Q9 lattice, BGK version
c        0=rest particles, 1-4, nearest-neigh (nn), 5-8 (nnn)
```

```
c ================================
          program bgk2d
c ================================
          implicit double precision(a-h,o-z)
          include 'bgk2.par'
c --------------------------------------------------------------
c --- input parameters

          call input

c --- initialisation

          call inithydro
          call equil
          call initpop

c ------- MAIN LOOP
          iconf=0
          do 10 istep = 1,nsteps

c periodic boundary conditions
css       call pbc
c mixed boundary conditions c (Poiseuille flow)
          call mbc

          call move

          call hydrovar

          call equil

          call colli

          if (iforce) then
           call force(istep,frce)
          endif
c Obstacle ?
          if (iobst.eq.1) call obst
c 0-dim diagnostic
          if (mod(istep,ndiag).eq.0) then
             call diag0D(istep)
          endif
c 1d movie for gnuplot
          if(mod(istep,500).eq.0) then
```

```
          call  movie
          endif
c  1d  profiles
          if  (mod(istep ,nout).eq.0)  then
             call  profil(istep ,frce)
          endif
c  2d  configs
          if  (mod(istep ,nout).eq.0)  then
             call  config(istep ,iconf)
             iconf-iconf+1
          endif


c-------- end  of  main  loop

10        continue
          end
c ========================
          subroutine  Input
c ========================
          implicit  double  precision(a-h,o-z)
          include 'bgk2.par'
c-------------------------------------------------------
          print*,' Number  of  steps '
          read(5,*)nsteps
          print*,' Number  of  steps  between  printing  profile '
          read(5,*)nout
          print*,' Number  of  steps  between  performing  diagnostics '
          read(5,*)ndiag
          print*,' Relaxation  frequency  omega'
          read(5,*)omega
          print*,'Applied  force    (.TRUE.  or .FALSE.)  ?'
          read(5,*)iforce
          print*,' Initial  density  and  velocity  for  the  Poiseuille  force '
          read(5,*)rho0 ,u0,v0
          print*,' Final  velocity  for  the  Poise  force '
          read(5,*) uf
          print*,' Linear  obstacle  (T  or  F?'
          read(5,*) iobst
          print*,' Obstacle  height?'
          read(5,*) nobst
          print*,' Obstacle  id ',iobst
          print*,' Length  of  the  obstacle  (multiple  of  2)',nobst
          print*,' File  for  output: 5  chars '
          read(5,'(A)')fileout
```

```
      open (10, file = fileout // '.uy')
      open (11, file = fileout // '.vy')
      open (12, file = fileout // '.rhoy')
      open (14, file = fileout // '.uvx')
      open (16, file = fileout // '.pop')
      open (50, file = fileout // '.probe')
      open (35, file = 'porous')

      print*,'*****************************************'
      print*,' Lattice BGK model, 2D with 9 velocities '
      print*,'*****************************************'
      print*,'Number of cells :',nx,'*',ny
      print*,'Nsteps :',nsteps
      print*,'Relaxation frequency :',omega
      print*,'Initial velocity for this Poiseuille force :',u0
      print*,'Initial density :',rho0
      print*,'Applied force :',iforce
      if (iobst.eq.1) then
          print*,' Linear Obstacle with length :',nobst
      endif
      write (6,'(A)')'Output file :',fileout
c lattice weights
      w0  = 4.0d0/9.0d0
      w1  = 1.0/9.0d0
      w2  = 1.0/36.0d0
c sound-speed and related constants
      cs2  = 1.0d0 / 3.0d0
      cs22 = 2.0d0 * cs2
      cssq = 2.0d0 / 9.0d0
c viscosity and nominal Reynolds
      visc = (1.0d0 / omega - 0.5d0) * cs2
      rey  = u0*ny/visc
      print*,' Viscosity and nominal Reynolds:',visc,rey

      if(visc.lt.0) stop "OMEGA OUT of (0,2) interval!!'"

c Applied force (based on Stokes problem)

      fpois = 8.0d0 * visc * uf / dfloat(ny) / dfloat(ny)
      fpois = rho0*fpois/6.  ! # of biased populations
      print*,' Intensity of the applied force ',fpois

      return
      end
```

```fortran
c ===============================
      subroutine Inithydro
c ===============================
      implicit double precision(a-h,o-z)
      include 'bgk2.par'
c-------------------------------------------------------
      write(6,*) 'u0',u0
      do j = 0, ny+1
        do i = 0, nx+1
          rho(i,j)  = rho0
          u(i,j)    = u0
          v(i,j)    = v0
        enddo
      enddo
      return
      end
c ============================
      subroutine Initpop
c ============================
      implicit double precision(a-h,o-z)
      include 'bgk2.par'
c----------------------------------------------------
      iseed = 15391
      ramp  = 0.01         ! random amplitude
      do j = 0, ny+1
        do i = 0, nx+1
          do ip=0,npop-1
            f(ip,i,j)=feq(ip,i,j)
            rr=2.*rand(iseed)-1.
            f(ip,i,j)=(1.+ramp*rr)*rho0/float(npop)
            iseed = iseed+1
          end do
        end do
      end do

      return
      end
c =========================
      subroutine Move
c =========================
      implicit double precision(a-h,o-z)
      include 'bgk2.par'
c---------------------------------------------
      do j = ny,1,-1
```

```fortran
      do i = 1, nx
         f(2,i,j) = f(2,i,j-1)
         f(6,i,j) = f(6,i+1,j-1)
      enddo
   enddo
   do j = ny,1,-1
      do i = nx,1,-1
         f(1,i,j) = f(1,i-1,j)
         f(5,i,j) = f(5,i-1,j-1)
      enddo
   enddo
   do j = 1,ny
      do i = nx,1,-1
      f(4,i,j) = f(4,i,j+1)
      f(8,i,j) = f(8,i-1,j+1)
      enddo
   enddo
   do j = 1,ny
      do i = 1,nx
      f(3,i,j) = f(3,i+1,j)
      f(7,i,j) = f(7,i+1,j+1)
   enddo
   enddo

   return
   end
c =========================
      subroutine Hydrovar
c =========================
      implicit double precision(a-h,o-z)
      include 'bgk2.par'
c----------------------------------------------
c hydro variables
      do j = 1,ny
       do i = 1,nx
        rho(i,j)=f(1,i,j)+f(2,i,j)+f(3,i,j)
     .          +f(4,i,j)+f(5,i,j)+f(6,i,j)
     .          +f(7,i,j)+f(8,i,j)+f(0,i,j)
        rhoi=1./rho(i,j)

        u(i,j)=(f(1,i,j)-f(3,i,j)+f(5,i,j)-
     .          f(6,i,j)-f(7,i,j)+f(8,i,j))*rhoi
```

```
      v(i,j)=(f(5,i,j)+f(2,i,j)+f(6,i,j)
   .          - f(7,i,j)-f(4,i,j)-f(8,i,j))*rhoi

         enddo
       enddo

       return
       end
c =========================
       subroutine Equil
c =========================
       implicit double precision(a-h,o-z)
       include 'bgk2.par'
c-----------------------------------------------------
c equils are written explicitly to avoid multiplications by zero
       do j = 0,ny+1
         do i = 0,nx+1
           rl = rho(i,j)
           ul = u(i,j)/cs2
           vl = v(i,j)/cs2
           uv = ul*vl
           usq = u(i,j)*u(i,j)
           vsq = v(i,j)*v(i,j)
           sumsq  = (usq+vsq)/cs22
           sumsq2 = sumsq*(1.0d0-cs2)/cs2
           u2 = usq/cssq
           v2 = vsq/cssq

           feq(0,i,j) = w0*(1.0d0 - sumsq)

           feq(1,i,j) = w1*(1.0d0 - sumsq  + u2 + ul)
           feq(2,i,j) = w1*(1.0d0 - sumsq  + v2 + vl)
           feq(3,i,j) = w1*(1.0d0 - sumsq  + u2 - ul)
           feq(4,i,j) = w1*(1.0d0 - sumsq  + v2 - vl)

           feq(5,i,j) = w2*(1.0d0 + sumsq2 + ul + vl + uv)
           feq(6,i,j) = w2*(1.0d0 + sumsq2 - ul + vl - uv)
           feq(7,i,j) = w2*(1.0d0 + sumsq2 - ul - vl + uv)
           feq(8,i,j) = w2*(1.0d0 + sumsq2 + ul - vl - uv)
         enddo
       enddo

       return
       end
```

```
c =========================
        subroutine  Colli
c =========================
        implicit double precision(a-h,o-z)
        include 'bgk2.par'
c----------------------------------------------------------
        do k = 0,npop-1
         do j = 1,ny
          do i = 1,nx
           f(k,i,j)=f(k,i,j)*(1.0d0-omega)+omega*feq(k,i,j)
          enddo
         enddo
        enddo

        return
        end
c ===================================
        subroutine  Force(it,frce)
c ===================================
        implicit double precision(a-h,o-z)
        include 'bgk2.par'
c----------------------------------------------------------
        frce = fpois
        do j = 1,ny
         do i = 1,nx
           f(1,i,j) = f(1,i,j) + frce
           f(5,i,j) = f(5,i,j) + frce
           f(8,i,j) = f(8,i,j) + frce

           f(3,i,j) = f(3,i,j) - frce
           f(6,i,j) = f(6,i,j) - frce
           f(7,i,j) = f(7,i,j) - frce
         enddo
        enddo

        return
        end
c =========================
        subroutine  Pbc
c =========================
        implicit double precision(a-h,o-z)
        include 'bgk2.par'
```

```
c-----------------------------------------------------------
c EAST
        do j = 1,ny
          f(1,0,j) = f(1,nx,j)
          f(5,0,j) = f(5,nx,j)
          f(8,0,j) = f(8,nx,j)
        enddo
c WEST
        do j = 1,ny
          f(3,nx+1,j) = f(3,1,j)
          f(6,nx+1,j) = f(6,1,j)
          f(7,nx+1,j) = f(7,1,j)
        enddo
c NORTH
        do i = 1,nx
          f(2,i,0) = f(2,i,ny)
          f(5,i,0) = f(5,i,ny)
          f(6,i,0) = f(6,i,ny)
        enddo
c SOUTH
        do i = 1,nx
          f(4,i,ny+1) = f(4,i,1)
          f(7,i,ny+1) = f(7,i,1)
          f(8,i,ny+1) = f(8,i,1)
        enddo

        return
        end
c ========================
        subroutine Mbc
c ========================
        implicit double precision(a-h,o-z)
        include 'bgk2.par'
c-----------------------------------------------------------
c WEST inlet
        do j = 1,ny
          f(1,0,j) = f(1,nx,j)
          f(5,0,j) = f(5,nx,j)
          f(8,0,j) = f(8,nx,j)
        enddo
c EAST outlet
        do j = 1,ny
          f(3,nx+1,j) = f(3,1,j)
          f(6,nx+1,j) = f(6,1,j)
```

```
              f(7,nx+1,j) = f(7,1,j)
          enddo
c NORTH solid
          do i = 1,nx
              f(4,i,ny+1) = f(2,i,ny)
              f(8,i,ny+1) = f(6,i+1,ny)
              f(7,i,ny+1) = f(5,i-1,ny)
          enddo
c SOUTH solid
          do i = 1,nx
              f(2,i,0) = f(4,i,1)
              f(6,i,0) = f(8,i-1,1)
              f(5,i,0) = f(7,i+1,1)
          enddo
c corners bounce-back
          f(8,0,ny+1)     = f(6,1,ny)
          f(5,0,0)        = f(7,1,1)
          f(7,nx+1,ny+1) = f(5,nx,ny)
          f(6,nx+1,0)     = f(8,nx,1)


          return
          end
c ==========================
          subroutine Obst
c ==========================
          implicit double precision(a-h,o-z)
          include 'bgk2.par'
c-------------------------------------------------------------
          i    = nx / 4
          jbot = ny/2-nobst/2
          jtop = ny/2+nobst/2+1
          do j = ny/2-nobst/2,ny/2+nobst/2+1
            f(1,i,j) = f(3,i+1,j)
            f(5,i,j) = f(7,i+1,j+1)
            f(8,i,j) = f(6,i+1,j-1)
            f(3,i,j) = f(1,i-1,j)
            f(7,i,j) = f(5,i-1,j)
            f(6,i,j) = f(8,i-1,j)
          enddo
c top
          f(2,i,jtop)=f(4,i,jtop+1)
          f(6,i,jtop)=f(8,i-1,jtop+1)
c bot
          f(4,i,jbot)=f(2,i,jbot-1)
```

```
      f(7,i,jbot)=f(5,i-1,jbot-1)

      return
      end
c =============================
      subroutine Movie
c =============================
      implicit double precision(a-h,o-z)
      include 'bgk2.par'
c-----------------------------------------------------------
      open(99,file='movie.out')
      do j = 1,ny
         write(99,*) j,u(nx/4,j),u(nx/2,j),u(3*nx/4,j)
      enddo
      write(99,'(bn)')
      write(99,'(bn)')

      return
      end
c ===================================
      subroutine Profil(it,frce)
c ===================================
      implicit double precision(a-h,o-z)
      include 'bgk2.par'
c-----------------------------------------------------------
      write(6,*) 'ucenter,force ',u(nx/2,ny/2),frce
      do j = 1,ny
         write(10,*) j,u(nx/4,j),u(nx/2,j),u(3*nx/4,j)
         write(11,*) j,v(nx/4,j),v(nx/2,j),v(3*nx/4,j)
         write(12,*) j,rho(nx/2,j)
         write(99,*) j,u(nx/2,j)
      enddo
      write(10,'(bn)')
      write(11,'(bn)')
      write(12,'(bn)')

      do i = 1,nx
         write(14,*) i,u(i,ny/2),v(i,ny/2)
         write(16,*) i,f(1,i,ny/2),f(3,i,ny/2)
      enddo
      write(14,'(bn)')

      write(50,*) it,u(nx/2,ny/2)
```

```
          return
          end
c ================================
          subroutine Diag0D(istep)
c ================================
          implicit double precision(a-h,o-z)
          include 'bgk2.par'
c-------------------------------------------------------------
          densit = 0.0d0
          do k = 0,npop-1
            do j= 1,ny
              do i = 1,nx
                densit = densit + f(k,i,j)
              enddo
            enddo
          enddo
          densit = densit / dfloat(nx*ny) /dfloat(npop)
          umoy = 0.0d0
          vmoy = 0.0d0

          do j = 1,ny
            do i = 1,nx
              umoy = umoy + u(i,j)
              vmoy = vmoy + v(i,j)
            enddo
          enddo

          umoy = umoy / dfloat(nx*ny)
          vmoy = vmoy / dfloat(nx*ny)

          print*,'diagnostic 0D : istep density umoy and vmoy',
     .          istep ,densit ,umoy,vmoy

          return
          end
c ======================================
          subroutine Config(istep ,iconf)
c ======================================
          implicit double precision(a-h,o-z)
          include 'bgk2.par'
c -----------------------------------------
          iout=60+iconf
          do j = 2,ny-1
            do i = 2,nx-1
              vor = u(i,j+1)-u(i,j-1)-v(i+1,j)+v(i-1,j)
```

```
              write(iout,*) i,j,u(i,j),v(i,j),vor
          enddo
          write(iout,'(bn)')
        enddo
        write(6,*) 'configuration at time and file >>>>>> ',istep,iout

        return
        end
c  ***********************************************
c        the bgk2.f codlet ends here
c  ***********************************************

c  ================================================
c  parameter file: bgk2.par
c  ================================================
        parameter (nx =128, ny = 64, npop = 9)
        character*5 fileout
        logical iforce

        common /constants/ cs2,cs22,cssq,omega,fpois,den,visc,
       .                   w0,w1,w2
        common /phys/      rho0,u0,v0,uf,fom
        common /arrays/    iflag(nx,ny),
       .                   u(0:nx+1,0:ny+1),v(0:nx+1,0:ny+1),
       .                   rho(0:nx+1,0:ny+1),
       .                   feq(0:npop-1,0:nx+1,0:ny+1),
       .                   f(0:npop-1,0:nx+1,0:ny+1)

        common /count/ iobst,nout,ndiag,nsteps,nobst
        common /ile/       fileout
        common /logic/ iforce
c  ================================================
c  input file: bgk2.inp
c  ================================================
50001           n. of steps
2000            n. of profiles
2000            n. of global diagnos
1.5             omega
T               force yes/no
1.0  0.10  0.0  rho,u,v at t=0
0.10            u forcing
0               obstacle 0=free, 1=plate
8               obstacle height
OUP18           label
```

Index